Time-Rock Units of the Geologic Column	Time Units of the Geologic Time Scale (Numbers are absolute dates in millions of years before the present)					Time Range of Several Groups of Plants and Animals
	Eon	Era	Period		Epoch	
	Phanerozoic Eon (*Phaneros* = "evident"; *Zoon* = "life")	Cenozoic Era	Quaternary		Recent or Holocene	
					Pleistocene — 2	
			Tertiary	Neogene	Pliocene — 5	
					Miocene — 23	
				Paleogene	Oligocene — 37	
					Eocene — 55	
					Paleocene — 65	
		Mesozoic Era	Cretaceous — 140			
			Jurassic — 195			
			Triassic — 230			
		Paleozoic Era	Permian — 280			
			Carboniferous	Pennsylvanian — 310		
				Mississippian — 345		
			Devonian — 400			
			Silurian — 435			
			Ordovician — 500			
			Cambrian — 600			
	Precambrian or Cryptozoic Eon	Proterozoic — 2600			Cryptozoic Comprises About 87% of the Geologic Time Scale	
		Archean — 3800				
		No Record			Origin of Earth About 4.6 Billion Years Ago	

Birds

Mammals

Reptiles

Amphibians

Land Plants

Fishes

Invertebrates

The Earth Through Time

Second Edition

Harold L. Levin
Professor of Geology
Washington University
St. Louis

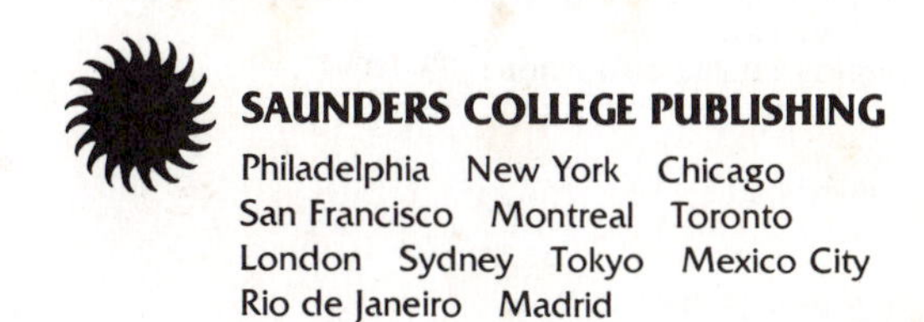

Address orders to:
383 Madison Avenue
New York, NY 10017

Address editorial correspondence to:
West Washington Square
Philadelphia, PA 19105

This book was set in Korinna by York Graphic Services.
The editors were Don Jackson, John Butler, Patrice L. Smith, Bonnie Boehme.
The art & design director was Richard L. Moore.
The text design was done by Nancy E. J. Grossman.
The cover design was done by Richard L. Moore.
The artwork was drawn by Tom Mallon.
The production manager was Tom O'Connor.

Front cover credit: Checkerboard Mesa, Zion, Utah. Photograph by © Peter Kresan.
Back cover credit: Castle Geyser, Yellowstone National Park. Copyright © 1982 by William T. Moore, Jr.

Library of Congress Cataloging in Publication Data

Levin, Harold L.
The earth through time.

Includes bibliographies and index.
1. Historical geology. I. Title.
QE28.3.L48 1983 551.7 82-10367
ISBN 0-03-058354-3

THE EARTH THROUGH TIME ISBN 0-03-058354-3

Library of Congress catalog card number 82-10367.

3456 032 987654321

CBS COLLEGE PUBLISHING
Saunders College Publishing
Holt, Rinehart and Winston
The Dryden Press

Preface

This is a book about the history of the earth. It is written for all those who have an interest in the planet on which they live. The frontiers of knowledge about the earth are advancing with exceptional speed. Our understanding of causes of past changes in geography, climate, and life are being revolutionized by fascinating new concepts of drifting continents, and of ocean floors that are endlessly in motion. In these chapters, the development and significance of these dynamic new ideas will be examined, as will the basic tenets of geology established by over three centuries of scientific observation.

The Earth Through Time is primarily an introductory textbook for a course in Historical Geology. In many colleges, students will already have had a previous course in Physical Geology and may therefore use the discussion of earth materials as a review. For the student having no previous acquaintance with geology, the chapters on earth materials and sedimentary rocks provide the background needed to proceed with the study of the earth's history.

Any history, whether geological or cultural, necessarily includes information about the nature and time of occurrence of important events. However, to comprehend geological history, which must be read from fossils and rocks, it is also important to understand the scientific reasoning by which events and conditions of the past can be interpreted. Therefore, cause and effect relationships and the bases for inferences about the past will be frequently discussed in these chapters.

The preparation of the second edition of a textbook provides benefits similar to those that occur when one gives a lecture for a second time. There is an opportunity to state some things better, to refashion an explanation, to find a better illustration, and to constructively expand and delete. Nearly always, second lectures are improved, and I am hopeful that this is true for second editions as well. In this second edition I have expanded the coverage of several topics, updated others, and provided some entirely new information all within the organization of the original edition. There is, for example, additional material on trace fossils, paleoecology, geologic activity within and near subduction zones, and the significance of alien terranes and microcontinents. The development of concepts fundamental to historical geology, the differentiation of the planet earth, stratigraphic correlation, and geochronology are all topics that have been expanded, and there is much new data on Paleozoic geography, hominid evolution, and spacecraft exploration of the solar system. The second edition also provides an updated treatment of the physical history of the western North American craton and adjacent orogenic belts. Recently published global paleogeographic maps have been added.

Nearly all of the above modifications result from comments provided by reviewers and individuals who have used the first edition in their historical geology courses. I extend my sincere thanks to all of these earth scientists, including reviewers William Ausich, (Wright State University), Michael Bikerman (University of Pittsburgh), Stanley Fagerlin (Southwest Missouri State University), John R. Huntsman (University of North Carolina), Allen Johnson (West Chester State College), Peter B. Leavens (University of Delaware), Joseph Lintz (University of Nevada at Reno), Dewey M. McLean (Virginia Polytechnic Institute), Thomas Roberts (University of Kentucky), and Thomas T. Zwick (Eastern Montana College). Reviewers of the first edition were Roger J. Cuffey (Pennsylvania State University), Larry Knox (Tennessee Technical University), Kenneth O. Stanley (Ohio State University), Calvin H.

Stevens (San Jose State University), and Eugene J. Tynan (University of Rhode Island). William H. Easton (University of Southern California) and Kenneth Van Dellen (Macomb Community College) also provided many useful suggestions derived from their classroom experience with the text. For the friendly persuasion that led to the original manuscript, I am indebted to Ms. Joan Garbutt, formerly of Saunders College Publishing.

I am, of course, deeply grateful to the staff of Saunders College Publishing for their vitality and professionalism. In particular, I benefited from the enthusiastic support and prudent advise of Donald C. Jackson, Patrice L. Smith, Richard L. Moore, and John P. Butler.

My deepest appreciation goes to Kay Helen Levin who provided persistent encouragement, and to whom this book and this author are dedicated.

HAROLD L. LEVIN

Contents Overview

Contents

Development of the Earth's Major Features 170

8

The Earth's Dynamic Crust 191

9

The Oldest Rocks 229

10

The Antique World of the Early Paleozoic 263

11

The Late Paleozoic 315

12

The Mesozoic Era 359

13

The Mesozoic Biosphere 387

14

The Cenozoic Era 423

The research vessel *Glomar Challenger*.

1

A Science Named Historical Geology

The face of places and their forms decay;
And that is solid earth that once was sea;
Seas, in their turn, retreating from the shore,
Make solid land, what ocean was before.
Ovid, *Metamorphoses*, XV

CHANGING VIEWS OF THE THIRD PLANET

We live on the third planet from the sun (Fig. 1–1). It was born nearly 5 billion years ago and since that time has circled continuously around the sun like a small spaceship in heavy traffic. The sun, in turn, carries the earth and its companion planets around the center of our galaxy while the great galactic spiral that contains the solar system moves silently through the universe.

Scarcely 100,000 years ago, primates called humans appeared on earth. Unlike earlier animals, these creatures of oversized brains and clever fingers asked questions about themselves and their surroundings. Their questioning has continued to this very hour. How was the earth formed? Why does the earth sometimes tremble? What lies beneath the seas and beyond the stars? Timidly, ancient people probed the limits of their world, fearing that they might tumble from its edge. Their descendants came to know the planet as an imperfect sphere and began an examination of every obscure corner of its surface. In harsher regions, exploration proceeded slowly. It has been only within the last 100 years that humans have penetrated the deep interior of Antarctica. Today, except for a few areas of great cold or dense forest, the continents are well charted. The new frontiers for exploration lie beneath the oceans and outward into space.

The advance of science along the oceanic frontiers began less than three decades ago. Crisscrossing the seas, ships equipped with precision depth recorders that employed the principle of echo sounding (Fig. 1–2) plotted continuous topographic profiles of the sea floors. Related methods utilizing stronger energy sources provided an image of the layers of rock and sediment beneath the ocean bottom. A panoramic view of undersea geology emerged that was more magnificent and complex than anyone had imagined (Fig. 1–3). The magnetic characteristics of the ocean floors

Figure 1–1 The Solar System. (From J. Turk and A. P. Turk, *Physical Science*. 2nd ed. Philadelphia, Saunders College Publishing, 1981.)

were also examined, and the scientific community was startled to learn that the earth's magnetic polarity had occasionally reversed itself. A unique deep sea drilling ship, the *Glomar Challenger* (Fig. 1–4), was constructed. With the help of the National Science Foundation, it was quickly put into operation. With this splendid vessel, geologists were able to penetrate deeply into the sea floor and bring on deck the sediment and rock needed to decipher the history of the ocean basins. Samples from beneath the sea as well as long strips of graph paper imprinted with jagged lines (Fig. 1–5) nourished exciting new hypotheses of drifting crustal segments, splitting continents, and changing world geography.

Exploration of the cosmic "ocean" has also contributed to our new view of the earth. Intricately engineered spacecraft circled the earth and set down on the moon. Two members of our inquisitive species reached down and held lunar pebbles in their gloved hands. Analysis of those pebbles yielded an astonishing number of clues to the early history of both the moon and the earth. Within the past decade, space vehicles have briefly experienced the hellish atmosphere of Venus, sampled the soil of Mars, and fixed electronic eyes on Mercury and Jupiter. Small satellites continue to circle the earth, transmitting repeating images of the planet. An unprecedented, synoptic view of a dynamic, delicately adjusted planet has emerged from these transmissions. On a day-to-day basis, data supplied by satellites have been used to find mineral deposits, to assess worldwide agricultural productivity, and to provide unexcelled meteoro-

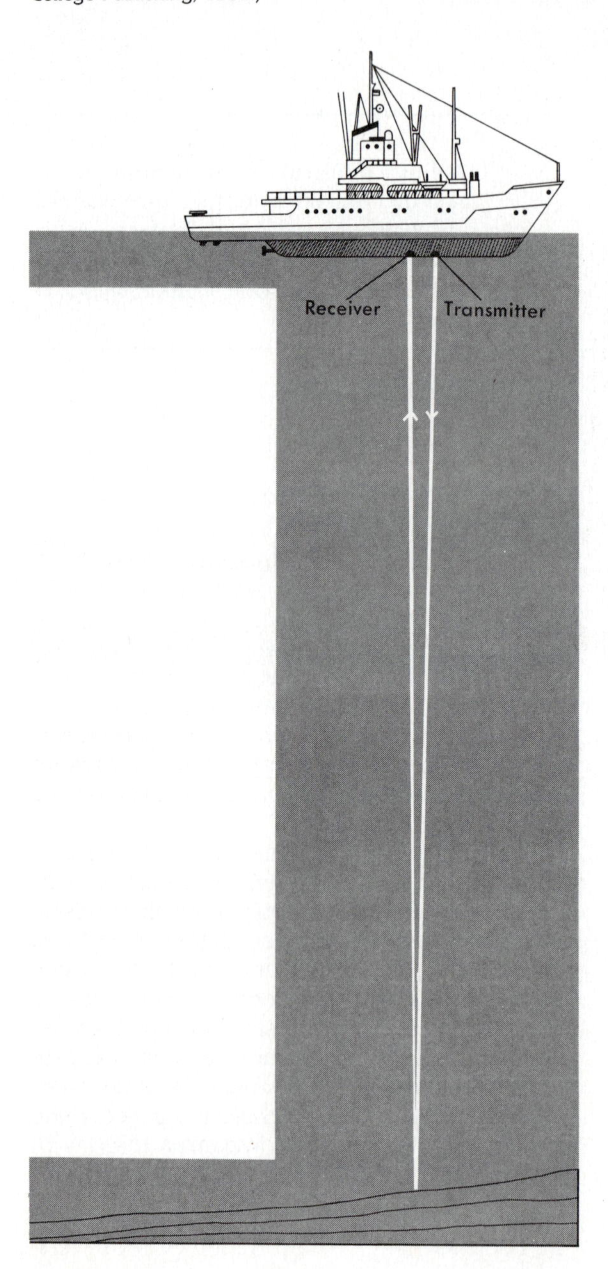

Figure 1–2 The principle of echo sounding. A transmitter sends a sound wave, which is reflected back to the surface by the ocean bottom and is picked up by a receiver. By knowing the total time involved and the speed of sound in the ocean (1500 meters per second), water depth can be determined. (From J. M. McCormick and J. V. Thiruvathukal. *Elements of Oceanography*. Philadelphia, Saunders College Publishing, 1981.)

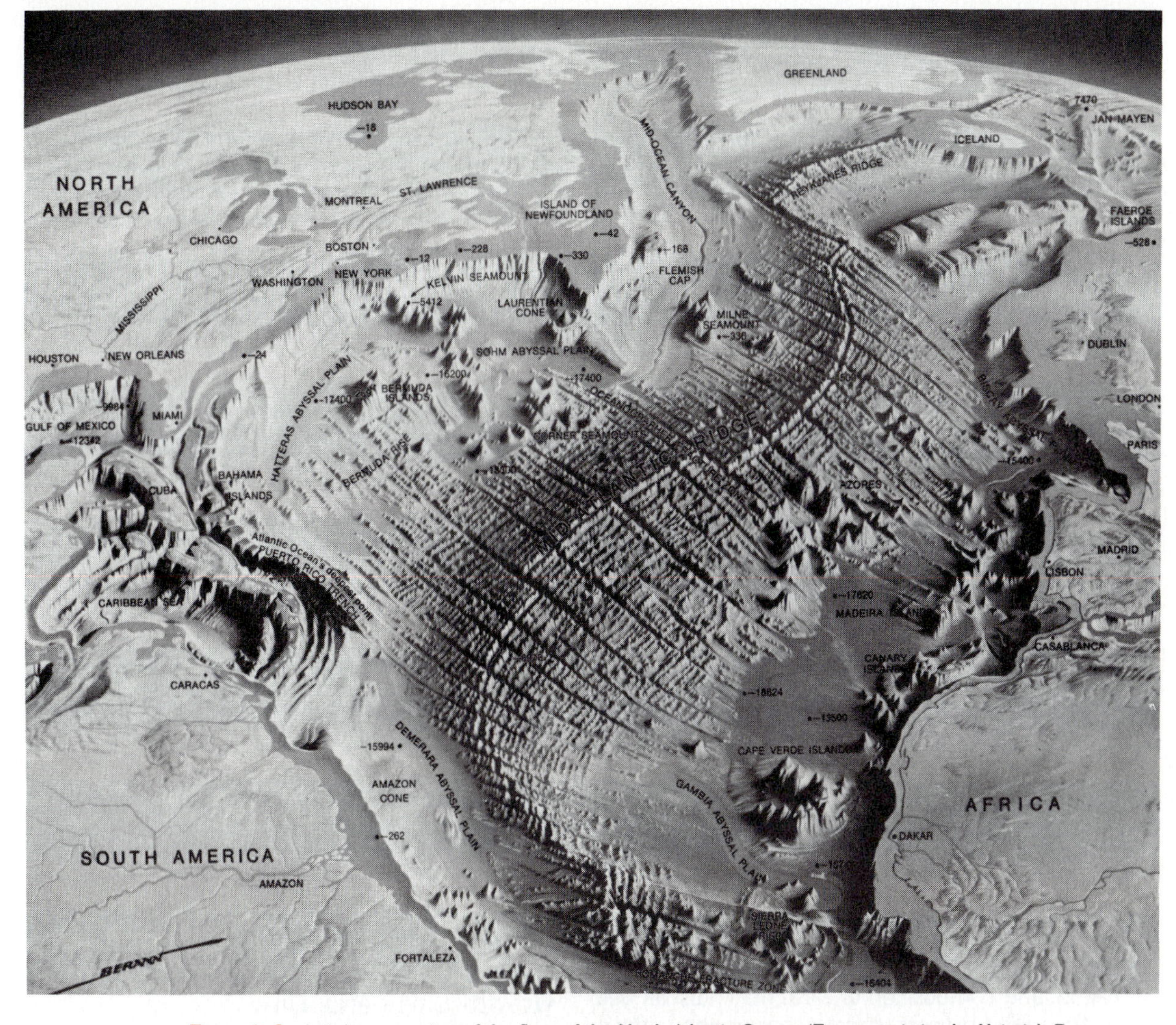

Figure 1–3 Artist's conception of the floor of the North Atlantic Ocean. (From a painting by Heinrich Berann, Courtesy of Alcoa; from J. M. McCormick and J. V. Thiruvathukal. *Elements of Oceanography*. 2nd ed. Philadelphia, Saunders College Publishing, 1981.)

logic information. Perhaps more important, however, these data have helped to make all the residents on the third planet more aware of the beauty and fragility of their home.

AN ECLECTIC SCIENCE

Geologists concern themselves with an exceptional variety of scientific tasks and must employ knowledge from diverse fields. Some examine the composition and texture of meteorites and moon rocks. With magnifiers and computers, others scrutinize photographs of planets to understand the features that characterize their surfaces. Still other geologists are busily unraveling the structure of mountain ranges (Fig. 1–6) in an attempt to understand the behavior of glaciers, underground water, volcanoes, and streams. Large numbers of geologists search for the fossil fuels and ores we require. They worry, as do you and I, about the fate of humans in a world of diminishing resources. To help in their work, geologists draw upon the knowledge of astronomy, physics, chemistry, mathematics, and biology. For example, petroleum geologists must understand the physics of moving fluids, the chemistry of hydrocarbons, and the biology of the invertebrate fossils they use to reconstruct the subsurface geology. Because geology incorporates the information of so many other scientific disciplines, it can be termed an **eclectic science.** "Eclectic" is a word useful in describing a body of selected information drawn from a variety of sources. All sciences are eclectic to some degree, and geology is decidedly so.

For convenience of study, the body of knowl-

Figure 1–4 The *Glomar Challenger,* an oceanographic research vessel designed for taking drill cores from the floor of the deep ocean. The vessel is shown here in Okinawa Harbor. (Photograph courtesy of D. J. Echols.)

edge called "geology" can be divided into **physical geology** and **historical geology.** The origin, classification, and composition of earth materials, as well as the varied processes that occur on the surface and in the deep interior of the earth, are the usual subjects of physical geology. Historical geology addresses itself to the earth's evolution, changes in the distribution of lands and seas, growth and destruction of mountains, succession of animals and plants through time, and chronologic changes in other planets in our solar system. Clearly, the division between physical and historical geology is somewhat arbitrary, and it is necessary to understand the physical aspects of our planet if we are to fully know its history.

THE TASK OF THE HISTORICAL GEOLOGIST

Simply stated, the task of historical geologists is to examine planetary materials and structures and to discover how they came into existence. Geologists have in the world about them the tangible *result* of past events, and they must work backward in time to discover the causes of those events. In their work, they use the usual procedures of science—namely, the collection of observations, the formulation of hypotheses to explain those observations, and the validation of the hypotheses by means of further observations. Thus, observed facts serve both as the basis for building hypotheses and as the ultimate check on their accuracy.

One way to better understand how geologists think and work is to describe a couple of interesting geologic research projects. One such project provided information about the number of days in a year 370,000,000 years ago. Another resulted in a description of the Mediterranean region as it looked 6 million years ago.

Growth Lines and Time

In the early 1960's, Professor John Wells of Cornell University was musing over the fact that,

Figure 1–5 Continuous seismic profiles showing abyssal hills and abyssal plains along the edge of the Mid-Atlantic Ridge. (From D. E. Hayes and A. C. Pimm, Initial Reports of Deep Sea Drilling Project. 14:341–376, 1972.)

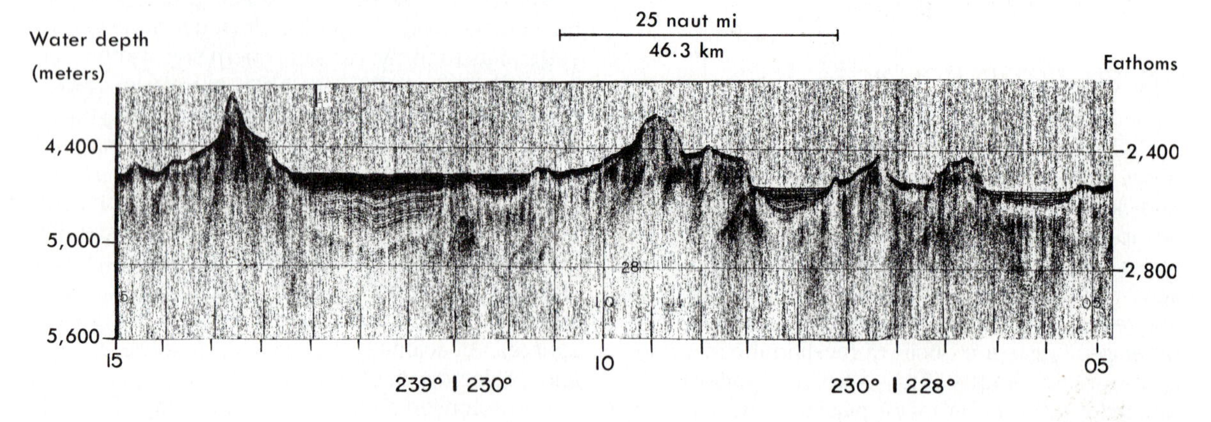

Figure 1–6 Geologist examining rock exposure for clues to the formation's transportational and depositional history. (Courtesy of Standard Oil Company of California. Photograph by Carl Iwasaki.)

although paleontology (the study of fossils and ancient life forms; from the Greek *palaios,* meaning "ancient") provided a way to determine *relative* geologic age, fossils could not be used in absolute **geochronology** (the chronology of the earth's history as determined by geologic events). His thoughts led him to consider the astronomic basis for time and to search for some detectable effect of movement in the sun-earth-moon system on the fossil remains of organisms. On theoretic grounds, geophysicists and astronomers had determined that the present 24-hour period of the earth's rotation on its axis has not been constant throughout geologic time. The earth has been slowing down because of tidal friction at a rate of about 2 seconds per 100,000 years. Thus, the length of the day has been increasing, and the number of days in a year has been steadily decreasing. On the basis of work completed by earlier researchers, Wells knew that the fine growth bands on the exoskeletons of coral organisms might represent daily growth increments (Fig. 1–7). Thus, a coral might secrete one thin band of calcium carbonate each day. Wells counted the growth bands on several species of living corals and found that the count "hovers around 360 in the space of a year's growth." Proceeding next to fossil corals of successively older geologic age, Wells counted correspondingly larger numbers of growth bands in the yearly segment of the exoskeleton. For example, on corals that had been determined to be about 370 million years old, Wells found 398 daily growth bands in a yearly increment. This evidence suggesting that there were nearly 400 days in a year 370 million years ago correlated astonishingly well with the calculations provided by the astronomers. Historical geologists could now state with considerable confidence that long ago when the first land vertebrates began to appear, the days were shorter, but there were more of them in a year. Furthermore, once the relationship between number of days in the year and absolute geologic age had been plotted on a graph (Fig. 1–8), then it might be possible to derive the age of a stratum by counting the growth bands on the fossil corals collected from that stratum. Of course, the method would be valid only if the slowdown of the earth's rotation has been uniform.

As in all forms of geochronology, the Wells method has its limitations. Only exceptionally well-preserved fossils are suitable, and large numbers must be subjected to growth band counts in order

Figure 1–7 Growth banding displayed by a specimen of the extinct coral *Heliophyllum halli.* The finer lines may represent daily growth increments. Together with the annual bands, the growth increments can be used to estimate the number of days in a year at the time the animal lived. (Photograph courtesy of George R. Clark, II.)

to obtain statistically valid results. Many specimens show variations caused by factors in the environment unrelated to ordinary daily changes. Nevertheless, the method has provided an interesting correlation between astronomic and biologic phenomena.

A Study of Mediterranean History

A second research effort that exemplifies the operation of the scientific method is embodied in the ongoing work of Kenneth J. Hsu of the Swiss Federal Institute of Technology and of William B. F. Ryan of the Lamont-Doherty Geological Observatory. These two geologists were chief scientists aboard the 1970 *Glomar Challenger* cruise to investigate the floor of the Mediterranean.

Long before the cruise, paleontologists had discovered evidence that the invertebrate fauna of the Mediterranean had changed abruptly about 6 million years ago. The older organisms of the region had nearly suffered annihilation, although a few species had managed to migrate into the Atlantic, and a still smaller group of hardy species had survived. Somewhat later, the refugees returned, bringing new species with them. What was the reason for the migrations and extinctions?

A second question concerned the origin of an enormous buried gorge that lay beneath—and nearly parallel to—the present course of the Rhone River, which flows from Switzerland to France. The buried canyon is nearly 1000 meters deep. What had happened in the geologic past to give a stream sufficient erosive powers to cut so deeply?

The third task for the *Challenger's* scientists was to try to determine the origin of the domelike masses buried deep beneath the Mediterranean sea floor. These structures had been detected

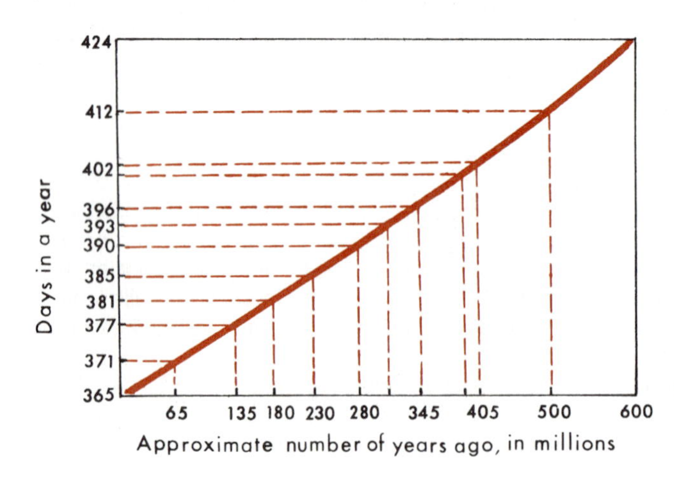

Figure 1–8 Changing length of the day through geologic time. (From J. W. Wells. *Nature,* 197:948–950, 1963.)

years earlier by echo-sounding instruments, but they had never been penetrated in the course of drilling. Were they salt domes such as are common along the United States Gulf Coast, and if so, why should there have been so much solid crystalline salt beneath the floor of the Mediterranean?

Finally, there was the presence—again detected by seismic echo-sounding instruments—of a hard layer of sedimentary rock 100 meters or so below the present sea floor. This hard layer was a strong acoustic reflector (layer of sediment or rock capable of reflecting sound waves) and was easily recognized nearly everywhere in the Mediterranean. Most deep sea deposits are soft, poorly reflective oozes. What was the composition and origin of the hard layer?

With such questions as these clearly before them, the scientists aboard the *Glomar Challenger* proceeded to the Mediterranean to search for the answers. On August 23, 1970, they recovered a sample from the surface of the hard layer. The sample was an unusual gravel, composed of pebbles of strangely hardened sediment that had once been ooze, as well as of pieces of oceanic volcanic rock and granules of gypsum. Not a single pebble was found that might have indicated that the gravel came from the nearby continent. This observation was properly noted. In the days following, samples of solid gypsum were repeatedly brought on deck as drilling operations penetrated the hard layer at its seismically predicted depth. Clearly, the "strong acoustic reflector" was a layer of gypsum. Furthermore, the gypsum was found to possess peculiarities of composition and structure that suggested it had formed on desert flats. Sediment above and below the gypsum layer contained tiny marine fossils, indicating open-ocean conditions. As they drilled into the central and deepest part of the Mediterranean Basin, the scientists took solid, shiny, crystalline salt from the core barrel. Interbedded with the salt were thin layers of what appeared to be windblown silt.

The time had come to formulate a hypothesis. The investigators theorized that about 20 million years ago the Mediterranean was a broad seaway linked to the Atlantic by two narrow straits. Crustal movements then closed the rather tenuous straits, and the now land-locked Mediterranean began to evaporate. The increasing salinity caused the extermination of scores of invertebrate species. Only a few organisms especially tolerant of hypersaline conditions remained. As evaporation continued, the remaining brine became so dense that the calcium sulfate of the hard layer was precipitated. In the central deeper part of the basin, the last of the brine evaporated to precipitate more soluble sodium chloride. Later, under the weight of overlying sediments, this salt flowed plastically upward to form salt domes. Before this happened, however, the Mediterranean was a vast "Death Valley" 3000 meters deep. Then, about $5\frac{1}{2}$ million years ago came the deluge. As a result of crustal adjustments and faulting, the Strait of Gibraltar opened, and water cascaded spectacularly back into the Mediterranean. Turbulent waters tore into the hardened salt flats, broke them up, and ground them into the pebbles observed in the first sample taken by the *Challenger.* As the basin was refilled, normal marine organisms returned. Soon layers of oceanic ooze began to accumulate above the old hard layer.

The salt and gypsum, the faunal changes, and the unusual gravel provided abundant evidence that the Mediterranean was once a desert. But how did this knowledge relate to the buried gorge beneath the Rhone Valley? The answer was clear. If the Mediterranean had been emptied, the surrounding lands would have stood high above the floor of the basin. The gradients of streams would have markedly increased, and their swiftly flowing waters would have incised rapidly, eroding deep canyons. Such canyons should also have been cut by other streams that once flowed into the Mediterranean. As news of the *Glomar Challenger's* findings spread, geologists who had worked in the region alerted the investigators to the presence of gorges in North Africa. A Russian geologist recalled a 240-meter chasm that had been encountered at the time the foundations had been prepared for the Aswan Dam in Egypt. Oil company geologists in Libya sent word of many similar gorges that they had detected from seismograph records. A final piece of the puzzle seemed to fall into place.

In the pages ahead, the methods of historical geology will be further demonstrated. Many of these methods enable geologists to place past events in their proper chronologic order and to discern changes in the earth's past climate, atmosphere, and geography.

THE FOUNDERS OF HISTORICAL GEOLOGY

As practiced today, geology is characterized by the use of wonder-working instruments and computerized techniques. The modern approach to geologic investigation seems extremely remote from the way the study was pursued a century or two ago. Yet, the basic style of thinking used by a geologist working with computer print-outs and electron probes has not changed much from that of a seventeenth century naturalist armed only with compass and hammer. Also, the progress of science is cumulative, and we build our new theories by using observations and postulates of those who preceded us. Four centuries ago Leonardo da Vinci (1452–1519) recognized the true significance of fossils and understood the impermanence of mountains. Robert Hooke (1635–1703), who was also intrigued by fossils, perceived that "it would not be impossible to raise a chronology out of them." Around the year 1669, three very important concepts were clearly enunciated by Nils Steensen.

Nicolaus Steno

Nils Steensen, who Latinized his name to Nicolaus Steno, was a Danish physician who settled in Florence, Italy. There he became physician to the Grand Duke of Tuscany. Since the duke was a generous employer, Steno had ample time to tramp across the countryside, visit quarries, and examine strata. His investigations led him to formulate such basic principles of historical geology as *superposition, original horizontality,* and *original lateral continuity.*

The **principle of superposition** states that in any sequence of undisturbed strata, the oldest layer is at the bottom, and successively higher layers are successively younger. It is a rather obvious axiom, which probably had been understood by many naturalists even before Steno. Yet Steno, on the basis of his observations of strata in northern Italy, was first to explain the concept formally. The fact that it is self-evident does not diminish the principle's importance in deciphering earth history. Furthermore, the superpositional relationship of strata is not always apparent in regions where layers have been steeply tilted or overturned (Fig. 1–9). In such instances, the geologist must examine the strata for clues useful in recognizing their uppermost layer. The way fossils lie in the rock and the evidence of mudcracks and ripple marks are particularly useful clues when one is trying to determine "which way was up."

The observation that strata are often tilted led Steno to his **principle of original horizontality.** He reasoned that most sedimentary particles settle from fluids under the influence of gravity. The sediment then must have been deposited in layers that were nearly horizontal and parallel to the surface on which they were accumulating (Fig. 1–10). Hence, steeply inclined strata indicate an episode of crustal disturbance after the time of deposition (Fig. 1–11).

The **principle of original lateral continuity** was the third of Steno's stratigraphic axioms. It pertains to the fact that, as originally deposited, strata either extend in all directions until they thin to a feather edge, or they end abruptly against the edges of the basin or area in which they were laid down. This observation is significant in that whenever one observes the exposed edge of strata in a cliff or valley

Figure 1–9 Steeply inclined strata that have been eroded to form a natural arch. Devon, England. (Photograph courtesy of the Institute of Geological Sciences, London.)

rpositional sequence of strata at Book Cliffs in northeastern Utah. (Photograph courtesy of

the strata continue
be determined by
ologists stand on a
a canyon, it is the
ntinuity that leads
ge of sandstone on
realize that the two
us.

Today, we recognize that Steno's principles are basic to the geologic specialty known as **stratigraphy,** which is the study of layered sedimentary rocks, including their texture, composition, arrangement, and correlation from place to place. Because stratigraphy enables geologic events as recorded in rocks to be placed in their correct sequence, it is the key to the history of the earth.

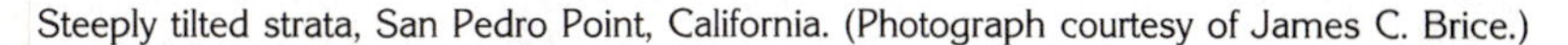

Steeply tilted strata, San Pedro Point, California. (Photograph courtesy of James C. Brice.)

Decoders of the Geologic Succession

The stratigraphic principles formulated by Steno in the seventeenth century were rediscovered several decades later by other European scientists. Among the most prominent of these early geologists were John Strachey (1671–1743), Giovanni Arduino (1714–1795), Johann G. Lehmann (1719–1767), Georg Füchsel (1722–1776), and Peter Simon Pallas (1741–1811). John Strachey is best remembered for his use of the principles of superposition and original lateral continuity in deciphering the stratigraphic succession of coal-bearing formations in Somerset and Northumberland, England. He clearly illustrated the sequence of formations encountered at the surface and in mines and described the manner in which horizontal strata rested upon the eroded edges of inclined older formations. Years later this type of stratigraphic relationship would be termed unconformable.

Whereas John Strachey was particularly interested in the local stratigraphic succession, other naturalists developed a broader, more global view of the geologic succession. In Italy, Giovanni Arduino wrote that there were basically three kinds of mountains, which he designated as Primary, Secondary, and Tertiary. Primary mountains were composed of crystalline rocks of the kind later to be named igneous and metamorphic. Arduino recognized that rocks of the Primary group were likely to be the oldest in a mountain system and were usually exposed along the central axis of the ranges. Secondary mountains were constructed of layered, well-consolidated, fossiliferous rocks. Such rocks were later to be named sedimentary. Arduino's Tertiary designation was reserved for mountains composed of unconsolidated gravel, sand, and clay, as well as lava beds.

Classifications similar to that of Arduino also appeared in the works of the German scientists Lehmann and Füchsel. These men were not "rocking-chair theorists," for both devoted long hours to making careful observations in the field. Füchsel worked chiefly in the mountains of Thuringia, whereas his contemporary Lehmann tramped on the rocks of the Harz and Erz Gebirge. They prepared excellent summaries of the stratigraphic succession in these mountains and further developed a remarkably perceptive understanding of some of the events involved in the making of mountain ranges.

This insight into the history of mountains was improved as a result of the work of an indefatigable field geologist named Peter Simon Pallas. Under the patronage of Catherine II of Russia, Pallas traveled across the whole of Asia and made careful studies of the Ural and Altai Mountains. He recognized the threefold division of mountains formulated by his predecessors. In addition, Pallas was able to construct a general geologic history of the Urals, and he provided a lucid description of how the rock assemblages change as one travels from the center to the flanks of mountain systems.

Abraham Gottlob Werner

One of the most influential geologists working in Europe near the close of the eighteenth century was Professor Abraham Gottlob Werner (1750–1817). Werner's eloquent and enthusiastic lectures at the Freiberg Mining Academy in Saxony transformed that school into an international center for geologic studies. Werner was a competent mineralogist, and many geologists of his day used his scheme for the identification of minerals and ores. Werner is not remembered, however, as much for his contributions to mineralogy as for the role he played in an interesting controversy. He believed that all rocks, including granite and basalt, were formed in water. Werner, and those that believed as he did, came to be known as Neptunists (from Neptune, the Roman god of the sea). According to the Neptunists, all bedrock had been precipitated from a universal ocean that once covered the entire earth. The fossils found in sedimentary rocks were cited as evidence for this ocean. Unfossiliferous rocks like granite and basalt were simply represented as the initial deposits of the great ocean and were thought to have precipitated and crystallized much like certain crystalline sedimentary rocks. Lava flows were poorly understood at the time, so their presence between layers of sediment was taken as evidence that the basalt must have been laid down in much the same way as the adjacent sedimentary rocks. Perhaps, reasoned Werner, Noah's Flood was the source of the widespread waters.

Werner's views did not go unchallenged for long. An opposing school of thought developed among those who believed correctly that igneous rocks congealed from molten material. Those having this view were called Plutonists, and they waged a war of words against the Neptunists. The debate began to subside when, in France, Werner's former student J. F. D'Aubisson de Voisins clearly demonstrated the volcanic origin of basalt with indisputable field evidence. D'Aubisson's interpretation was supported by James Hutton in Scotland, who stated that basalt "formed in the bowels of the earth of melted matter poured into rents and openings of the strata."

James Hutton

James Hutton, an Edinburgh physician and geologist, is remembered as a staunch opponent of Neptunism, but more importantly for his development of the **principle of uniformitarianism.** In simple terms, uniformitarianism stipulates that the past history of the earth is best interpreted in terms of what is known about the present. The principle is often paraphrased as "the present is the key to the past." Thus, geologic features formed long ago were produced by processes that are still at work today. These processes—whether deposition, erosion, or volcanism—are governed by unchanging natural laws. Today, geologists are careful to point out that although processes long ago were the same as those today, they varied in rate and intensity. At times, continents may have stood relatively higher above the oceans. This higher elevation would have resulted in higher rates of erosion and relatively harsher climatic conditions compared with intervening episodes, when the lands were low and partially covered with inland seas. Similarly, volcanism at one time or another in the geologic past may have been more frequent than it is now. However, ancient volcanoes should have disgorged gases and deposited lava and ash just as present-day volcanoes do when they erupt. Modern glaciers are more limited in area than are those of the recent geologic past; nevertheless, they form erosional and depositional features in much the same manner as their larger and more ancient counterparts.

Like many "rules," uniformitarianism is not without its exceptions. For example, before the earth had evolved an atmosphere like that existing today, processes acting upon the earth must have been unlike those now at work. Therefore, it is not at all unreasonable to ask "what *is* uniform?" The answer is that the physical and chemical laws governing geologic processes are uniform. In order to eliminate the errors inherent in Hutton's form of uniformitarianism and to stress validly the uniformity not of processes but of natural laws, many geologists prefer to abandon the term uniformitarianism and to substitute the term **actualism.** Actualism is the principle that the same processes and natural laws prevailed in the past as those we can now see in operation or can infer from observations.

The concept of uniformitarianism was not the only contribution James Hutton made to geology. In his *Theory of the Earth,* published in 1785, he brought together many of the formerly separate thoughts of the naturalists who preceded him. He showed that rocks recorded events that had occurred over immense periods of time and that the earth had experienced many episodes of upheaval separated by quieter times of denudation and sedimentation. In his own words, there had been a "succession of former worlds." Hutton saw a world of cycles in which water sculptured the surface of the earth and carried the erosional detritus from the land into the sea. The sediment of the sea was compacted into stratified rocks, and then by the action of enormous forces, the layers were cast up to form new lands. In this endless process, Hutton found "no vestige of a beginning, no prospect of an end." In this phrase, we see Hutton's intoxicating view of the immensity of geologic time. No longer could geologists compress earth history into the short span suggested by the Book of Genesis.

At the Isle of Arran (an island of Scotland) and at Jedburgh, Scotland, Hutton came across an exposure of rock where steeply inclined older strata had been beveled by erosion and covered by flat-lying younger layers (Fig. 1–12). It was clear to Hutton that the older sequence was not only tilted but also partly removed by erosion before the younger rocks were deposited. The erosional surface meant that there was a gap or hiatus in the rock record. In 1805, Robert Jameson named this relationship an **unconformity.** More specifically, Hutton's Jedburgh exposure was an **angular unconformity** because the lower beds were tilted at an angle to the upper. This and other unconformities provided Hutton with evidence for periods of denudation in his "succession of worlds." Although he did not use the word "unconformity," he was the first to understand and explain the significance of this concept.

Frequently, Hutton was able to recognize distinctive strata in different parts of Scotland—provided that their composition and texture did not change within his area of investigation. What he was unable to do was to recognize rocks of the same age but of different appearance. In other words, he was unable to determine the chronologic equivalence of strata. This problem was resolved in England around 1800 by William Smith.

William "Strata" Smith

William Smith was an English surveyor and engineer who devoted 24 years to the task of tracing out the strata of England and representing them on a map. Small wonder that he acquired the nickname "Strata." He was employed to locate the routes of canals, to design drainage for marshes, and to restore springs. In the course of this work, he independently came to understand the principles of stratigraphy, for they were of immediate use

Figure 1–12 Unconformity at Siccar Point, Scotland. It was here that the historical significance of an unconformity was first realized by James Hutton in 1788.

to him. By knowing that different types of stratified rocks occur in a definite sequence and that they can be identified by their composition and texture (lithology), the soils they form, and the fossils they contain, he was able to predict the kinds and thicknesses of rock that would have to be excavated in future engineering projects. His use of fossils was particularly significant. Prior to Smith's time, collectors rarely noted the precise beds from which fossils were taken. Smith, on the other hand, carefully recorded the occurrence of fossils and quickly became aware that certain rock units could be identified by the particular assemblages of fossils they contained. He used this knowledge first to trace strata over relatively short distances and then to extend over great distances his "correlations" to strata of the same age but of different lithology. Ultimately, this knowledge led to the **principle of biologic succession,** which stipulates that the life of each age in the earth's long history was unique for particular periods, that the fossil remains of life permit geologists to recognize contemporaneous deposits around the world, and that fossils could be used to assemble the scattered fragments of the record into a chronologic sequence.

"Strata" Smith did not know why each unit of rock had a particular fauna. This was 60 years before the publication of Darwin's *Origin of Species.* Today, we recognize that different kinds of animals and plants succeed one another in time because life has evolved continuously. Because of this continuous change, or evolution, only rocks formed during the same age could contain similar assemblages of fossils.

News of Smith's success as a surveyor spread widely, and he was called to all parts of England for consultation. On his many trips, he kept careful records of the types of rocks he saw and the fossils they contained. Armed with his notes and observations, he prepared a geologic map of England that has remained substantially correct even today. His work clearly demonstrated the validity of the principle of biologic succession.

Cuvier and Brongniart

The use of fossils for the correlation and recognition of formations was not exclusively William Smith's discovery. At the same time that Smith was making his observations in England, two scientists across the English Channel in France were diligently advancing the study of fossils. They were Baron Georges Léopold Cuvier (1769–1832) and his close associate Alexander Brongniart (1770–1847). Cuvier and Brongniart established the foundations of vertebrate paleontology. They validated Smith's findings that fossils display a definite succession of types within a sequence of strata and that this succession remains more or less constant wherever found. In the course of their studies, Cuvier and Brongniart recognized that great groupings of strata were separated by distinct unconformities. As one would pass from such a group or sequence into the overlying one, a dramatic change in the total fossilized fauna was apparent. From this observation the Frenchmen erroneously concluded that the history of life was marked by frightful catastrophes, each of which resulted in the extinction of old faunas and the appearance of new ones. When, however, geologic studies were carried to parts of the earth outside Europe, it became apparent that the supposed catastrophes, if they occurred at all, were local. The fossil records and rocks missing in Europe were found preserved on other continents. Further, there was clear evidence of a biologic continuum, in which the beginnings of each new group of organisms could be found among the older group.

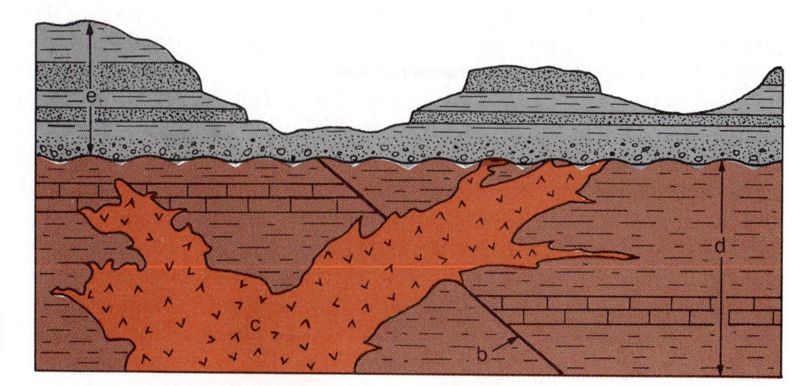

Figure 1–13 Time sequence of events from spatial relations of strata, faults, and intrusions.

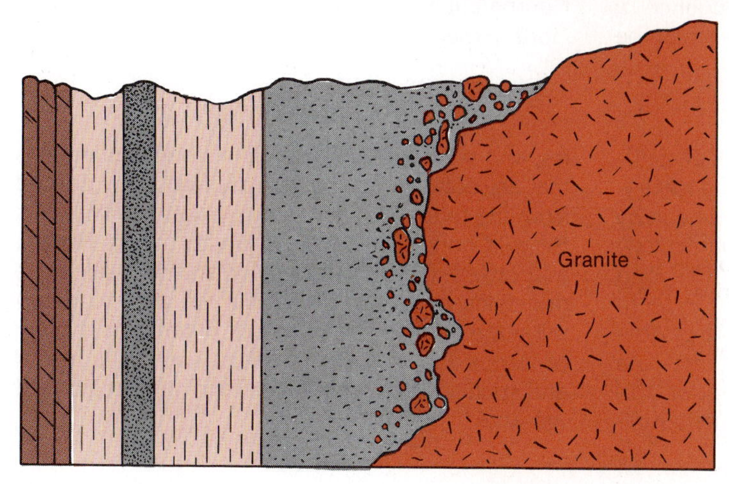

Figure 1–14 Granite inclusions in sandstone indicate that granite is the older unit. (Originally horizontal strata have been tilted to a vertical orientation.)

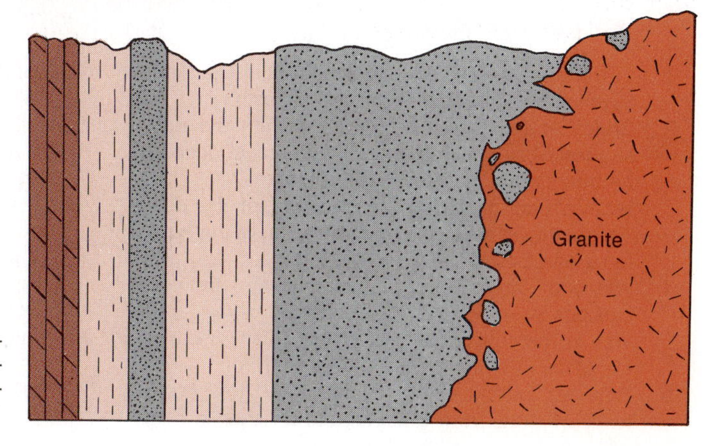

Figure 1–15 Inclusions of sandstone in granite indicate that sandstone is the older unit. (After being deposited, strata were tilted to their present vertical orientation.)

Simultaneous worldwide catastrophism was shown to be an untenable hypothesis.

Charles Lyell

In the early nineteenth century, a geologist wrote a book that would both amplify the ideas of Hutton and present under one title the most important geologic concepts of the day. His name was Charles Lyell, and he authored the classic *Principles of Geology*. The first volume of this work was printed in 1830. It grew to four volumes and became immensely important in the Great Britain of Queen Victoria. In this work can be found additional postulates useful in establishing relative ages of rock units. For example, Lyell discussed the general principle that a geologic feature that cuts across or penetrates another body of rock must be younger than the rock mass penetrated. In other words, the feature that is cut is older than the feature that crosses it. The generalization applies not only to rocks but also to geologic stuctures like faults and unconformities. Thus, in Figure 1–13 fault *b* is younger than stratigraphic sequence *d;* the intrusion of igneous rock c is younger than the fault, since it cuts across it; and by superposition, rock sequence *e* is youngest of all.

Another generalization to be found in Lyell's *Principles* relates to **inclusions.** Lyell logically discerned that fragments within larger rock masses are older than the rock masses in which they are enclosed. Thus, whenever two rock masses are in contact, the one containing pieces of the other will be the younger of the two.

In Figure 1–14, the pebbles of granite (a coarse-grained igneous rock) within the sandstone tell us that the granite is older and that the eroded granite fragments were incorporated into the sandstone. In Figure 1–15, the granite was intruded as a melt into the sandstone. Because there are sandstone inclusions in the granite, the granite must be the younger of the two units.

By the middle of the nineteenth century, the major concepts needed to unravel earth history were well understood by geologists. Armed with the self-evident and universal rules of superposition, initial horizontality, biologic succession, cross-cutting relationships, and inclusions, geologists proceeded to decipher the geologic history, not only of Europe but also of other regions, including western North America.

Summary

Simply stated, geology is the study of all naturally occurring processes, phenomena, and materials of the past and present earth. Historical geology is that branch of the science concerned particularly with decoding the rock record of the earth's long history. In the last few decades, advances in technology have added immensely to the store of geologic knowledge. Interpretation of the new data requires an exceptional understanding not only of geology but also of physics, chemistry, mathematics, and biology. However, the historical inferences that are drawn from the data are frequently derived from the fundamental geologic axioms of Nicolaus Steno, Charles Lyell, William Smith, James Hutton, and others. These early geologists formulated the principles by which geologists are able to determine the relative age of a rock outcrop, its history of deposition and deformation, and its temporal and spatial relationship to strata in other regions of the world.

Questions for Review

1. Summarize the evidence indicating that the Mediterranean was once a deep desert basin.
2. Under what circumstances, other than the ones suggested by Hsu and Ryan, might gravel deposits be transported into deep oceanic areas?
3. What events preceded the filling up of the deep gorges that in turn led to the formation of the Mediterranean Basin?
4. Why is the reporting of methods, results, and conclusions so important to scientists?
5. Why is the concept of uniformitarianism useful in understanding the history of the earth?
6. How did William Smith demonstrate the equivalency of strata from one place to another in England?
7. What criteria might a geologist seek in distinguishing the top from the bottom of a sequence of strata that had been turned into vertical alignment by mountain-building forces?
8. If a granitic intrusion cuts into or across several strata, which is older—the granite or the strata?
9. If an erosional surface (unconformity) cuts across or truncates folded strata, did the erosion occur before or after the strata were folded?
10. In coming upon the eroded edge of a stratum, one may correctly surmise that the stratum continues laterally in all directions from the exposure. How may the stratum terminate?

Terms to Remember

actualism
angular unconformity
geochronology
historical geology
inclusions
physical geology
principle of cross-cutting relationships
principle of biologic succession
principle of original horizontality
principle of original lateral continuity
principle of superposition
principle of uniformitarianism
stratification
stratigraphy
unconformity

Supplemental Readings and References

Adams, F. D. 1938. *The Birth and Development of the Geological Sciences.* Baltimore, The Williams & Wilkins Co. (Reprinted by Dover Press, 1954.)

Albritton, C. C., Jr. (ed.). 1963. *The Fabric of Geology.* San Francisco, Freeman, Cooper and Co.

Bailey, E. 1967. *James Hutton—The Founder of Modern Geology.* New York, American Elsevier Publishing Co., Inc.

Chamberlin, T. C. 1965. The method of multiple working hypotheses. *Science* 148:754–759. (Reprint of 1890 paper.)

Eiseley, L. C. 1959. Charles Lyell, *Sci. Am.* 201(8):98–106. (Offprint No. 846. San Francisco, W. H. Freeman Co., Publishers.)

Geikie, A. 1905. *The Founders of Geology* (2d ed.). New York, Macmillan. (Reprinted by Dover Press, 1954.)

Hsu, K. J. 1972. When the Mediterranean dried up. *Sci. Am.* 227(6):26–36.

Hudson, J. H., Shinn, E. A., Halley, R. B., and Lidz, B. 1976. Sclerochronology: a tool for interpreting past environments. *Geology* 4:361–364.

Siever, R. 1968. Science—observational, experimental, historical. *Am. Sci.* 56:70–77.

QUARTZ CRYSTALS
(Photograph courtesy of Wards Natural Science Establishment, Inc., Rochester, N.Y.)

2

Earth Materials

> One of the most generally useful sciences, and nearly indispensable to civilized society, is mineralogy or the natural history of minerals.
>
> Abraham Gottlob Werner, *On the External Characters of Minerals*, 1774

ATOMS AND MOLECULES

The Structure of Atoms

Rocks are composed of minerals, minerals are composed of atoms, and atoms are the basic building blocks of all matter. Knowledge of atoms is therefore necessary to understand the nature of all geologic materials. Such knowledge also later helps us to understand the methods by which rocks containing radioactive elements can be dated.

Atoms are the smallest particles of an element that can enter into a chemical reaction. They are also the smallest chemically indivisible particles of an element. An individual atom consists of an extremely minute but heavy nucleus surrounded by rapidly moving **electrons.** The electrons are relatively farther apart than are the planets surrounding our sun; consequently, the atom consists primarily of empty space. However, the electrons move so rapidly around the nucleus that they effectively fill the space within their orbits, giving size to the atom and repelling other atoms that may approach.

In the nucleus of the atom are closely compacted particles called **protons,** which carry a unit charge of positive electricity equal to the unit charge of negative electricity carried by the electron. Associated with the protons in the nucleus are electrically neutral particles having the same mass as protons. These are called **neutrons.** Modern atomic physics has made us aware of still other particles in the nucleus. For our understanding of the atom, however, knowledge of protons and neutrons is sufficient. The number of protons in the nucleus of an atom establishes its number of positive charges and is called its **atomic number.** Each chemical element is composed of atoms having a particular atomic number. Thus, every element has a different number of protons in its nucleus. In naturally occurring elements, the number may range from 1 proton in hydrogen (Table 2–1) to 92 in uranium.

Isotopes

The **mass** of an atom is approximately equal to the sum of the masses of its protons and neutrons. (The mass of electrons is so small that it need not be considered.) Carbon 12 is used as the standard for comparison of mass. By setting the atomic mass of carbon at 12, the atomic mass of hydrogen, which is the lightest of the elements, is just a bit greater than 1 (1.008, to be precise). The nearest whole number to the mass of a nucleus is called its **mass number.** Some atoms of the same

Element and Symbol	Atomic Number (Number of Protons in Nucleus)	Number of Neutrons in Nucleus	Atomic Mass	Electrons in Various Levels	Total Number of Electrons
Hydrogen (H)	1	0	1	1	(1)
Helium (He)	2	2	4	2	(2)
Carbon 12 (C)*	6	6	12	2–4	(6)
Carbon 14 (C)	6	8	14	2–4	(6)
Oxygen (O)	8	8	16	2–6	(8)
Sodium (Na)	11	12	23	2–8–1	(11)
Magnesium (Mg)	12	13	25	2–8–2	(12)
Aluminum (Al)	13	14	27	2–8–3	(13)
Silicon (Si)	14	14	28	2–8–4	(14)
Chlorine 35 (Cl)*	17	18	35	2–8–7	(17)
Chlorine 37 (Cl)	17	20	37	2–8–7	(17)
Potassium (K)	19	20	39	2–8–8–1	(19)
Calcium (Ca)	20	20	40	2–8–8–2	(20)
Iron (Fe)	26	30	56	2–8–14–2	(26)
Barium (Ba)	56	82	138	2–8–18–18–8–2	(56)
Lead 208 (Pb)*	82	126	208	2–8–18–32–18–4	(82)
Lead 206 (Pb)	82	124	206	2–8–18–32–18–4	(82)
Radium (Ra)	88	138	226	2–8–18–32–18–8–2	(88)
Uranium 238 (U)	92	146	238	2–8–18–32–18–12–2	(92)

Table 2–1 STRUCTURE OF SOME GEOLOGICALLY IMPORTANT ELEMENTS

* When two isotopes of an element are given, the most abundant is starred; for other elements, only the most abundant isotope is given. Note carefully that ordinary chemical atomic weights are not given; these are mixtures of isotopes and are therefore not whole numbers.

substance have different mass numbers. Such variants are called **isotopes.** Isotopes are two or more varieties of the same element that have the same atomic number and chemical properties but differ in mass numbers because they have a varying number of neutrons in the nucleus. For example, there are at least three different isotopes of oxygen (Table 2–2). These have atomic masses of 16, 17, and 18, respectively; oxygen 16 is the most abundant isotopic form in nature.

Energy Levels

The number of electrons surrounding the nucleus equals the number of protons in the nucleus. An atom, therefore, is electrically neutral. The electrons do not orbit randomly but are distributed in a way that determines how atoms combine to form molecules. This arrangement is sometimes visualized as consisting of a central nucleus surrounded by a number of electrons moving in definite orbits. It must be remembered, however, that this is just a way of thinking about the atom and is not a true picture of one. It has been possible to deduce the arrangement of the orbital electrons from x-ray studies of the atom, from atomic spectra, and from the atom's chemical behavior. Such deductions show that electrons are arranged so that specific numbers of them occur in each of various energy levels or "shells" (Fig. 2–1). The simplest of all atoms is hydrogen, which has one proton in the nucleus and one planetary electron. The helium atom has two planetary electrons close to the nucleus—a condition also present in the inner shell of heavier elements.

The greatest number of energy levels of even the most complex atoms is seven. Each of the first four levels may contain no more than (but may contain less than) the following number of electrons: 2 in the first level, 8 in the second level, 18 in

Table 2–2 STABLE ISOTOPES OF OXYGEN*

Isotope	Number of Protons	Number of Neutrons	Atomic Mass
^{16}O	8	8	16
^{17}O	8	9	17
^{18}O	8	10	18

* Isotopes of oxygen ("oxygen 16," "oxygen 17," and "oxygen 18") are written ^{16}O, ^{17}O, ^{18}O, respectively. The superscript designates the sum of the protons and neutrons.

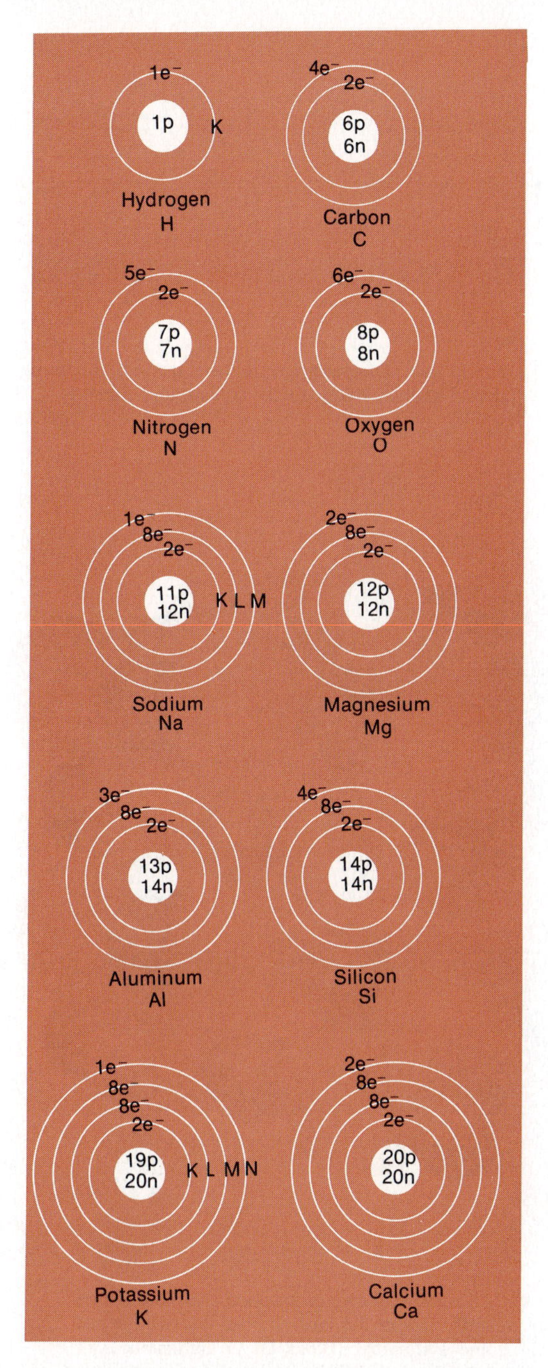

Figure 2–1 Distribution of electrons in some geologically important atoms. (These are not "pictures" of atoms, but merely diagrammatic representations showing the number of electrons in each level of the atom.)

the third level, and 32 in the fourth level. The relationship between the number of the level and the maximum number of electrons that a level can contain is thus 2 × (level number)^2. Although this relationship might hold true for the levels beyond the fourth, it cannot be established, because no element is known that has these outer levels entirely filled. Uranium, the heaviest naturally occurring element, has the following number of electrons in each level, beginning with the first: 2–8–18–32–18–12–2.

Another fact that has emerged from the study of the electronic structure of atoms is that those elements whose outermost electron shell contains eight electrons are chemically the most stable. Of course, for an atom like helium, which has only two electrons, stability is attained with those two electrons. In all cases of an outer ring containing fewer than eight electrons, the atom tends to gain or lose electrons to form stable structures. For example, the diagram of the chlorine atom shows that it has only seven electrons in its outer ring (Fig. 2–2). Chlorine is inclined to borrow one electron from sodium, thereby giving both atoms a stable configuration. Atoms like neon, argon, and krypton are called **noble** or **inert gases.** Their outer rings already contain the maximum quota of eight electrons, and they therefore tend to neither lend nor borrow electrons.

Combining Atoms

Ionic Bonding

The tendency for atoms to form a complete outer ring of electrons results in the combination of atoms into **molecules**—distinct groups of two or more atoms tightly bound together. The atoms tend to achieve a complete outer ring by the smallest possible change. Thus, an element having fewer than four electrons in its outer ring will tend to give up those electrons, whereas an element having more than four will tend to borrow electrons to complete its outer ring. In the combination of sodium and chlorine, it is easier for sodium to lend the single electron in its outer ring than to borrow seven. Upon lending this electron, the sodium atom becomes positively charged. Chlorine, which borrowed an electron, is now negatively charged. A charged atom—that is, an atom in which the number of protons is either more or less than the number of electrons—is called an **ion.** Because unlike charges of electricity attract, the positively charged sodium ion is drawn tightly to the negatively charged chlorine ion to form a molecule of common salt, NaCl (Fig. 2–3). This type of chemical union is referred to as **ionic bonding.** The example points up the fact that metals (like sodium) tend to be lenders of electrons, whereas nonmetals (like chlorine) are borrowers.

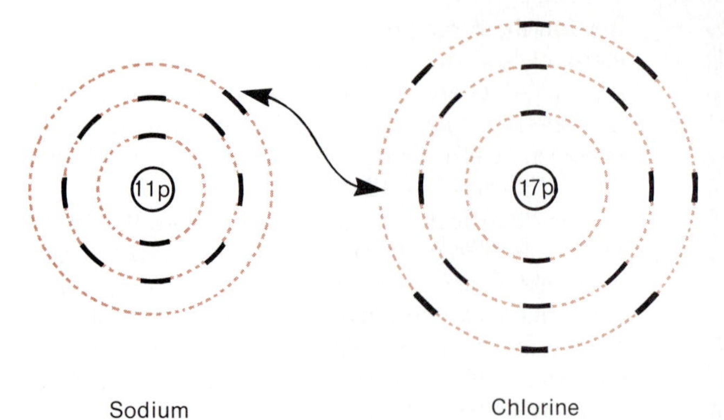

Figure 2–2 Diagrammatic representation of the formation of sodium chloride from sodium and chlorine.

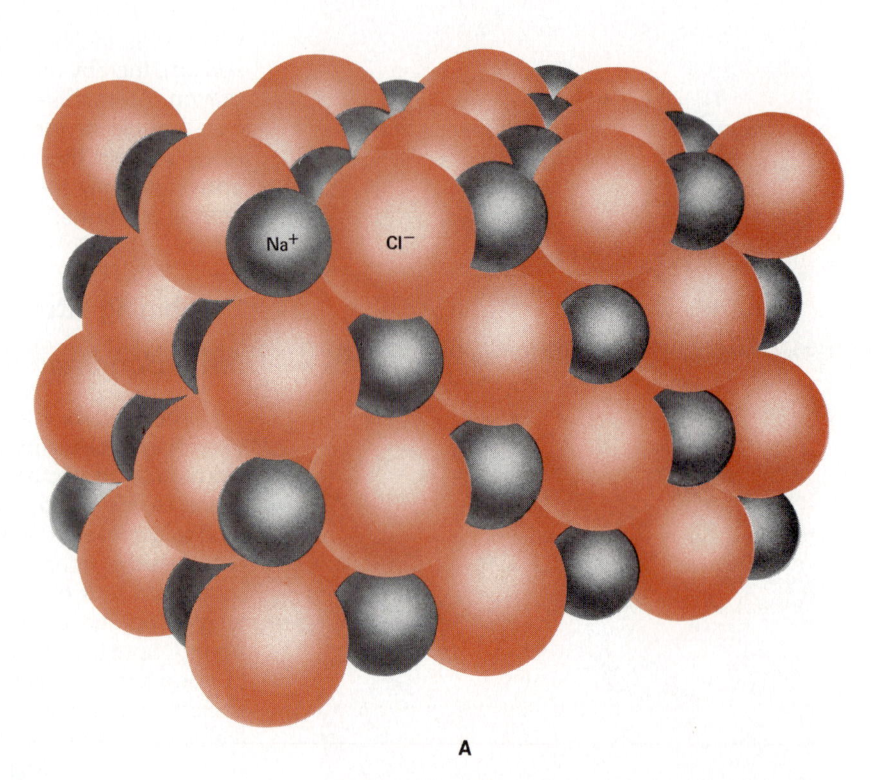

Na^+ ion

Cl^- ion

B

Figure 2–3 Structure of sodium chloride crystal. (a) Model showing relative sizes of the ions. (b) Ball-and-stick model showing cubic symmetry. (From D. O. Johnston, J. T. Netterville, J. L. Wood, and M. M. Jones. *Chemistry and the Environment.* Philadelphia, W. B. Saunders Co., 1973.)

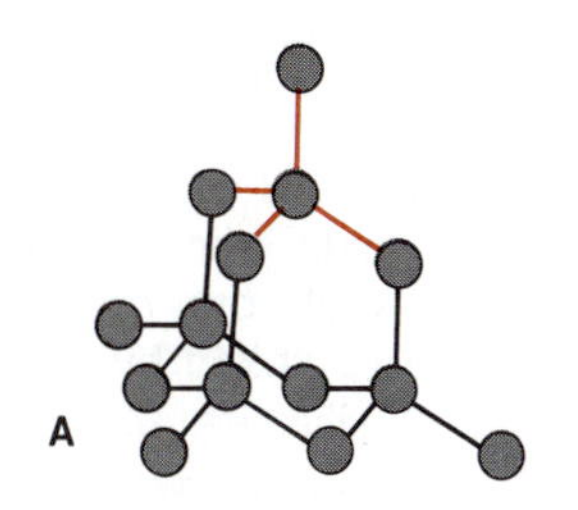

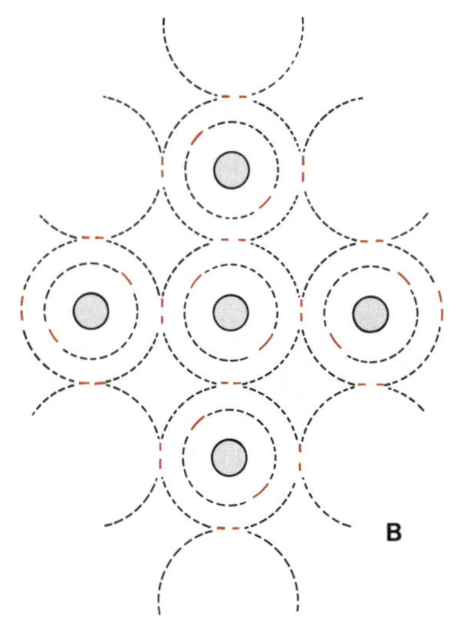

Figure 2–4 (A) The carbon atoms in a diamond are arranged in tetrahedrons consisting of four carbon atoms surrounding a central carbon atom. (B) A plan view of the atoms in a diamond. Note that each carbon atom, which individually has four electrons in its outer shell, shares one outer electron with each of its four neighbors. A stable outer shell of eight electrons is achieved by sharing pairs of electrons, and this type of bonding is termed covalent.

Covalent Bonding

Some elements contain half of, or nearly half of, the required number of electrons in their outer ring. Two examples are carbon (atomic number 6) and silicon (atomic number 14). Such atoms do not form ions and therefore act as electrical insulators. They tend to achieve the stable configuration of eight outer-shell electrons by *sharing* pairs of their electrons. In short, these atoms unite, not by borrowing or lending single electrons, but by sharing pairs of electrons that revolve around both nuclei and serve both atoms simultaneously. This manner of combining atoms is termed **covalent bonding.** A simple example of this type of combination is the bonding in diamond, in which each carbon atom is surrounded by four others (Fig. 2–4).

Size of Ions

We have seen that the electrical charge of ions is one factor in determining which atoms can be combined. Another is the *size* of the ions involved. The size, or **ionic radius,** of an ion depends upon the number of electron shells surrounding its nucleus and upon the charge on the nucleus. If an atom loses an electron and becomes a positive ion, the excess positive charge on the nucleus pulls the electron shells inward. The reverse occurs if an electron is gained. The radii of atoms and ions have been determined by using x-ray techniques to measure the distances between successive rows of atoms or ions in crystals.

Table 2–3 ABUNDANCES OF CHEMICAL ELEMENTS IN THE EARTH'S CRUST*

Element and Symbol	Percentage by Weight	Percentage by Number of Atoms	Percentage by Volume
Oxygen (O)	46.6	62.6	93.8
Silicon (Si)	27.7	21.2	0.9
Aluminum (Al)	8.1	6.5	0.5
Iron (Fe)	5.0	1.9	0.4
Calcium (Ca)	3.6	1.9	1.0
Sodium (Na)	2.8	2.6	1.3
Potassium (K)	2.6	1.4	1.8
Magnesium (Mg)	2.1	1.9	0.3
All other elements	1.5		
	100.0	100.0†	100.0†

* Based on B. Mason 1966. *Principles of Geochemistry.* New York, John Wiley & Sons, Inc.
† Includes only the first eight elements.

The number of ions that can be packed perfectly around another ion depends upon their respective ionic radii. For example, if an ion were the size of an orange, it could be surrounded by more grapefruit-sized ions than if it were the size of a cherry. Together, the electrical charges and the ionic radii will determine which elements can combine into molecules and minerals. Ionic radii will be influential in determining the internal atomic arrangement of molecules and the external form of crystals.

Order Among the Silicates

About 75 per cent by weight of the earth's crust is composed of the two elements oxygen and silicon (Table 2–3). For the most part, these elements occur in combination with such abundant metals as aluminum, iron, calcium, sodium, potassium, and magnesium to form an important group of minerals called the **silicates.** A single family of silicate minerals, the feldspars, comprises about one half of the material of the crust, and a single mineral species called **quartz** (see Fig. 2–8) represents a sizable portion of the remainder.

The fundamental unit in the crystal structure of silicates is called a **silicon tetrahedron** (Fig. 2–5), which consists of a compact tetrahedral arrangement of four oxygen ions around a central silicon ion. This pyramidal shape is a consequence of the very small size and high charge of the silicon ion and of the relatively large size of the oxygen ions. The silicon tetrahedron, however, is not an electrically neutral unit. The combining of four oxygen ions (each with two negative charges) and one silicon ion (with four positive charges) leaves the resultant tetrahedron with four unsatisfied negative charges. To correct the imbalance, the tetrahedral unit must either bond to one or more additional positive ions (such as magnesium or iron) or share oxygen atoms with neighboring tetrahedra. Of the abundant silicate minerals, only one, called *olivine,* has all of its tetrahedra linked by metallic ions. In the other abundant silicates, the tetrahedra are strongly linked by sharing oxygens. The pattern developed by the connected tetrahedra not only forms the atomic structure of the mineral but also determines many of its properties. The various silicate structural types are as follows:

1. SINGLE TETRAHEDRA STRUCTURE. These minerals are constructed of individual tetrahedra, the corners of which are linked by positive (metallic) ions. **Olivine,** a mineral common to the lavas of the Hawaiian Islands, has this structure.

2. SINGLE- OR DOUBLE-CHAIN STRUCTURE. In single-chain structure, each tetrahedron shares two corners of its base with two other tetrahedra; in the double chains, two single chains are combined by sharing mutual oxygen atoms (Fig. 2–6). Either kind of chain is bonded by positive metallic ions. The fibrous texture of the mineral **asbestos** is an obvious reflection of that mineral's basic chain structure.

3. SHEET STRUCTURE. Every tetrahedron shares three corners of its base with three other tetrahedra to form continuous, flat sheets of tetrahedra (Fig. 2–7). **Mica** is such a mineral, and it can be readily split into paper-thin sheets, thus reflecting its basic structure.

4. FRAMEWORK STRUCTURE. Each tetrahedron shares all four corners with other tetrahedra to form a continuous three-dimensional framework. **Quartz** is a mineral having a framework structure. Because the bonds are so strong in

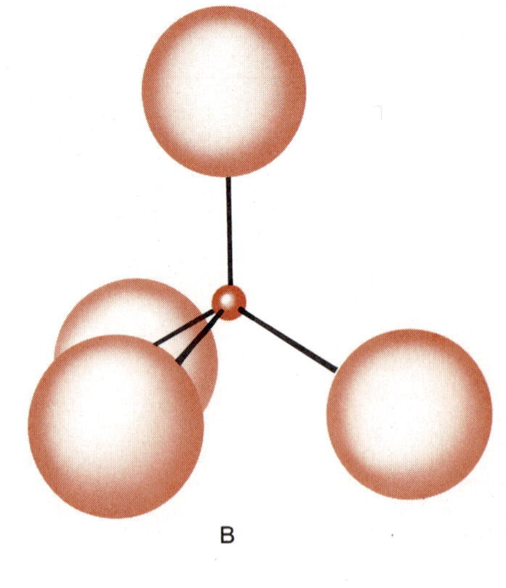

Figure 2–5 Two ways of representing the silicon-oxygen tetrahedron $(SiO_4)^{4-}$. The *packing model* in (A) represents in true scale the relative sizes and geometry of the tetrahedron. The *nuclear model* in (B) shows only the nuclei of the atoms but represents more clearly the angle of bonding and the relative positions of the nuclei.

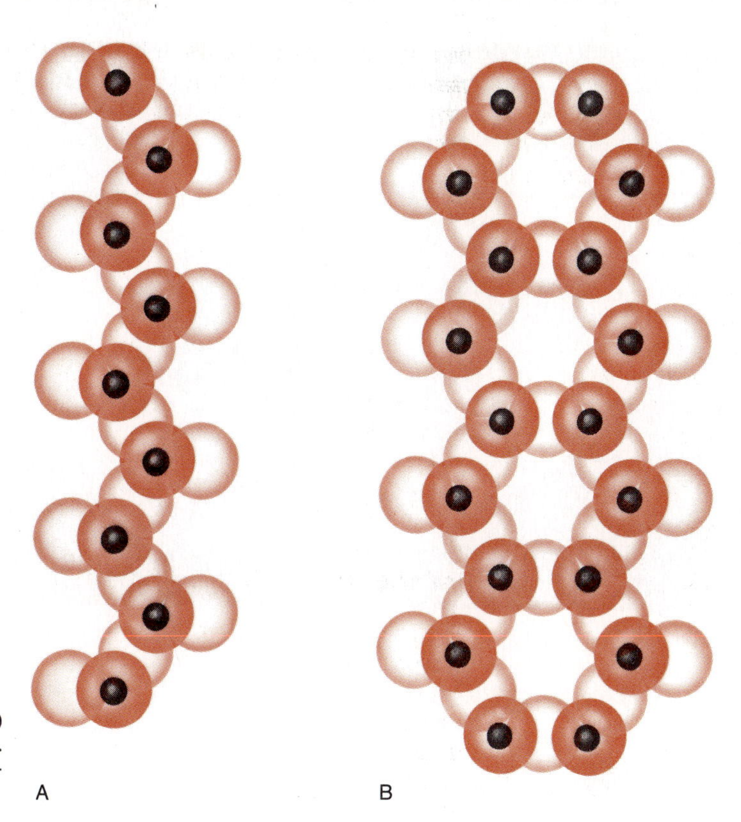

Figure 2–6 Single chain (A) and double chain (B) silicon-oxygen textrahedra viewed from above. The small silicon ions are located beneath the topmost oxygen atoms.

every direction in quartz, they produce a mineral that is exceptionally hard, with no tendency to break along preferred directions.

The silicate structures described here illustrate the way atoms are arranged in an orderly fashion to produce minerals and suggest that some of the more visible physical properties of minerals are reflections of their atomic structure. Thus, scaly minerals like mica split along parallel planes because they have tetrahedra arranged in sheets; quartz will not cleave smoothly in any direction because its tetrahedra form a continuous three-dimensional network. Olivine weathers more rapidly than quartz because if the iron that links the tetrahedra is chemically removed during weathering, the tetrahedra fall apart. The reason that diamond is heavier than graphite is that the carbon ions are more closely packed; in addition, graphite is opaque because the ions are spaced so far apart that light striking the graphite surface tends to be absorbed. No matter what the property of a mineral, its explanation can be found among its atoms.

MINERALS

A **mineral** may be defined as a naturally occurring element or compound formed by inorganic processes that has a definite chemical composition or range of compositions as well as distinctive properties and form that reflect its characteristic atomic structure. Minerals can usually be identified by such physical properties as color, hardness,

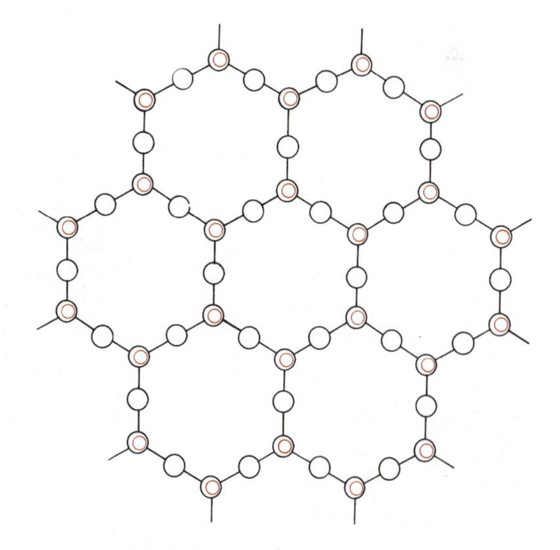

Figure 2–7 Lattice diagram of silicate sheet structure viewed from above. The small spheres represent silicon ions beneath oxygen ions.

Figure 2–8 Cavity in a rock that has become lined with quartz crystals. Such hollowed-out rocks are called *geodes.* (Photograph courtesy of Wards Natural Science Establishment, Inc., Rochester, N.Y.)

Figure 2–9 Quartz crystal. (Photograph courtesy of Wards Natural Science Establishment, Inc., Rochester, N.Y.)

Figure 2–10 Agate. (Photograph courtesy of Wards Natural Science Establishment, Inc., Rochester, N.Y.)

Table 2–4 COMMON ROCK-FORMING SILICATE MINERALS

Silicate Mineral	Composition	Physical Properties
Quartz	Silicon dioxide (Silica, SiO_2)	Hardness of 7 (on scale of 1 to 10);* will not cleave (fractures unevenly); specific gravity: 2.65
Potassium feldspar group	Aluminosilicates of potassium	Hardness of 6.0–6.5; cleaves well in two directions; pink or white; specific gravity: 2.5–2.6
Plagioclase feldspar group	Aluminosilicates of sodium and calcium	Hardness of 6.0–6.5; cleaves well in two directions; white or gray; may show striations on cleavage planes; specific gravity: 2.6–2.7
Muscovite mica	Aluminosilicates of potassium with water	Hardness of 2–3; cleaves perfectly in one direction, yielding flexible thin plates; colorless; transparent in thin sheets; specific gravity: 2.8–3.0
Biotite mica	Aluminosilicates of magnesium, iron, potassium, with water	Hardness of 2.5–3.0; cleaves perfectly in one direction, yielding flexible, thin plates; black to dark brown; specific gravity: 2.7–3.2
Pyroxene group	Silicates of aluminum, calcium, magnesium, and iron	Hardness of 5–6; cleaves in two directions at 83 and 93°; black to dark green; specific gravity: 3.1—3.5
Amphibole group	Silicates of aluminum, calcium, magnesium, and iron	Hardness of 5–6; cleaves in two directions at 56° and 124°; black to dark green; specific gravity: 3.0–3.3
Olivine	Silicate of magnesium and iron	Hardness of 6.5–7.0; light green; transparent to translucent; specific gravity: 3.2–3.6
Garnet group	Aluminosilicates of iron, calcium, magnesium, and manganese	Hardness of 6.5–7.5; uneven fracture, red, brown, or yellow; specific gravity: 3.5–4.3

* The scale of hardness used by geologists was formulated in 1822 by Frederich Mohs. Beginning with diamond as the hardest mineral, he arranged the following table:

10 Diamond	5 Apatite
9 Corundum	4 Fluorite
8 Topaz	3 Calcite
7 Quartz	2 Gypsum
6 Feldspar	1 Talc

density, crystal structure, and cleavage. There are more than 2000 mineral species that have been discovered and scientifically described, but most of these are rarely encountered. For our present purposes, it is important to consider only those minerals that compose the bulk of common rocks or that are particularly useful in interpreting ancient environments. These minerals fall into the following general categories.

Important Rock-forming Minerals

Quartz

The principal silicate minerals and some of their properties are listed in Table 2–4. The mineral **quartz** (SiO_2) is one of the most familiar and important of all the silicate minerals (Fig. 2–8). It is common in many different families of rocks. As mentioned earlier, quartz represents the ultimate in cross-linkage of silica tetrahedra; it therefore will not break along smooth planes. However, the tetrahedra are joined only at the corners and in a relatively open arrangement. It is thus not a *dense* mineral, but it is quite *hard* because of its framework structure. Indeed, it will easily scratch glass. When quartz crystals are permitted to grow in an open cavity, they frequently develop the hexagonal prisms topped by pyramids that are prized by crystal collectors (Fig. 2–9). More frequently, the crystal faces cannot be discerned because orderly addition of atoms is interrupted by contact with other growing crystals and results in an aggregate that exhibits crystalline texture. Such common minerals as **chert, flint, jasper,** and **agate** (Fig. 2–10) are sedimentary varieties of quartz. *Chert* is a dense, hard, white mineral composed of submicrocrystalline quartz that forms as a result of the precipitation of silicon dioxide by either biologic or chemical means. *Flint* is the popular name for a dark gray or black variety of chert in which the dark color results from inclusions of organic matter. *Jasper* is an opaque variety of quartz and is reddish, greenish, brown, or yellow in color. *Agate* is a variety of quartz that exhibits distinct banding.

The origin of chert is a complex problem made even more difficult by the fact that different varieties are formed by somewhat different processes. Some cherts are replacements of earlier carbonate rocks. Others appear to have formed as a result of the solution and reprecipitation of silica from the siliceous skeletal remains of organisms. Small amounts of chert may also be precipitated directly from concentrated aqueous solutions.

The Feldspars

The **feldspars** are the most abundant constituents of rocks, composing about 60 per cent of the total weight of the earth's crust. There are two major families of feldspars: the **potassium feldspar group,** which constitutes the potassium aluminosilicates, and the **plagioclase group,** which comprises the aluminosilicates of sodium and calcium (Fig. 2–11). Members of the plagioclase group exhibit a wide range in composition—from a calcium-rich end member called *anorthite* ($CaAl_2Si_2O_8$) to a sodium-rich end member called *albite* ($NaAl_2Si_3O_8$). Between these two extremes, plagioclase minerals containing both sodium and calcium occur. The substitution of sodium for potassium, however, is not random but rather is governed by the temperature and composition of the parent material. Thus, by examining the feldspar content of a once molten rock, it is possible to infer

Figure 2–11 Rock composed of intergrown plagioclase crystals (anorthite), which are white, and black pyroxene crystals. (Photograph courtesy of Wards Natural Science Establishment, Inc., Rochester, N.Y.)

Figure 2–12 Mica, showing characteristic cleavage along one directional plane. (Photograph courtesy of Wards Natural Science Establishment, Inc., Rochester, N.Y.)

the physical and chemical conditions under which it originated. Feldspars are nearly as hard as quartz and range in color from white or pink to bluish-gray. They have good **cleavage** (break along smooth planes) in two directions, and the resulting flat, often rectangular, surfaces are useful in identification. Those plagioclase feldspars with abundant sodium tend to characterize silica-rich rocks like granite, whereas calcium-rich plagioclases occur in rocks like the Hawaiian lavas (called *basalts*).

Mica

Mica (Fig. 2–12) is a silicate mineral easily recognized by its perfect and conspicuous cleavage along one directional plane. The two chief varieties are the colorless or pale-colored **muscovite** mica, which is a hydrous potassium aluminum silicate, and the dark-colored **biotite mica,** which also contains magnesium and iron. Muscovite is a common rock-forming silicate.

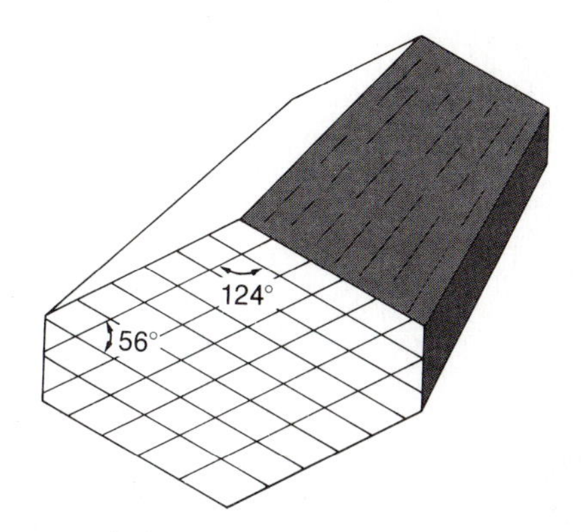

Figure 2–13 Typical shape of a hornblende crystal. Planes of cleavage intersect at 56° and 124°.

Amphiboles and Pyroxenes

The **amphibole** group comprises silicates of aluminum, calcium, magnesium, and iron. Silica tetrahedra in this mineral family are linked into double chains. The most common amphibole is *hornblende.* Hornblende is a black to dark green mineral. Crystals tend to be long and narrow. Two good cleavages are developed parallel to the long axis, and these intersect each other at angles of 56° and 124° (Fig. 2–13).

Another family of silicate minerals resembling the amphiboles in physical and chemical properties is the **pyroxenes.** These are silicates with abundant calcium, magnesium, and iron, as well as aluminum, titanium, and sodium. A representative pyroxene is *augite.* Typically, augite crystals are rather stubby in shape, with good cleavages developed along two planes that are nearly at right angles (87° and 93°) to one another (Fig. 2–14). Thus the cross-section of the crystal appears nearly square. Unlike hornblende, which has a double-chain silicate structure, augite is constructed of single chains. This difference accounts for augite's having differently shaped cleavage pieces.

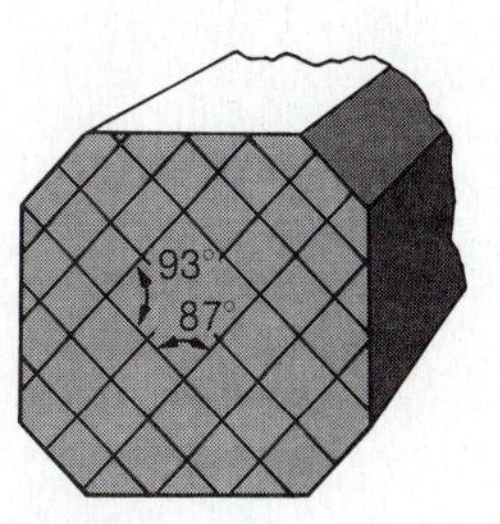

Figure 2–14 Typical shape of an augite crystal. Planes of cleavage intersect at 87° and 93°.

Olivine

Olivine (Fig. 2–15) has been mentioned earlier as a silicate mineral in which oxygen ions in the tetrahedra are not shared. Instead, tetrahedra are strongly bonded by iron or magnesium ions. Olivine can range in composition from iron silicate to magnesium silicate; in nature, however, most olivines contain both magnesium and iron and are designated by the formula $(MgFe)_2SiO_4$. The substitution of magnesium for iron in the olivine structure is not random but is controlled by temperature and the availability of iron and magnesium. Most commonly, olivine appears as a greenish, glassy, granular mass. Olivines, pyroxenes, amphiboles, and biotite micas together constitute a large group of blackish or greenish, iron- and magnesium-rich minerals called **ferromagnesians.**

The Clay Minerals

The **clay minerals** are silicates of hydrogen and aluminum with additions of magnesium, iron, and potassium. Like the micas, they have sheet structure (Fig. 2–16). However, because individual flakes are extremely small, they require the magnification of the electron microscope for adequate examination. The sheets of tetrahedra are weakly bonded, and the potassium, sodium, and magnesium ions may enter and leave exchangeable sites in the atomic structure. Unlike many of the silicates discussed previously, clays form as a result of weathering of other aluminum silicate minerals. As a group, clays are known by their amorphous form, softness, low density, and plasticity when wet. In order to identify individual species of clay minerals, however, it is necessary to use x-ray diffraction techniques.

Nonsilicate Minerals

Approximately 3 per cent of the earth's crust is composed of nonsilicate minerals. Among the most widespread of these are the carbonates and evaporites.

Calcite and **dolomite** are the two most important members of the carbonate mineral group. Calcite ($CaCO_3$), which is the main constituent of limestone and marble, forms in many ways. It is secreted as skeletal material by certain invertebrate animals, precipitated directly from sea water, or formed as dripstone in caverns. Calcite is easily recognized by the rhombohedron-shaped cleaved fragments (Fig. 2–17) and by the fact that an application of hydrochloric acid on its surface will cause effervescence. Dolomite [$CaMg(CO_3)_2$] is a carbonate of calcium and magnesium that occurs in extensive layers of a rock called **dolostone.** In the field, geologists often distinguish dolomite by the fact that it will not effervesce in dilute hydrochloric acid unless it is powdered. Many ancient dolostones are thought to have been formed by recrystallization of still older limestones. Because calcium and magnesium ions are nearly the same

Figure 2–15 Granular mass of olivine (Photograph courtesy of Wards Natural Science Establishment, Inc., Rochester, N.Y.)

Figure 2–16 Electron micrograph of the clay mineral kaolinite. The flaky, stack-of-cards character of the clay crystals is a manifestation of their silicate sheet structure. Magnified about 2000 times. (Courtesy of Kevex Corporation.)

Figure 2–17 Calcite, showing the characteristic rhombohedral shape of cleavage pieces. (Photograph courtesy of Wards Natural Science Establishment, Inc., Rochester, N.Y.)

Figure 2–18 Halite. (Photograph courtesy of Wards Natural Science Establishment, Inc., Rochester, N.Y.)

size, it is possible for the magnesium to replace calcium in the atomic structure.

Evaporites are minerals that have precipitated from bodies of water subjected to intense evaporation. They include common rock salt (Fig. 2–18), or **halite** (NaCl), and **gypsum** ($CaSO_4 \cdot 2H_2O$).

We have noted that minerals are composed of atoms. Particularly abundant in the crust are the atoms of only eight elements (Table 2–1): oxygen, silicon, aluminum, iron, calcium, sodium, potassium, and magnesium. The minerals that commonly occur in the crust are also few in number. They include the seven silicate minerals or mineral groups, the carbonates, and the evaporites.

Minerals as Clues to Rock History

Minerals are studied by geologists not as mere constituents of the crust but as clues to the history of the rocks in which they occur. Some minerals develop exclusively in ocean water and provide the geologist with evidence that a particular sediment was deposited in the sea rather than in a freshwater lake. A thick bed of *halite* indicates aridity and evaporation so extreme that the brine had become ten times saltier than ordinary sea water. The magnetic properties of *magnetite,* an oxide of iron, can in certain situations provide clues to the position of continents relative to the earth's magnetic poles. Certain minerals form within a narrow range of conditions and can therefore be used to diagnose the pressures and temperatures involved in the formation of crustal rocks and mountains. *Diamonds,* for example, form only at high temperatures and extremely high pressures. Like diamond, *graphite* is a crystalline variety of carbon. However, it forms at lower temperatures and pressures. Other minerals contain radioactive isotopes that permit us to know the age of the parent rocks. By their size, crystals of feldspar give the petrologist information about how some ancient molten mass congealed. By their shape, grains of *quartz* contain a history of what has happened to them since they were eroded from some ancient granitic terrain. The buried mineral products of weathering may provide information about past climatic conditions in a region.

THE FAMILIES OF ROCKS

Rock Conversions

We have noted that minerals can provide information about the environment in which they formed. With study it is therefore possible to know the origin of the rocks containing those minerals. Geologists are agreed on a fundamental division of rocks into three great families according to difference of origin. **Igneous rocks** are those that have cooled from a molten state. **Sedimentary rocks** consist of materials derived from other rocks and deposited by water, ice, or wind. **Metamorphic rocks** are any that have been changed from previously existing rocks by the action of heat, pressure, and associated chemical activity. The changes may include a recrystallization of the previous minerals or growth of entirely new minerals.

In regard to the three major groupings of rocks, it is important to remember that they are not immutable. The earth's crust is dynamic and ever-changing. Any sedimentary or metamorphic rock may be partially melted to produce igneous rocks, and any previously existing rock of any category can be compressed and altered during mountain-building to produce metamorphic rocks. The weathered and eroded residue of any family of rock can be observed today being transported to the sea for deposition and conversion into sedimentary rocks. These changes can be incorporated into a schematic diagram that is designated the "rock cycle" (Fig. 2–19).

Although rocks are classified into groups that have had similar origin, the identification of rocks is not based on origin, but on description. For identification it is necessary to know general appearance as well as chemical and physical properties. **Texture** (size, shape, and arrangement of constituent minerals) and **mineral composition** are essential for identification. Inferences regarding the origin of the rock are based on geologic observations and experimentation.

Igneous Rocks

Igneous rocks constitute over 90 per cent by volume of the earth's crust, although their great abundance may go unnoticed because they are extensively covered by sedimentary rocks. Much of our mountain scenery is sculptured in igneous rocks formed long ago. Current volcanic activity (Fig. 2–20) provides an often spectacular reminder of the processes that produce igneous (from the Latin *ignis,* meaning "fire") rocks (Fig. 2–21). It is an appropriate name for rocks that develop from cooling masses of molten material derived from exceptionally hot parts of the earth's interior. **Magma** is the term used to describe this mixture of molten silicates and gases while it is still beneath the surface. If it should penetrate to the surface, it loses most of its gases and vapors and becomes **lava** (Fig. 2–22).

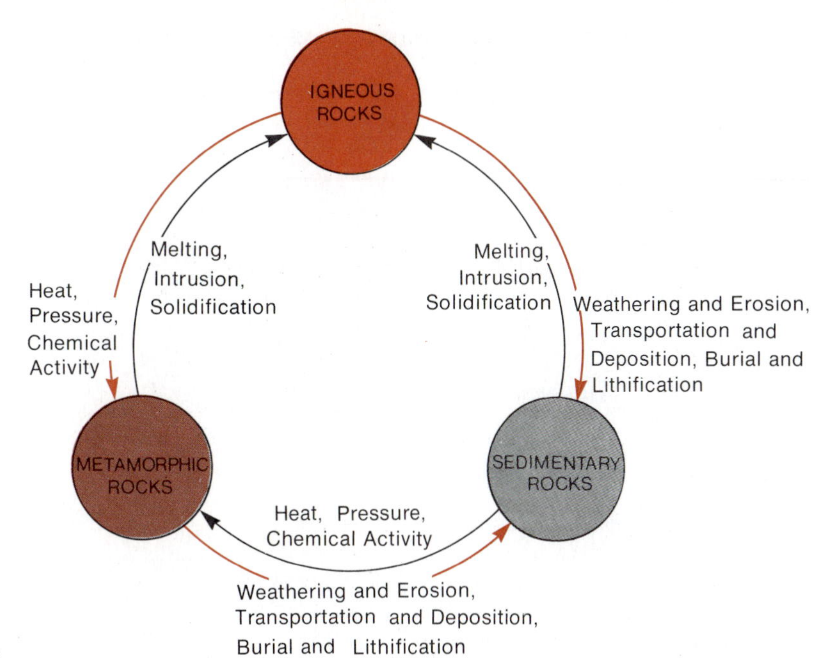

Figure 2–19 The rock cycle.

Cooling History of Igneous Rocks

Intrusive, or **plutonic, igneous rocks** are those that have congealed from magma once located deep beneath the surface. Large masses of such rocks are sometimes called *plutons.* Their presence at the surface of the earth results from crustal uplift and erosional removal of the overlying rocks. **Extrusive,** or **volcanic, igneous rocks** are derived from lava and harden at the surface of the earth. The grain size of igneous rocks is an index to their history of cooling. Magmas lose heat slowly and retain water, tending to inhibit profusive formation of crystal nuclei. Thus, there is time and space for the growth of larger crystals around fewer nuclei. In typical intrusive rocks like granite, diorite, and gabbro (Table 2–5), the intergrown crystals are large enough to be readily seen without magnification. In contrast, extrusive igneous rocks have a **finer** texture in which crystals are too small to be seen with the unaided eye. The structure of such rocks reflects sudden chilling of molten silicates as they were ejected at the surface of the earth. In lavas, water vapor is quickly lost to the atmosphere,

Table 2–5 COMMON IGNEOUS ROCKS

Coarsely Crystalline Intrusive Igneous Rocks	Approximate Extrusive Equivalents	Common Silicate Mineral Components	Average Specific Gravity
Granite	Rhyolite	Quartz Potash feldspar Sodium plagioclase (Minor biotite, amphibole, magnetite)	2.7
Diorite	Andesite	Sodium-calcium plagioclase Amphibole, biotite (Minor pyroxene)	2.8
Gabbro	Basalt	Calcium plagioclase Pyroxene Olivine (Minor amphibole, ilmenite)	3.0

Figure 2–20 Aerial view of erupting Mount St. Helens volcano in southwestern Washington State, May 18, 1980. The plume of steam and ash rose over 20,000 meters above the crater rim. (Photograph courtesy of U.S. Department of the Interior, Geological Survey.)

crystal nucleation is rapid, and the melt begins to form a solid before there is sufficient time to grow larger crystals. Such an extrusive rock is *basalt,* composed of ferromagnesian minerals and tiny rectangular grains of plagioclase feldspars. *Obsidian* is an extrusive rock that cooled so rapidly that there was insufficient time for crystallization; the melt therefore froze into a glass.

If coarsely crystalline igneous rocks indicated slow cooling, and finely crystalline ones indicate rapid cooling, then what would be the cooling history of a rock with large crystals immersed in a very fine-grained matrix? Such rocks are said to have **porphyritic texture.** The large crystals *(phenocrysts)* were formed slowly at depth and were then swept upward and incorporated in the lava as it hardened at the surface.

Composition of Igneous Rocks

The mineral composition of an igneous rock provides insight into the amounts and kinds of different ions in the parent magma. Altogether, in igneous rocks there are only about eight elements that are abundant: oxygen, silicon, aluminum, calcium, iron, magnesium, sodium, and potassium. These combine in specific ways to form the feldspars, ferromagnesian minerals, micas, and quartz minerals that are the constituents of igneous rocks.

For many years, geologists have recognized that intrusive igneous rocks tend to be richer in silica than are extrusive rocks. Later studies indicated that this relatively greater amount of silica served to keep the melt below the surface by increasing *viscosity* (resistance to flow). Even while still within

Figure 2–21 Lava flow entering ocean. Kilauea Volcano, 1970. (Photograph courtesy of D. A. Swanson, U.S. Geological Survey.)

Figure 2–22 Solidified ropy lava, or "pahoehoe," forms a drapery-like structure. Hawaii, Kilauea Volcano. (Photograph courtesy of D. A. Swanson, U.S. Geological Survey.)

Figure 2–23 Granite. (Photograph courtesy of Wards Natural Science Establishment, Inc., Rochester, N.Y.)

the melt, silica tetrahedra develop and tend to align themselves into chains and rings that thicken the melt and retard its upward progress.

Granite (Fig. 2–23) is the most abundant silica-rich intrusive rock. It is derived from melts so rich in silica that after all linkages with metallic ions are satisfied, there is still enough silicon and oxygen remaining to form quartz grains within the rock. It is true that there are silica-rich extrusive rocks as well (i.e., *rhyolite*), but they are far less abundant than extrusives containing lower percentages of silica. Most high-silica rocks are light-colored.

Basalt (Fig. 2–24) is a fine-grained extrusive rock derived from a low-silica melt. Basaltic lavas have been observed to flow with the approximate consistency of motor oil, a fact accounting for the frequency with which it is found at the earth's surface. Since the silica percentage is low, quartz grains are rarely in basalt.

There is one final important relationship in the chemistry of magmas and lavas. The silica-rich melts also tend to include ample quantities of potassium and sodium and therefore yield crystals of potassium feldspar, sodium plagioclase, and mica, along with the ubiquitous quartz. Calcium, magnesium, and iron are rather minor constituents of silica-rich magmas, but they increase in abundance in rocks deficient in silica. The low-silica rocks utilize these elements in the formation of calcium plagioclases and ferromagnesian minerals. Because of the ferromagnesians and grayish plagioclase minerals, low-silica rocks tend to be dark gray, black, or green.

Figure 2–24 Basalt dike cutting across a rhyolitic rock mass. Silver Mines, Missouri.

Volcanic Activity

Although deciphering the history of a granitic mass is certainly intellectually stimulating, it is unlikely to evoke the feelings of awe and excitement one experiences when viewing volcanic activity. Volcanoes are basically vents in the earth's surface through which hot gases and molten rock flow from the earth's interior. The extrusions may be quiet or explosive. Quiet eruptions are exemplified by the Hawaiian volcanoes and are frequently characterized by truly enormous outpourings of low-viscosity lava (Fig. 2–25). The lava spreads widely to form the gentle slopes of a "shield volcano." Explosive eruptions are caused by the sudden release of molten rock driven upward by large pockets of compressed gases. Such explosive eruptions can literally blow the volcanic cone to bits. The catastrophic eruption of Krakatoa in 1883 was heard 5000 kilometers away and was responsible for the death of 36,000 humans. Fortunately for our species, most eruptions are not so violent.

There are, of course, all stages of volcanic activity between quiet and violently explosive. Perhaps in response to changes in the composition of the parent melt, some volcanoes have even been known to shift from one type of activity to another. Volcanic activity also includes successive outpourings of lava from great fissures so as to form lava plateaus that extend over thousands of square kilometers. Such fiery floods long ago produced the Columbia and Snake River Plateau as well as the Deccan Plateau of India.

By far, the most abundant kind of volcanic rock is basalt. It underlies the ocean basins, has been built into midoceanic ridges, and has accumulated sufficiently in places like Hawaii and Iceland to have produced substantial land area.

How and at what depth did this great volume of basalt originate? To answer this question it is necessary to refer briefly to a model of the earth's interior that has been formulated by the study of earthquake waves. The model depicts the earth's basaltic **crust** as a thin zone about 6 km thick and overlying the **mantle** of denser olivine- and pyroxene-rich rocks. The boundary between the crust and the mantle is recognized by an abrupt change in the velocity of earthquake waves as they travel downward into the earth. For many years geologists believed that basaltic lavas originated from the lower part of the basaltic crust. However, several recent lines of evidence suggest that the basaltic lavas may have come from molten pockets of upper mantle material. For example, present-day volcanic activity is closely associated with deep earthquakes that occur within the mantle far be-

Figure 2–25 Photograph taken at night of lava falls cascading over a ridge during the eruption of Kilauea. (Photograph courtesy of D.A. Swanson, U.S. Geological Survey.)

neath the crust. It is quite likely that fractures produced by these earthquakes could serve as passages for the escape of molten material to the surface. A detailed study of earthquake shocks from particular volcanic eruptions in Hawaii indicates that the erupting lavas were derived from pockets of molten material within the upper mantle at depths of about 100 km. A weak plastic zone (called the "low velocity zone") in the mantle appears to represent the level at which the lavas originated. The mechanism by which they developed is called partial melting. **Partial melting** is that general process by which a rock subjected to high temperature and pressure is partly melted and the liquid component is moved to another location. At the new location, the separated liquid may solidify into rocks having different composition from the parent mass. The word "partial" in the expression "partial melting" refers to the fact that some minerals melt at lower temperatures than others, and so for a time the material being melted resembles a hot slush composed of liquid and still solid crystals. The molten fraction is usually less dense than the solids from which it was derived and thus tends to separate from the parent mass and work its way toward the surface. In this way melts of basaltic composition separated from denser rocks of the upper mantle and eventually made their way to the surface to form volcanoes.

Many complex and interrelated factors control where in the mantle partial melting may occur or even if it will occur at all. Generally, heat in excess of 1500°C is required, but the precise temperature for melting is also influenced by pressure and the water content of the rock. As pressure increases, the temperature at which particular minerals melt also rises. Thus a rock that would melt at 1000°C near the surface will not melt in deeper zones of higher pressure until it reaches far greater temperatures. Water has an effect opposite to that of pressure, for its presence will allow a rock to start melting at lower temperatures and shallower depths than it would have otherwise. Laboratory experiments indicate that the melting of "dry" mantle rock can occur at depths of about 350 km but that the presence of only a little water can cause partial melting and yield basaltic liquid from depths as shallow as 100 km.

Not all lavas found at the earth's surface are basaltic. Volcanoes of the more explosive type that are located at the edge of continents around the Pacific and in the Mediterranean extrude a lava called **andesite.** Andesite contains more silica than basalt, and its lava is thus more viscous. This greater resistance to flow contributes to the gas containment that precedes explosive volcanic activity. Andesites are considered to be intermediate in silica content between the rocks of the continental crust and those of the oceanic crust of the earth.

Andesitic rocks may originate in more than one way. Some emplacements result from originally basaltic magmas in which minerals like olivine and pyroxene form early and settle out, thus leaving the remaining melt relatively richer in silica. This process is called **fractional crystallization.** Evidence for this mode of origin is provided by Iceland's volcanoes. Iceland is a volcanic island formed on oceanic basaltic crust. It has been observed that the longer the quiet period between eruptions of Icelandic volcanoes, the more siliceous is the lava that is extruded. Apparently, longer periods of quiescence provide time for fractional crystallization and settling.

Andesitic melts are also believed to form by partial melting of mantle materials in the presence of some water. They may also result when a silica-rich older rock is assimilated by a basaltic magma. In addition to these theories, it has been suggested that andesites may result from the melting of oceanic crust and siliceous marine sediments as they descend into hot zones of the mantle. The wet, silica-enriched melts of andesitic composition might then rise buoyantly and erupt along volcanic island arcs. In the chapter dealing with plate tectonics (Chapter 8), we will examine this theory more fully.

Metamorphic Rocks

All Things Change

Sir Charles Lyell recognized that igneous or sedimentary rocks, if subjected to high temperature, pressure, and the chemical action of solutions and gases, can be altered to quite different kinds of rocks. Lyell borrowed the term *metamorphism* (from the Latin *metamorphosis,* meaning "change of form") to describe this process. It is still used today to describe alterations in rocks brought about by physical or chemical changes in the environment that are intermediate between those that result in igneous rocks and those that produce sedimentary rocks. Any previously existing rock may be converted to a metamorphic rock, and the changes primarily involve recrystallization of minerals in the rock while it remains in the solid state. In the process of recrystallization, the textural characteristics of the parent rock may be changed while at the same time new minerals develop that are stable under the new conditions of pressure and temperature. New elements need not be introduced; instead, those that are already present are incorporated into different and often denser minerals. Variations in heat and pressure may result in different kinds of metamorphic rocks, even from the same parent material.

Most metamorphic rocks exhibit a layering called **foliation,** which results from the parallel alignment of mineral grains. Whether this foliation is very fine or coarse depends upon the size and shapes of the constituent minerals. A few metamorphic rocks (*marble* is a familiar example) do not develop foliation.

Metamorphism

Alterations of rock immediately adjacent to igneous intrusions constitute **contact metamorphism.** The changes that occur in the intruded rock are largely the result of high temperatures and the emanation of chemically active vapors that accompany igneous intrusions. Such factors as the size of the magmatic body, its composition and fluidity, and the nature of the intruded rock also influence the kind and degree of contact metamorphism. Important ore deposits are commonly situated in the metamorphosed rock surrounding the intrusives. Examples of such deposits include magnetite and copper ores in metamorphic zones around granite intrusives in the Urals, central Asia, the Appalachian Mountains, Utah, and New Mexico.

Regional or **dynamothermal metamorphism** is a type of rock alteration that is areally extensive and occurs under the conditions of great confining pressures and heat accompanying deep burial and mountain building. In a subsequent chapter, we will discuss how rocks deposited in crustal troughs adjacent to continents may be compressed into mountain systems and thus be regionally metamorphosed. Metamorphic "index minerals" known to form under specific temperature and pressure conditions are used to decipher the history of growth of these ancient mountainous regions, even when only the roots of the ranges remain.

Kinds of Metamorphic Rocks

As any rock can be metamorphosed in a number of different ways, there are hundreds of different kinds of metamorphic rocks. However, for our purposes, we need only consider several that occur extensively at the earth's surface. It is convenient to divide metamorphic rocks into two groups based on the presence or absence of foliation.

FOLIATED METAMORPHIC ROCKS

Slate. In slate, the foliation is microscopic and caused by the parallel alignment of minute flakes of silicates with sheet structure (Fig. 2–26). The planes of foliation are quite smooth, and the rock may be split along these planes of "slaty cleavage." The planes of foliation may lie at any angle to the bedding in the parent rock. Slate is derived from the regional metamorphism of shale.

Phyllite. The texture in phyllite is also very fine, although some grains of mica, chlorite, garnet, or quartz may be visible. Phyllite surfaces often develop a wrinkled aspect and are more lustrous than slate. Phyllite represents an intermediate degree of metamorphism between slate and schist. The parent rocks are commonly shale or slate.

Schist. The platy or needle-like minerals in schist are sufficiently large to be visible to the unaided eye; the minerals tend to be segregated into distinct layers. Schists are named according to the most conspicuous mineral present. Thus, there are mica schists, amphibole schists, chlorite schists, and many others. Shales are the usual parent rocks for schists, although some are derived from fine-grained volcanic rocks.

Gneiss. This is a coarse-grained, evenly granular rock. Foliation results from segregation of minerals into bands rich in quartz, feldspar, biotite, or amphibole. Foliation is coarse and appears less distinct than in schist. High-silica igneous rocks and sandstones are the usual parent rocks for gneisses.

NONFOLIATED METAMORPHIC ROCKS

Marble. A fine to coarsely crystalline rock, marble is composed of calcite or dolomite and therefore is relatively soft. (It can be scratched with steel.) Marble is derived from limestone or dolostone.

Quartzite. A fine-grained, often sugary-textured rock, quartzite is composed of intergrown quartz and therefore is very hard. Rock will break through constituent grains; it may be any color. Quartzite is derived from quartz sandstone.

Greenstone. A dark green rock, greenstone has a texture so fine that mineral components, except for scattered larger crystals, cannot be seen without magnification. It is derived by the low-grade metamorphism of low-silica volcanic rocks.

Figure 2–26 Specimen of slate from outcrop near Bangor, Pennsylvania (Photograph courtesy of Wards Natural Science Establishment, Inc., Rochester, N.Y.)

Granitization

Before leaving the subject of metamorphic rocks, we must mention a metamorphic aspect of **granite.** In the previous section, we regarded granite as having formed from a silicate melt in the usual igneous manner. For a good many years, geologists were fully satisfied with that explanation. They could find many places where the contacts between the granite and the older rock were sharp, showed the effects of being baked by a hot intruding mass, and contained pieces or inclusions of the intruded rock that appeared to have "fallen into" the granite. Clearly, such granites were once molten. However, not all granite contacts were so clearly intrusive. Rock exposures were found in which true granites seemed to merge imperceptibly into rocks of metamorphic character and ultimately into sedimentary rocks. The granites seemed to have been converted from these older rocks in the course of an episode of intense regional compression. Recent studies of the isotopic composition of granites, as well as detailed field work, may provide an explanation for the two kinds of granite occurrence. One begins with a thick wedge of sediment that is compressed by mountain-building forces. Near the base of the wedge, sedimentary rocks are converted to rocks of gneissic character. Enough partial melting occurs that a film of silicate melt moves upward, surrounds older grains, and gradually converts the entire body into granite without ever completely melting it. The result is a granite with gradational contacts. This process of converting solid rocks to granites without causing them to become a magma has been termed **granitization.** Higher in the deformed wedge, the films and streams of silicate liquid come together to form the classic intrusive kinds of granite with sharp contact relationships.

Historical Significance of Metamorphic Rocks

We have noted that the conditions for metamorphism are developed in regions that have been subjected to intense compressional deformation. Such regions of the earth's crust either now have, or once had, great mountain ranges. Thus, where large tracts of low-lying metamorphic terrain are exposed at the earth's surface, geologists conclude that crustal uplift and long periods of erosion leveled the mountains. Metamorphic rock exposures at many localities across the eastern half of Canada represent the truncated stumps of ancient mountain systems.

From studies of the mineralogic composition of metamorphic rocks, it is often possible for geologists to reconstruct the conditions under which the rocks were altered and then to make inferences about the directions of compressional forces, pressures, temperatures, and the nature of parent rocks. Investigators are aided in these studies by the knowledge that specific metamorphic minerals form and are stable within finite limits of temperature and pressure. Maps of **metamorphic facies,** or zones of rocks that formed under specific conditions, can be constructed. Commonly, such maps delineate broad bands of metamorphic rocks, each of which formed under sequentially more intense conditions of pressure and temperature. Imagine a terrane that was once underlain by a thick sequence of calcareous shales, was subjected to compression so as to produce mountains, and then experienced loss of those mountains by erosion. One might then begin a traverse across this eroded surface on unmetamorphosed shales that were not involved in the mountain building (Fig. 2–27). These shales would contain only unaltered sedimentary minerals. Progressing farther, toward the area of most intense metamorphism, one might see that the shales had given way to slates bearing the green metamorphic mineral chlorite. Still farther along the traverse, schists containing such intermediate-grade metamorphic minerals as biotite and garnet would appear. Finally, one might come upon coarsely foliated schists containing kyanite, staurolite, and sillimanite—minerals that develop under high temperature and pressure.

Metamorphic minerals do not always appear in the orderly fashion indicated in the preceding example. Depending on the nature and depth of metamorphism, temperature may increase at a faster or slower rate than does pressure, resulting in the growth of different index minerals. It is also possible for a previously metamorphosed terrane to experience a second, less severe episode of metamorphism. In such cases, a lowering of metamorphic grade might result. Yet another factor that influences the kinds of minerals produced is the mineralogic composition of the parent rock and the amount of water present during metamorphism.

Metamorphic rocks may contain a wealth of historical information. A marble containing flakes and veins of chlorite that is dated as 1 billion years old tells the geologist many things. It records the existence of an ancient, somewhat clayey limestone that experienced a relatively low level of metamorphism. Because limestone was the parent rock, the geologist may infer that conditions on earth at that early date were suitable for the precipitation of carbonate rocks. Such conditions would

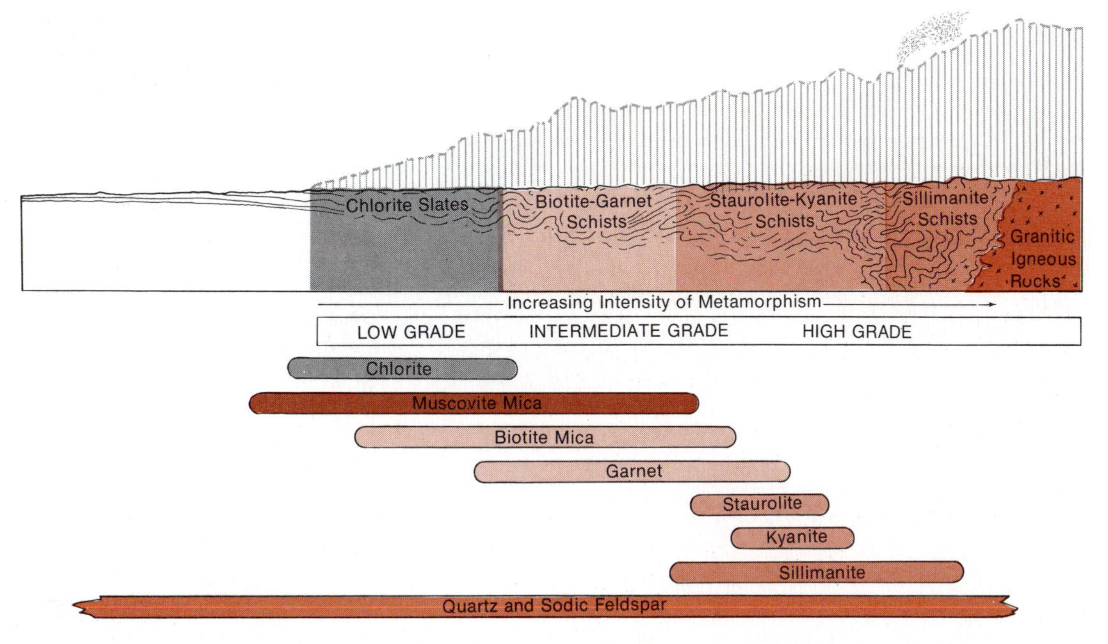

Figure 2–27 Changes in the mineralogic composition of a terrain originally underlain by shales following regional metamorphism.

include an atmosphere and hydrosphere similar to that existing today and considerably unlike the nonoxygenic environment that characterized the earth's most ancient ages.

Sedimentary Rocks

Introduction

Sedimentary rocks are simply rocks composed of consolidated sediment—particles that are the product of weathering and erosion of any previously existing rock or soil. The components of sedimentary rocks may range from large boulders to the molecules dissolved in water. Sediment is deposited through such agents as wind, water, ice, or mineral-secreting organisms. The loose sediment is converted into coherent solid rock by any of several processes: precipitation of a cementing material around individual grains, compaction, or crystallization. These processes constitute **lithification.**

The most obvious feature of sedimentary rocks is their occurrence in beds or layers called **strata.** Stratification is commonly the result of changes in the conditions of deposition that cause materials of somewhat different nature to be deposited for a period of time. For example, the velocity of a stream might decrease, causing particles to settle out that might otherwise have stayed in suspension. In another situation, the kind of materials brought into a given depositional site by streams might change, and there would then be a corresponding change in composition of the accumulating layers.

Sandstone, shale, and carbonate rocks (such as limestone) constitute the most abundant sedimentary rocks. **Sandstones** are composed of grains of quartz, feldspar, and other particles that are cemented or otherwise consolidated. **Shale** consists largely of very fine particles of quartz and abundant clay. The **carbonates** are rocks formed when the carbon dioxide contained in water combines with oxides of calcium and magnesium.

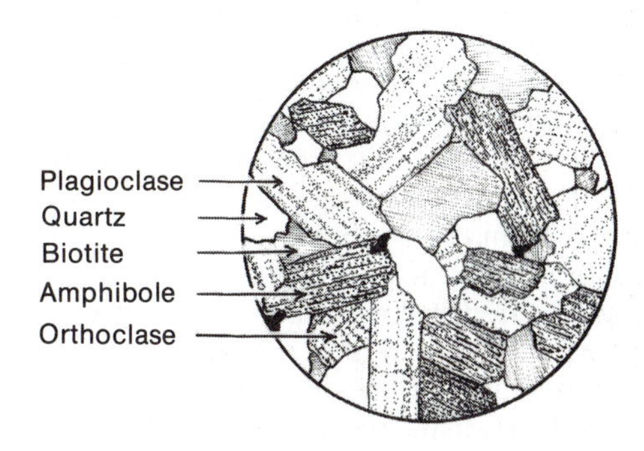

Figure 2–28 Sketch of a thin section of granodiorite as viewed through the petrographic microscope. (Diameter is about 3 mm.)

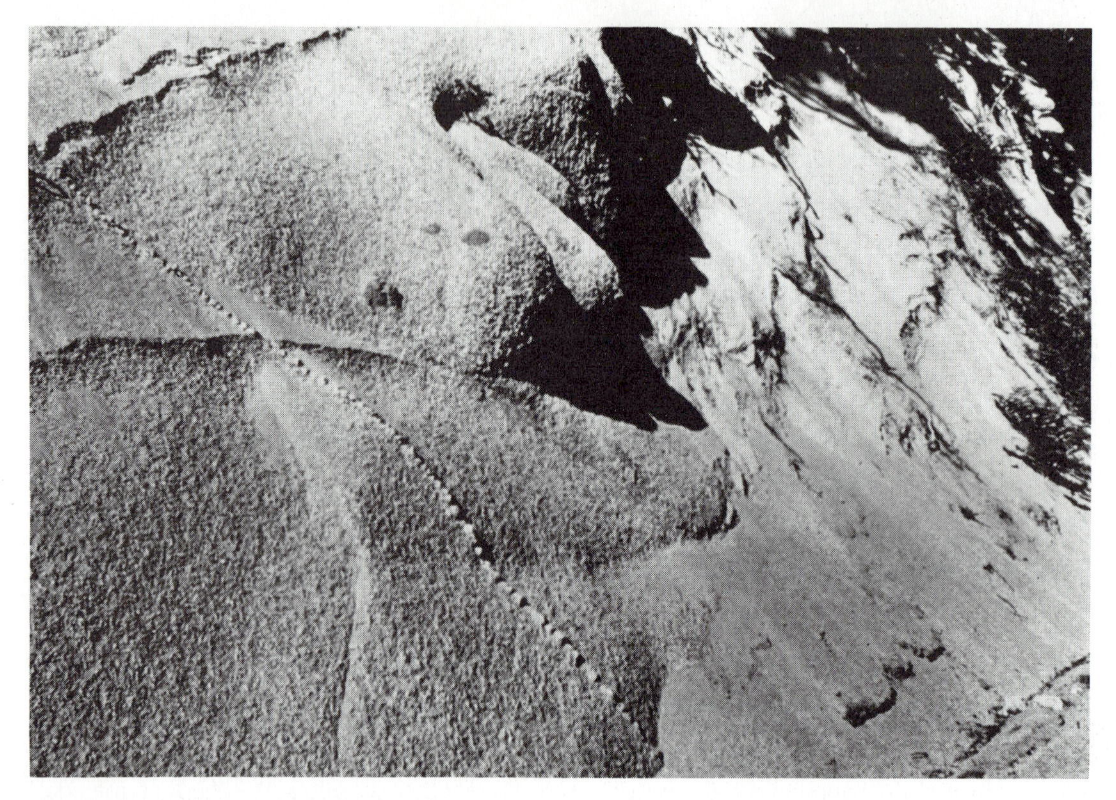

Figure 2–29 Exposure of weathered granodiorite showing accumulation of clastic grains resulting from disintegration of the parent rock. Sierra Nevada, California.

Derivation of Sedimentary Materials

Sedimentary rocks must have originally come from the decomposition of older rocks. Commonly, the older rocks are igneous; indeed, these were once the only rocks on earth. It is therefore instructive to review the manner in which the common components of sedimentary rocks might be derived from an abundant kind of igneous rock. One such igneous rock is called **granodiorite** (Fig. 2–28). It is an ordinary intrusive rock consisting of quartz, calcium and sodium feldspar, potassium feldspar, biotite, and amphibole. Can such a rock, if chemically decomposed in a temperate climate, yield the materials required for the formation of sandstones, shales, and limestones?

Consider, first, the quartz in the granodiorite. Quartz will persist almost unchanged during weathering. It is one of the most chemically stable of all the common silicate minerals. As the parent rock is gradually decomposed, quartz grains tend to be washed out and carried away to be deposited as sand that will one day become sandstones (Fig. 2–29).

The feldspars decay far more readily than quartz. They are primarily aluminum silicates of potassium, sodium, and calcium. In the weathering process, the last three elements are largely dissolved and carried away by solutions as bicarbonate ions (although some may remain in soils within clay minerals). Ultimately, they reach the sea, where they may stay in solution, or they are deposited as layers of limestone. If large quantities of lake or sea water are evaporated, **evaporites** like *halite* ($NaCl$) or *gypsum* ($CaSO_4 \cdot 2H_2O$) may be formed. Of course, not all the feldspars and micas in the granodiorite necessarily decay. Some may persist as detrital grains that become incorporated into sandstones and other sediments.

The decomposition of the plagioclase feldspar in granodiorite can be expressed by the equation:

$$CaAl_2Si_2O_8 \cdot 2NaAlSi_3O_8 + 4H_2NO_3 + 2(nH_2O)$$

Plagioclase Feldspars + Carbonic Acid (Water plus Carbon Dioxide) + Water

yields

$$Ca(HCO_3)_2 + 2NaHCO_3 + 2Al_2(OH)_2Si_4O_{10} \cdot nH_2O$$

Soluble Calcium Bicarbonate + Soluble Sodium Bicarbonate + Clay Mineral

The equation shows that the principal nonsoluble product of the decomposition of feldspars (and other aluminum silcates) is clay. Here, then, is the greatest primary source for the clay of shales, claystones, and soils.

Biotite is another mineral in the granodiorite source rock. Decomposition of biotite, which is a potassium, magnesium, and iron alumino silicate, yields soluble potassium and magnesium carbonates, small amounts of soluble silica, and iron oxides. The iron oxides serve to color many sedimentary rocks in tints of brown and red.

Variety Among Sedimentary Rocks

Sedimentary rocks are classified according to their **composition** and **texture.** The term "texture" refers to the size and shape of the individual grains and to their arrangement in the rock. A rock that has a **clastic** texture is composed of grains of sand, silt, or parts of rocks or fossils. Most sedimentary rocks have a clastic texture, but many others are formed by the intergrowth of crystals and are therefore crystalline. In very simple classifications, sedimentary rocks are divided into clastic and nonclastic categories.

CLASTIC ROCKS

Clastic (or detrital) rocks are those composed of individual fragments of mineral or rock (Fig. 2–30). The fragments may range in size from huge boulders to microscopic particles. The group includes such common sedimentary rocks as conglomerate, sandstones, and shales; the materials of these rocks are derived, at least initially, from weathering and erosion of pre-existing rocks on land.

Texture is the key to naming the major clastic rocks. **Conglomerate,** for example, is composed of water-worn, rounded particles larger than 2 mm in diameter (Fig. 2–31). **Breccias** are composed of fragments that are angular but similar in size to conglomerates. In **sandstones,** grains range between 0.0625 and 2.0 mm. The varieties of sandstones are then subdivided largely according to composition. **Siltstones** are finer than sandstones (0.004 to 0.0625 mm), and **shales** are composed of particles finer than 0.004 mm. Shales may contain abundant clay minerals, which are flaky minerals that align parallel to bedding planes. As a result, shales characteristically split into thin slabs parallel to bedding planes. This property is termed **fissility.** Rocks lacking fissility but composed of clay-sized particles are called **claystones** or **mudstones.**

NONCLASTIC ROCKS

Nonclastic rocks are formed from materials either directly precipitated from solution or secreted by the activities of plants or animals. In contrast to the clastic group, such sediments have undergone little if any transportation. Dissolved substances are brought to the oceans by rivers and under suitable conditions may be precipitated or extracted by organisms to form the minerals of nonclastic rocks. The most prevalent of the nonclastic group are the **carbonate rocks,** which are composed of the carbonate minerals calcite

Figure 2–30 Alternating beds of sandstone (lighter strata) and shale (dark strata) exposed along the coast of California, west of San Francisco Bay. (Photograph courtesy of J. C. Brice.)

Figure 2–31 Conglomerate. Sespe formation of Tertiary Age. San Joaquin Hills, California. (Photograph courtesy of J. G. Vedder, U.S. Geological Survey.)

($CaCO_3$, which crystallizes in the hexagonal crystal system), aragonite ($CaCO_3$, which crystallizes in the orthorhombic crystal system), and dolomite ($CaMg(CO_3)_2$, which, like calcite, forms hexagonal crystals). Of these rocks, calcite is the predominant mineral in limestones (see Fig. 2–17), although the other carbonate minerals, as well as clay and varieties of quartz, may also be present in variable amounts. Indeed, limestones (Fig. 2–32) show a wide variation in texture and composition.

Limestones. The most abundant limestones are of marine origin and have formed as a result of precipitation of calcite or aragonite by organisms and the incorporation of skeletons of those organisms into sedimentary deposits. Inorganic precipitation of carbonate minerals may also form deposits of limestone. The importance of this process is questionable, however, because the precipitation is nearly always closely associated with photosynthetic and respiratory activities of organisms or with the release of tiny particles of aragonite upon the decay of green algae. Strictly speaking, it appears very few marine limestones are the result of direct chemical precipitation.

After the calcium carbonate has accumulated, it becomes recrystallized or otherwise consolidated into indurated rock that may be variously colored—from white, through tints of brown, to gray. Limestones tend to be well stratified, frequently contain nodules and inclusions of chert, and are often highly fossiliferous (containing fossils). The rock may range in texture from coarsely granular to very fine-grained and aphanitic.

In general, limestones consist of one or more of a combination of such textural components as **micrite, carbonate clasts, oolites,** or **carbonate spar.** Micrite is a uniformly fine-grained, "muddy" texture in which individual particles cannot be discerned without considerable magnification. Micritic texture is apparently the result of consolidation of carbonate mud and ooze. Carbonate clasts are sand- or gravel-sized pieces of carbonate. The most common clasts are either *bioclasts* (skeletal fragments of marine invertebrates) or *oolites* (Fig. 2–33), which are spherical grains formed by the precipitation of carbonate around a nucleus. Sparry carbonate is a clear crystalline carbonate that is normally deposited between the clasts as a cement or has developed by replacement of calcite. These textural categories of limestone as seen through the microscope are shown in Figure 2–34. They permit classification of particular samples as micritic limestone, clastic limestone, oolitic limestone, or sparry (crystalline) limestone.

Figure 2–32 Limestone of Mississippian Age exposed in an abandoned quarry, Columbia, Missouri. (Photograph courtesy of Stephen D. Levin.)

Figure 2–33 Oolites. (Diameter of field is 2 mm.) (Photograph courtesy of J. C. Brice.)

There are many varieties of limestone. **Chalk** is a soft, porous variety that is composed largely of extremely minute calcareous skeletal elements called *coccoliths* (Fig. 2–35). Coccoliths are secreted by marine golden brown algae. **Lithographic limestone** is a dense, micritic limestone once widely used as an etching surface in printing illustrations. Some limestones consist almost entirely of skeletal remains of reef corals and other frequently skeletonized marine invertebrates.

Dolostone. This is a nonclastic rock composed largely of the mineral *dolomite,* which is a calcium-and-magnesium carbonate. As found in exposures, dolostone is not easily distinguished from limestone. The usual field test for distinguishing dolostone from limestone is to apply cold dilute hydrochloric acid. Unlike limestone, which bubbles readily, dolostone will effervesce only slightly, if at

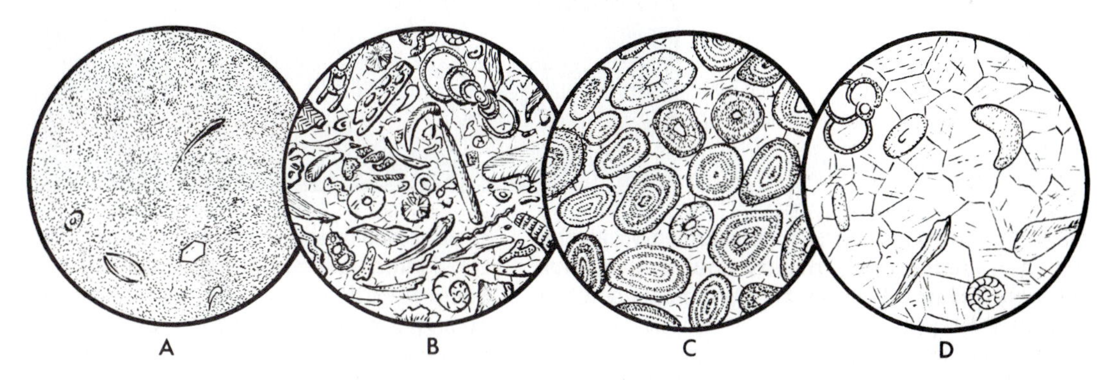

Figure 2–34 Textures of limestones as seen in thin section under the microscope. (A) Aphanitic limestone or micrite. (B) Bioclastic limestone with fine-grained sparry calcite as cement. (C) Oölitic limestone. (D) Sparry or crystalline limestone.

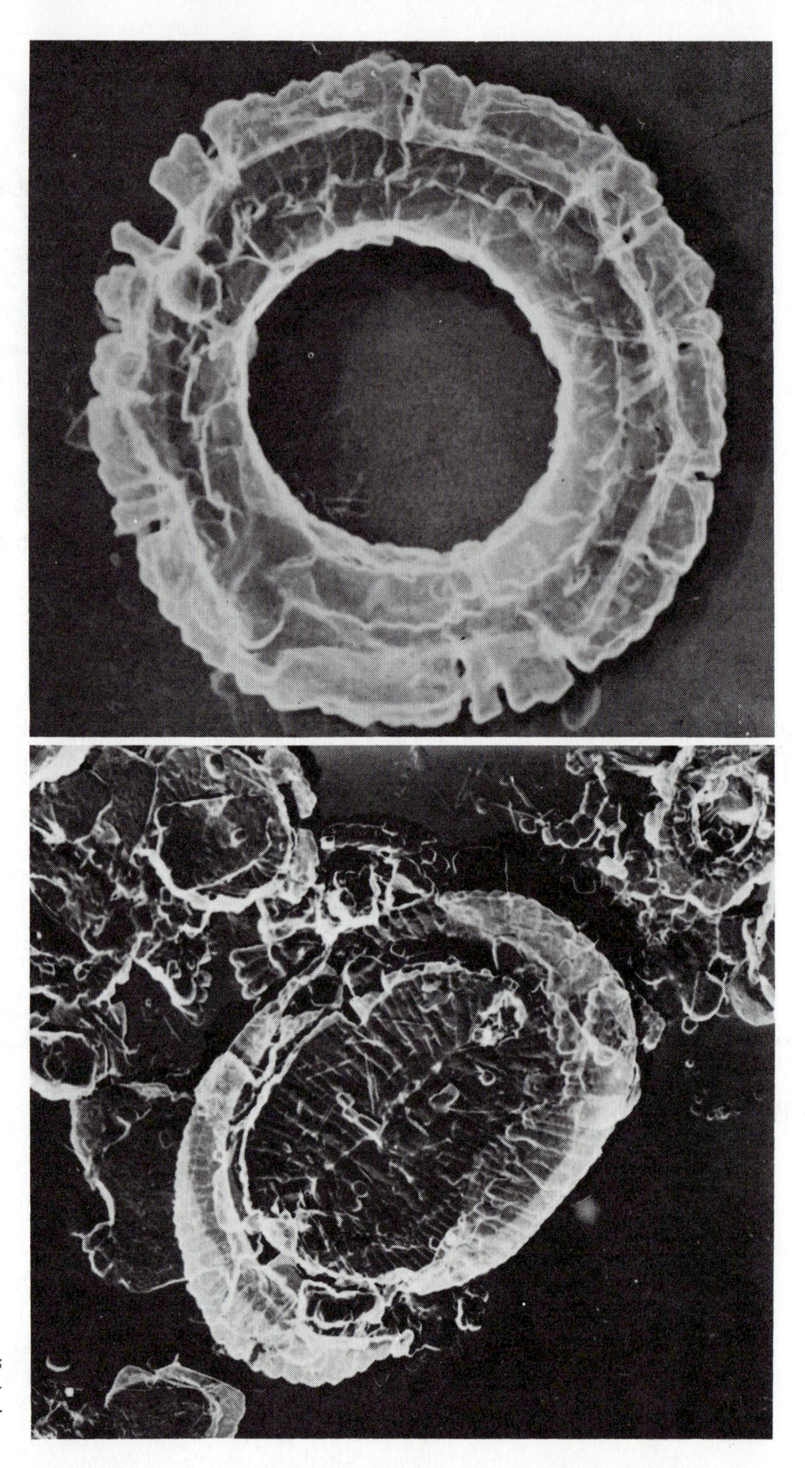

Figure 2–35 Coccoliths. *Coccolithus doronicoides* (above); *Helicosphaera carteri* (below). (Largest dimension is 22 microns.)

all. In thin sections of the rock examined with the aid of a petrographic microscope, the uniform rhombic grains of dolomite are a trait useful in identification (Fig. 2–36).

The origin of dolostones is somewhat problematic. The mineral dolomite is not secreted by organisms in shell-building. Direct precipitation from sea water does not normally occur today, except in a few environments where the sediment is steeped in abnormally saline water. Such an origin is not considered adequate to explain the thick sequences of dolomitic rock commonly found in the geologic record. The most widely believed theory for the origin of dolostones is that they result from partial replacement of calcium by magnesium in the original calcareous sediment. However, it is not known how long or at what time in the history of the rock this "dolomitization" occurs.

Figure 2–36 Rhomb-shaped crystals of dolomite. (Diameter of field is 2 mm.) (Photograph courtesy of J. C. Brice.)

Figure 2–37 The chert nodules in this limestone are the darker bodies, which, because of their greater resistance to weathering, stand out in relief. Mississippian limestone. Hardin County, Illinois. (Photograph courtesy of C. Butts and the U.S. Geological Survey.)

Figure 2–38 Vertical beds of chert. Claremont Chert, Berkeley Hills, California. (Photograph courtesy of J. C. Brice.)

Chert. We have previously mentioned a form of microcrystalline quartz called chert (SiO_2), noting its occurrence as nodules in limestones (Fig. 2–37). The origin of these nodules is still being debated among petrologists, although the majority believe that such nodules form as replacements of carbonate sediment by silica-rich sea water trapped in the sediment. Some cherts occur in areally extensive layers and thus qualify as monomineralic rocks. These so-called bedded cherts (Fig. 2–38) are thought to have formed from the accumulation of the siliceous remains of diatoms and radiolaria and from subsequent reorganization of the silica into a microcrystalline quartz. Silica from the dissolution of volcanic ash is believed to enhance the process; indeed, many bedded cherts are found in association with ash beds and submarine lava flows.

Evaporites. In the previous chapter, we noted that evaporites are chemically precipitated rocks that are formed as a result of evaporation of saline water bodies. Only about 3 per cent of all sedimentary rocks consist of evaporites. Evaporite sequences of strata are composed chiefly of such minerals as gypsum ($CaSO_4 \cdot 2H_2O$), anhydrite ($CaSO_4$), halite (NaCl), and associated calcite and dolomite. Extensive ancient deposits of evaporites are currently being commercially worked in Michigan, Kansas, Texas, New Mexico, Germany, and Israel. The conditions required for precipitation of thick sequences of evaporites include warm, relatively arid climates and a physiographic situation that would provide periodic additions of sea water to the evaporating marine basin. In the Gulf Coastal Region of the United States, as a result of the pressure of overlying rocks, deeply buried deposits of salt have flowed plastically upward to form underground domes of salt. In the process, the salt arched overlying strata, thereby producing structures in which petroleum could collect (Fig. 2–39).

Coal. Coal is a carbonaceous rock resulting from the accumulation of plant matter in a swampy environment combined with alteration of that plant tissue by both biochemical and physical processes until it is converted to a consolidated carbon-rich material. The biochemical and physical changes may produce a series of products ranging from peat and lignite to bituminous and anthracite coal. For coal to form, plant tissue must be accumulated under water or be quickly buried, because vegetable matter, if left exposed to air, is readily oxidized to water and carbon dioxide. With underwater accumulation or quick burial of plant material, a major part of the carbon can be retained.

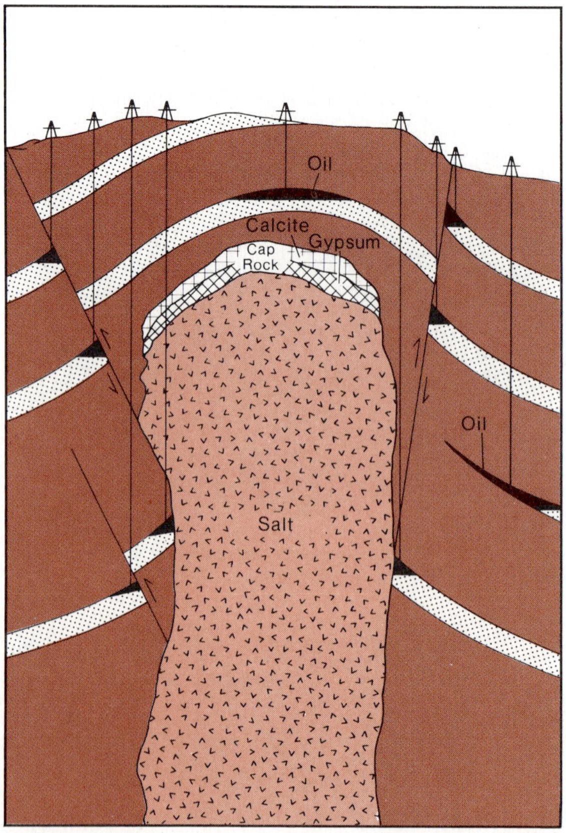

Figure 2–39 Salt dome, illustrating possibilities for oil entrapment in domelike structures (top center) and by faults, and the pinchout of oil-bearing strata.

Summary

Rocks are the very obvious materials of which the crust of the earth is composed. The rocks are themselves composed of minerals, and minerals in turn are constructed of the atoms of chemical elements. Centuries ago, a Greek philosopher named Democritus speculated that water and rocks were made of invisibly tiny pieces of matter. Democritus coined the term "atom" to describe these particles. The present understanding of atoms is that they have a central nucleus containing one or more positively charged protons and electrically neutral neutrons. Circling the nucleus are negatively charged electrons. The arrangement of electrons into levels, the number of atomic particles, the size of the atoms, and their electrical characteristics determine how the atoms will combine into molecules and ultimately into minerals.

Silicate minerals are by far the most important of the rock-forming minerals. The fundamental unit in silicate mineral structures is a tetrahedral arrangement of four oxygen ions around a central silicon ion. By the sharing of oxygen ions between neighboring tetrahedra or positively charged ions such as calcium, magnesium, or iron, the tetrahedra form chains, sheets, or three-dimensional frameworks; these structures are the underlying reasons for the specific properties of a mineral.

Of the principal rock-forming mineral families, quartz, feldspar, mica, amphiboles, pyroxenes, and olivine are silicates that were initially crystallized from molten material. Clay minerals, calcite, dolomite, and certain varieties of quartz (such as chert and flint) are formed by processes of weathering and precipitation at temperatures that prevail at the earth's surface.

Rocks are aggregates of minerals. Igneous rocks are those that have cooled from a molten silicate material. Consolidated materials derived from other rocks by weathering, erosion, transport, and deposition constitute sedimentary rocks. Metamorphic rocks are those altered from older rocks that have been subjected to heat, pressure, and chemically active solutions.

Rocks and minerals are the materials from which geologists must make their interpretations of ancient environments and past geologic happenings. Thus, the brief survey of earth materials presented in this chapter contains considerable background material that will be helpful in later sections of this book.

Questions for Review

1. What is the difference between an ion and an atom?
2. How does a crystal differ from a crystalline solid?
3. Distinguish between covalent bonding and ionic bonding.
4. Why are the silicate minerals so important in geology?
5. What is the probable cooling history of an igneous rock that is for the most part very finely crystalline but also contains large crystals that appear sporadically within the finely crystalline material?
6. In terms of atomic structure, explain why asbestos is fibrous and mica is scaly. Why does the mineral quartz not break into fragments with smooth surfaces?
7. What type of igneous rock best approximates the composition of the crust beneath the oceans? The continental crust?
8. Of the mineral groups discussed in this chapter, which are particularly characteristic of sedimentary rocks?
9. What inferences can be drawn from the color of igneous rocks?
10. Account for the observation that most intrusive igneous rocks tend to be rich in silica, whereas extrusive rocks contain relatively less silica.
11. What is the origin of foliation in metamorphic rocks? List the foliated metamorphic rocks in order of increasing coarseness.
12. What historical interpretation can be made about regions in which extensive tracts of metamorphic and igneous rocks are exposed at the earth's surface?
13. Formulate a definition of *stratification*. What causes stratification in a sedimentary rock?
14. What are the differences between limestone and dolostone? How might limestone be altered to dolostone?

Terms to Remember

amphibole
amphibolite
andesite
asbestos
atom
atomic number
basalt
biotite mica
breccias
calcite
carbonates
chalk
chert
clastic rocks
conglomerate
contact metamorphism
covalent bonding
dolomite
dolostone
dynamothermal metamorphism
electron
evaporite
extrusive igneous rocks
fissility
flint
foliation
fractional crystallization
gneiss
granite
granodiorite
greenstone
gypsum
halite
igneous rocks
intrusive igneous rocks
ion
ionic bonding
isotope
lithographic limestone
magma
mantle
mass
mass number
mica
micrite
mineral
molecule
muscovite
neutron
olivine
oolites
partial melting
phyllite
plagioclase feldspar
porphyritic texture
potassium feldspar group
proton
quartz
quartzite
regional metamorphism
sandstone
schist
shales
silicon tetrahedron
siltstone
slate
strata
texture

Supplemental Reading and References

Blatt, H., Middleton, G., and Murray, R. 1972. *Origin of Sedimentary Rocks.* Englewood Cliffs, N.J., Prentice-Hall, Inc.

Dietrich, R. V. 1980. *Stones.* San Francisco, W. H. Freeman.

Dietrich, R. V., and Skinner, B. J. 1979. *Rocks and Minerals.* New York, John Wiley & Sons.

Dunbar, C. O., and Rodgers, J. 1957. *Principles of Stratigraphy.* New York, John Wiley & Sons.

Ernst, W. G. 1969. *Earth Materials.* Englewood Cliffs, N.J., Prentice-Hall, Inc.

Garrels, R. M., and Mackenzie, F. T. 1971. *Evolution of Sedimentary Rocks.* New York, W. H. Norton & Co., Inc.

Holden, J. C. 1980. Mount St. Helens. *Science 80* 1(6): 49–57.

Pough, F. H. 1960. *A Field Guide to Rocks and Minerals.* Cambridge, Mass., Houghton Mifflin Co.

Tennissen, A. C. 1974. *The Nature of Earth Materials.* Englewood Cliffs, N.J., Prentice-Hall, Inc.

Geologist working in Colorado Rocky Mountains
(Photograph courtesy of Exxon, U.S.A.)

3

The Historical Significance of Sedimentary Rocks

In the high mountains, I have seen shells. They are sometimes embedded in rocks. The rocks must have been earthy materials in days of old, and the shells must have lived in water. The low places are now elevated high, and the soft material turned into hard stone.

Chu-Hsi, AD 1200

THE SEDIMENTOLOGIC ARCHIVES

Ever since the earth has had an atmosphere and hydrosphere, sediments have been accumulating on its surface. The sediments, now formed into sedimentary rocks, contain features that tell us about the environment in which they were deposited. By interpreting these bits of evidence in successively higher strata, one can decipher the geologic history of a part of the earth.

The Tectonic Setting

The type of rock eroded to produce sediment, the climate under which processes of weathering and erosion take place, the method of transportation of sedimentary materials, and the "tectonics" of a region all determine the type of sedimentary rock to be formed at a particular location. **Tectonics** may be defined as the study of the deformation or structural behavior of a large area of the earth's crust over a long period of time. For example, a region may be tectonically stable, subsiding, rising gently, or more actively rising to produce mountains and plateaus. Where a source area has recently been compressed and uplifted, an abundance of coarse clastics derived from the rugged upland source area will be supplied to the basin. In the geologic past, such a tectonic setting has resulted in the accumulation of great "clastic wedges" of sediment that thickened and became coarser toward the former mountainous source area. In other tectonic settings of the past, the source area has been stable and topographically more subdued, so that finer particles and dissolved solids became the most abundant components being carried by streams.

The tectonic setting influences not only the size of clastic particles being carried to sites of deposition, but also the thickness of the accumulating deposit. For example, if a former marine basin of deposition had been provided with an ample supply of sediment and was experiencing tectonic subsidence, enormous thicknesses of sediments might accumulate. Over a century ago, James Hall (an early New York geologist) recognized that the thick accumulations of shallow water sedimentary rocks in the Appalachian region required that crustal subsidence accompany deposition. His reasoning was quite straightforward. It was easy to visualize filling a basin that was 40,000 ft deep with 40,000 ft of sediment. However, if the basin was only several hundred feet deep, the only way to get tens of thousands of feet of sediment into it would be to have subsidence occurring simultaneously with sedimentation.

In a marine basin of deposition that is stable or subsiding very slowly, the surface on which sedimentation is occurring is likely to remain within the zone of wave activity for a long time. Wave action and currents will wear, sort, and distribute the sediment into broad, blanket-like layers. If the supply of sediment is small, this type of sedimentation will continue indefinitely. Should the supply of sediment become too great for currents and waves to transport, however, the surface of sedimentation would rise above sea level, and deltas would form.

It is possible to view the tectonic framework of entire continents as well as of particular areas. The principal tectonic elements of a continent are **cratons** and **mobile belts.** The craton consists of the central stable region of a continent. Such regions have a generally subdued topography and are composed of ancient igneous and metamorphic rocks that may be either exposed or covered by flat-lying younger sedimentary strata.

Mobile belts are long, relatively narrow, often arcuate crustal zones of great tectonic instability. Today such belts are recognized by a high frequency of earthquakes and volcanic eruptions. Mobile belts known as geosynclines have been extremely important in the evolution of continents. **Geosynclines** develop along the margins of cratons that have subsided over long periods of time. They serve as collecting troughs for tens of thousands of meters of sedimentary and volcanic material. Geosynclines tend to become increasingly more unstable through time. As this instability increases, the geosynclinal wedge of sediment is compressed, metamorphosed, intruded, and eventually deformed into great mountain ranges.

The tectonic setting of deposition largely determines the nature of sedimentary deposits. Conversely, the kind of tectonic setting can often be inferred from a rock's textural and structural features and from its color, composition, and fossils. The tectonic and historical significance of some of these characteristics of sedimentary rocks will be frequently examined in the pages to follow.

COLOR IN SEDIMENTARY ROCKS

We have seen that color in igneous rocks can be used to indicate the approximate amount of ferromagnesian minerals. Color in sedimentary rock can also provide useful clues to identification. For example, varieties of chert can be identified as flint if they are gray or black, or as jasper if they are red. Color is also useful in providing clues to the environment of deposition of sedimentary rocks. Of the sedimentary coloring agents, carbon and the oxides and hydroxides of iron are clearly the most important.

Black Coloration

Black and dark gray coloration in sedimentary rocks—especially shales—usually results from the presence of organic carbon compounds and iron sulfides. The occurrence of an amount of organic carbon sufficient to result in black coloration implies an abundance of organisms in or near the depositional areas as well as environmental circumstances that kept the remains of those organisms from being completely destroyed by oxidation or bacterial action. These circumstances are present in many marine, lake, and estuarine environments today. In a typical situation, the remains of organisms that lived in or near the depositional basin settle to the bottom and accumulate. In the quiet bottom environment, dissolved oxygen needed by aerobic bacteria to attack and break down organic matter may be lacking. There may also be insufficient oxygen for scavenging bottom-dwellers that might feed on the debris. Thus, organic decay is limited to the slow and incomplete activity of anaerobic bacteria; consequently, incompletely decomposed material rich in black carbon tends to accumulate. In such an environment, iron combines with sulfur to form finely divided iron sulfide (pyrite, FeS_2), which further contributes to the blackish coloration. Such environments of deposition are likely to yield toxic solutions of hydrogen sulfide (H_2S). The lethal solutions rise to poison other organisms and thus contribute to the process of accumulation. Black sediments do not always form in restricted basins. They may develop in relatively open areas, provided the rate of accumulation of organic matter exceeds the ability of the environment to cause its decomposition.

Red Coloration

Hues of brown, red, and green are frequently formed in sedimentary rocks as a result of their iron oxide content. Few, if any, sedimentary rocks are free of iron, and less than 0.1 per cent of this metal can color a sediment a deep red. The iron pigments not only are ubiquitous in sediments but also are difficult to remove in most natural solutions.

Iron forms two sorts of ions: **ferrous** iron has two positive charges, whereas **ferric** iron has three positive charges. Thus, iron may form two oxides: FeO (ferrous oxide) and Fe_2O_3 (ferric oxide). In air, ferrous iron is slowly oxidized to ferric iron. When oxygen is in short supply, ferric iron may be similarly reduced to ferrous iron. Ferric minerals like **hematite** tend to color the rock red, brown, or purple, whereas the ferrous compounds impart hues of gray and green. Hydrous ferric oxide **(limonite)** is often yellowish in color.

Red Beds

Strata colored in shades of red, brown, or purple by ferric iron are designated **red beds** by geologists. Oxidizing conditions required for the development of ferric compounds are more typical of nonmarine than marine environments; most red beds are flood plain, alluvial fan, or deltaic deposits. Some, however, are originally reddish sediment carried into the open sea. Electron microscope studies (Walker, 1967) of red beds forming today in Baja, California, indicate that the red coloration developed long after the sediment was deposited. After burial, the decay of clastic ferromagnesian minerals released iron that was oxidized by the oxygen in underground water circulating through the pore spaces. Thus, red coloration may be imparted in the subsurface and may be independent of climate. The paleoenvironmental interpretations one can draw from red beds should be based to a large degree on the associated rocks and sedimentary structures. Red beds interspersed with evaporite layers indicate warm and arid conditions.

Although red beds are more likely to represent nonmarine than marine deposition, occasionally the reddish strata are interbedded with fossiliferous marine limestones. In such cases, the color may be inherited from red soils of nearby continental areas. Lands located in warm, humid climates often develop such reddish soils. When the soil particles arrive at the marine depositional site, they will retain their red coloration if there is insufficient organic matter present to reduce the ferric iron to the ferrous state. Otherwise, they will be converted to the gray or green colorations of ferrous compounds.

In summary, sedimentary rocks of red coloration may be a product of the source materials, may have developed after burial as a result of a lengthy period of subsurface alteration, or may be the result of subaerial oxidation. Geologists are suspicious of the last possibility, because most modern desert sediments are not red unless composed of materials from nearby outcroppings of older red beds.

HISTORICAL SIGNIFICANCE OF TEXTURE

The size, shape, and arrangement of mineral grains in a rock constitute its **texture.** In addition to the larger grains themselves, the textural appearance of a rock is influenced by the materials that hold the particles together. **Matrix** is one category of bonding material that consists of finer clastic particles (often clay) that were deposited at the same time as the larger grains and that fill the crevices between them. **Cement,** on the other hand, is a chemical precipitate that crystallizes in the voids between grains following deposition. *Silica* (SiO_2) and *calcium carbonate* ($CaCO_3$) are common natural cements.

Texture can provide many clues to the history of a particular rock formation. In carbonate rocks, extremely fine-grained textures such as developed in lithographic limestones probably indicate deposition in quiet water. Fine carbonate muds, which are the parent sediment of such rocks, are not likely to settle to the bottom in turbulent water. Whole unbroken fossil shells confirm the quiet water interpretation. Limestones containing the worn and broken fragments of fossil shells are likely to be the products of reworking by wave action. They are turbulent water deposits. Because the original textures of carbonates are often altered by recrystallization, historical interpretations are more frequently derived from analysis of texture in clastic sedimentary rocks.

Size and Sorting of Clastic Grains

Geologists universally use a scale of particle sizes known as the **Wentworth Scale** to categorize clastic sediments (Table 3–1). After disaggregation of a rock in the laboratory, the particles can be passed through a series of successively finer sieves

Table 3–1 SIZE RANGE OF SEDIMENTARY PARTICLES

Wentworth Scale (*in millimeters*)	Particle Name
	Boulders
256	
128	Cobbles
64	
32	
16	Pebbles
8	
4	
	Granules
2	
	Very coarse sand
1.0	
	Coarse sand
0.5 (1/2)	
	Medium sand
0.25 (1/4)	
	Fine sand
0.125 (1/8)	Very fine sand
0.0625 (1/16)	
0.0313 (1/32)	
0.0156 (1/64)	
	Silt
0.0078 (1/128)	
0.0039 (1/256)	
	Clay

and the weight percentage of each size range in the rock may be determined. It is obvious that a stronger current of water (or wind) is required to move a big particle than a small one. Therefore, the size distribution of grains tells the geologist something about the turbulence and velocity of currents. It can also be an indicator of the mode and extent of transportation. If sand, silt, and clay are supplied by streams to a coastline, the turbulent inshore waters will winnow out the finer particles, so that gradations from sandy nearshore deposits to offshore silty and clayey deposits frequently result (Fig. 3–1). Sandstones formed from such inshore sands may retain considerable porosity and provide void space for petroleum accumulations. For this reason, one often finds petroleum geologists assiduously making maps showing the grain size of deeply buried ancient formations to determine areas of coarser and more permeable clastic rock.

One aspect of a clastic rock's texture that involves grain size is sorting. **Sorting** is an expression of the range of particle sizes deviating from the average size. Rocks composed of particles that are all about the same average size are said to be "well sorted" (Fig. 3–2), and those that include grains with a wide range of sizes are termed "poorly sorted" (Fig. 3–3). Sorting often provides clues to conditions of transportation and deposition. Wind, for example, winnows the dust particles from sand, producing grains that are all of about the same size. Wind also sorts the particles that it carries in suspension. Only rarely is the velocity of winds sufficient to carry grains larger than 0.2 mm. While carrying grains of that size, winds sweep finer particles into the higher regions of the atmosphere. When the wind subsides, well-sorted silt-sized particles drop and accumulate. In general, windblown deposits are better sorted than deposits formed in an area of wave action, and wave-washed sediments are better sorted than stream deposits. It must be kept in mind however, that if a source sediment is already well sorted, the resulting deposit will be similarly well sorted. Poor sorting occurs when sediment is rapidly deposited without being selectively separated into sizes by currents. Poorly

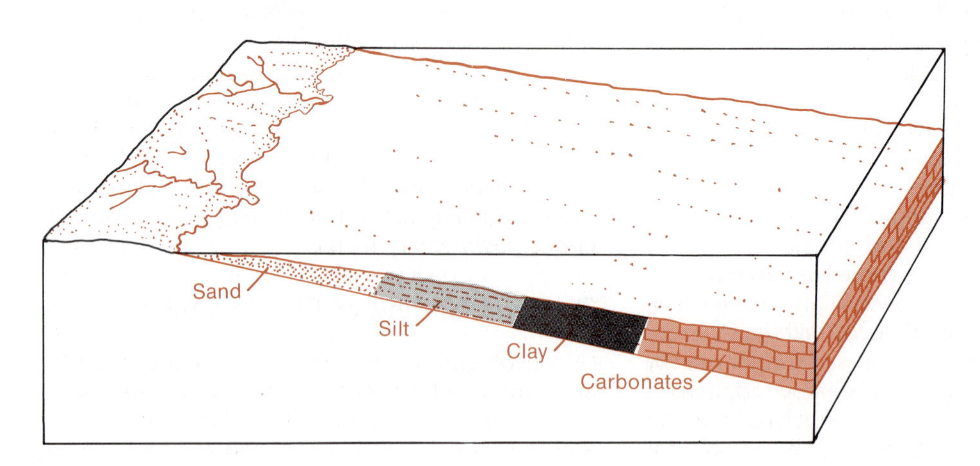

Figure 3–1 Gradation of coarser nearshore sediments to finer offshore deposits.

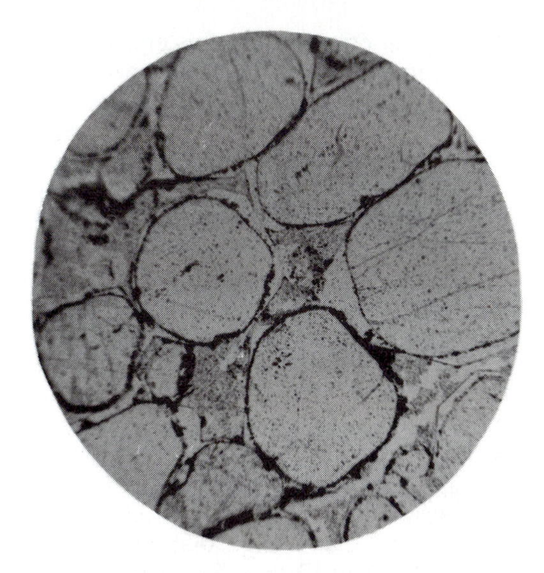

Figure 3–2 Thin section of a well-sorted sandstone as seen under the microscope. Quartz grains have developed silica overgrowths, which are visible as ghostly crystal outlines. (Diameter of area is 1.0 mm.)

Figure 3–3 Thin section of a poorly sorted sandstone (graywacke) observed under the microscope with crossed polarizers. (Diameter of area is 1.2 mm.)

sorted conglomerates and sandstones are deposited at the foot of mountains, where stream velocity is suddenly checked. Another example of a poorly sorted conglomerate is **tillite,** a rock deposited by glacial ice containing all particle sizes in a heterogeneous mixture.

Figure 3–4 Well-rounded grains of quartz viewed under the microscope. From the St. Peter Formation near Pacific, Missouri.

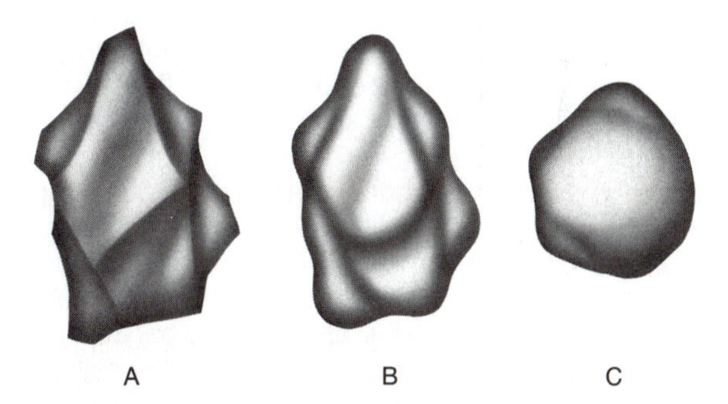

Figure 3–5 Shape of sediment particles. (A) An angular particle (all edges sharp). (B) A rounded grain that has little sphericity. (C) A well-rounded, highly spherical grain. *Roundness* refers to the smoothing of edges and corners, whereas sphericity measures the degree of approach of a particle to a sphere.

Shape of Clastic Grains

The **shape** of particles in a clastic sedimentary rock can also be useful in determining its history. Shape can be described in terms of **rounding** of particle edges and **sphericity** (how closely the grain approaches the shape of a sphere) (Figs. 3–4 and 3–5). A particle becomes rounded by having sharp corners and edges removed by impact with other particles. The relatively heavy impacts between pebbles and granules being transported by water cause rapid rounding. Lighter impacts occur between sand grains in water transport; the water provides a cushioning effect. The result is far slower rounding. Indeed, most of the rounding of sand grains may occur during wind transport. Clearly, the roundness of a particle can be used to infer the history of abrasion. It is a reflection of the distance the particle has traveled, the transporting medium, and the rigor of transport. It can sometimes be used as evidence for recycling of older sediments.

Arrangement of Clastic Grains

The third element in our definition of texture is the **arrangement** of the grains in the clastic rock. Geologists examine the rock to ascertain whether grains are the same size and if they are clustered into zones or heterogeneously mixed. These observations may help to determine whether the sediment had been winnowed and sorted by currents or had been dumped rapidly. Such factors as the medium of transport, surface of deposition, and direction and velocity of currents control grain orientation. Geologists are particularly interested in studying grain orientation as a means of determining the current direction of the transporting medium. In general, sand grains deposited in moving water tend to acquire a preferred orientation in which the long axes of most of the elongate grains are aligned parallel to the direction of flow. Such particles also tend to be slightly tilted in the "upstream" direction.

To determine whether preferred orientation exists in a rock unit, geologists must collect many carefully oriented samples (Fig. 3–6) and subject the samples to a statistical analysis of the orientation of constituent grains. Many times, because of weakness or variability of currents, there may be little or no preferred orientation. Also, orientation in windblown sands is almost always less well developed than in water-carried sands. Several grain orientation studies of modern beach sands show that grain orientation in this environment is mainly controlled by backwash; in general, grains become aligned parallel to the shoreline. In stream-deposited sands, grain orientation is usually parallel to the elongation of the sand body. In glacial sediment, both elongate sand grains and pebbles show a longitudinal orientation parallel to the direction of ice movement. Grain orientation analyses are useful in determining sediment distributional patterns in the geologic past and have occasionally provided important clues to the subsurface location and trend of petroleum-bearing sandstone strata.

INFERENCES FROM PRIMARY SEDIMENTARY STRUCTURES

Primary Sedimentary Structures

A **primary sedimentary structure,** such as bedding or ripple marks, is one that forms during the deposition of sediment. In contrast, secondary sedimentary structures, such as folds and faults, develop after deposition. Primary structures are extremely useful to geologists interested in reconstructing ancient environments. For example, **mud cracks** indicate drying after deposition. These conditions are common on valley flats and in tidal zones. Mud cracks (Figs. 3–7 and 3–8) develop by shrinkage of mud or clay on drying and are most abundant in the marine environment. **Cross-bedding** is an arrangement of beds or laminations in which one set of layers is inclined relative to the others (Figs. 3–9 and 3–10). The cross-bedding units can be formed by the advance of a delta (Fig. 3–11) or a dune (Fig. 3–12). A depositional envi-

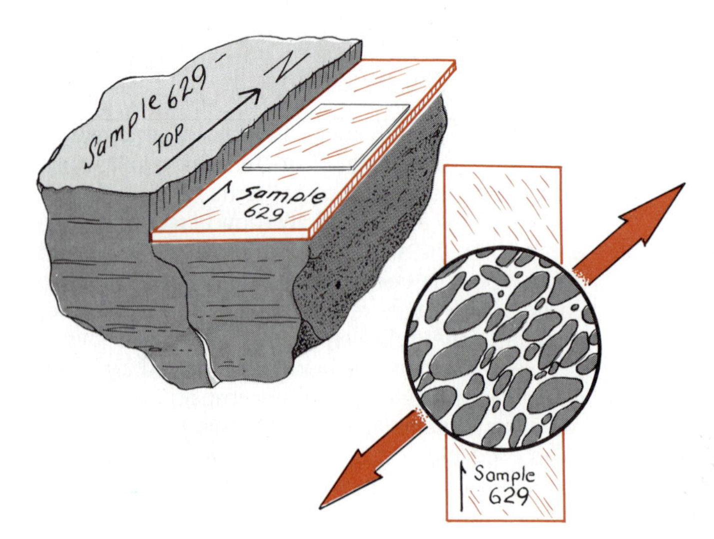

Figure 3–6 Grain orientation study. One method of studying grain orientation is to prepare an oriented thin section of a rock whose field orientation has been recorded. Grain orientations are then measured under a microscope equipped with a rotating stage. The angle of the long axis of each elongated grain from the north line is determined.. From many individual measurements, a mean orientation—in this example, about N 45° E or S 45° W—is determined and its statistical significance evaluated. A thin section cut perpendicular to bedding might reveal the tilt of the grains and might then be used to determine that the transportation medium flowed northeast rather than southwest.

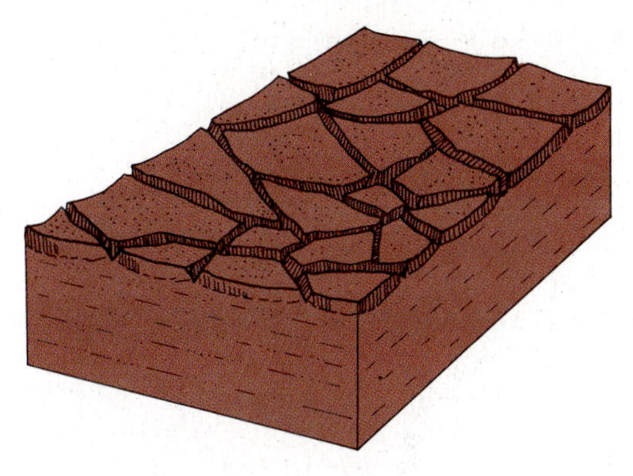

Figure 3–7 Mud cracks forming irregular polygonal pattern in clayey sediment. Crack system is caused by shrinkage following loss of water. Mud cracks are an indication of the terrestrial origin of the beds in which their casts are preserved.

Figure 3–8 Ancient mudcracks in silty shale, Isle Royale National Park, Keweenaw County, Michigan. Polygonal crack system was caused by shrinkage following loss of water. Cracks were subsequently filled with the darker sediment. Compare with similar features in Figure 4–35. (Photograph courtesy of N. K. Huber, U.S. Geological Survey.)

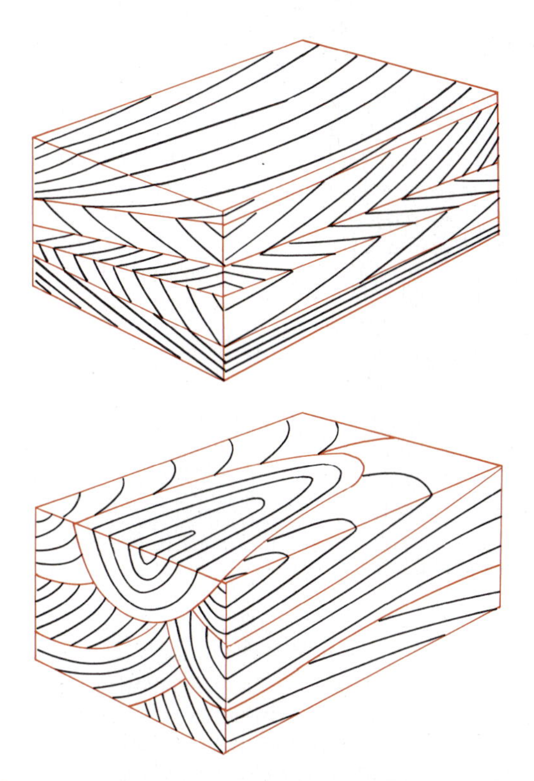

Figure 3–9 Two types of cross-bedding. The upper block shows planar laminations, as seen in beach deposits and dunes; the lower block represents trough laminations, as often formed in river channels. The orange line at the base of each set of laminations represents a surface of erosion that truncates older sets below.

Figure 3–10 Cross-bedded waterlaid sandstone. Tar Springs Formation of Mississippian Age, southern Illinois. This is an example of planar cross-bedding in that units are neither wedge-shaped nor very thick.

ronment dominated by currents is inferred from cross-bedding. The currents may be wind or water. In either medium, the direction of the inclination of the sloping beds is a useful indicator of the direction taken by the current. By plotting these directions on maps, geologists have been able to determine the pattern of prevailing winds at various times in the geologic past (Fig. 3–13).

Graded bedding consists of repeated beds, each of which has the coarsest grains at the base and successively finer grains nearer the top (Fig. 3–14). Although graded bedding may form simply as the result of faster settling of coarser, heavier grains in a sedimentary mix, it appears to be partic-

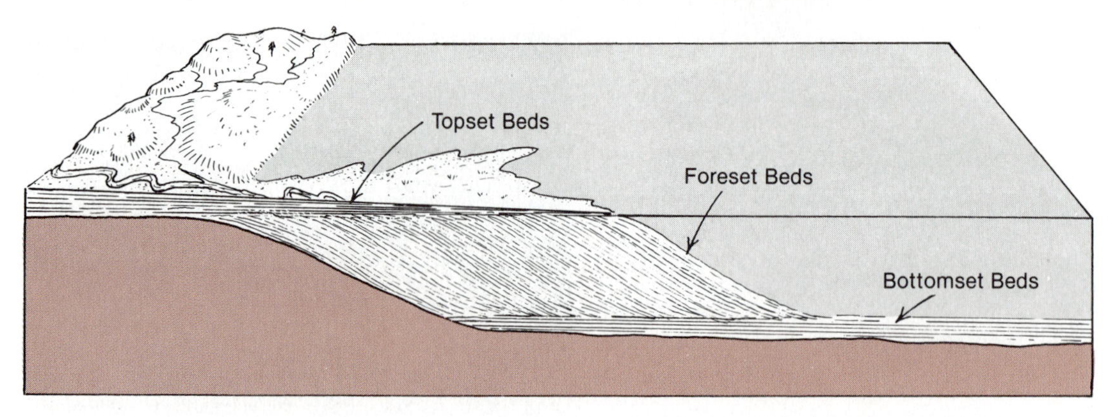

Figure 3–11 Cross-bedding in a delta. The succession of inclined foreset beds are deposited over bottomset beds that were laid down earlier. Topset beds are deposited by the stream above the foreset beds.

Figure 3–12 The dunes of Great Sand Dunes National Monument in south central Colorado (Photograph courtesy U.S. Department of the Interior, National Park Service.)

ularly characteristic of deposition by **turbidity currents.** Turbidity currents are masses of water containing large amounts of suspended material; the currents, having been made more dense than surrounding water by suspended muddy sediment, flow turbulently down slopes below relatively clearer water. Such currents are characteristically triggered by submarine earthquakes and landslides that may occur along steeply sloping regions of the sea floor. The forward part of the turbidity current contains coarser debris than does the tail. As a result, the sediment deposited at a given place on the sea bottom grades from coarse to fine as the "head" and then the "tail" of the current passes over it. The presence of graded beds may indicate former turbidity currents. Geologists believe that turbidity currents frequently characterize unstable, tectonically active environments.

Ripple marks are commonly seen sedimentary features that developed along the surfaces of bedding planes (Fig. 3–15). *Symmetric ripple marks* (Fig. 3–16) are formed by the oscillatory motion of water beneath waves. *Asymmetric ripple marks* are formed by air or water currents and are useful in indicating the direction of movement of currents. For example, ripple marks form at right angles to current directions; the steeper side of the asymmetric variety faces the direction in which the medium is flowing. Although there are instances of ripple marks developed at great depths on the sea floor, more frequently these features occur in shallow water areas.

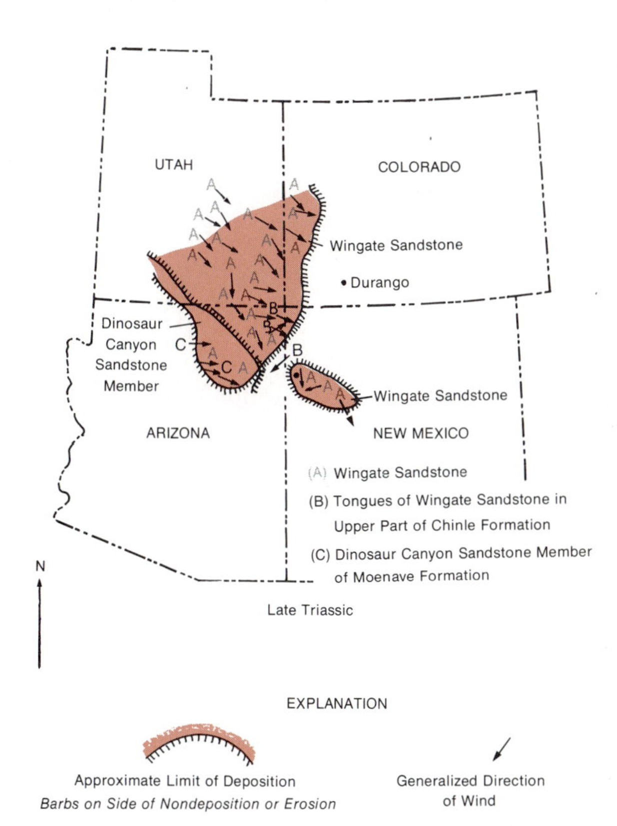

Figure 3–13 Wind directions determined from cross-stratified Late Triassic sandstones of the Colorado Plateau. The dune beds indicate, on the average, paleo winds blowing from the northwest to the southeast during the Triassic Period. (After F. G. Poole, U.S. Geological Survey Prof. Paper No. 450-D, 1962, pp. 147–151.)

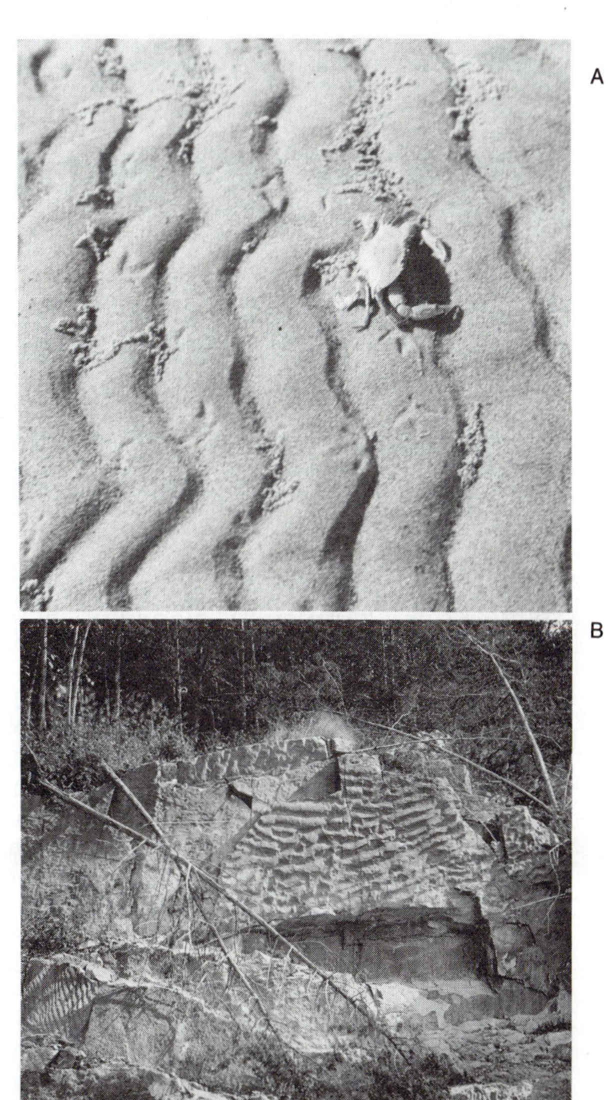

Figure 3–15 (A) Ripple marks formed in sand along modern beach. (B) Ripple marks in Baraboo Formation, Precambrian of Wisconsin. The Baraboo beds, now tilted vertically, also represent a beach of a Precambrian sea. (Upper photograph courtesy of E. D. McKee, U.S. Geological Survey.)

Figure 3–14 Graded bedding.

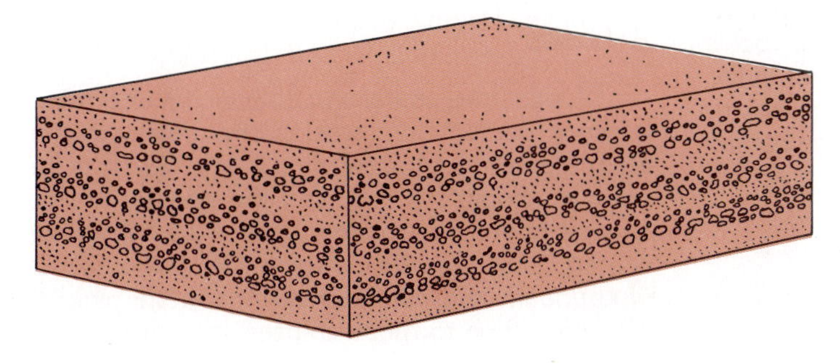

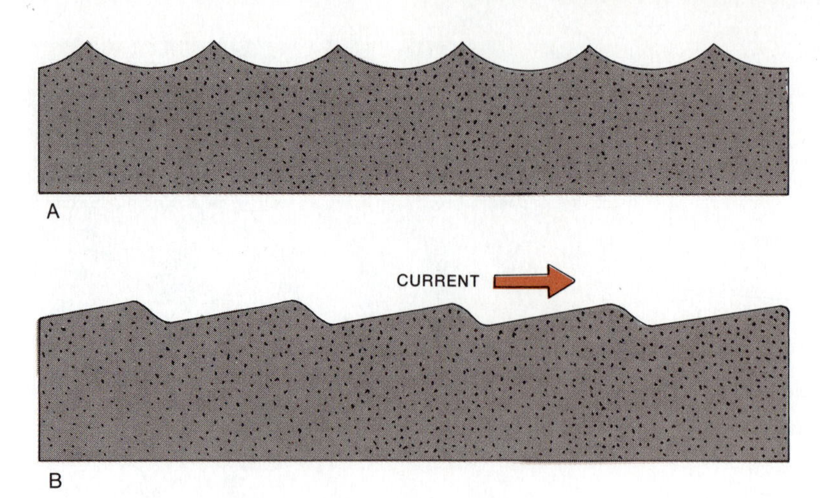

Figure 3–16 Profiles of ripple marks. (A) Oscillatory ripples. (B) Aqueous current ripples.

THE MEANING OF SANDSTONES

"Mature" and "Immature" Sandstones

Of all the clastic sedimentary rocks, sandstones have been studied the most completely and provide the greatest amount of information about ancient environmental conditions. The mineral composition of the grains in sandstones provides important information about the source areas for the sediment as well as the history of the sediment prior to deposition. Often by studying the grains closely, one can ascertain whether the source material was granitic, metamorphic, volcanic, or sedimentary. The mineral content also provides a clue to the compositional maturity of sandstones. Rigorous weathering and long transport tend to reduce the less stable feldspars and ferromagnesian minerals to clay and iron compounds and to cause rounding and sorting of grains. Hence, one can assume that a sandstone rich in these less durable and angular components underwent relatively less transport and other forms of geologic duress. Such sediments are termed **immature** and are most frequently deposited close to their source areas. On the other hand, quartz is an especially durable component of sandstone. Quartz can be used as an indicator of a sandstone's maturity; the higher the percentage of quartz, the greater the maturity. In addition to providing an indication of a rock's maturity, composition is an important factor in the classification of sandstones into *graywackes, subgraywackes, arkoses,* and *orthoquartzites* (Fig. 3–17).

Graywackes

Graywackes (from the German term *wacken,* meaning "waste or barren") are immature sandstones consisting of significant quantities of dark, very fine-grained material. Normally, this fine matrix consists of clay, chlorite, micas, and silt. There is little or no cement, and the sand-sized grains are not in close contact because they are separated by the finer matrix particles. The matrix constitutes approximately 30 per cent of the rock, the remaining coarser grains consisting of quartz, feldspar, and rock particles. Graywacke has a dirty, "poured in" appearance. The poor sorting, angularity of grains, and heterogeneous composition of graywackes indicate an unstable source and depositional area in which debris resulting from accelerated erosion of highlands is carried rapidly to subsiding basins. Graded bedding, interspersed layers of volcanic rocks, and cherts (which may indirectly derive their silica from volcanic ash) attest further to dynamic conditions in the area of deposition. The inferred tectonic setting is very unstable, with deposition occurring offshore from an actively rising mountainous region (Fig. 3–18). Graywackes and associated shales and cherts may contain fossils of deep-water organisms, indicating deposition at great depth. Such shallow water sedimentary structures as cross-bedding and ripple marks are rarely found. To the experienced petrologist, graywackes clearly indicate dynamic, unstable conditions.

Arkoses

Sandstones containing 25 per cent or more of feldspar (derived from erosion of a granitic source area) are called **arkoses** (Fig. 3–17). Quartz is the most abundant mineral, and the angular-to-subangular grains are bonded together by calcareous cement, clay minerals, or iron oxide. The presence of abundant feldspars and iron imparts a pinkish-gray coloration to many arkoses. In general, arko-

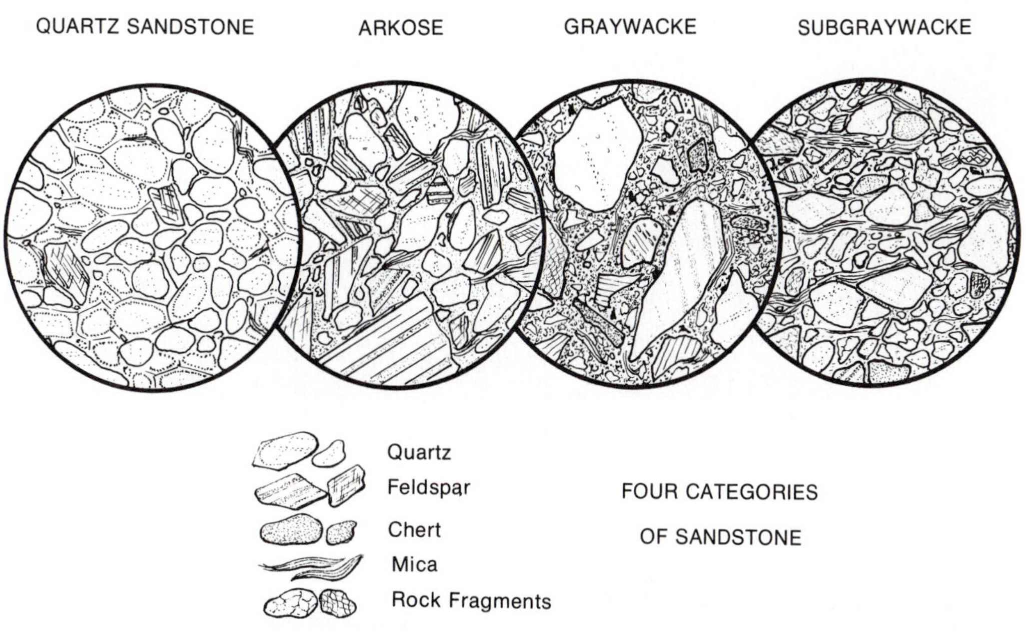

QUARTZ SANDSTONE ARKOSE GRAYWACKE SUBGRAYWACKE

Quartz
Feldspar
Chert
Mica
Rock Fragments

FOUR CATEGORIES OF SANDSTONE

Figure 3–17 Four categories of sandstone as seen in thin section under the microscope. (Diameter of field is about 4 mm.)

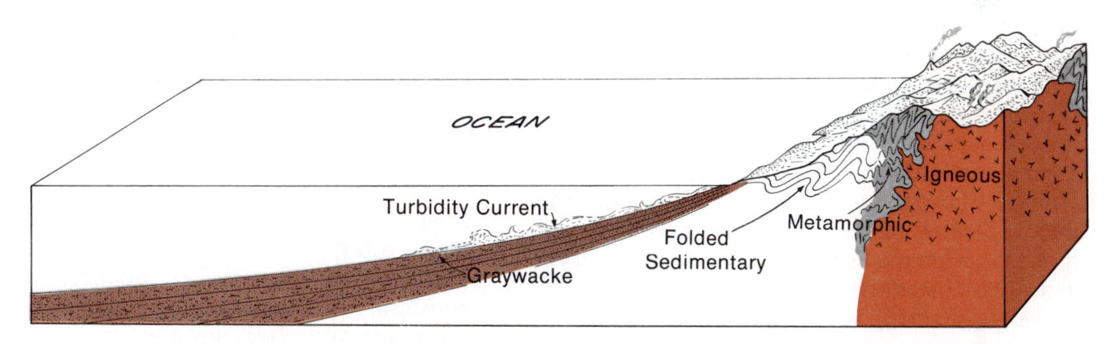

Figure 3–18 Tectonic setting in which graywacke is deposited. Frequently, graywackes are transported and deposited by masses of water highly charged with suspended sediment. Because of the suspended matter, the mass is denser than surrounding water and moves along the sea floor as a "turbidity current."

ses tend to be coarse sandstones that are moderately well sorted. They may develop as basal sandstones derived from the erosion of a granitic coastal area experiencing a marine transgression. Such blanket-like deposits may exhibit cross-bedding and ripple marks. Arkosic sandstones are more frequently nonmarine in origin, accumulating as feldspathic sands in **fault troughs or sedimentary basins** adjacent to granitic highlands (Fig. 3–19). A sedimentary basin is any broad, low continental or marine catchment area for sediments eroded from nearby uplands. (A sedimentary basin differs from a structural basin, in which rocks have been bent into a basin-like shape.)

Orthoquartzites

Because of the incorporation of "quartzite" into the term "orthoquartzite," students might infer the rock is metamorphic. It is not! **Orthoquartzites** are rocks characterized by dominance of quartz with little or no feldspar, mica, or fine clastic matrix. The quartz grains are well sorted and well rounded. They are most commonly held together by such cements as calcite and silica. Chemical cements such as these tend to be more characteristic of "clean" sandstones like orthoquartzites than of "dirtier" rocks like graywacke. The presence of a dense, clayey matrix seems to retard the formation

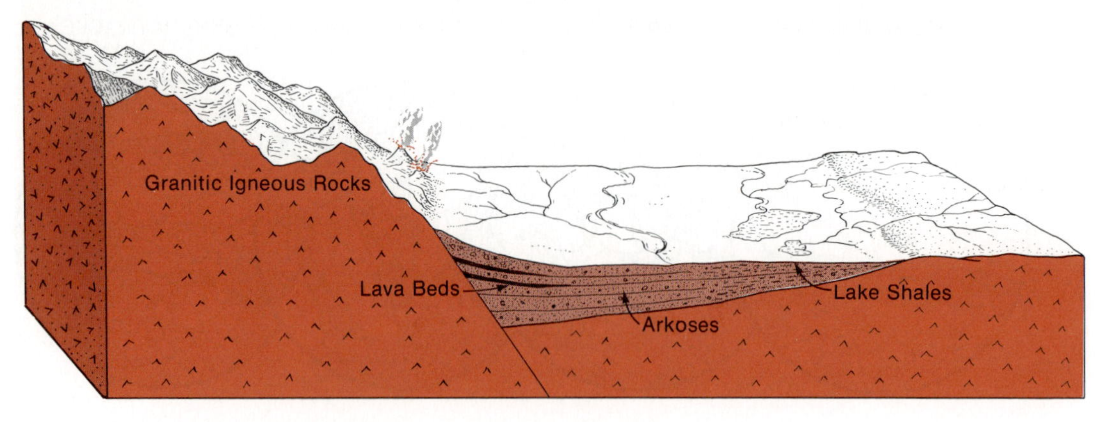

Figure 3–19 Geologic environment in which arkose may be deposited.

of chemical cement, perhaps because fine material fills pore openings where crystallization might occur.

Calcite cement may develop between the grains as a simple, finely crystalline filling, or large crystals may form and each may incorporate hundreds of quartz grains. Silica cement in orthoquartzites commonly develops as overgrowths on the original quartz grains (Fig. 3–20). During this process, ions are added in crystallographic continuity with the host grain even when the host grain is well rounded. The crystal faces of the overgrowths are clearly visible when the rocks are examined with the microscope, and the boundary that separates the original grain and the overgrowth is often marked by a thin zone of impurities that once coated the host grain.

Cross-bedding and ripple marks characterize these mature sandstones. Orthoquartzites occur as widespread but thin "blankets" of sand. They are frequently associated with, or grade into, limestone or dolostone beds. A geologist may infer that the tectonic setting for such associations of orthoquartzites and carbonate rocks is a slowly subsiding, relatively stable, shallow water depositional area that is a great distance from mountainous regions and is covered by extensive but shallow seas on the craton or near its border (Fig. 3–21).

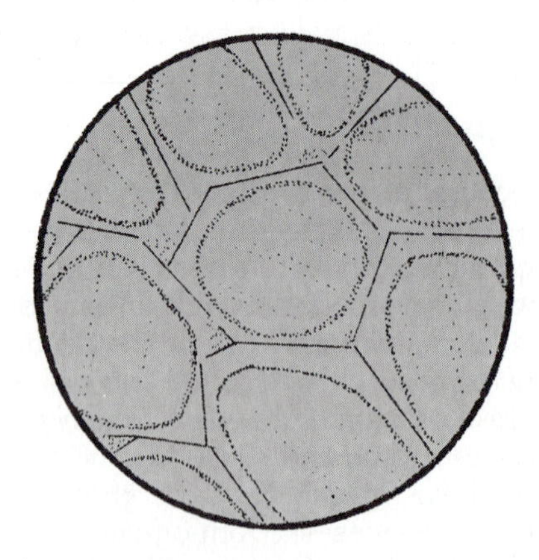
Figure 3–20 Orthoquartzite with silica cement formed as overgrowths in crystallographic continuity with host grains.

Subgraywackes

Arkoses, orthoquartzites, and graywackes are rather distinct kinds of sandstones. A sandstone that is more of a transitional type is termed a **subgraywacke** (Fig. 3–22). Subgraywackes are sandstones in which feldspars are relatively scarce and quartz and chert are abundant; they have a fine-grained detrital matrix that rarely exceeds 15 per cent. The voids are filled with mineral cement and clay. In many respects, subgraywacke is a rock intermediate between graywacke and orthoquartzite. For example, in subgraywackes the quartz grains are better rounded and more abundant, the sorting is better, the quantity of matrix material is lower, and the porosity is greater than in graywackes. These differences are the result of more reworking and transporting of the subgraywacke sediment. Subgraywackes often exhibit cross-bedding and ripple marks. The most frequent environment for such rocks are deltaic coastal plains, where they may be deposited subaerially or as marine deposits. Coal beds and micaceous shales are frequently associated with subgraywackes. These sandstones really warrant a more distinctive name, for they are the most abundant of all the sandstone groups.

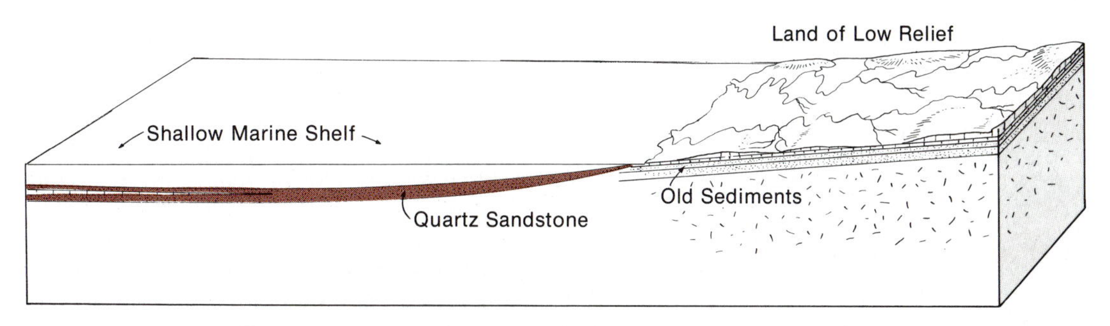

Figure 3–21 Idealized geologic conditions under which quartz sandstone may be deposited. There is little tectonic movement in this environment. Water depth is shallow, and the basin subsides only very slowly.

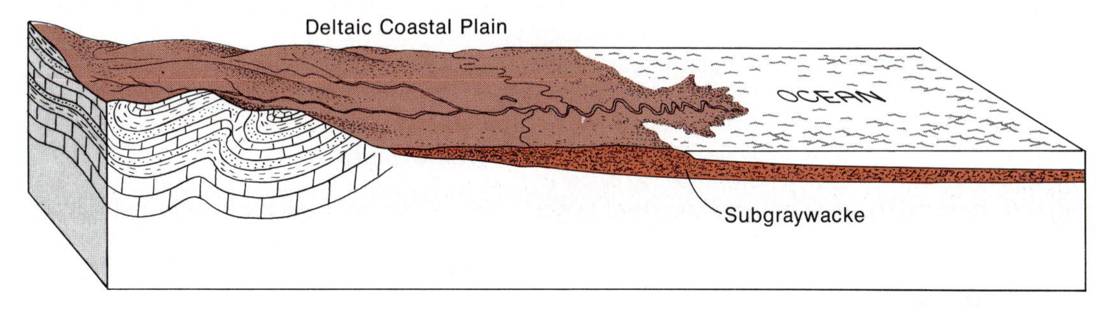

Figure 3–22 Deltaic environment in which subgraywackes may be deposited.

THE MEANING OF LIMESTONES

Most carbonate sedimentation today occurs relatively near the equator in warm, clear, oceanic waters. In such areas, calcareous algae and corals thrive, and marine animals construct thicker shells of calcite and aragonite. Clayey, turbid water inhibits carbonate formation, so that marine areas muddied by an influx of sediment-laden river water are not likey to be sites of carbonate accumulations.

Tropical seas today are essentially saturated with calcium carbonate. Therefore, any process that removes carbon dioxide from the water will tend to precipitate **lime** ($CaCO_3$). The mechanisms for removing carbon dioxide may include increases in temperature, evaporation, an influx of supersaturated water into an area of tiny $CaCO_3$ nuclei or "seeds," various biologic processes in marine animals, ammonia production in sea water during bacterial decay, and photosynthetic activity of marine plants. Of these mechanisms, photosynthesis is of greatest importance. Because photosynthetic green algae thrive at depths of less than about 15 meters, carbonate sedimentation is most rapid in similarly shallow marine locations.

The lime "muds" that are prevalent in modern carbonate-depositing environments may develop in more than one way. Some are derived from the death and decay of calcareous algae that grow on the sea floor. Additional fine carbonate material is composed of the microscopic shells of unicellular marine organisms. Some result from the precipitation of tiny crystals of lime from sea water that has been chemically altered by the biologic processes of marine plants. Coarser particles result from the abrasion of the shells of invertebrates or consist of fecal pellets produced by burrowing organisms.

The requirement of warm, clear, shallow seas for the accumulation of modern carbonates seems to apply equally well to ancient deposits. Major sequences of ancient limestones are relatively free of clay and frequently contain an abundance of fossils thought to be representative of shallow, warm seas. Ancient carbonate rocks have developed in a variety of tectonic settings. Thick sections of limestones and dolostones have formed in subsiding basins in west Texas, Alberta, and Michigan. In such areas, optimum conditions for carbonate sedimentation resulted in a rate of accumulation that approximately equaled subsidence. Thick deposits of limestones are also known to have occurred along the cratonic margins of geosynclines. Thinner, but very extensive blankets of limestones have accumulated on the stable cratonic regions of the continents when those regions were subjected to marine inundations. Limestones generally are not characteristic of the more dynamic parts of geosynclinal belts, probably because of the influx of clay and other siliceous clastics in such environments, as well as rapid changes in water depth.

THE MEANING OF SHALES

Shale is a general term for very fine-textured, fissile (that is, capable of being split into thin layers) rock composed of clay, mud, and silt. In general, the environmental significance of shales parallels that of the sandstones with which they are associated. Frequently, the silt-size particles in shales are similar in composition and shape to the associated sandstone beds. These silty components can frequently be extracted from the shale sample by disaggregating the rock in water and repeatedly pouring off the muddy liquid, retaining the larger particles as a residue.

In shales associated with orthoquartzite strata, the silt fraction often consists predominantly of rounded quartz grains. Such **"quartz shales"** result from the reworking of older residual clays by transgressing shallow seas. Their association with thin, widespread limestones and orthoquartzites provides evidence for their deposition under stable tectonic conditions.

"Feldspathic shales" contain at least 10 per cent feldspar in the silt size and tend to be rich in the clay mineral kaolinite. Feldspathic shales are frequent associates of arkoses and are presumed to have formed in a similar environment. Such shales are representative of the finer sediment winnowed from coarser detritus and deposited in quieter locations.

"Chloritic shales" are geosynclinal deposits and are usually associated with graywackes. As implied by their name, flakes of chlorite are common among the silt-sized components. The less flaky particles tend to be angular. Fissility in these rocks is less well developed than in other shales. The clay and silt particles in chloritic shales are generally derived from mountainous, unstable source areas nearby.

A shale type that is the approximate equivalent of a subgraywacke can be designated a **"micaceous shale."** Mica flakes, quartz, and feldspar are all among its silt-sized components. Micaceous shales are deposited under conditions somewhat less stable than the environment for quartz shales. They are particularly characteristic of ancient deltaic deposits.

The clay minerals that occur in shales are complex hydrous aluminosilicates with constituent atoms arranged in silicate sheet structures. **Kaolinites, montmorillonites,** and **illites** are the three major groups of clay minerals. Kaolinites are the purest and seem to have a preferred occurrence in terrestrial environments. Montmorillonites may contain magnesium, calcium, or sodium or any combination of these three, whereas potassium is an essential constituent of illites. Illites are the predominant clay mineral in more ancient shales. Unfortunately, positive identification of clay minerals in hand specimens or under the microscope is not possible and can be accomplished only by means of x-ray, thermal, or chemical analyses.

THE SEDIMENTARY ROCK RECORD

Rock Units

Rock units are bodies of rock that can be subdivided and recognized wherever they occur on the basis of observable objective criteria. They can be identified in the field and mapped as units that are distinctive and different from neighboring units. The fundamental rock unit is the **formation,** which may be defined as a mappable, lithologically distinct body of rock with recognizable contacts with other formations (Fig. 3–23).

Maps that record the distribution and nature of rock units and the occurrence of geologic features such as folds and faults are called **geologic maps.** To prepare a geologic map, formations must first be matched or correlated from place to place. In this way, the areal or lateral extent of each rock unit can be determined. If the formation is continuously exposed, as are formations along the walls of the Grand Canyon, widely separated exposures can be correlated by tracing them continuously from place to place. More frequently, exposures are not continuous but may be covered or eroded away at various places. In such cases, lithologic attributes or other special properties of the unit serve to identify it at isolated exposures. The formations can also be correlated by their position in the sequence of strata. For example, three shale units in an area might be superficially similar in appearance, but if one is underlain by fossiliferous limestone, another by red beds, and a third by dolostone, each becomes mappable as a distinct unit recognizable by its relationship to subjacent formations.

Naming Rock Formations

Formations are given two names: (1) a geographic name that refers to a locality where the

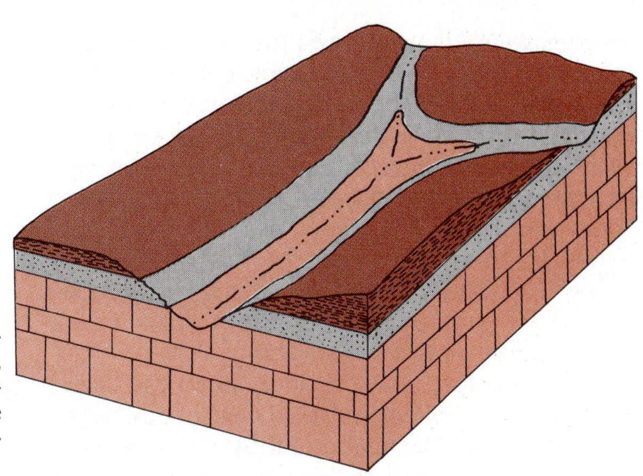

Figure 3–23 Formations. The diagram shows three formations. In practice, these formations would be formally named, often after a geographic location near which they are well exposed. For example, the three formations shown here might be designated the "Cedar City Limestone," "Big Springs Sandstone," and "Plattsburgh Shale."

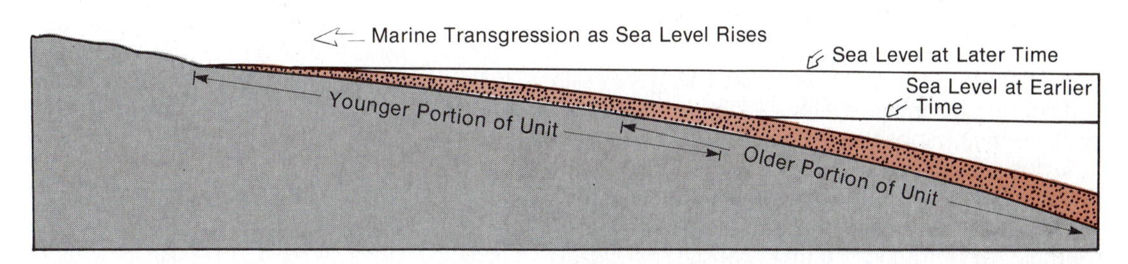

Figure 3–24 Diagram showing how the original deposits of a formation may vary in age from place to place.

formation is well exposed or where it was first described and (2) a rock name if the formation is primarily of one lithologic type. For example, the "St. Louis Limestone" is a carbonate formation named after exposures at St. Louis. When formations have several lithologic types within them, the locality name may be followed by the word "formation." Distinctive smaller units within formations may be split out as "members," and formations may be combined into groups because of related lithologic characteristics (or by their position between distinct stratigraphic breaks). In mapping formations and other rock units, geologists recognize that the unit may or may not be the same age everywhere it is encountered. The nearshore sands deposited by a sea slowly transgressing across a coastal plain may deposit a single blanket of orthoquartzite. However, it will be older where the transgression began and younger where it ended (Fig. 3–24).

Time-Rock Units

Historically, geologists attempting to reconstruct the biologic and physical environment of a particular region for a particular time interval quickly recognized that rock units, like formations, were not adequate. As we have noted, a given rock unit might be of different ages at different localities. Thus, it was necessary to formulate the term **time-rock unit** (also known as a chronostratigraphic unit) for sedimentary layers deposited during a particular time interval. The fundamental time-rock unit is the **system.** A system corresponds to a unit of pure time called a **period** in the geologic calendar (Table 3–2). Both terms have the same "first name" so that it is correct to refer, for example, to the "Devonian Period" (Table 3–3) as a certain length of time in geologic history, whereas the "Devonian System" refers to the rocks that are the tangible record of that period of time. Geologic periods may be divided into **epochs** and **ages.** Similarly, a system is divisible into **series,** which in turn may include yet smaller units termed **stages.** As depicted in Table 3–2, geologists have also established time and time-rock names for units above the level of the period and system. These include the time term **era,** with its time-rock equivalent **erathem,** and **eon,** for which the time-rock term **eonothem** is employed.

Time-rock units may include more than a single lithologic type. Unlike rock units, however, they are consistently the same age throughout their lateral extent. They permit the geologists to prepare maps of ancient geography for particular times. The areal extent of a time-rock unit can be determined by examination of surface exposures and records of rocks penetrated by deep drill holes. Upper and lower boundaries are intangible time planes determined largely on the basis of fossil evidence. The determination of these time boundaries and the correlation of time-rock units is of cardinal importance to the historical geologist. In this work, a vast store of information about the "geologic life spans" of particular species that occur as fossils is used. A decription of these paleontologic methods for determining age equivalency of strata appears in the next chapter.

Facies

There is yet another term used by geologists in reference to those characteristics of rocks that relate to their depositional environments. The term is **facies.** A sedimentary facies is the general appearance or aspect of a rock from which its environment of deposition can be determined. For example, a formation might consist of clastic limestone along one of its lateral margins and lithographic limestone elsewhere. Geologists might then delineate a "clastic limestone facies" and interpret it as a nearshore part of the formation, whereas they might interpret the micritic limestone facies as a former offshore deposit. In this case, the distinguishing characteristics are lithologic (rather than biologic); therefore, the facies can be further

Table 3–2 TERMS COMMONLY USED FOR THE TIME AND TIME-ROCK DIVISIONS

Time Divisions	Time-Rock Divisions
EON	EONOTHEM
ERA	ERATHEM
PERIOD	SYSTEM
EPOCH	SERIES
AGE	STAGE
	ZONE

Table 3–3 THE GEOLOGIC TIME SCALE

Time Units of the Geologic Time Scale
(*Numbers are absolute dates in millions of years before present*)

Eon	*Eras*	*Periods*	*Epochs*	
Phanerozoic	*Cenozoic*	Quaternary	Recent or Holocene	
			Pleistocene	2.0
		Tertiary	Pliocene	5
			Miocene	23
			Oligocene	37
			Eocene	55
			Paleocene	
		65		
	Mesozoic	Cretaceous		
		140		
		Jurassic		
		195		
		Triassic		
		225		
	Paleozoic	Permian		
		280		
		Pennsylvanian (Upper Carboniferous)		
		320		
		Mississippian (Lower Carboniferous)		
		360		
		Devonian		
		400		
		Silurian		
		435		
		Ordovician		
		500		
		Cambrian		
		570		

Precambrian Time
or
Cryptozoic Eon

designated as **lithofacies.** In other cases, the rock unit may be lithologically uniform, but the fossil assemblages differ and permit recognition of different **biofacies** that reflect differences in the environment. A limestone unit, for example, might contain abundant fossils of shallow water reef corals along its thinning edge and elsewhere be characterized by remains of deep-water sea urchins and snails. There would thus be two biofacies developed—one reflecting deeper water than the other. They could, therefore, be designated the "coral" and the "echinoid-gastropod biofacies."

The term "facies" can also be applied to sediment accumulating today. For example, bottom sediments are systematically sampled by means of devices lowered to the sea floor by ships. The location and sediment type for each sample is then plotted on a map.

Lithofacies maps (which are described in more detail in a later section) show the areal distribution of the different types of sediments. They are useful in depicting the areal variations in sedimentary environments for a stratigraphic unit that covers a particular area. Such maps provide a view of different facies of essentially the same age. If one were able to make such maps of successively different times, it would become apparent that ancient facies have shifted their localities as the seas **advanced** or **retreated,** or as environmental conditions changed.

Consider for a moment an arm of the sea slowly transgressing (advancing over) the land. The sediment deposited on the sea floor may idealistically consist of a nearshore sand facies, an offshore mud facies, and a far offshore carbonate facies. As the shoreline advances inland, the boundaries of these facies also shift in the same direction, thereby developing an overlap sequence (Fig. 3–25) in which coarser sediments are covered by finer ones. Should the sea subsequently begin a withdrawal, the facies boundaries will again move in the same direction as the shoreline, creating as they do so an **offlap** sequence of beds (Fig. 3–26). In offlap situations, coarser nearshore sediment tends to lie above finer sediments. Also, because offlap units are deposited during marine regressions, recently deposited sediment is exposed to erosion, and part of the sedimentary sequence is lost. Study of sequential vertical changes in lithology, such as those represented by offlap and overlap relationships, is one method by which geologists recognize ancient advances and retreats of the seas and chart the positions of shorelines.

Correlation

When examining an isolated exposure of rock in a road cut or the bank of a stream, a geologist is aware that the rock may continue laterally beneath the cover of soil and loose sediment and that the same stratum or rock body, or its equivalent, is likely to be found at other localities. The determina-

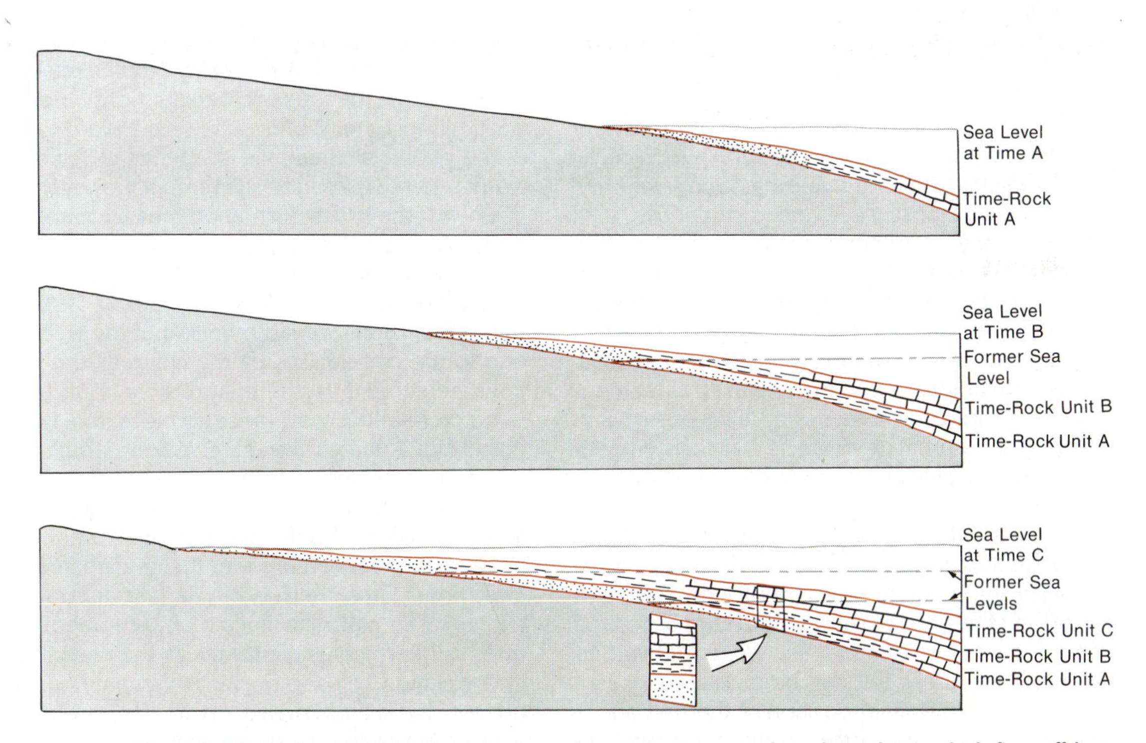

Figure 3–25 Sedimentation during a transgression produces an *overlap* relationship in which finer offshore lithofacies overlie coarser nearshore facies (see inset), nearshore facies are progressively displaced away from a marine point of reference, and older beds are protected from erosion by younger beds.

Walther's law of succession of facies

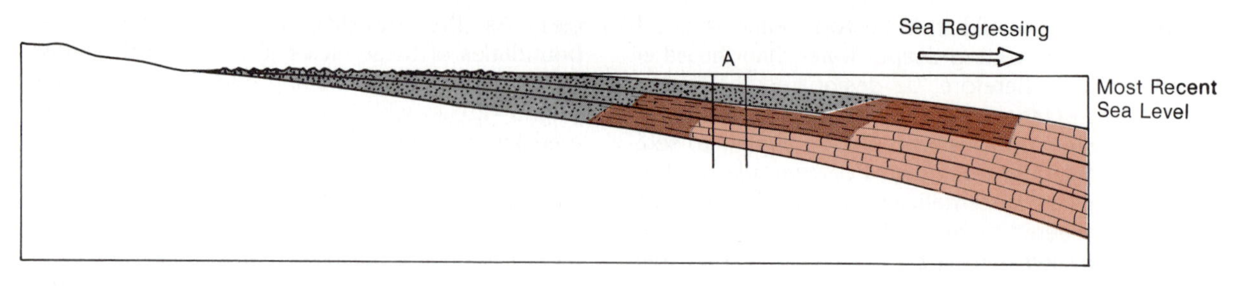

Figure 3–26 Sedimentation during a regression produces an offlap relationship in which coarser nearshore lithofacies overlie finer offshore lithofacies as shown in A. The sandy nearshore facies is progressively displaced toward the marine point of reference. Older beds are subjected to erosion as the regression of the sea proceeds.

tion of the equivalence of bodies of rock in different localities is called **correlation.** The rock bodies may be equivalent in their lithology (composition, texture, color, and so on), in their age, in the fossils they contain, or in a combination of these attributes. Because there is more than one meaning for the term, geologists are careful to indicate the kind of correlation used in solving a particular geologic problem. In some cases it is only necessary to trace the occurrence of a lithologically distinctive unit, and the age of that unit is not critical. Other problems can be solved only through the correlation of rocks that are of the same age. Such correlations involve time-rock units and are of the utmost importance in geology. They are the basis for the geologic time scale and are essential in working out the geologic history of any region.

The correlation of strata from one locality to another may be accomplished in several ways. If the strata are well exposed at the earth's surface, as in arid regions, where soil and plant cover is thin, then it may be possible to trace distinctive rock units for many kilometers across the countryside by actually walking along the exposed strata. In using this straightforward method of correlation, the geologist may sketch the contacts between units directly on topographic maps or aerial photographs. The notations can then be used in the construction of geologic maps. It is also possible to construct a map of the contacts between correlative units in the field by using appropriate surveying instruments.

In areas where bedrock is covered by dense vegetation and a thick layer of soil, geologists must rely upon intermittent exposures found here and there along the sides of valleys, in stream beds, and in road cuts. Correlations are more difficult to make in these areas but can be facilitated by recognizing the similarity in position of the bed one is trying to correlate with other units in the total sequence of strata. A formation may have changed somewhat in appearance between two localities, but if it always lies above or below a distinctive stratum of consistent appearance, then the correlation of the problematic formation is confirmed (Fig. 3–27).

A simple illustration of how correlations are used to build a composite picture of the rock record is provided in Figure 3–28. A geologist working along the sea cliffs at location 1 recognizes a dense oölitic limestone (formation F) at the lip of the cliff. The limestone is underlain by formations E and D. Months later, the geologist continues the survey in the canyon at location 2. Because of its distinctive character, the geologist recognizes the oölitic limestone in the canyon as the same formation seen earlier along the coast and makes this correlation. The formation below F in the canyon is somewhat more clayey than that at locality 1 but is inferred to be the same because it occurs right under the oölitic limestone. Working upward toward location 3, the geologist maps the sequence of formations from G to K. Questions still remain, however. What lies below the lowest formation thus far found? Perhaps years later an oil well, such as that at location 3, might provide the answer. Drilling reveals that formations C, B, and A lie beneath D. Petroleum geologists monitoring the drilling of the well would add to the correlations by matching all the formations penetrated by the drill to those found earlier in outcrop. In this way, piece by piece, a network of correlations across an entire region is built up.

For correlations of time-rock units, one cannot depend upon similarities in lithology to establish equivalence. Rocks of similar appearance have been formed repeatedly over the long span of geologic time. Thus, there is the danger of correlating two apparently similar units that were deposited at quite different times. Fortunately, the use of fossils in correlation helps to prevent mismatching. Methods of correlation based on fossils are fully described in the next chapter. They are based on the fact that animals and plants have undergone change through geologic time, and therefore the

Figure 3–27 If the lithology of a rock is not sufficiently distinctive to permit its correlation from one locality to another, its position in relation to distinctive rock units above and below may aid in correlation. In the example shown here, the limestone unit at locality A can be correlated to the lowest of the four limestone units at locality B because of its stratal position between the gray shale and sandstone units.

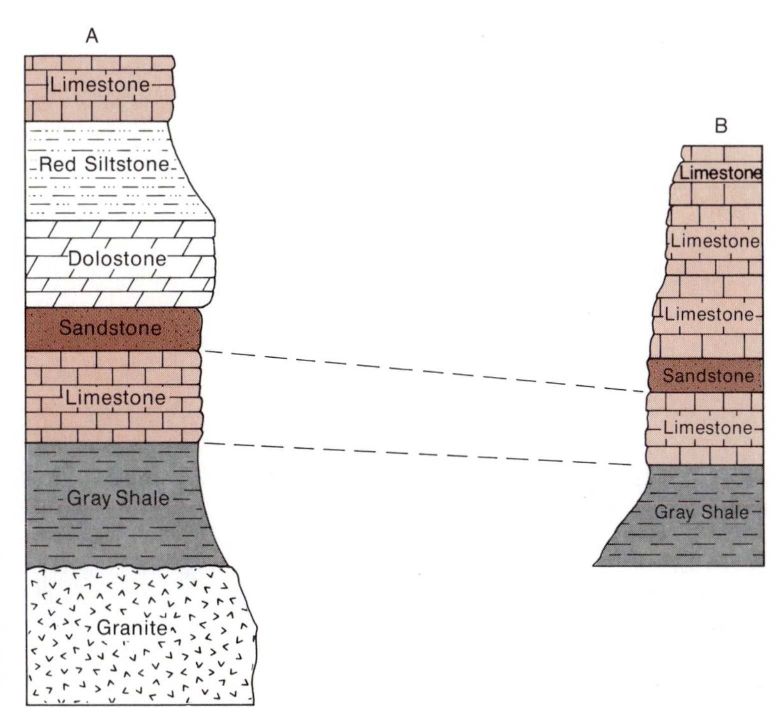

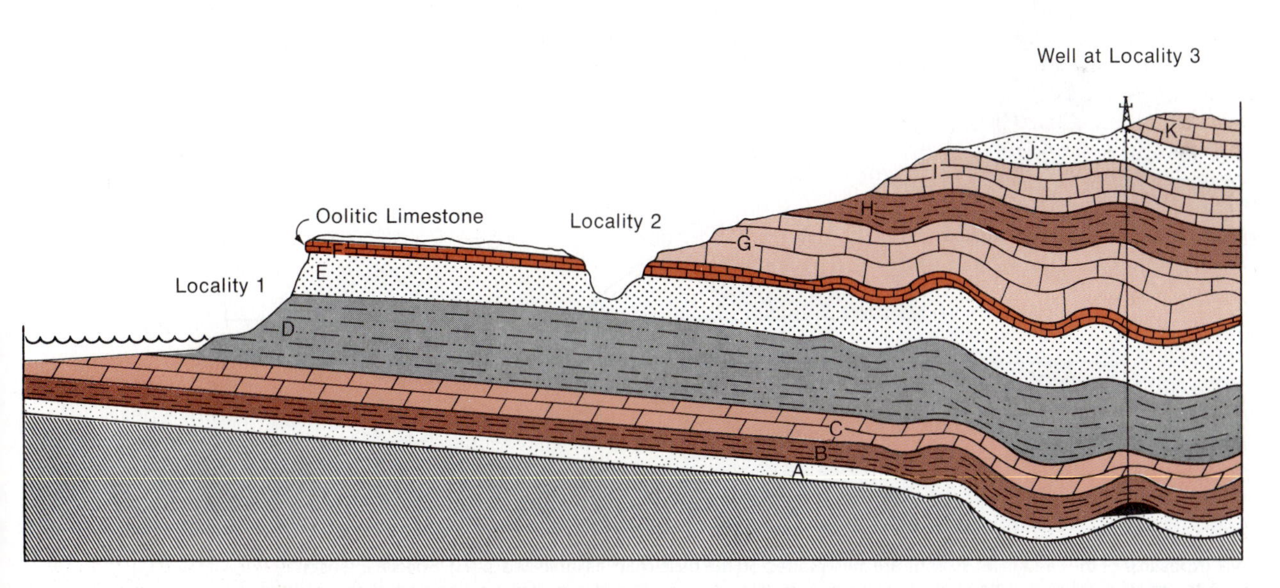

Figure 3–28 An understanding of the sequence of formations in an area usually begins with examination of surface rocks and correlation between isolated exposures. Study of samples from deep wells permits the geologist to expand the known sequence of formations and to verify the areal extent and thickness of both surface and subsurface formations.

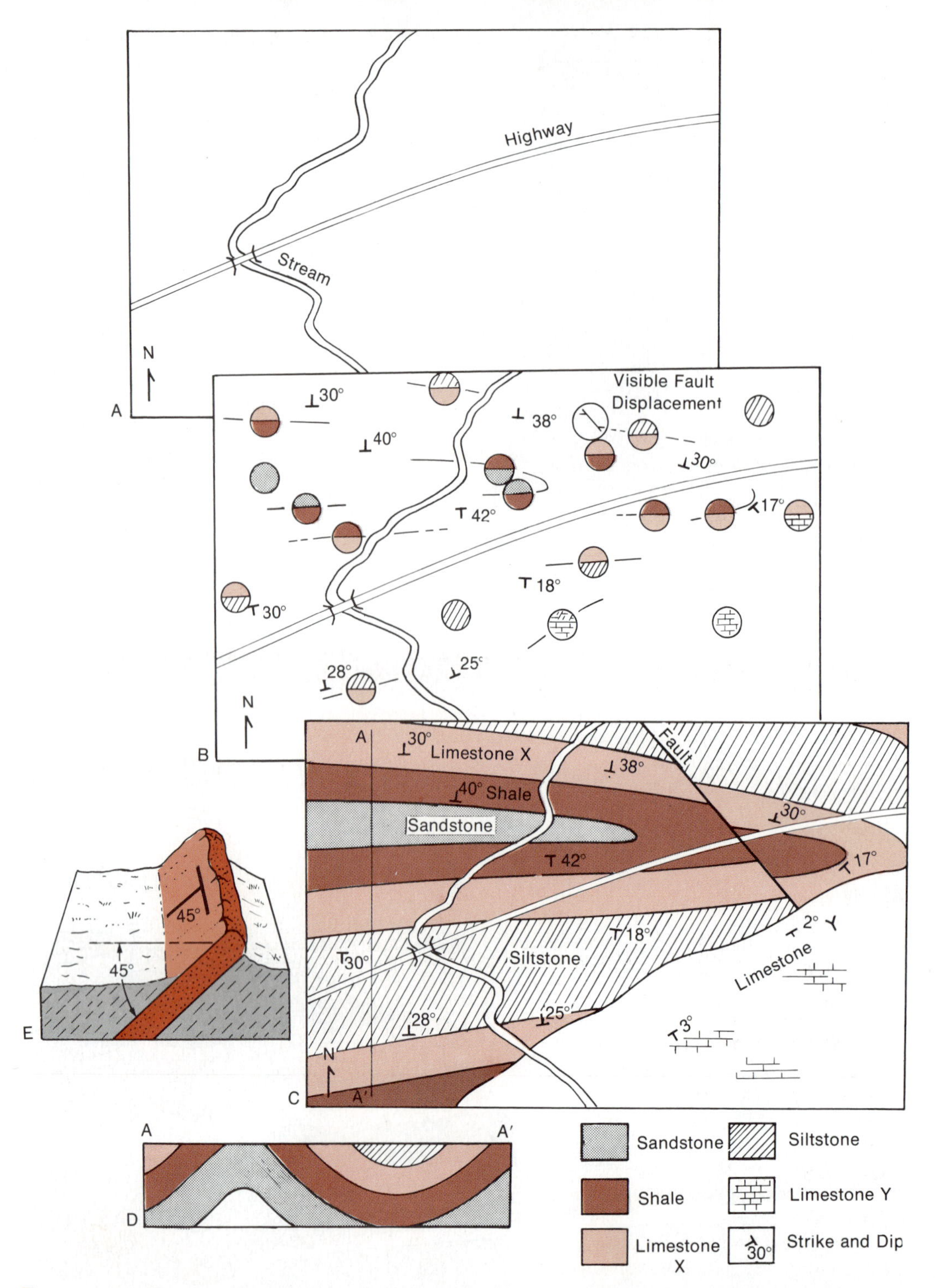

Figure 3–29 Steps in the preparation of a geologic map. A suitable base map (A) is selected. The locations of rock exposures of the various formations are then plotted on the base map, as indicated in (B). Special attention is given to exposures that include contacts between formations; where they can be followed horizontally, they are traced onto the base map also. *Strike* (the compass direction of a line formed by the intersection of the surface of a bed and a horizontal plane) and *dip* are measured wherever possible and added to the data on the base map. After careful field study and synthesis of all the available information, formation boundaries are drawn to best fit the data. On the completed map (C), color patterns are used to show the appearance of the area if there were no soil cover. A cross-section along line A–A′ is shown in (D); (E) is a block diagram illustrating strike and dip.

fossil remains of life are recognizably different in rocks of different ages. Conversely, rocks of the same age but from widely separated regions can be expected to contain similar assemblages of fossils.

Unfortunately, there are complications to these generalizations. In order for two strata to have similar fossils, they would have to have been deposited in rather similar environments. A sandstone formed on a river floodplain would have quite different fossils than one formed at the same time in a nearshore marine environment. How might one go about establishing that the floodplain deposit could be correlated to the marine deposit? In some cases this might be done by physically tracing out the beds along a cliff or valley side. Occasionally, one is able to find fossils that actually do occur in both deposits. Pollen grains, for example, could have been wafted by the wind into both environments. Possibly, both deposits occur directly above a distinctive, firmly correlated stratum such as a layer of volcanic ash. Ash beds are particularly good time markers because they are deposited over a wide area during a relatively brief interval of time. Such key beds are exceptionally useful in establishing the correlation of overlying strata. Finally, the geologist may be able to obtain the actual age of the strata using radioactive methods, and these values can then be used to establish the correlation.

MAPPING THE PAST

Geologic Maps

The study of the geologic history of an area ordinarily begins with the construction of a geologic map. Assume, for a moment, that all the loose material and vegetation were miraculously removed from your home state, so that bedrock would be exposed everywhere. Imagine, further, that the surfaces of the formations now exposed were each painted a different color and photographed vertically from an airplane. Such a photograph would constitute a simple geologic map. In actual practice, a geologic map is prepared by locating contact lines between formations in the field and then plotting these contacts on a base map. Symbols are added to the colored areas to indicate formations and lithologic regions, mineral deposits, and structures such as folds and faults. Once the geologic map is completed, a geologist can tell a good deal about the geologic history of an area. The formations depicted represent sequential "pages" in the geologic record. From the simple geologic map shown as Figure 3–29, the geologist is able to deduce that there was an ancient period of compressional folding, that the folds were subsequently faulted, and that an advance of the sea resulted in deposition of younger sedimentary layers above the more ancient folded strata.

Paleogeographic Maps

A map showing the geography of a region or area at some specific time in the geologic past is termed a **paleogeographic map.** Such maps are really interpretations based on all available paleontologic and geologic data. The majority of such maps show the distribution of ancient lands and seas (Fig. 3–30). Paleogeographic maps are, at best, of limited accuracy, because as seas advance and retreat endlessly through time, the line drawn at the sea's edge may represent an average of several shoreline positions. They are nevertheless useful for showing general geographic conditions within regions or continents. To prepare a paleogeographic map, one would plot all occurrences of rocks of a given time interval on a map and enclose the area of occurrence in boundary lines. Areas of nonoccurrence may be places of no

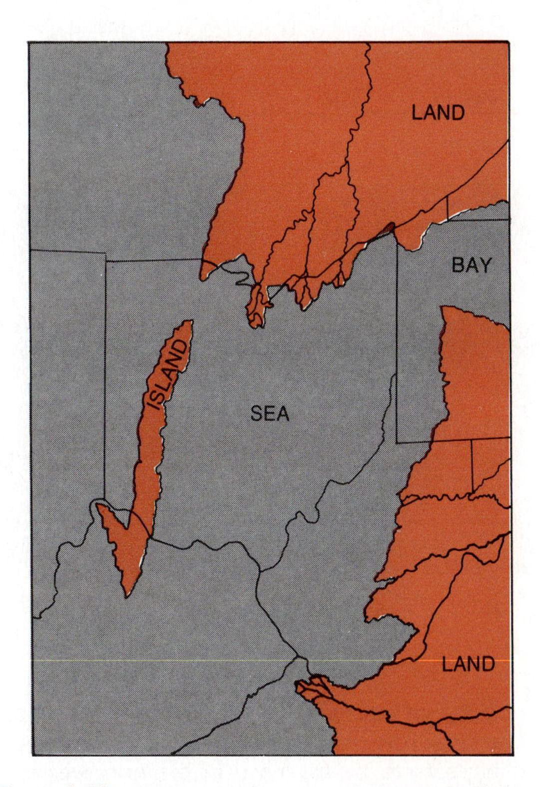

Figure 3–30 Paleogeographic map of Ohio and adjoining states during an early part of the Mississippian Period. The data for this study were obtained from outcrops and over 40,000 well records. (After J. F. Pepper, W. J. de Witt, and D. F. Demarest, U.S. Geological Survey Prof. Paper 259, 1954.)

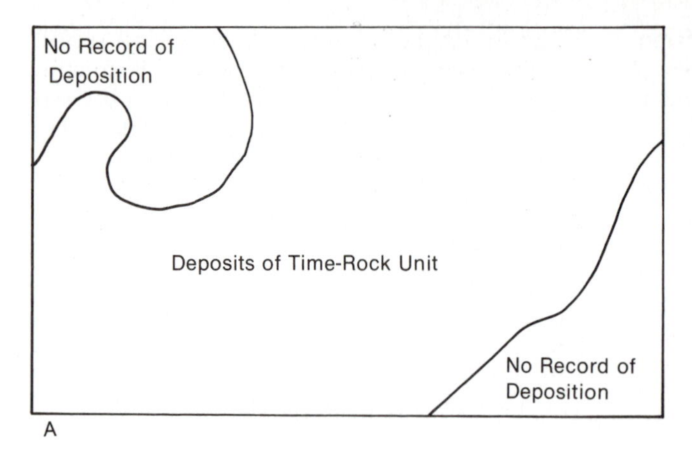

Figure 3–31 Stages in the construction of a paleogeographic map. (A) Area of occurrence of time-rock unit. (B) Plot of rock types within the time-rock unit. (C) Paleogeographic reconstruction.

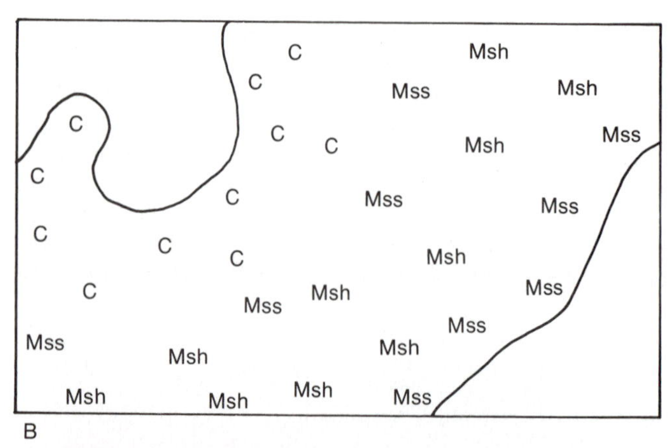

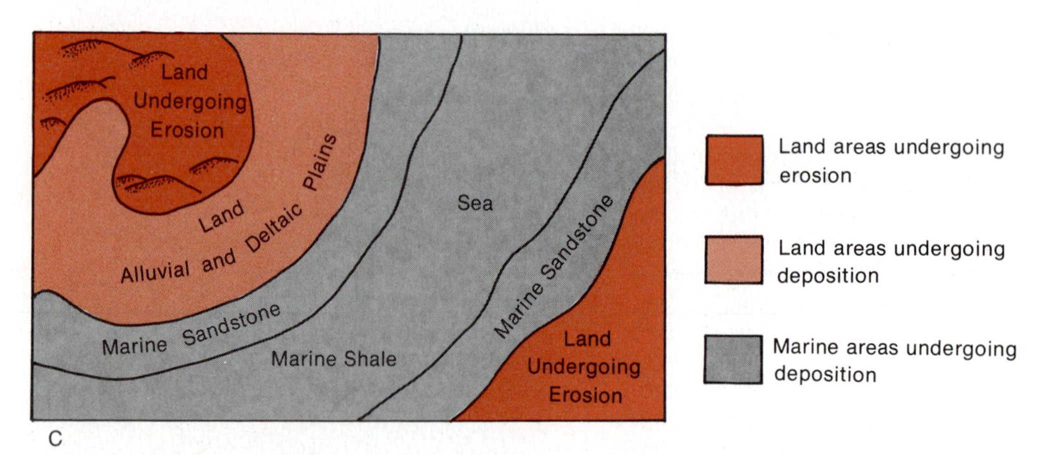

deposition or places where deposits once existed but were subsequently eroded away. Nondeposition appears to be the case in the northwestern corner of Figure 3–31, for that area is nicely encircled by sandstones that grade outward to shales. (If the time-rock unit thins toward areas of nondeposition, this interpretation would be strengthened.) With the help of fossils, the nature of the sediments—that is, whether marine or nonmarine—is determined and plotted on the map. The final step is to complete the paleogeographic reconstruction (Fig. 3–31C).

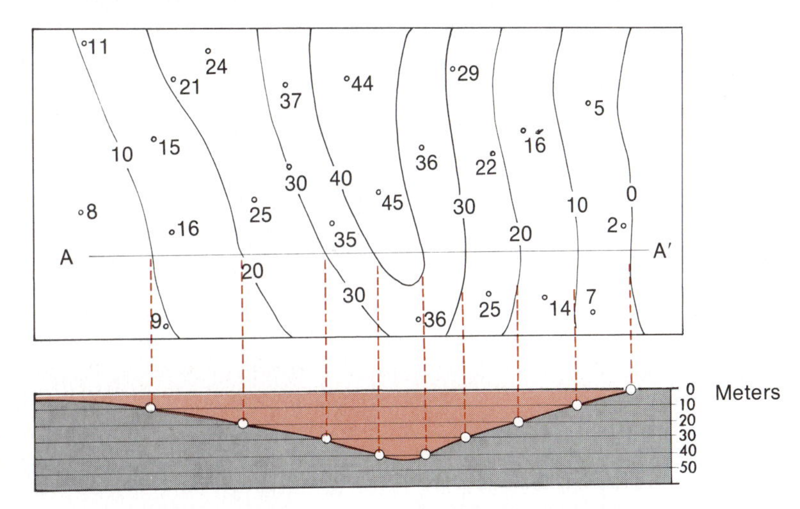

Figure 3–32 Diagram illustrating the construction of a simple isopach map in an area of undeformed strata.

Isopach Maps

Isopach maps are prepared by geologists in order to illustrate changes in the thickness of a formation or time-rock unit. The lines on an isopach map (Fig. 3–32) connect points at which the unit is of the same thickness. On a base map, the geologists plot the thickness of units as they are revealed in drilling or in measured surface sections. Isopach lines are then drawn to conform to the data points as perfectly as possible. Ordinarily, the upper surface of the unit being mapped is used as the datum from which thickness measurements are made. An isopach map may be very useful in determining the size and shape of a depositional basin, the position of shorelines, and areas of uplift. Figure 3–33 is an isopach map of Upper Ordovician formations in Pennsylvania and adjoining states. The map indicates a semicircular center of subsidence in southern New York and Pennsylvania in which over 2000 ft of sediment accumulated. The isopach pattern further indicates a highland source area to the southeast.

Lithofacies Maps

Maps constructed to show areal variations in facies can provide additional details and validity to paleogeographic interpretations. Such graphic representations are called **lithofacies maps.** Figure 3–34 is a hypothetical base map of an area subjected to exploratory drilling by oil companies. The logs for the wells are shown below their locations. Geologists first correlate the formations. Then, assuming that the unconformity represents one time plane and the ash bed another, they define the time-rock unit as "X." Paleontologic study of the rocks between the time planes confirms the validity of the time-rock unit. Geologists may now prepare the lithofacies map. Time-rock unit X is missing at well Number 11; this may be the result of its not being deposited there or, having been deposited, of its being eroded away. It is logical that the sandy facies was deposited adjacent to a north-south trending shoreline.

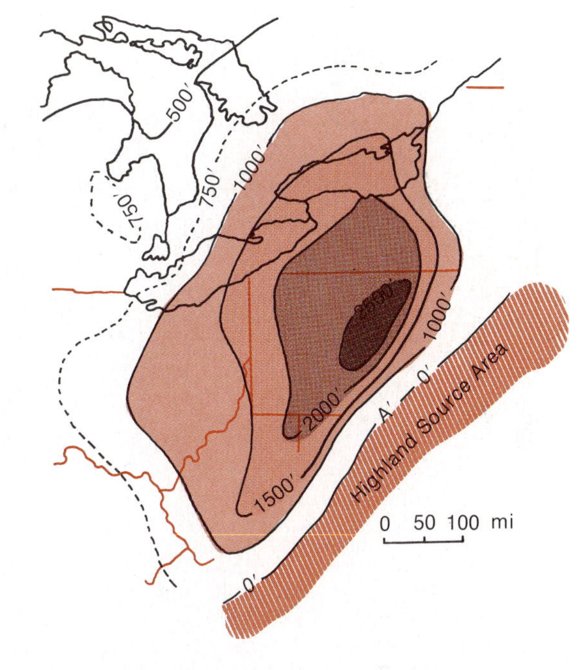

Figure 3–33 Isopach map of Upper Ordovician formations in Pennsylvania and adjoining states. (After M. Kay. *Geol. Soc. Am. Mem.* No. 48, 1951.)

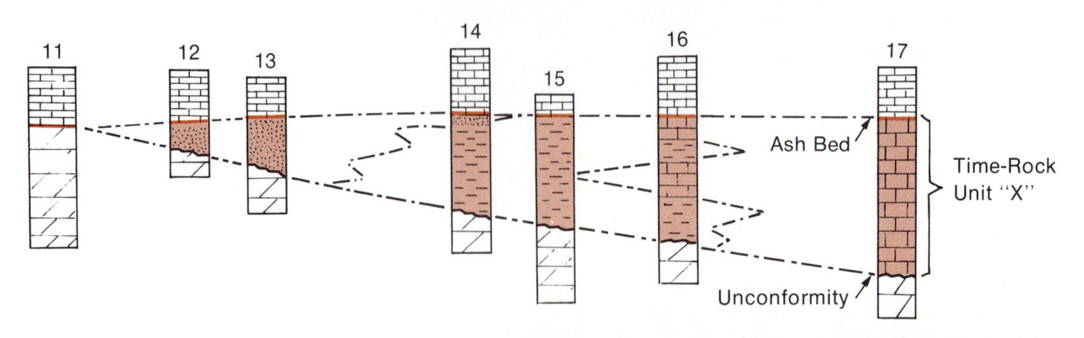

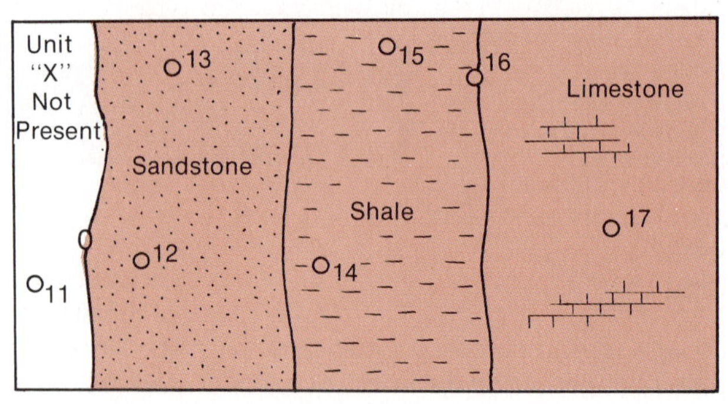

Figure 3–34 Diagram illustrating the preparation of a lithofacies map from a subsurface time-rock unit. Well locations are indicated by small circles. Correlation of rock units between wells is indicated by dashed lines. Because of the few control points, the exact position of lithofacies boundaries on this map is somewhat arbitrary.

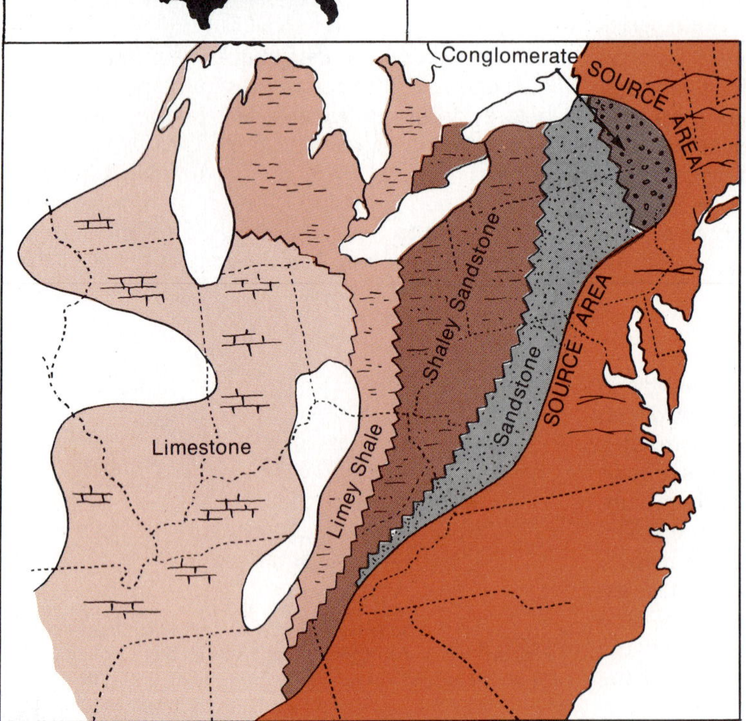

Figure 3–35 Lithofacies map of Lower Silurian rocks in the eastern United States. (After T. W. Amsden. *Bull. Am. Assoc. Petrol. Geol.*, 39:60–74, 1955.)

Figure 3–35 is a lithofacies map of rocks deposited over 400 million years ago in the eastern United States. From this map, one can infer the existence of a highland area that existed at that time along our eastern seaboard and that supplied the coarse clastics. Detrital sediments from the source area become fine, and the section thins as one proceeds westward from the source highlands. Finally, as far west as Indiana, the map indicates that only carbonate precipitates were laid down. The conglomerates were probably the deposits of great alluvial fans built out from the ancient mountain system. Because the northern border of the time-rock unit cuts across the bands of facies, it is considered to be an erosional border and does not represent the original extent of the unit mapped.

The lithofacies maps just described provide a qualitative interpretation of areal changes in rock bodies. Quantitative lithofacies maps can also be constructed and are frequently used in the study of subsurface formations that are known primarily from well records. By means of contour lines, such maps show the areal distribution of some measurable characteristic of the unit being mapped. For example, contours may be drawn on the percentage of one lithologic component compared to the total unit or on the ratio of one rock type to the others within the unit. An isopach map is ordinarily the base map for any of the quantitative maps, since one must know the total stratigraphic thickness of the unit with which individual components are compared.

Perhaps the most important fact to remember about the mapping techniques used in deciphering earth history is that they are limitless in variety. New and inventive schemes for mapping the past appear regularly in geologic journals as geologists continue to probe into the details of ancient geologic events.

IMPORTANCE OF SEDIMENTARY ROCKS TO SOCIETY TODAY

From caveman to astronaut, humans have learned to use sedimentary rocks to improve their lot. Men and women of antiquity fashioned tools from chert and flint, and they later learned the art of manufacturing useful containers from clay. Their successors developed the means of extracting copper and iron from the rocks and from these metals made more sophisticated implements. In addition to the metals they sometimes contain, sedimentary formations include coal seams and contain the oil, gas, and groundwater important to our present way of life.

Mineral Fuels

The mineral fuels—petroleum, gas, and coal—are our most important sedimentary resources. They are essential for heat and power and metal refining; they are also the source of many chemicals useful in the manufacturing of plastics and fertilizers. **Coal** is a brownish-black to black combustible rock that forms in beds from a few inches to many feet in thickness and is interbedded with shales, sandstones, and other sedimentary rocks. Extensive coal-bearing sequences characterize the Pennsylvanian and Cretaceous Systems. Pennsylvanian sequences of strata in the Appalachians include over 100 individual beds of coal, each composed of the compressed and altered remains of land plants. Coal is abundant in the United States and will become increasingly more important over the next three decades as our reserves of petroleum are consumed.

Commercial accumulations of oil and gas require rather special geologic conditions. These conditions are found almost exclusively in sedimentary rocks. First, there must be a **source rock** for the petroleum. Ordinarily, this is a series of beds rich in the organic remains of unicellular organisms that accumulated along with the particles of sediment. Second, there must be a permeable **reservoir rock** such as sandstone or porous limestone to provide passage and storage for the gas and oil. The reservoir rock is covered by an impermeable **cap rock** to prevent the upward escape of hydrocarbons. Finally, there must be an oil-trapping structure (Fig. 3–36) so that the hydrocarbons can be concentrated in one place. Petroleum geologists seek out these structures in many ways. Long before drilling programs are begun, geologists conduct gravity surveys and seismic "reflective shooting" surveys (Fig. 3–37) to reveal clues to underground structures. Later, exploratory wells are drilled, and by careful analyses of electrical, lithologic, and paleontologic logs of these wells, a picture of the configuration of the underground layers is gradually developed (Fig. 3–38).

Construction Materials

Other sedimentary materials useful to humans include clays for use in ceramics and bricks, limestones for building stones and cement, sand and gravel for concrete and glass, and evaporites for use in the chemical industry. Some sedimentary strata are rich in iron oxide. Those near Birmingham, Alabama, have been mined and smelted for several decades. Because minerals such as gold, chromite, diamond, cassiterite (tin oxide), and magnetite are heavier than quartz and

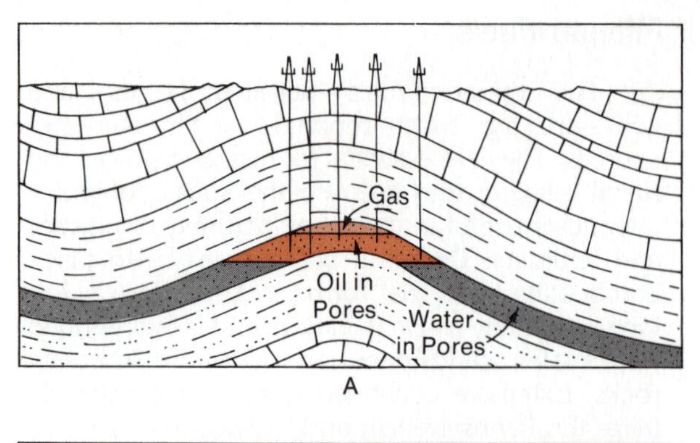

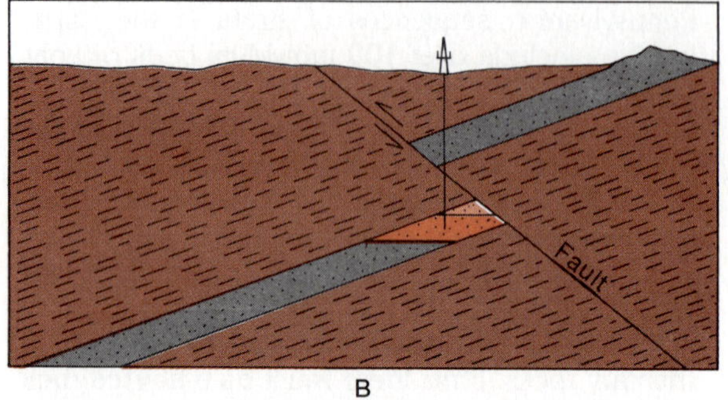

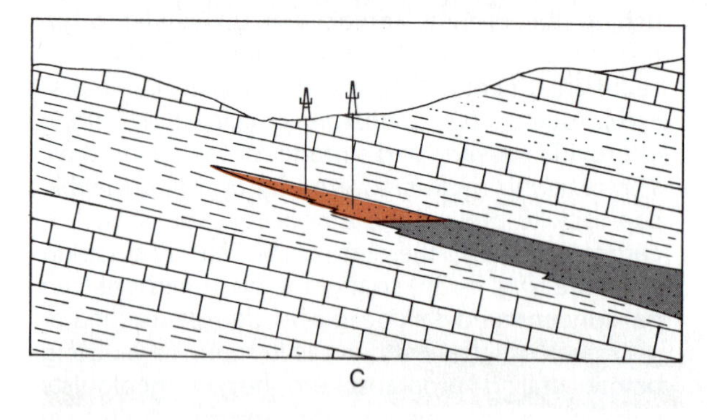

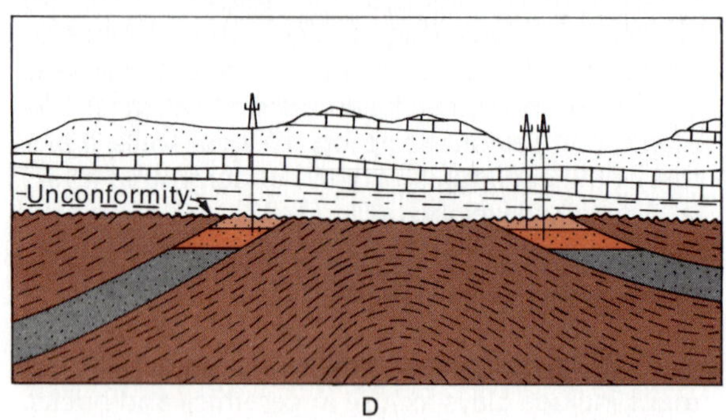

Figure 3–36 Idealized cross-sections depicting geologic conditions that may lead to the accumulation of petroleum and natural gas. (A) Anticlinal trap. (B) Fault trap. (C) Stratigraphic trap. (D) Entrapment beneath an unconformity.

other common silicates, they tend to accumulate in quiet areas of streams or to be concentrated by waves as **placer** deposits. Finally, there are sedimentary ores that form in place by the deep chemical decay of a parent rock. The most important metal concentrated in this manner is aluminum. Bauxite, an almost pure hydrous aluminum oxide, is the ore mineral of aluminum. Bauxites devel-

Figure 3–37 Seismograph party sets off seismic blast near the Great Divide, Colorado. Shooting, recording, and cable trucks on the road. (Photograph courtesy of Standard Oil Company of California.)

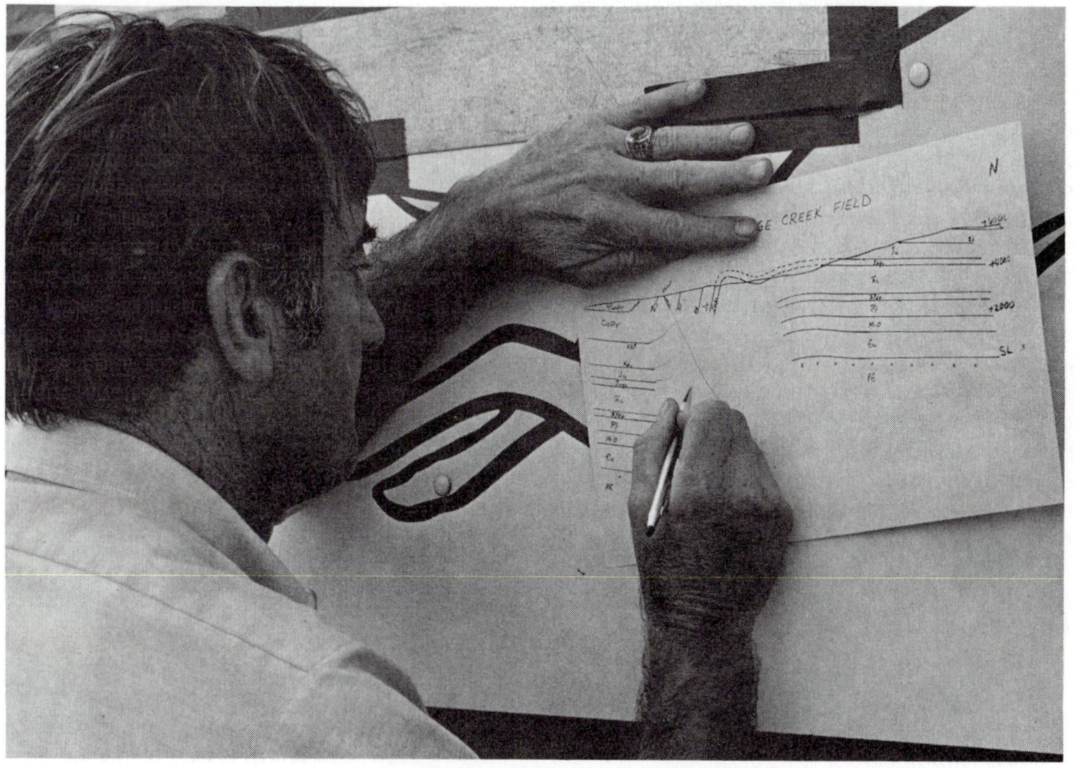

Figure 3–38 Geologist constructing stratigraphic cross-section of subsurface formations. (Photograph courtesy of C. Iwasaki, Standard Oil Company of California.)

oped in once humid, tropical regions as a result of the weathering of source rocks rich in aluminum silicates.

Limited Supply of Mineral Resources

Any commentary on mineral resources would be incomplete without noting that ore bodies and oil fields are limited in extent and are the result of unusual associations of geologic conditions. Furthermore, it is not theory but fact that they are exhaustible and irreplaceable. Their effect upon the standards of living and society is so enormous that every citizen should become involved in political decisions relating to the management of our natural resources.

Summary

Sedimentary rocks represent the material record of environments that once existed on the earth. For this reason, they are of great importance to the science of historical geology. All sedimentary rocks are formed by the accumulation and consolidation of the products of weathering derived from older rock masses as well as by accumulation of organic debris. Because of their mode of formation, the composition of sedimentary rocks can frequently provide information about source areas. The materials of sedimentary rocks are often transported by wind, water, or ice or are carried in solution to be precipitated in a particular environment of deposition. The transporting media often impart characteristics of texture or composition that can be used to reconstruct the depositional history and tectonic setting. Fossils in sedimentary rocks are splendid environmental indicators. They tell us if strata are marine or nonmarine, if the water was deep or shallow, or if the climate was cold or warm.

Color in a rock provides clues to the chemistry of the depositional medium. In the case of clastic rocks, the size, shape, and arrangement of grains can provide data about the energy of the transporting agent, the distance the grains had traveled, erosional recycling, and the degree to which movements of the earth's crust had disturbed a basin of deposition. Similar kinds of information are elucidated by such primary sedimentary structures as graded bedding, cross-bedding, and current ripple marks. Sandstones are particularly useful in paleoenvironmental studies. Orthoquartzites, graywackes, arkoses, and subgraywackes accumulate in particular paleogeographic and tectonic situations. It is the geologist's task to discover the details of those situations by examining the rocks.

Geologists usually divide successions of sedimentary rocks into rock units that are sufficiently distinctive in color, texture, or composition to be recognized easily and mapped. Such rock units are called formations and are not necessarily of the same age throughout their areal extent. A time-rock unit differs from a rock unit in that it is an assemblage of strata deposited within a particular interval of time. The examination of variations in the distinctive aspects or facies of a time-rock unit is of great importance in formulating a valid picture of conditions long ago. Geologists employ various graphic methods to record variations in facies and other attributes of sedimentary rocks. These methods include the preparation of lithofacies, biofacies, geologic, and isopach maps. If examined in chronologic sequence, such maps are useful not only in reconstructing ancient geography but also in providing a picture of the changing patterns of lands and seas to which parts of the earth have been subjected.

Sedimentary rocks are the source of a great many materials necessary to modern civilization. Some of these materials, like clay and building stones, are rather commonplace. Others, like placer deposits of diamonds, chromite, and gold, occur only in limited amounts at a few localities. Aluminum, a metal derived from a residual mineral named bauxite, is another extensively exploited sedimentary resource. However, the fossil fuels—coal and petroleum—are the most important of all sedimentary materials today. Indeed, oil, coal, and gas power our industrialized world. These resources are present on the earth in finite quantities and must be carefully managed.

Questions for Review

1. What are geosynclines? What evidence suggests that most of the world's great mountain ranges were once occupied by elongate marine basins of deposition?
2. Why are sandstones and siltstones of nonmarine desert environments rarely black or dark gray in color?
3. Under what circumstances might a reddish siltstone containing fossils of marine invertebrates have originated?

4. How does matrix in a rock differ from cement? What kinds of cements can bind a sedimentary rock together? Which cement is most durable?
5. What properties of a sedimentary rock can be used to determine its pre-existing parent rock, its mode of transportation, and its depositional environment?
6. What sort of history might you infer for a sandstone composed of very angular grains, poorly sorted and indurated with a 30 per cent matrix of mud?
7. What is the difference between a "rock unit" and a "time-rock unit"? How is each defined and recognized?
8. Distinguish between an "immature" and a "mature" sandstone.
9. Compare the probable tectonic environments in which a graywacke and an orthoquartzite might be deposited.
10. Why, in general, are fine-grained sediments deposited farther from a shoreline than coarse-grained sediments?
11. A geologist studying a sequence of strata discovers that limestone is overlain by shale, which in turn is overlain by sandstone. What might this signify with regard to the advance or retreat of a shoreline with time?
12. What features or properties of sedimentary rocks are useful in determining the direction of currents of the depositing medium? What features indicate relatively shallow water deposition?
13. What are turbidity currents? How might turbidity current deposits be recognized?
14. What sort of paleogeographic information can be derived from lithofacies maps? From isopach maps?
15. What are the essential natural requirements for the occurrence of an oil "pool"?

Terms to Remember

arkoses
biofacies
cap rock
cement
cratons
cross-bedding
epoch
facies
formation
geologic map
geosynclines
graded bedding
graywackes
hematite
immature sandstone
isopach map
limonite
lithofacies
lithofacies map
matrix
mature sandstone
mobile belts
montmorillonites
mud cracks
offlap
orthoquartzite
paleogeographic map
period
placer deposit
red beds
reservoir rock
ripple marks
rounding
series
sorting
source rock
sphericity
stage
subgraywackes
system
tectonics
texture
time-rock unit
Wentworth Scale

Supplemental Readings and References

Blatt, H., Middleton, G., and Murray, R. 1972. *Origin of Sedimentary Rocks.* Englewood Cliffs, N.J., Prentice-Hall, Inc.

Dunbar, C. O., and Rogers, J. 1957. *Principles of Stratigraphy.* New York, John Wiley & Sons.

Garrels, R. M., and MacKenzie, F. T. 1971. *Evolution of Sedimentary Rocks.* New York, W. W. Norton & Co., Inc.

Krumbein, W. C., and Sloss, L. L. 1963. *Stratigraphy and Sedimentation* (2nd ed.). San Francisco, W. H. Freeman & Co.

Laporte, I. F. 1968. *Ancient Environments.* Englewood Cliffs, N.J., Prentice-Hall, Inc.

Matthews, R. K. 1974. *Dynamic Stratigraphy.* Englewood Cliffs, N.J., Prentice-Hall, Inc.

Reineck, H. E., and Singh, I. B. 1973. *Depositional Sedimentary Environments.* New York, Springer-Verlag.

Rigby, J. K., and Hamblin, W. K. (eds.). 1972. Recognition of ancient sedimentary environments. *Soc. Econ. Paleontal. Mineral. Spec. Publ.* No. 16, 1–340.

Selley, R. C. 1970. *Ancient Environments.* Ithaca, N.Y., Cornell University Press.

Walker, T. R. 1967. Formation of red beds in ancient and modern deserts. *Geol. Soc. Am. Bull.* 78:353–368.

Wilson, J. L. 1975. *Carbonate Facies in Geologic History.* New York, Springer-Verlag.

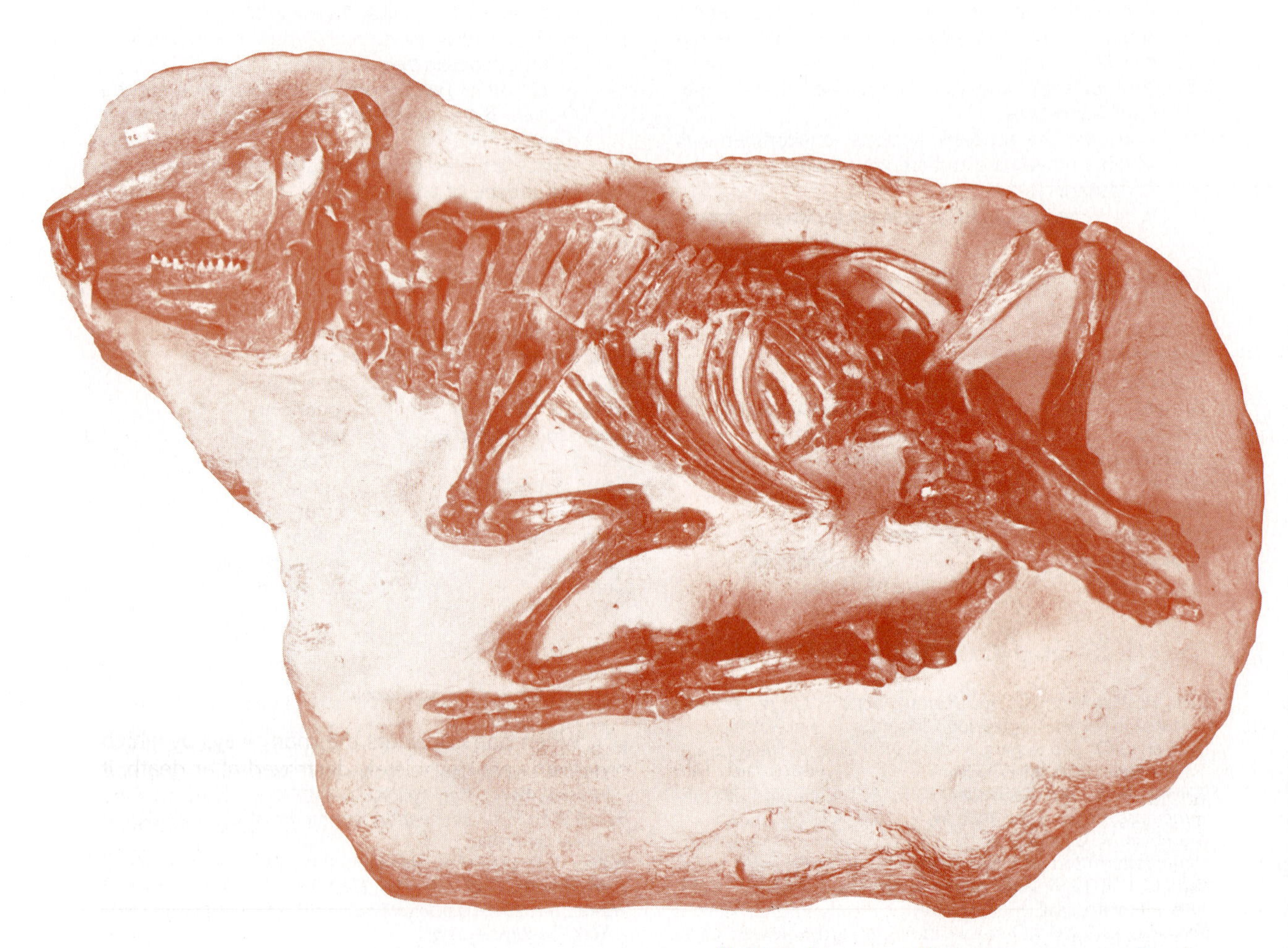

Skeleton of a peccary, *Platygonus compressus,* from ice age deposits in a Mississippi River bluff. (Photograph courtesy of the U.S. Geological Survey.)

4

Life Through Time

The search for a fossil may be considered at least
as rational as the pursuit of a hare.
William Smith (1769–1839)

THE NATURE OF THE FOSSIL RECORD

The Meaning of Fossils

The earth is about 4.6 billion years old. Life has been present on it for about 3.2 billion years. The science that seeks to understand all aspects of the succession of plants and animals over the great span of time is called **paleontology.** It is a science based on the study of **fossils,** the remains or traces of ancient life. The work of **paleontologists** is far more complicated (and exceedingly more interesting) than merely describing and cataloguing ancient creatures. They must determine how these animals and plants of long ago lived and how they grew. They seek out the ancestors and descendants of these life forms, attempt to know their true "fleshed-out" appearances, and try to determine the extremes of temperature, salinity, or moisture they could tolerate. Paleontologists must determine how the past inhabitants of the earth fit into the total web of life, how they interacted with each other and the environment, and how they can now be utilized in deciphering the age of strata.

Preservation

Types of fossilization

When one considers the many ways by which organisms are completely destroyed after death, it is remarkable that fossils are as common as they are. Attack by scavengers and bacteria, chemical decay, and destruction by erosion and other geologic agencies make the odds against preservation very high. However, the chances of escaping complete destruction are vastly improved if the organism happens to have a mineralized skeleton and dies in a place where it can be quickly buried by sediment. Both of these conditions are met especially well in the sea, where shelled invertebrates flourish and are covered by the continuous rain of sedimentary particles. Although most fossils are found in marine sedimentary rocks, they also are found in terrestrial deposits left by streams and lakes. On occasion, animals and plants have been preserved after becoming immersed in tar or quicksand, trapped in ice or lava flows, or engulfed by rapid falls of volcanic ash.

The term "fossil" often implies **petrification—**

literally, a transformation into stone. After the death of an organism, the soft tissue is ordinarily consumed by scavengers and bacteria. The empty shell may be left behind, and if it is sufficiently durable and resistant to dissolution, it may remain basically unchanged for a long period of time. Indeed, unaltered shells of marine invertebrates are known from deposits over 100 million years old. In many marine creatures, however, the skeleton is composed of a kind of crystalline calcium carbonate called aragonite. Although aragonite has the same composition as the more familiar mineral known as calcite, its ions are packed differently, and it crystallizes in a different crystal system. Aragonite is also relatively unstable and in time changes by recrystallization to the more stable calcite.

Many other processes may alter the shell of a clam or snail and enhance its chances for preservation. Water containing dissolved silica, calcium carbonate, or iron may circulate through the enclosing sediment and be deposited in cavities once occupied by veins, canals, nerves, or tissues. In such cases, the original composition of the bone or shell remains, but the fossil is made harder and more durable. This addition of a chemically precipitated substance into pore spaces is termed **permineralization.**

Petrification may also involve a molecular exchange of the original substance via substitution of mineral matter of a different composition. This process is termed **replacement,** because as each original molecule is removed, it is replaced by a molecule of silica or calcium carbonate, or less frequently by sulfides and oxides of iron. Replacement can be marvelously precise, so that even the growth lines on shells and structures of wood may be perfectly preserved (Fig. 4–1).

Figure 4–1 Petrified wood. The original wood has been replaced by silica. (Photograph courtesy of Wards Natural Science Establishment, Inc.)

Figure 4–2 Carbonized leaf from ash beds near Florissant, Colorado.

Another type of fossilization, known as **carbonization,** occurs when soft tissues are preserved as thin films of carbon. Leaves and tissue of soft-bodied organisms such as jellyfish or worms may accumulate, become buried and compressed, and lose the volatile constituents. The carbon often remains behind as a blackened silhouette (Fig. 4–2).

Fossils may also take the form of molds, imprints, and casts. Any organic structure may leave an impression of itself if it is pressed into a soft material and if that material is capable of retaining the imprint. Commonly, among shell-bearing invertebrates, the shell is dissolved after burial and lithification, leaving a vacant **mold** bearing the surface features of the original shell. In many instances, molds are later filled, forming **casts** that faithfully show the original form of the shell (Fig. 4–3).

Although it is certainly true that the possession of hard parts enhances the prospects of preservation, organisms having soft tissues and organs are also occasionally preserved. Insects have been found preserved in the hardened resins of coniferous trees (Fig. 4–4). X-ray examination of thin slabs of rock sometimes reveals the ghostly outlines of tentacles, digestive tracts, and visual organs of a variety of marine creatures. Soft parts,

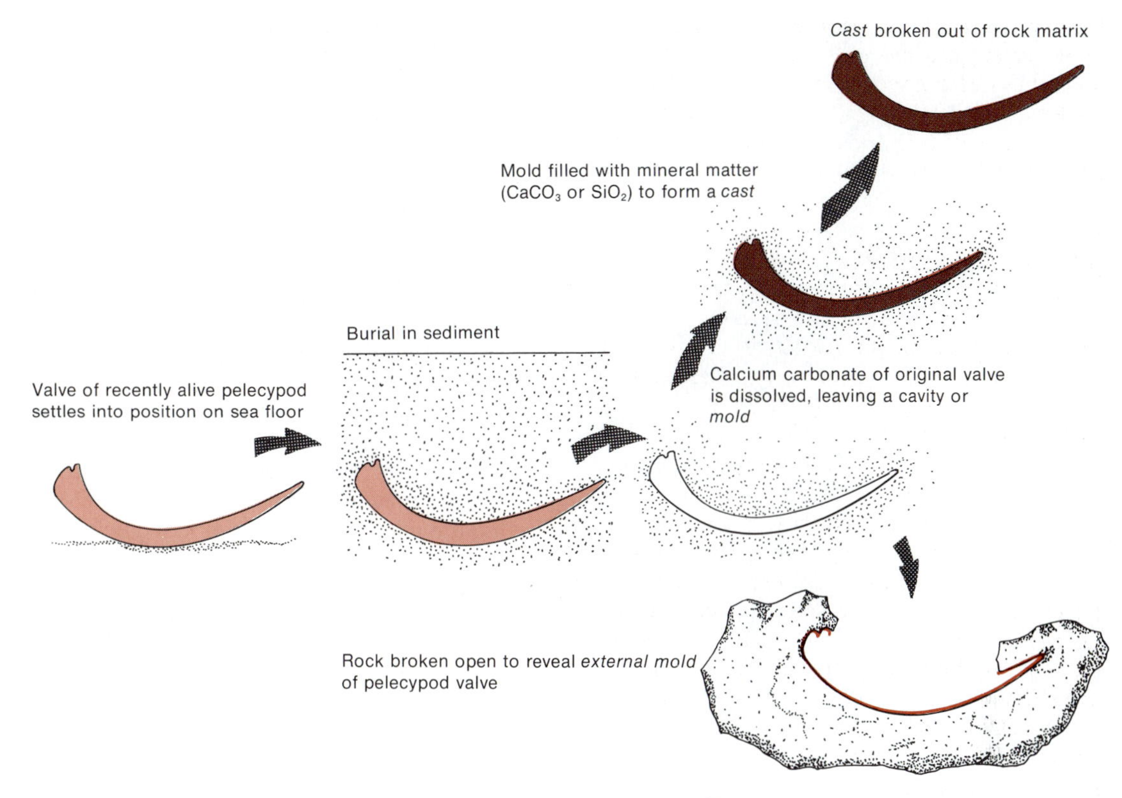

Figure 4–3 Origin of fossil molds and casts.

such as skin, hair, and viscera of ice age mammoths have been preserved in frozen soil (Fig. 4–5) or have been "pickled" in the oozing tar of oil seeps. The delicate remains of the world's oldest unicellular organisms are found embedded in chert that hardened over them 3 billion years ago.

Evidence of ancient life does not consist solely of petrifications, molds, and casts. Sometimes the paleontologist is able to obtain clues to an animal's appearance and how it lived by examining tracks, burrows, and borings. Such markings are called **trace fossils,** and the study of trace fossils is termed **ichnology.** The tracks of ancient vertebrate animals may indicate whether the animals that made them were bipedal (walked on two legs) or quadrupedal (walked on four legs), digitigrade (walked on toes) or plantigrade (walked on the flat of the foot), whether it had an elongate or a short body, if it was lightly built or ponderous, and sometimes whether it was aquatic (with webbed toes) or possibly a flesh-eating predator (with sharp claws).

Trace fossils of invertebrate animals are more frequently found than the tracks of vertebrates, and they are also useful indicators of the habits of ancient creatures. One can sometimes infer if the trace-making invertebrate was crawling, resting, grazing, feeding, or simply living within a relatively permanent dwelling. For example, crawling traces, as might be expected, are linear and show directed movement (Fig. 4–6A). Shallow depressions that more or less reflect the shape of the animal may be resting traces (Fig. 4–6B). Simple or U-shaped structures more or less perpendicular to bedding are often dwelling traces (Fig. 4–6C). Grazing traces (Fig. 4–6D) occur along bedding planes and are characterized by a systematic meandering or concentric and parallel patterns that represent the animal's effort to cover the area containing food in an efficient manner. The three-dimensional counterparts of grazing traces are called feeding traces. Feeding traces consist of systems of branched or unbranched burrows, as shown in Figure 4–6E.

The Incomplete Record of Life

The record of life on earth is incomplete. If it were a finished record, it would have to include information on all past forms of life for every increment of time as well as documentation of their abundances and geographic distribution. Clearly, this goal is unattainable. Only a limited number of animals and plants have survived the hazards of preservation. Many that were preserved have never

Figure 4–4 Worker ant *(Sphecomylma freyi)* preserved in Cretaceous amber. (Photograph courtesy of Professor Frank M. Carpenter.)

been exposed to our view by erosion or drilling. Others have simply not yet been discovered. The record is rather more complete for marine life with hard external skeletons—and less complete for land life lacking bone or shell. Because of erosion, there are many gaps in the record; these are not unexpected.

Figure 4–7, from an article by Joseph Barrell, illustrates the intermittent nature of the rock record and, hence, the contained fossil record as well. The figure depicts the accumulation of a sequence of strata during a period of general marine transgression (advance of the sea over the land) at a particular locality. As one might expect, the major transgressions have minor regressions superimposed upon them. These lesser comings and goings of the sea may result from a variety of factors, including the release or entrapment of glacial ice on continents, changes in the shape of ocean basins, or movements of the earth's crust. Because the lands are exposed to erosion during the regression phases, there is a loss of previously deposited sediment or at least a lack of deposition. The rock record is built up only during times when the sea level rises again. Thus, the rock column shown on the

Figure 4–5 In the summer of 1977, the carcass of this baby mammoth was dug from frozen soil (permafrost) in northeastern Siberia. The mammoth stood about 104 cm tall at the shoulders, was covered with reddish hair, and was judged to be only several months old at the time of death. Dating by the radiocarbon method indicates death occurred 44,000 years ago. (Photograph courtesy of Klavdija Novikova, Biologopoczvennyj Institut, Vladivostok, USSR.)

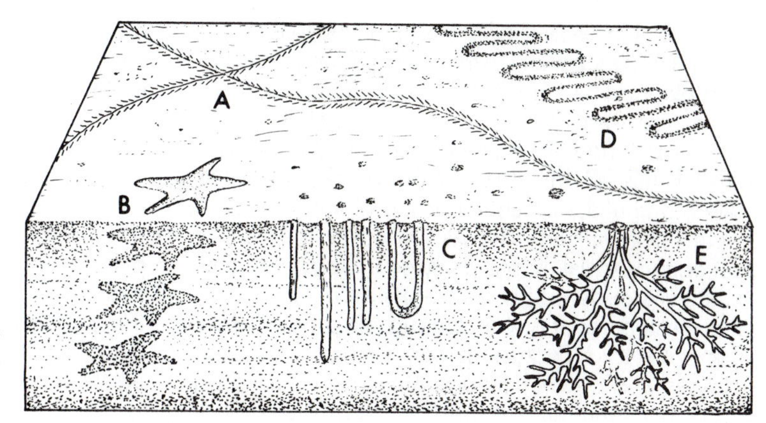

Figure 4–6 Traces that reflect animal behavior. (A) Crawling traces, (B) resting traces, (C) dwelling traces, (D) grazing traces, (E) feeding traces.

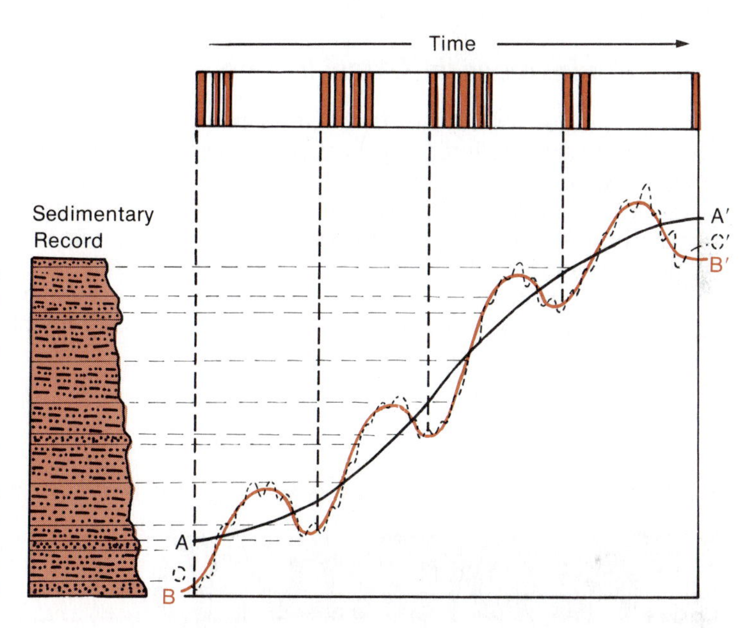

Figure 4–7 Diagram showing the sedimentary record resulting from oscillations in sea level and the time value of the sedimentary record. Curve A–A′ represents a major trend of rising sea level with resulting marine transgression. The two smaller curves (B–B′ and C–C′) represent shorter frequency oscillations (minor transgressions and regressions). During a rise in sea level, sediments accumulate; during a regression, part of the previously deposited sediment is removed by erosion. Thus, the sedimentary record represents only that fraction of geologic time shown at the top by the colored bars. (After J. Barrell. *Geol. Soc. Am. Bull.,* 28:745, 1917.)

left side of the figure represents only a small fraction of time—namely, the parts of the time graph that are shown in color.

Of course, in nature there will be other factors, including those causing changes in the supply of sediment, that would affect the sedimentary sequence. The important facts, however, are that the rock record is incomplete and that where there are no rocks of a particular age there can be no fossils for that period. Even when the sedimentary sequence is relatively representative of a given time span, there may be numerous rock intervals that are barren of fossils. In spite of these difficulties, by piecing together the data from many localities around the world, paleontologists have managed to uncover a history of life that merits most serious attention.

THE RANK AND ORDER OF LIFE

Paleontology and biology are closely related. Both are basically concerned with organisms. It is therefore appropriate that both of these sciences employ the same methods for naming and classifying organisms.

Naming and Classifying Organisms

The Linnaean System of Classification

Because of the large number of living and fossil animals and plants, random naming would be confusing and inefficient. Realizing this, the Swedish naturalist **Carl von Linné** (also known as

Carolus Linnaeus) (1707–1778) established a carefully conceived system for naming animals and plants. The Linnaean system utilizes morphologic structure as the basis of classification and employs binomial nomenclature at the species level. In this scheme, the first, or **generic,** name was used to indicate a general group of creatures that were visibly related, such as all doglike forms. The second, or **trivial,** name denoted a definite and restricted group—a **species.** *Canis lupus,* for example, designates the wolf among all canids. Linné went on to recognize larger divisions such as classes and orders. His groupings were based on traits that seemed most basic or natural; the modern nomenclatural system, however, is based on an attempt to indicate evolutionary relationships.

Concepts Involved in Classification

The species, of course, is the fundamental category in the modern classification of living things. A **species** may be defined as a population or group of populations composed of individuals that are basically alike in their structural and functional characteristics and that are actually or potentially able to interbreed. The ability to interbreed successfully is interpreted as evidence that the individuals of a population are closely related genetically. Species exist during a comparatively short period of geologic time, are derived from ancestral populations, and in turn are the population from which new species are derived.

Of course, it is not always possible to determine if individuals of an extinct population actually were able to interbreed successfully. Therefore, paleontologists must rely heavily on morphologic traits and the overall characteristics of the grouping they are tentatively considering to be a species. In deciding on the validity of a proposed species, biologists have the advantage of being able to view success in interbreeding. Furthermore, because they work in the present, biologists often see populations as isolated, well-defined units. Life, however, is a continuum. The isolated species observed today represent only a slice of that biologic continuum—a momentary time horizon (Fig. 4–8). Traced backward in time, species merge into their ancestral populations. The decision about where a species began in a lineage and where it terminated is necessarily subjective. Thus, in deciding if a group of similar fossilized organisms is to be considered a species, the paleontologist attempts to gather evidence that there is no more variation among individuals of the same species in the "ver-

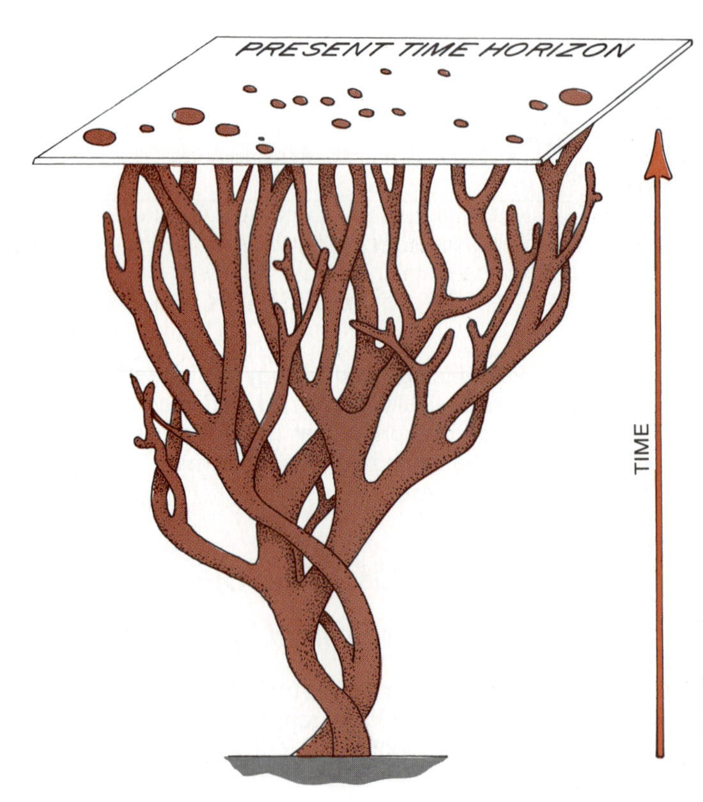

Figure 4–8 Diagram of a "family tree" of organisms. Living species appear as more or less isolated groups. However, "boundaries" between species are difficult to determine in the vertical (time) sense, because the "boundaries" are often morphologically transitional.

tical," or time, sense than there is among individuals that lived at about the same time.

TAXONOMY. Although there are many problems associated with the meaning of the word "species," this unit is necessarily basic in the modern system of classification called **taxonomy.** In this system, the various categories of living things are arranged in a hierarchy that expresses levels of kinship. For example, a **genus** (pl. *genera*) is a group of species that have close ancestral relationships; a **family** is a group of related genera; an **order** is group of related families; a **class** is a group of related orders; a **phylum** (pl. *phyla*) is a group of related classes; and a **kingdom** is a large group of related phyla. To use an example familiar to all: Individual humans are members of the Kingdom **Animalia,** Phylum **Chordata,** Class **Mammalia,** Order **Primates,** Family **Hominidae,** Genus ***Homo,*** and species ***sapiens.***

Biologic classification is more than a mere system for cataloguing various organisms. Because it is based on organic structures, it also depicts the broad outlines of ancestral evolutionary relationships. In a sense, the classification is a depiction of the tree of life. (For reference, a simplified classification of the major groups of frequently fossilized living and extinct organisms is provided in Appendix A.)

RULES OF NOMENCLATURE. Because of the incredible variety of living and once living organisms, scientists at the turn of the century recognized the need for a set of rules governing scientific nomenclature. To meet this need, the international congresses of zoologists and biologists established "Commissions on Nomenclature" and assigned to them the responsibility for drawing up the rules (*International Code of Biological Nomenclature*) governing the names that could be applied to categories of organisms. The code stipulates that no two genera can have the same name. It states that scientific names must be either Latin or Latinized and be printed, preferably in italics. Further, the name of the genus must be a single word (nominative singular) and must begin with a capital letter. The species name begins with a small letter and must agree grammatically with the generic name. To prevent duplication of names, the **Law of Priority** was established; this stipulated that the generic or specific name that is first published is the only valid one (provided it is published along with a detailed description and in an acceptable publication). In the event that a species is inadvertently described a second time and given another name, the second name is suppressed as a synonym. Conversely, when the same name is used for two different species, only the earlier designation is valid. These are, of course, only a few of the more familiar rules in the International Code of Biological Nomenclature.

ORGANIC EVOLUTION

The **doctrine of organic evolution,** as it is sometimes called, is based on a body of scientific observations indicating that organisms of today are descendants of those that lived in the past; further, that the modern, complex organic world developed from older and simpler forms of life. In another sense, evolution is an expression of nature's way of changing organisms in response to changes in their environments. Fossils provide the tangible evidence for evolutionary change. Sir Charles Lyell and Charles Darwin knew quite well that modifications in lineages of related organisms become apparent in the fossils of successively younger strata. It is one of the tasks of paleontologists to understand the mechanisms and environmental factors that resulted in these modifications.

Evolution's Ways and Means

The concept that animals and plants have been continuously changing through time is not a new one. Anaximander (611–547 BC) alluded to the idea 25 centuries ago. This famous Greek taught that life arose from mud warmed in the sun; that plants came first, then animals, and finally humans. Unfortunately, during the Dark and Medieval Ages, complete faith in the scriptural account of special creation of "species" effectively stifled imaginative thinking about evolution. It was not until the eighteenth century in Europe that intellectuals like Jean Lamarck, Georges Buffon, and Carl von Linné began to challenge the concept of special creation. Buffon and Linné were convinced that evolution had occurred, but they offered no explanation of its cause. Nevertheless, Linné provided scientists with a workable scheme for classifying animals and plants, and Buffon set down a careful definition of species, noting that species were separate entities that were not able to interbreed successfully with other dissimilar species. The first general theory of evolution to excite considerable public attention was developed by the celebrated naturalist Jean Baptiste de Lamarck in the period between 1801 and 1815. Lamarck's writings were followed a half century later by those of Charles Darwin and Alfred R. Wallace.

Lamarck's Theory of Evolution

Lamarck's theory of evolution stipulated that all species, including man, are descended from other species. His theory was based on a belief that new structures in an organism appear because of needs or "inner want" of the organism and that structures once acquired in this way are somehow inherited by later generations. In similar but reverse fashion, little used structures would disappear in succeeding generations. For example, Lamarck believed that snakes evolved from lizards that had a strong preference for crawling. Because of this inner need to have long, thin bodies, certain lizards developed such bodies, and because legs became less and less useful in crawling, these structures gradually disappeared. Lamarck's ideas were challenged almost immediately. There was no way to prove by experimentation that such a thing as "inner want" existed. More important, Lamarck's belief that characteristics acquired during the life span of an individual could then be inherited was tested and shown to be invalid. We know that some organisms can "adapt" to environmental conditions in a narrowly limited way within a life span. For example, a fair-skinned woman on the beach can be made protectively tan. However, no amount of sunbathing will result in her first child being born with a suntan. There is no way that somatic, or body, cells can pass characteristics over to reproductive cells and thereby on to the next generation. Thus, the Lamarckian concept of evolution based on use or disuse of organs was discredited.

Darwin's Theory of Natural Selection

Charles Darwin (1809–1882) praised Lamarck for his courage in perceiving that lineages of organisms do change and that the changes require many generations and great lengths of time. Building on some of the ideas expressed by Lamarck, Lyell, and others, Darwin provided the key concept of organic evolution by means of **natural selection.** A colleague, Alfred R. Wallace, shared in the formulation of this important new idea. The biologically important aspects of the Darwinian theory are embodied in the following three observations: (1) There is always a certain amount of variation among the individuals of the same kind of organism. (2) Organisms are potentially able to produce more offspring than required for simple maintenance of their numbers and, indeed, more than can be supported by the environment. Darwin and Wallace observed that populations do not grow in size indefinitely, so they must be limited by the survival of only a small portion of the offspring. (3) Competition for food and living space inevitably results in the elimination of the weaker or less well-adapted individuals and in the survival of those individuals that vary in some way that makes them better fitted to their environments. The survivors breed and provide the offspring from which the less well-adapted members of the subsequent generation are again culled. Thus, over many generations, the variations useful as adaptations to an environment will tend to prevail and to be maintained until such time as the environment itself is changed.

Darwin's *The Origin of Species by Means of Natural Selection* appeared in 1859. It was published at a time when at least part of the European intellectual atmosphere was more liberal and less satisfied with the theologic doctrine that every species had been independently created. Although the ideas of Darwin and Wallace continued to disturb the religious feelings of some, they nevertheless increasingly acquired adherents among nineteenth and twentieth century scientists. Darwin did not consider his views impious. He saw a "grandeur in this view of life with its several powers, having been originally breathed by the Creator into a few forms or into one; and that . . . . from so simple a beginning endless forms most beautiful and most wonderful have been and are being evolved."

Mendelian Principles of Inheritance

Every theory has its strong and weak points. In Darwin's theory, the weakness lay in an inability to explain the *cause* of variability in a way that could be experimentally verified. The cause of at least a part of that variability was discovered in the course of elegant experiments on garden peas conducted by a Moravian monk named **J. Gregor Mendel** (1822–1884). Mendel discovered the basic principles of inheritance. His findings, printed in 1865 in an obscure journal, were unknown to Darwin and unheeded by the scientific community until 1900, when the article was rediscovered. Mendel described the mechanism by which traits are transmitted from adults to offspring. In his experiments with graden peas, he demonstrated that heredity in plants is determined by "character determiners" that divide in the pollen and ovules and are recombined in specific ways during fertilization. Mendel called these hereditary regulators "factors." They have since come to be known as *genes.*

As currently understood, genes are chemical units or segments of a nucleic acid—specifically deoxyribonucleic acid (DNA). As suggested by careful chemical and x-ray studies, the DNA molecule is conceived as two parallel strands twisted

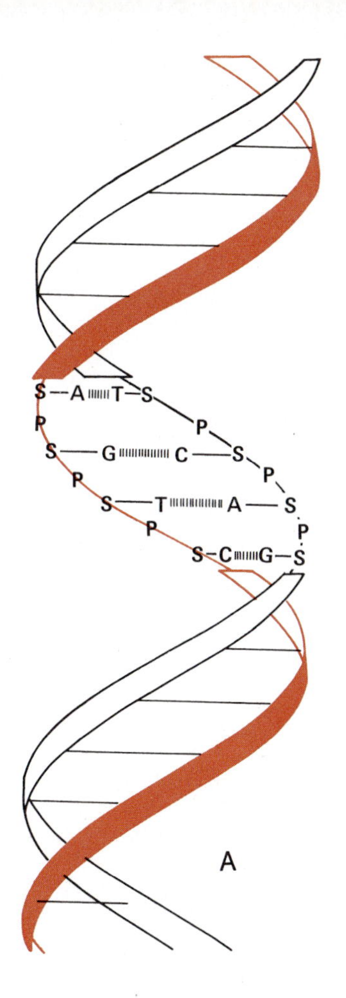

Figure 4–9 The double helix structure of the DNA molecule. (A) Schematic representation of part of the molecule. S–sugar; P–phosphate; A–adenine; T–thymine; G–guanine; C–cytosine. (B) F. H. C. Crick (right) and J. D. Watson (left) working in the Cavendish Laboratory at Cambridge, England, built scale models of the double helical structure of DNA. They received the Nobel Prize in 1962 for their work on the structure of the DNA molecule.

somewhat like the handrails of a spiral staircase (Fig. 4–9). The twisted strands are made up of phosphate and sugar compounds and are linked with cross-members composed of specific nitrogenous bases.

THE GENE. Each **gene** is equivalent to a portion of the DNA molecule. Any given gene is believed to direct the formation of a specific enzyme, which in turn passes out of the nucleus of a cell into the surrounding portion to regulate some specific aspect of the cell's growth, metabolism, or reproduction. In almost all organisms, genes are linked together to form larger units called **chromosomes,** the central axis of which consists of a very long DNA molecule comprising hundreds of genes.

Cell Division and Reproduction

MITOSIS, MEIOSIS, AND FERTILIZATION. In the formation of sex cells and in reproduction, chromosomes are recombined and redistributed to produce a variety of gene combinations in the offspring. The result is **variability,** some of which is apparent in the appearance of members of succeeding generations. The mechanism for this variability is, of course, associated with cell division and reproduction. The kind and number of chromosomes are constant for species and differ between species. Except in bacteria and blue-green algae, the chromosomes are located in the nucleus of the cells and occur in duplicate pairs. Thus, each chromosome has a homologous mate. In humans, for example, there are 46 chromosomes, or 23 pairs. Cells with paired homologous chromosomes are designated **diploid** cells. In all living things, new cells are being produced constantly to replace worn-out or injured cells and to permit growth. In asexual organisms, and in all the **somatic,** or body, cells of sexual organisms, the process of cell division that produces new diploid cells with exact replicas of the chromosomal components of the parent cells is called **mitosis** (Fig. 4–10).

In most organisms with sexual reproduction, a second type of division, called **meiosis,** takes place when **gametes** (egg cells or sperm cells) are

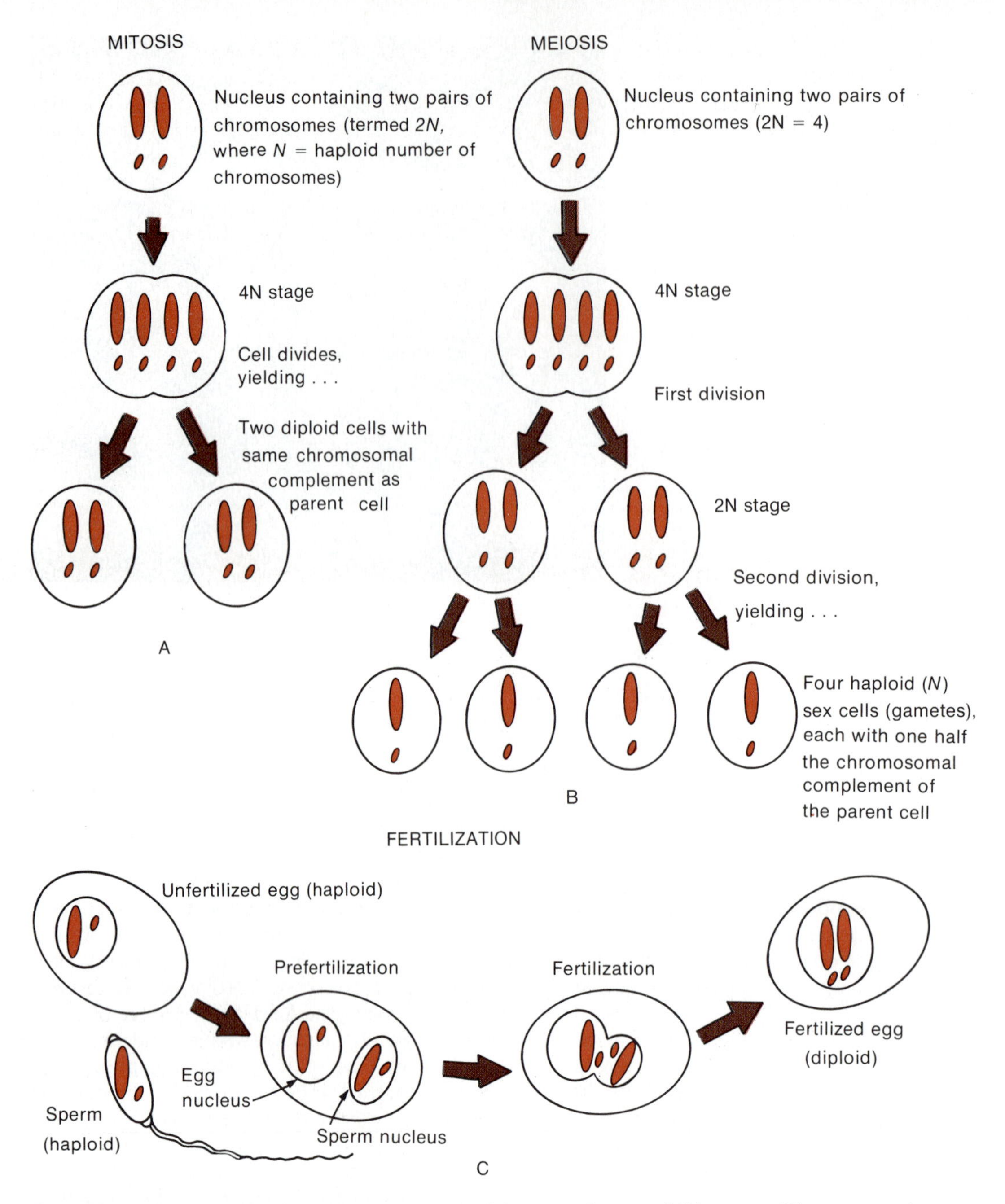

Figure 4–10 Greatly simplified summary comparison of the major features of (A) mitosis; (B) meiosis; and (C) fertilization.

formed. Meiosis may occur in unicellular organisms, but in multicellular forms it takes place only in the reproductive organs (testes or ovaries). Meiosis consists of two cell divisions. In the initial division, the chromosomal pairs are divided in random fashion so that each of the potential new cells receives one member of each original pair. These daughter cells are termed **haploid,** because they do not have paired chromosomes. The haploid cells then reproduce themselves exactly as in mitosis to form a total of four haploid cells, which are the gametes, or reproductive cells. When two gametes meet during sexual reproduction, the sperm enters the egg to form a single cell, which can now be called the **fertilized egg.** Because two gametes have been combined into one cell, there is now a full complement of chromosomes and genes representing a mix from both parents. The fertilized egg now begins a process of growth by mitotic cell division that eventually leads to a complete organism.

MUTATIONS. Recombination of genes alone, however, would not result in evolution as we understand it, because the possibilities for variation within an unchanging collection of genes are finite. However, another source of variation, called mutation, permits entirely new traits to appear and spread. A **mutation** is an inheritable change in genetic material within cells. The altered, or mutant, gene produces an effect different from that possible with the original DNA configuration. As a result of mutations, new characters may appear that were not present in previous generations.

Mutation as a basis for evolution was first proposed in 1901 by Hugo DeVries. From his experiments on primrose plants, he found that of 50,000

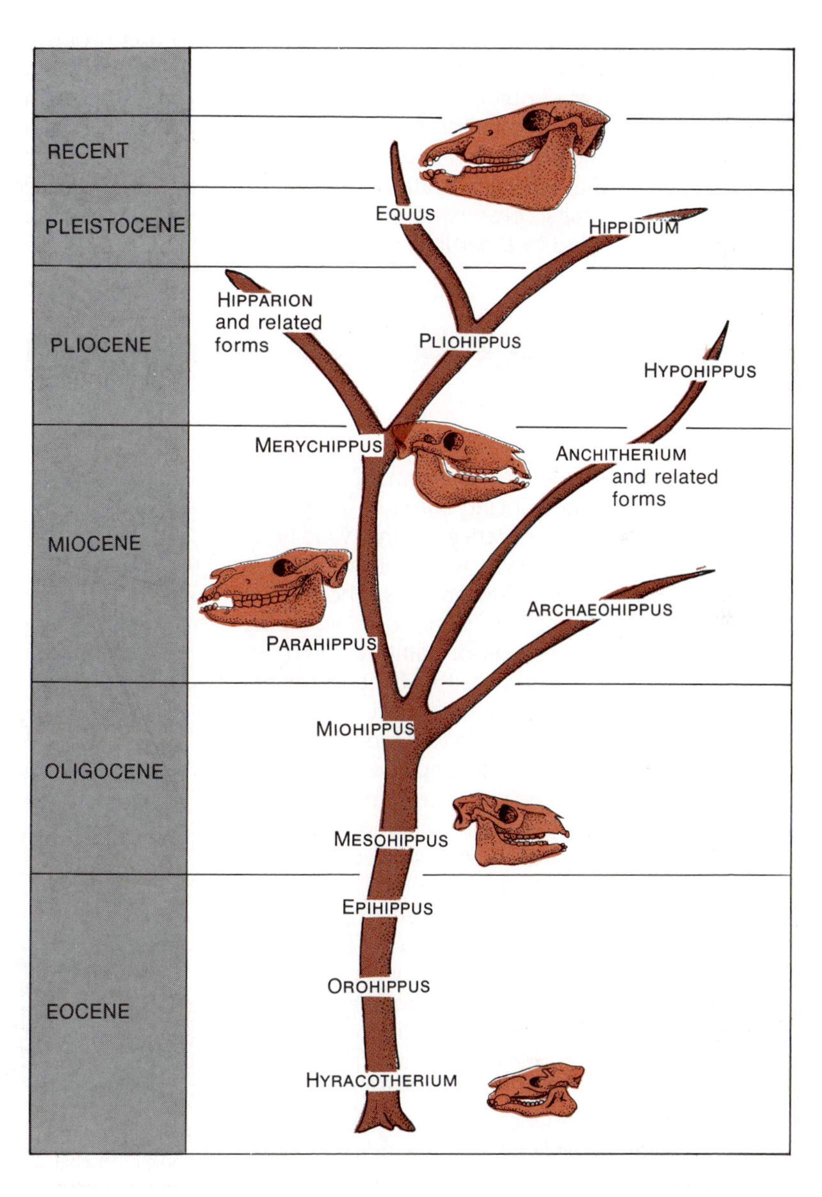

Figure 4–11 Simplified diagram to show the evolutionary relations of major genera of horses. (Data from E. H. Colbert. *Evolution of the Vertebrates.* New York, John Wiley & Sons, Inc., 1969.)

specimens, at least 800 showed striking new (that is, noninherited) variations that were in turn transmitted to offspring.

Later experimentors showed that mutations may occur at different levels within the genetic material. At the lower, or molecular, level, the building block sequence of DNA in one or more genes may be altered. At a higher level, the mutation may result from a structural alteration of the chromosome in which the DNA is arranged. At a still higher level, a mutation may result from an increase in the number of chromosomes.

Mutations can be caused by ultraviolet light, cosmic and gamma rays, chemicals, and certain drugs. They may also occur spontaneously without a specific causative agent.

Clearly, organic evolution requires change from at least three different sources: mutations, gene recombinations, and natural selection. Mutations are the ultimate source of new and different genetic material. Recombination spreads the new material through the population and mixes the new with the old. Natural selection sorts out the multitude of varying traits, preserving those that by chance are best fitted to a particular environment.

Evidence for Evolution

Paleontologic Clues

The proponents of the concept of evolution had a difficult time convincing some of their nineteenth century contemporaries of the validity of Darwin's theory. One reason was that evolution is an almost imperceptibly slow process. The human life span is too short to witness evolutionary changes across generations of plants or animals. Fortunately, we can overcome this difficulty by examining the remains of organisms left in rocks of successively younger age. If life has evolved gradually, the fossils preserved in consecutive formations should exhibit those changes. Indeed, many examples are known of sequential morphologic changes among related creatures during successive intervals of geologic time. The most famous example is provided by fossil horses recovered from formations of the Cenozoic Age (Fig. 4–11). The oldest horses thus far discovered as fossils were about the size of fox terriers. Unlike modern horses, they had four toes on their front feet and three on the rear. As one examines the many branching lineages of the horse "family tree" in successively higher (hence, younger) formations, one is able to see the results of evolutionary change. The animal shows an increase in size, a reduction of side toes with emphasis on the middle toe, an increase in the height and complexity of teeth, and a deepening and lengthening of the skull.

The change in horse dentition through time provides an interesting example of the important link between environment and organic evolution. During the time that the horse family was evolving, there is paleontologic evidence that grasslands were becoming increasingly widespread in North America and Eurasia. Grass is a rather harsh food for an animal that must feed upon plants. The blades contain silica, and because grass grows closely to the ground, it is usually coated with abrasive dust. Early members of the horse family lived in forested areas and fed upon the less abrasive leaves of trees and shrubs. They were browsers and had the low-crowned teeth of leaf-eaters. Later members of the horse family were affected by selection processes that favored variants better able to cope with the problem of tooth wear from eating grasses. They had, over many generations, evolved the high-crowned teeth with complex patterns of enamel that characterize grazing animals (Fig. 4–12). These horses were likely to have been well

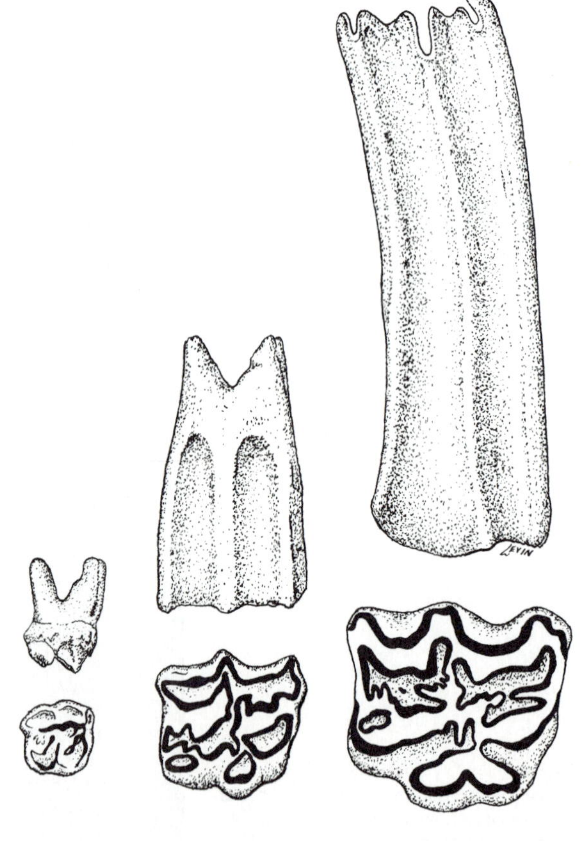

Figure 4–12 Development of high-crowned grinding molars in horses. Enamel shown in black. From left to right, *Hyracotherium, Merychippus,* and *Equus.* (From H. L. Levin, *Life Through Time.* Dubuque, Iowa, William C. Brown Co., 1975.)

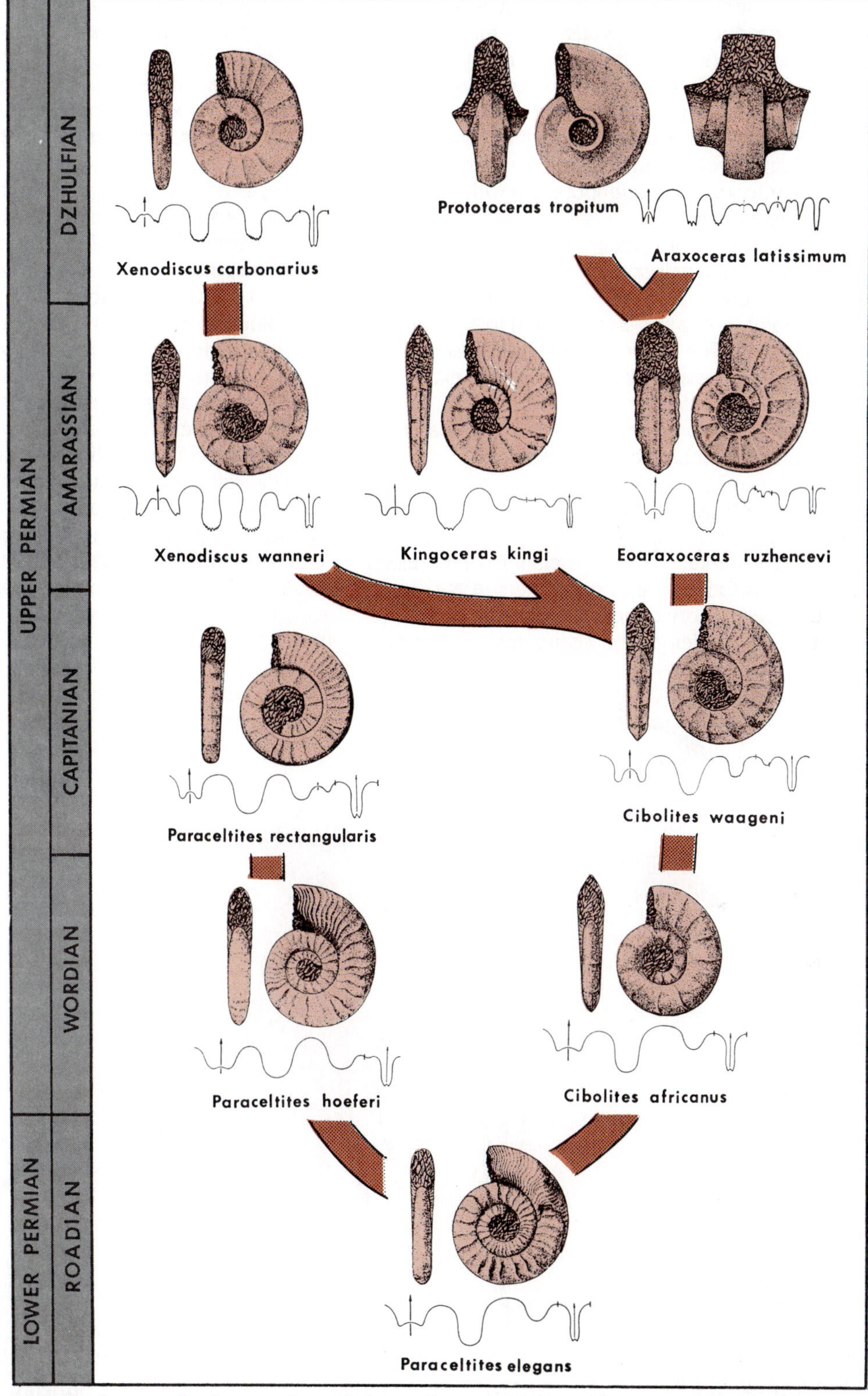

Figure 4–13 An example of progressive evolutionary change in a group of Permian ammonoid cephalopods. According to this interpretation, two evolutionary lineages originated from *Paraceltites elegans,* one terminating in *P. rectangularis* and a second producing *Cibolites waageni.* The latter was the root stock for three additional lineages. The curved lines beneath each drawing are tracings of the suture lines; the terms along the left border are the names of Permian Stages. (From C. Spinosa, W. M. Furnish, and B. F. Glenister. J. Paleontol. Vol. 49, No. 2, 1975.)

nourished and hence lived longer lives and were able to produce more progeny. Their evolution must also have affected the evolution of the predators that pursued them and may even have contributed to the evolution of species of grasses that were resistant to damage by grazing. Evolution is an intricate process in which every animal and plant interacts with its neighbors and with the physical environment.

Although fossil remains of horses provide a fine illustration of paleontologic evidence for evolution, hundreds of other examples representing every major group of animals and plants would serve as well. The marine invertebrates known as cephalopods provide exceptionally fine examples of progressive evolutionary changes (Fig. 4–13), and they are also one of the fossil groups most useful for paleontologic correlation.

Biologic Clues

Supplemental to the paleontologic clues favoring evolution are several persuasive arguments that are more directly biologic in nature. For example, in studies of the comparative morphology of organisms, it is not at all unusual to find body parts that are evidently of similar origin, structure, and development, even though they may be adapted for different functions in various species. For example, in seed plants, leaves are found as petals, tendrils, and thorns. In four-limbed vertebrates, the bones of the limbs may vary in size and shape, but they are fundamentally similar and in similar relative positions in birds, horses, whales, and humans (Fig. 4–14). Such basically similar structures in superficially dissimilar organisms are referred to as **homologous.** The differences in homologous structures are the result of variations and adaptations to particular environmental conditions, but the similarities indicate genetic relationship and common ancestry.

As another line of evidence, biologists point to the existence in animals and plants of useless and usually reduced structures that in other related species are well developed and functional. These **vestigial organs** are the "vestiges" of body parts that were utilized in earlier ancestral forms. Genetic material inherited from those ancestors still produces the vestigial organs, but the importance of these structures to the well-being of the organism

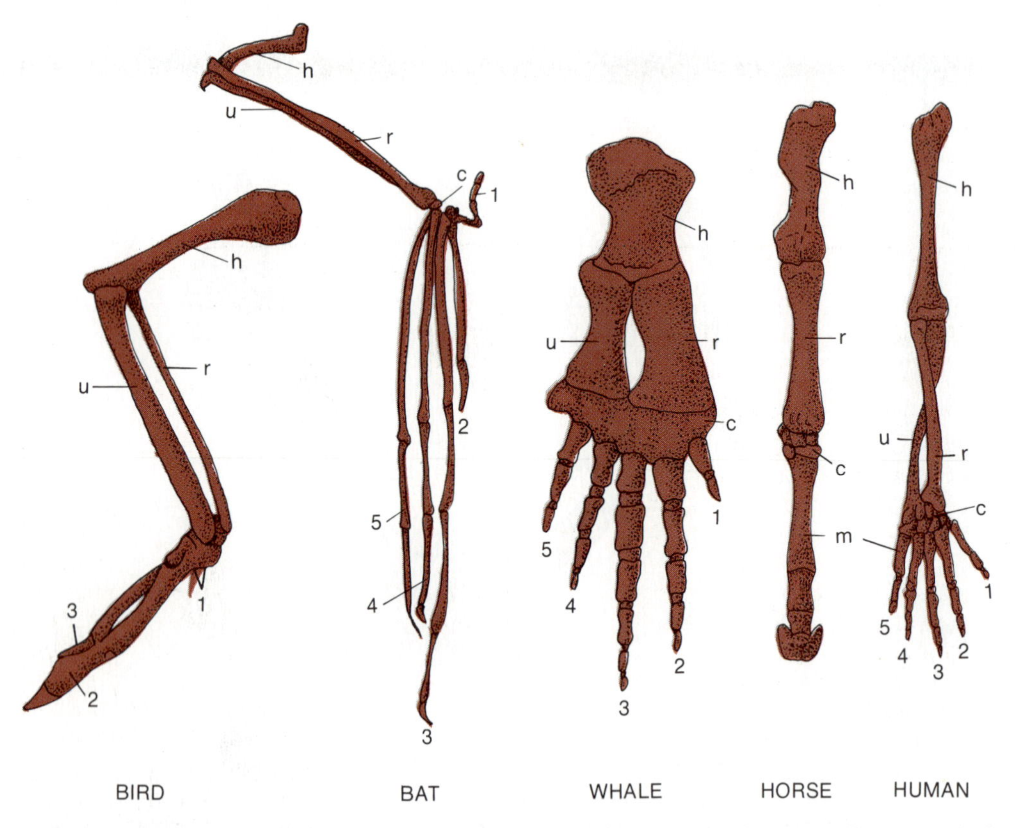

Figure 4–14 Skeleton of right forelimb of several vertebrates to show similarity of structure. Key: *c*, carpals; *h*, humerus; *m*, metacarpals; *r*, radius; *u*, ulna; 1–5, digits.

has diminished with changes in environment and habits. Humans have over 100 of these vestigial structures, including the appendix, the ear muscles, and the coccyx (tail vertebrae). The vestigial pelvic bones in such different animals as the boa constrictor and whale clearly suggest that they were evolved from four-legged animals.

Biologists have provided further evidence of evolution through comparative studies of the embryos of vertebrate animals. In their early stages of development, embryos of fish, birds, and mammals are strikingly similar. It would seem that all these animals received basic sets of genes from remote common ancestors. These genes control embryologic development for a time. Later in the developmental process, other genes begin to assume control and to cause each different species to develop in its own unique way.

Many plants and animals resemble one another, not only in structure and embryologic development but in biochemistry as well. Present knowledge indicates that all organisms produce nucleic acids—especially DNA—and that all use a compound called **adenosine diphosphate (ADP)** in life processes that involve energy storage and transfer. Digestive enzymes and hormone secretions are similar in many forms of life. Proteins extracted from corresponding tissues of closely related animals show definite similarities. The antigenic reactions of blood from the various groups of humans have been found to be practically identical or rather similar to such reactions in the blood of anthropoid apes.

FOSSILS AND STRATIGRAPHY

Establishing Age Equivalence of Strata with Fossils

One of William Smith's major contributions to geology was his recognition that individual strata contain definite assemblages of fossils. Because of the continuous change in life through time, superposition, and the observation that once species have become extinct they do not reappear in later ages, fossils can be used to recognize the approximate age of a unit and its place in the stratigraphic column. (Such a method for judging the age of a unit would not be possible with inorganic characteristics of strata, because they frequently recur in various parts of the geologic column.) Further, because organic evolution occurs all around the world, rocks formed during the same age in identical environments but diverse localities should contain similar fauna and flora if there has been an opportunity for genetic exchange. This fact permits geologists to match chronologically or correlate strata from place to place. It provides a means for establishing the age equivalence of strata in widely separated parts of the globe.

The Geologic Range

Before fossils could be used as indicators of age equivalence, it was first necessary to determine the relative ages of the major units of rock on the basis of superposition. Geologists began by working out the superpositional sequences locally, then they added sections from other localities around the world, fitting the segments together into a composite geologic column (Fig. 4–15). The next step was to determine the fossil assemblage from each time-rock unit and to identify the various genera and species. This work began well over a century ago and is still very much in progress. Gradually, it became possible to recognize the oldest, or first appearance of, particular species as well their youngest occurrence (last to appear) stratigraphically. The interval between the first and last appearance of a species constitutes the **geologic range.** Clearly, the geologic range of any ancient organism is not known *a priori* but is determined only by recording its occurrence in numerous stratigraphic sequences from hundreds of locations. Figure 4–16 illustrates just one such study that records the geologic ranges of nine genera of large fossil protozoans. Today, ranges are well known for some species and groups of species but relatively poorly known for others. Fortunately, there is now enough data that isolated sections of strata in geologically unexplored areas can be located in the composite geologic column by use of their contained fossils.

Identification of Time-Rock Units

The method for identifying time-rock units is illustrated by Figure 4–17. Geologists working in Region 1 come upon three time-rock systems of strata designated *O*, *D*, and *M*. Perhaps years later in Region 2, they again find units *O* and *D*, but in addition, they recognize an older unit—namely, *C*, below and hence older than *O*. Finally, while working in Region 3, they find a "new" unit, *S*, sandwiched between units *O* and *D*. The section, complete insofar as can be known from the available evidence, consists of four units that decrease in age as one progresses upward from *C* to *M*. The geologists may then plot the ranges of fossil species found in *C* to *M* alongside the geologic column. If they should next find themselves in an un-

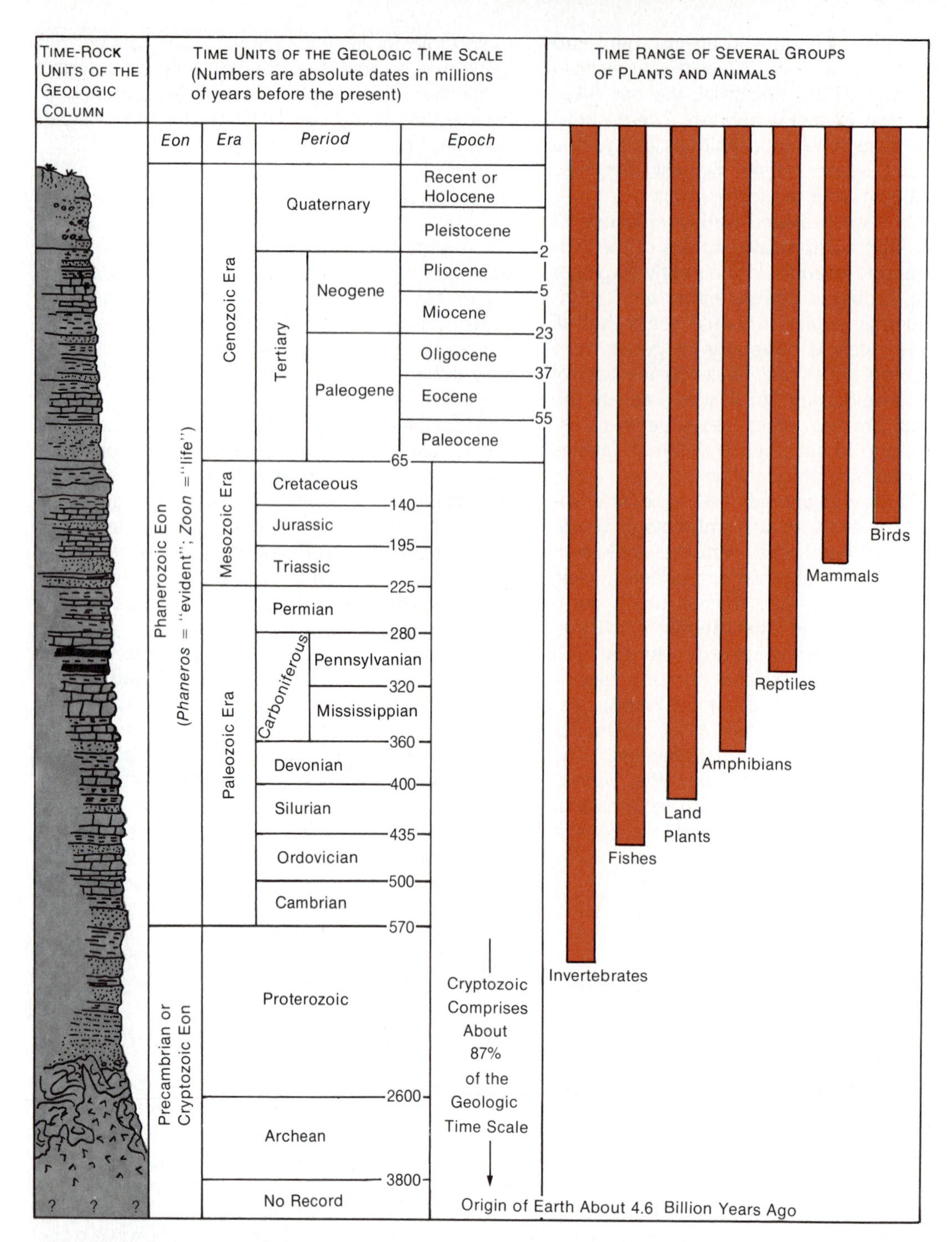

Figure 4–15 Geologic time scale.

explored region, they might experience difficulty in attempting to locate the position of this rock sequence in the standard column, especially if the lithologic traits of the rocks had changed. However, on discovering a bed containing species *A*, they are at least able to say that the rock sequence might be *C*, *O*, or *S*. Should they later find fossil species ***B*** in association with ***A***, they might then state that the outcrop in the unexplored region correlates in time with unit *S* in Region 3. In this way, the strata around the world are incorporated into an accounting of global stratigraphy.

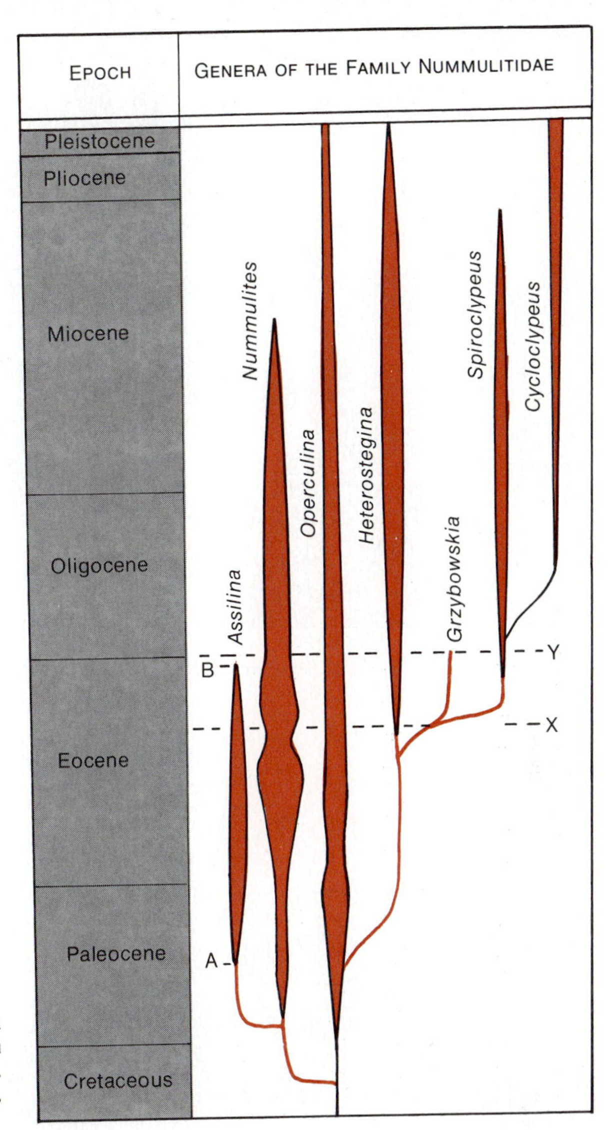

Figure 4–16 Phylogeny and time range of several genera within the family Nummulitidae, here used to illustrate how a range zone and concurrent range zone might be delimited. (Simplified from R. S. Barnett. *Paleontology,* 48(6):1253, 1974.)

Paleontologic Correlation

Of course, the preceding illustration is greatly simplified, and future discoveries may extend known fossil ranges. Geologists must always keep in mind that the chance inability to find key fossils might lead to erroneous interpretations. Fortunately, the chances of making such mistakes are diminished by the practice of using entire assemblages of fossils. Two or three million years from now, geologists might have difficulty in firmly establishing on fossil evidence that the North American opossum, the Australian wallaby, and the African aardvark lived during the same episode of geologic time. However, if they found fossils of *Homo sapiens* with each of these animals, it would indicate their contemporaneity. In this example, *Homo sapiens* can be considered the **cosmopolitan** species, for it is not restricted to any single geographic location within the terrestrial environment. The aardvark and wallaby are said to be **endemic,** in that they are confined to a particular area.

In the case of fossilized marine animals, cosmopolitan species have been especially useful in establishing the contemporaneity of strata, whereas endemic species are generally good indicators of the environment in which strata were deposited. Endemic species may slowly migrate from one locality to another. For example, the peculiar screw-shaped marine fossil bryozoan known as

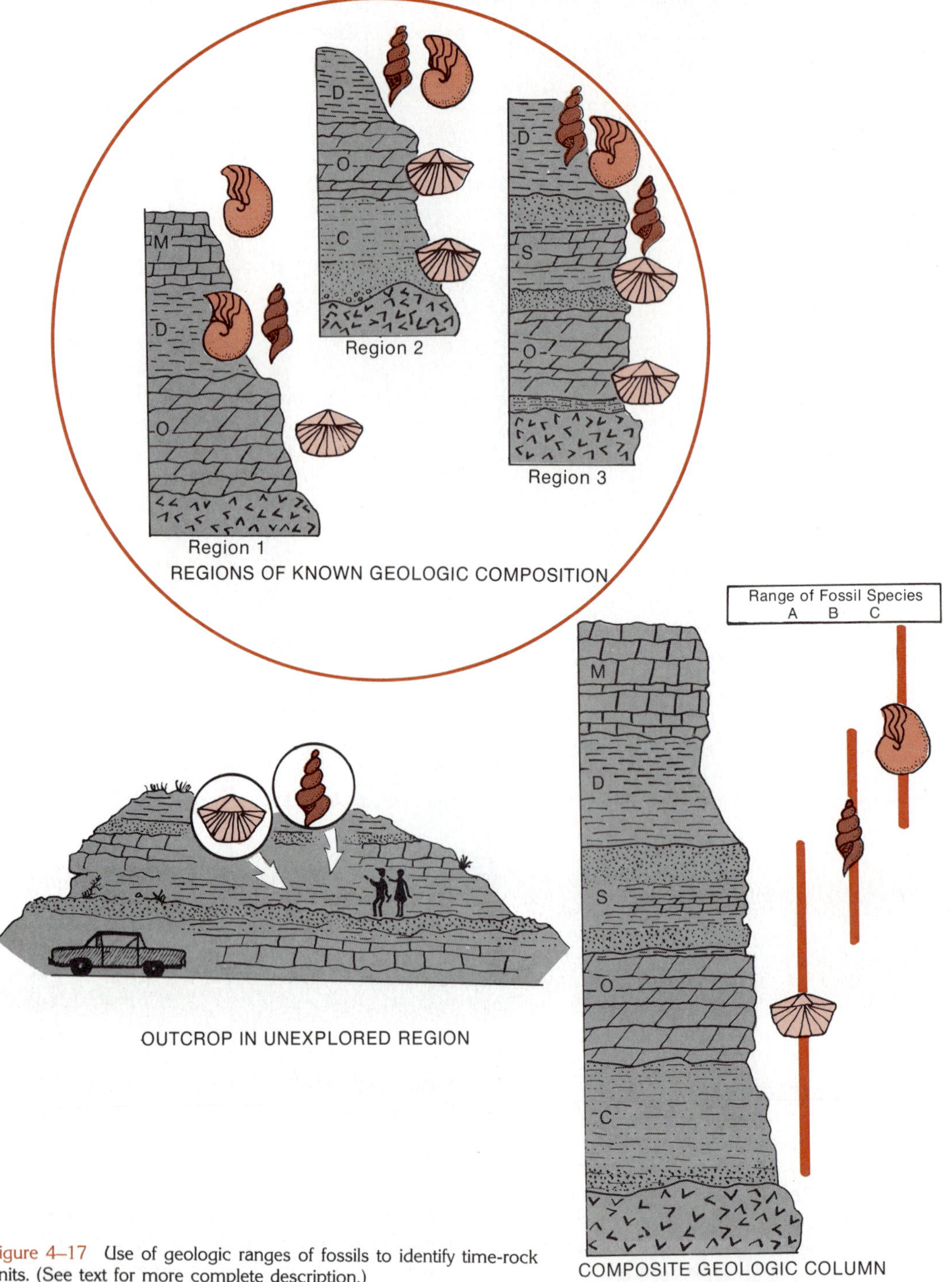

Figure 4–17 Use of geologic ranges of fossils to identify time-rock units. (See text for more complete description.)

Archimedes (Fig. 4–18) was endemic to central North America during the Mississippian Period but migrated steadily westward, finally reaching Nevada by the Pennsylvanian, and Russia by the Permian Period.

One must assume a tentative attitude in making correlations based on fossils. The validity of a correlation increases each time the sequence of faunal changes is found again at different locations around the world. Geologists must also be aware

Figure 4–18 The bryozoan *Archimedes.* This specimen shows only the screwlike axis. A fragile, lacy colony was attached to the sharp, revolving edge. (Width of specimen is 5 cm.)

that all changes need not reflect evolution but, instead, may indicate faunal migrations or shifts in flora that accompanied ancient environmental changes. In this regard, the sudden disappearance of a fossil need not mean that it became extinct, but rather that it moved elsewhere. One might also note that the earliest appearance of a fossil in rocks of a given region might mean it had evolved there; however, it might also signify only that an older species had come into the new locality.

In addition to being on the alert for faunal changes that are related to shifting of ancient environments (as opposed to changes resulting from evolution), paleontologists must also be concerned about the possibility of **reworked fossils.** At various times in the geologic past, weathering and erosion has freed fossils from their host rock. Much as would be the case with any clastic particle, these fossils might then be reworked into younger beds, and the younger strata might then be mistakenly assigned to an older geologic time. Some fossils are particularly resistant to erosion and chemical decay and therefore are more susceptible to reworking. Among such resistant fossils are conodonts (Fig. 4–19), tiny, toothlike structures of phosphatic composition and uncertain origin, which are abundant in some Paleozoic strata.

GUIDE FOSSILS. Many fossils are rare and restricted to a few localities. Others are abundant, widely dispersed, and derived from organisms that lived during a relatively short span of geologic time. Such short-lived but widespread fossils are called **guide fossils,** because they are especially useful in identifying time-rock units and correlating them from one area to another. A guide fossil with a short geologic range is clearly more useful than one with a long range. A fossil species that lived during the total duration of a geologic era would not be of much use in identifying the rocks of one of the subdivisions of lesser duration within that era. It is apparent that rate of evolution is an important factor in the development of guide fossils. Simply stated, **rate of evolution** is a measure of how much biologic change has occurred with respect to time. Figure 4–20 provides an example of a change in the rate of evolution among horned

Figure 4–19 Conodonts (50×).

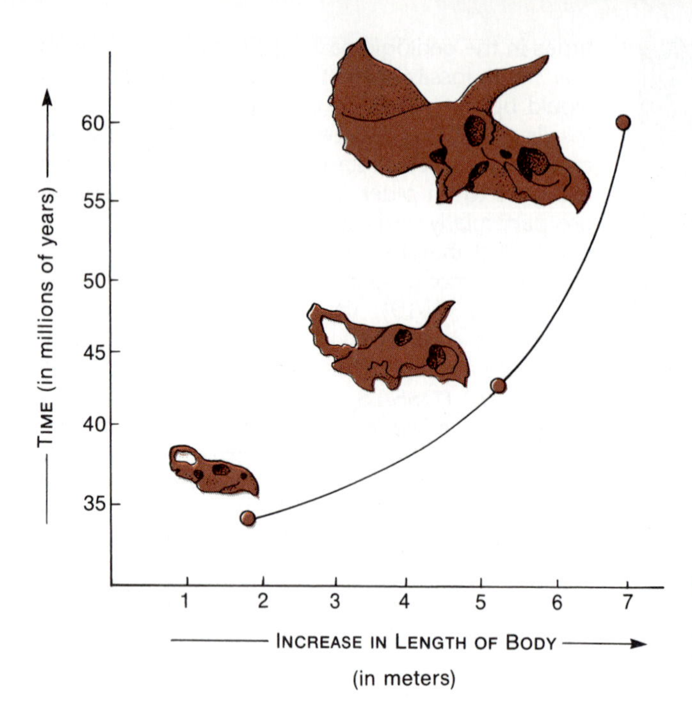

Figure 4–20 Example of a study on rate of evolutionary change. In this lineage of horned dinosaurs, the rate of increase in size was rapid at first and then appeared to decrease. (Adapted from E. H. Colbert. *Evolution,* 2:145–163, 1948.)

dinosaurs. The data indicate that, although increase in size was a persistent evolutionary trend, the rate of that change diminished with time.

Obviously, different animals have evolved at different rates. Human beings and their ancestors, for example, are considered to have had a relatively rapid rate of evolution. In general, lineages with rapid rates of evolution provide greater numbers of guide fossils. Usually, rates of evolution can be related to changes in the environment or to the inherent reproductive or genetic characteristics of the evolving populations.

Although guide fossils are of great convenience to geologists, correlations and interpretations based on assemblages of fossils are often more useful and less susceptible to error or uncertainties caused by undiscovered, reworked, or missing individual species. Also, by using the **overlapping geologic ranges** of particular members of the assemblage, it is often possible to recognize the deposits of smaller increments of time. The advantage of overlapping ranges can be illustrated with the help of Figure 4–21. Here, rather than using actual ranges of species, larger animal categories are used. It is apparent from this chart that a rock containing both stromatoporoids and goniatites can only be considered Devonian, whereas the occurrence of members of only one of the fossil groups would not provide as narrow a limit to the age of the rock.

ZONES. Geologists frequently employ the term **zone** when describing a body of rock that is identified strictly on the basis of its contained fossils. Paleontologic zones may vary in thickness or lithology and may be either local or global in lateral extent. Without formally naming them, we have already described the three major types of zones: range zone, assemblage zone, and concurrent-range zone. A **range zone** is simply the rock body representing the total geologic life span of a distinct group of organisms. For example, in Figure 4–16, the *Assilina* Range Zone is marked by the first (lowest) occurrence of that genus at point A and its extinction point at B. Geologists may also designate **assemblage zones** selected on the basis of several coexisting taxa and named after an easily recognized and usually common member of the assemblage. In certain areas, however, even though the guide fossil for the assemblage zone may not be present, the other members permit recognition of the zone. The third type of zone is the **concurrent-range zone.** It is recognized by the overlapping ranges of two or more *taxa* (sing. *taxon*). (The term **taxon** refers to a group of organisms that constitute a particular taxonomic category, such as species, genus, and family.) For example, the interval between X and Y in Figure 4–16 might be designated the *Assilina-Heterostegina* concurrent-range zone.

Zones are important in stratigraphy, for they

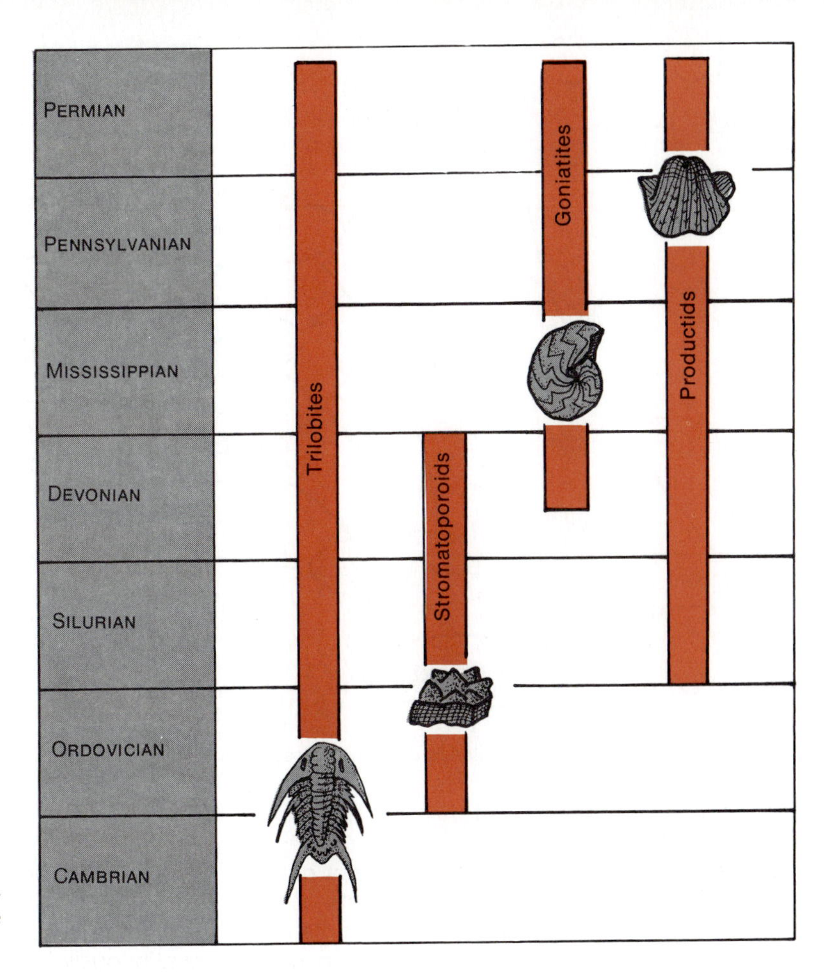

Figure 4–21 Advantage of using overlapping ranges of fossil taxa to recognize smaller increments of geologic time.

permit one to recognize correlative parts of the geologic column even when the lithologic characteristics of the rock are different.

FOSSILS AS CLUES TO ANCIENT ENVIRONMENTS

Paleoecology

Fossils, cold and mute as they may appear to be, are vestiges of once lively animals and plants that nourished themselves, grew, reproduced, and interacted in countless ways with their surroundings. The study of the interaction between ancient organisms and their environment is called **paleoecology.** Paleoecologists attempt to ascertain where ancient creatures lived and why they lived there. This is accomplished in various ways. One can sometimes compare the fossil species with a still living similar organism and assume that both organisms had the same needs, habits, and tolerances. One can also examine the anatomy of the fossil and determine which structures were likely to have developed in response to particular biologic and physical conditions in the environment. Such modifications for living in a certain way and certain kind of place are called adaptations. The fins of fish, claws of tigers, and blubber of whales are all adaptations that are useful for survival in particular environments. Not all adaptations are morphologic. Biochemical adaptations are of equal importance, although more difficult to recognize in fossils.

The language of paleoecology is derived from **ecology,** the study of present relationships between organisms and their environments. The largest unit in the study of ecology, and paleoecology as well, is called an **ecosystem.** Any selected part of the physical environment, together with the animals and plants in it, constitutes an ecosystem. An ecosystem may be as large as the earth or as small as a garden pond. Paleoecologists are keenly interested in ancient ecosystems, especially those of

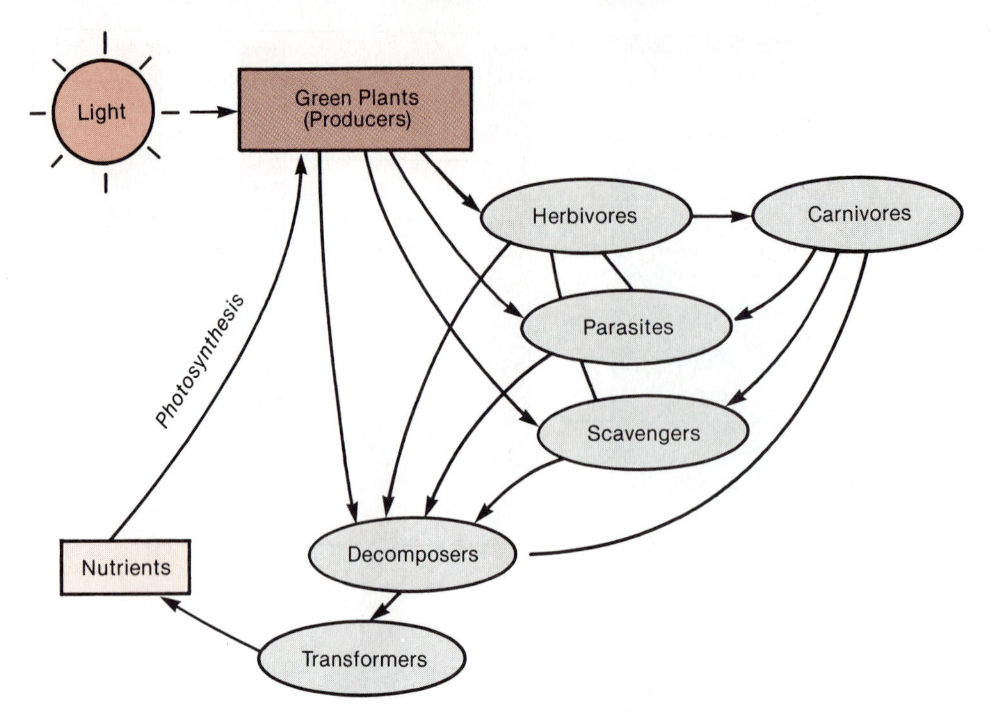

Figure 4–22 The movement of materials through an ecosystem. Components within ellipses are consumers.

the ocean. This is because of the richness of the marine fossil record. The physical aspects of the ocean ecosystem include the water itself, its dissolved gases (especially oxyen and carbon dioxide), salts (phosphates, nitrates, chlorides, and carbonates of sodium, potassium, and calcium), various organic compounds, turbidity, pressure, light penetration, and temperature. The biologic components of the ocean ecosystem are commonly divided according to so-called trophic, or feeding, levels (Fig. 4–22). There are, for example, **producer organisms** such as green plants, which, by means of photosynthesis, manufacture compounds from simple inorganic substances. These producer organisms in the sea are mostly very small in size (less than 0.1 mm) and include diatoms and other forms of algae. Producer organisms are eaten by **consumer organisms** such as mollusks, crustaceans, and fishes. The primary consumers commonly feed directly upon unicellular plants and thus may be further designated **herbivores.** The secondary consumers, which eat the herbivores, are called **carnivores.** Tertiary and even quaternary consumers feed on carnivores from lower trophic levels. The ecosystem also contains decomposers and transformers, including bacteria and fungi, which are able to break down the organic compounds in dead organisms and waste matter and produce simpler materials that can be utilized by the producers. Thus, in an ecosystem, the basic chemical components of life are continuously being recycled. The marine ecosystems also contain **parasites,** which feed on other organisms without necessarily killing them, and **scavengers,** which derive their nourishment from dead organisms.

Within any given ecosystem one finds the specific environments or habitats where certain organisms live. In the ocean, habitats range from the cold and dark realm of the abyss to the warm, illuminated, turbulent areas of a tropical coral reef. Within habitats one may find a number of **ecologic niches,** in which particular organisms make their living. An ecologic niche really represents the sum of all the biologic and physical conditions that permit the organism occupying the niche to survive. In the coral reef habitat, there are numerous niches. Some are occupied by the tiny, tentacled coral animals that build the framework of the reef and feed on tiny, suspended microorganisms. Marine snails occupy a niche along the surface of algal mats, where with snail-like slowness they graze on films of algae. In areas of soft sediment, plump lugworms unfastidiously consume mud for its content of organic nutrients. The niche for worms in the coral reef is quite different from that of the reef-dwelling predatory fishes, which dart from dark cavities among the coral heads to snare unwary victims.

An important task of paleoecologists is to use the fossils they encounter to determine the characteristics of the environment in which the enclosing sediment was deposited. To facilitate these studies, it is useful to have a simple classification of environments. In the ocean, the classification begins with a twofold division into pelagic and benthic realms. The **pelagic** realm consists of the water mass lying above the ocean floor and can be further divided into a **neritic zone,** which overlies the continental shelves, and an **oceanic zone,** which extends seaward from the shelves (Fig. 4–23). Within the pelagic realm one finds myriads of mostly tiny animals and plants that drift, float, or feebly swim. They are called **plankton. Phytoplankton** are plants and include algae such as the diatoms and coccolithophorids, which are to be discussed in a subsequent chapter. Planktonic animal life is designated **zooplankton** and includes protozoan creatures such as radiolarians and foraminifers, as well as larvae of an immense number of invertebrate animals that live as adults on the sea floor.

The pelagic realm is also the home of **nekton,** or true swimming animals. Nekton are able to travel where they choose under their own power, and this is a very advantageous adaptation. A swimming creature can search for its food and does not have to depend upon chance currents to bring it a morsel to eat. It can sometimes escape predators and can move to more favorable areas when conditions become difficult. The nekton are a diverse group. For the most part, they are members of the Chordata and include a vast array of fishes and fewer numbers of reptiles and mammals. There are, however, nonchordate (having no backbone) nektonic animals as well, the most numerous being squid, shrimp, and krill.

The second great division of the ocean is the **benthic** realm. It begins with a narrow zone above high tide called the **supralittoral zone.** Several kinds of marine plants, insects, and crustaceans have adapted to this difficult marine environment in which spray from waves provides necessary moisture. Seaward of the supralittoral zone is the area between high and low tide. It is called the **littoral zone.** Organisms living in the littoral zone must be able to tolerate the alternately wet and dry conditions that occur there. Some avoid the drying influence of the air by migrating back and forth with the tides or by burrowing into the wet sand. Others, like insects and certain crabs, have adapted themselves to exposure to air.

The largest number of benthic animals and plants live seaward of the littoral zone in the continuously submerged **sublittoral zone.** This zone extends from low tide levels down to the edge of the continental shelf (about 200 meters). In most parts of the benthic littoral zone, light penetrates all the way to the sea floor. Various kinds of algae thrive, as well as abundant protozoans, sponges, corals, worms, mollusks, crustaceans, and sea urchins. These animals are never subjected to desiccation. Their adaptations are mainly associated with food gathering and protection from predators. Some of the benthic animals live on top of the sediment that carpets the sea floor and are therefore dubbed

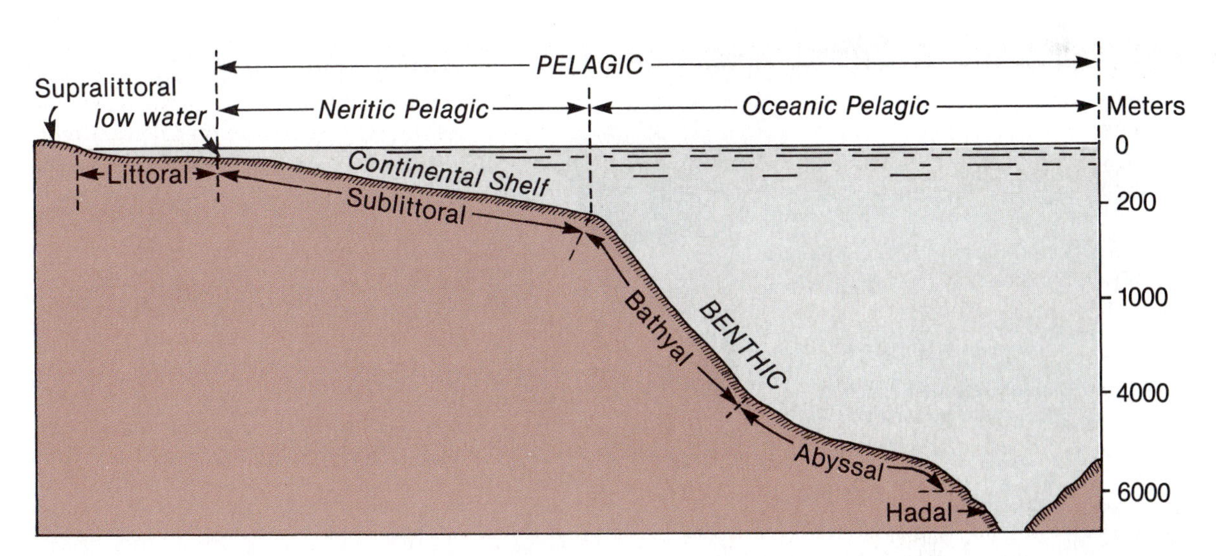

Figure 4–23 Classification of marine environments.

epifaunal. Others, termed **infaunal,** burrow into the sediment. Those bottom-dwellers capable of locomotion are further designated **vagile,** in contrast to immobile creatures, which are **sessile.**

Beyond the continental shelves, the benthic environment is subjected to colder temperatures, little or no light penetration, and high pressures. Plants are unable to live at these depths. One encounters the **bathyal environment** from the edge of the shelf to a depth of about 4000 meters. Still deeper levels constitute the **abyssal environment.** The term **hadal** is reserved for the extreme depths found in oceanic trenches. As might be expected, animals are less abundant in the abyssal and hadal environments. Mostly they are scavengers and carnivores that depend upon the slow fall of food from higher levels.

Use of Fossils in Reconstructing Ancient Geography

The geographic distribution of present-day animals and plants is closely controlled by environmental limitations. Any given species has a definite range of conditions under which it can live and breed, and it is generally not found outside that range. Ancient organisms, of course, had similar restrictions on where they could survive. If we note the locations of fossil species of the same age on a map and correctly infer the environment in which they lived, we can produce a paleogeographic map for that particular time interval. One might begin by plotting on a simple base map the locations of marine fossils that lived at a particular time. This provides an idea of which areas were occupied by seas and might even suggest locations for ancient coastlines. Figure 4–24 shows major land and sea regions during the mid-Carboniferous Period. The locations of marine protozoans called *fusulinids* are plotted as open circles. Notice that the lofty ranges of the Rocky Mountains were non-existent, their present locations occupied by a great North-South seaway.

Having obtained a fair idea of the marine regions, one might next give attention to the locations of fossils diagnostic of land areas. The fossilized bones of land animals such as dinosaurs or

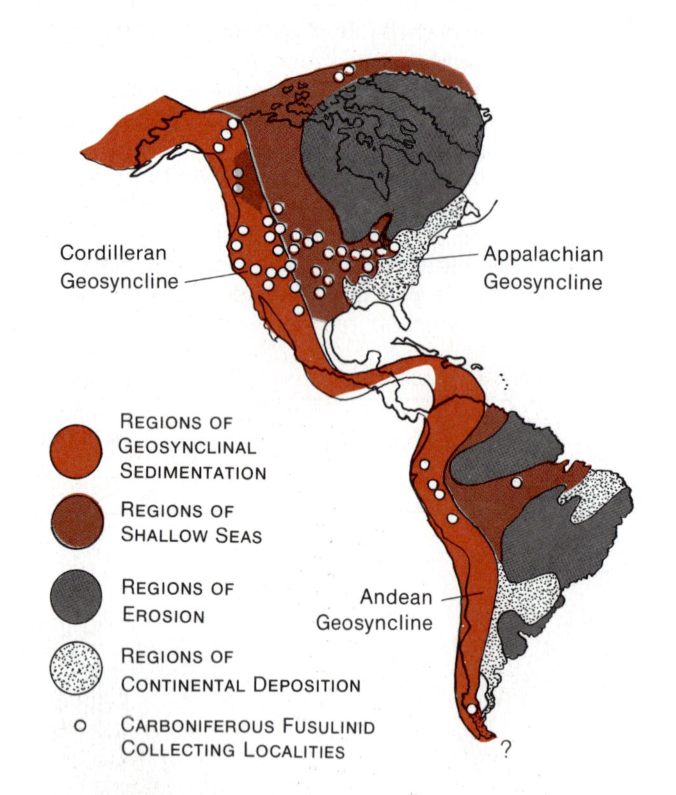

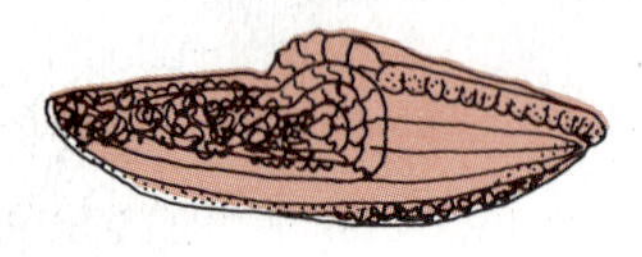

Enlarged sketch of a fusulinid with section cut out to show internal structure.

Figure 4–24 Major land and sea regions in North and South America during the Carboniferous Period. (Adapted from C. A. Ross. *J. Paleontol.,* 41(6):1341–1354, 1967.)

mastodons would suggest a terrestrial environment, as would known occurrences of remains of land plants. If the rocks contain no diagnostic land fossils, but do contain carbon, geologists are sometimes able to tell if deposition occurred in marine or freshwater bodies by the ratio of the isotope carbon 13 to carbon 12 in the sediment.

By an analysis of the fossils and the nature of the enclosing sediment, it is often possible to recognize deeper or shallower parts of the marine realm or to discern particular land environments such as ancient floodplains, prairies, deserts, and lakes. River deposits may yield the remains of freshwater clams and fossil leaves. A mingling of the fossils of land organisms and sea organisms might be the result of a stream entering the sea and perhaps building a delta in the process.

The migration and dispersal patterns of land animals, as indicated by the fossils they left behind, is one important indicator of former land connections as well as mountainous or oceanic barriers that once existed between continents. Today, for example, the Bering Strait between North America and Asia prevents migration of animals between the two continents. The approximately 80 km between the two shorelines (Fig. 4–25), however, is covered by less than 50 meters of water; this might lead one to wonder whether Asia was once connected to North America. The fossil record shows that a land bridge did connect these two continents on several occasions during the Cenozoic Era. The earliest Cenozoic strata on either continent have a fossil fauna uncontaminated by foreign species. Somewhat higher and younger rocks contain fossil remains of animals that heretofore had been found only on the opposite continent. These remains mark a time of land connection. The last such connection may have existed only 14,000 to 15,000 years ago, when Stone Age hunters used the route to enter North America.

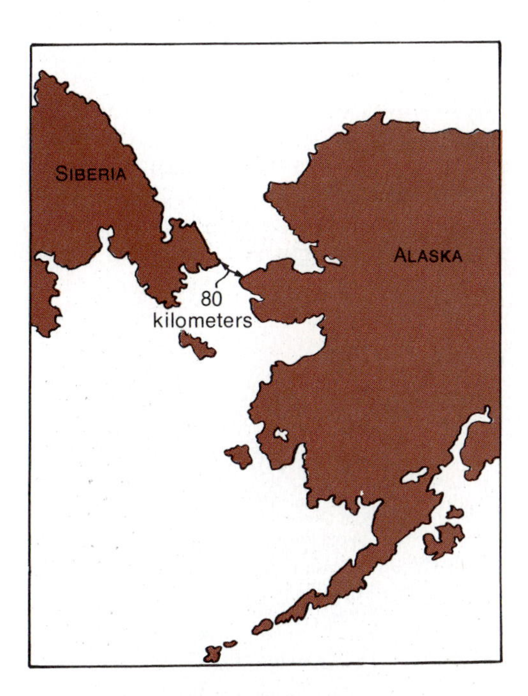

Figure 4–25 The Bering Strait between North America and Asia. The area between the two continents would become land if sea level were lowered by only 45 meters. Such a lowering of sea level occurred at various times during the Cenozoic Era and permitted migrations across the land bridge.

Another familiar example of how fossils aid in paleogeographic reconstructions is found in South America. Careful analyses of fossil remains indicate that, in the Early Cenozoic Era, South America was isolated from North America; as a result, a uniquely South American fauna of mammals evolved over a period of 30 to 40 million years. The establishment of a land connection between the two Americas is recognized in strata only a few million years old (Late Pliocene) by the appearance of a mixture of species formerly restricted to either North or South America. (The migrations were decidedly detrimental to South American species.)

In addition to deciphering the positions of shorelines or locations of land bridges, paleontologists can also provide data that help to locate the equator, parallels of latitude, and pole positions of long ago. It has been observed that in the higher latitudes of the globe, one is likely to find large numbers of individuals but that these are members of relatively few different species. In contrast, equatorial regions tend to develop a large number of species, but with comparatively fewer individuals within each species. Stated differently, the variety or **species diversity** for most higher categories of plants and animals increases from the poles toward the equator. This is probably related to the fact that relatively few species can adapt to the rigors of polar climates. Conversely, there is a stable input of solar energy at the equator, less duress caused by seasons, and a more stable food supply. These warmer areas place less stress on organisms and provide opportunity for continuous, uninterrupted evolution. Of course, in particular areas, generalizations about species diversity can be upset by special circumstances of altitude, upwelling, or other vagaries of air and water currents.

Another way to locate former equatorial regions (and therefore also the polar regions that lie 90° of latitude to either side) is by plotting the locations of fossil coral reefs of a particular age on a world map. Nearly all living coral reefs lie within 30° of the equator (Fig. 4–26). It is not an unreasona-

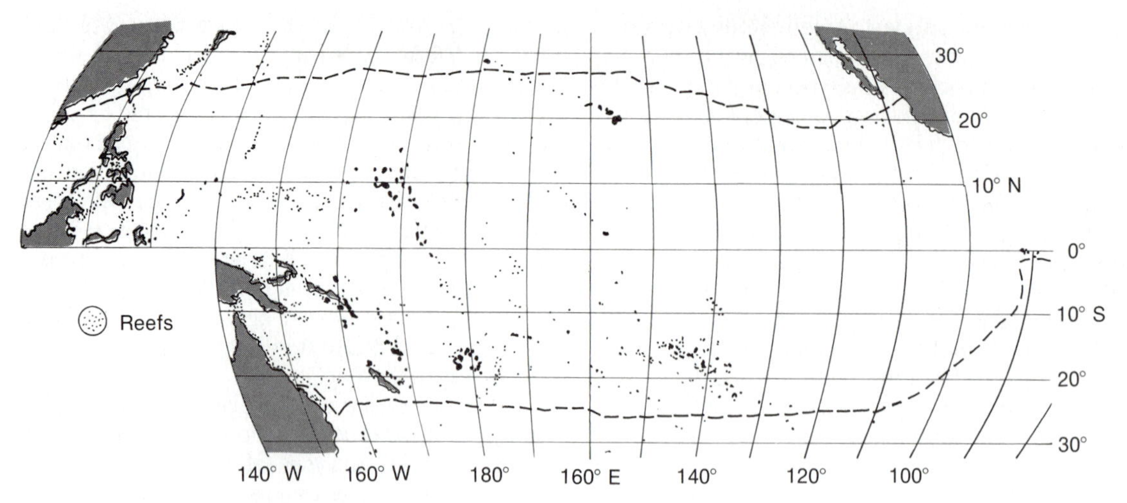

Figure 4–26 Latitudinal limits of living coral reefs in the Pacific Ocean shown by broken lines. (After J. W. Wells. *Geol. Soc. Am. Mem. 67,* 1:609–631, 1957.)

ble assumption that the ancient reef corals had similar geographic preferences.

Use of Fossils in the Interpretation of Ancient Climatic Conditions

Among the many climatic factors that limit the distribution of organisms, **climate** (especially the temperature component of climate) is of great importance. There are many ways by which paleoecologists gain information on ancient climates. An analysis of fossil spore (Fig. 4–27) and pollen grains (Fig. 4–28) often provides outstanding evidence of past climatic conditions. Living organisms with known tolerances can often be directly compared with fossil relatives. The coelenterates of modern coral reefs, for example, thrive in regions where temperatures rarely fall below 18′C (Celsius). It is a reasonable assumption that ancient reefs were thus limited in a similar way to the warmer parts of the oceans. One is aided in making such inferences by the study of all associated fossils as well as by textural or compositional features of the enclosing rock. Even when a close living analogue is not available, plants and animals may exhibit morphologic traits useful in determining paleoclimatology. Plants with aerial roots, lack of annular rings, and large wood cell structure indicate tropical or subtropical climates lacking in strongly contrasting seasons. Marine mollusks (e.g., clams, oysters, snails) with well-developed spinosity and thicker shells tend to occur in warmer regions of the oceans. In the case of particular species of marine protozoans, variation in the average size of individuals or in the direction of coiling can provide clues to cooler or warmer conditions.

An example of environmentally induced changes in shell coiling directions is provided by the planktonic (floating) foraminifer *Globorotalia truncatulinoides* (Fig. 4–29). In the Pacific Ocean, left-coiling tests (shells) of these foraminifers have dominated in periods of relatively cold climate, whereas right-coiling tests dominated during warmer episodes. Another foraminifer, *Neoglobquadrina pachyderma,* exhibits similar changes in coiling directions, not only in the Pacific Ocean but the Atlantic as well. Such reversals in coiling may occur quickly and over broad geographic areas. For this reason, they are exceptionally useful in correlation studies.

Aside from morphologic changes within species, entire assemblages of foraminifers are widely

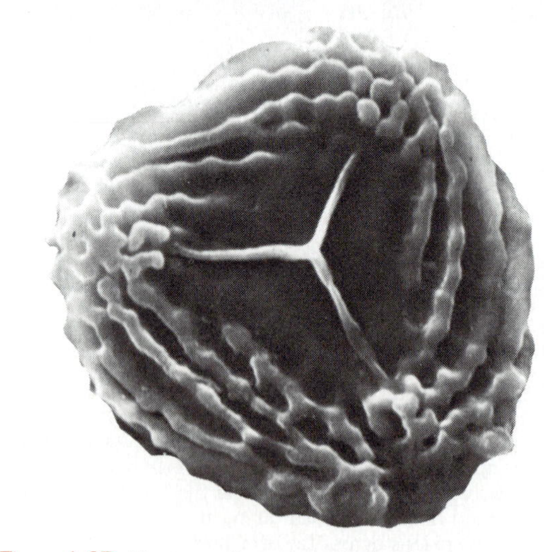

Figure 4–27 Electron micrograph of a fossil spore grain, magnified 1000 times. (Courtesy of Standard Oil Company of California.)

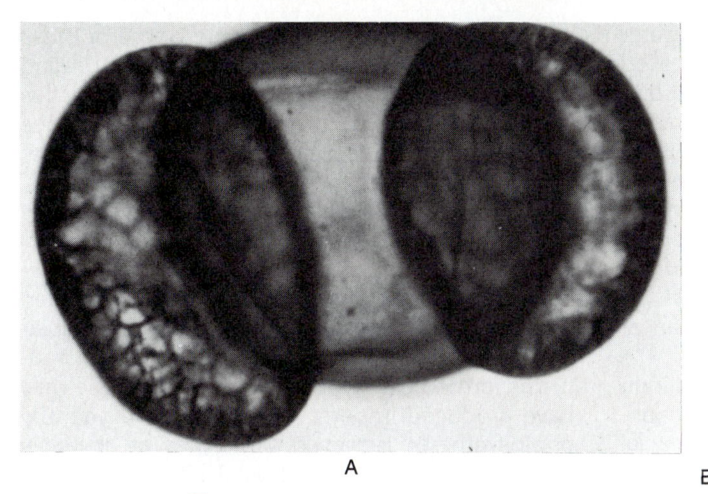

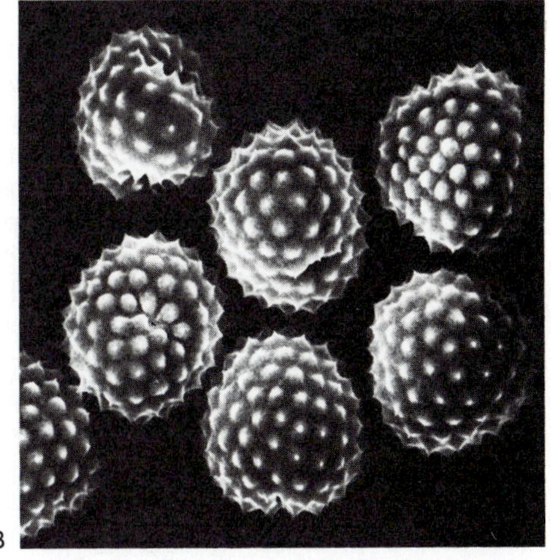

Figure 4–28 Pollen grains. (A) Pollen grain of a fir. Note the inflated wings, which make the pollen grain more bouyant. The specimen is about 125 microns in its longest dimension. (B) The spiny pollen grains of ragweed, as viewed with the aid of the electron microscope. These grains are 12 to 15 microns in diameter. (Electron micrograph courtesy of Walter Lewis.)

used in paleoecologic studies. Today, as in the past, the benthic (bottom-dwelling) species have inhabited all major marine environments. As a result of the accumulated data on living foraminifers, inferences about salinity and water depths can be made. Frequently, fossil foraminifers have provided the means for recognition of ancient estuaries, coastal lagoons, and nearshore or deep oceanic deposits.

Sometimes, when the overall morphology of a fossil does not provide clues to temperature or climate, the compositions of the skeleton can be used. Magnesium, for example, can substitute for calcium in the calcium carbonate of shelled invertebrates. For particular groups of living marine invertebrates, it has been found that those living in warmer waters have higher magnesium values than do those residing in colder areas. This knowledge has been used to interpret the climate at the times fossil forms were living.

In another kind of biogeochemical study of shell matter, investigators have found that certain organisms will build their skeleton of calcium carbonate in both of its two mineral forms: *aragonite* and *calcite.* Moreover, in colder realms, particular species will have higher calcite-to-aragonite ratios, whereas their fellow species living in warmer waters have lower calcite-to-aragonite ratios. Secretion of calcium carbonate in the form of aragonite appears to be "easier" at higher temperatures.

There are, however, problems when geochemical means are used to deduce temperature. Aragonite is an unstable mineral that reverts to the more stable calcite. It is nearly absent in Paleozoic rocks. High-magnesium calcite is also unstable. Magnesium may be leached out of the sediment after burial, causing the conversion of high-magnesium calcite to its more stable low-magnesium counterpart.

An elegant method for learning about relative temperature in the ancient marine environment involves analyses of the ratios of isotopes of oxygen. In addition to the common oxygen 16, there are two heavier oxygen isotopes: oxygen 17 and oxygen 18. In the late 1940's, the distinguished chemist Harold Urey reasoned that because the ratio of oxygen 16 to oxygen 18 in sea water is dependent on temperature, the ratios of those isotopes in the calcium carbonate ($CaCO_3$) of shells should reflect the enclosing water's temperature at the time that carbonate was secreted. In other words, changes in water temperature long ago should be reflected (and recorded) by corresponding changes in the $^{18}O/^{16}O$ ratio in calcareous shells. It was found that the ratio between the ^{18}O and ^{16}O in shell matter decreases 0.02 per cent for each 1'C lowering of water temperature. Urey and his coworkers, in one of the pioneering studies, used extinct squidlike animals called belemnites (Fig. 4–30) as subjects for their experiments. They

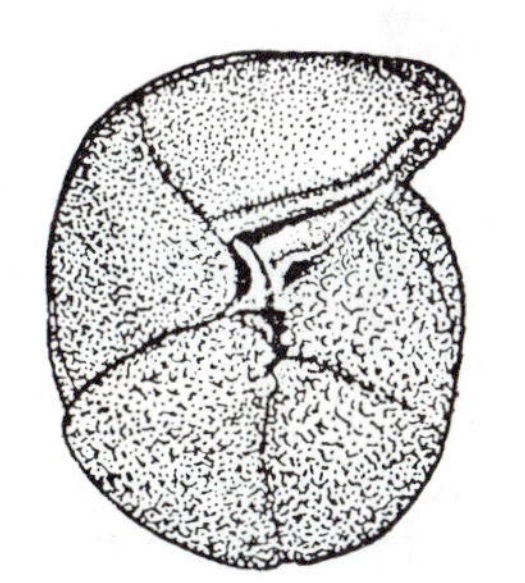

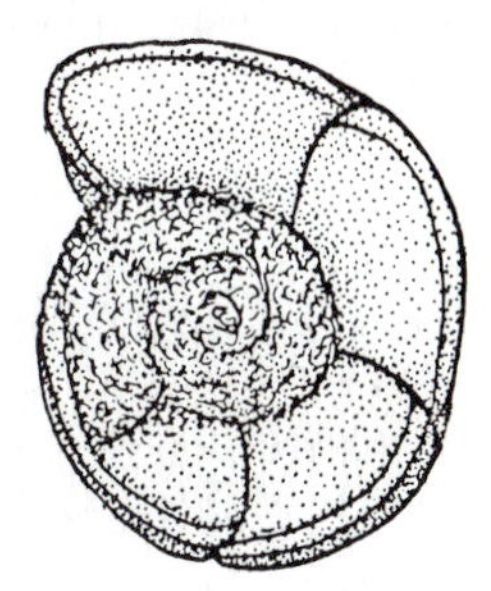

Figure 4–29 Shell coiling in the foraminifer *Globorotalia truncatulinoides.* Sketch of both sides of a single left-coiling specimen. (Diameter is about 0.9 mm.) (From F. L. Parker. *Micropaleontology.* 8(2):219–254, 1962.)

were able to determine that the mean sea water temperature in which the Jurassic belemnites lived was 17.6′C (± 6.0′ maximal seasonal variation). Subsequent studies by Robert Bowen on Jurassic and Cretaceous belemnites from around the world provided paleotemperature interpretations for extended periods of time (Fig. 4–31), permitted checks on the quite different locations of past geographic poles (determined by geophysical studies), and indicated that tropical and semitropical climatic belts were once more widespread than at present.

Oxygen isotope studies are also used as indicators of post-Miocene episodes of extensive continental glaciation. This is possible because glacial ice locks up the lighter isotope, causing an increase in the heavier isotope in sea water. Thus the oxygen isotope method provides a direct measure of water temperature only for sediments deposited at times when ice caps were not actively growing or melting.

AN OVERVIEW OF THE HISTORY OF LIFE

According to our best evidence, the earth was formed about 4.6 billion years ago. For a time, it was devoid of any form of life. Our earliest direct indications of ancient life are remains of bacteria and the most primitive of algae discovered in rocks over 3.2 billion years old. These early fossils are usually considered evidence that the long evolutionary march had begun. However, they also stand at the end of another long and remarkable period during which living things presumably evolved from nonliving chemical compounds. We have no direct geologic evidence to tell us how and when the transition from nonliving to living occurred. What we do have are reasonable hypotheses supported by careful experimentation. Some of these hypotheses will be examined in the chapter dealing with the Precambrian Period. It will suffice for this brief overview to note that life came into existence early in that great 4-billion-year span of time prior to the beginning of the Cambrian Period.

For most of Precambrian time, living things left only occasional traces. Here and there, primarily in siliceous rocks as old as 3.2 to about 0.7 billion years old, paleontologists have been rewarded with finds of bacteria and filamentous or unicellular blue-green algae. Occasionally, filamentous organisms formed extensive algal mats, and these may have produced important quantities of oxygen by photosynthesis. However, all life was at or below the unicellular level until about 1 billion years ago, when the world's first multicellular organisms left their trails and burrows in rocks that were someday to be called the *Torrowangee Group of Australia.* In rocks deposited about 0.7 billion years ago, fossil metazoans recognizable as worms, coelenterates, and arthropods have been found—albeit rarely—in scattered spots around the world. Thus, near the end of the Precambrian Eon, the stage was set for the evolution of a wide range of Paleozoic plants and animals.

The Evolutionary History of Animals

Some rather general observations are possible after a brief scanning of the history of life following the Precambrian. One notes that the principal groups of invertebrates appear early in the Paleozoic (Fig. 4–32). Less advanced members of each phylum characterized the earlier geologic periods, whereas more advanced members came along later. Also, most of the principal phyla are still represented by animals today. In our review of fossil creatures, we are not startled by sudden appearances of bizarre or exotic animals and plants. Fur-

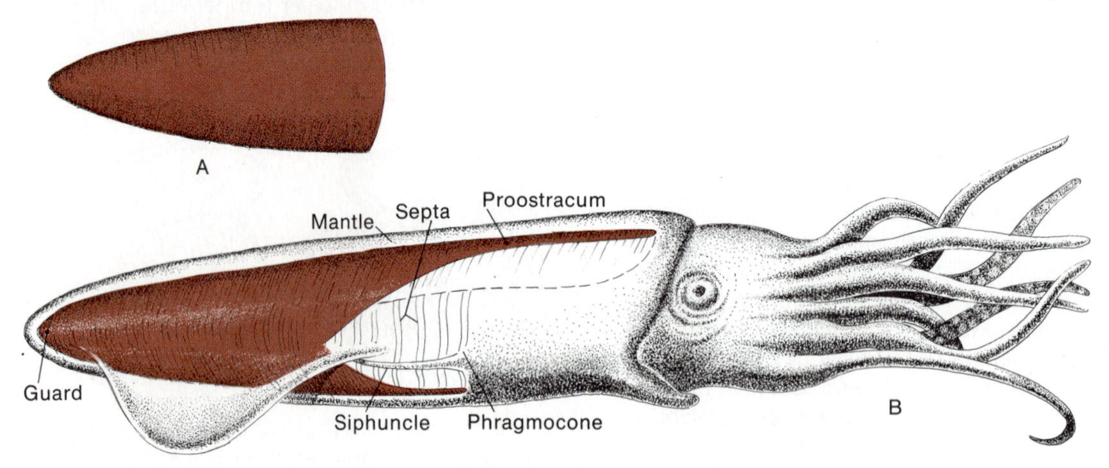

Figure 4–30 The usual belemnite fossil consists only of the guard, shown in A. An interpretation of the appearance of a belemnite as it would appear if living is shown in B.

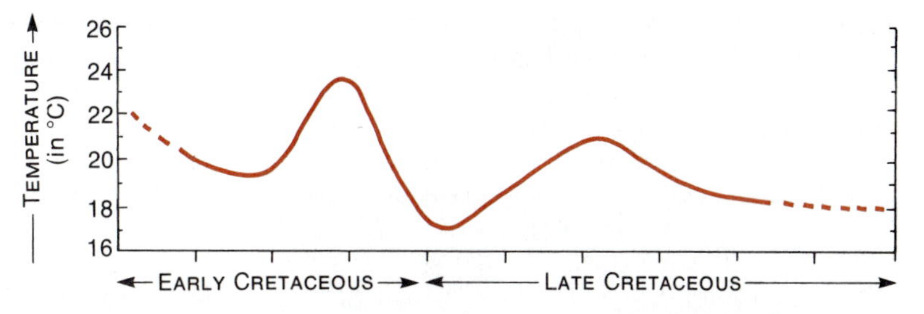

Figure 4–31 Paleotemperature data for the Cretaceous obtained by oxygen isotope analysis of belemnite fossils. The curve shows two episodes of relatively warm conditions with a decline in temperatures during the early to late Cretaceous transition. (After R. Bowen. *J. Geology,* 69(1): 1961.)

ther, we are able to recognize periods of environmental adversity that caused extinctions. Such episodes were usually followed by much longer intervals of recovery and more or less orderly evolution. One final observation is that there has been a persistent gain in the overall diversity of life down through the ages.

After the Precambrian Era, life became so varied that it is difficult to describe it both briefly and adequately. We can, however, make some generalizations. Such marine invertebrate animals as trilobites, brachiopods, nautiloids, horn corals, honeycomb corals, and twiglike bryozoans were abundant during the Paleozoic Era. The Paleozoic was also the era when fishes, amphibians, and reptiles appeared and left behind a fascinating record of the conquest of the lands. Remains of more modern corals, diverse pelecypods, sea urchins, and ammonoids characterize the marine strata of the Mesozoic Era. However, the Mesozoic is best known as the **"age of reptiles,"** when dinosaurs and their kin dominated the continents. No less important than these big beasts, however, were the rat-sized primitive mammals and earliest birds that skittered about perhaps unnoticed by the "thunder beasts." The Cenozoic faunas often superficially resembled those of the Mesozoic. However, there were no ammonoid cephalopods in this most recent era. Rather, rocks of the Cenozoic are recognized by distinctive families of protozoans (foraminifera) and a host of modern-looking snails, clams, sea urchins, barnacles, and encrusting bryozoa. Because the Cenozoic saw the expansion of warm-blooded creatures like ourselves, it is appropriately termed the "age of mammals."

The Evolutionary History of Plants

Long before the first animal appeared on earth, there were plants. Plants also experienced a complex and marvelous evolutionary history. That history began with the origin of the primitive, unicellular aquatic algae of the Precambrian Eon. Except for the blue-green algae and bacteria noted

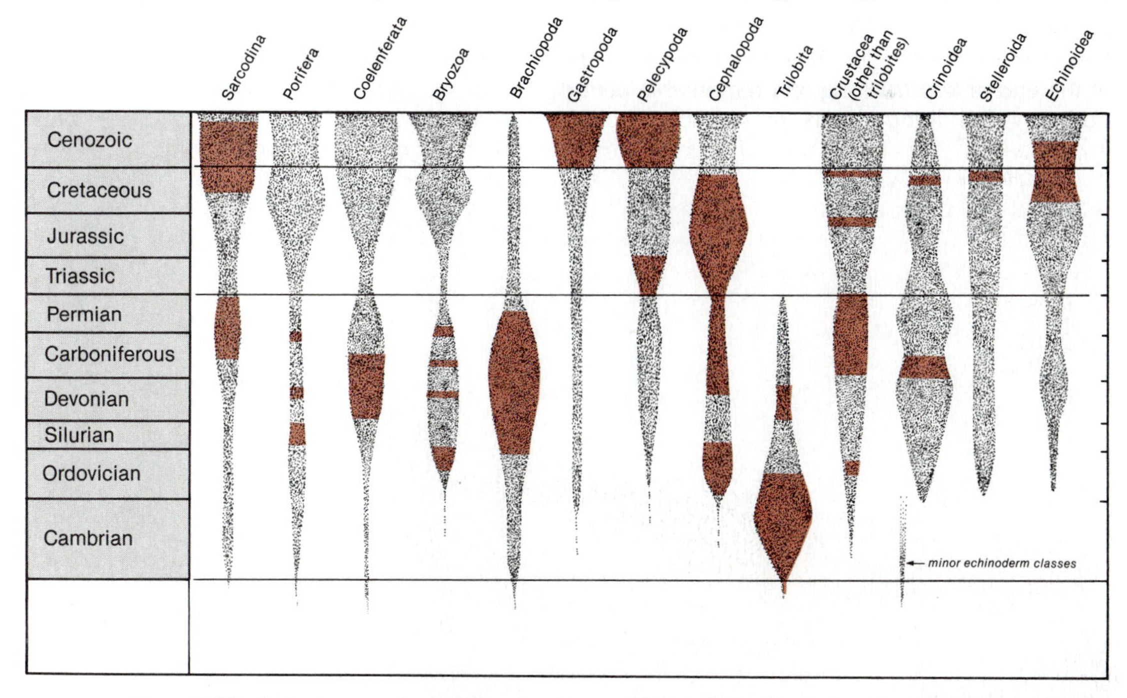

Figure 4–32 Geologic ranges and relative abundances of frequently fossilized categories of invertebrate animals. Width of range bands indicates relative abundance. Colored areas indicate where fossils of a particular category are widely used in zoning and correlation.

Figure 4–33 Foraminifera obtained from the Monterey Formation (Miocene) of California.

previously, the fossil record of this earliest, but important, part of plant evolution is extremely poor. It improves during the second stage in plant history with the origin in the Silurian Period of seedless (spore-bearing) vascular land plants. These plants proliferated in the coal forests of the Late Paleozoic Era. A third episode in plant evolution is marked by the advent of nonflowering pollen- and seed-producing plants. Although this was also a Late Paleozoic event, the seed plants expanded and diversified during the Mesozoic Era. Finally, in the last period of the Mesozoic, plants having both seeds and flowers evolved. Grasses were added to the flowering plant floras during the Cenozoic, and the lands in the age of mammals took on a decidedly modern appearance.

FOSSILS AND THE SEARCH FOR MINERAL RESOURCES

Paleontologists and Their Role

One might be surprised to learn that the majority of paleontologists in the world are not based in museums and universities but, rather, are industrial scientists using their knowledge of fossils in the search for economically valuable minerals. Most of these paleontologists are engaged in the search for oil and gas. In their investigations, they generally utilize only very small fossils, because these are not likely to be broken by drilling tools. During the drilling process, samples of the rock being penetrated are returned to the surface in the drilling mud or in rock cores. The rock is then broken down, and the microfossils are extracted for study. By far, the most useful microfossils are the protozoans known as *foraminifers* (Fig. 4–33). Other microfossils that have been used frequently in petroleum exploration include tiny, toothlike structures called *conodonts,* small crustaceans known as *ostracodes* (Fig. 4–34), *pollen grains* and *spores,* and the calcareous remains of marine algae. **Micropaleontologists** (Fig. 4–35), specialists in the study of microscopic fossils, prepare subsurface logs that depict the depth from which each species of fossil was obtained. From the accumu-

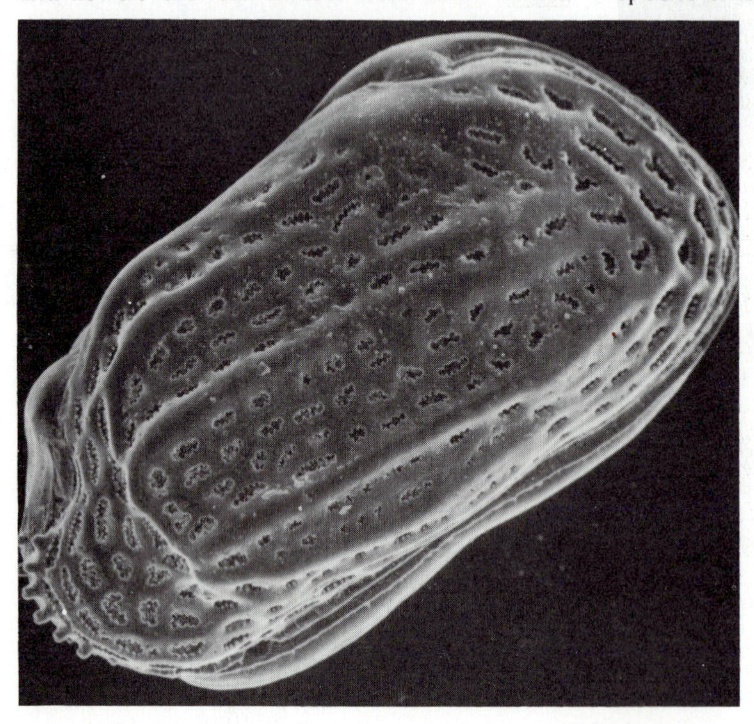

Figure 4–34 Electron micrograph of an ostracode (130×).

lated information of many such paleontologic logs, one is able to identify the geologic period of the formations through which the drill penetrates. Particular time-rock units can be recognized by their unique index fossils. Knowledge of the location of such key fossils relative to known oil-bearing strata is useful when wells are drilled in unproved territory. Indeed, the microfossils permit correlation from well to well and from area to area across entire petroliferous (oil-containing) provinces. The vertical successions of fossil assemblages in well samples often reflect changes in the depth of water in which the sediment was deposited and can therefore be used to infer tectonic conditions within the petroleum province. Along a single stratigraphic zone, knowledge of the depth of water preferences for microfossils may help in locating the porous reef or strand line sediments that are occasionally saturated with petroleum.

Paleontologists also play a role in the exploration of resources other than oil and gas. Mining geologists use fossils to date strata above and below mineralized beds and to help determine directions of movements along faults that offset the ore-bearing strata. With the aid of fossils and careful field work, they are often able to pinpoint the geologic time in which valuable metals were emplaced and then to use that knowledge to seek out ores in unexplored areas. Paleontologic mapping of ancient algal reefs has shown how these structures control the emplacement of metal sulfides. Whenever the origin of a mineral body is related to particular environmental conditions at the earth's surface, it is likely that paleontology will provide useful information for exploration.

Figure 4–35 Micropaleontologists employing the electron microscope in the examination of microfossils. (Courtesy of Standard Oil Company of California. Photographed by Dick Tolbert.)

SPECULATIONS ABOUT LIFE ON OTHER PLANETS

While reflecting upon the long and complex history of life on earth, one might drift on to questions of whether or not other planets had—or now have—life as we know it. Is organic evolution a process that is unique to this planet? What properties of the earth have made it a suitable place for the origin and evolution of life? Could those conditions exist elsewhere in the universe?

To begin with, the earth is large enough that its gravitational attraction is sufficient to retain an atmosphere. The temperature of most of the earth's surface is low enough to provide an abundance of water in its liquid form. In addition, that temperature is suitable for chemical reactions required for life processes and is itself a consequence of the size of our sun and its distance from the earth. Our sun-star also has a life span sufficiently long to permit time for the emergence and evolution of life. Finally, the earth has always had all of the chemical elements required for life processes.

Life in Our Solar System

We may first consider the possibilities for extraterrestrial life here in our own solar system. All evidence to date indicates that because of either size or distance from the sun, conditions are currently too harsh on neighboring planets to permit the evolution of *higher* forms of life. Scientists are still evaluating the rather ambiguous results of experiments conducted as part of the Viking mission to Mars. Thus far, they have been unable to discern any unequivocal evidence for life on the red planet.

Scientists had predicted that no life could exist on the moon because of the lack of an atmosphere and water as well as the extremes of temperature. The predictions were correct. The moon appears to be sterile.

Life in the Universe

Thus, we see that the earth is indeed biologically special when compared with other planetary bodies in our solar system. In contrast, however, it may not be so peculiar at all in the vast realms of the universe. Astronomers estimate that there are about 150 billion stars in our galaxy, and the number of galaxies now appears to be nearly limitless. Of those 150 billion stars in our galaxy, a not unreasonable estimate stipulates that at least 1 billion have planets with size and temperature conditions similar to the earth's. These are potentially habitable planets. For the entire known universe (of which our galaxy is but a small part), reputable scientists have estimated that there are as many as 100,000,000,000,000,000,000 planetary systems similar to our solar system. Such calculations indicate that it is probable that life does exist out there somewhere. Indeed, the universe may be rich in suitable habitats for life—but what kind of life? If we assume such life was formed from the same universal store of atoms and under physical conditions not too dissimilar to those that existed on earth, then we might very well recognize it as a living thing. But it is highly unlikely that duplicates of humans, cows, or butterflies exist on other planets. There are many variables in the evolutionary interactions of genetics, environment, and time involved in the making of a particular species. To say that the very same mutations, genetic recombinations, and environmental conditions producing a sparrow could occur in precisely the same sequential steps in time on a distant planet seems most improbable.

Summary

Fossils are the remains, or traces, of life of the geologic past. The processes that are important in fossilization are varied and include the precipitation of chemical substances into pore spaces (permineralization), molecular exchange for substances that were once part of the organism with inorganic substances (replacement), or compression of the animal and plant into a thin film of carbonized remains (carbonization). Tracks, trails, molds, and casts are additional varieties of fossils. Paleontologists, the specialists that study fossils, are aware that the fossil record is not complete for all forms of life that have existed on earth. It is more complete for organisms that had hard parts such as teeth, bone, or shell and for organisms that lived on the floors of shallow seas, where the deposition of sediment is relatively continuous and rapid.

It has been repeatedly demonstrated that particular lineages of fossil organisms show progressive changes in form. These observations are a part of the evidence for Darwin's conclusion that life has experienced a continuous development from simple forms to more complex forms. Because of this continuous change through time, rock layers from different periods can be recognized and correlated on the basis of the kinds of fossils they contain.

The concept of organic evolution received worldwide attention following the statement of the concept by Charles Darwin in 1859. Darwin is credited with demonstrating the principle of natural selection as a key factor in evolution. Our modern concept of evolution combines natural selection with more recently discovered hereditary mechanisms for changes by means of mutations and recombinations of hereditary materials.

An important goal for paleontologists is to understand the complex relationships between ancient plants and animals and their habitats. This area of paleontology is termed paleoecology. Paleoecology can, in turn, provide information about the distribution of ancient lands and seas, ancient climates, depth of water, natural barriers to migration, and former locations of continents.

Because fossils are so useful to persons inferring environments of deposition, as well as in correlating and determining relative geologic age of strata, they are widely employed by commercial paleontologists engaged in the search for fossil fuels. It is likely that this aspect of paleontology will continue in importance as the search for economically essential minerals intensifies. Paleontologists will continue their work in the future not only in such areas of exploration but possibly also in the detection of once living things in planets other than the earth. Statistics suggest the strong probability of present or past life somewhere in the universe.

Questions for Review

1. What is a fossil? When are fossils most valuable as index, or guide, fossils?
2. Account for the observation that there are more fossils of marine invertebrates than of any other group of organisms.
3. How has paleontology contributed to the concept of organic evolution?
4. How may fossils aid in deciphering ancient climates? Give an example.
5. Which would probably become better guide fossils: organisms with slow rates of dispersal or organisms with rapid rates of dispersal? Give reasons for your answer.
6. Fossil *A* occurs in rocks of the Cambrian and Ordovician Periods. Fossil *B* occurs in rocks that range from Early Ordovician through Permian in

age. Fossil *C* occurs in Mississippian through Permian rocks.

a. What is the maximum possible range of age for a rock containing only fossil *B*?
b. What is the maximum possible range for a rock containing both fossil *A* and fossil *B*?
c. Which is the better guide fossil, *A* or *C*? Support your answer.

7. A time-rock unit contains a different fossil assemblage at one locality than it does at another one located 50 mi away. Suggest a possible cause for the dissimilarity.
8. Does the rate of evolution of a particular group of related organisms affect their usefulness as guide fossils? Substantiate your answer.
9. Name the kinds of fossils that have been extensively studied and utilized by commercial paleontologists in the exploration for petroleum.
10. Who was the scientist largely responsible for developing our modern system of biologic nomenclature? Why is it called a "binomial" system?
11. Explain why our knowledge of evolution suggests that although there may be life outside the earth, there will probably not be human beings precisely like ourselves.
12. What are the major stages in the evolutionary history of terrestrial plants?
13. What were the contributions of Darwin, Mendel, and DeVries to our modern concepts of organic evolution?
14. In fossilization by replacement, what are the usual, or most common, replacing substances?
15. Why is the determination of a fossil species more subjective than the determination of a still-living (not extinct) species?
16. Using fossils for age correlation is dependent upon *a priori* knowledge of their time ranges. How is this information obtained? What might cause the information to be in error?

Terms to Remember

abyssal
bathyal
benthic
carbonization
carnivore
cast
class (taxonomic)
consumer organism
cosmopolitan species
Darwin
deoxyribonucleic acid (DNA)
diploid
ecologic niche
ecology
endemic species
epifaunal
evolution
extinct
family (taxonomic)
fossil
gamete
gene
genus
geologic range
guide fossils
hadal
haploid
herbivore
homologous
ichnology
infaunal
kingdom (taxonomic)
Lamarckian theory of evolution
Linnaean system of classification
Linné
littoral zone
meiosis
micropaleontologist
mitosis
mutation
natural selection
nekton
neritic zone
oceanic zone
order (taxonomic)
organic evolution
natural selection
paleoecology
paleontology
parasite
pelagic
permineralization
petrification
phylum
phytoplankton
plankton
primary consumer
rate of evolution
replacement
scavenger
sessile
species
species diversity
sublittoral zone
supralittoral zone
taxon
taxonomy
vagile
vestigial organs
trace fossil
zone (fossil)
zooplankton

Supplemental Readings and References

Barnett, R. S. 1974. Application of numerical taxonomy to the classification of the *Nummulitidae. J. Paleontol.* 48(6): 1249–1263.

Beerbower, J. R. 1968. Search for the Past (2d ed.). Englewood Cliffs, N.J., Prentice-Hall, Inc.

Bowen, R. 1961. Oxygen isotope paleotemperature measurements on Cretaceous Belemnoidea from Europe, India, and Japan. *J. Paleontol.* 35(5):1077–1084.

Clarkson, E. N. K. 1979. *Invertebrate Paleontology and Evolution.* London, Geo. Allen & Unwin, Ltd.

Cowen, R. 1976. *History of Life.* New York, McGraw-Hill Book Co.

Darwin, C. 1859. The Origin of Species (1963 ed.). Introduction by H. L. Carson. New York, Washington Square Press.

Delevoryas, T. 1962. *Morphology and Evolution of Fossil Plants.* New York, Holt, Rinehart, and Winston.

Dott, R. H., Jr., and Batten, R. L. 1981. *Evolution of the Earth* (3d ed.). New York, McGraw-Hill Book Co.

Dunbar, C. O., and Waage, K. M. 1969. *Historical Geology* (3d ed.). New York, John Wiley & Sons.

Eiseley, L. 1961. *Darwin's Century.* New York, Doubleday & Co.

Laporte, L. F. 1968. *Ancient Environments.* Englewood Cliffs, N.J., Prentice-Hall, Inc.

Lowenstam, H. A. 1954. Factors affecting aragonite-calcite ratios in carbonate-secreting marine organisms. *J. Geol.,* 69:241–260.

McAlester, A. L. 1968. *The History of Life.* Englewood Cliffs, N.J., Prentice-Hall, Inc.

Schopf, J. M. 1975. Modes of fossil preservation. *Rev. Paleobol.* 20:27–53.

Shklovski, I. S., and Sagan, C. 1966. *Intelligent Life in the Universe.* San Francisco, Holden-Day, Inc.

Spinosa, C., Furnish, W. M., and Glenister, B. F. 1975. The Xenodiscidae, Permian ceratitoid ammonoids. *J. Paleontol.,* 56(2):239–283.

Valentine, J. W. 1973. *Evolutionary Paleoecology of the Marine Biosphere.* Englewood Cliffs, N.J., Prentice-Hall, Inc.

The Grand Canyon of the Colorado River. This is a view down Kaibab Trail below Yaki Point. (Courtesy of U.S. National Park Service, photograph by G. A. Grant.)

5

Time and Geology

> Each grain of sand, each minute crystal in the rocks about us is a tiny clock, ticking off the years since it was formed. It is not always easy to read them, and we need complex instruments to do it, but they are true clocks or chronometers. The story they tell numbers the pages of earth history.
>
> Patrick M. Hurley, *How Old is the Earth?*
> Doubleday Anchor Books, 1959

INTRODUCTION

We humans have a fascination with the concept of time. Geology instructors are particularly aware of this fascination, for they are regularly asked the age of various rock and mineral specimens brought to the university by students, inquisitive artisans, and returning vacationers. If informed that the samples are tens or even hundreds of millions of years old, the collectors are often pleased, but they are also perplexed. "How can this fellow know the age of this specimen by just looking at it?" they think. If they insist on knowing the answer to that question, they may next receive a short discourse on the subject of geologic time. It is explained that the rock exposures from which the specimens were obtained have long ago been organized into a standard chronologic sequence based largely on superposition, evolution as indicated by fossils, and actual rock ages in years obtained from the study of radioactive elements in the rock. The geologist's initial estimate of age is based on *experience.* He may have spent a few hours on his knees at those same collecting localities and had a background of information to draw upon. Thus, at least sometimes, he can recognize particular rocks as being of a certain age. The science that permits him to accomplish this feat is called **geochronology.** It is a science that was born over 4 centuries ago when Nicholas Steno described how the position of strata in a superpositional sequence could be used to show the **relative geologic age** of the layers. This simple, but important, idea was expanded and refined much later by William "Strata" Smith and some of his contemporaries. These practical geologists showed how it was also possible to correlate strata. Outcrop by outcrop the rock sequences with their contained fossils were pieced together, one above the other, until a standard geologic time scale based on relative ages had been constructed.

THE STANDARD GEOLOGIC TIME SCALE

The early geologists had no way of knowing how many time units would be represented in the completed geologic time scale, nor could they know which fossils would be useful in correlation or which new strata might be discovered at a future time in some distant corner of the globe. Consequently, the time scale grew piecemeal, in an unsystematic manner. Units were named as they were discovered and studied. Sometimes the name for a unit was borrowed from local geography, from a mountain range in which rocks of a particular age were well exposed, or from an ancient tribe of Welshmen; sometimes the name was suggested by the kind of rocks that predominated.

Divisions in the Geologic Time Scale

The two major divisions in the geologic time scale are termed **eons.** Approximately seven eighths of all of earth history was expended in the first eon—the Cryptozoic Eon (informally termed "Precambrian"). To a geologist, the biblical phrase "In the beginning" alludes to this long interval of time that began about 4.6 billion years ago.

It was during the Cryptozoic Eon (literally meaning "hidden life") that the earth had completed the process of gathering together most of the rocky substance, possibly from what was part of a much older cloud of turbulent cosmic dust. In addition, it was the interval during which life on earth first appeared. Cryptozoic sequences around the world include great tracts of igneous and metamorphic rocks. The antiquity of these rocks was recognized in the mid-1700's by Johann Lehman, a professor of mineralogy in Berlin, who referred to them as the "primary series." One frequently finds the term in the writing of French and Italian geologists who were contemporaries of Lehman. In 1833, the term appeared again when Sir Charles Lyell used it in his formulation of a surprisingly modern geologic time scale. Lyell and his predecessors recognized these "primary" rocks by their crystalline character and took their uppermost boundary to be an unconformity that separated them from the overlying—and therefore younger—fossiliferous strata.

All of the remainder of geologic time is included in the second, or **Phanerozoic Eon.** As a result of careful study of superposition accompanied by correlations based on the abundant fossil record of the Phanerozoic, geologists have divided it into three major subdivisions, termed **eras.** The oldest era is the **Paleozoic,** which we now know lasted about 370 million years. Following the Paleozoic is the **Mesozoic Era,** which continued about 170 million years. The **Cenozoic Era,** in which we are now living, began about 60 million years ago.

As shown in Figure 4–15, the eras are divided into shorter time units called **periods;** periods may in turn be divided into **epochs.** In Chapter 3 we noted that eras, periods, epochs, and divisions of epochs, called **ages,** all represent intangible increments of pure time. They are geologic time units. The rocks formed during a specified interval of time are called **time-rock units.** For example, strata laid down during a given *period* compose a time-rock unit called a **system.** As indicated in Chapter 3, one may properly speak of climatic changes during the "Cambrian Period" as indicated by fossils found in rocks of the "Cambrian System." Each of the geologic systems is recognized largely by its distinctive fauna and flora of fossils. The fossils are different in stage of evolution from other fossils in both older and younger systems. **Series** is the time-rock term used for rocks deposited during an epoch, whereas **stage** represents the tangible rock record of an age (see Table 3–2).

Recognition of Time Units

Units of geologic time bear the same names as the time-rock units to which they correspond. Thus, we may speak of the "Jurassic System" or the "Jurassic Period" according to whether we are referring to the rocks themselves or to the time during which they accumulated.

Time terms have come into use as a matter of convenience. Their definition is necessarily dependent upon the existence of tangible time-rock units. The steps leading to the recognition of time-rock units began with the use of superposition in establishing age relationships. Local sections of strata were used by early geologists to recognize beds of successively different age and, thereby, to record successive evolutionary changes in fauna and flora. (The order and nature of these evolutionary changes could be determined because higher layers are successively younger.) Once the faunal and floral succession was deciphered, fossils provided an additional tool for establishing the order of events. They could also be used for correlation, so that strata at one locality could be related to the strata of various other localities. No single place on earth contains a complete sequence of strata from all geologic ages. Hence, correlation to standard sections of many widely distributed local sections was necessary in constructing the geologic time

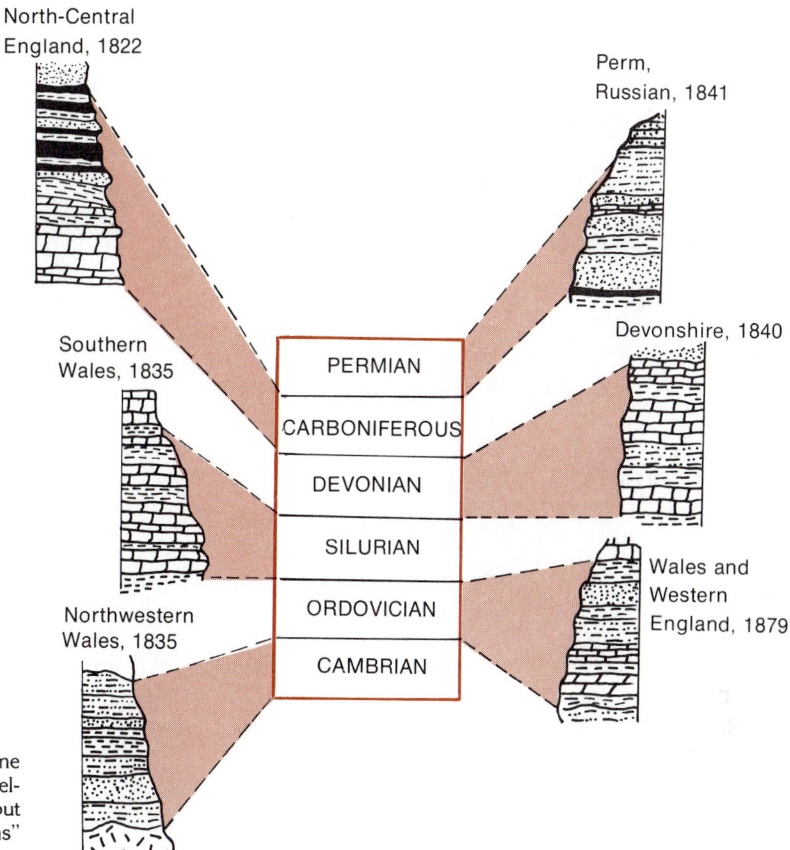

Figure 5–1 The standard geologic time scale for the Paleozoic and other eras developed without benefit of a grand plan, but rather by the compilation of "type sections" for each of the systems.

scale (Fig. 5–1). Clearly, the time scale was not conceived as a coherent whole but rather evolved part by part as a result of the individual studies of many earth scientists. Indeed, for some units at the series and stage level, the process continues even today. The fact that the time scale developed in piecemeal fashion is apparent when one reviews its growth and development.

The Cambrian System

The rocks of the **Cambrian System** take their name from *Cambria,* the Latin name for Wales. Exposures of strata in Wales (Fig. 5–2) provide a standard section with which rocks elsewhere in Europe and on other continents can be correlated. The standard section in Wales is named Cambrian *by definition.* All other sections deposited during the same time as the rocks in Wales are recognized as Cambrian *by comparison* to the standard section.

Adam Sedgewick, a Yorkshire clergyman and professor of geology at Cambridge, named the *Cambrian* in the 1830's for outcrops of poorly fossiliferous graywackes and dark siltstones and sandstones. The area in north Wales that Sedgewick studied was noted for its complexity, yet he was able to unravel its geologic history on the basis of spatial relationships and lithology.

The Silurian and Ordovician Systems

At about the same time that Sedgewick was laboring with outcrops that were to become the Cambrian System, another geologist, Sir Roderick Impey Murchison, had begun studies of fossiliferous strata outcropping in the hills of south Wales. Murchison named these rocks the **Silurian** (Fig. 5–3), taking the name from early inhabitants of western England and Wales know as the *Silures.* In 1835, Murchison and Sedgewick jointly presented a paper, *On the Silurian and Cambrian Systems, Exhibiting the Order in Which the Older Sedimentary Strata Succeed Each Other in England and Wales.* With this publication, the two geologists introduced the basis of the modern time scale. In the years that followed, a controversy arose between the two men that was to sever their

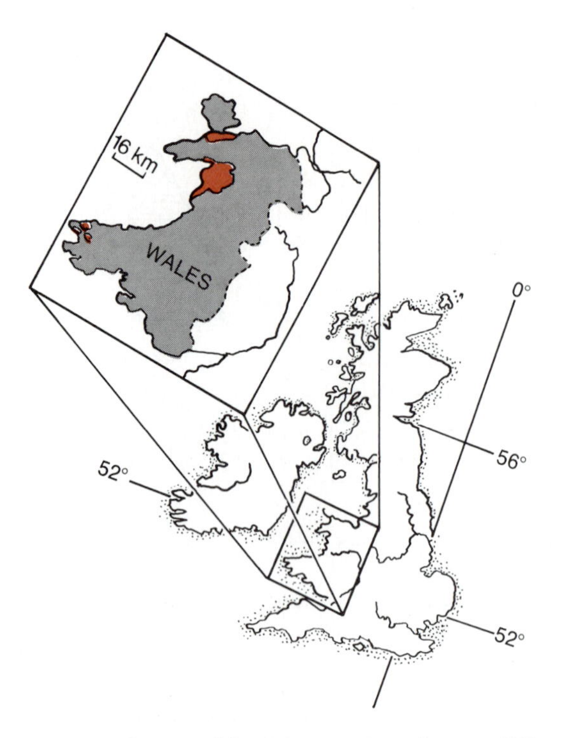

Figure 5–2 Outcrop of Cambrian strata in northwestern Wales, where these rocks were named.

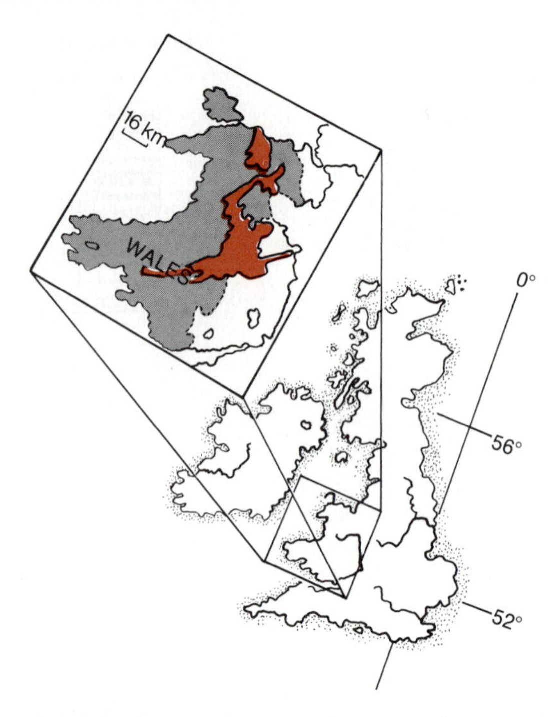

Figure 5–3 Outcrop of strata of the Silurian System in the classic region of Wales and western England, where Sir Roderick Impey Murchison named them.

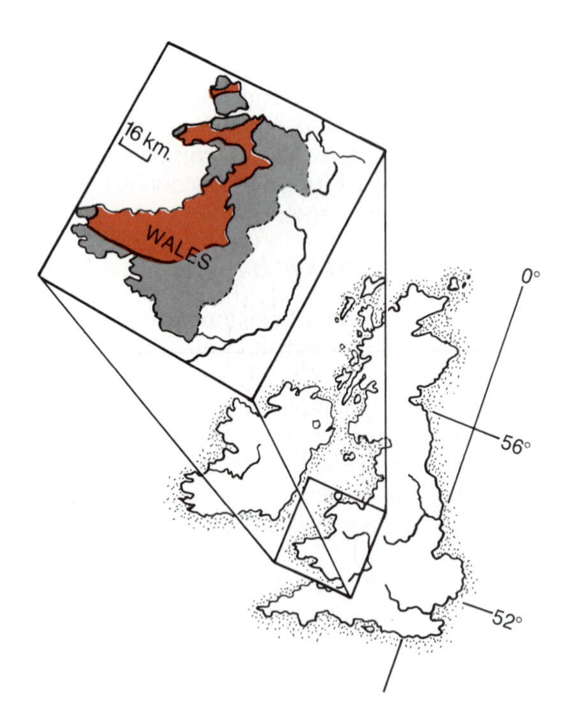

Figure 5–4 Outcrop pattern of strata of the Ordovician System in the type region of Wales.

friendship. Because Sedgewick had not described fossils distinctive of the Cambrian, the unit could not be recognized in other countries. Murchison argued, therefore, that the Cambrian was not a valid system. During the 1850's, he maintained that all fossiliferous strata above the "Primary Series" (the old name for Precambrian) and below the Old Red Sandstone (of Devonian age) belonged within the Silurian System. Sedgewick, of course, disagreed, but his opinion that the Cambrian was a valid system did not receive wide support until fossils were described from the upper part of the sequence. The fossils proved to be similar to faunas in Europe and North America. Hence, the Cambrian did meet the test of recognition outside England. Using these fossils as a basis for reinterpretation, the English geologist Charles Lapworth proposed combining the upper part of Sedgewick's Cambrian and the lower part of Murchison's Silurian into a new system. In 1879, he named the system **Ordovician** after the *Ordovices,* an early Celtic tribe (Fig. 5–4). The first three systems of the Paleozoic were now established (Fig. 5–5).

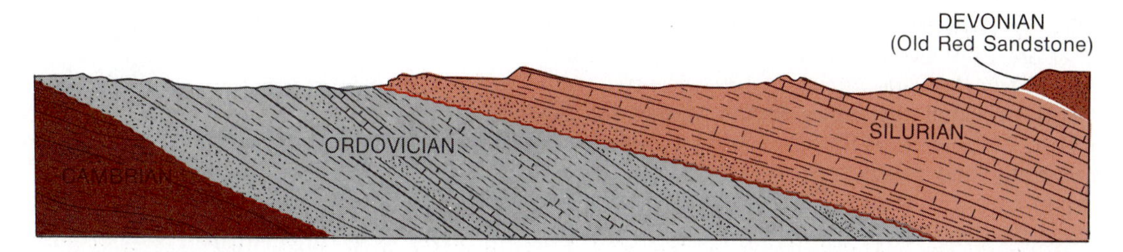

Figure 5–5 Generalized geologic cross-section for the Silurian "type region." Unconformities separate the Ordovician from the Cambrian and Silurian Systems. Silurian strata are inclined toward the east, with more resistant rocks forming escarpments that face toward the west.

The Devonian System

The **Devonian System** was proposed for outcrops near Devonshire, England, (Fig. 5–6) by Sedgewick and Murchison in 1839 (prior to the years of their bitter debate). They based their proposal on the fact that the rocks in question lay beneath the previously recognized Carboniferous System and contained a fauna that was distinctive and different from that of the underlying Silurian and overlying Carboniferous. In their interpretation of the intermediate nature of the fauna, they were aided by the studies of William Lonsdale, a retired army officer who had become a self-taught specialist on fossil corals. Further evidence that the new unit was a valid one came when Murchison and Sedgewick were able to recognize it in the Rhineland region of Europe. The Devonian rocks of Devonshire were also found to be equivalent to the widely known *Old Red Sandstone* of south Wales.

The Carboniferous System

The term **Carboniferous** was coined in 1822 by the English geologists William Conybeare and William Phillips to designate strata that included beds of coal in north-central England. Subsequently, it became convenient in Europe and Britain to divide the system into a *Lower Carboniferous* and *Upper Carboniferous*—the latter containing most of the workable coal seams. Two systems in North America, the **Mississippian** and **Pennsylvanian,** are broadly equivalent to these subdivisions. The American geologist Alexander Winchell formally proposed the name Mississip-

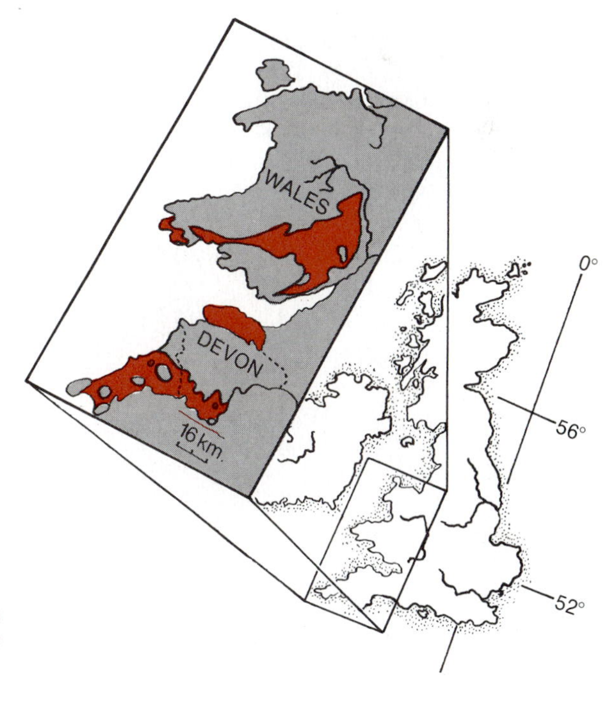

Figure 5–6 Outcrops of strata of the Devonian System in Wales and southwestern England, where these rocks were named.

pian in 1870 for the Lower Carboniferous strata that are extensively exposed in the Upper Mississippi River drainage region. In 1891, Henry S. Williams provided the name Pennsylvanian for the Upper Carboniferous System. Although both Pennsylvanian and Mississippian Systems are recognized by most United Stated geologists, neither term has been employed outside North America.

The Permian System

The **Permian System** takes its name from the small Russian town of Perm on the western side of the Ural Mountains (Fig. 5–7). In 1840 and 1841, Murchison, in company with the French paleontologist Edouard de Verneuil and several Russian companions, traveled extensively across western Russia. To his delight, Murchison found he was able to recognize Silurian, Devonian, and Carboniferous rocks by the fossils they contained. As a result he became even more convinced that groups of fossil organisms succeed one another in a definite and determinable order (this finding is now labeled the *principle of biologic succession*). Murchison established the new Permian System for rocks that overlay the Carboniferous System and contained fossils similar to those in German strata (the Zechstein beds), which had the same stratigraphic position as the Magnesian Limestone in England. Field studies had previously shown that the Magnesian Limestone rested upon Carboniferous strata. Thus Murchison was able to include the Magnesian Limestone within the Permian by correlation. The fossils of the new system appeared distinctly intermediate between those of the Carboniferous below and the Triassic above. Murchison's establishment of the Permian system provides a fine example of the logic employed by early geologists in putting together the pieces of the standard time scale.

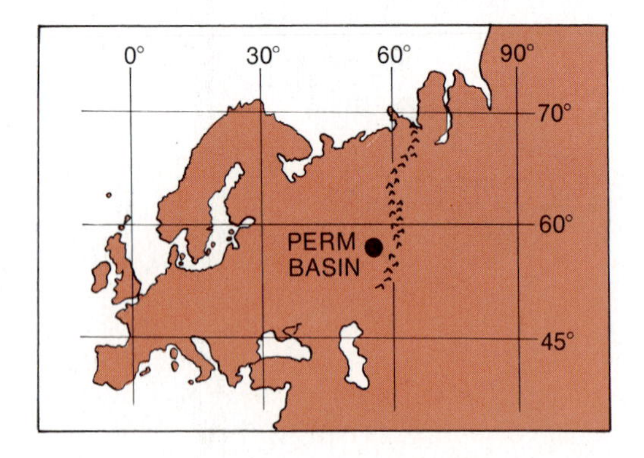

Figure 5–7 Location of the basin near Perm (Molotov), where Murchison established the Permian System.

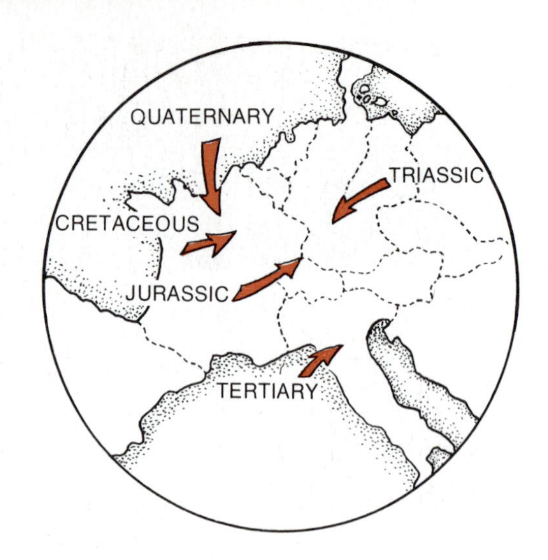

Figure 5–8 Type areas for the systems of the Mesozoic and Cenozoic.

The Triassic System

The influence of British geologists in providing names for the systems of the Paleozoic is by now obvious. However, their presence is not as evident in the development of Mesozoic nomenclature (Fig. 5–8). The **Triassic,** for example, was applied in 1834 by a German geologist named Frederich von Alberti. The term refers to a threefold division of rocks of this age in Germany. However, because the German strata in the type area are poorly fossiliferous, the standard of reference has been shifted to richly fossiliferous marine strata in the Alps.

The Jurassic System

Another German scientist, Alexander von Humboldt, proposed the term **Jurassic** for strata of the Jura Mountains between France and Switzerland. However, in 1795 when he used the term, the concept of systems had not been developed. As a result, the Jurassic was redefined as a valid geologic system in 1839 by Leopold von Buch.

The Cretaceous System

During the same year that Conybeare and Phillips were defining the Carboniferous, a Belgian geologist named Omalius d'Halloy proposed the term **Cretaceous** (from the Latin, *Creta,* meaning "chalk") for rock outcrops in France, Belgium, and Holland. Although chalk beds are prevalent in some Cretaceous exposures, the system is actually recognized on the basis of fossils. Indeed, some thick sections of Cretaceous rocks contain no chalks whatsoever.

The Tertiary System

The name **Tertiary** leads us back to the time when geology was just beginning as a science. Giovanni Arduino suggested a classification with four major divisions: Primary, Secondary, Tertiary, and Quaternary. The Tertiary was derived from his 1759 description of unconsolidated "montes tertiarii" sediments at the foot of the Italian Alps. Later, the Tertiary was more precisely defined, and standard sections for series of the Tertiary were established in France. The **Eocene, Miocene,** and **Pliocene,** for example, were proposed by Charles Lyell in 1832 on the basis of the proportions of species of living marine invertebrates in the fossil fauna. By definition, 3 per cent of the fossil fauna of the Eocene still live, whereas the Miocene contained 17 per cent, and the Pliocene contained 50 to 67 per cent. The term **Oligocene** was proposed by August von Beyrich in 1854, and the **Paleocene** was proposed 20 years later by Wilhelm Schimper. Other system names are also used in place of the Tertiary. Some geologists prefer the terms **Paleogene** (for the Paleocene, Eocene, and Oligocene) and **Neogene** (for the Miocene and Pliocene).

The Quaternary System

In 1829, the French geologist Jules Desnoyers proposed the term **Quaternary** for certain sediments and volcanics exposed in northern France. Although these deposits contained few fossils, Desnoyers was convinced on the basis of field studies that they were younger than Tertiary rocks. In the decade following Desnoyer's establishment of the Quaternary, the unit was divided by Charles Lyell into an older **Pleistocene Series,** composed primarily of deposits formed during the glacial ages, and the younger **Recent Series.**

This brief review describing how geologists drew up a table of geologic time clearly shows a lack of any grand and coherent design. These geologic pioneers were influenced by conspicuous changes in assemblages of fossils from one sequence of strata to another. In many places in Europe they found that such changes frequently occurred above and below an unconformity. The success of their methods is apparent from the fact that, by and large, the systems have persisted and found wide use even to the present day.

ABSOLUTE AGE OF THE EARTH

Early Attempts at Absolute Geochronology

After having constructed a geologic time scale on the basis of relative age, it is understandable that geologists would seek some way to assign absolute ages in millions of years to the various periods and epochs. From the time of Hutton, leaders in the scientific community were convinced that the earth was indeed very old, and certainly it was much older than the approximately 6000 years estimated by biblical scholars from calculations involving the ages of post-Adamite generations. But how old was the earth? And how might one quantify the geologic time scale?

To geologists of the 1800's it was apparent that to determine the absolute age of the earth or of particular rock bodies, they would have to concentrate on natural processes that continue in a single recognizable way and that also leave some sort of tangible record in the rocks. Evolution is one such process, and Charles Lyell recognized this. By comparing the amount of evolution exhibited by marine mollusks in the various series of the Tertiary System with the amount that had occurred since the beginning of the Pleistocene ice age, Lyell estimated that 80 million years had elapsed since the beginning of the Cenozoic. He came astonishingly close to the mark. However, for older sequences, estimates based on rates of evolution were difficult, not only because of missing parts in the fossil record, but also because rates of evolution for many taxa were not well understood.

In another attempt, geologists reasoned that if rates of deposition could be determined for sedimentary rocks, then they might be able to estimate the time required for deposition of a given thickness of strata. Similar reasoning suggested that one could estimate total elapsed geologic time by dividing the average thickness of sediment transported annually to the oceans into the total thickness of sedimentary rock that had ever been deposited in the past. Unfortunately, such estimates did not adequately account for past differences in rates of sedimentation or losses to the total stratigraphic section during episodes of erosion. Also, some very ancient sediments were no longer recognizable, having been converted to igneous and metamorphic rocks in the course of mountain building. As a result of these uncertainties, estimates of the earth's total age based on sedimentational rates ranged from as little as a million to over a billion years.

Some types of sediment do permit one to know the rate of deposition quite accurately. For example, shales deposited in lakes often exhibit summer and winter layers that differ in color and texture and can be counted much like growth rings in trees. The regularly repeated layers are called **varves.** Each summer-winter couplet represents deposition during 1 year, and an entire sequence of varves provides a record of the exact number of years that a particular environment existed. An illustration of the use of varves is provided by a study completed in 1929 by W. H. Bradley. By counting the varves in the shales of Wyoming's Green River Formation, Bradley was able to show that 6½ million years were required to deposit a thickness of 790 meters of shale.

In yet another geochronologic scheme, investigators attempted to determine the total age of the oceans. They speculated that the oceans' basins had been filled very shortly after the origin of the planet and thus would be only slightly younger in age than the earth itself. The best known of the calculations for the age of the oceans were made by the distinguished Irish geologist John Joly. From information provided by gauges placed at the mouths of streams, Joly was able to estimate the annual increment of salt to the oceans. Then, knowing the salinity of ocean water and the approximate volume of water, he calculated the amount of salt already held in solution in the oceans. An estimate of the age of the oceans was derived from the following formula:

$$\frac{\text{Total Salt Content in Ocean (in grams)}}{\text{Rate of Salt Added Each Year (grams per year)}} \text{ equals } \text{Age of Ocean (in years)}$$

Beginning with essentially nonsaline oceans, it would have taken about 90 million years for the oceans to reach their present salinity, according to Joly. The figure, however, was off the mark by a factor of 50, largely because there was no way to account accurately for recycled salt and salt incorporated into clay minerals deposited on the sea floors. Vast quantities of salt once in the sea had become extensive evaporite deposits on land; some of the salt being carried back to the sea had been dissolved, not from primary rocks, but from eroding marine strata on the continents. Even though in error, Joly's calculations clearly supported those geologists who insisted on an age for the earth far in excess of a few million years. The belief in the earth's immense antiquity was also supported by Darwin, Huxley, and other evolutionary biologists who saw the need for time in the hundreds of millions of years to accomplish the organic evolution apparent in the fossil record.

The opinion of the geologists and biologists that the earth was immensely old was soon to be challenged by the physicists. Spearheading this attack was Lord Kelvin, considered by many the outstanding physicist of the nineteenth century. Kelvin calculated the age of the earth on the assumption that it had cooled from a molten state and that the rate of cooling followed ordinary laws of heat conduction and radiation. Kelvin estimated the number of years it would have taken the earth to cool from a hot mass to its present condition. His assertions regarding the age of the earth varied, over 2 decades of debate, but in his later years he confidently believed that 24 to 40 million years was a reasonable age for the earth. The biologists and geologists found Kelvin's estimates difficult to accept. But how could they do battle against his elegant mathematics when they were themselves armed only with inaccurate dating schemes and geologic intuition? For those geologists unwilling to capitulate, however, new discoveries showed their beliefs to be correct and Kelvin's to be unavoidably wrong.

The correct answer to the question "how old is the earth?" was provided only after the discovery of **radioactivity,** a phenomenon unknown to Kelvin during his active years. With the detection of natural radioactivity by Henri Becquerel in 1896, followed by the isolation of radium by Marie and Pierre Curie 2 years later, the world became aware that the earth had its own built-in source of heat. It was not inexorably cooling at a steady and predictable rate, as Kelvin had suggested.

Radiometric Methods of Dating the Earth

Radioactivity

The radioactivity discovered by Henri Becquerel was a consequence of the fact that some elements, such as uranium and thorium, are unstable. Such elements will decay to form other elements or other isotopes of the same element. To understand what is meant by "decay," let us consider what happens to a radioactive element like uranium 238. Uranium 238 has an atomic weight of 238. The "238" represents the sum of the atom's protons and neutrons (each proton and neutron having a "weight" of 1). Uranium has an atomic number (number of protons) of 92. Such atoms with specific atomic number and weight are sometimes termed **nuclides.** Sooner or later (and entirely spontaneously) the uranium 238 atom will fire off a particle from the nucleus called an **alpha particle.** Alpha particles are positively charged ions of helium. They have an atomic weight of 4 and an atomic number of 2. Thus, when the alpha particle is emitted, the new atom will have an atomic weight of 234 and an atomic number of 90. From the decay of the parent nuclide, uranium 238, the daughter of the nuclide, thorium 234, is obtained (Fig. 5–9). A shorthand equation for this change is written:

$$^{238}_{92}\text{U} \rightarrow {}^{234}_{90}\text{Th} + {}^{4}_{2}\text{He}$$

This change is not, however, the end of the process, for the nucleus of thorium 234 is not stable. It eventually emits a **beta particle** (an electron discharged from the nucleus when a neutron splits into a proton and an electron). There is now an extra proton in the nucleus but no loss of atomic weight because electrons are essentially weightless. Thus, from $^{234}_{90}\text{Th}$ the daughter element $^{234}_{91}\text{Pa}$ (protactinium) is formed. In this case, the atomic number has been increased by one. In other instances, the beta particle may be captured by the nucleus, where it combines with a proton to form a neutron. The loss of the proton would decrease the atomic number by one.

A third kind of emission in the radioactive decay process is called **gamma radiation.** It consists of a form of invisible electromagnetic waves having even shorter wavelengths than do x-rays.

As alpha and beta particles, as well as gamma radiation, move through the surrounding materials, their energy is transformed into increased activity of the electrons in the atoms of the surrounding medium. The result is *heat.* This radiogenic source of heat was the unknown entity in Lord Kelvin's calculations of the earth's thermal history.

The Clocks in the Rocks (Radiometric Dating)

Nuclear adjustments such as those described previously occur many times before a final, stable daughter element, such as lead, is formed. The rate at which the steps in the process take place is unaffected by changes in temperature, pressure, or the chemical environment, since these do not involve the nucleus. Indeed, one can confidently assume that the rate of decay of long-lived isotopes has not varied since the earth came into existence. Therefore, once a quantity of radioactive nuclides has been incorporated into a growing mineral crystal, that quantity will begin to decay at a steady rate with a definite percentage of the radiogenic atoms undergoing decay in each increment of time. Each radioactive element has a particular mode of decay and a unique decay rate. As time passes, the quan-

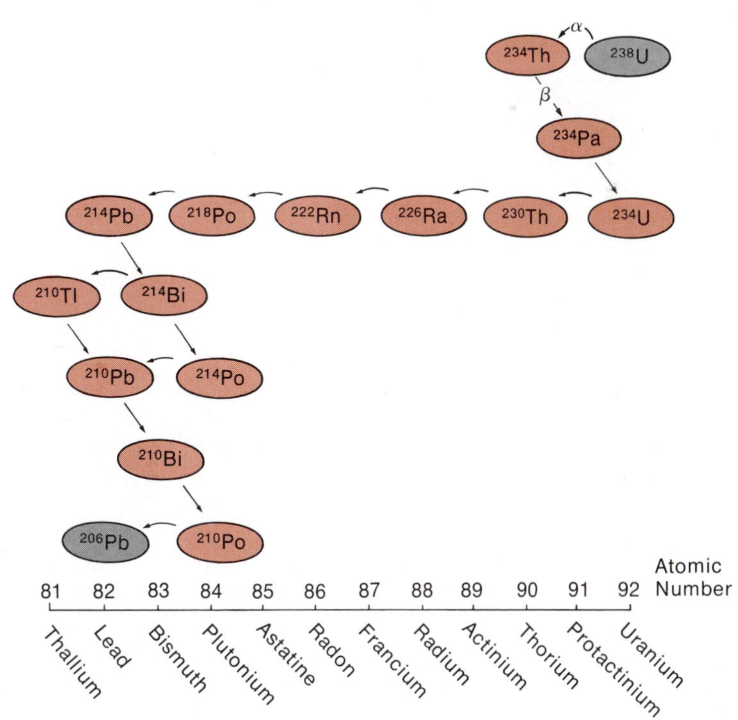

Figure 5–9 Radioactive decay series of uranium 238 (^{238}U) to lead 206 (^{206}Pb).

tity of the parent nuclide diminishes and the amount of daughter atoms increases, thereby indicating how much time has elapsed since the clock began its time-keeping. The "beginning," or "time zero," for any mineral containing radioactive nuclides would be the moment when the radioactive parent atoms became part of a mineral from which daughter elements could not escape. The retention of daughter elements is essential, for they must be counted to determine the original quantity of the parent nuclide (Fig. 5–10).

The determination of the ratio of parent to daughter nuclides is usually accomplished with the use of a **mass spectrometer,** an analytical instrument capable of separating and measuring the proportions of minute particles according to their mass differences. In the mass spectrometer, samples of elements are vaporized in an evacuated chamber, where they are bombarded by a stream of electrons. This bombardment knocks electrons off the atoms, leaving them positively charged. A stream of these positively charged ions is deflected as it passes between plates that bear opposite charges of electricity. The degree of deflection is proportional to the masses of the atoms (Fig. 5–11).

Not all radioactive decays are measured by means of a mass spectrometer. In the case of carbon 14, which decays by beta particle emission, the measurement of nuclides is accomplished indirectly by the use of a very sensitive **geiger counter.**

Of the three major families of rocks, the igneous clan lends itself best to radiometric dating. The dates obtained from such rocks indicate the time that a silicate melt containing radioactive elements solidified. In contrast to igneous rocks, sedimentary rocks can only rarely be dated radiometrically. Some dates for sedimentary strata have been obtained from a mineral called *glauconite,* which is believed to form "in place" at the time of deposition. This greenish mineral contains radioactive potassium 40, which decays to argon 40 and can be used in geochronology. Because of possible losses in the daughter element argon, care must be taken in interpreting dates, however; in most instances, potassium-argon dates derived from glauconite are considered minimal ages for the enclosing strata. As for classic sedimentary rocks that contain radioactive elements in their detrital mineral grains, the ages obtained refer to the parent rock that was eroded and is older than the sedimentary layer.

Dates obtained from metamorphic rocks may also require special care in interpretation. The age of a particular mineral may record the time the rock first formed or any one of a number of subsequent metamorphic recrystallizations.

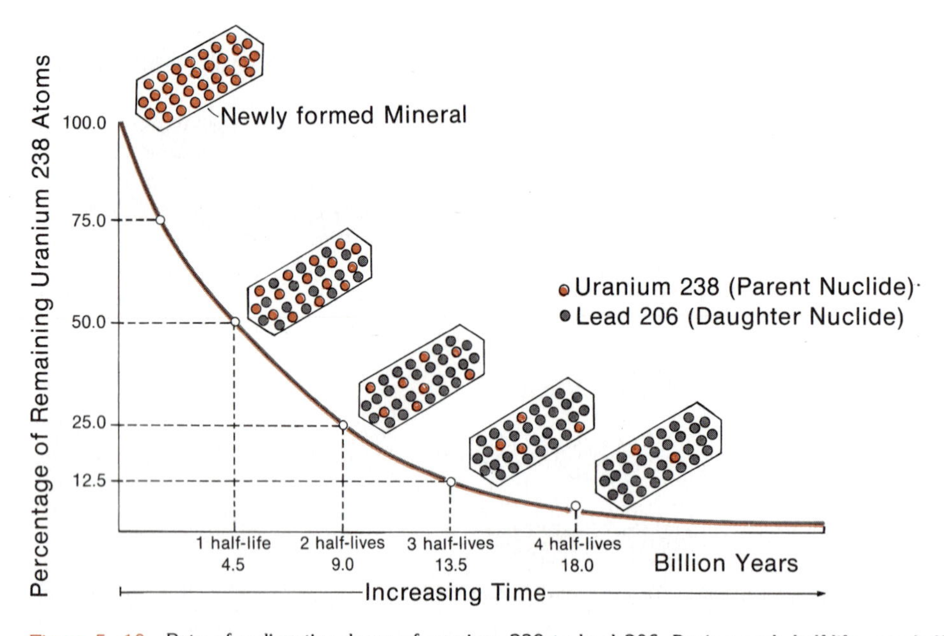

Figure 5–10 Rate of radioactive decay of uranium 238 to lead 206. During each half-life, one half of the remaining amount of the radioactive element decays to its daughter element. In this simplified diagram only the parent and daughter nuclides are shown, and the assumption is made that there was no contamination by daughter nuclides at the time the mineral formed.

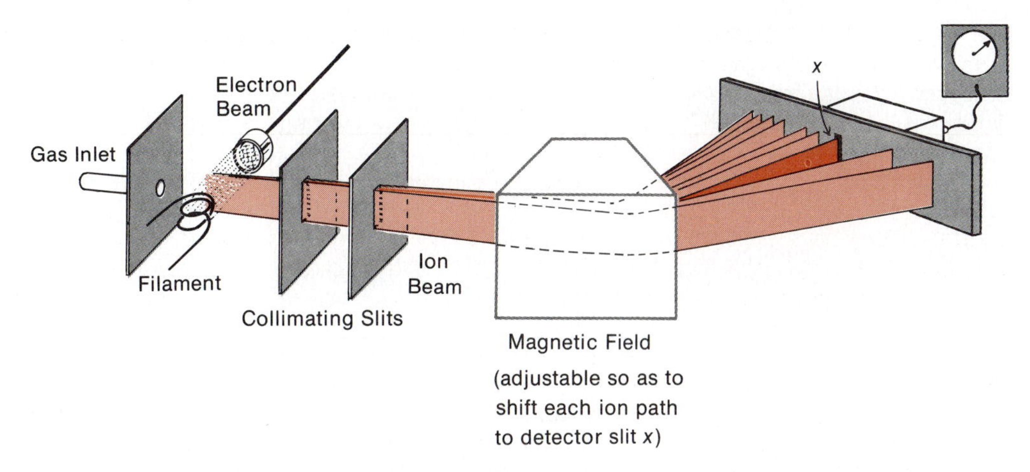

Figure 5–11 Schematic drawing of a mass spectrometer. In this type of spectrometer the intensity of each ion beam is measured electrically (rather than recorded photographically) to permit determination of the isotopic abundances required for radiometric dating.

There are many other problems that can affect the validity of a radiometric age. If some of the daughter products are removed from the sample by weathering or leaching, its age would be underestimated. If the element being analyzed was a gas, some of that gas (as with argon in glauconite) might have diffused out of the rock. The heat accompanying burial or mountain building might enhance such losses. There is also the possibility that at a *later time* older rocks may be partially remelted so that the age would be that of the second, rather than the initial, melting event. Clearly, great care must be taken in understanding the field relationships of the rock masses under investigation and in selecting samples.

Once an age has been determined for a particular rock unit, it is sometimes possible to use that date to approximate the age of adjacent rock masses. A shale lying below a lava flow that is 110 million years old and above another flow dated at 180 million years old must be between 110 and 180 million years of age (Fig. 5–12). Fossils within the shale might permit one to assign it to a particular geologic system or series and, by correlation, to extend the age data around the world.

Half-Life

There is no way that one can predict with certainty the moment of disintegration for any individual radioactive atom in a mineral. We do know that it would take an infinitely long time for all of the atoms in a quantity of radioactive elements to be entirely transformed to stable daughter products. Experimenters have also shown that the decline in the number of atoms is rapid in the early stages but becomes progressively slower in the later stages (see Fig. 5–10). One can statistically forecast what percentage of a large population of atoms will decay in a certain amount of time.

Because of these features of radioactivity, it is convenient to consider the number of years needed for half of the original quantity of atoms to decay. This span of years is termed the **half-life.** Thus, at the end of the years constituting one half-life, one half of the original quantity of radioactive element still has not undergone decay. After another half-life, one half of what was left is halved, so that one fourth of the original quantity remains. After a third half-life, only one eighth would remain, and so on.

Every radioactive nuclide has its own unique half-life. Uranium 235, for example, has a half-life of 704 million years. Thus, if a sample contains 50 per cent of the original amount of uranium 235 and 50 per cent of its daughter product, lead 207, then that sample is 704 million years old. If the analyses indicate 25 per cent of uranium 235 and 75 per cent of lead 207, two half-lives would have elapsed, and the sample would be 1408 million years old (Fig. 5–13).

The Principal Geologic Timekeepers

At one time, there were many more radioactive nuclides present on earth than there are now. Many of these had short half-lives and have long since decayed to undetectable quantities. Fortunately, for those interested in dating the earth's most ancient rocks, there remain a few long-lived radioactive nuclides. The most useful of these are uranium 238, uranium 235, rubidium 87, and po-

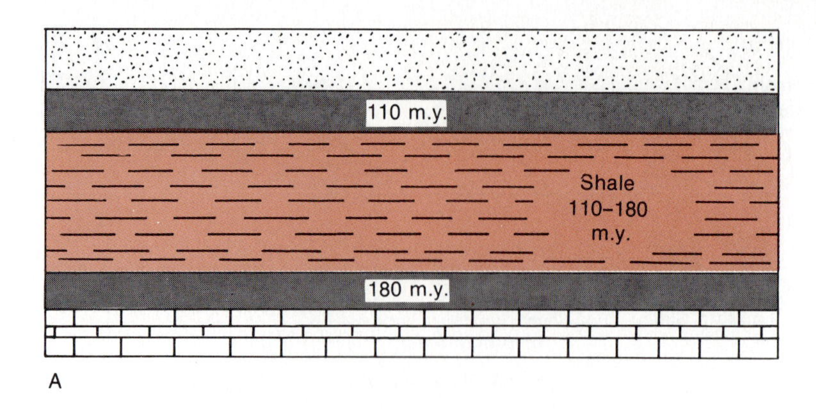

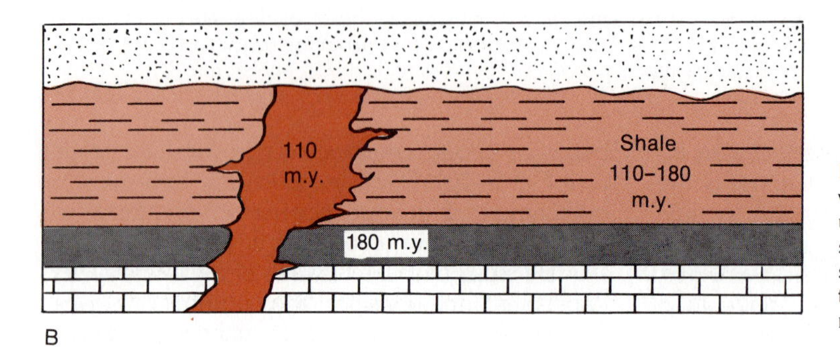

Figure 5–12 Igneous rocks that have provided absolute radiogenic ages can often be used to date sedimentary layers. In A, the shale is bracketed by two lava flows. In B, the shale lies above the older flow and is intruded by a younger igneous body. (Note: m.y. = million years.)

tassium 40 (Table 5–1). There are also a few short-lived radioactive elements that are used for dating more recent events. Carbon 14 is an example of such a short-lived isotope. There are also short-lived nuclides that represent segments of a uranium or thorium decay series.

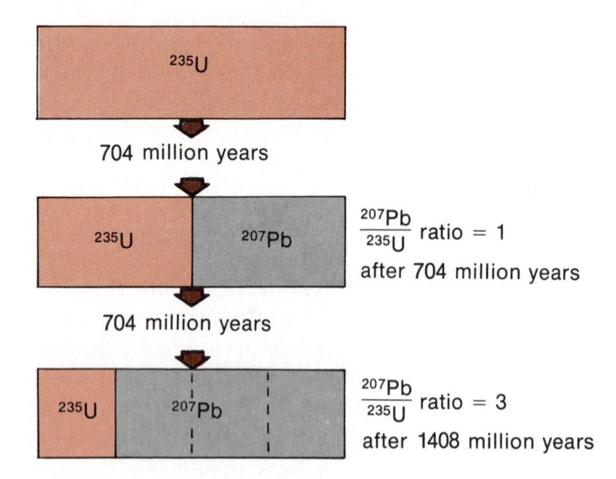

Figure 5–13 Radioactive decay of uranium 235 to lead 207.

Timekeepers That Produce Lead

Dating methods involving lead require radioactive nuclides of uranium or thorium that were incorporated into the earth's crust when it congealed. To determine the age of a sample of mineral or rock, one must know the original quantity of parent nuclides as well as the quantity remaining at the present time. The original number of parent atoms should be equal to the sum of the present quantity of parent atoms and daughter atoms. This raises the question of whether or not some of the lead may not have already been in the mineral and, if not detected, cause its radiometric age to exceed its true age. Lead 204, which is never produced by decay, provides a means of detecting original lead. All common lead contains a mixture of four lead isotopes. In most minerals used for dating, the proportions of the lead isotopes are nearly constant, so that lead 204 can be used to calculate the quantities of *original* lead 206 and lead 207. These quantities can then be subtracted from the total to give the amount due to radioactivity.

As we have seen, different isotopes decay at different rates. Geochronologists take advantage of this fact by simultaneously analyzing two or three

Table 5–1 SOME OF THE MORE USEFUL NUCLIDES FOR RADIOACTIVE DATING

Parent Nuclide*	Half-Life (in years)†	Daughter Nuclide	Materials Dated
Carbon 14	5730	Nitrogen 14	Organic materials
Uranium 235	704 million (7.04×10^8)	Lead 207 (and helium)	Zircon, uraninite, pitchblende
Potassium 40	1251 million (1.25×10^9)	Argon 40 (and calcium 40‡)	Muscovite, biotite, hornblende, volcanic rock, glauconite, K-feldspar
Uranium 238	4468 million (4.47×10^9)	Lead 206 (and helium)	Zircon, uraninite, pitchblende
Rubidium 87	48,800 million (4.88×10^{10})	Strontium 87	K-micas, K-feldspars, biotite, metamorphic rock, glauconite

*A *nuclide* is a convenient term for any particular atom (recognized by its particular combination of neutrons and protons.)
†Half-life data from Steiger, R. H., and Jäger, E. 1977. Subcommission on geochronology: convention on the use of decay constants in geo- and cosmochronology. *Earth and Planetary Science Letters,* 36:359–362.
‡Although potassium 40 decays to argon 40 and calcium 40, only argon is used in the dating method because most minerals contain considerable calcium 40, even before decay has begun.

isotope pairs as a means to cross-check ages and detect errors. For example, if the $^{235}U/^{207}Pb$ radiometric ages and the $^{238}U/^{206}Pb$ ages agree, then they are said to be **concordant;** there then exists a high probability that the radiometric age is valid. Uranium-lead ages that vary widely are said to be **discordant.** However, even the best of concordant dates do not agree perfectly but are expected to vary within reasonable limits. Unavoidable losses or gains of isotopes by interactions with surrounding solutions or from the heat accompanying geologic processes are the usual causes for variance. Radiometric ages should be considered reasonable approximations of true age.

Radiometric ages that depend upon uranium/lead ratios may also be checked against ages derived from lead 207 to lead 206. Because the half-life of uranium 235 is much less that the half-life of uranium 238, the ratio of lead 207 (produced by the decay of uranium 235) to lead 206 will change regularly with age and can be used as a radioactive timekeeper (Fig. 5–14).

The Potassium-Argon Method

Potassium and argon are another radioactive pair widely used for dating rocks. By means of **electron capture** (causing a proton to be transformed into a neutron), about 11 per cent of the potassium 40 in a mineral decays to argon 40, which may then be retained within the parent mineral. The remaining potassium 40 decays to calcium 40 (by emission of a beta particle from a neutron, thereby transforming it into a proton). The decay of potassium 40 to calcium 40 is not useful for obtaining radiometric ages, because radiogenic calcium cannot be distinguished from original calcium in a rock. Thus, geochronologists concentrate their efforts on the 11 per cent of potassium 40 atoms that decay to argon. One advantage of using argon is that it is inert—that is, it does not combine chemically with other elements. Argon 40 found in a mineral is very likely to have originated there following the decay of adjacent potassium 40 atoms in the mineral.

Another advantage to the potassium-argon scheme for dating rocks is that potassium 40 is an

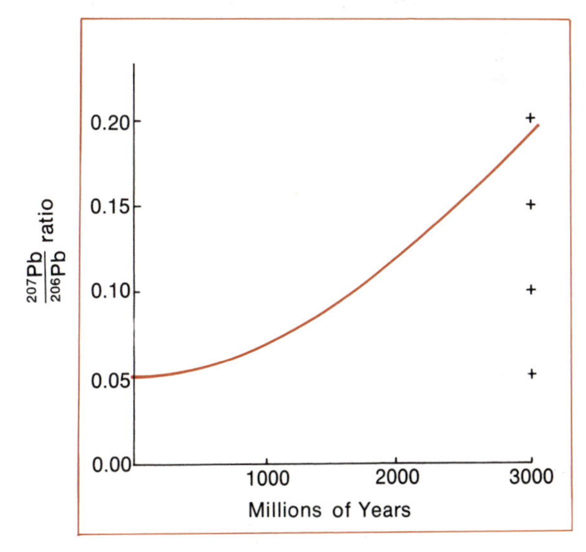

Figure 5–14 Graph showing how the ratio of lead 207 to lead 206 can be used as a measure of age.

abundant constituent of many common minerals, including micas, feldspars, and hornblendes. However, like all radiometric methods **potassium-argon dating** is not without its limitations. A sample will yield a valid age only if none of the argon has leaked out of the mineral being analyzed. Leakage may indeed occur if the rock has experienced temperatures above about 125°C. In specific localities, the ages of rocks dated by this method reflect the last episode of heating rather than the time of origin of the rock itself. A less serious problem is mechanical entrapment of atmospheric argon in flowing lavas.

The half-life of potassium 40 is 1251 million years. As illustrated in Figure 5–15, if the ratio of potassium 40 to argon 40 is found to be 1 to 1, then the age of the sample is 1251 million years. If the ratio is 3 to 1, then yet another half-life has elapsed, and the rock would have a radiogenic age of two half-lives, or 2502 million years.

Potassium-argon is widely used in deciphering various types of geologic problems. Geologists are now using the method in studies relating to sea floor movements. For many years, scientists have been curious about the alignment of the major Hawaiian Islands and the adjacent seamounts. With the advent of the theory of sea floor spreading, scientists developed the concept that these volcanic islands were built over a relatively fixed "hot spot" deep in the upper mantle. Conduits from the "hot spot" brought lavas up to the sea floor, where eruptions periodically occurred. Volcanoes that developed over the "hot spot" were then conveyed along by sea floor movement, and new volcanoes were produced over the vacated position. Geologists reasoned that if this process had taken place in the Hawaiian Islands, then potassium-argon radiometric ages should change in sequence along the island-seamount chain. The dates do indeed support the theory, and they even suggest that the direction of movement changed from a more northerly trend to northwesterly trend about 40 million years ago (see Figure 8–43).

The Rubidium-Strontium Method

The dating method based on the disintegration by beta decay of rubidium 87 to strontium 87 can sometimes be used as a check on potassium-argon dates, because rubidium and potassium are often found in the same minerals. The rubidium-strontium scheme has a further advantage in that the strontium daughter nuclide is not diffused by relatively mild heating events, as is the case with argon.

In the rubidium-strontium method, a number of samples are collected from the rock body to be dated. With the aid of the mass spectrometer, the amounts of radioactive rubidium 87, its daughter product strontium 87, and strontium 86 are calcu-

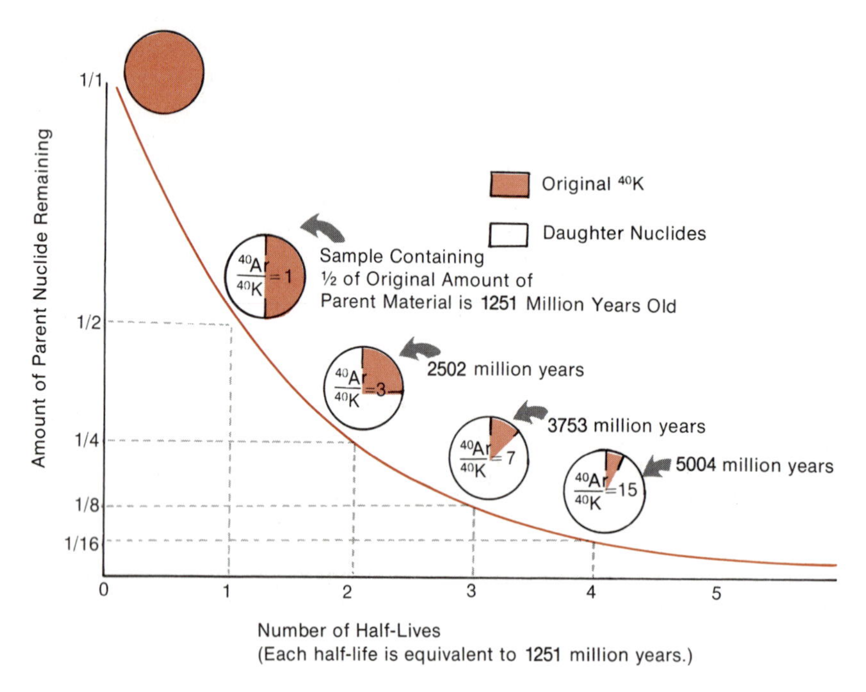

Figure 5–15 Decay curve for potassium 40.

lated for each sample. Strontium 86 is an isotope not derived from radioactive decay. A graph is then prepared in which the $^{87}Rb/^{86}Sr$ ratio in each sample is plotted against the $^{87}Sr/^{86}Sr$ ratio (Figure 5–16). From the points on the graph, a straight line is constructed that is termed an **isochron.** The slope of the isochron results from the fact that, with the passage of time, there is continuous decay of rubidium 87, which causes the rubidium 87/strontium 86 ratio to decrease. Conversely, the strontium 87/strontium 86 ratio increases as strontium 87 is produced by the decay of rubidium 87. The older the rocks being investigated, the more the original isotope ratios will have been changed, and the greater will be the inclination of the isochron. The slope of the isochron permits a computation of the age of the rock.

The rubidium-strontium and potassium-argon methods need not always depend upon the collection of discrete mineral grains containing the required isotopes. Sometimes the rock under investigation is so finely crystalline and the critical minerals so tiny and dispersed that it is difficult or impossible to obtain a suitable collection of minerals. In such instances, large samples of the entire rock may be used for age determination. This method is called **whole-rock analysis.** It is useful not only for fine-grained rocks but also for rocks in which the yield of useful isotopes from mineral separates is too low for analysis. Whole-rock analysis has also been useful in determining the age of rocks that have been so severely metamorphosed that the potassium-argon or rubidium-strontium radiometric clocks of individual minerals have been reset. In such cases, the age obtained from the minerals would be that of the episode of metamorphism, not the total age of the rock itself. The required isotopes and their decay products, however, may have merely moved to nearby locations within the same rock body, and therefore analyses of large chunks of the whole rock may provide valid radiometric age determinations.

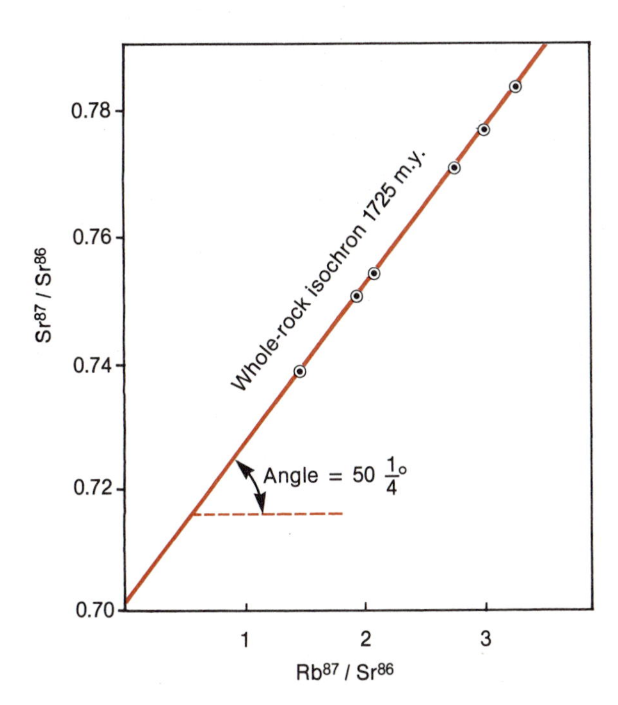

Figure 5–16 Whole-rock rubidium-strontium isochron for a set of samples of a Precambrian granite body exposed near Sudbury, Ontario. (Modified from T. E. Krogh et al., *Carnegie Institute Washington Year Book,* Vol. 66, 1968, p. 530.)

The Carbon 14 Method

Techniques for age determination based on content of radiocarbon were first devised by W. F. Libby and his associates at the University of Chicago in 1947. It has become an indispensable aid to archaeologic research and frequently is useful in deciphering the very recent events in geologic history. Because of the short half-life of carbon 14—a mere 5730 years—organic substances older than about 40,000 years no longer contain carbon 14 in measurable amounts.

Unlike uranium 238 and rubidium 87, carbon 14 is created continuously in the earth's upper atmosphere. The story of its origin begins with **cosmic rays,** which are extremely high-energy particles (mostly protons) that bombard the earth continuously. Such particles strike atoms in the upper atmosphere and split their nuclei into smaller particles, among which are neutrons. Carbon 14 is formed when a neutron strikes an atom of nitrogen 14. As a result of the collision, the nitrogen atom emits a proton and becomes carbon 14 (Fig. 5–17). Radioactive carbon is being created by this process at the rate of about two atoms per second for every square centimeter of the earth's surface. The newly created carbon 14 combines quickly with oxygen to form CO_2, which is then distributed by wind and water currents around the globe. It soon finds its way into photosynthetic plants, because they utilize carbon dioxide from the atmosphere to build tissues. Plants containing carbon 14 are ingested by animals, and the isotope becomes a part of their tissues as well.

Eventually, carbon 14 decays back to nitrogen 14 by emission of a beta particle. A plant removing CO_2 from the atmosphere should receive a share of carbon 14 proportional to that in the atmosphere. A state of equilibrium is reached in which the gain in newly produced carbon 14 is balanced by the decay loss. The rate of production of carbon

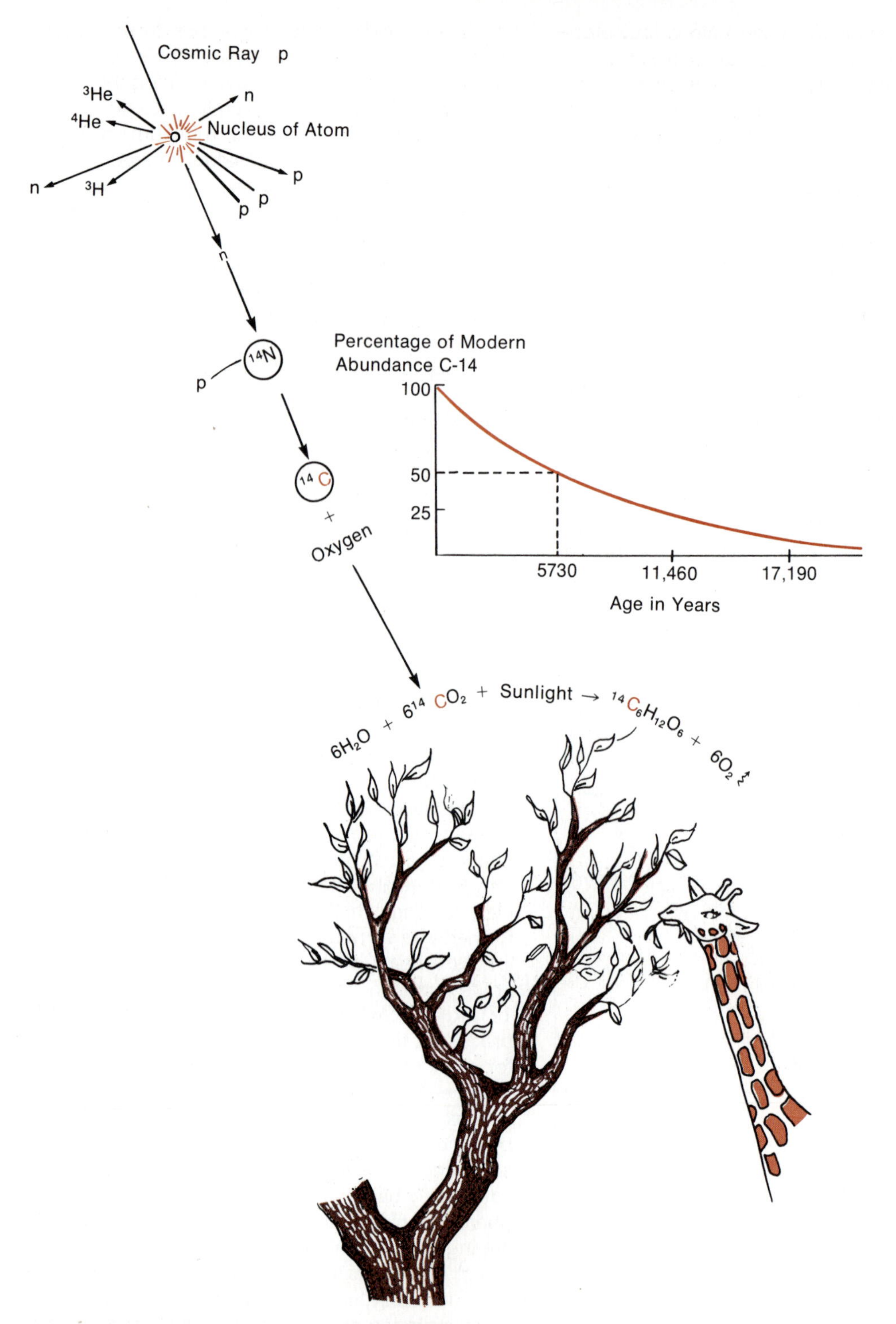

Figure 5–17 Carbon 14 is formed from nitrogen in the atmosphere. It combines with oxygen to form radioactive carbon dioxide and is then incorporated into all living things.

14 has varied somewhat over the past several thousand years (Fig. 5–18). As a result, corrections in age calculations must be made. Such corrections are derived from analyses of standards such as wood samples, whose exact age is known.

The age of some ancient bit of organic material is not determined from the ratio of parent to daughter nuclides, as is done with previously discussed dating schemes. Rather, the age is estimated from the ratio of carbon 14 to all other carbon in the sample. After an animal or plant dies, there can be no further replacement of carbon from atmospheric CO_2, and the amount of carbon 14 already present in the once living organism begins to diminish in accordance with the rate of carbon 14 decay. Thus, if the carbon 14 fraction of the total carbon in a piece of pine tree buried in volcanic ash were found to be about 25 per cent of the quantity in living pines, then the age of the wood (and the volcanic activity) would be two half-lives, or 11,460 years. To allow for unavoidable error, the age of the wood might be expressed as 11,460 $\pm$ 250 years.

The carbon 14 technique has considerable value to geologists studying the most recent events of the Pleistocene ice age. Prior to the development of the method, the age of sediments deposited by the last advance of continental glaciers was surmised to be about 25,000 years. Radiocarbon dates of a layer of peat beneath the glacial sediments provided an age of only 11,400 years. The method has also been found useful in dating the geologically recent uppermost layer of sediment on the sea floors. For the deeper and older marine sediments, however, the thorium method described further on is employed.

Methods Involving Thorium 230

The past 2 decades of intensive exploration of the sea floor has prompted the development of yet other dating methods that are especially useful for oceanic sediments too old to be dated with carbon 14. These new techniques utilize isotopes that are produced in the intermediary stages of the uranium decay series. Scientists who developed these methods recognized that most of the uranium brought to the oceans by streams remains in solution. While in solution it decays, eventually producing thorium 230. The thorium isotope is precipitated and becomes a component of ocean floor sediments. Thorium 230 itself decays with a half-life of 75,000 years. Because lower levels of the sediment have been undergoing decay longer, geochronologists are able to detect the expected decrease in quantity of thorium 230 at greater depths in a cored column of sediment. The thorium 230 concentration of each measured interval is compared to the quantity in the surface layer, and a time scale relating age to quantity of remaining parent nuclides is then formulated.

Deep sea sediments can also be dated by means of a procedure based upon thorium and protactinium. Thorium 230 has a half-life of 75,000 years and is in the decay series from uranium 238. Protactinium 231 has a half-life of only 34,300 years and is in the line of descent from uranium 235. Both parent nuclides are precipitated in the same proportions, but because of their different rates of decay, the ratio of the two changes regularly with time. Thus, the greater the differences in the quantity of undecayed parent isotopes, the older the sediment.

Nuclear Fission Track Timekeepers

Nuclear particle fission tracks were discovered about 20 years ago when scientists using the electron microscope were able to examine the areas around presumed locations of radioactive particles that were embedded in mica. Closer examination showed that the tracks were really small tunnels—like bullet holes—that were produced when high-energy particles of the nucleus of uranium were fired off in the course of **spontaneous fission** (spontaneous fragmentation of an atom into two or more lighter atoms and nuclear particles). The particles speed through the orderly rows of atoms in the crystal, tearing away electrons from atoms located along the path of trajectory and ren-

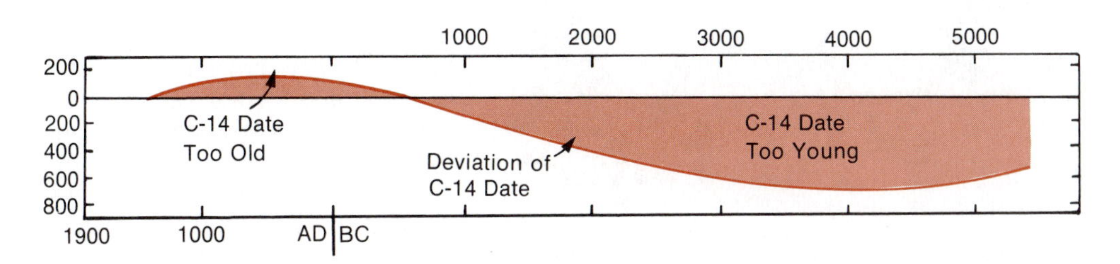

Figure 5–18 Deviation of carbon 14 ages from true ages from the present back to about 5000 BC. Data are obtained from analysis of bristle cone pines from the western United States. Calculations of ^{14}C deviations are based on half-life of 5,730 years. (Adapted from E. K. Ralph, H. N. Michael, and M. C. Han. Radiocarbon dates and reality. *MASCA Newsletter,* 9:(1), 1973.)

dering them positively charged. Their mutual repulsion produces the track (Fig. 5–19). The tracks are only a few atoms in width and are impossible to see without an electron microscope. Therefore, the sample is immersed for a short period of time in a suitable solution (acid or alkali), which rushes up into the tubes, enlarging the track tunnel so that it can be seen with an ordinary microscope.

The natural rate of track production by uranium atoms is very slow and occurs at a constant rate. For this reason, the tracks can be used to determine the number of years that have elapsed since the uranium-bearing mineral solidified. One first determines the number of uranium atoms that have already disintegrated. This number is obtained with the aid of the microscope by counting the etched tracks. Next, one must find the original number of uranium atoms. This quantity can be determined by bombarding the sample with neutrons in a reactor and thereby causing the remaining uranium to undergo fission. A second count of tracks reveals the original quantity of uranium. Finally, one must know the spontaneous fission decay rate for uranium 238. This information is determined by counting the tracks in a piece of uranium-bearing synthetic glass of known date of manufacture.

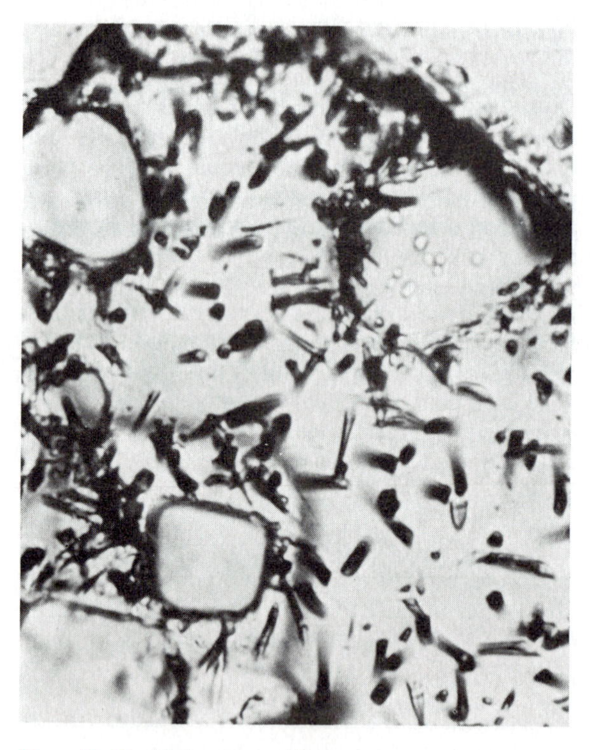

Figure 5–19 Fission tracks. These tracks were produced by plutonium 244 in a melilite crystal that was extracted from a meteorite. The small crystals are spinel inclusions. Melilite is a calcium-magnesium-alumino silicate. (Photograph courtesy of Frank Podosek, Department of Earth and Planetary Sciences, Washington University.)

Fission track dating is of particular interest to geochronologists because it can potentially be used to date specimens only a few centuries old as well as to date rocks billions of years in age. The method helps to date the period between about 40,000 and 1 million years ago—a period for which neither carbon 14 nor potassium-argon methods are suitable. As with all radiometric techniques, however, there can be problems. If rocks have been subjected to high temperatures, tracks may heal and fade away.

THE AGE OF THE EARTH

Anyone interested in the total age of the earth must decide what event constitutes its "birth." Most geologists assume "year 1" commenced as soon as the earth had collected most of its present mass and had developed a solid crust. Unfortunately, rocks that date from those earliest years have not been found on earth. They have long since been altered and converted to other rocks by various geologic processes. The oldest rocks on earth that have been dated thus far include 3.4-billion-year-old granites from the Barberton Mountain Land of South Africa, 3.7-billion-year-old granites of southwestern Greenland, and metamorphic rocks of about the same age from Minnesota. These dates permit us to say the planet is *at least* about 3.7 billion years old.

Meteorites, which many consider to be remnants of a disrupted planet that originally formed at about the same time as the earth, have provided uranium-lead and rubidium-strontium ages of about 4.6 billion years. From such data, and from estimates of how long it would take to produce the quantities of various lead isotopes now found on the earth, geochronologists feel that the 4.6-billion-year age for the earth can be accepted with confidence. Substantiating evidence for this conclusion comes from returned moon rocks. The ages of these rocks range from 3.3 to about 4.6 billion years. The older age determinations are derived from rocks collected on the lunar highland, which may represent the original lunar crust. Certainly, the moons and planets of our solar system originated as a result of the same cosmic processes and at about the same time.

Summary

In no other science does time play as significant a part as in geology. Time provides the frame of reference necessary to the interpretation of events, processes, and inhabitants of the earth. In the earlier stages of its development, geology was dependent upon relative dating of events. James Hutton helped scientists visualize the enormous periods needed to accomplish the events indicated in sequences of strata, and the geologists who followed him pieced together the many local stratigraphic sections, using fossils and superposition. A scale of relative geologic time gradually emerged. Initial attempts to decide what the rock succession meant in terms of years were made by estimating the amount of salt in the ocean, the average rate of deposition of sediment, and the rate of cooling of the earth. However, these early schemes did little more than suggest that the planet was at least tens of millions of years old and that the traditional concept of a 6000-year-old earth did not agree with what could be observed geologically.

An adequate means of measuring geologic time was achieved only after the discovery of radioactivity at about the turn of the twentieth century. Scientists found that the rate of decay by radioactivity of certain elements is constant and can be measured and that the proportion of parent and daughter elements can be used to reveal how long they had been present in a rock. Over the years, continuing efforts by investigators as well as improvements in instrumentation (particularly of the mass spectrometer) have provided many thousands of age determinations. Frequently, these radiometric dates have shed light on difficult geologic problems, provided a way to determine rates of movement of crustal rocks, and permitted geologists to date mountain building or to determine the time of volcanic eruptions. In a few, but highly important, regions radiometric dates have been related to particular fossiliferous strata and have thereby helped to quantify the geologic time scale and to permit estimation of rates of organic evolution. Radiometric dating has also changed the way humans view their place in the totality of time.

The radiometric transformations most widely used in determining absolute ages are uranium 238 to lead 206, uranium 235 to lead 207, thorium 232 to lead 208, potassium 40 to argon 40, rubidium 87 to strontium 87, and carbon 14 to nitrogen 14. Methods involving uranium/lead ratios are of importance in dating the earth's oldest rocks. The short-lived carbon 14 isotope that is created by cosmic ray bombardment of the atmosphere provides a means to date the most recent events in earth history. For rocks of intermediate age, schemes involving potassium/argon ratios, those utilizing intermediate elements in decay series, or those employing fission fragment tracks are most useful. A figure of 4.6 billion years for the earth's total age is now supported by ages based on meteorites and on lead ratios from terrestrial samples.

Improvements in radiometric geochronology are being made daily and will provide further calibration of the standard geologic time scale in the future. Some of the time boundaries in the scale, such as that between the Cretaceous and Tertiary Systems, are already well validated. Others, like the boundary between the Paleozoic and Mesozoic, require additional refinement. Additional efforts to further incorporate radiometric ages into sections of sedimentary rocks are among the continuing tasks of historical geologists. The usual methods for determining the age of strata involve the dating of intrusions that penetrate these sediments or the dating of interbedded volcanic layers. Less frequently, strata can be dated by means of radiogenic isotopes incorporated within sedimentary minerals that formed in place at the time of sedimentation. At present, the best radiometric age estimates indicate that Paleozoic sedimentation began about 570 million years ago, the Mesozoic Era began about 225 million years ago, and the Cenozoic commenced about 65 million years ago.

Questions for Review

1. By what methods did geologists attempt to determine the age of the earth before the discovery of radioactivity? Why were these methods inadequate?
2. What types of radiation accompany the decay of radioactive isotopes?
3. In making an age determination based upon the uranium-lead method, why should an investigator select an unweathered sample for analysis?
4. How do the isotopes carbon 12 and carbon 14 differ from one another in regard to the following: (a) number of protons, (b) number of electrons, and (c) number of neutrons? What is the origin of carbon 14?
5. Define "half-life." Why is this term used instead of an expression like "whole-life" in radiometric dating? What are the half-lives of uranium 238, potassium 40, and carbon 14?
6. How do fission tracks originate? What geologic conditions might destroy these tracks?
7. Pebbles of basalt within a conglomerate yield a radiometric age of 300 million years. What can be said about the age of the conglomerate? Several miles away, the same conglomerate strata is bisected by a 200-million-year-old dike. What now can be said about the age of the conglomerate in this location?
8. Has the amount of uranium in the earth increased, decreased, or remained about the same over the past 4.5 billion years? What can be stated about changes in the amount of lead?
9. State the estimated age of a sample of mummified skin from a prehistoric human that contained 12.5 per cent of an original quantity of carbon 14.
10. How are dating methods involving decay of radioactive elements unlike methods for determining elapsed time by the funneling of sand through an hour glass?
11. State the effect on the radiometric age of a zircon crystal being dated by the potassium-argon method if a small amount of argon 40 escaped from the crystal.
12. What is the advantage of having both uranium and thorium present in a mineral being used for a radiometric age determination?
13. What are *varves,* and how might they be used to determine the time span of existence of an ancient lake?
14. Minerals suitable for radiometric age determinations are usually components of igneous rocks. How, then, can absolute ages be obtained for sedimentary formations?
15. Discuss the contributions to the development of the geologic time scale made by Charles Lapworth, Frederick von Alberti, and Charles Lyell.
16. What observations were used by Sir Roderick I. Murchison in the establishment of the Permian System?
17. What is the essential difference between a "time-rock unit" and a "time-unit?" Give an example of each.
18. What characteristics of the rocks of north Wales caused Adam Sedgewick difficulty in establishing them as a valid geologic system?

Terms to Remember

alpha particle
beta particle
Cambrian System
Carboniferous System
Cenozoic Era
concordant
cosmic rays
Cryptozoic Eon
Devonian System
discordant
electron capture
epoch
era
gamma radiation
geiger counter
geochronology
half-life
mass spectrometer
Mesozoic Era
Mississippian System
nuclear fission tracks
nuclide
Ordovician System
Paleozoic Era
Phanerozoic Eon
Pennsylvanian System
period
Permian System
potassium 40
potassium-argon method
Precambrian
Quaternary System
radioactivity
relative geologic age
series
Silurian System
stage
system
time-rock unit
Triassic System
whole-rock radiometric analysis

Supplemental Readings and References

Berry, W. B. N. 1968. *Growth of a Prehistoric Time Scale.* San Francisco, W. H. Freeman Co.

Eicher, D. L. 1976. *Geologic Time* (2d ed.). Englewood Cliffs, N.J., Prentice-Hall, Inc.

Faul, H. 1966. *Ages of Rocks, Planets, and Stars.* New York, McGraw-Hill Book Co.

Fleischer, R. L., Price, P. B., and Walker, R. M. 1965. Tracks of charged particles in solids. *Science* 149(3682):383–393.

Hamilton, E. I. 1965. *Applied Geochronology.* New York, Academic Press, Inc.

Harbaugh, J. W. 1969. *Stratigraphy and Geologic Time.* Dubuque, Iowa, William C. Brown Co.

Hurley, P. M. 1959. *How Old Is the Earth?* New York, Doubleday & Co., Inc. (Anchor Books).

Lancelot, V., and Larson, R. L. 1975. Sedimentary and tectonic evolution of the northwestern Pacific. *Initial Reports of the Deep Sea Drilling Project* XXXII:925–939.

Larson, R. L., and Chase, C. G. 1972. Late Mesozoic evolution of the western Pacific Ocean. *Geol. Soc. Am. Bull.* 83(12):3627–3644.

Schaffer, O. A., and Zahringer, J. (eds.). 1966. *Potassium-Argon Dating.* New York, Springer-Verlag.

Steiger, R. H., and Jäger, E. 1977. Subcommission on geochronology: convention on the use of decay constants in geo- and cosmochronology. *Earth and Planetary Science Letters* 36:359–362.

Toulmin, S., and Goodfield, J. 1965. *The Discovery of Time.* New York, McGraw-Hill Book Co.

An Apollo 17 astronaut examining samples from a boulder believed to have rolled 22 million years ago from a higher terrain near the landing site. (Photograph courtesy of National Aeronautics and Space Administration; photograph 17-140-2149.)

6

Planetary Beginnings

> Give me matter and motion, and I will construct the universe.
>
> Rene Descartes, 1640

THE EARTH'S PLACE IN THE COSMIC ENVIRONMENT

The earth is one of nine planets that revolve around a rather average star, our sun. The planets, in order of increasing distance from the sun, are Mercury, Venus, Earth, Mars, Jupiter, Saturn, Uranus, Neptune, and Pluto (Pluto is considered by many astronomers to be a large satellite that escaped from Neptune). A belt of asteroids orbits the sun in the region between the paths of Mars and Jupiter. This grouping of planets and asteroids around the sun constitutes our solar system (Fig. 6–1). Certainly, ours is not the only such system in the universe. It is true that the suspected planets of other systems are too distant to be detected directly, but we know that they are out there because of the wobbles their orbital movements and gravitational pull cause in the stars they circle.

Our solar system is a small part of a much larger aggregate of stars, planets, dust, and gases called a **galaxy.** Galaxies are also numerous in the universe, and their constituents are arranged into

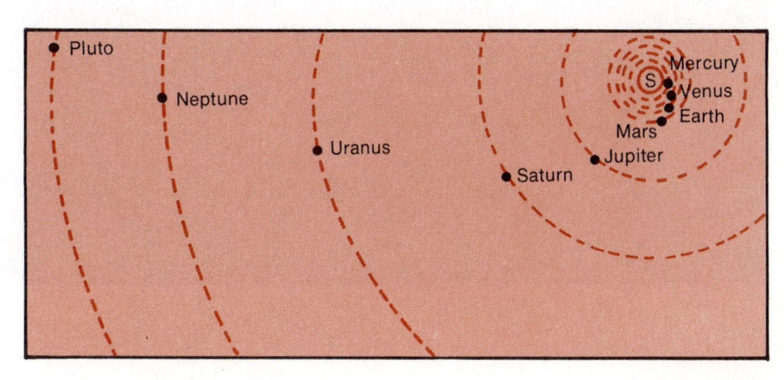

Figure 6–1 Relative distances of the planets from the sun.

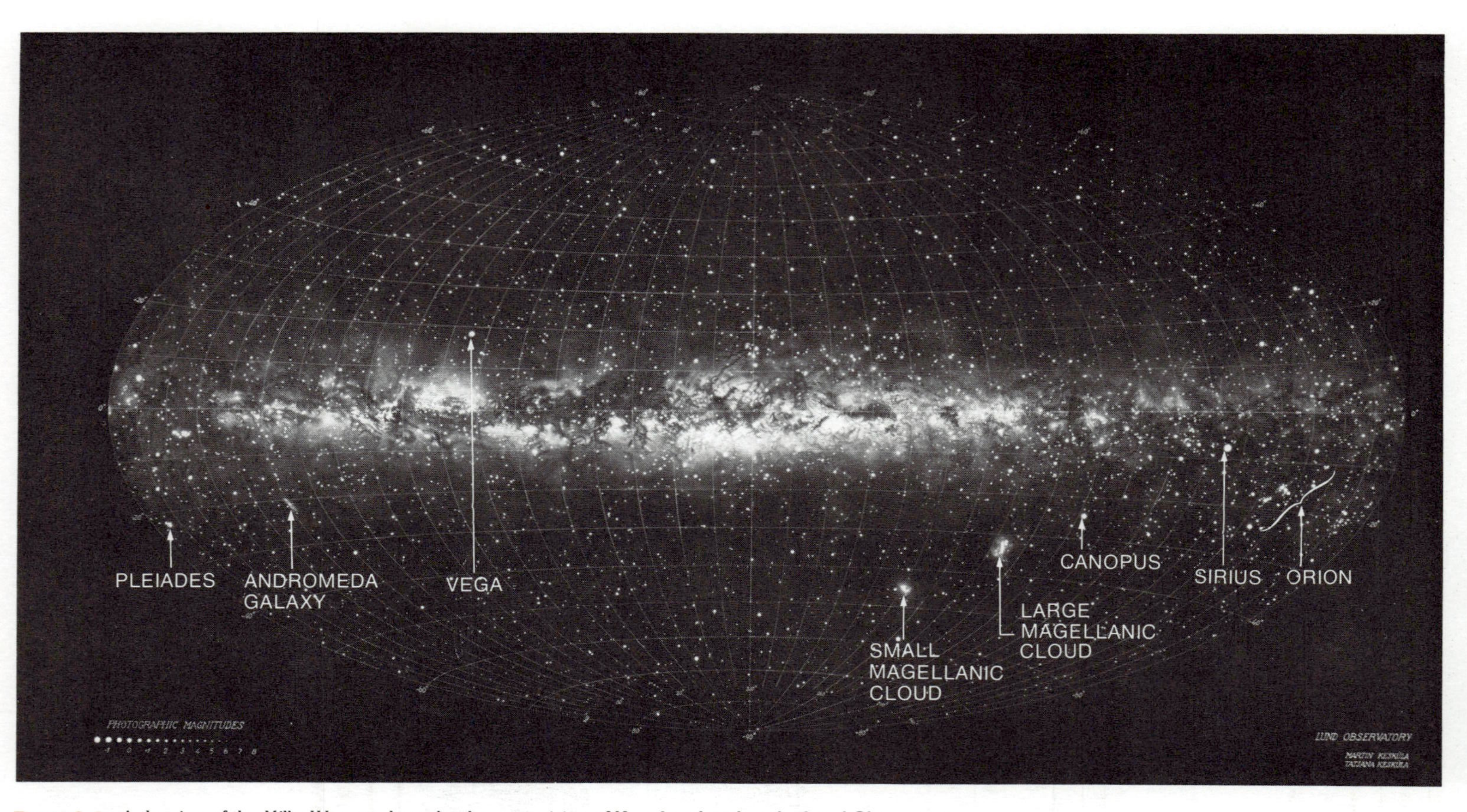

Figure 6–2 A drawing of the Milky Way, made under the supervision of Knut Lundmark at the Lund Observatory in Sweden. Seven thousand stars plus the Milky Way are shown in this panorama, which is in coordinates such that the Milky Way falls along the equator. (Courtesy of Lund Observatory, Sweden; from J. M. Pasachoff. *Contemporary Astronomy.* 2nd ed. Philadelphia, Saunders College Publishing, 1981.)

several different general forms: tightly packed elliptical galaxies, irregular galaxies, and the more familiar discoidal spiral galaxies with their glowing central bulge and great curving arms.

The galaxy in whiich our solar system is located is called the **Milky Way Galaxy** because, as we look toward its dense central bulge, we see a great milky haze of light in the heavens. The Milky Way Galaxy is of the spiral category (Fig. 6–2). It rotates slowly in space, completing one rotation about every 240 million years. Our sun is located about two thirds of the distance (or 26,000 light years) outward from the center of the galaxy to its edge.

Figure 6–3 Relative sizes and approximate orbits of the planets. (From J. M. Pasachoff. *Contemporary Astronomy.* 2nd ed. Philadelphia, Saunders College Publishing, 1981.)

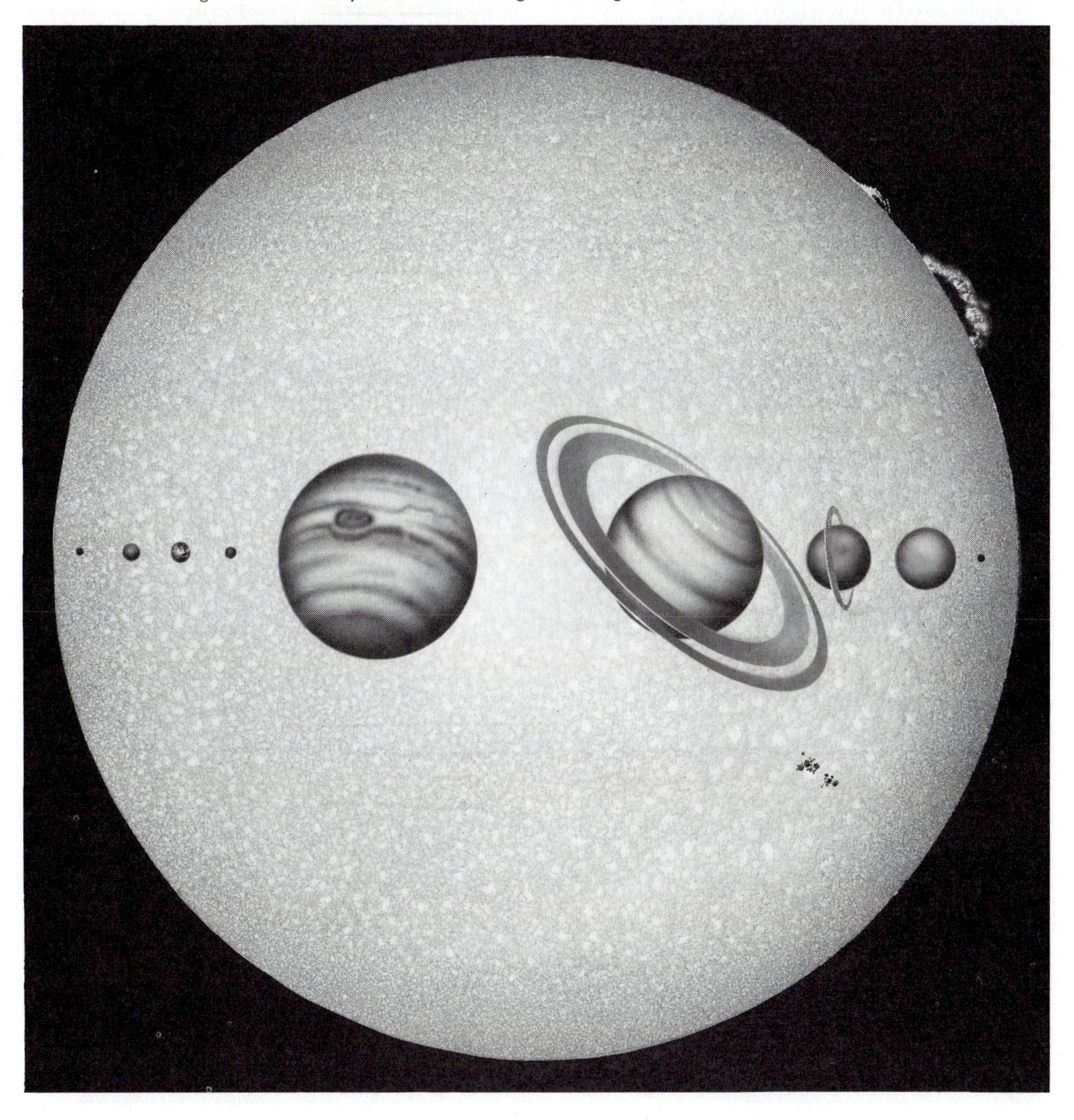

THE SUN

The largest, and for many reasons the most important, member of the solar system is the sun. Although only a modest star in comparison to others in our galaxy, it nevertheless has a mass 333,000 times that of the earth and is 109 times larger (Fig. 6–3). It is composed mostly of hydrogen (about 70 per cent) and helium (about 27 per cent). Remaining heavier elements exist as gases in the hot interior of the star.

The surface of the sun, and the part from which light reaching us emanates, is called the **photosphere** (Fig. 6–4). Above the photosphere lies the solar atmosphere. It consists first of a layer of cooler gas that absorbs certain wavelengths of light. This layer is called the **"reversing layer."** Lying above the reversing layer is an atmospheric zone termed the **chromosphere.** The hot hydrogen forming the chromosphere is visible as a red glow during a solar eclipse. The uppermost atmospheric layer is the **corona,** a seething, tenuous shroud of fiery gases extending millions of kilometers outward from the sun. Beneath these three layers is the sun's interior, where temperatures exceeding 20 million degrees Celsius exist.

Although an enormous amount of solar energy is intercepted by the earth, our planet is not roasted but is able to maintain a range of temperatures roughly between −50 and +60°C. The maintenance of this vital temperature range is made possible by three factors. First, because of rotation, the earth receives energy from the sun on one hemisphere, whereas it returns heat to space over its entire surface. Second, some of the incoming radiation is reflected off the atmosphere and clouds and is directed back into space without ever reaching ground level. Finally, a part of the intercepted radiation is absorbed by the atmosphere and radiated back into space without warming the earth's surface.

There are other forms of radiation besides sunlight that are intercepted by the earth (Fig. 6–5). Three per cent of the incoming rays are ultraviolet. Because of their destructive effects on life, it is

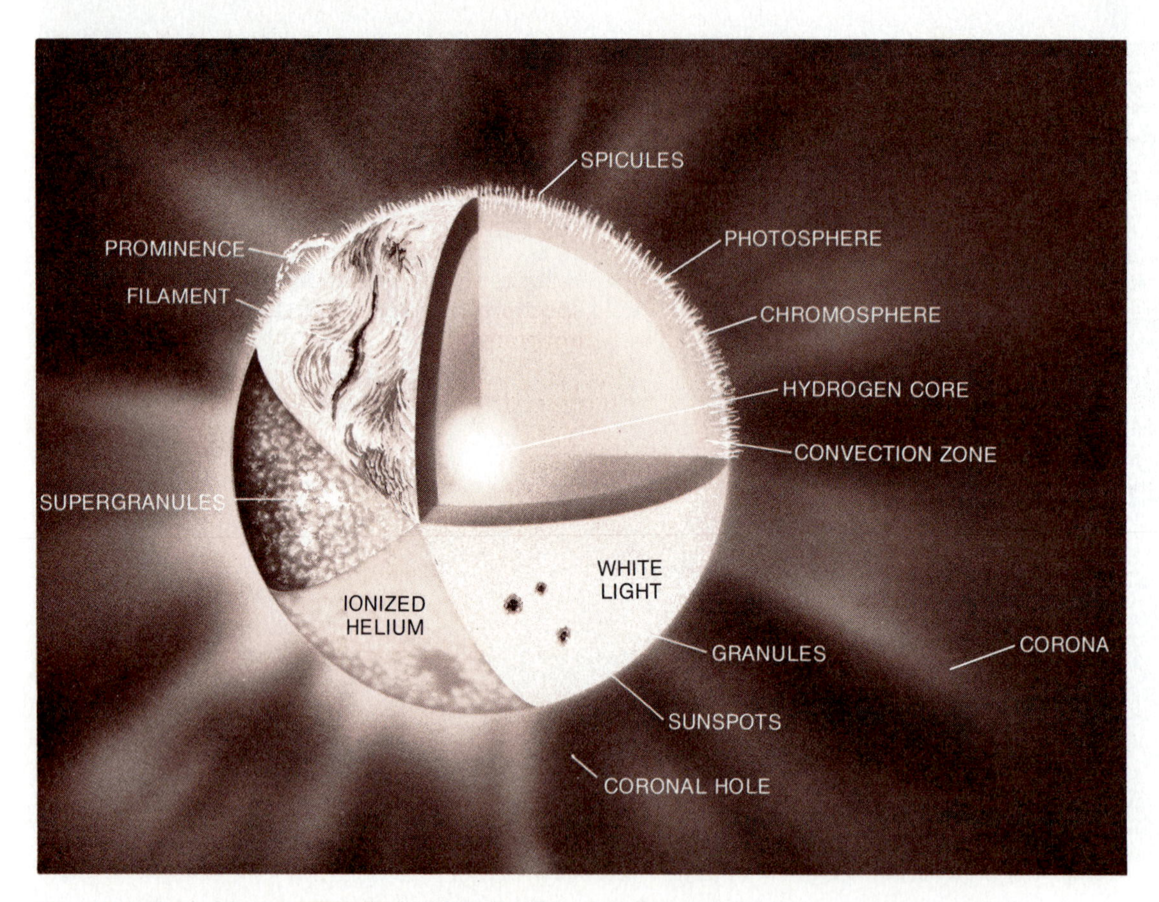

Figure 6–4 Parts of the solar atmosphere and interior. (From J. M. Pasachoff. *Contemporary Astronomy*. 2nd ed. Philadelphia, Saunders College Publishing, 1981.)

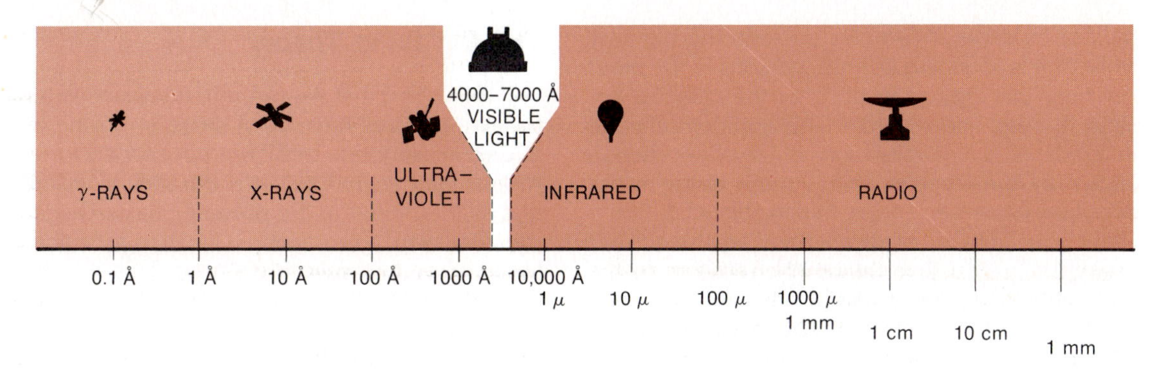

Figure 6–5 The electromagnetic spectrum. In the scale, an angstrom (Å) is 10^{-8} cm (0.000 000 01 cm) and a micron (μ) is 10^{-4} cm (0.000 1 cm). (From J. M. Pasachoff. *Contemporary Astronomy*. 2nd ed. Philadelphia, Saunders College Publishing, 1981.)

fortunate that most of this short-wavelength radiation is absorbed in the ozone layer of the atmosphere. X-rays and gamma rays are also absorbed in the upper atmosphere. Radio frequency waves are reflected. Much of the infrared radiation is absorbed by water vapor and carbon dioxide, causing the lower zones of the atmosphere to be warmed.

The energy that maintains the sun as a great glowing sphere of gases is derived from a continuous thermonuclear reaction called **fusion.** In the fusion process, hydrogen is changed to helium, and excess mass is converted to energy (Fig. 6–6). Each second, the sun transmits an amount of energy equivalent to that which would result from the burning of 25 billion billion lb of coal. This energy from the sun is the ultimate force behind the many geologic processes that sculpture the earth's surface. For example, the sun's rays aid in the evaporation of surface waters, which in turn results in clouds that provide the precipitation required for erosion. Along with the earth's rotation, the sun's radiation results in winds and ocean currents. Some scientists believe that protracted variations in the heat received from the sun may trigger episodes of continental glaciation or may reduce lush forests to barren wastelands. In company with the moon, the sun helps move the tides. Even primitive people knew the sun's importance and recognized it as the fountainhead and sustainer of life.

SPECULATIONS ABOUT THE UNIVERSE

All theories that attempt to describe the development of the universe are forced into conformity with an important astronomic observation called the **red shift.** To comprehend the red shift, one must first recall a few characteristics of light and color. A band of colors, like that formed when a beam of light is spread out by a prism, is called a **spectrum.** The spectra of stars are studied by astronomers. They reveal not only the star's composition but also whether the star is moving toward or away from the earth and at what speed.

To understand how this information can be obtained, recall listening to the apparent change in pitch from a moving source of sound as it passes

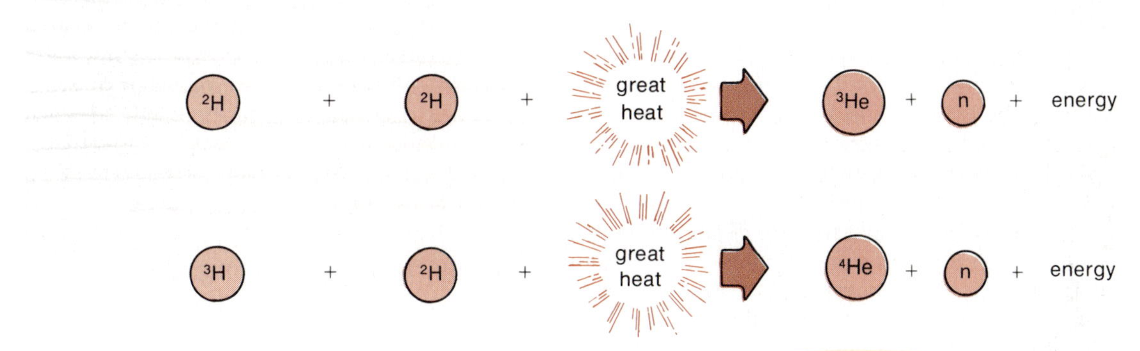

Figure 6–6 Examples of two fusion reactions (n=neutron).

by. For example, if two cars pass on the highway while one is sounding its horn, persons in the other car hear the change in pitch as the vehicles approach, pass, and separate. This apparent change in pitch is a manifestation of a phenomenon known as the **Doppler shift.** As the distance between the listener and noise-making object is diminishing, the number of waves entering the ear per second is greater than the number emitted by the approaching sounding object. It is as if the wavelength is shortened, and so the sound rises in pitch. As listener and noise source move apart, the reverse occurs. There is an "apparent" lengthening of waves and lowering of pitch.

An analogy might be imagined in which a toy boat is directed in toward the point where a large rock had been thrown into the center of a pond, causing widening circles of waves. The boat would bob up and down frequently while moving toward the source of the waves. If, however, the boat moved away from the center of the pond, there would be longer intervals between each up-and-down motion. The waves would appear to have increased in length.

A similar effect occurs when the waves are light waves rather than sound or water waves. An approaching star is recognized by an apparent shortening of wavelength, and spectral lines are shifted toward the blue end of the spectrum. If the star is moving away from us, the wavelengths are increased, and the spectral lines shift toward the red end of the spectrum. Furthermore, the greater the velocity of separation, the wider the observed red shift. By 1914, the astronomer W. M. Slipher had determined the spectra of 14 entire galaxies and found that the light from 12 of those exhibited the now famous red shift. Slipher reasoned that most of the galaxies within our range of observation were moving away from us at speeds of hundreds of miles per second. In 1929, Edwin Hubble made the further discovery that the red shift increased with increasing distances of the galaxies and that the more distant the galaxy, the higher its receding velocity.

Some astronomers speculate that the red shift may not be related to the Doppler effect but rather is caused by light that undergoes some sort of fatigue in traveling over such vast distances. Another possible explanation is that it is caused by some sort of ongoing physical changes in the light sources or by factors that may be related in a relativistic way to the curvature of space. Such ideas are being seriously considered, but they are not at all as widely accepted as the notion that the red shifts indicate that the universe is expanding.

The Big Bang Theory

Assuming that the red shift does indeed tell us that the galaxies have been steadily moving apart at their present rate of 32 km per second for each million light years, then it is possible to calculate that all the matter in the present galaxies began to move outward from a single location in space between 10 and 15 billion years ago. This knowledge has led to a hypothesis for the birth of the universe that is appropriately termed the **"big bang theory."** According to this idea, all the material now in the universe was assembled into a huge agglomeration of individual protons, neutrons, and electrons that existed at temperatures and pressures so extreme that they prevented the formation of atoms as we know them. Then came an explosion that destroyed the agglomeration and propelled its matter outward at enormous velocities. The sudden cooling that accompanied expansion would have favored the combination of atomic particles into atoms of the lighter elements, which might then have condensed into clouds of gases and eventually incorporated into the bodies of galactic systems.

The Steady State Theory

Many astronomers were uneasy about a cosmic evolution that simply starts with a big bang and ends with galaxies disappearing somewhere out in the farthest reaches of the universe. The theory does not adequately account for the origin of the particles in the initial agglomerate, and some scientists are uncomfortable with a system that simply seems to run down. As a result, a new theory was proposed that contained the red shift observation but depicted a universe with no beginning and no end. It was termed the **"steady state theory,"** and although it is no longer a serious contender, it is still frequently discussed in astronomy journals.

As implied by the name, the steady state theory suggests that conditions within the universe have always been and always will be as they are now. As expansion occurs, new matter, initially in the form of hydrogen, is created from combinations of atomic particles in the space between galaxies at the same approximate rate that older matter is receding. In this way, a uniform density of matter in the universe is maintained.

Theory of an Oscillating Universe

A third theory of the universe stipulates that for every big bang there is a preceding "big squeeze."

According to this idea, after the great explosion that causes the initial expansion, gravitational forces begin to prevail and cause all the matter to be drawn back to its place of origin. The expanding universe would thus become a contracting one. In the last stages of contraction, material would be returning at enormous velocities, resulting in compression of all the returned matter into an enormously dense mass in which temperatures and pressures would be so extreme that stars and their component atoms would be disassembled into their component atomic particles. From this mass would come the next big bang and the initiation of yet another episode of expansion and contraction. It is estimated each cycle might be about 100 billion years in length.

THE BIRTH OF THE SOLAR SYSTEM

Scientists now have fairly reliable estimates of *when* the earth originated. The question of *how* it originated is a subject of continuing investigation. Current programs in space exploration, including the landing of spacecraft on the moon, Venus, and Mars, and the flyby missions to Mercury, Saturn, and Jupiter have vastly improved our ability to speculate with some validity about the earth's origin. The new observations have added many refinements to older theories, all of which were obliged to account for the following characteristics of the solar system.

1. The orbital paths of planets occur within approximately the same plane, so that the solar system has the shape of a flat disk.
2. The orbits are nearly circular, and all lie near the plane of the sun's rotation (Fig. 6–7).
3. The direction of planetary revolution is counterclockwise (called **prograde**) when viewed from a hypothetic point high in space above the solar system. The direction of rotation of the planets (with the exception of Venus and Uranus) on their axes is also counterclockwise.
4. With few exceptions, the angle between the axis of rotation and the pole of orbit of each planet is small (Fig. 6–8). This angle of inclination is 23.5° for the earth. The angle of inclination determines the nature of seasons on planets (Fig. 6–9).
5. The sun is overwhelmingly the most massive body in the solar system. It contains 99.9 per cent of all the mass in the system. However, the sun is rotating so slowly that it has only 2 per cent of the total angular momentum in the solar system. The sun turns slowly on its axis while the planets race around the sun relatively rapidly.
6. Planetary distances from the sun can be described by a simple relationship called **Bode's Law.** Bode, a German astronomer, noted that each planet tends to be roughly twice as far from the sun as its next sunward neighbor.
7. With few exceptions, satellite systems tend to mimic the movements of the larger planetary bodies that they orbit.
8. As indicated by their differing densities, as well as by moon, earth, and meteorite samples, the planets have different compositions. Mercury, Venus, Earth, and Mars have mean densities of 3.8, 4.8, 5.5, and 3.3, respectively. They constitute the small, dense inner, or **terrestrial, planets.** Jupiter, Saturn, Uranus, and Neptune have lesser densities of 1.34, 0.71, 1.27, and 1.58, respectively. They are termed the outer, or **Jovian, planets.**
9. Hydrogen and helium constitute about 98 per

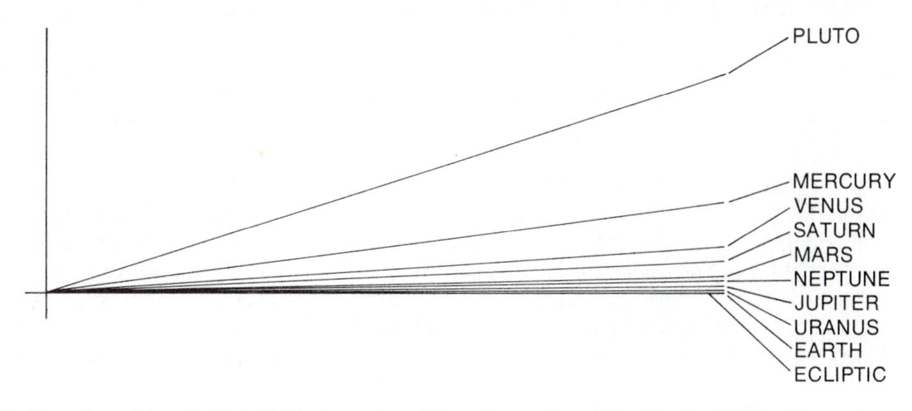

Figure 6–7 The orbits of the planets, with the exception of Pluto, have only small inclinations to the ecliptic plane. (From J. M. Pasachoff. *Contemporary Astronomy.* 2nd ed. Philadelphia, Saunders College Publishing, 1981.)

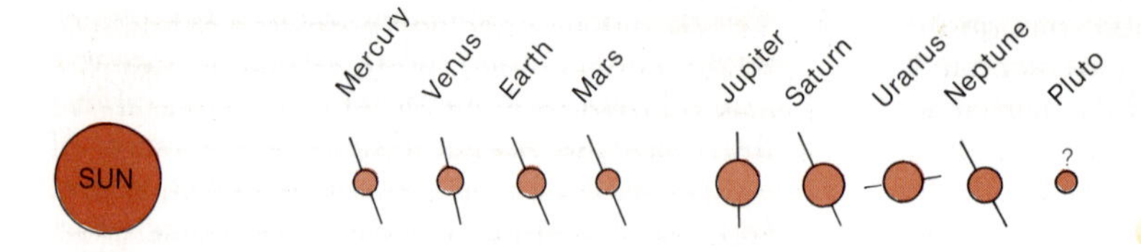

Figure 6–8 Planetary angles of inclination.

cent of the sun's mass. In contrast, the terrestrial planets have large quantities of such elements as oxygen, silicon, and iron. Once again, the Jovian planets are observed to contain enormous quantities of hydrogen and helium.

The Nebular Hypothesis

One of the earliest scientific attempts to explain the origin of the solar system was made in 1755 by the German philosopher Immanuel Kant. About 40 years later, the brilliant French mathematician Pierre Laplace independently developed a theory much like that of Kant, but he described it in a more comprehensive and scientific manner. This early account of the earth's origin was titled the **nebular hypothesis.** Kant and Laplace speculated that the solar system was once a great cloud of slowly rotating, hot gaseous material. As it cooled, it shrank; and as it shrank, it spun ever more rapidly, much as an ice skater spins faster when she pulls in her arms. The spinning motion would cause the nebula to flatten to a disklike shape. Eventually, centrifugal force around the margin of the disk would become sufficient to cause a ring of material to separate and be left behind as the rest of the nebula continued to contract. Subsequently, as the shrinking parent disk continued to spin faster, smaller rings would separate. In the next stage, matter in each ring would begin to condense, particles would collect into large aggregates, and aggregates would eventually collide and collect into planets and satellites. Laplace had neatly accounted for the flattened shape of the solar system, the circular orbits of the planets, and their spacing outward from the sun. Unfortunately, the theory had a serious flaw. It predicted that the sun should be the fastest rotating body of all (have most of the angular momentum), and, as we have noted, this is not the case.

The Planetesimal Hypothesis

Recognition of the inability of the nebular hypothesis to account for the slow rotation of the sun led to other theories that were based on a chance encounter between the sun and another body. One such theory, called the **planetesimal hypothesis,** was proposed in 1900 by Forest R. Moulton, an astronomer, and Thomas C. Chamberlin, a geologist. These two University of Chicago professors proposed that another star passed so near the sun that it raised enormous tides and pulled outward great gaseous filaments of matter. The material of

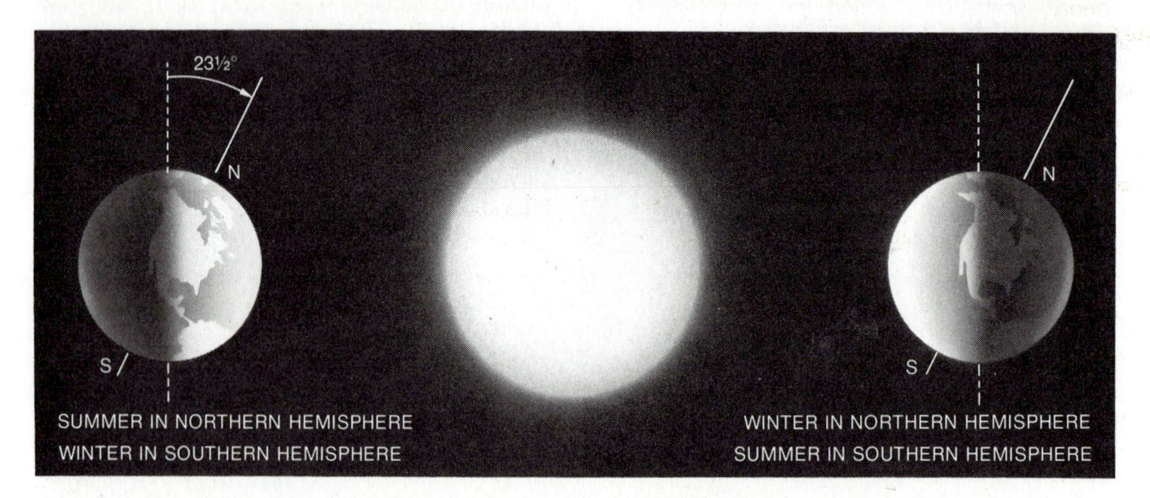

Figure 6–9 Seasons occur because a planet's axis is inclined with respect to the plane of its revolution around the sun. As shown here, when the earth's northern hemisphere is tilted toward the sun, it has its summer. At the same time the southern hemisphere is experiencing winter. (From J. M. Pasachoff. *Contemporary Astronomy*. 2nd ed. Philadelphia, Saunders College Publishing, 1981.)

the filaments condensed and formed into gases and particles, and the particles in turn accreted into larger masses (planetesimals) that ultimately evolved into planets and satellites. The passing star was considered to have imparted a strong oblique thrust to the ejected materials. Thus, the sun was not caused to spin rapidly, but considerable momentum was transferred to the outer material that was to become the planets.

In the two decades that followed presentation of the planetesimal hypothesis, several modifications of the basic encounter concept were published. The best known of these was developed by James H. Jeans and Harold Jeffries. Their theory was based on the grazing collision of a passing star with the sun and did not depend on tidal disturbances.

The encounter hypotheses were not without weaknesses, and they soon came to be viewed with skepticism. Collisions of stars appear to be rare events in our galaxy, and calculations suggested that the filamentous matter torn from the sun would simply dissipate into space rather than be drawn together into planets.

The Protoplanet Hypothesis

From Turbulent Eddies to Protoplanets

As a result of dissatisfaction with the encounter theories, astronomers began to re-evaluate the old nebular concepts in the light of new information about the characteristics of the sun. The revised theory, sometimes called the **protoplanet hypothesis,** was first proposed by C. F. Von Weizacker and later modified by Gerard P. Kuiper.

The hypothesis begins with a cold, rarefied cloud of dust particles and gases. The cloud gradually contracts and flattens as it takes on a counterclockwise rotation. About 90 per cent of its mass remains concentrated in the thicker central part of the disklike cloud. While rotation and contraction are occurring, the phenomenon of **turbulence** (not considered in the earlier nebular hypothesis) begins to affect the cloud system. Any departure from the smooth flow of a gas or fluid constitutes turbulence. In the dust cloud it consisted of chaotic swirling and churning movements that were superimposed on the grander primary motion of the entire cloud. Each turbulent eddy was affected by the movement of adjacent eddies, and some served as collecting sites for matter brought to their location by neighboring swirls. When the disk had shrunk to a size somewhat larger than the present solar system, its now denser condition permitted condensation of the concentrated knots of matter. Smaller particles merged to build chunks up to meters in diameter, and the dense swarms swept up finer particles within their orbital paths, thereby growing in size. These swarms of solids and gases were designated **protoplanets** by Kuiper. They were enormously larger than present-day planets. Each rotated somewhat like a miniature dust cloud, and each eventually swept away most of the debris in its orbital path and was able to revolve around the central mass without collision with other protoplanets (Fig. 6–10). Their formation from the original dark, cold cloud required an estimated 10 million years.

Within the protoplanets, condensation of ions into solid particles and accretion of the particles into larger chunks led eventually to true planetary bodies. The process is presumed to have involved gravitational settling of the heavier particles (like iron) toward the center of the protoplanet; contraction of the initially rarefied cloud; and, with that contraction, an increase in the rate at which the mass rotated.

The Birth of the Sun and Planets

While the protoplanets were in the process of condensation and accretion, material that had been pulled into the central region of the nebula condensed, shrank, and was heated to several million degrees by gravitational compression. The sun was born. Later, when pressures and temperatures within the core of the sun became sufficiently high, thermonuclear reactions began, pouring out additional energy derived from matter by the nuclear fusion of hydrogen atoms into atoms of helium. (Today, the sun converts about 596 million metric tons of hydrogen into 532 million metric tons of helium each second. The difference represents matter that has been converted to energy.)

As the sun began its thermal history, its radiation ionized surrounding gases in the nebula. These ionized gases in turn interacted with the sun's magnetic lines of force, causing the star's magnetic field to adhere to the surrounding nebula and drag it around during rotation. At the same time, the nebula caused a drag on the magnetic field and slowed down the sun's rotation. In this way, the old problem of why the sun rotates slowly was explained.

The stream of radiation from the sun sometimes called **solar wind,** drove enormous quantities of the lighter elements and frozen gases outward into space. (This solar force is what causes a comet's tail to show wavy streaming (Fig. 6–11) or to

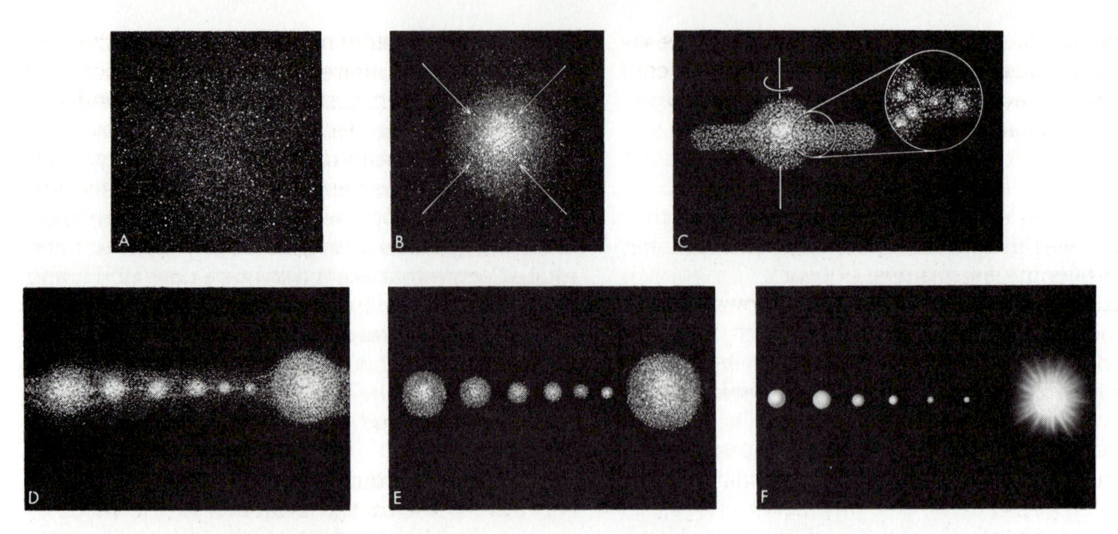

Figure 6–10 An illustration of the protoplanet hypothesis. A and B show the dust cloud taking form and beginning to rotate counterclockwise as it contracts and flattens. In C the nebula is distinctly disklike with a central bulge representing the protosun. Materials in the thin portion of the nebula condense and accrete to form protoplanets (E), which in turn contract to become planets (F). (From J. M. Pasachoff. *Contemporary Astronomy*. 2nd ed. Philadelphia, Saunders College Publishing, 1981.)

be bent away from the sun.) As would be expected, the planets nearest the sun lost enormous quantities of lighter matter and thus have smaller masses but greater densities than the outer planets. They are composed primarily of rock and metal that could not be moved away by the solar wind. The Jovian planets were the least affected and retained their considerable volumes of hydrogen and helium.

Asteroids and Comets

Proponents of the protoplanet hypothesis assume that a major protoplanet failed to develop in the belt now occupied by the asteroids. Possibly a number of small eddies evolved in this zone and eventually produced the larger asteroids. Some of these may have collided to produce smaller asteroids and meteorites.

The origin of comets is far more speculative. Astronomers examining the spectra of comets find indications that they are composed of frozen methane, ammonia, and water, in which particles of heavier elements may be embedded, somewhat like sand grains in a snowball. These comets may be composed in part of the flux driven spaceward by solar wind, or they may be made of matter that once was too far out on the periphery of the nebula to have been drawn into the evolving protoplanets. Whatever their origin, comets are truly denizens of interstellar space. Some comets approach sufficiently close to earth to be almost within range of our present space technology. It is not unreasonable to expect that in a decade or two, space scientists may send a probe to comets to analyze them chemically and thereby obtain new insights into solar system history.

The protoplanet hypothesis should not be considered the final word in cosmology. Many questions relating to the role of gravitation and the

Figure 6–11 Photograph of the Comet Mrkos. The wavy streaming in the tail was caused by solar wind. (Hale Observatory photograph; from J. M. Pasachoff. *Contemporary Astronomy*. 2nd ed. Philadelphia, Saunders College Publishing, 1981.)

distribution of certain of the heavy elements remain to be answered. Even if the hypothesis continues to be accepted, it can be expected to undergo considerable change in the years ahead. As the concept now stands, it is probably the most widely favored of hypotheses for solar system origin. Large telescopes have revealed the existence of true nebulae between stars, and some of these great swirls of gas and dust appear to be forming new stars. Planetary systems like our own may be forming even now. The hypothesis also accounts for the known spacing of planets, their directions of rotation and revolution, and the distribution of angular momentum within the system.

For the historical geologist, the protoplanet hypothesis presents a foundation on which to build the history of the earth. One may begin with a solid earth that originated in the cold state by the collection of particles that were derived from a cloud of interstellar matter and view the solar system as having formed from rather ordinary cosmic processes. The formation of the earth was probably not a freakish occurrence.

METEORITES AS SAMPLES OF THE SOLAR SYSTEM

Because meteorites are masses of mineral or rock that have reached the earth from space, they are of great importance to scientists interested in the origin and history of planets. They are the only objects from the universe beyond the moon and the earth that can currently be held in hand and scrutinized in the laboratory. Besides revealing something of the composition of the solar system, they provide important clues to the earth's age.

Meteors (Fig. 6–12) are the objects familiarly known as "shooting stars," a misleading term, because they are not stars at all. They are thought to be pieces of materials originally left in space during the formation of the solar system, later captured by the earth's gravity, and then completely vaporized before reaching the earth's surface. If a portion of the meteor survives its fall through the atmosphere and crashes into the earth, the object is then termed a **meteorite.** About 500 meteorites as large as or larger than a baseball reach the earth's surface each year. Weathering and erosion on the earth have erased most of the craters made by these impacts, but there remain about 70 partially preserved craters or clusters of craters. In some of these, all that remains is the root zone of densely shattered and shocked rocks. Geologists refer to these remnants as astroblems. A large number of astroblems are known from the area of ancient rocks that surround Hudson's Bay in Canada.

Craters caused by meteorite bombardment are clearly evident on the moon. The moon is also affected by micrometeorites (less than 1 mm in diameter), for these smaller infalling particles are considered a major cause of lunar erosion. Most of

Figure 6–12 A meteor crossing the field of view of the Palomar telescope while it was taking a 15-minute exposure of the Comet Kobayashi-Berger-Milon, August, 1975. (Hale Observatory photograph/John Huchra; from J. M. Pasachoff. *Contemporary Astronomy*. 2nd ed. Philadelphia, Saunders College Publishing, 1981.)

the micrometeorites come from comets, whereas the larger meteorites are probably derived from collisions in the asteroid belt.

Upon entering the atmosphere, meteors are heated to incandescence by the friction created during collision with air molecules, and for a few moments they appear as glowing orbs with bright, gaseous tails. At this point, they can be photographed and their spectrum analyzed. These analyses indicate that the larger meteors are composed chiefly of iron, calcium, silicon, aluminum, and sodium. Meteors are frequent visitors. Over the whole earth, many thousands can be seen each night. Fortunately, most of those that enter the earth's atmosphere are vaporized before they strike the earth's surface. More meteors are seen at certain times than at others. When the number of sightings is above average, the earth is experiencing a **meteor shower.** Showers tend to occur at the same time each year and represent movement of the earth through the orbits of defunct comets. Also, the rate at which meteors are seen increases after midnight on the night of a shower, because that side of the earth is facing the oncoming interplanetary debris.

Meteorites can be classified according to their composition as ordinary chondrites, carbonaceous chondrites, achondrites, irons, and stony-irons (Fig. 6–13). The most abundant of these types are the **ordinary chondrites** (Fig. 6–14), which are crystalline stony bodies composed of high-temperature ferromagnesian minerals. Their mineralogy indicates they formed in a closed system at temperatures of about 500 to 1000°C. Ordinary chondrites can be dated by the uranium-lead, strontium-rubidium, and potassium-argon methods, and they are found to be about 4600 million years old. Many ordinary and carbonaceous chondrites contain spherical bodies called **chondrules** (Fig. 6–15). Their spherical shape suggests that chondrules may have formed as droplets of molten material that was then rapidly chilled to preserve the rounded form of the original droplet. Perhaps the melting occurred during the early heating of the sun, and the chondrules solidified as they moved outward toward cooler regions of space.

Figure 6–13 The major categories and proportions of meteorites.

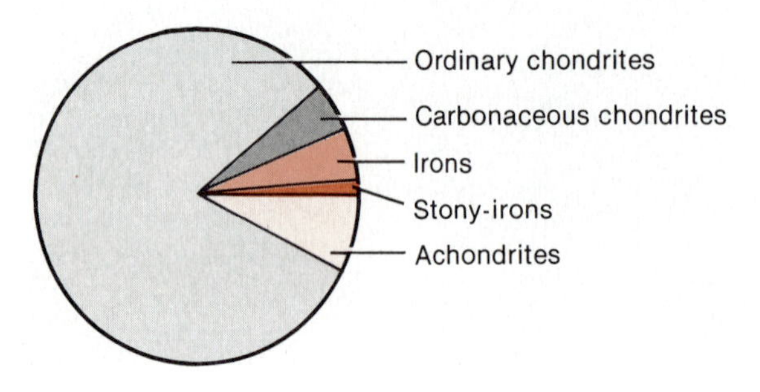

Less abundant than the ordinary chondrites are the **carbonaceous chondrites.** In these meteorites one finds the same abundance of metallic elements as in ordinary chondrites, with the addition of hydrous minerals, more volatile trace elements, and about 5 per cent of organic compounds, including inorganically produced amino acids. Suspended in the blackish, earthy matrix of carbonaceous chondrites are both chondrules and irregular pieces of crystalline, high-temperature minerals that may have condensed from a cooling vapor. These bodies are referred to as calcium-aluminum–rich inclusions.

The matrix material of most carbonaceous chondrites consists of relatively low-temperature minerals, whereas the chondrules and calcium-aluminum–rich inclusions are composed of minerals that form at higher temperatures. It appears that these two components originated under different conditions and were subsequently brought together.

One of the most interesting findings about carbonaceous chondrites is that their overall composition is very similar to that of the sun. Indeed, if it were possible to extract some solar material, cool it down, and condense it, the condensate would be chemically similar to carbonaceous chondrites. This similarity suggests that carbonaceous chondrites are samples of primitive planetary material that formed when the sun formed and since that time have never experienced melting and the kind of mineral fractionation that characterizes igneous rocks.

A small percentage of stony meteorites do not contain chondrules and therefore are termed **achondrites.** Most of these have compositions similar to terrestrial basalts, and many exhibit angular, broken grains. These fragmental textures indicate that achondrites may be the products of collisions of larger bodies. One can imagine, for example, the collision of two asteroids having metallic cores enveloped in silicates. The fragments resulting from the collision would consist of a range of compositions from metallic to stony.

The **iron meteorites** (Fig. 6–16) comprise about 6 per cent of all meteorites found on earth. Most iron meteorites are intergrowths of two varieties of nickel-iron alloy. The large size of the crystals and metallurgic calculations indicate that some iron-nickel meteorites cooled as slowly as 1°C per million years. Such a slow rate of cooling would be possible only in objects at least as large

Figure 6–14 Meteorite of the type termed an ordinary *chondrite.* Chondrites are stony meteorites that contain BB-sized round spherules (chondrules) composed of olivine and certain pyroxenes. (Photograph courtesy of The Institute of Meteoritics of the University of New Mexico.)

Figure 6–15 Chondrules in a carbonaceous chondrite. (Photograph courtesy of the Institute of Meteoritics of the University of New Mexico.)

Figure 6–16 Iron meteorites. The upper specimen has been cut and polished, then etched with acid to show the interesting pattern of interlocking crystals called the "Widmanstätten Pattern." (Photograph courtesy of the Institute of Meteoritics of the University of New Mexico.)

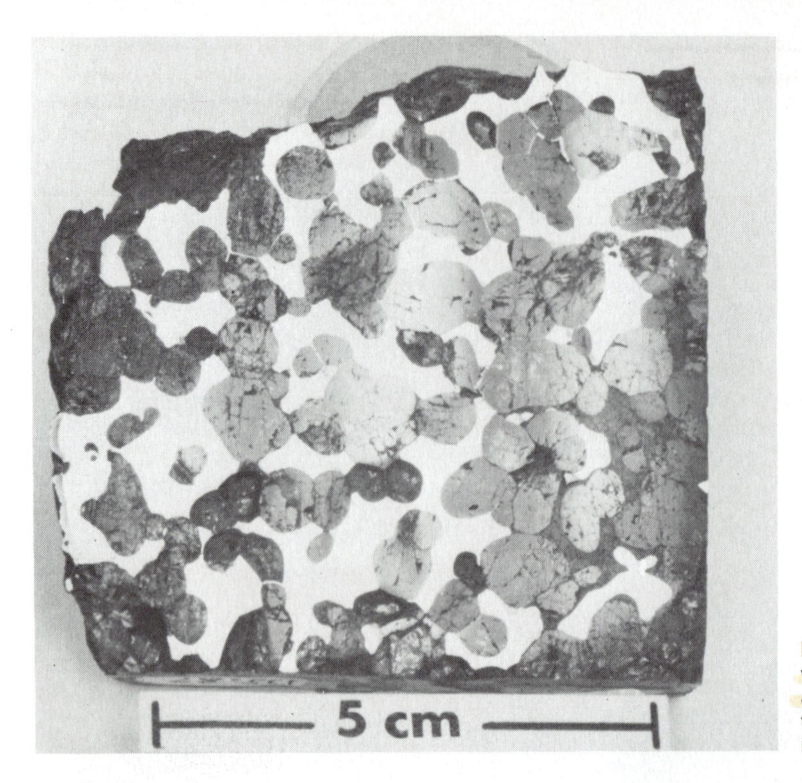

Figure 6–17 Stony-iron meteorite. The white areas are composed of iron-nickel, and the gray areas are mostly olivine. (Photograph courtesy of the Institute of Meteoritics of the University of New Mexico.)

as asteroids. Thus the history of the iron meteorites probably involved a long episode of slow crystallization within a parent body sufficiently large to provide insulation for the hot interior, followed by violent disruption of that body by a collision to produce the meteorites.

The least abundant of all the categories of meteorites are the **stony-irons** (Fig. 6–17). As indicated by their name, they are composed of silicate minerals and iron-nickel metal in about equal amounts. The stony-iron meteorites are generally considered to represent fragments from the interfaces between the silicate and metal portions of asteroidal bodies. In some cases, they may have originated at the boundary between an iron core and the silicate mantle of a planetary body destroyed by collision.

THE MOON

Moon Features

As satellites go, our moon is rather large in relation to the size of its parent planet. It has a diameter of over one fourth of the earth, and its density of about 3.3 is the same as that of the upper part of the earth's mantle. Gravity on the moon is insufficient to maintain an atmosphere. The moon and the earth revolve around their mutual center of gravity as a sort of "double planet." The moon orbits the earth and rotates on its axis at the same rate. As a result, we always see the same side of the moon and must depend on transmissions from space vehicles for images of the far side (Fig. 6–18). These photographs have revealed a densely cratered surface interrupted only by few and relatively small, level areas.

Telescopic observation of the near side of the moon was first made in 1609 by a rather irascible professor of mathematics at the University of Padua. The mathematician, of course, was Galileo Galilei. Galileo learned how Dutch lens grinders had combined a series of lenses into a tube to magnify images of distant objects. He quickly set to work and managed to build a telescope capable of magnifying objects by about 30 times. With his telescope, Galileo was able to see that the surface of the moon was not smooth, but rather "uneven, rough, full of cavities and prominences . . ." He recognized "small spots," which we now know are craters, and dark patches that children imagine to be the eyes and mouth of the "man in the moon." These dark patches lie within large basins. Galileo named them maria ("seas") and incorrectly suggested they were filled with water. Today, we know the dark areas have been flooded, not with water but with basaltic lavas. These lavas are believed to have been produced during the impact of large meteorites rather than by subsurface heating. In

Figure 6–18 The far side of the moon as photographed by Apollo 16 (Photograph courtesy of National Aeronautics and Space Administration.)

many maria, the lava has been contained within depressions that are hundreds of kilometers across and 10 to 20 km deep. The depressions are termed **mare basins** (Fig. 6–19). Mare basins had to have developed prior to the extrusion of the dark lavas that form the maria, for these lavas are contained within the basins.

Galileo also provided the name for the lighter-hued, rougher terrains of the moon. He called these regions terrae, although modern space scientists refer to them as **lunar highlands.** The highlands are the oldest parts of the moon, having already formed long before the earliest mare lavas appeared. The highlands are heavily cratered regions reminiscent of a battlefield following a heavy artillery barrage. They provide stark evidence of the moon's early episodes of intense meteoritic bombardment.

On the moon, as on the earth, craters are roughly circular, steep-sided basins normally caused by either volcanic activity or meteoritic impact (Fig. 6–20). Although a few craters in the moon show evidence of volcanic origin, the great majority are clearly the result of meteorite impact. Unlike the earth, the moon has insufficient atmosphere to burn up approaching meteorites, and thus the frequency of impact is high. Lunar geologists are able to determine the relative ages of certain areas on the moon's surface by the density of craters. Younger areas have fewer craters than do older ones. Another aid to recognizing differences in the age of lunar features is provided by the rays of material splashed radially outward from the crater at the time of impact. Younger rays and other impact features cross and partially cover older features. Light-colored impact rays com-

Figure 6–19 In order to show the whole moon, but still show the detail that does not show up well at full moon, the Lick Observatory has put together this composite of first and third quarters. Note the dark maria and the lighter, heavily cratered highlands. Two young craters, Copernicus and Kepler, can be seen to have rays of light material emanating from them. This is a ground-based view of the moon, and the satellite is shown inverted, as it would be when viewed through a telescope. (Lick Observatory photograph; from J. M. Pasachoff. *Contemporary Astronomy.* 2nd ed. Philadelphia, Saunders College Publishing, 1981.)

posed of finely crushed rock are exceptionally well developed in the large lunar crater known as *Copernicus* (see Fig. 6–19). This familiar feature is about 90 km across and is rimmed by concentric ridges of hummocky material blasted out by the impact of the crater-forming body. The center of the crater floor is raised and thought to represent an upward adjustment following impact. Not all craters have such spectacular rays. Impact rays are sometimes obliterated by the rather continuous rain of tiny meteorites that strike the moon. For this reason, younger craters may still be adorned with impact rays, whereas older craters have lost these features.

There is ample evidence of former volcanic activity on the moon. We have already noted the extensive lava flooding of mare basins. In addition to these flood basalts, dome-shaped volcanoes measuring about 10 km in basal diameter have been located. At least 1 per cent of the lunar craters are believed to have resulted from volcanic activity. Meandering channelways called **sinuous rilles** occur on the surface of our satellite. Most of these are thought to have been produced by the turbulent flow of lava or by the collapse of lava tunnels. These sinuous rilles are distinctly different from straight **linear rilles,** which probably result from faulting. In the maria, astronomers have also observed features named **wrinkle ridges,** thought to develop by the extrusion of lava. Tall volcanic

Figure 6–20 This view from the orbiting Apollo 15 Command Module shows a smaller crater, Krieger B, superimposed on a larger crater, Krieger. Obviously, the smaller crater is younger than the larger one. Several rilles (clefts along the lunar surface that can be hundreds of kilometers in length) and ridges are also visible. (Photograph courtesy of National Aeronautics and Space Administration.)

peaks are not found on the moon. Perhaps this is because lunar lavas, like lavas associated with shield volcanoes on the earth, are highly fluid and spread laterally rather than piling up around a vent.

Are lunar volcanoes still active today? In 1958, Russian astronomers observed evidence of gas emission from the crater *Alphonsus,* and on two occasions in 1963, observers at the Lowell Observatory saw glowing red spots in and near the crater *Aristarchus.* These observations suggest that the moon still experiences sporadic volcanic activity.

Moon Rocks

Investigations of samples returned from the moon as well as instrumental data gathered by orbiting spacecraft indicate that three kinds of moon rocks are most prevalent. The lunar maria are covered primarily with basaltic rocks rich in pyroxenes and plagioclase (Figs. 6–21, 6–22). Although there are differences in proportions of particular minerals, these moon basalts resemble earth basalts and were probably derived from a silicate melt similar in composition to the rocks of the earth's mantle. On the lunar highlands, rocks composed almost entirely of calcium plagioclase feldspar are abundant (Fig. 6–23). These gabbro-like rocks are named **anorthosites.** The third abundant lunar rock also resembles gabbro but more specifically would be termed **norite.** In norite, the pyroxene minerals lack calcium and aluminum. The distribution of norite on the moon is rather random.

The samples brought to earth from the moon were all obtained from the blanket of rock fragments and dust that covers the moon's surface. Within this loose material, called **regolith** (Fig. 6–24), it is possible to find fragments of rocks of all sizes, ranging from microscopic grains to huge boulders. Some of the boulders and cobbles consist of aggregates of smaller rock fragments welded together as lunar breccias.

At Tranquillity Base, site of the first moon landing, the regolith was estimated to be between three and four meters thick. It is believed to have formed as a result of meteorite impacts, each of which would dislodge a mass of debris many hundred times greater than its own mass. Calculations indicate that meteorite impacts amply account for the amounts of loose regolith on the moon.

Radiometric dating of lunar rocks has been of great importance in an understanding of the sequence of events in the moon's history. Most mare basalts are between 3.2 and 3.8 billion years old

Figure 6–21 A basalt collected from the lunar maria by Apollo 15 astronauts. This lava solidified so rapidly that bubbles formed by escaping gas were trapped, forming vesicles. The specimen is 14 inches across. (Photograph courtesy of National Aeronautics and Space Administration.)

Figure 6–22 Photomicrograph of a polished thin section of a moon rock, about 2 mm long. Mineral grains are predominantly olivine, pyroxene, and plagioclase. The dark matrix is largely brown glass. (Photograph courtesy of T. Bernatowicz, Department of Earth and Planetary Sciences, Washington University.)

and were extruded during the great lava floods that produced the mare basins (Fig. 6–25). For the most part, the lunar breccias were formed during impact events that occurred prior to these floodings. The oldest rocks on the moon were taken from the lunar highlands, and they have been found to be between 4.5 and 4.6 billion years old. Thus, they are about the same age as the oldest meteorites and indeed probably crystallized at about the time the moon originated.

Samples of moon rocks now being studied in the United States have been found to retain weak magnetism imparted to the rocks at the time they solidified. Lunar geologists have suggested that this so-called remanent magnetism indicates the existence of a once liquid but now solidified core and that the moon had a significant magnetic field between about 4.2 and 3.2 billion years ago. The moon has no general magnetic field today.

Mascons

Prior to sending a manned spacecraft to the moon, the United States had launched five spacecraft designed to circle the moon and transmit high-resolution pictures of the moon's surface back to earth. The speed and location of these craft (named "orbiters") were precisely monitored by radio tracking devices. Much to the surprise of the scientists monitoring the orbiters, their speed was not constant but increased as the craft approached one of the dark, circular mare and decreased as they passed over the mare. Evidently, there were mass concentrations, dubbed **mascons,** beneath the maria. Gravitational attraction of the mascon caused approaching orbiters to speed up and then slowed the orbiter down as it passed overhead. Very likely, mascons are caused by dense layers of basalt within mare basins. Calculations indicate that a layer 5 to 10 km thick and about 10 per cent

Figure 6–23 Rock from the lunar highlands. As indicated by the larger crystals, these rocks may be among the earliest to solidify when the moon formed. The specimen shown here is 4.6 billion years old. It is composed primarily of olivine and feldspar. (Photograph courtesy of National Aeronautics and Space Administration.)

Figure 6–24 Geologist-astronaut H. H. Schmitt samples lunar regolith with a specially designed rake. (Photograph courtesy of National Aeronautics and Space Administration.)

denser than lunar highland rocks would produce the effect noticed in the orbiters. Scientists are intrigued, however, by the fact that these concentrations of mass have not sunk into the lunar interior. This indicates that the inside of the moon is not weak and plastic but rather is sufficiently rigid to support mascons for billions of years.

Lunar Processes

In the first chapter of this book, we alluded to earth processes capable of raising mountain ranges, as well as such opposing processes as erosion and weathering that are constantly at work wearing away the lands. Geologic processes also operate on the moon, but they are slower and of a different kind. The moon is a desolate world in which there is no air or surface water. Whenever gases or liquids escape to the lunar surface, they are dissipated into space because the moon lacks sufficient mass, and hence gravitational attraction, to hold them. There can be no stream-cut canyons and valleys, no glacial deposits, and no accumulations of wave-sorted sand. Rocks brought back to earth from the moon seem bright and newly formed and lack the discoloration common to weathered earth rocks. Yet erosion does occur on the moon, primarily through the continuous bombardment of the lunar surface by large and small meteorites. The larger meteorites, of course, produce craters and rough terrain, but it is the rain of small meteorites and **micrometeorites** (those with diameters of less than a millimeter) that subtly alter the moon's surface and are important in the formation of the lunar soil. The upper or exposed surfaces of lunar rocks brought back to earth from the moon were riddled with tiny, glass-lined pits produced by the impact of fast-moving micrometeorites. The micrometeorite barrage continues today, as was evident on examination of the Apollo spacecraft after its return to earth. Ten tiny microcraters were chipped into its windows.

Bombardment of the lunar surface by objects from space is largely responsible for the development of lunar soil. Beginning with a newly solidified lava, incoming particles blast out a multitude of small craters. As if struck by bullets, large particles are pulverized to produce fine fragments that pile up around the new crater and cover older ones. The impacts of larger meteorites are capable of scattering debris over a wide area.

Cosmic radiation is another process that plays a less vigorous but persistent role in weathering the lunar surface. Cosmic rays are fast-moving atomic particles, mostly consisting of protons of hydrogen and helium nuclei. On earth, we are largely protected from cosmic rays by the atmosphere, but the moon is subjected to a continuous onslaught. Cosmic rays penetrate moon minerals and damage their atomic lattices. If these minerals are dipped in a corrosive chemical bath, the chemical will enlarge the damage paths made by the passage of cosmic particles. On examination of these minerals with a microscope, they are frequently found to be riddled with cosmic ray particle tracks. Even the plastic space helmets worn by the Apollo 11 astronauts contained tiny cosmic ray damage tracks. When closing their eyes, the astronauts could "see" bright streaks of light that were actually produced by the flight of a cosmic ray particle through the eyeball. Fortunately, the astronauts were not exposed to this radiation long enough to endanger their health.

Temperature variations on the moon may also contribute to the production of loose surface materials. Lunar day and night are each about 2

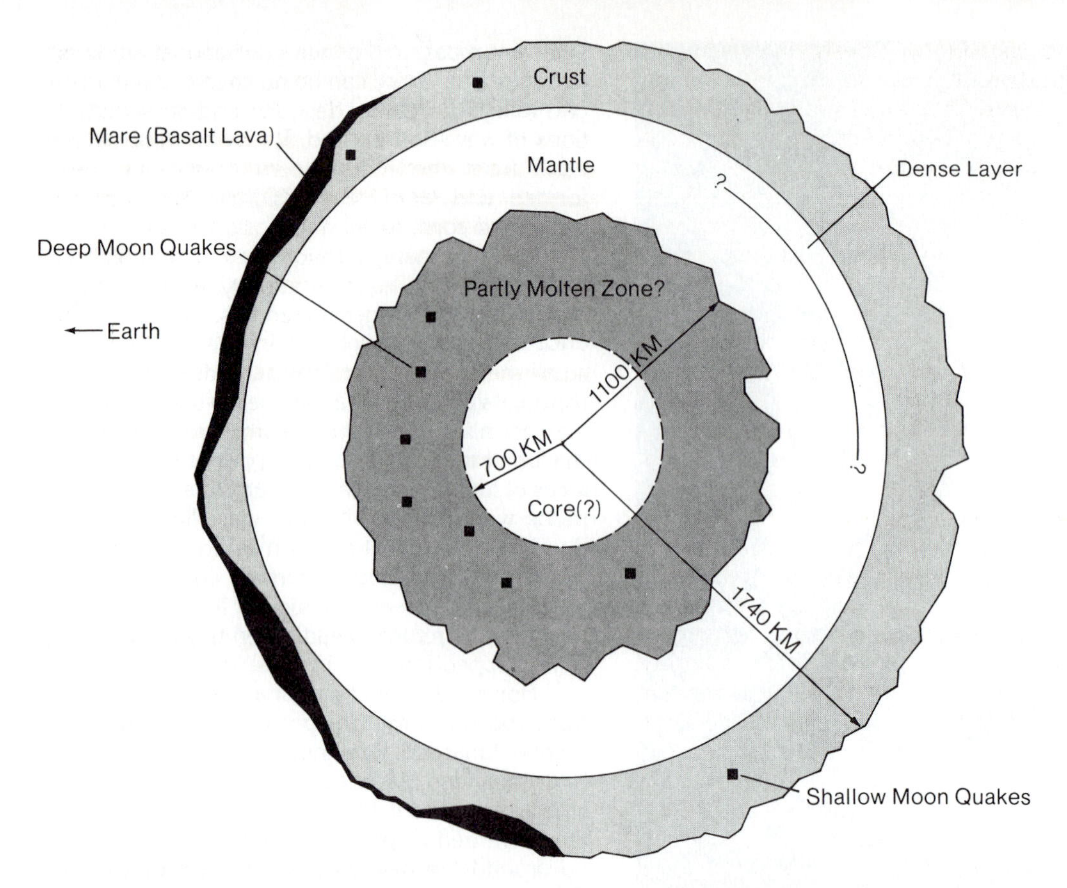

Figure 6–25 Cross-section through the moon showing an outer crust, an inner mantle, and an innermost zone that may still be partly molten. Within the innermost zone, there may be a small metallic core (dashed circle). The earth-facing side is relatively smooth and shows larger accumulations of the mare basin lavas. The far side is more rugged and has almost no lava. The drawing is not to scale, and the ruggedness of the far side is exaggerated. Data for the drawing were provided by A. M. Dainty, N. R. Goins, and M. N. Toksoz. (Courtesy of National Aeronautics and Space Administration. From B. M. French. *What's New on the Moon?*)

weeks long. During daytime, temperatures at the surface rise to 134°C, whereas at night they drop to −170°C. Although not yet proven, the alternate expansion with heating and contraction with cooling weakens grain boundaries and could result in fracturing of rock.

Although gravity on the moon is weaker than on earth, it is nevertheless an effective agent in moving lunar materials from high to low places. Whenever loose material on slopes is unable to resist the pull of gravity, it will break away and produce masses of slumped debris. Steplike slump masses are recognized as a common feature along the rims of some lunar craters. Elsewhere, photographs reveal tracks in lunar soils that record the rolling and sliding of large boulders down slopes.

The Origin of the Moon

Even with the explosion of scientific data from the space program, scientists are unable to support confidently a single theory for the origin of the moon. As a result, there are three hypotheses now being actively examined. The first of these stipulates that the moon accreted along with the earth as an integrated two-planet system. The second concept suggests that both bodies accreted independently and that the moon was then captured by the earth. The third hypothesis proposes that the moon broke off from the earth before either had completely solidified.

If both the earth and the moon formed together from the same part of the solar nebula, then they should have similar composition (as indicated in part by specific gravity). However, the average specific gravity of the earth is about 5.5, whereas that of the moon is only 3.3. This difference, together with the absence of a dipolar magnetic field on the moon, indicates that it lacks an iron core. These considerations have caused many scientists to view with skepticism the double-planet concept for the origin of the moon.

The compositional difficulties seem to be resolved in the second hypothesis, in which the moon is considered to have formed in a distant part of the solar system but was subsequently captured when it approached the earth's gravitational field. Unfortunately, there are some severe problems relating to the mechanics of such a capture, and long-term measurements of the moon's movements required for verification of the concept are lacking. However, the capture hypothesis will be periodically re-examined in the light of new, highly accurate measurements made with the aid of laser reflectors left behind on the moon's surface.

In 1898, George Darwin, the son of Charles Darwin, speculated that the moon was composed of a large mass of material that spun off the earth at an early time when its rotation was much faster than now. The earth's centrifugal force, together with the sun's tide-raising ability, first caused a huge tidal bulge on the earth. The bulge then separated to form the moon. When the moon's composition was found to be much like that of the earth's mantle, scientists speculated that the development of the earth had first proceeded to the point at which the earth's core of iron and other heavy elements had formed. After these heavy elements (which are substantially depleted in the moon's composition) had settled inward, a great bulge of mantle material formed, then pinched off, and began to orbit in widening circles around the earth. This interesting theory will continue to be examined over the next several years. It cannot be accepted as yet, largely because of calculations indicating that if the moon spun off the mantle, it would subsequently be drawn back again. Geophysicists are also uncertain that there would be sufficient force available to lift the mass of the moon from the earth and throw it into orbit.

As some scientists continue to debate the possibility that the moon was derived from the earth's mantle, a larger number of investigators have come to favor the hypothesis that states that the moon, along with the other solar system bodies, was born about 4.6 billion years ago by the accumulation of matter and smaller bodies from the original dust cloud. Theoretically, the infall of small objects released so much energy that the outer part of the moon became molten to a depth of several hundred kilometers, and minerals then separated out to produce the crust of anorthosite.

The anorthositic primordial surface was next subjected to a barrage of meteorites that lasted nearly 1.5 billion years and that produced the intricately scarred and cratered landscapes of the lunar highlands. Then internal heat generated by the steady decay of such radioactive minerals as uranium and thorium began to melt the moon again, especially at rather shallow depths of 100 to 250 km. The result of this melting event was massive extrusions of lava that spread out over parts of the lunar surface during the period from about 3.8 to 3.1 billion years ago. The moon was much quieter after the last of these eruptions. Meteorite impacts continued to sculpture its surface, sometimes forming spectacular craters like *Copernicus* (see Fig. 6–19). This was the time also when most of the lunar soil layer accumulated.

Today, except for minor moonquakes associated with tidal stresses, the moon is a relatively quiet satellite. On its bleak, airless surface, forms of life comparable to those on earth will probably never be found. The rocks that have been brought to the earth for study have not provided the slightest trace of organic activity. It is unlikely that life ever existed on our satellite.

THE EARTH'S NEIGHBORING INNER PLANETS

Mercury

Mercury has the distinction of being the smallest and swiftest of the inner planets. It revolves rapidly, making a complete journey around the sun every 88 earth days. Mercury makes a complete rotation on its axis every 59.65 days, which means that the planet rotates three times while encircling the sun twice.

Mariner 10, launched in 1973, provided our first close look at Mercury (Fig. 6–26). Following its launching, Mariner 10 flew by Venus and, using the gravity of Venus to deflect its trajectory, went on to take a look at Mercury. Indeed, the spacecraft's orbit permitted visits over Mercury on three occasions. Mariner 10 data and photographs transmitted back to earth revealed a planet with a rather moonlike surface. Densely cratered terrains as well as smooth areas resembling maria were quite apparent. Some of the craters exhibited rays of light-colored impact debris, just as do craters on the moon. Long, linear scarps were discerned, and these are thought to be fracture zones along which crustal adjustments occurred. Perhaps tidal forces, or shrinkage of the planet as it cooled, may have produced these elongate scarps.

Mercury has a weak magnetic field, a thin and all but negligible atmosphere composed mostly of helium, a lightweight crust covered with fine dust, and an iron core that comprises nearly 75 per cent of its radius.

Figure 6–26 Photomosaic of Mercury as photographed by the Mariner spacecraft. (Photograph courtesy of the National Aeronautics and Space Administration.)

Venus

Although Venus is similar to the earth in size and shape, its surface conditions are quite different from those on earth. Indeed, our planet is a paradise compared to Venus. Maximum surface temperature on our nearest planetary neighbor is a searing 475°C. The planet is shrouded in a thick layer of clouds that extend 40 km above the surface (Fig. 6–27). Carbon dioxide is the principal gas in the atmosphere of Venus, although there are lesser amounts of oxygen, nitrogen, and sulfuric acid vapors. The large amount of carbon dioxide serves as an insulating blanket in trapping solar energy by the greenhouse effect (the heating of an atmosphere by the absorption of infrared energy remitted by the planet as it receives light energy in the visible range from the sun). Surface atmos-

Figure 6–27 Photomosaic of Venus, made in ultraviolet light from the Mariner 10 spacecraft in 1974. (Photograph courtesy of the National Aeronautics and Space Administration.)

pheric pressure on Venus is 90 times greater than it is on earth. The light yellow coloration of the planet as seen from the earth is the result of droplets of sulfuric acid in the clouds that shroud Venus.

Venus rotates through one complete turn each 243 days. Its direction of rotation is retrograde with respect to the stars, that is, opposite to the paths of revolution of the planets around the sun. Its period of revolution around the sun is 225 days. From its size and density, planetary scientists infer an interior not unlike that of the earth. Most of what we know about the surface of Venus is derived from radio telescopes and radar that attempt to "see" through the cloud cover. Such devices have detected both rough and smooth terrain, impact craters, volcanoes, and lava flows. A quick closer look at the surface of Venus was provided in 1982, when the Russian spacecraft Venera 13 and Venera 14 actually succeeded in landing on the planet. In spite of the crushing atmospheric pressure and intense heat (sufficient to melt tin), the spacecraft survived for over an hour. Venera 13 transmitted a view of a rather level field of scattered, flattish rocks on a solid surface. The photographs from Venera 14 showed rounded, apparently eroded boulders. Robot chemical analysis on the spacecraft provided data that suggested the boulders might have a basaltic composition.

Additional information about Venus was provided early in 1979 by the American Pioneer Venus orbiter spacecraft. Pioneer Venus confirmed many of the findings of the Venera 9 and 10 missions and provided substantiating data for the intense electrical activity that characterizes the atmospheric envelope of Venus. The geologists monitoring the data transmitted by Pioneer Venus were particularly interested in images of the planet's solid surface, as revealed by the spacecraft's ground-scanning radar. A huge canyon was discovered, which extended for a distance of 1400 km and which may be a great tectonic crack in the planet's surface. In places, the canyon is 280 km wide and over 4.6 km deep. The spacecraft's radar imagery also revealed huge impact craters, some of which measured 700 km in diameter. A magnificent range of mountains rising 10 km above the planet's average surface and broad, level plateaus were added to the inventory of known topographic features on Venus.

Unlike the earth, Venus has no oceans. This is probably because it is too close to the sun for water vapor to condense and form oceans. Without oceans, in which CO_2 can be combined with calcium and magnesium to form carbonate rocks, there is no apparent mechanism to take and remove CO_2 from the atmosphere. The gas would persist, locked to its planet by gravity, and contribute through the greenhouse effect to the oppressive heat that characterized Venus. It is a sobering thought that if our planet had formed about 10 million km closer to the sun, it would be a similarly hostile place.

Mars

General Characteristics

The planet Mars has always been of special interest to humans because of speculation that some form of life, however humble and microscopic, may now exist or at one time have existed there. It is a planet that is only a little farther from the sun than is our own; it has an atmosphere that includes clouds; it has developed white polar caps; and it has seasons and a richly varied landscape. That landscape (Fig. 6–28) has been splendidly revealed to us by the Mariner and Viking missions. It includes magnificent craters, colossal volcanic peaks, deep gorges, sinuous channels, and an extensive fracture system. Evidence of wind erosion is clearly seen on photographs of the Martian surface, and there are indications of the work of ice as well. The planet has a diameter of only about 50

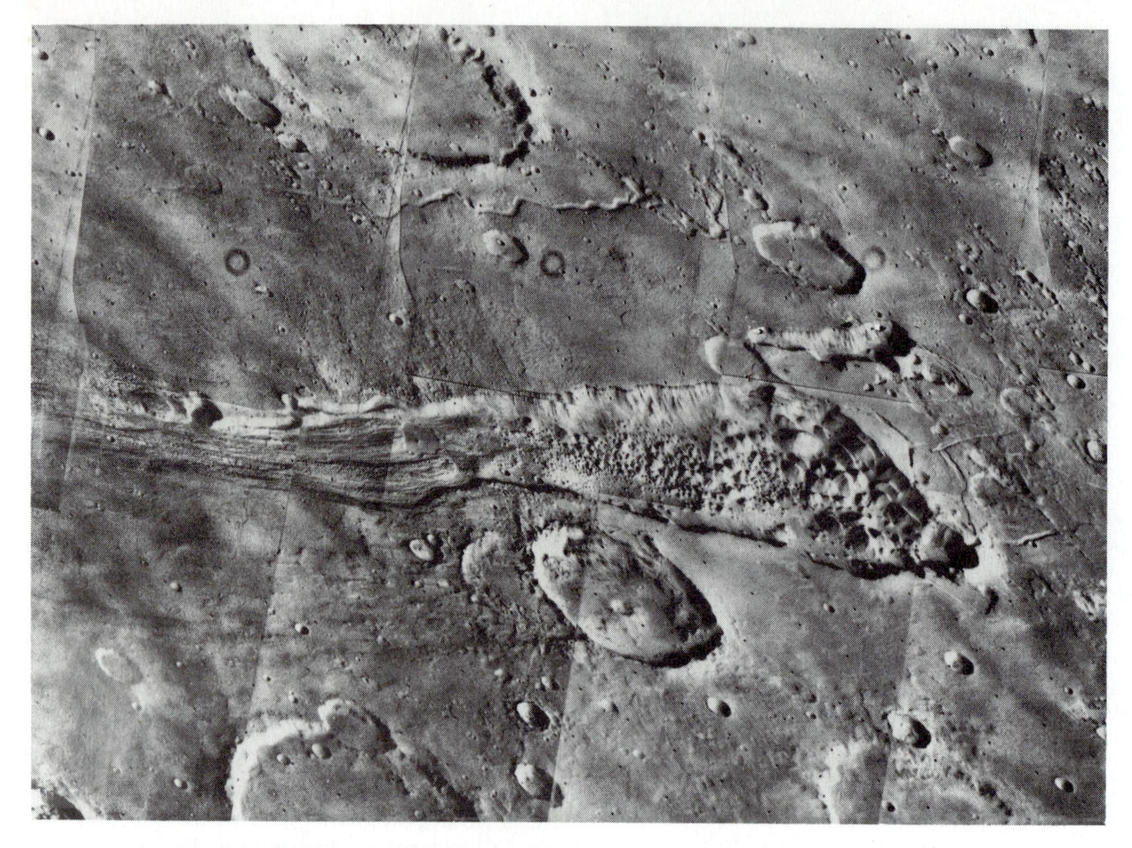

Figure 6–28 The Martian surface as revealed in a mosaic of photographs taken by Viking 1 on July 3, 1976. The large valley in the center was probably caused by downfaulting, possibly in association with the melting of subsurface ice. The hummocky topography on the valley floor may have also resulted from this process. (Photograph courtesy of National Aeronautics and Space Administration.)

per cent that of the earth and has only 10 per cent the mass of the earth.

A year on Mars is 687 earth days long, and its axial tilt is similar to that of the earth. Carbon dioxide is the principal gas in the Martian atmosphere, although small amounts of nitrogen, oxygen, and carbon monoxide are also present. The atmosphere is nearly 150 times thinner than the earth's atmosphere. The planet has virtually no greenhouse effect and temperatures vary at the equator from about +21 to −85°C.

Mars has two moons. Their names are *Phobos* (meaning fear) and *Deimos* (meaning panic). They are large chunks of rock about 27 and 15 km across, respectively.

The somewhat lower density of Mars, compared with other inner planets, and its lack of a detectable magnetic field suggest that its iron may be largely scattered throughout the planet. This interpretation is strengthened by the color photographs transmitted from the Viking Landers. Those photographs show the rusty red and orange hues typically developed on earth by limonite and hematite. Indeed, the sky often takes on a pinkish hue, probably because of the rust-colored dust. Apparently, iron at the Martian surface has reacted with water vapor and oxygen to provide the color that so intrigued early planetary observers.

Martian Landscape Features

As a result of the work accomplished by the Mariner and Viking Orbiters, nearly the entire surface of Mars had been photographed by 1976. Crisp, high-resolution photographs revealed craters, dunes, volcanoes, and canyons in unsurpassed clarity. On a grand scale, the planet's physiography is divisible into two rather different halves. The southern hemisphere is mostly heavily cratered rough terrain. This surface is inferred to be older than the smoother, only lightly cratered surface of the northern hemisphere. The boundary between the two halves occurs along a great circle inclined about 30° from the equator.

The Viking spacecraft that reached Mars in the

Figure 6–29 Photograph of Martian dune field remarkably similar to dunes seen in deserts on earth. Photograph taken by camera on *Viking Lander* on August 3, 1976. The sharp dune crests indicate that the most recent wind storms transported sediment particles from upper left to lower right. Large boulder at left is about 3 meters long. (Photograph courtesy of the National Aeronautics and Space Administration.)

summer of 1976 each contained two parts, namely, a lander and an orbiter. The landers, of course, provided scientists with the first ground view of Mars. The photographs transmitted from the landers to the orbiters and thence to the earth show a sandy-looking terrain littered with cobbles and boulders (Fig. 6–29). Here and there, undisturbed bedrock was visible. Many of the rocks strewn about were clearly of volcanic origin. Some had vesicular texture, whereas others were finely crystalline, as in an aphanitic igneous rock. The devices on the lander that analyzed the soil indicated that the fine, loose material was rich in iron, magnesium, and calcium. Very likely, it was derived from parent rocks rich in ferromagnesian minerals.

Craters and Dunes

The density of craters in the southern half of Mars is similar to the crater density in the lunar highlands. Most of these craters were probably excavated between 4 and 4.5 billion years ago by the same torrential barrage of meteorites that pelted the lunar highlands.

Photographs clearly show row upon row of dunes on the floors of some of the larger craters. If the thinness of the Martian atmosphere is taken into account, then the size and spacing of these dunes can be used to estimate the velocity of the winds that formed them. Calculations indicate that the dune-forming winds blew at speeds of up to 100 meters per second. Dunes and other evidence of the work of wind are especially common near the poles. The incorporation of dust in the polar ice itself is inferred from the stratified appearance, interpreted to be alternate layers of dust and ice (Fig. 6–30). Dust storms on Mars are prodigious events. As Mariner 9 began its encirclement of the planet, a dust storm was in progress, and useful photographs of the Martian surface could not be transmitted until the dust subsided.

Volcanic Features

Although most of the craters on Mars have been formed by the impact of meteorites, a fewer number are certainly volcanic. Easily the most spectacular of the volcanic craters is *Olympus Mons* (see Fig. 6–31). This gigantic feature is more than 500 km across at its base and rises 30 km above the surrounding terrain. It is equal in volume to the total mass of all the lava extruded in the entire Hawaiian Island chain. Around its summit are lava flows and the narrow rilles interpreted to be old lava channels or possibly collapsed lava tubes. Flood lavas are also evident, not only in the smoother northern hemisphere but also among the craters of the rugged southern half of Mars.

Channels and Canyons

There is so very little water today in the Martian

Figure 6–30 The frosty Martian North Pole. In this view one can see the perennial ice cap that overlies stratified deposits thought to be alternating layers of dust and ice. (Photograph courtesy of the National Aeronautics and Space Administration and Michael Botts.)

atmosphere that if all of it were to be condensed in one place, it would probably provide only enough water to fill a community swimming pool. Yet the surface of the planet is dissected with channels and canyons that grow wider and deeper in the downslope direction. Some of these channels have braided patterns, sinuous form, and tributary branches such as are characteristically developed by streams on earth (Fig. 6–32). These features suggest that flowing water was an important geologic agent at some earlier time in Martian history. There may once have been a considerable reserve of water frozen in the subsurface as a kind of permafrost. Permafrost is ground that remains below the freezing temperature and contains ice throughout the year. Many investigators theorize that the larger channels, some of which are over 5 km wide and 500 km long, were excavated by water gushing out onto the surface as the subsurface layer melted. Large tongues of debris closely resembling mudflows may also have received their water from such a process. Many channels on Mars dwarf our own Grand Canyon in size and, in order to form, would have required torrential floods so spectacular as to be hard to visualize by earth standards.

Ice Caps

Long before spacecraft landed on Mars, telescopic observations revealed the presence of polar ice caps. The ice caps on Mars, however, differ from the earth's ice caps in that the Martian polar caps are solid carbon dioxide (like commercial dry ice). Carbon dioxide precipitates out of the Martian atmosphere during the cooler seasons and partially evaporates during warmer times. The result is alternate growth and shrinkage of the ice caps. Melting is never complete, and a small area of frozen material remains. Some investigators believe this residual cap may consist partly of water ice.

Martian History

Like the earth, Mars was transformed from a protoplanet by accretion of a multitude of smaller objects. This accretion was followed by an episode of differentiation during which the once homogeneous body became partitioned into masses of dif-

Figure 6–31 The giant Martian volcano, Olympus Mons. This enormous volcano has a diameter of 600 km (375 mi), and the rim rises 24 km (15 mi) above the surrounding terrain. (Photograph courtesy of R. Arvidson, Department of Earth and Planetary Sciences, Washington University.)

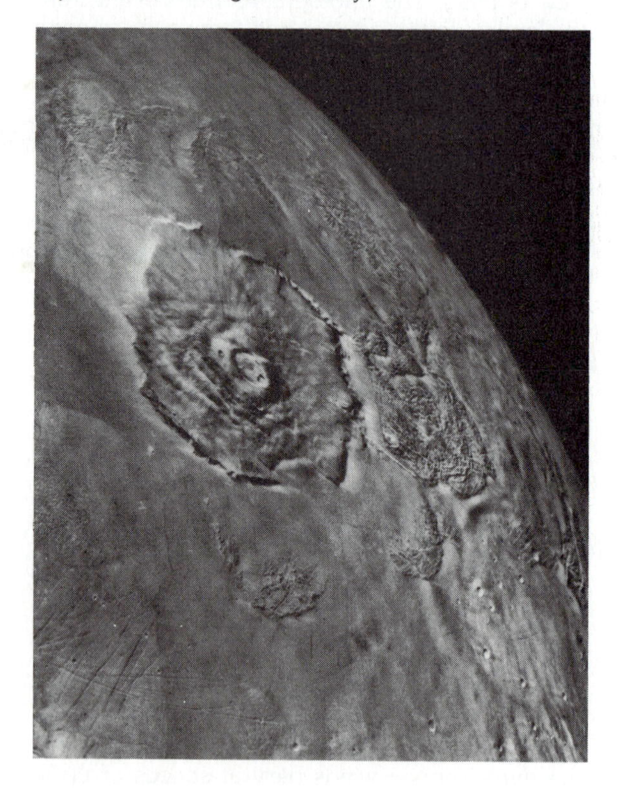

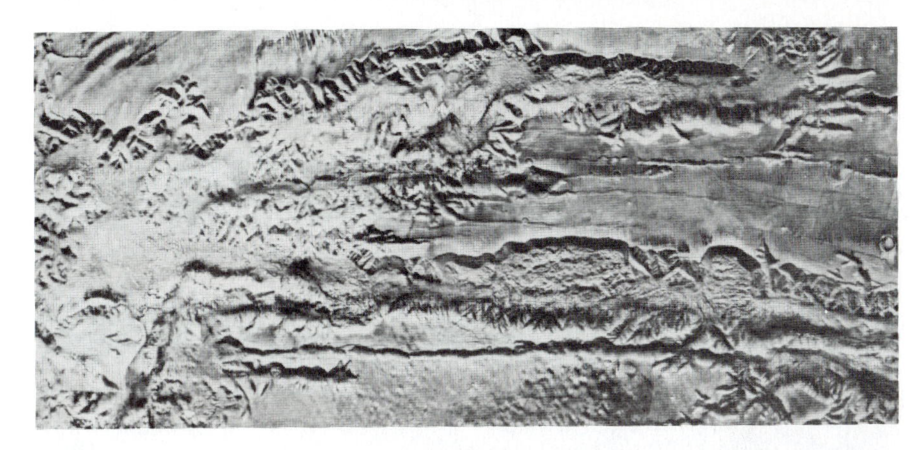

Figure 6–32 The immense Martian canyon system named Valles Marineris, as photographed by Viking orbiters. It has been suggested that these canyons are in part the result of slumping and fluvial erosion as ground ice melted and gushed onto the surface. (Photograph courtesy of the National Aeronautics and Space Administration.)

ferent chemical composition and physical properties. Very likely the differentiation process involved partial or complete melting so that fluids could move from one region of the planet to another.

During the initial billion years or so of Martian history, the planet experienced heavy bombardment of meteorites and asteroids, with the result that the exposed crust became densely cratered. The craters have been modified by relatively recent wind erosion. The crust was disturbed not only by cratering but also by the faulting and fracturing that accompanied volume changes in the mantle.

It has been postulated that a dense atmosphere formed on Mars as a result of volcanic outgassing (removal of embedded gas by heating) and that the production of water and carbon dioxide far exceeded its loss into space. The planet was warmer at that time, perhaps because of a greenhouse effect. For a time, running water was an important geologic agent, as evidenced by the sinuous furrows incised throughout the cratered terrain. In certain regions of the planet, a small decline in surface temperatures may have caused the water contained in surface materials to freeze, thus trapping still liquid subsurface water in pockets and channels. With local melting, the subsurface water may have gushed upward and eroded large channels while simultaneously causing entire areas of surficial material to collapse in mudflows.

Eventually, the supply of atmospheric water and carbon dioxide became so depleted that the planet began to cool. Remaining carbon dioxide and water migrated to polar caps and to subsurface reservoirs, where it has remained to the present. Processes other than water erosion continued to modify the planet, however. Winds on Mars of 110 meters per second in velocity are not unusual. They are responsible for fields of Martian sand dunes and erosional stripping of Martian soils.

Because Mars was once warmer, had more water, and possessed the elemental raw materials thought to be required for the development of organisms, scientists have long speculated on the possibility of finding primitive forms of life on the planet. For this reason, a series of experiments aboard the Viking Lander was designed to search for signs of present or former life. The results were disappointing. Not a trace of organic material could be detected in the Martian soil. However, it is still too early to rule out the possibility that continuing research in the decades ahead may provide evidence of organisms.

THE OUTER PLANETS

Beyond the orbit of Mars lies the ring of asteroids, and beyond the asteroids are the orbits of Jupiter, Saturn, Uranus, Neptune, and Pluto. The first four of these planets are also called the **Jovian** (for Jupiter) **planets.** They are similar to one another in having large size and relatively low densities. Pluto is not a Jovian planet. It is so far away and small that it is difficult to observe, but it appears to be a stony mass that was once a satellite of another planet.

Jupiter

Jupiter, named for the leader of the ancient Roman gods, is the largest planet in our solar system. The diameter of this huge planet is 11 times

Figure 6–33 Jupiter. Voyager 1 took this photograph of Jupiter in 1979 at a range of 32.7 million km.

greater than the earth's diameter, and its volume exceeds that of all the other planets combined. Jupiter spins rapidly on its axis, making one full rotation in slightly less than 10 earth hours. This rapid rotation results in the formation of the encircling colored atmospheric bands for which the planet is famous (Fig. 6–33). Hydrogen, helium, and lesser quantities of ammonia and methane are the predominant gases in Jupiter's atmosphere. On the basis of experiments with these gases, some investigators believe that the reddish and orange colors in the atmosphere are derived from methane-based compounds. It is impossible to discern the surface of the solid interior of Jupiter with optical telescopes. The atmosphere, which is several hundred kilometers thick, passes gradually to liquid and eventually solid matter. The solid interior of the giant planet may consist of highly compressed hydrogen, possibly surrounding a rocky core. Jupiter has at least 13 satellites. The large planet with its many moons rather resembles a miniature solar system.

We have learned about Jupiter not only from earth-based telescopes but from close encounters by unmanned spacecraft. The most successful of the robot explorations occurred in 1979 when the two Voyager spacecraft swept past Jupiter and transmitted magnificently clear images of the planet's brightly colored atmosphere, the tempestuous Great Red Spot (Fig. 6–34), five of Jupiter's moons, and a ring of debris less than 0.6 km thick that circles the planet about 55,000 km above the tops of the clouds. The discovery of the ring on Jupiter makes that planet the third in the solar system known to possess such a feature (Saturn and Uranus are the other planets known to have rings).

Jupiter's Great Red Spot, which is twice the size of the earth, is an intriguing feature known to astronomers for centuries. As clearly revealed by the Voyager mission, this many-hued disturbance is a gargantuan atmospheric storm that extends deep into the cloud cover and turns in a complete counterclockwise revolution every six earth days.

In addition to the Great Red Spot, the com-

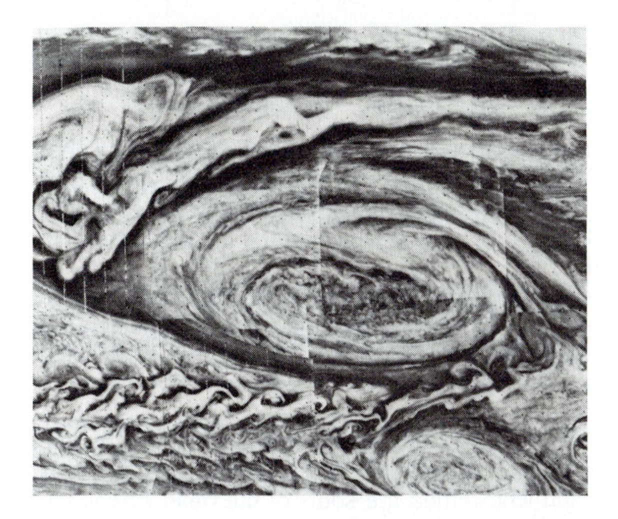

Figure 6–34 Photomosaic of Jupiter's Great Red Spot, assembled from pictures taken by Voyager 1 in 1979 at a distance of 1.8 million km. The smallest clouds visible in this mosaic are 33 km across. (Photograph courtesy of the National Aeronautics and Space Administration.)

puter-controlled devices aboard the Voyager spacecraft radioed back the distinct images of great bolts of lightning, whose occurrence had only been suspected before the flyby. Another discovery was an auroral display far brighter than any northern lights ever seen on earth. Voyager also confirmed the presence of an immense magnetic field surrounding Jupiter. First detected by the Pioneer 10 spacecraft in 1973, the magnetic field extends outward from the planet for 16 million km, encompasses all of the larger satellites, and traps enormous quantities of charged particles from the solar wind. The presence of the field indicates that Jupiter may have a magnetic core, possibly resulting from fluid movements within its spinning interior.

Saturn

Galileo, with the aid of his primitive telescope, was the first person to observe the distinctive ring around Saturn (Fig. 6–35). Actually, there are three concentric rings, with the brightest and broadest in the center. The rings are composed of billions of orbiting particles of dust and ash, as well as fewer large fragments. All revolve around Saturn in the approximate plane of the planet's equator. Saturn's 10 moons also revolve in this plane. Although the area of the rings is enormous, most of the material is concentrated only several meters thick. Sunlight reflected from the myriads of particles gives the illusion that the ring is solid. Scientists are uncertain about the precise origin of Saturn's rings. One group of astronomers suggest that the particles in the rings are the remains of an exploded satellite. Other argue the rings are a remnant of the particle cloud from which satellites form.

If Saturn were dropped into an ocean big enough to hold it, it would float. Its density is only 70 per cent that of water. Saturn's mass, however, is equivalent to 95 earth masses. Measurements of Saturn's size and density suggest that the planet may have a central core of heavy elements (probably mostly iron), surrounded by an outer core of hot compressed volatiles such as methane, ammonia, and water. These two core regions, however, constitute only a small part of the total vol-

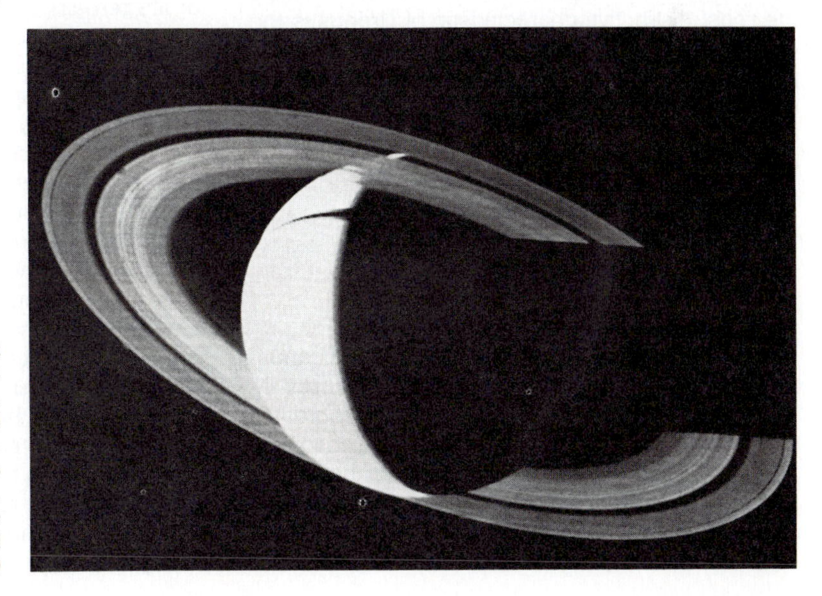

Figure 6–35 Voyager 1 looked back at Saturn on November 16, 1980, four days after the spacecraft flew past the planet, to observe the appearance of Saturn and its rings from this unique perspective. Voyager 1 took this image when it was 5.3 million kilometers from the planet. Saturn's shadow falls across part of the rings. (Photograph courtesy of National Aeronautics and Space Administration.)

ume of the planet, in which hydrogen and helium are overwhelmingly the predominant elements.

As was true for Jupiter, our knowledge of Saturn has been greatly improved as a result of data collected by unmanned spacecraft. In particular, the trajectory of the Pioneer 11 spacecraft was programmed so that it would sweep by Saturn in September of 1979. Pioneer 11's journey from the earth to Saturn took 6 years, but as it came close to the great planet, it provided us with our most impressive views of Saturn, its rings, and several of its moons. The data transmitted back to the earth helped planetary scientists confirm that Saturn has a magnetic field, trapped radiation belts, and an internal source of heat.

Spacecraft exploration of Saturn has not ended. Voyager 2, having departed from the vicinity of Jupiter in 1979, is continuing its journey through the solar system. Its trajectory will carry it past Saturn and thence on to distant Uranus. Voyager 2 is scheduled to approach Uranus, a planet 1.6 billion miles from the earth, during January of 1986.

Uranus

Beyond the orbit of Saturn lies the planet Uranus. Like Saturn, Uranus also has rings, but they are too faint to be observed telescopically. Uranus has a low density (1.6 gm per cubic centimeter) and a frigid surface temperature of about −185°C. Five moons circle Uranus. Thus far, earth-based instruments have only been able to detect hydrogen and methane on Uranus, although helium may also be present. The methane may be responsible for the greenish appearance of Uranus. Perhaps the most distinctive characteristic of Uranus is the orientation of its axis of rotation, which lies in the plane of orbit. The planet's direction of rotation is in the reverse direction to that of all other planets except Venus. This retrograde direction is also characteristic of the orbits of the five satellites around Uranus.

Neptune

Through the telescope, Neptune, like Uranus, appears as a greenish orb. Neptune's density of 2.21 gm per cubic centimeter is just slightly greater than the density of Uranus. Neptune's atmosphere may also resemble that of Uranus. Both hydrogen and helium have been detected spectroscopically, and probably methane is also present. Two moons circle the planet.

Pluto

Far out on the outer limits of our solar system lies the orbit of Pluto. It is so small and so distant that less is known about it than the other planets. We do observe that Pluto's orbit is tilted an unusual 17° to the ecliptic plane and that its orbit crosses that of Neptune. Because of the high inclination of Pluto's orbit, however, there is no danger of a collision with Neptune. Its small size and orbital peculiarities suggest that Pluto is not a true planet, but rather a former satellite of Neptune that escaped the gravitational pull of that planet and now occupies a separate orbit. In 1978, scientists of the U.S. Naval Observatory found that Pluto has a moon. They named it Charon, after the mythologic boatman who ferried the souls of the dead across the river Styx to Hades.

DOOMSDAY

Anyone interested in the history of the earth is naturally led to a consideration of its future. Clearly, all that has happened, is happening, and will happen to this planet is inevitably linked to the history and destiny of the sun. The sun is a star, and like other stars, it will run its course from birth to death. By observing other stars that are at various stages in their histories, astronomers have pieced together a story of what happens when these bodies begin to deplete their store of nuclear fuel. Our star has been "burning" for about 5 billion years. In this span of time, it has consumed about half of its nuclear fuel and so has about another 5 billion years of life remaining.

What will "doomsday" for the sun and earth be like? Most astronomers believe that once the sun's original supply of hydrogen has been largely converted to helium by fusion, the deep core will begin to contract. In the process of contracting, it will generate great amounts of heat that will in turn trigger thermonuclear reactions capable of producing incredibly high temperatures. The seething nuclear inferno will then cause the sun to puff up into an enormous sphere and to turn blood-red. Our sun will have become a "red giant" (Fig. 6–36). As the great ball of rarefied gases continues to expand, it will engulf and vaporize Mercury and Venus. What will happen to the earth can be readily imagined. The searing heat will boil away the oceans and heat the crust even to the melting point of some surface rocks. This purgatorial state of affairs may continue for a billion years or so, until finally the sun has dissipated its energy. The dying sun may then begin to shrink into the form typified by stars known as "white dwarfs." Such dwarf stars have exhausted their supply of nuclear fuels and remain incandescent only because of their residual internal heat. Eventually, even this light will be extinguished, and the sun will collapse inward under its own gravity. It will become a small, cold, inert planet. The scorched and frazzled earth will continue as a barren, frigid sphere unable to free itself from the gravitational grasp of the dead sun.

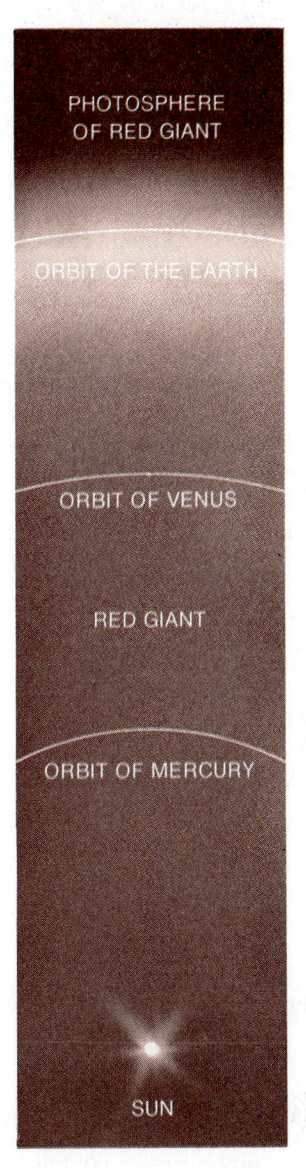

Figure 6–36 A red giant swells to a size that would incorporate everything within the diameter of the earth's orbit. (From J. M. Pasachoff. *Contemporary Astronomy*. 2nd ed. Philadelphia, Saunders College Publishing, 1981.)

Summary

The planet earth is one of a group of nine planets revolving around a modest-sized star, our sun. Planets and sun constitute our solar system, which, together with myriads of other stars, planets, dust, and gas, make up the Milky Way Galaxy. Our sun provides energy to the earth, which drives geologic processes and which has ultimately determined the character of terrestrial life. The sun is composed mostly of hydrogen and helium. The helium results from the fusion of hydrogen nuclei with the conversion of a part of the hydrogen into solar energy.

The study of light from stars has revealed that spectral lines shift toward the red end of the spectrum, indicating that currently the universe is expanding. This knowledge has been incorporated into three hypotheses for the origin of the universe. The "big bang" concept suggests that the universe began when a gigantic dense mass of material exploded, blasting particles outward in all directions. Proponents of the "steady state" hypothesis postulate that matter is continually being created in the empty realms of space at a rate equal to the depletion of matter as it expands outward. As suggested by its name, the "oscillating universe" concept is based on an alternation of a "big bang," which propels matter outward, followed by a reversal in which the exploded materials return and subsequently explode again.

Theories for the origin of our solar system must conform to the known facts of the system's distribution of mass and angular momentum, as well as the known orbital and rotational characteristics of the planets. Although for a period in the early part of the present century, theories based on encounters of stars were popular, more recent ideas for solar system origin postulate condensation and accretion of planets from a nebula of gases and dust particles. Currently, the most popular of such concepts is the "protoplanet hypothesis." It suggests that the earth was formed by the cold accretion of particles collected within a turbulent eddy located on a disk-shaped rotating cloud of dust and gas.

Meteorites are important to planetary geologists because they are truly extraterrestrial samples of the solar system. They can be classified as ordinary chondrites, carbonaceous chondrites, achondrites, irons, and stony-irons and are believed to be fragments resulting mostly from collisions of one or more small planetary bodies that were once present in the asteroid belt.

The moon, which has been studied by telescope for centuries, was first visited by humans in July, 1969. Its general features include the relatively smooth, dark mare basins and the rugged lunar highlands. Samples of the loose regolith on the moon have been brought back to earth. They include basaltic rocks from the maria and specimens of anorthosite from the highlands. Study of the age and distribution of these rocks suggests that lunar history began about 4.6 billion years ago, when the moon had developed its surficial skin of anorthositic igneous rock. This surface was then flooded in various regions by the basalts of the maria about 1 billion years later.

Mars is a planet of particular interest because of the possibility that life may have once existed there. It has a cratered topography reminiscent of the moon and at the same time has features that appear to have been formed by processes similar to those operating on earth. Dunes, stratified sediments, and sinuous rills suggest erosion and deposition by wind, and, perhaps at some earlier time, by water and ice.

The earth's environment and history is inexorably tied to the history of the sun. Our sun will continue to shine with a hot yellow light for about another 5 billion years as its hydrogen is slowly converted to helium. When the hydrogen fuel nears exhaustion and helium becomes the principal solar material, further nuclear reactions will release still more energy. The sun will then become a red giant, and its heat will sear the earth's surface. After this phase of its history, the sun will gradually fade and die.

Questions for Review

1. What are the names of the planets in our solar system? In what galaxy is our solar system located?
2. What is the source of the sun's heat? Given the amount of solar radiation intercepted by the earth, why is the earth's surface not much hotter than it is? What will happen to the sun when it has expended its supply of nuclear fuel?
3. What evidence indicates that the universe is expanding? What theories for the origin of the universe include the concept of an expanding universe?
4. What regularities or uniformities exist in the solar system that indicate that it is truly a coherent system?
5. In general, how do the terrestrial planets differ from the Jovian planets?
6. What is the relationship between the tilt of a planet's axis of rotation and a planet's seasons? Which planets have an unusual axial angle of inclination?
7. What was the principal objection to the original nebular hypothesis for the origin of the solar system? How is this problem apparently resolved in the protoplanet hypothesis?
8. What are the principal kinds of meteorites that have been found on the earth? In what way are they thought to have originated? What are chondrites?
9. What evidence indicates that, unlike the earth, the moon does not have an iron-nickel core?
10. How do the rocks of the lunar maria and lunar highlands differ in composition and age?
11. What is the derivation of the reddish coloration of the planet Mars?
12. How does Mars differ from the earth in density, internal structure, and average range of equatorial temperatures?
13. Why does one side of the moon have more craters than the other? In what two ways have lunar craters been developed?
14. Without resorting to radiometric dating, how do lunar geologists determine the relative ages of craters on the moon?
15. What is lunar regolith? How is it thought to have originated?

Terms to Remember

achondrite
anorthosite
big bang theory
Bode's Law
chondrite
chondrule
differentiation
Doppler shift
fusion
galaxy
iron meteorites
Jovian planet
linear rilles
lunar highland
maria
mascon
meteor
meteorite
meteoroid
Milky Way Galaxy
nebular hypothesis
norite
photosphere
planetesimal hypothesis
protoplanet
red shift
regolith
sinuous rilles
solar wind
steady state theory
stony meteorite
stony-iron meteorite
terrae
terrestrial planet
turbulence
wrinkle ridges

Supplemental Readings and References

Abell, G.O. 1975. *Exploration of the Universe* (3d ed.). New York, Holt, Rinehart and Winston.

Arvidson, R.E., Binder, A.B., Jones, K.L. 1978. The surface of mars. *Sci. Am.* 238:3, 76–91.

Beatty, J.K., O'Leary, B., and Chaikin, A. (eds.). 1981. *The New Solar System.* Cambridge, Mass., Sky Publishing Corp.; London, Cambridge University Press.

French, B.M. 1977. *The Moon Book.* Middlesex, England, Penguin Books, Ltd.

Gamow, G. 1964. *A Star Called the Sun.* New York, Viking Press, Inc.

Hartmann, W.K. 1973. *Moons and Planets: An Introduction to Planetary Science.* Belmont, Calif., Wadsworth Publishing Co., Inc.

King, E.A. 1976. *Space Geology: An Introduction.* New York, John Wiley & Sons, Inc.

Mutch, T.A., Arvidson, R.E., Head, J.W., Jones, K.L., and Saunders, R.S. 1976. *The Geology of Mars.* Princeton, N.J., Princeton University Press.

Wood, J.A. 1979. *The Solar System.* Englewood Cliffs, N.J., Prentice-Hall, Inc.

York, D. 1975. *Planet Earth.* New York, McGraw-Hill Book Co.

The meeting of atmosphere, lithosphere, and hydrosphere. Here at Sea Arch Overlook in Volcanoes National Park, Hawaii, waves attack recently solidified lava.

Development of the Earth's Major Features

The world is the geologist's great puzzle box.
Louis Agassiz, 1894

THE EARTH TAKES FORM

As described in the preceding chapter, the earth began to take form when particles orbiting the sun aggregated to form a body sufficiently large to be termed a planet. The planet earth continued to grow over a span of 50 to 90 million years as a result of the ongoing shower of particles and meteorites that impacted upon its surface. Eventually, however, most of the material in or near the earth's orbital path had accreted onto the planetary surface, and the earth's active growth stage ended. At this time in its early history, the earth had neither a dense metallic core surrounded by a heavy mantle of silicates nor a thin crust of relatively low-density rocks. These features developed in the next stage of the earth's early history as a result of differentiation. **Differentiation** refers to the many processes by which a planet becomes internally zoned as heavy materials move toward its center and light materials accumulate near the surface. Differentiation produced the core, mantle, and crust.

THE NATURE OF THE DEEP INTERIOR

Much of what we know about the earth's deep interior has been indirectly determined from the study of earthquake waves (seismic waves). Such waves are generated by sudden ruptures or failures in rocks of the solid earth that have been subjected to stress. Actual breakage of coherent rock masses need not always occur. Periodic slippage along fracture zones may also produce earthquakes and the seismic waves that accompany them.

Messages from Inner Space

The shock generated by an earthquake moves outward in waves that travel at different speeds through materials differing in physical properties. These waves can be designated **primary, secondary,** and **surface waves.** Primary waves, or P-waves, are the speediest of the three kinds of earthquake waves and therefore are the first to arrive at a recording station after there has been a distant earthquake (Fig. 7–1). They travel through the earth at velocities of 5 to 15 km per second. In primary waves, pulses of energy are transmitted in such a way that the movement of rock particles is parallel to the direction of transmission of the waves. Thus, a given particle of rock set in motion by the shock of an earthquake is driven into its neighbor and bounces back. The neighbor strikes the next particle and rebounds, and subsequent particles continue the motion as a series of compressions and expansions that speed away from the source of shock (Fig. 7–2). Thus, P-waves are similar to sound waves in that they are longitudinal waves that travel by compression and rarefaction of the earth, somewhat like the coiled spring of the toy "Slinky."

When earthquake waves pass through rocks that differ in such properties as density, rigidity, or compressibility, they are bent or refracted. For example, P-waves are successively refracted as they traverse from less dense to more dense zones of the interior. The result is a pronounced curvature of P-wave paths (Fig. 7–3).

Secondary waves, which are also termed S-waves or transverse waves, are able to travel at velocities of about 4 to 7 km per second. The movement of rock particles in secondary waves is at right angles to the direction of propagation of the energy (Fig. 7–4). A demonstration of this type of wave is easily managed by tying a length of rope to a hook and then shaking the opposite end. A series of undulations will develop in the rope and move toward the hook—that is, in the direction of propagation. Yet any given "particle" or point along the rope will move up and down in a direction perpendicular to the path of propagation. Both P- and S-waves are sometimes termed **body waves** because of their ability to penetrate into the depths, or body, of the earth's interior.

Surface waves are seismic waves that are propagated only through the earth's outermost layers and are guided in their propagation by the earth's surface. There are actually an assortment of different types of surface waves having various complicated motions. The Rayleigh surface waves, for example, have an elliptical particle motion that is opposite in direction to that of propagation, whereas "Love waves"' vibrate horizontally and perpendicularly to wave propagation (Fig. 7–5). Surface waves are the last to arrive at the seismograph station. They are the cause of the enormous destruction that can result from severe earthquakes near densely populated areas. This destruction results because they are channeled through the thin, outer region of the earth and their energy is less rapidly dissipated into the large volumes of rock traversed by body waves. Indeed, surface waves may encircle the earth several times before friction causes them to fade.

The Divisions of Inner Space

Our knowledge of the earth's interior is derived largely from investigations of primary and secondary waves. In addition to the observation that the velocities of these waves increase regularly with depth, it has also been discovered that there is a sharp jump, or discontinuity, in wave velocities at two particular levels. One of these is called the **Mohorovičić discontinuity** (or, simply, "Moho"), after its discoverer. This discontinuity lies at about 30 to 40 km below the surface of continents and at lesser depths beneath the ocean floors. Another major discontinuity occurs nearly halfway to the center of the earth at a depth of 2900 km. This is called the **Gutenberg discontinuity** in honor of the geophysicist whose careful work defined its location. These discontinuities make it possible to think of the earth as divided into the **crust,** which is the thin surface layer extending downward to the Mohorovičić discontinuity; the **mantle,** which extends from the base of the crust to the Gutenberg discontinuity; and the **core,** which extends from 2900 km to the very center of the earth (Fig. 7–6).

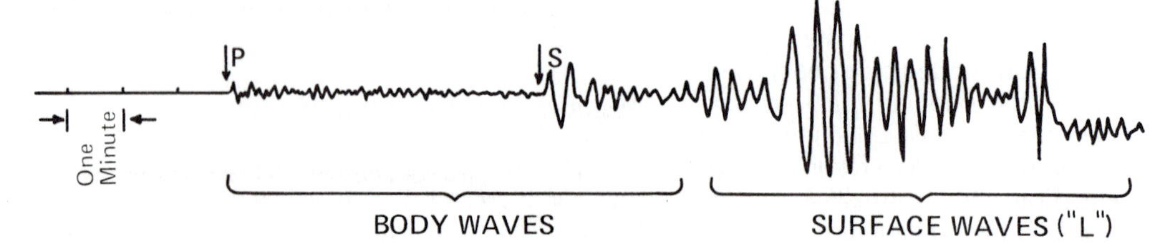

Figure 7–1 A typical seismogram. Time runs from left to right. P, S, and L waves are indicated. (From J. S. Watkins, M. L. Bottino, and M. Morisawa. *Our Geological Environment.* Philadelphia, Saunders College Publishing, 1975, p. 30.)

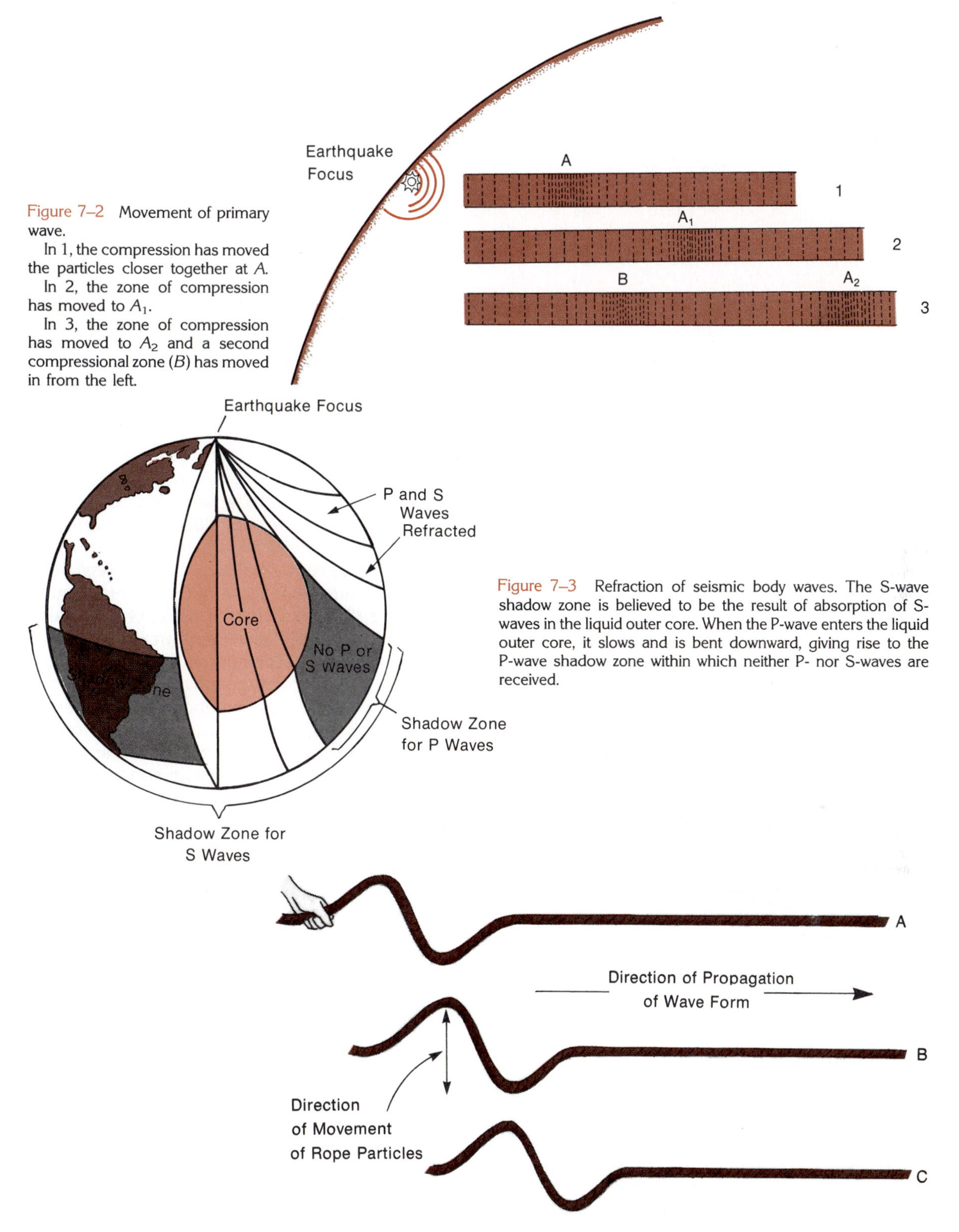

Figure 7–2 Movement of primary wave.
In 1, the compression has moved the particles closer together at *A*.
In 2, the zone of compression has moved to A_1.
In 3, the zone of compression has moved to A_2 and a second compressional zone (*B*) has moved in from the left.

Figure 7–3 Refraction of seismic body waves. The S-wave shadow zone is believed to be the result of absorption of S-waves in the liquid outer core. When the P-wave enters the liquid outer core, it slows and is bent downward, giving rise to the P-wave shadow zone within which neither P- nor S-waves are received.

Figure 7–4 Analogy of propagation of S-waves by displacement of a rope. In rocks, as in this rope, particle movement is at right angles to the direction of propagation of the wave. A, B, and C show the displacement of the crest from left to right at successive increments of time.

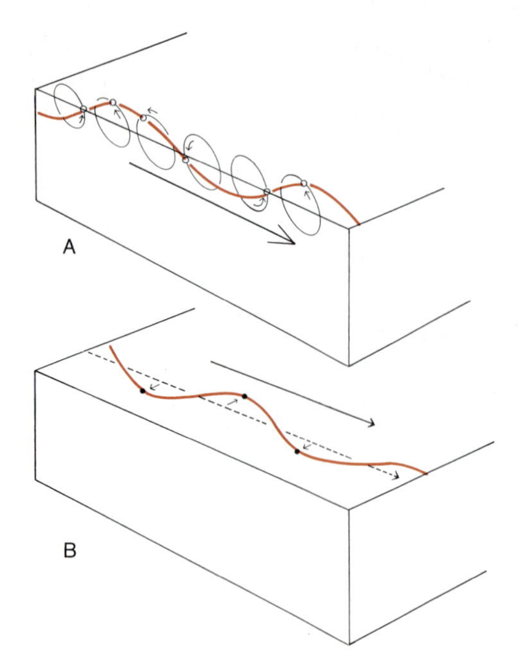

Figure 7–5 (A) Elliptical particle motion in Rayleigh surface waves (ground surface shown without displacement). (B) Particle motion in Love wave is in horizontal plane and perpendicular to direction of wave propagation.

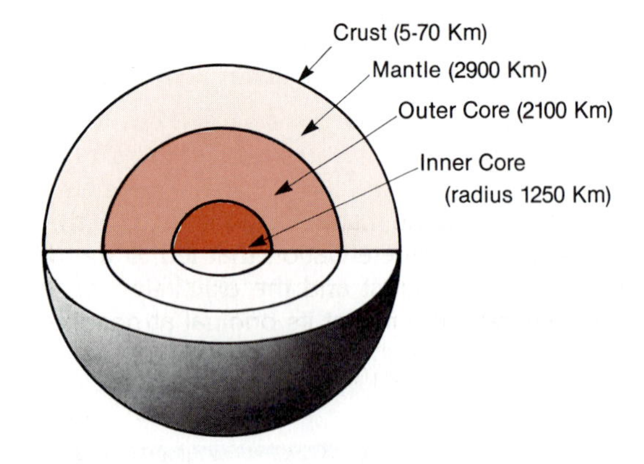

Figure 7–6 The interior of the earth. The crust appears as a line at this scale.

The Earth's Core

Travel Paths of Body Waves

Much of our present understanding of the nature of the earth's core is derived from knowledge of the behavior of earthquake waves. Secondary waves, for example, are unable to travel through fluids. Thus, if S-waves were to enter a molten region of the earth's interior, they would be absorbed there and would not be able to continue. It was the observation that S-waves generated by earthquakes do not appear on the opposite side of the earth that led to the theory that the material of the outer core is fluid. Thus, if an earthquake occurred in Tasmania, only P-waves would be detected in New York, because the core would prevent S-waves from passing through. This phenomenon gives rise to an **S-wave shadow zone**—a region on the side of the earth opposite an earthquake that receives no S-waves. Such a zone extends beyond about 105° from the source, or **focus,** of an earthquake.

Unlike S-waves, primary waves are not absorbed by fluids. They are, however, abruptly slowed and refracted when they enter a fluid medium. Therefore, as the primary waves travel toward the molten outer core of the earth, their velocity is checked and they are refracted downward. The result is a P-wave shadow zone that extends from about 105 to 143° from the focus.

The core itself may be divided into an inner and outer part, although the boundary may not be a sharp one. Evidence for subdividing the core comes from the observation that the shadow zone is not totally devoid of P-wave arrivals. Very weak P-waves are detected. Some of these shadow zone arrivals are thought to be the result of the earth's having an inner core in which P-waves are transmitted at higher speeds and deflected outward to the shadow zone. Most geophysicists believe the inner core is of the same composition as the outer core but that it exists as a solid because of the enormous pressures at the center of the planet. The boundary between the inner and outer core may be drawn at a depth of about 5120 km.

Clues to Core Composition

Seismic data not only permit division of the earth's interior into zones but also provide our most reliable evidence for the composition of those zones. Logical choices of earth materials having appropriate density and seismic characteristics are assigned to the internal parts and validated by comparison with other relevant data. Molten iron that contains nickel is the theoretic choice for the material of the core. Seismic evidence suggests that the core ranges in density from 10 gm per cubic centimeter near the outer boundary of the core to 12 gm per cubic centimeter at the center. Molten iron at the earth's surface has a density of only 7.5 gm per cubic centimeter and would not seem an ideal choice. Calculations, however, indicate that the extreme pressure in the core would so crowd the iron atoms that the density of the metal would be raised even higher than indicated by seismology. The admixture of nickel and perhaps lesser amounts of sulfur or silicon would then bring the density into conformity with the seismic information.

Support for the theory that the core is composed mostly of iron (85 per cent) with a lesser amount of nickel has come not only from seismic investigations but also from the study of the metallic meteorites mentioned in the previous chapter. Such meteorites are alloys of iron and nickel. They appear to have been formed under conditions of extremely high temperature and pressure, such as exist in some planetary interiors, and are very likely fragments from the cores of planets shattered by collisions in space.

Differentiation and Density Layering

How does the existence of a dense metallic core relate to theories on the origin of the earth? The currently favored theory would have the earth and other planets created from the atoms, molecules, and particles circulating within a turbulent cosmic dust cloud. According to this concept, the earth would have acquired most of its materials while relatively cool, and its various elements would have been mixed and dispersed from its surface to its center. How, then, did layering develop? The answer that comes immediately to mind is that after the earth had accumulated most of its matter, it became partially or entirely molten. Somewhat like a great blast furnace, much of the excess iron and nickel percolated downward to form the core. Remaining iron and other metals combined with silicon and oxygen and separated into the overlying less dense mantle. Still lighter components may have separated out of the mantle to form an uppermost crustal layer. Geologists do not insist on complete melting for this differentiation of materials into layers. Partial melting was very likely also occurring on a vast scale, as well as solid diffusion of ions mobilized by high temperatures and pressures. Indeed, it would be difficult to account for the present abundance of volatile elements on earth if the planet had melted completely, for under such circumstances these materials would be driven off. Even for partial melting and solid diffusion to have occurred, however, an enormous amount of heat would have been required.

The heat needed for differentiation did not come from a single source. Some of it was accretionary heat derived from the bombardment of particles and meteoritic materials when the planet was still actively growing. A large part of the heat generated by the great infall of impacting bodies would have been preserved because additional showers of debris would have formed an insulating layer over the scorched impact areas. Also, as more and more material accumulated at the surface of the planet, the weight of the overburden on underlying zones resulted in a temperature rise caused by gravitational compression. It has been estimated that the combined effects of accretion and compression may have given the earth an average internal temperature of about 1000°C at this time in its early history. Compression also served to reduce the size of the planet.

Another extremely important source of heat for differentiation was derived from the decay of radioactive isotopes incorporated within the materials of the planet. Much of this radioactively generated heat accumulated deep inside the earth, where the poor thermal conductivity of surrounding rocks insured that it would not be lost. There was more heat derived from radioactivity in the early days of the earth's development than today, because many of the isotopes having relatively short half-lives had not yet disintegrated.

Geophysicists estimate that about 1 billion years after the solid earth had formed, the interior had accumulated enough heat to melt iron at depths of 400 to 800 km. Myriads of heavy molten droplets of iron slowly gravitated toward the planetary center, displacing lighter materials that had been there (Fig. 7–7). This movement of iron toward the earth's center released an enormous amount of gravitational energy, which, when converted to heat, further accelerated the process of differentiation. As the mantle began to take form, it, too, experienced differentiation that led to the development of the crust and the exhalation of the gases that gave the planet its original atmosphere and hydrosphere. Thus, ultimately the planet was changed from a primordial aggregate of stony and metallic minerals to an internally zoned sphere having a dense core, a surficial layer of light crustal rocks, and an intervening mantle.

The Mantle

Materials of the Mantle

As was the case with the earth's core, our understanding of the composition and structure of the mantle is based on indirect evidence. Because the mantle's average density is about 4.5 gm per cubic centimeter, it is believed to have a stony, rather than metallic, composition. Oxygen and silicon probably predominate and are accompanied by iron and magnesium as the most abundant metallic ions. The iron-rich rock **peridotite** approximates fairly well the kind of material inferred for the

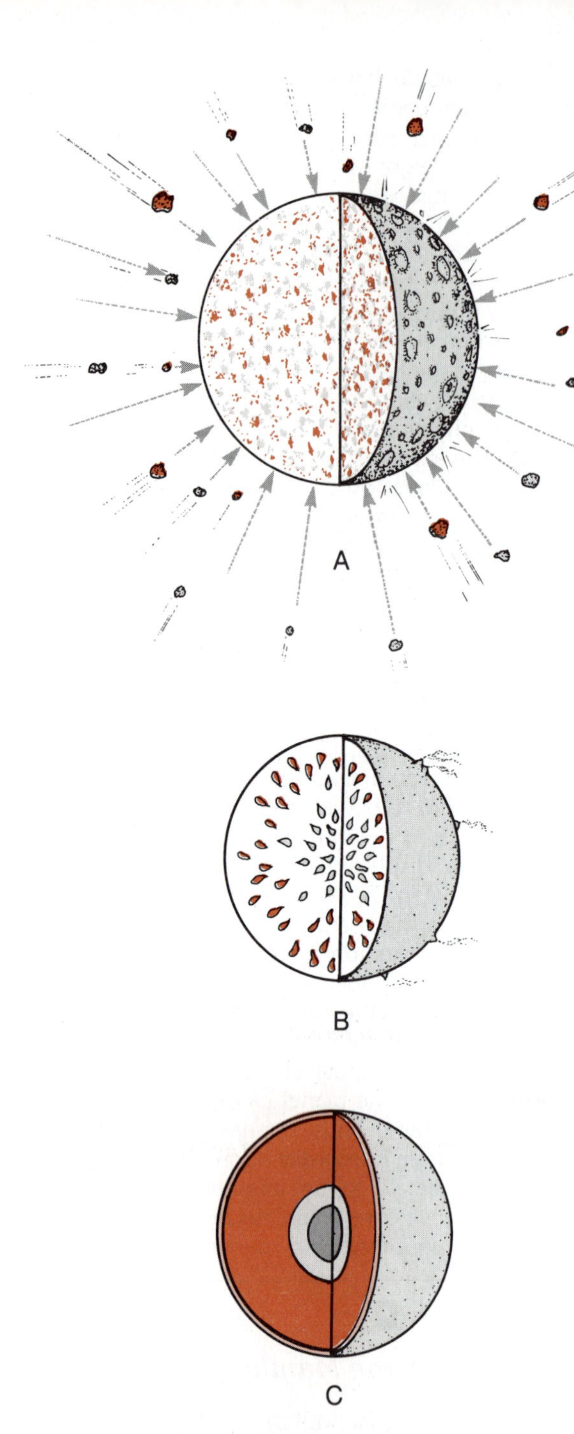

Figure 7–7 Conceptual diagrams of stages in the earth's early history. (A) represents growth of the planet by the aggregation of particles and meteorites that bombarded its surface. At this time, the earth was composed of a homogeneous mixture of materials. In (B), the earth has experienced a loss of volume because of gravitational compression. Temperatures in the interior have reached a level where differentiation has begun. Iron (gray drops) sink toward the interior to form the core, whereas lighter silicates move upward. The result of the differentiation of the planet is evident in (C) by the formation of core, mantle, and crust.

mantle. A peridotitic rock not only would be appropriate for the mantle's density but also is similar in composition to stony meteorites as well as rocks that are thought to have reached the earth's surface from the upper part of the mantle itself. Such suspected mantle rocks are indeed rare. They are rich in olivine and pyroxenes and contain small amounts of certain minerals, including diamonds, that can form only under pressures greater than those characteristic of the crust.

Layers of the Mantle

The mantle is not merely a thick, homogeneous layer surrounding the core but is itself composed of several concentric layers that can be detected by studying earthquake data. In Figure 7–8, note that within the mantle there are three zones of rapid increase in wave velocity. These sudden increases cannot be explained as simply the result of pressure increases with depth. Such increases would be too gradual and insufficient to cause seismic wave velocity to increase so abruptly. There must, therefore, be some change in the physical nature of the material. Geophysicists have demonstrated that if there is an increase in the speed of seismic waves with depth, it is likely to be caused by the tremendous increases in the elastic properties of the rocks.

The first zone of rapid increase in seismic wave velocity begins at a depth of about 160 km. The second is encountered at about 400 km and is taken to mark the base of the **upper mantle.** Beneath the upper mantle is the **transition zone,** which extends downward to about 650 km. The **lower mantle** lies beneath the transition zone and above the core.

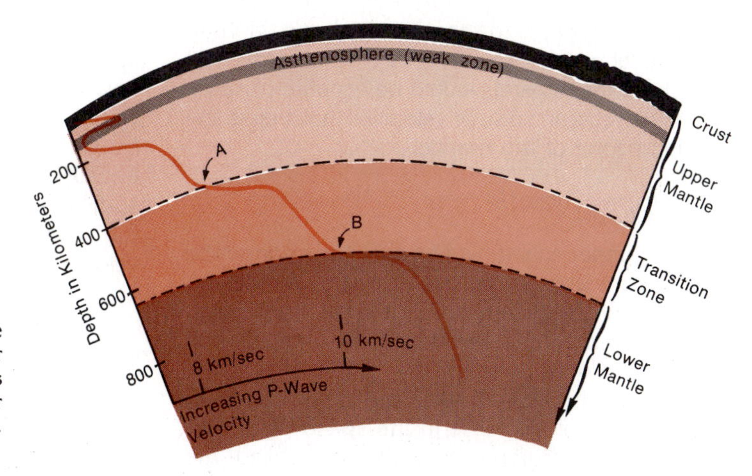

Figure 7–8 Generalized graph of average P-wave velocities vs. depth in the upper 1000 km of the mantle, showing two zones of velocity increase at *A* and *B*. (Data for graph primarily from L. R. Johnson. *J. Geophys. Res.,* 72:6318, 1967.)

The composition of the lower mantle is difficult to infer. A plausible guess, and one that is not contradicted by seismic data or calculated temperatures and pressures, is that it consists mostly of oxides and silicates of magnesium and iron. The overlying transitional zone may be the result of rearrangement of iron, magnesium, silicon, and oxygen atoms into denser and more compact crystals. For example, near the earth's surface, olivine and pyroxene are relatively stable minerals. However, in the high-temperature and high-pressure environment below a depth of about 400 km, olivine is likely to be converted to minerals whose atoms are more closely packed (Fig. 7–9). Similar dense

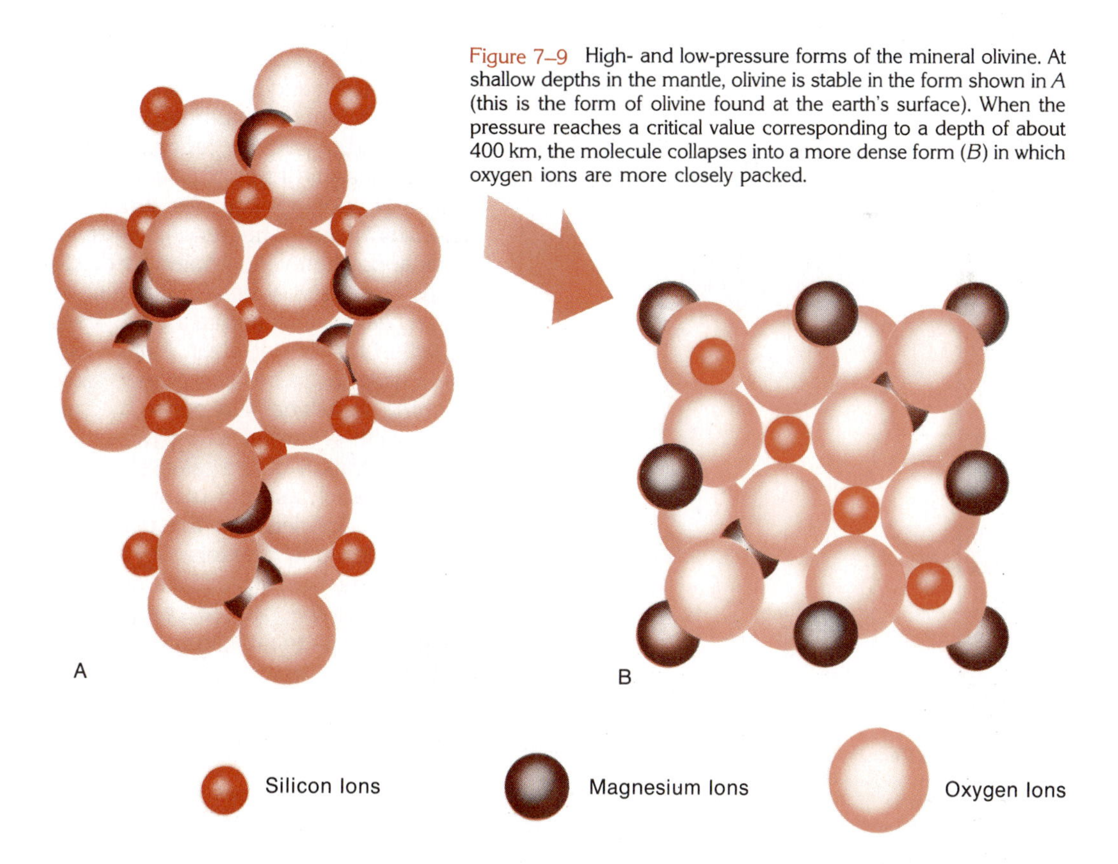

Figure 7–9 High- and low-pressure forms of the mineral olivine. At shallow depths in the mantle, olivine is stable in the form shown in *A* (this is the form of olivine found at the earth's surface). When the pressure reaches a critical value corresponding to a depth of about 400 km, the molecule collapses into a more dense form (*B*) in which oxygen ions are more closely packed.

crystals have been formed from olivine in high pressure laboratory experiments. Rocks composed of such dense minerals would be capable of causing the increase in seismic velocities that marks the transition zone of the mantle.

The Upper Mantle

The upper mantle is of particular importance to the historical geologist because its evolution and internal movements affect the geology of the crust. The most remarkable feature of the upper mantle is the **low-velocity zone** (Fig. 7–10). As suggested by its name, this is a region in the upper mantle in which there is a decrease in the speed of S- and P-waves. This region, which is also called the **asthenosphere,** lies at depths of about 60 to 120 km and is approximately 100 km in thickness. Another term useful in designating regions of the mantle is **mesosphere,** a zone beneath the asthenosphere where pressures are sufficiently great to impart greater strength and rigidity to the rock.

It is likely that seismic waves are slowed in the asthenosphere because they enter a region that is in the state of a crystalline-liquid mixture. In such a mineral "slush," 1 to 10 per cent of the material would consist of droplets and pockets of molten silicates. This interpretation is strengthened by the observation that in certain regions of the asthenosphere, S-waves are absorbed as if by large bodies of magma. If the interpretation is correct, then there is a layer within the mantle—the asthenosphere—which is capable of considerable motion and flow. Conceivably, such a slippery mobile layer would markedly influence movements in the overlying, more rigid upper mantle and crust. Indeed, because of the potential for causing changes in surficial layers of the earth, and as a primary source for basaltic magmas, the properties of the low-velocity layer must be considered in any theory of crustal evolution.

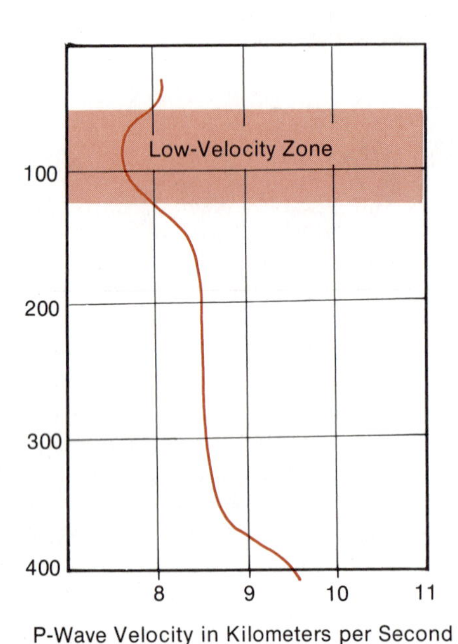

Figure 7–10 Profile of P-wave velocity for part of the upper mantle showing low-velocity zone.

THE CRUST OF THE EARTH

The crust of the earth is seismically defined as all of the solid earth above the Mohorovičić discontinuity. It is the thin, rocky veneer that constitutes the continents and the floors of the oceans. The crust is not a homogeneous shell in which low places were filled with water to make oceans and higher places formed continents. Rather, there are two distinct kinds of crust, which, because of their distinctive compositions and physical properties, determine the very existence of separate continents and ocean basins.

The Oceanic Crust

Beneath the varied topography of the ocean floors lies an oceanic crust that is approximately 5 to 12 km thick and has an average density of about 3.0 gm per cubic centimeter. Three layers of this oceanic crust can be recognized. On the upper surface is a thin layer of unconsolidated sediment that rests upon the irregular surface of the igneous basement layer. The second layer is about 4 km thick and, wherever sampled by dredge or core, is found to consist of basalts that had been extruded under water. The nature of the deepest layer of basaltic crust is not clear. Many suspect that it is metamorphosed basaltic mantle material that has become somewhat less dense by chemically combining with sea water. In the oceanic crust, we find a concentration of the relatively heavier common elements such as iron, magnesium, and calcium. They are included into the plagioclase feldspars, amphiboles, and pyroxenes of basalts. It is apparent that the oceanic basalts are rather similar in chemistry to the inferred composition of rocks of the uppermost mantle, from which they were probably derived.

The Continental Crust

At the boundaries of the ocean basins, the Mohorovičić discontinuity plunges sharply beneath

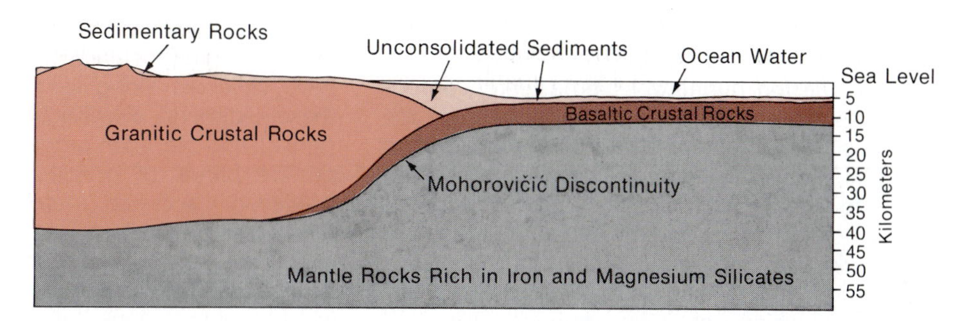

Figure 7–11 Generalized cross-section of a segment of the earth's crust showing location of the Mohorovičić discontinuity.

the thicker continental crust (Fig. 7–11). The depth to the Moho beneath the continents averages about 35 km, although the crust may be considerably thicker or thinner in particular regions. The continental crust is not only thicker than its oceanic counterpart but also less dense, averaging about 2.7 gm per cubic centimeter. As a result, continents "float" higher on the denser mantle than do the adjacent oceanic crustal segments. Somewhat like great stony icebergs, the roots of continents extend downward into the mantle. The concept of light crustal rocks "floating" on denser mantle rocks has been given the name **isostacy.** Were it not for isostacy, mountain ranges would gradually subside, for there are no rocks having sufficient strength to bear the heavy load of mountain ranges. Thus, mountains are not supported by the strength of the crust but rather are in a state of flotational equilibrium with denser underlying rocks.

Although continental crust is referred to as being "granitic," it is really an assortment of a variety of rocks, which, if all melted together, could be converted to granites or, more specifically, granodiorites. Igneous continental rocks are richer in silicon and potassium and poorer in iron, magnesium, and calcium than are igneous oceanic rocks. Also, extensive regions of the continents are blanketed by sedimentary rocks. On the average, continental sedimentary sequences rarely exceed a few kilometers in thickness, except in narrow, deeply subsiding tracts, where they may accumulate to thicknesses in excess of 15 km.

In terms of basic geologic character, the continental crust can be divided into three kinds of regions—namely, shields, platforms, and orogenic belts. **Shields** are great, broadly convex, relatively immobile regions usually constructed of Precambrian metamorphic and igneous rocks. Although great mountain ranges once traversed parts of shields, the mountains have long since been lost to erosion, and only the roots of the old ranges remain. In most continents, **platforms** are located adjacent to shields. Platforms are also stable regions of the crust. Their basic structure consists of an underlying basement of ancient crystalline rocks covered by essentially horizontal sedimentary strata. Platforms and shields together compose the **craton,** or stable nucleus of the continent. The third division of continental crust consists of great linear or arcuate tracts that have been subjected to severe deformation and mountain building. These are the **orogenic belts** typically found near the edges of continents. The reason for their existence will be explored in the pages that lie ahead.

THE BIRTH AND GROWTH OF CONTINENTS

The Embryonic Continents

How did the lighter granitic masses that constitute the continents originate? Answers to this question are found in two hypotheses. The first suggests that when the earth was in a molten state, granitic melts collected near the surface as a discontinuous light silicate slag. At intervals, slabs of this thin and unstable granitic crust might have sunk and been partially engulfed by upwelling basalts. Eventually the crust and upper mantle would have become sufficiently thick and cool to resist the instability. Separate regions of basaltic and granitic crust would settle into flotational equilibrium on the peridotitic mantle material. It is not known precisely when such a process may have begun. However, it is known from radiometric dates of ancient rocks that both types of crust already existed on earth somewhat over 3½ billion years ago.

The second hypothesis for the origin of the continents begins with a continuous thin crust of lava derived from the partially molten interior of the earth. Very likely, this primordial crust experienced periodic melting, during which time lighter components were separated out and distributed near the earth's surface. Wherever uplands existed, as along volcanic island arcs, the solidified lavas were subjected to erosion and the ordinary processes of oxidation, carbonation, and hydration that accompany chemical weathering. The products of this weathering and erosion were the earth's earliest sediments, which were then altered by rising hot gases and silica-rich solutions from below. Recycling and melting of this well-cooked and now lighter mix of earth materials led ultimately to rocks of granitic character that formed the nucleus of a continent. The new continent might then have provided a source for additional sediment that would have collected along its margins. Subsequently, these sediments also might have been metamorphosed and melted in orogenic events, so that successive bands of granitic crust would become welded or "accreted" onto the initial continental nucleus (Fig. 7–12).

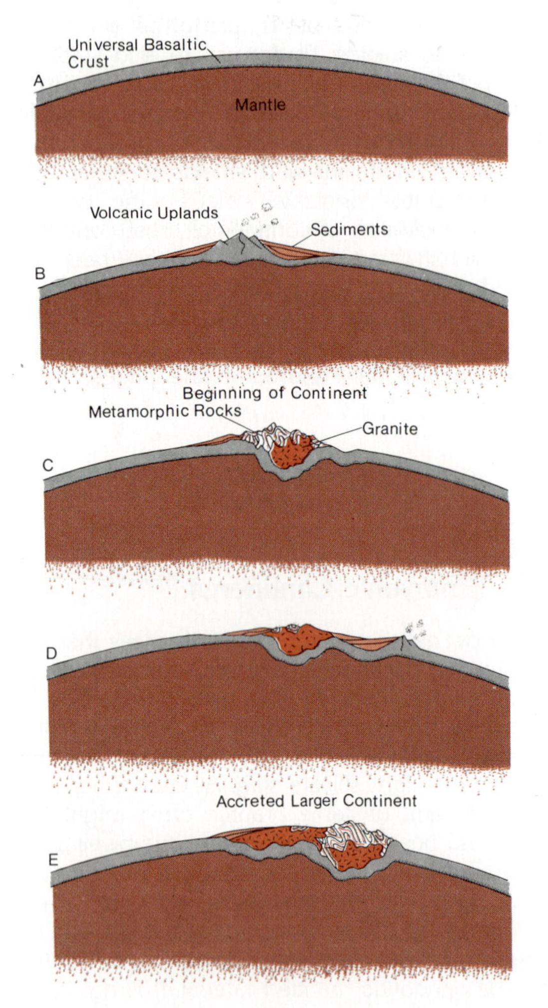

Figure 7–12 Model for the gradual evolution of continents by orogeny and conversion of sedimentary to granitic igneous rocks. (A and B) Original basaltic crust with volcanic uplands as source for sediments. (C) Orogeny of sedimentary accumulation to produce metamorphic and granitic rocks. (D) New sediments accumulate along margins of continental nucleus. (E) The process is repeated.

Proponents of the concept that the original solid crust was entirely basaltic support their view with the observations that (1) active volcanic arcs are developed today along zones of crustal weakness; and (2) compressional mountain ranges tend to be located, as if accreted, along the margins of the stable granitic core of the continents. Those opposed to the concept have geochemical questions. In conversion of the weathered products of basalt into granite, what happened to the excess iron and magnesium? Also, one needs additions of potassium and sodium to make granite. These elements are rare in sediments that have formed from basalts. Thus, the problem of the origin of the continents will continue to be examined in the decade ahead.

Geosynclines and Continent Building

The precise form and locations of the earth's first continents will probably never be determined. Only tiny patches of crust as old as 3.6 millon years have been left for us to examine. Whatever their original condition, continents have certainly undergone complex changes, growth, and fragmentation since they were first formed. Frequently, these changes have been associated with a cycle of geologic events involving large-scale crustal downwarpings along features called geosynclines.

The Classic Geosynclinal Model

As considered briefly in Chapter 3, a **geosyncline** is a great, elongate trough in the earth's crust that receives a thick sequence of sediments over a long span of geologic time and may ultimately be compressed into a major system of mountains. Geosynclines are truly global features and may extend over hundreds or thousands of kilometers. Typically, they are developed in the mobile zone between the continents and ocean basins, but they have also appeared within the ocean basins or in the central parts of continental masses. In the 1930's a German geologist, Hans Stille, suggested that geosynclines bordering continents can sometimes be distinguished by the nature of the sediments they contain. Stille identified the shallower part of the downfold that lay nearest the continent as the **miogeosyncline.** The miogeosyncline may actually represent only a transitional zone along the subsiding edge of a continent. Within it are deposited thick sequences of sandstone, shale, and limestone that are for the most part erosional products derived from the **craton,** a term applied to the broad, stable interior of continents. Indeed, sediments accumulating in the miogeosyncline resemble cratonic sediments but accumulate to much greater thicknesses.

The more actively subsiding and dynamic portion of the geosyncline lies parallel to, and seaward of, the miogeosyncline (Fig. 7–13). It was named the **eugeosyncline** by Stille. This outer eugeosynclinal belt typically contains a rock sequence rich in graywacke sandstones, siliceous shales, cherts, tuffs, volcanic ash, and submarine lavas. Rock assemblages such as this suggest an adjacent volcanic source region, perhaps in the form of a large system of volcanic islands parallel to the eugeosyncline.

By studying broad, elongate belts of erosionally truncated ancient rocks that are believed to have originated in geosynclines and by thoroughly investigating folded mountain ranges, geologists have been able to discern a historical sequence of events for geosynclines. Stages in the sequence can be recognized in all of the new and old mountain ranges of the world, although their magnitude and duration may vary, and stages may be interrupted or missing (Fig. 7–13).

Stages of Geosynclinal History

The cycle theoretically begins with the **depositional stage,** in which sediments are poured into the newly formed trough. A volcanic island arc may develop on the outer margin of the downfold. Differences in sedimentation may then become apparent as craton-derived material fills the inner miogeosynclinal belt while beds of graywacke, shale, chert, and volcanics accumulate in the outer eugeosynclinal tract. Enlargement of the volcanic ridge, sedimentation, and contemporaneous sagging and flexing may continue for tens of millions of years before the geosyncline enters into the **orogenic stage.** As a result of compressional movements from one or both sides, the sediments within the trough are squeezed, deformed, and crumpled. The oldest and bottom-lying sediments, already forced downward by the enormous burden of overlying layers, begin to respond to the pressure and heat. Partial or total melting occurs in some of these deep regions, and in others, the mineral content of the original sediment is reorganized during metamorphism and converted to rocks of granitic character. These great granitic masses form the batholithic roots of rising moun-

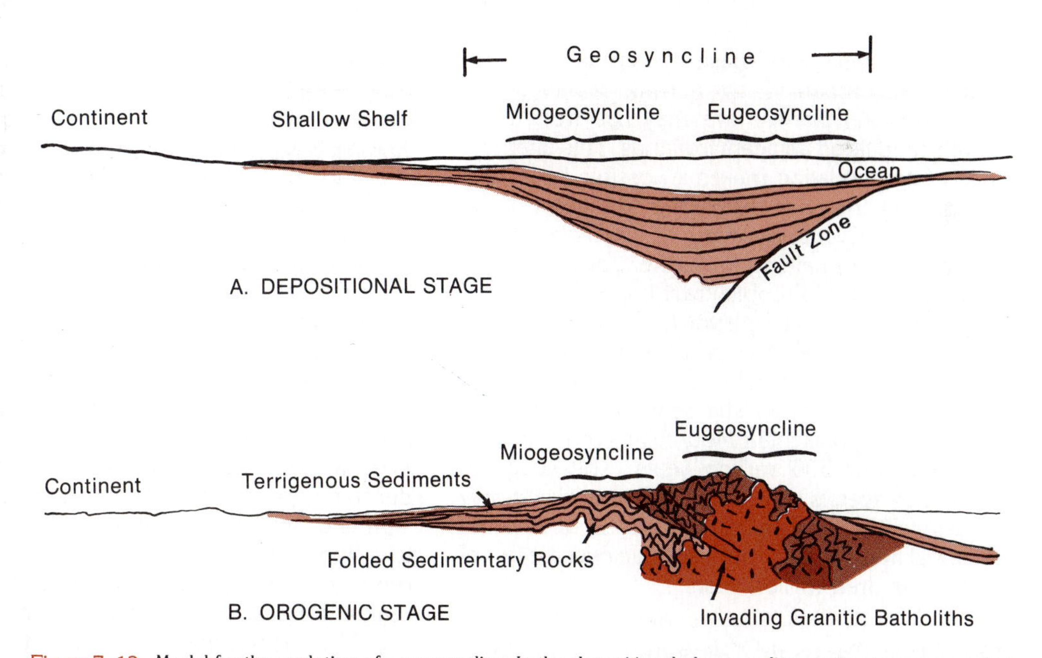

Figure 7–13 Model for the evolution of a geosyncline. In the depositional phase, sedimentation is accompanied by subsidence. Crustal shortening, igneous intrusions, as well as folding and faulting occur during the deformational, or orogenic, phase. In the plate tectonics model, stage A would occur after separation of plates and opening of an ocean basin. Sedimentation would occur on the trailing margins of continents. The deformational stage shown in B would be the result of plate collision and closing of the ocean basin.

tain ranges. In addition to the intrusions and metamorphism, the orogeny is accompanied by concertina-like folding of layers of sediment. Great reverse or thrust faults develop and move sections of rock over and onto the miogeosynclinal belt and craton. The geosynclinal tracts are squeezed to a fraction of their former widths, and great mountains, such as the Alps, Himalayas, Appalachians, and Western Cordillera of the Americas, are heaved upward from what were once tranquil seaways. The younger of these ranges still preserve the sediments that once accumulated in the geosynclinal troughs.

The final episode in this model of geosynclinal evolution may be appropriately designated the **erosional stage.** In this stage, two forces interact in such a way that they cause sporadic rejuvenation of the mountain system. The old cliché "what goes up must come down" applies very well to mountain ranges. Particle by particle, the substance of the highlands heaved upward during the orogenic stage is dissolved or loosened and carried to lower altitudes and ultimately to the oceans. Eventually, the mountains may be reduced to low-lying plains on which are exposed rocks that once existed deep within the bowels of the compressed geosyncline. As a result of the erosion, a great weight is lifted from the geosynclinal belt. In much the same way that a boat rises in the water as it is unloaded, the mountains periodically rise to assume a new position of gravitational equilibrium with surrounding portions of the crust. These movements are frequently accompanied by normal faulting, and along the fault planes, lava may rise from deep sources and flood adjacent lowlands. This alternate uplift and denudation makes it possible for geologists to view the once deeply buried interior of the geosyncline. Eventually, there is insufficient lightweight "root" remaining for further upward adjustments, and the old mobile tract becomes a stable part of the continental platform.

Of course, the passage of each stage in the history of a geosynclinal tract may require an incredible length of time. The geosyncline that gave birth to the Appalachian Mountains received sediment for over 300 million years. The orogenic phase, however, is relatively short, comprising only a few tens of millions of years. Longest of all is the erosional finale, which may continue over a span of several hundred million years.

The Plate Tectonic Geosynclinal Model

Geologists can observe ample evidence that ancient sediments frequently accumulated to great thicknesses in geosynclinal tracts and that these deposits then underwent compression. But what was the cause of the subsidence that permitted such great thicknesses of sediment to accumulate? It has been demonstrated that the weight of the sediment alone would not sufficiently depress the earth's crust. Also, what mechanism caused the ultimate compression of the wedge of sediments during the orogenic phase?

In recent years, tentative answers to these questions have been provided by a new model of geosynclinal evolution that involves the concept of **plate tectonics,** which will be discussed in detail in Chapter 8. At this point, it is sufficient to know the central idea of plate tectonics: The earth's crust and upper mantle are divided into a shifting mosaic of relatively brittle slabs in which continents may be embedded. In such a system of plates moving about over the surface of a sphere, there will be places where plates slide past one another, places where they move away from one another, and places where they move toward one another and collide. Each moving plate must have a *leading edge,* which is likely to develop compressional structures as it moves against other plates, and a *trailing edge,* which may experience sagging under the influence of tensional stresses. It is primarily along this trailing edge that the great geosynclinal prisms of sediments are deposited upon the sagging margin of the craton (Fig. 7–14). If, however, patterns of plate motion should change (as they apparently have many times in the geologic past), a geosynclinal prism of sediments may experience a direct encounter with the leading margin of another plate. The result would be deformation of the thick sedimentary wedge.

If the opposing plate is composed of relatively dense oceanic crust, its forward margin is likely to slide beneath the plate bearing the marginal geosyncline. According to the plate tectonic model for geosynclinal evolution, when the oceanic plate reaches depths of about 100 km, basaltic magmas are generated, rise toward the surface, and erupt as submarine volcanoes. As subduction continues, magmas of more granitic composition are formed at depths by the melting of sea floor sediments and oceanic crust. The upward movement of the less dense magmas and general heating causes expansion and uplift of the surface. Mountains begin to take form. Strata adjacent to the intrusions are metamorphosed, and on the flanks of the uplifted region, folds and faults form by gravitational sliding. The new granites and associated metamorphics become welded to the edge of the craton as an addition to the continent.

The American geologist Robert Dietz believes that a plate tectonic model can adequately explain

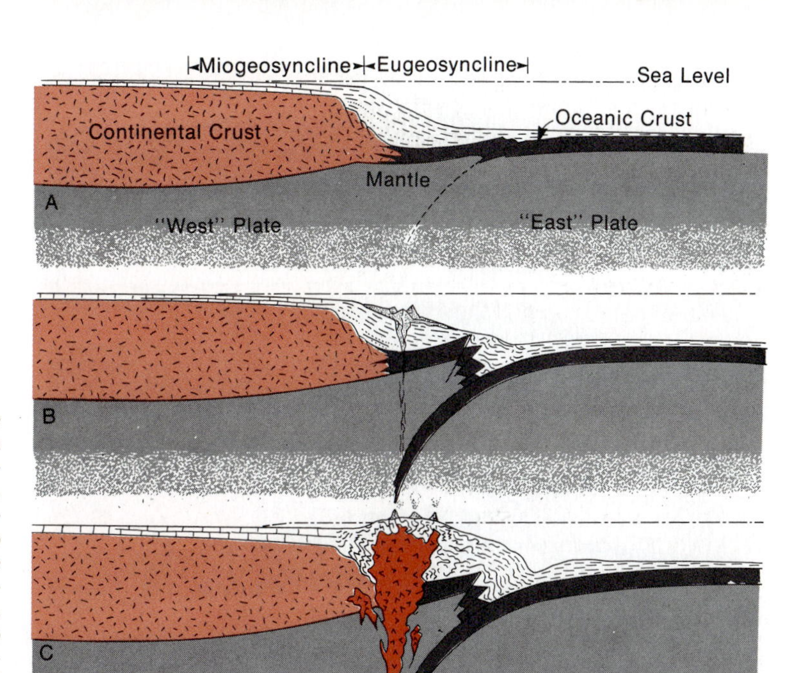

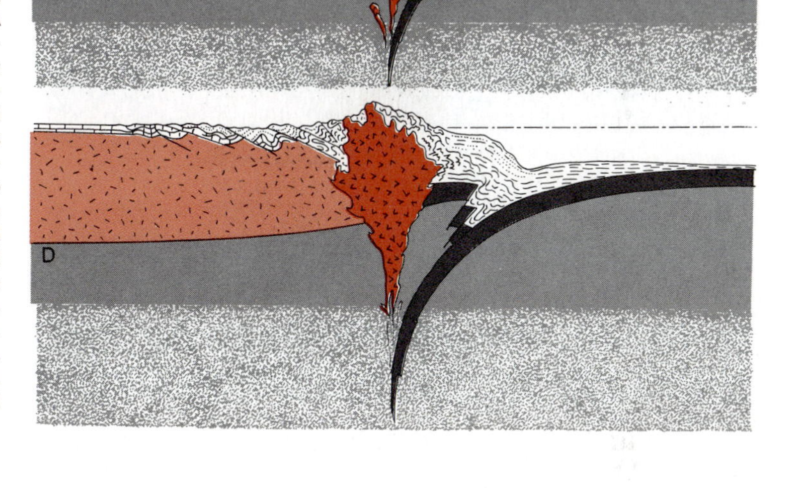

Figure 7–14 Plate-tectonic model for the origin of a mountain belt. (A) Trailing margin of a continent sags and receives thick accumulation of sediments as it drifts to the left. Carbonates on continental shelf grade seaward to graywackes and shales and finally to deep sea sediments. (B) Lithospheric plate bearing oceanic crust collides and is subducted beneath plate bearing continent. When subducted plate has descended to about 100 km, basaltic-type lavas, generated by melting of the lithosphere, rise and are extruded as submarine volcanics. A trench begins to form in the zone of subduction. (C) As subduction continues, granitic type magmas, formed from molten subducted materials, rise. Their buoyancy and heat cause doming at the surface, and this is accompanied by metamorphism and deformation. Graywackes accumulate adjacent to the rising mountain chain. (D) Subduction and orogeny continue. Sheets of folded and metamorphosed rocks are thrust faulted toward the craton, and new granites are welded into the older continental crust. (Modified from J. F. Dewey and J. M. Bird. Mountain belts and the new global tectonics. *J. Geophys. Res.,* 75:2625–2647, 1970. Copyright, The American Geophysical Union.)

most of the structural and sedimentologic features of ranges like the Appalachians. Figure 7–15 illustrates the model proposed by Dietz. The sequence of events begins with thick (geosynclinal) bodies of sediment accumulating on the rear, or trailing, edges of diverging plates. The ocean basin is presumably widened as the continents drift apart. At a later period, the movement of the crustal plates is reversed, the oceans narrow, and they are ultimately squeezed out of existence as the plates collide and crumple the accumulated sediment. As suggested by the geology of eastern North America, the plates may again separate, re-establishing oceanic tracts.

According to these new concepts, the physical setting for the sedimentational phase of geosynclinal evolution would be similar to that now present off the Atlantic Coast of the United States. The submarine physiographic divisions in this region consist of the continental shelf, the continental slope, and the continental rise. The **continental shelf** is the submerged, very gently sloping (about 0.1°) edge of the continent. Seaward of the shelf, the imposing **continental slope** drops off toward the deep ocean floors. At the base of the slope, and banked up against it, is a thick sedimentary fill that forms the **continental rise.**

Over the past 150 million years, sediments derived from the interior of North America have accumulated on the continental shelf. Eventually, a great wedge of sandstones, shales, and marls has been deposited on the shelf (Fig. 7–16). These deposits bear an easily recognized similarity to the rocks found in the miogeosynclinal tracts of an-

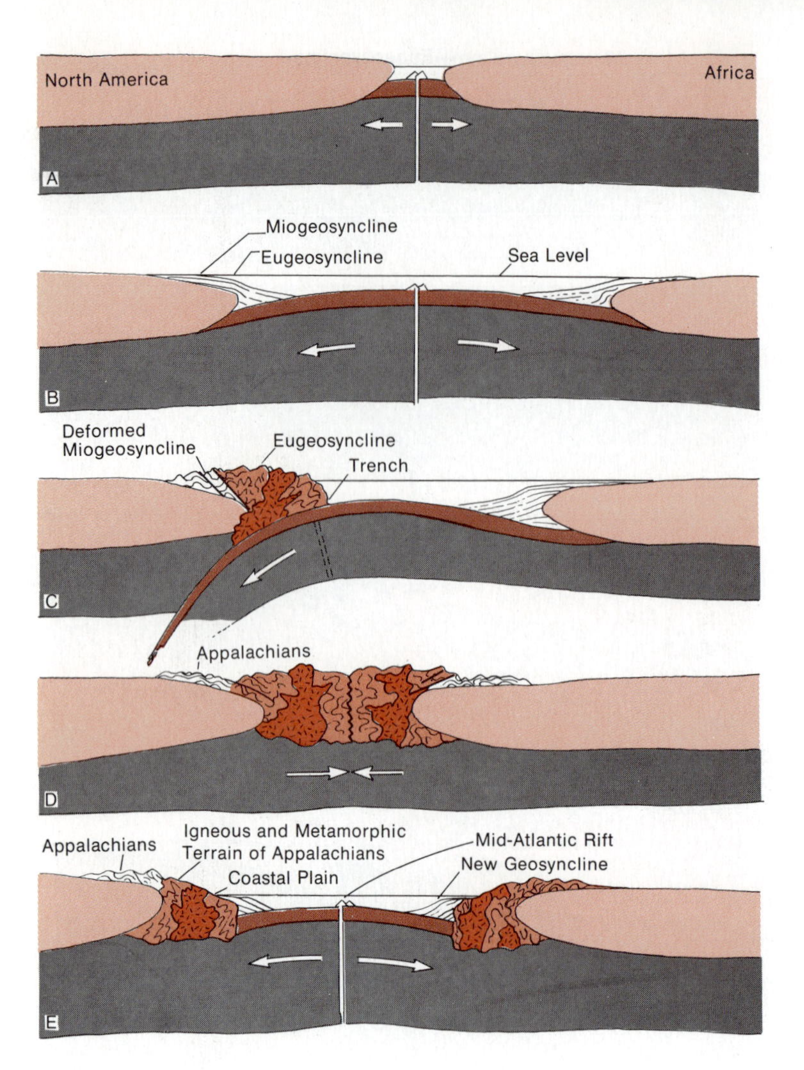

Figure 7–15 Plate tectonic model for events leading to the formation of the Appalachian Mountains and present Atlantic Ocean, as proposed by Robert S. Dietz. (A, B) Continents separate, and the ancestral Atlantic Ocean is inserted; geosynclinal couplets form on the margins of continents. (C, D) Plates converge, the subduction zone is formed, and the eugeosyncline is collapsed. Continents are sutured together. (E) Rifting occurs again, and new geosynclines form along continental margins.

cient geosynclines, and they reach similar thicknesses as well. Thus, continental shelves located along the inactive, trailing edges of continents may be modern miogeosynclines in their depositional stage. One would look for the modern counterpart of the eugeosyncline seaward of the continental shelf, if only because studies of ancient geosynclines show this relationship. A suitable depositional site for eugeosynclinal sediments is the continental rise. Here sediment reaches a final resting place at the base of the continental slope. It has accumulated there to a thickness of over 10 km. The sediment is poorly sorted, exhibits graded bedding, and has many features suggesting it has been transported by turbidity currents. Geologists call such deposits **turbidites** and recognize their similarity to the thick graywacke assemblages found in old eugeosynclines. Indeed, if these turbidites were someday to be underthrust and compressed by the encounter with a converging tectonic plate, they would be converted to the severely metamorphosed and deformed eugeosynclinal tracts of mountain ranges.

An interesting feature of this new idea about the modern counterparts of mio- and eugeosynclines is that the development of an initial trough to receive the sediment is not really needed. This pleases geologists, who have always had some difficulty finding a present-day example of such a trough. Without the need for a trough, the "syncline" part of the terms miogeosyncline and eugeosyncline is no longer appropriate, and so many prefer to shorten these rather ponderous terms to miogeocline and eugeocline.

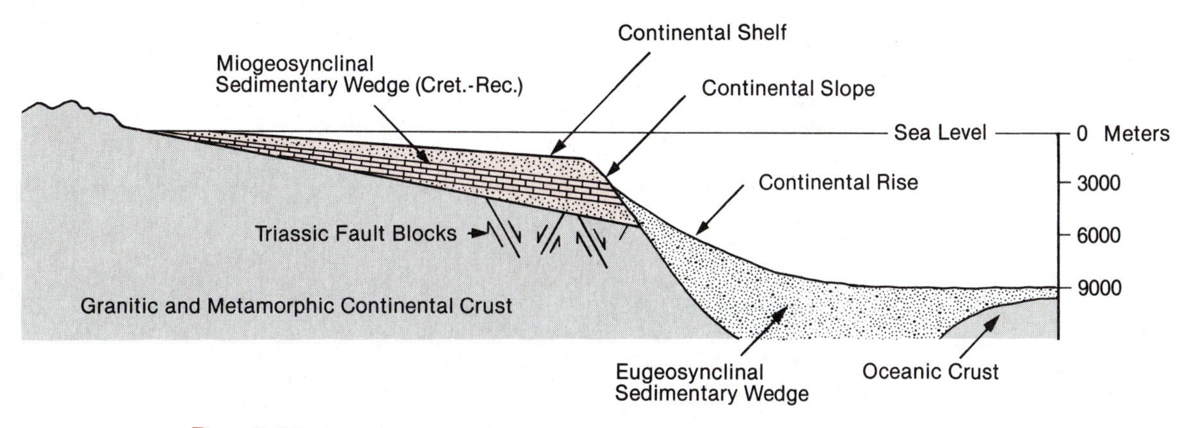

Figure 7–16 Geosynclinal sediments accumulating on the passive eastern coast of the United States (vertical scale exaggerated).

ATMOSPHERIC EVOLUTION

The Primitive Atmosphere

We have described how the earth formed by means of the aggregation of small particles and larger bodies that were themselves once part of a great disk of gas and dust. It is likely that the grains of dust and the meteorites incorporated into our growing planet contained a small but significant amount of gases as well, and these volatiles were thus distributed throughout the earth's interior. Water, for example, may have been present in certain hydrated minerals (minerals that contain water, like serpentine). Molecular nitrogen (N_2) and carbon dioxide (CO_2) could be derived from nitrogen and carbon atoms that were trapped in original meteoritic materials like carbonaceous chondrites. At a later stage in the earth's evolution, when it began to heat up and begin the process of differentiation, water vapor and other gases were released from the rocks that held them and were vented into the developing atmosphere. The precise chemical nature of the gases released in this way would have been dependent upon the temperatures of the rocks from which they were derived, as well as upon the state of oxidation of the rocks through which the gases passed. Geochemists believe the released gases consisted chiefly of water vapor, carbon dioxide, nitrogen, carbon monoxide, and hydrogen. It is unlikely that any significant amount of oxygen was released, as it would have been readily combined with iron and thereby efficiently reincorporated into the rocks.

Once at the earth's surface, the gases that had been vented (Fig. 7–17) were subjected to a variety of changes that determined their relative abundances in the primordial atmosphere. Water vapor was removed by condensation, and oceans began to take form. Through the weathering of crustal rocks, some of the carbon dioxide was converted to carbonate rocks. Because it is a very light element, hydrogen escaped rapidly into space. Thus it seems likely that nitrogen and carbon dioxide were the chief gases in the primordial atmosphere, with water vapor, carbon monoxide, and hydrogen present, but in lesser quantities. This early atmosphere was reducing and nonoxygenic in character. Within it the initial chemical steps were taken that led to the development of life on earth.

Oxygen-Rich Atmosphere

The change from an oxygen-poor to an oxygen-rich atmosphere occurred as a result of two processes. The first was dissociation of water molecules into hydrogen and oxygen. Termed **photochemical dissociation,** this process occurs in the upper atmosphere when water molecules are split by high-energy beams of ultraviolet light from the sun. Photochemical dissociation, however, does not produce free oxygen at a rate sufficient to balance loss of the gas by dissipation into space. Another far more important oxygen-generating mechanism came with the advent of life on earth and, more specifically, with life that had evolved the remarkable capability of separating carbon dioxide into carbon and free oxygen. We now know this process as **photosynthesis.** It has made the earth a truly unique planet in our solar system.

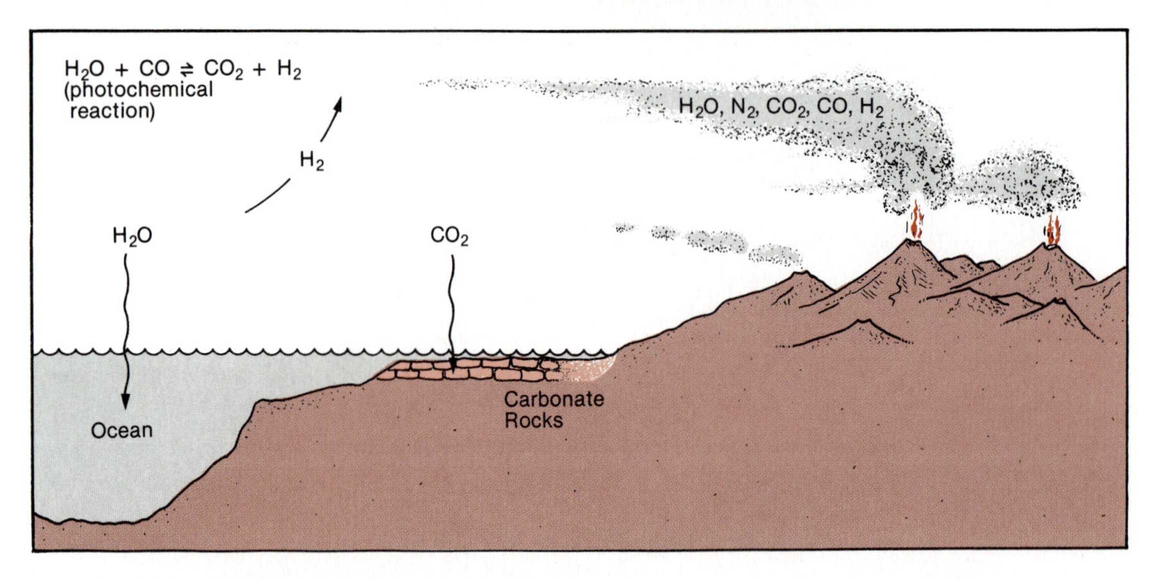

Figure 7–17 The relative amounts of gases in the primordial atmosphere varied from the abundances vented to the exterior of the earth during differentiation. Nitrogen tended to be retained in the atmosphere, whereas much of the water vapor was lost to the oceans by condensation, some CO_2 was combined with calcium and magnesium derived from weathering and extracted to form carbonate rocks, and hydrogen, being light, was lost to space.

Geologic Clues to the Nature of Ancient Atmospheres

The most ancient rocks providing evidence of the atmospheric changes described previously are about 3.5 billion years old. Although they are metamorphic rocks, many of them still contain depositional features indicating they were originally waterlaid sediments. Mineral grains within these old rocks were formed or modified by weathering, and weathering cannot occur without an atmosphere. Further, it is highly probable that the atmosphere was anoxygenic, for there is an obvious absence of oxidized iron such as hematite. Rather, the sulfide of iron called pyrite is the usual iron mineral found in these rocks. A carbon dioxide–rich atmosphere is a rather acidic one, in that carbon dioxide and water combine to form carbonic acid. In such an environment, alkaline rocks such as dolostone and limestone do not develop, and this may account for the absence of carbonate rocks from this early stage of earth history. Chert, a rock more tolerant of such conditions, is frequently found associated with these ancient rocks.

The advent of an oxygen–rich atmosphere is reflected by the state of oxidation of iron in the rocks overlying (and hence younger than) those described in the preceding paragraph. Beginning in rocks that are about 3 billion years old, one observes evidence of periodic abundance of oxygen. Many formations exhibit an alternation of rusty-red and gray bands. The former are colored by ferric iron oxide (Fe_2O_3) and indicate abundant free oxygen in the environment. The grayish layers contain oxygen-deficient iron compounds and suggest a lesser availability of oxygen. The alternation of color bands might have been caused by a fluctuating and perhaps season-related supply of oxygen. In the iron mining district north of Lake Superior, the ores above the banded iron deposits are uniformly oxidized. These rocks are about 1.8 billion years old and are evidence of the persistent presence of an oxygenic atmosphere by that time. In other parts of the world, one finds extensive outcrops of dolostone and limestone of about the same age. If carbon dioxide was still abundant in the atmosphere, these carbonate rocks would not have been deposited.

THE ORIGIN OF THE OCEANS

If the water in the atmosphere was outgassed from the interior, then the bodies of water that accumulated on the surface of the earth were derived from the atmosphere after it had cooled sufficiently to condense and allow rain to fall. Gradually, the great depressions in the primordial crust began to fill, and oceans began to take form. One cannot help but wonder if the enormous volume of oceanic water could have all come from the interior. Calculations provide a very strong "yes" answer to

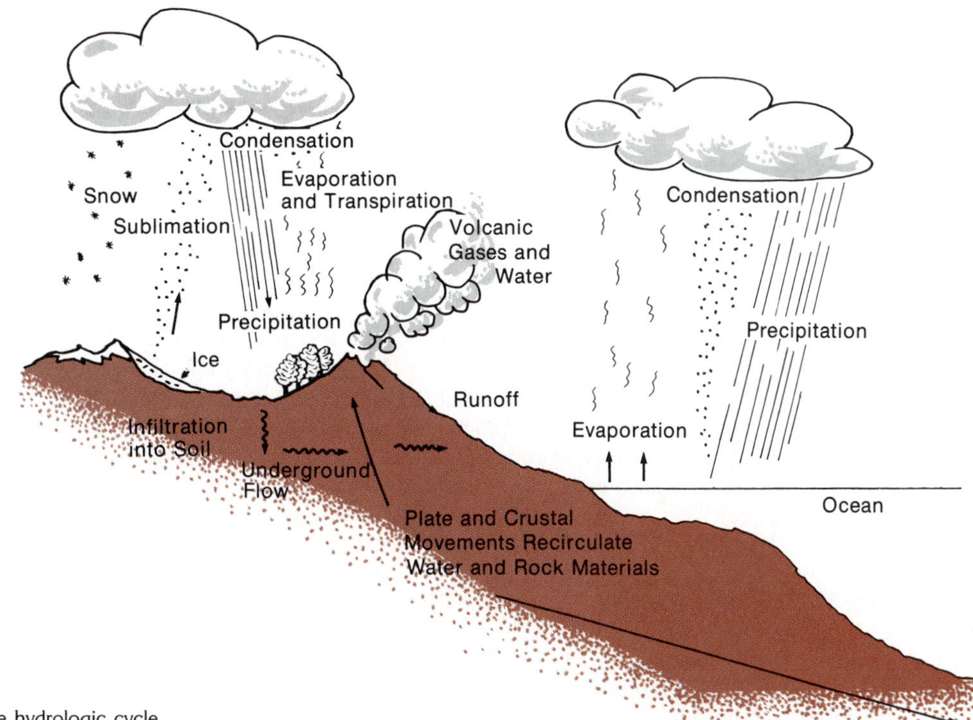

Figure 7–18 The hydrologic cycle.

this question. Vast amounts of water were locked within hydrous silicate minerals that were ubiquitous within the primordial earth. These waters, sweated from the earth's interior and precipitated onto the uplands, began immediately to dissolve soluble minerals and carry the solutes to the sea. In this way, the oceans quickly acquired the saltiness that is their most obvious compositional characteristic. They have maintained a relatively consistent composition by precipitating surplus solutes at about the same rate as they are supplied. Of course, because of its high solubility, sodium remains in sea water longer than do other common elements. However, the fossil record of marine organisms suggests that even this element has not varied appreciably in sea water for at least the past 600 million years.

The earth probably outgassed its present quantity of water rather early in its history and has been partially recycling it ever since. The heating that was required to melt the core, as well as possible stresses induced in the crust by the nearby moon, might have so disturbed the crust as to promote outgassing at a rapid rate. Soon thereafter, the oceans settled into a sort of "steady state" system and probably have not gained appreciably in volume. Water is continuously recirculated by evaporation and precipitation—processes powered by the sun and gravity. Some of the water in the oceans is temporarily lost by being incorporated into hydrous clay minerals that settle to the ocean floors. However, even this water has not permanently vanished, for the sediments may be moved to orogenic belts and melted into magmas that return the water to the surface in the course of volcanic eruptions (Fig. 7–18).

Summary

The nature of the earth's deep interior is derived from the study of earthquake waves. Among the various kinds of earthquake or seismic waves are primary, secondary, and surface waves. Primary and secondary waves (also called body waves) pass deep within the earth and therefore are the most instructive. Study of abrupt changes in the velocities of seismic waves at different depths provides the basis for a threefold division of the earth into a central core; a thick, overlying mantle; and a thin, enveloping crust. Sudden changes in earthquake wave velocities are termed discontinuities.

The core of the earth, as indicated by the Gutenberg discontinuity, begins at a depth of 2900 km. It is likely that the outer core is molten and that the entire core is composed of iron with small amounts of nickel and possibly either silicon or sulfur. The core probably originated during an episode of heating, when heavier constituents of protoplanet were drawn by gravity to the center and lighter components rose to the surface.Constituting about 80 per cent of the earth's volume, the mantle is composed of iron and magnesium silicates such as olivine and pyroxene. A warm, rather plastic zone in the mantle is recognized by lower seismic wave velocities. The presence of this low-velocity layer, or asthenosphere, is an important element in the modern theory of plate tectonics. The asthenosphere may function as a weak plastic layer upon which horizontal motions in more rigid surface layers can occur.

The seismic boundary that separates the mantle from the overlying crust is the Mohorovičić discontinuity. It lies far deeper under the continents than under the ocean basins. Thus, the continental crust is thicker than the oceanic crust. There are compositional and density differences as well. The continental crust has an overall granitic composition and is less dense than the oceanic crust, which is composed of basaltic rocks.

Hypotheses that attempt to account for the origin of the continents either begin with a thin universal granitic crust that subsequently broke apart and partially foundered or suggest that granitic crust developed from remelting and metamorphism of the weathered products of an original universal basaltic crust. According to the latter view, sediments from the erosion of basaltic highlands might be converted to initial continental masses during orogeny. Once the nucleus of a continent had formed, it might accumulate thick sequences of sediment along its margins. Orogeny would fold, metamorphose, and melt the sediments and weld the entire mass onto the continents. In this way, continental masses may have grown by accretion.

Elongate subsiding tracts in the earth's crust that accumulate sediment are called geosynclines. Marginal geosynclinal belts seem to pass through successive stages of deposition, orogeny, and erosion. These stages may be incorporated into the theory of plate tectonics. For example, the depositional phase may develop along crustal plate boundaries that do not abut against other plates. At some later time in geologic history, such plate edges may move against an opposing plate, resulting in compression of the bordering geosynclinal filling and the creation of a mountain range.

The earth's primitive atmosphere was rich in carbon dioxide and nitrogen, with lesser but significant amounts of water vapor, carbon monoxide, and hydrogen. This atmosphere and the hydrosphere were formed of gases vented from the earth's interior. The present atmosphere of the earth was initiated by additions of large amounts of uncombined oxygen, most of which originated as a by-product of photosynthesis.

Questions for Review

1. What major internal zones of the earth would be penetrated if one were able to drill a well from the North Pole to the center of the earth?
2. What are the three main kinds of earthquake waves? Describe their characteristics.
3. What does the presence of an S-wave shadow zone indicate about the interior of the earth?
4. What is the Mohorovičić discontinuity? The Gutenberg discontinuity? Where is each located?
5. How does continental crust differ from the oceanic crust in thickness, composition, and density?
6. What is the inferred composition of the mantle? Of the core? On what evidence are these inferred compositions based?
7. Explain why mountain areas may experience uplift after they have been largely eroded away.
8. Compare a eugeosyncline with a miogeosyncline in terms of thickness, characteristic rocks, and stability.
9. In what way does the activity of volcanoes support the hypothesis that the earth's atmosphere and hydrosphere were derived from the interior by degassing?
10. What is the origin of most of the free oxygen in the earth's atmosphere? What geologic evidence attests to the presence of oxygen in ancient rocks?
11. Account for the granite core that is characteristic of mountain ranges.
12. According to one theory, the original crust of the earth was entirely basaltic in composition. How does this theory account for the presence of granitic crust?
13. What geologic interpretation might one make about the source of a fissure or volcanic pipe filled with ultrabasic rocks containing diamonds?
14. What are the three main stages in the history of a geosynclinal tract?

Terms to Remember

asthenosphere
body wave
continental rise
continental shelf
continental slope
core
craton
crust
differentiation
eugeosyncline
focus
geosyncline
Gutenberg discontinuity
low-velocity zone
mantle
mesosphere
miogeosyncline
Mohorovičić discontinuity
oceanic crust
orogenic belt
peridotite
photochemical dissociation
photosynthesis
plate tectonics
platform
primary seismic waves
secondary seismic waves
shield
surface wave
seismic shadow zones
turbidite

Supplemental Readings and References

Bolt, B. A. 1973. The fine structure of the earth's interior. *Sci. Am.* 228:24–33.

Bott, M. H. P. 1971. *The Interior of the Earth.* New York, St. Martin's Press, Inc.

Clark, S. P., Jr. 1971. *The Structure of the Earth.* Englewood Cliffs, N.J., Prentice-Hall, Inc.

Dietz, R. S. 1972. Geosynclines, mountains, and continent building. *Sci. Am.* 228:24–33.

Jacobs, J. A. 1976. *The Earth's Core.* San Francisco, Academic Press, Inc.

Pollack, J. B. 1981. Atmospheres of the terrestrial planets. *In* Beatty, J. K., O'Leary, B., and Chaikin, A. (eds.). *The New Solar System.* Cambridge, Mass., Sky Publishing Corp.; London, Cambridge University Press.

Ringwood, A. E. 1975. *Composition and Petrology of the Earth's Mantle.* New York, McGraw-Hill Book Co.

Rubey, W. W. 1955. Development of the hydrosphere and atmosphere in the crust of the earth. *Geol. Soc. Am. Spec. Paper* 62:631–650.

York, D. 1975. *Planet Earth.* New York, McGraw-Hill Book Co.

Photograph taken from *Landsat* (satellite) at an altitude of 900 km. Near the center is the Dead Sea. The smaller body of water to the north is Lake Tiberias. To the northwest is the Mediterranean. The biblical cities of Sodom and Gomorrah were probably located at a site now flooded at the south end of the Dead Sea. Both Lake Tiberias and the Dead Sea are located along a north-south trending system of faults and occur in downfaulted blocks or grabens. The steep sides of the faults form the rugged terrain on either side of the Dead Sea rift valley. The fault system was apparently formed by stresses associated with the separation of Arabia from Africa.

8

The Earth's Dynamic Crust

What was solid earth has become the sea,
and solid ground has issued from the
bosom of the waters.
Ovid, *Metamorphoses*

A CHANGEABLE PLANET

Christopher Columbus was a visionary. It is easy to imagine him standing straddle-legged on the deck of the Santa Maria speculating about the unseen mysteries that lay beyond the horizon. Although an imaginative man, Columbus would not have guessed that the ocean floor far beneath him once did not exist and that the waters over which he sailed had filled the widening chasm formed where a continent had been torn apart.

Great ocean basins, continents, mountains, and plains appear to us to be everlasting. The human life span is far too short to permit a view of lands heaving and moving about. People are allowed time only to witness an occasional earthquake or volcanic eruption. Fortunately, the myopic view of the earth can be corrected with the immense perspectives provided by historical geology. Along the banks of streams and in the rocks exposed at the sides of our highways, it is possible to examine the cumulative effects of scores of geologic events. One may also observe the unrelenting degradation of the lands by water and ice, and it becomes apparent that, if continents were unchangeable, plateaus and mountains would have long ago been reduced to low-lying plains. That highlands remain at all attests to the fact that the degradational forces of weathering and erosion must be balanced by internal forces that uplift and rebuild. The occurrence of upheavals is further demonstrated by layers of fossiliferous marine sediments that are now poised in cloud-flanked mountain peaks. Crustal uplift is implicit in exposures of originally deep-seated rocks, such as granite. These granitic terrains are exposed to view only after removal of miles of overlying rocks beneath which the granite had congealed. This removal could not have been accomplished without a comparable amount of uplift. Elsewhere, one may observe broadly meandering streams with a pattern of sinuosity normally found in low-lying regions of the continents. They are now incising deep canyons, for the low-lying lands over which they once wandered have been elevated high above sea level. Such uplifts, along with the many other movements of the earth's crust, break and deform rocks and produce geologic structures such as faults, folds, and unconformities. A brief

Figure 8–1 Fault developed in a sequence of sandstones, shales, and siltstones. Arrows indicate directions of relative movement. (a) is a fault plane that dips about 40° and contains about 4 cm of clayey fault gouge; (b) is a gently dipping sandstone stratum; (c) is shale; (d) is siltstone. (Photograph courtesy of United States Geological Survey and W. L. Adkison.)

Figure 8–2 Fault scarp exposed north of Bitterwater, California. (Photograph courtesy of J. C. Brice.)

discussion of these structures will provide the necessary background for discussion of our principal topic, plate tectonics.

Geologic Structures

A **fault** is a break in the earth's crustal rocks along which there has been movement (Figs. 8–1, 8–2, and 8–3). According to the relative movement of the rocks on either side of the plane of breakage, faults can be classified as **normal, reverse,** or **lateral** (Fig. 8–4). Lateral faults can be further designated as right lateral or left lateral if one looks across the fault zone to see if the opposite block moved to the right or left. Reverse faults in which the shear plane is inclined only a few degrees from

Figure 8–3 A mountain range formed by faulting: the Grand Tetons of Wyoming viewed from Signal Mountain.

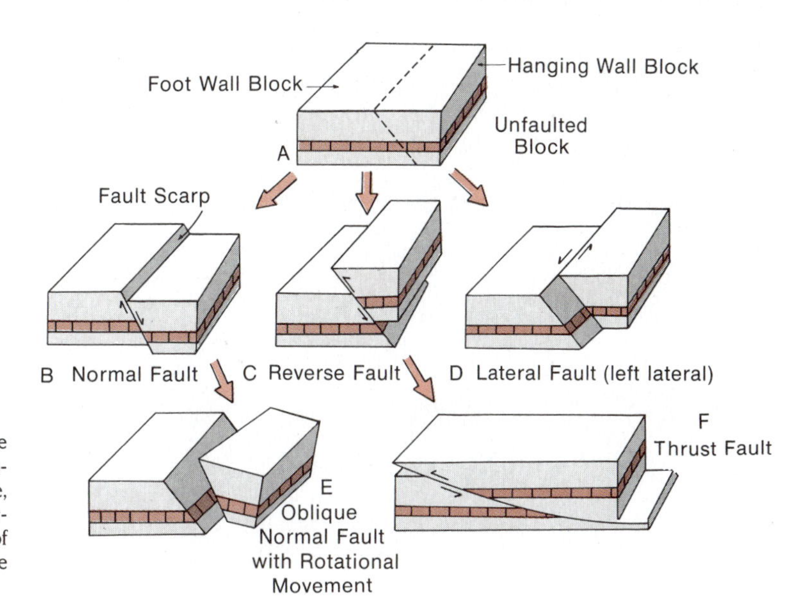

Figure 8–4 Kinds of faults. A shows the unfaulted block with the position of the potential fault shown by dashed line. In nature, movements along faults may vary in direction, as shown in E. A thrust fault is a type of reverse fault that is inclined at a low angle from the horizontal.

the horizontal are termed **thrust faults.** In normal faults, the mass of rock that lies above the shear plane and that is termed the **hanging wall** appears to move downward relative to the opposite side or foot wall. Such faults occur where rocks are subjected to tensional forces—forces that tend to stretch the crust. On the other hand, reverse faults exhibit a hanging wall that has moved up relative to the **foot wall.** If we imagine ourselves holding two blocks cut from wood like those of Figure 8–4, we find that they must be shoved together, or compressed, in order to cause reverse as well as thrust faulting. Thus, regions of the earth's crust containing numerous reverse faults (and folded strata as well) are likely to have experienced compression at some time in the geologic past. Although examples of all kinds of faults may be found in any mountain belt, compressional structures are decidedly the most prevalent in the world's great mountain ranges.

No less important than faults as evidence for the earth's instability are the bends in rock strata that are termed **folds** (Figs. 8–5, 8–6, and 8–7). Sediments are, of course, originally laid down in approximately level layers. As noted earlier, the perceptive naturalist Nicolaus Steno referred to this

Figure 8–5 Anticline in Early Paleozoic strata in the Neptune Range, Pensacola Mountains, Antarctica. (Photograph courtesy of J. R. Ege, United States Geological Survey.)

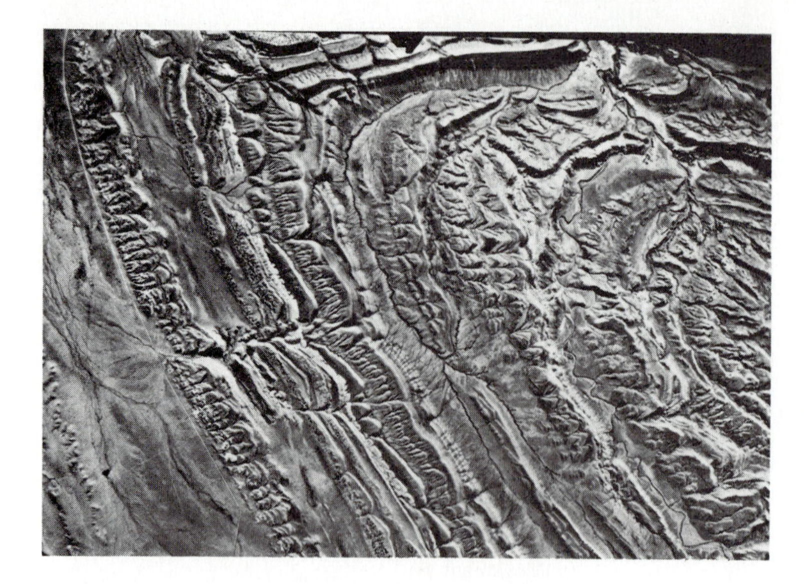

Figure 8–6 An anticlinal fold in Wyoming. Because of the sparseness of vegetation and soil cover, rock layers are clearly exposed and differences in resistance to erosion cause layers to be etched into sharp relief. (Photograph courtesy of the United States Geological Survey.)

observation as the "law of original horizontality." Steno correctly inferred that if strata were originally horizontal, then folds, flexures, or inclinations of those beds were direct evidence of crustal movement. The principal categories of folds are **anticlines, synclines, domes, basins,** and **monoclines** (Fig. 8–8).

Although most folds (and particularly those found in mountain belts) are formed by compression (Fig. 8–9), some may develop under a variety of other circumstances. For example, some relatively small folds may result from differential compaction as loose sediment is converted to sedimentary rock, from slumping of mounds of sediment at the time of its accumulation, from an upward protuberance of rock masses from below, from draping over uplifted fault blocks, and from crumpling as great blankets of strata slide over an older rock surface.

Other geologic structures, distinctly different from either folds or faults, also provide evidence for crustal movements. These structures are called **unconformities.** An unconformity is a buried erosion surface with older rocks lying beneath the surface of erosion and markedly younger rocks resting above that surface. Because erosion has removed an unknown thickness from the underlying strata and there has been no accumulation of sediment and fossils for a period of time, the unconformity represents a gap or break of significant

Figure 8–7 Ouachita Mountains of Arkansas. The landscape here has been developed on folded rocks. Because rainfall here is considerably greater than in the area shown in Figure 8–6, minor differences in rock resistance are obscured by soil and vegetation. (Photograph courtesy of the United States Geological Survey.)

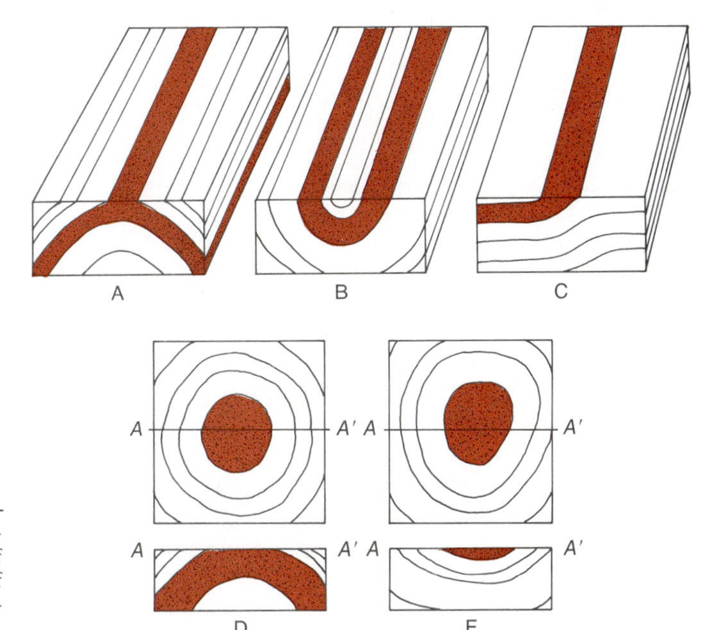

Figure 8–8 Kinds of folds. (A) Anticline. (B) Syncline. (C) Monocline. (D) Map view and cross section of a dome. Notice older strata are in center of outcrop pattern. (E) Map view and cross section of a basin. Younger rocks occur in central area of outcrop pattern.

duration in the geologic record of the region in which it is found. Indeed, the hiatus may amount to tens or even hundreds of millions of years.

The types of unconformities illustrated in Figure 8–10 differ regarding the orientation of the rocks beneath the erosional surface. Of the types shown, the **angular unconformity** provides the most readily apparent evidence for crustal deformation. The eighteenth century geologist James Hutton described what later became known as an

Figure 8–9 Complex folding within the Arkansas Novaculite formation at Caddo Gap, Arkansas. Note nearly vertical fault along left side of photograph. (Photograph courtesy of C. G. Stone, Arkansas Geological Commission.)

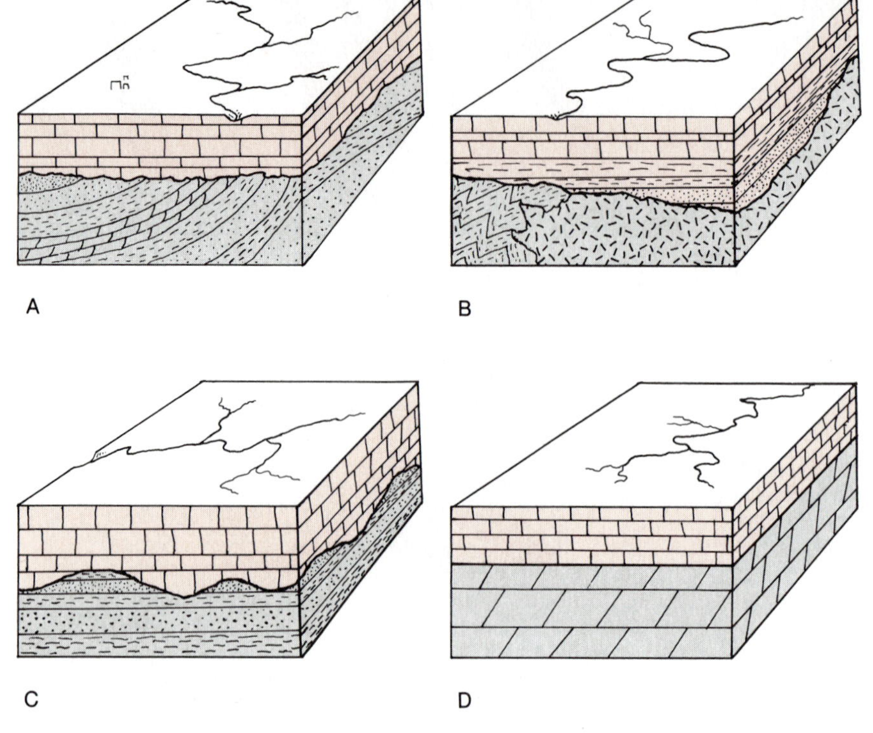

Figure 8–10 Types of unconformity. (A) Angular unconformity; (B) nonconformity; (C) disconformity; (D) paraconformity.

angular unconformity at Siccar Point on the Scottish coast of the North Sea. Using the rules of superposition and original horizontality, he was able to infer the sequence of events suggested in Figure 8–11. For example, Hutton realized that according to the concept of original horizontality, the once level Silurian beds below the surface of the unconformity must have buckled and folded after they were deposited, and then a lengthy period of erosion followed, during which the mountains formed by the folding were worn down. Subsequently, sediments of Devonian age were deposited on the truncated and tilted edges of the older Silurian rocks.

Examples of unconformities are abundant on every continent. Some do not reflect the degree of deformation apparent in the strata at Siccar Point but rather document the simple withdrawal of the sea for a period of time, followed by another marine transgression. The result may be a **disconformity,** in which parallel strata are separated by an erosional surface. The withdrawal and advance of the sea may be caused by fluctuations in the volume of ocean water, but more commonly it is the result of crustal uplift and subsidence. **Nonconformities** are surfaces at which stratified rocks rest on older intrusive igneous and metamorphic rocks. The crystalline rocks were emplaced deep within the roots of ancient mountain ranges that subsequently experienced repeated episodes of erosion and uplift. Eventually, the igneous and metamorphic core of the mountains lay exposed and provided the surface upon which the younger strata were deposited. The final type of unconformity is termed a **paraconformity** and consists only of a bedding plane between parallel strata. Paraconformities record a pause in deposition and therefore, like other unconformities, represent a gap in the geologic record. Paraconformities can be recognized only when fossils, or minerals that can be dated radiogenically, from the beds above and below indicate that there is an unrecorded segment of geologic time.

PLATE TECTONICS

Perhaps more than other scientists, geologists are accustomed to viewing the earth in its entirety. It is their task to assemble the multitude of observations about minerals, earthquakes, and landforms and then to weave them into a coherent view of the whole earth. Recently, a new theory has been developed that provides a marvelous base for as-

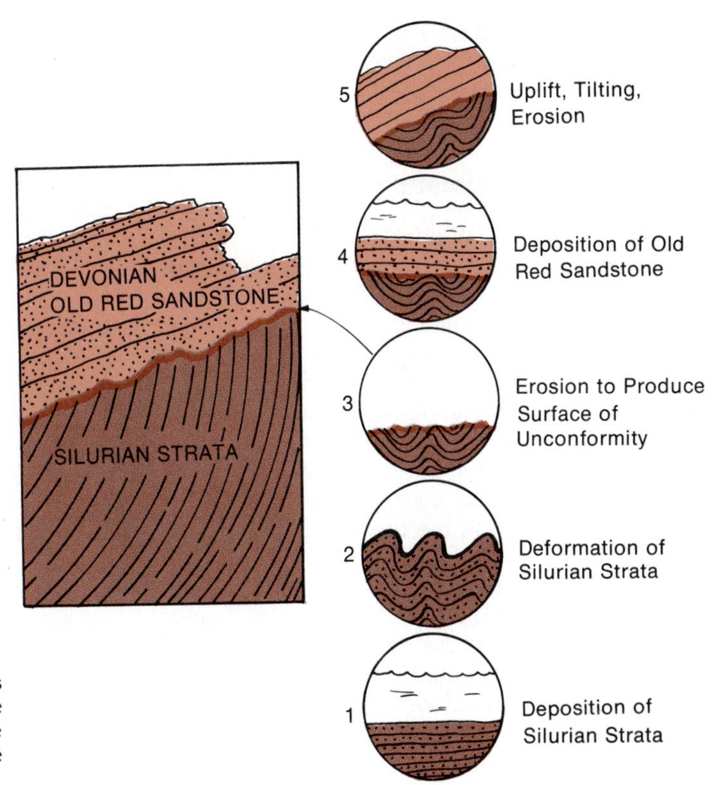

Figure 8–11 Simplified sketch of part of "Hutton's unconformity" (compare with photograph, Figure 1–12) and a diagrammatic representation of the historical sequence of events that produced the geologic structure at Siccar Point, Scotland.

sociating a great variety of geologic information. The new concept has been named **plate tectonics,** and it has taken on the dimensions of a scientific revolution. The new ideas were late in coming. Perhaps the most important reason for their tardy arrival was simply that, until recently, there were large regions of the globe that could not be adequately studied. This was particularly true of the vast realms beneath the oceans. For 2 centuries, the continents, which constitute only about 30 per cent of the earth's surface, received most of the attention. Technologic problems and formidable costs so limited exploration of the ocean floors that oceanographic research was possible only for relatively few earth scientists.

The lack of adequate information about the ocean floors began to be corrected in the years following World War II. Research related to naval operations produced submarine detection devices that also proved useful in measuring magnetic properties of rocks. In the mid-1940's to late 1950's, the need to monitor atomic explosions resulted in the establishment of a worldwide network of seismometers. This network provided the precise information about the global pattern of earthquakes. The magnetic field over large portions of the sea floor was soon to be charted by the use of newly developed and delicate *magnetometers.* Other technologic advances ultimately permitted scientists to examine rock that had been carefully dated by radiometric methods and then to determine the nature of the earth's magnetic field at the time those rocks had formed. Geologically recent reversals of the magnetic field were soon detected, correlated, and accurately dated. A massive, federally funded program to map the bottom of the oceans was launched, and depth information from improved echo depth sounding devices poured into data collecting rooms to be translated into maps and charts.

A new picture of the ocean floor began to emerge (see Figure 1–3). It was at once awesome, alien, and majestic. Great chasms, flat-topped, submerged mountains, boundless abyssal plains, and interminable volcanic ranges appeared on the new maps and begged an explanation. How did the volcanic midoceanic ridges and deep sea trenches originate? Why were both so prone to earthquake activity? Why was the Atlantic Ridge so nicely centered and parallel to the coastlines of the continents on either side? As the topographic, magnetic, and geochronologic data accumulated, the relationship of these questions became apparent. An old theory called **continental drift** was re-examined, and the new, more encompassing theory of plate tectonics was formulated. It was an idea whose time had come.

Fragmented Continents

It requires only a brief examination of the world map to notice the remarkable parallelism of the continental shorelines on either side of the Atlantic Ocean. If the continents were pieces of a jigsaw puzzle, it would seem easy to fit the great "nose" of Brazil into the re-entrant of the African coastline. Similarly, Greenland might be inserted between North America and northwestern Europe. It is not surprising, therefore, that earlier generations of mapgazers also noticed the fit and formulated theories involving the breakup of an ancient supercontinent.

In 1858 there appeared a work titled *La Création et ses Mystères Devoilés.* Its author, A. Snider, postulated that before the time of Noah and the biblical flood, there existed a great region of dry land. This antique land developed great cracks encrusted with volcanoes, and during the Great Deluge, a portion separated at a north-south trending crack and drifted westward. Thus, North America came into existence.

Near the close of the nineteenth century, the Austrian scientist Eduard Suess became particularly intrigued by the many geologic similarities shared by India, Africa, and South America. He formulated a more complete theory of a supercontinent that drifted apart following fragmentation. Suess called that great land mass **Gondwanaland** after Gondwana, a geologic province in east-central India.

The next serious effort to convince the scientific community of the validity of these ideas was made in the early decades of the twentieth century by the energetic German meteorologist Alfred Wegener. His book, *Die Entstehung der Kontinente und Ozeane* ("The Origin of the Continents and Oceans"), is considered a milestone in the historical development of the concept of continental drift.

Wegener's hypothesis was quite straightforward. Building on the earlier notions of Eduard Suess, he argued again for the existence in the past of a supercontinent that he dubbed **Pangaea.** That portion of Pangaea that was to separate and form North America and Eurasia came to be known as **Laurasia,** whereas the southern portion retained the earlier designation of Gondwanaland. According to Wegener's perception, Pangaea was surrounded by a universal ocean named **Panthalassa,** which opened to receive the shifting continents when they began to split apart some 200 million years ago (Fig. 8–12). The fragments of Pangaea drifted along like great stony rafts on the denser material below. In Wegener's view, the bulldozing forward edge of the slab might be expected to crumple and produce mountain ranges like the Andes.

The assiduous investigations of Wegener were not to go unchallenged. Criticism was leveled chiefly against his notion that the continents slid along through denser oceanic crust in the manner of giant granitic "icebergs." The eminent and scientifically formidable physicist Sir Harold Jeffries calculated that the ocean floor was far too rigid to allow for the passage of continents, no matter what the imagined driving mechanism. It is now known that Jeffries was correct in asserting that continents cannot—and do not—plow through oceanic crust. Shortly, it will be shown that continents do move, but they move only as passive passengers on large rafts of lithosphere that glide over a comparatively soft and plastic upper layer of the earth's mantle. Nevertheless, much of the evidence Wegener and others had assembled can be used to substantiate both the old and the newer concepts. We should have a look at this evidence.

The most convincing evidence for continental drift remains the geographic fit of the continents. Indeed, the correspondence is far too good to be fortuitous, even when one considers the expected modifications of shorelines resulting from erosion, deformation, or intrusions following the breakup of Pangaea some 200 million years ago. A still closer match results when one fits the continents together to include the continental shelves, which are really only submerged portions of the continents. Such a computerized and error-tested fitting of continents was carried out by Sir Edward Bul-

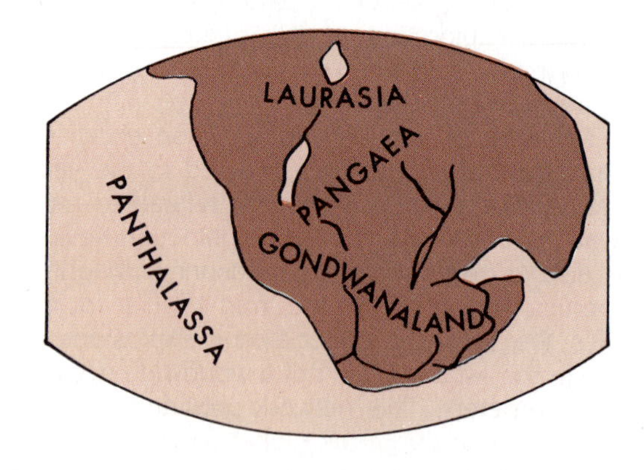

Figure 8–12 Map showing supercontinent Pangaea about 200 million years ago, as conceived by Alfred Wegener.

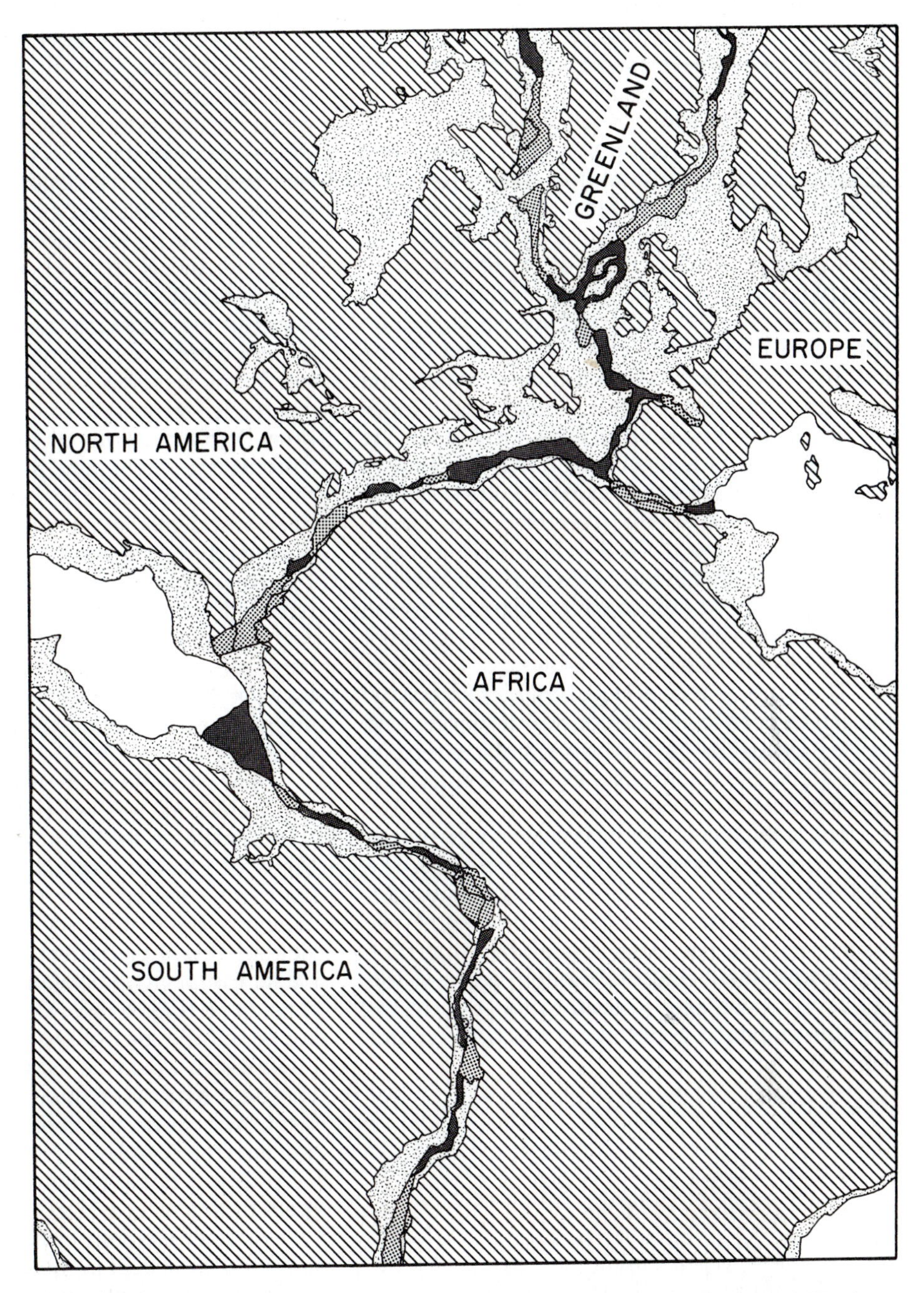

Figure 8–13 Fit of the continents as determined by Sir Edward Bullard, J. E. Everett, and A. G. Smith. The fit was made along the continental slope (light stippled pattern) at the 500-fathom contour line. Overlaps (shown in heavy dots) and gaps (shown in the darkest pattern) are probably the result of deformation and sedimentation after rifting. (Adapted from E. C. Bullard et al. *Philos. Trans. R. Soc. Lond.,* VA 258:41, 1965.)

lard, J. E. Everett, and A. G. Smith of the University of Cambridge (Fig. 8–13). Remarkably, this work showed that over most of the boundary the average mismatch was no more than a degree—a snug fit indeed.

Another line of evidence favoring the drift theory involves sedimentologic criteria indicating similarity of climatic conditions for widely separated parts of the world that were once closely adjacent to one another. For example, in such widely separated places as South America, southern Africa, India, Antarctica, Australia, and Tasmania, one finds glacially grooved rock surfaces and deposits of glacial rubble developed in the course of late Paleozoic continental glaciation. The deposits of poorly sorted clay, sand, cobbles, and boulders are called **tillites.** They, along with the grooves and scratches on rock surfaces that were apparently beneath the moving ice, attest to a great ice age that affected Gondwanaland at a time when it was as yet unfragmented and lying at or near the south polar region (Fig. 8–14). Furthermore, if the directions of the grooves on the bedrock are plotted on a map, they indicate the center of ice accumulation and the directions in which it moved. Unless the southern continents are reassembled into Gondwanaland, this center would be located in the ocean, and great ice sheets do not develop centers of accumulation in the ocean. Hence, the existence of Gondwanaland seems plausible. In a few instances, oddly foreign boulders in the tillites of one continent are found in the deposits of another continent now located thousands of miles across the oceans. Petrologists are able to trace their parent outcrops to the distant land masses.

There are additional clues to paleoclimatology that can be used to test the concept of moving land masses. Trees that have grown in tropical regions of the globe characteristically lack the annual rings resulting from seasonal variations in growth. Exceptionally thick coal seams containing fossil logs from such trees imply a tropical paleoclimate. The locations of such coal deposits should approximate an equatorial zone relative to the ancient pole position for that age. Also, it is evident that such coal seams now being exploited in northern latitudes must have been moved to those locations from the equatorial zones along which the source vegetation accumulated.

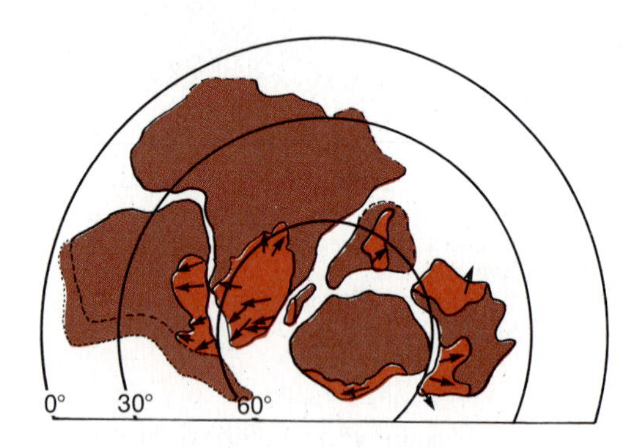

Figure 8–14 Reconstruction of Gondwanaland near the beginning of Permian time, showing the distribution of glacial deposits (shown in orange tint). Arrows show direction of ice movement as determined from glacial scratches on bedrock. (Modified from W. Hamilton and D. Krinsley. *Geol. Soc. Am. Bull.*, 78:783–799, 1967.)

Because of the decrease in solubility of calcium carbonate with rising temperatures, thick deposits of marine limestone also imply relatively warm climatic conditions. Arid conditions, such as those existing today on either side of the equatorial rain belt, can be recognized in ancient rocks by desert sandstones and evaporites. (Evaporites are chemical precipitates such as salt and gypsum that characteristically form when a body of water containing dissolved solids is evaporated.) Today such deposits are ordinarily formed in warm, arid regions located about 30° north or south of the equator. If one believes continents have always been where they are now, then it is difficult to explain the great Permian salt deposits now found in northern Europe, the Urals, and the southwestern United States (Fig. 8–15). The evaporites had probably been precipitated in warmer latitudes before Laurasia migrated northward.

A somewhat similar kind of evidence can be obtained by examining the locations of Permian reef deposits. Modern reef corals are restricted to a band around the earth that is within 30° of the equator. Ancient reef deposits are now found far to the north of the latitudes at which they had originated.

At least some of the paleontologic support for continental drift was well known to Suess and Wegener. Usually, in the Gondwana strata overlying the tillites or glaciated surfaces, there can be found nonmarine sedimentary rocks and coal beds containing a distinctive assemblage of fossil plants. Named after a prominent member of the assemblage, the plants are referred to as the ***Glossopteris* flora** (Fig. 8–16). Paleobotanists who have supported the idea of shifting continents have argued that it would be virtually impossible for this complex temperate flora to have developed in identical ways on the southern continents as they are separated today. The seeds of *Glossopteris* were far too heavy to have been blown over such great distances of ocean by the wind.

Another element of paleontologic corroboration for the concept of moving land masses is provided by the distribution in the Southern Hemisphere of fossils of a small aquatic reptile named ***Mesosaurus*** (Fig. 8–17). An interpretation, based on its skeletal remains, of this animal's habits, as well as the nature of the sediment in which its fossils are found, strongly suggests that it once inhabited lakes and estuaries and was not an inhabitant of the open ocean. The discovery of fossil remains of *Mesosaurus* both in Africa and in South America lends credence to the notion that these continents were once attached. Perhaps for a time, as they were just beginning to separate, the location

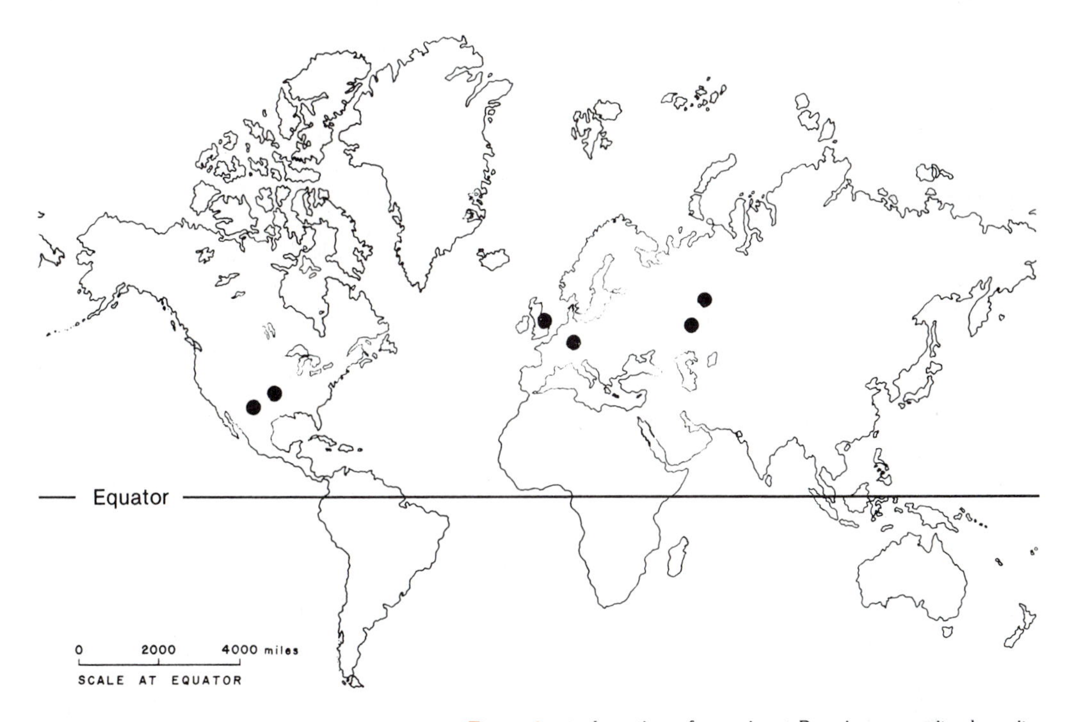

Figure 8–15 Location of prominent Permian evaporite deposits.

Figure 8–16 A bouquet of leaves from members of the *Glossopteris* flora. (A) *Sphenopteris.* (B) *Schizoneura.* (C) *Glossopteris.* (D) *Sphenophyllum.* (C) *Gangamopteris.*

Figure 8–17 *Mesosaurus.*

of the present coastlines became dotted with lakes and protected bodies of water that harbored *Mesosaurus.* This seems a more plausible explanation than one requiring the creature to navigate the South Atlantic or to wend its way across formidable latitudinal climatic barriers by following the shorelines northward and around the perimeter of the Atlantic.

Before fragmentation of a supercontinent such as Laurasia, one would expect to find numerous similar plants and animals living at corresponding latitudes on either side of the line of future separation. Such similarities do occur in the fossil record of continents now separated by extensive oceanic tracts. Faunas of Silurian and Devonian fishes, for example, are comparable in such now distant locations as Great Britain, Germany, Spitzbergen, eastern North America, and Quebec. These similarities, not only in fishes but also in amphibians, persist into the Carboniferous Period. In the Permian and periods of the Mesozoic Era, one again finds striking similarities in the reptilian faunas of Europe and North America. The fossil evidence implies not only the former existence of continuous land connections but also a uniformity of environmental conditions between the elements of Laurasia that once were at approximately similar latitudinal zones.

When one turns to Cenozoic mammalian faunas, however, one finds the situation to be quite different. Distinctive faunal elements are evident on separate continents and are especially evident in Australia, South America, and Africa. Apparently, as the continents became separated from each other by ocean barriers, genetic isolation resulted in morphologic divergence. The modern world's enormous biologic diversity is at least partially a result of evolutionary processes operating on more or less isolated continents. The faunas and floras during the periods prior to the breakup of Pangaea were less diverse.

The character, sequence, age, and distribution of rock units have also been examined for insights into concepts of drift. One might presume that locations close to one another on the hypothetic supercontinent would have environmental similarities that would result in resemblances in the kinds of rocks deposited. As indicated by the correlation chart of southern continents (Table 8–1), there is such a similarity in the geologic sections of now widely separated land masses. The sections begin with glacial deposits such as the Dwyka Tillite of Africa and the Itararé Tillites of South America. These cobbly layers are overlain by nonmarine sandstones and shales containing the *Glossopteris* flora and layers of coal. This overlaying may indicate a warming trend from boreal glacial to more temperate climates. However, one must be careful with such interpretations, for our most recent continental glaciations occurred under a cool temperate climatic regime. The next higher group of strata shows evidence of deposition under terrestrial conditions with an abundance of alluvial, eolian, and stream deposits that contain fossils of an interesting group of mammal-like reptiles. The uppermost beds of the sequences often include basalts (Fig. 8–18) and other volcanic rocks that are of similar age and composition in India and Africa but are somewhat younger in South America. The absence of the reptile-bearing Triassic strata and the Jurassic volcanics in Australia is believed to be the result of Australia's separating from Africa at an early date—probably early in the Permian—and long before the separation of Africa and South America. Australia must have retained just enough of a connection with Africa to permit the entry of dinosaurs and marsupials.

Yet another way to test the notion of a super-

Table 8–1 GONDWANA CORRELATIONS

System	*Southern Brazil*	*South Africa*	*Peninsular India*
Cretaceous	Basalt	Marine Sediments	Volcanics Marine Sediments
Jurassic	(Jurassic Rocks Not Present)	Basalt	Sandstone and Shale Volcanics Sandstone and Shale
Triassic	Sandstone and Shale with Reptiles	Sandstone and Shale with Reptiles	Sandstone and Shale with Reptiles
Permian	Shale and Sandstones with *Glossopteris* Flora Shale with *Mesosaurus*	Shale and Sandstones with Coal and *Glossopteris* Shale with *Mesosaurus*	Shale and Sandstones with Coal and *Glossopteris* Shale
Carboniferous	Sandstone Shale and Coal with *Glossopteris* Tillite	Sandstone Shale and Coal with *Glossopteris* Tillite	Tillite

continent is to see whether or not geologic structures, such as the trends of folds and faults, match up when now distant continents are hypothetically juxtaposed. Such correlative trends do exist. Folds and faults are often difficult to date, although if successful one can establish the contemporaneity of a fault lineage or fold axis now on separated continents. One may also examine the outcrop patterns of Precambrian basement rocks and discern correlative boundaries between now widely separated continents. A folded geosynclinal sequence of Precambrian strata in central Gabon, for example, can be traced into the Bahia Province of Brazil. Also, isotopically dated Precambrian rocks of West Africa can be correlated with rocks of similar age in northeastern Brazil.

Figure 8–18 Drakensberg Mountains of South Africa. The range is composed of a sequence of thick lava flows, some of the more prominent of which can be seen as bluffs near the top of the range. (From J. S. Watkins, M. L. Bottino, and M. Morisawa. *Our Geological Environment.* Philadelphia, Saunders College Publishing, 1975.)

The Testimony of Paleomagnetism

One can sympathize with Alfred Wegener in his losing battle with the imposing Sir Harold Jeffries. As previously noted, Wegener's ideas came along a half century too early. The scientific discoveries of the past 2 decades would have provided him with evidence that even Jeffries would have had difficulty in refuting. The new information came from the study of magnetism that had been imparted to ancient minerals and rocks and preserved down to the present time. To understand this **paleomagnetism,** as it is called, it is necessary to digress for a moment and consider the general nature of the earth's present magnetic field.

The Earth's Present Magnetism

It is common knowledge that the earth has a magnetic field. It is this field that causes the alignment of a compass needle. The origin of the magnetic field is still a question that has not been fully resolved, but many geophysicists believe it is generated as the rotation of the earth causes slow movements in the liquid outer core. The magnetic lines of force resemble those that would be formed if there were an imaginary bar magnet extending through the earth's interior. The long axis of the magnet would be the conceptual equivalent of the earth's magnetic axis, and the ends would correspond to the north and south geomagnetic poles (Fig. 8–19). Although today the geomagnetic poles are located about 11° of latitude from the rotational axis, they slowly shift position. When averaged over several thousand years, the geomagnetic poles and the geographic poles do coincide. If we do assume that this relationship has always held true, then by calculating ancient magnetic pole positions from paleomagnetism in rocks, we have coincidentally located the earth's geographic poles. It should be kept in mind, however, that such interpretations are based on the supposition that the rotational and magnetic poles have always been relatively close together. This seems a reasonable assumption based on the modern condition, as well as on paleontologic studies that have shown inferred ancient climatic zones in plausible locations relative to ancient pole positions. Another assumption is that the earth has always been dipolar. Paleomagnetic studies from around the world thus far support this supposition.

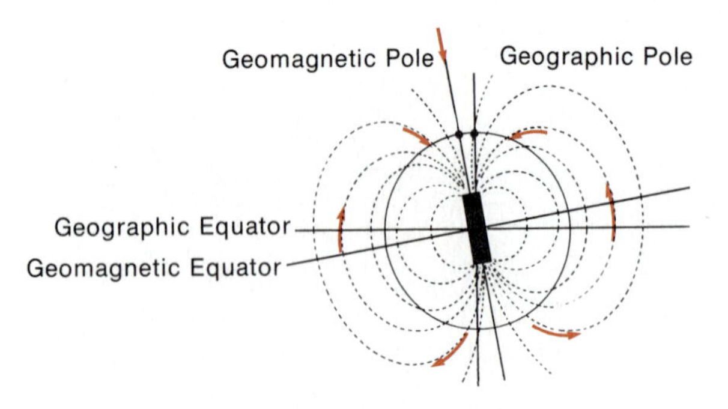

Figure 8–19 Magnetic lines of force for a simple dipole model of the earth's magnetic field.

Remanent Magnetism

The magnetic information frozen into rocks may originate in several ways. Imagine for a moment the outpouring of lava from a volcano. As the lava begins to cool, magnetic iron oxide minerals form and align their polarity with the earth's magnetic field. That alignment is then retained in the rock as its crystallization is completed. In a simple analogy, the magnetic orientations of the minerals responded as if they were tiny compass needles immersed in a viscous liquid. Because they are aligned parallel to the magnetic lines of force surrounding the earth, they not only point the way toward the poles (magnetic declination) but also become increasingly more inclined from the horizontal as the poles are approached (Fig. 8–20). This inclination, when detected in paleomagnetic analysis, can be used to determine the latitude at which an igneous body containing magnetic minerals cooled and solidified.

Another name for the magnetism frozen into ancient rocks is **remanent magnetism,** or **RM.** The type described previously, in which igneous rocks cool past the Curie temperature (also called the "Curie point") of its magnetic minerals, is further classified as *thermoremanent magnetism.* The Curie temperature is simply that temperature above which a substance is no longer magnetic. A few minerals, the most important of which is magnetite (Fe_3O_4), have the property of acquiring RM. Magnetite is widespread in varying amounts in virtually all rocks. The manner in which RM is acquired is complicated but can be explained in a general way. Remanent magnetism in a mineral is ultimately due to the fact that some atoms and ions (charged atoms) have so-called magnetic moments; this means that they behave like tiny magnets. The magnetic moment of a single atom or ion is produced by the spin of its electrons. When magnetite takes on its RM, the iron ions align themselves within the crystal lattice so that their magnetic moments are parallel. Now most igneous rocks crystallize at temperatures in excess of 900°C. At these extreme temperatures, the atoms in the minerals have a large amount of energy and

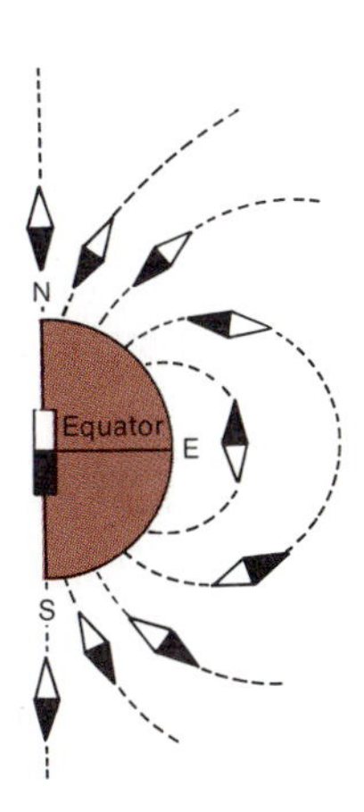

Figure 8–20 A freely suspended compass needle aligns itself in the direction of the earth's magnetic field. The inclination of the needle will vary from horizontal at the equator to vertical at the poles.

are being violently shaken. The vibrating, shaking atoms are unable to line up until more cooling occurs. Finally, when the Curie temperature is reached (578°C for magnetite), enough energy has been lost to allow the atoms to come into alignment. They will stay aligned at still lower temperatures because the earth's field is not strong enough to alter the alignment already frozen into the minerals.

Igneous rocks are not the only kinds of earth materials that can acquire remanent magnetism. In lakes and seas that receive sediments from the erosion of nearby land areas, detrital grains of magnetite settle slowly through the water and rotate so that their directions of magnetization parallel the earth's magnetic field. They may continue to move into alignment while the sediment is still wet and uncompacted, but once the sediment is cemented or compacted, the depositional remanent magnetism is locked in.

Over the past 2 decades, geophysicists have been measuring and accumulating paleomagnetic data for all the major divisions of geologic time. Their results are partly responsible for the recent revival of interest in drift hypotheses. For example, when ancient pole positions were located on maps, it appeared that they were in different positions relative to a particular continent at different periods of time in the geologic past. Either the poles had moved relative to stationary continents, or the poles had remained in fixed positions while the continents shifted about. If the poles were wandering and the continents "stayed put," then a geophysicist working on the paleomagnetism of Ordovician rocks in France should arrive at the same location for the Ordovician poles as a geophysicist doing similar work on Ordovician rocks from the United States. In short, the paleomagnetically determined pole positions for a particular age should be the same for all continents. On the other hand, if the continents had moved and the poles were fixed, then we should find that pole positions for a

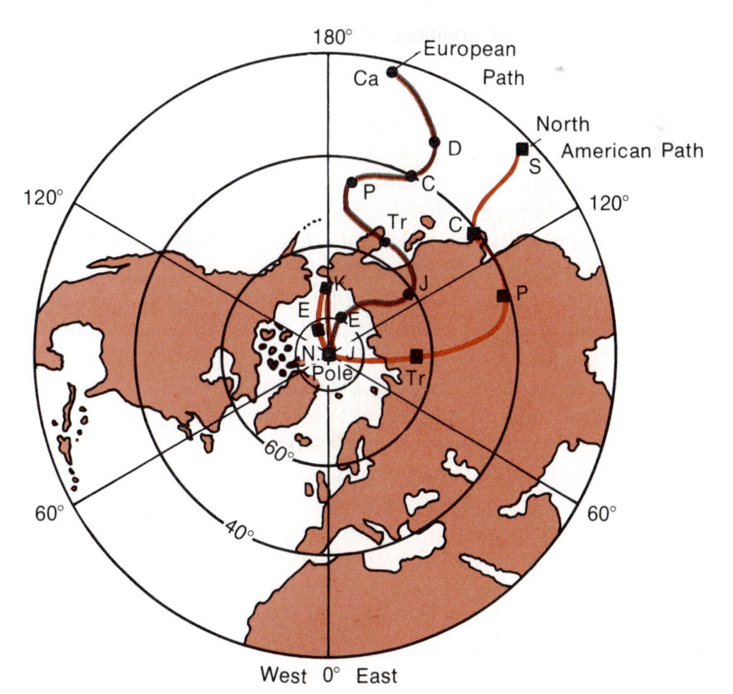

Figure 8–21 Apparent paths of polar wandering for Europe and North America. A scatter of points have been averaged to a single point for each geologic period: *Ca,* Cambrian; *S,* Silurian; *D,* Devonian; *C,* Carboniferous; *P,* Permian; *Tr,* Triassic; *J,* Jurassic; *K,* Cretaceous; *E,* Eocene. (After M. H. P. Bott. *The Interior of the Earth: Structure and Processes.* New York, St. Martin's Press, 1971.)

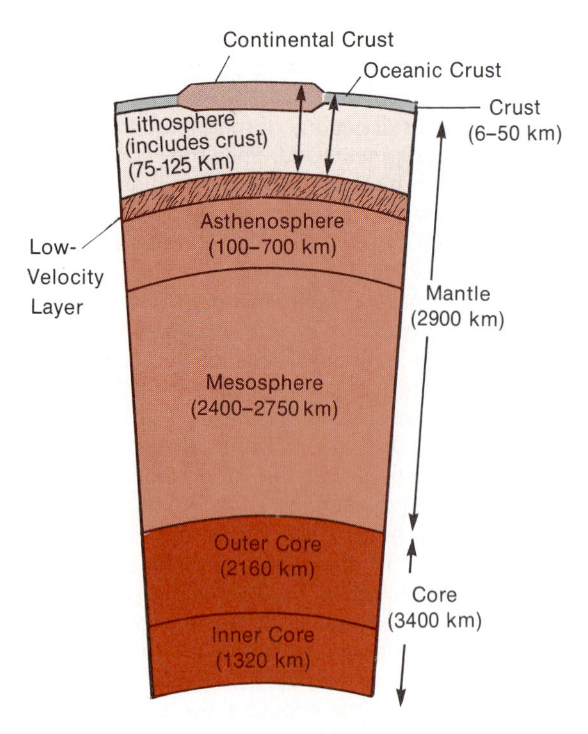

Figure 8–22 Divisions of the earth's interior. (For clarity, the divisions have not been drawn to scale.)

particular geologic time would be different for different continents. The data suggest that this latter situation is the more valid.

Another way to view the data of paleomagnetism is to examine what are called **polar wandering curves** (the name implies that the poles are moving, but as we have just noted, this is not likely). These curves are merely lines on a map connecting pole positions relative to a specific continent for various times during the geologic past. As shown in Figure 8–21, the curves for North America and Europe meet in recent time at the present North Pole. This means that the paleomagnetic data from recently formed rocks from both continents indicate the same pole position. A plot of the more ancient poles results in two similarly shaped but increasingly divergent curves. If this divergence resulted from a drifting apart of Europe and North America, then one should be able to reverse the movements mentally and see if the curves do not come together. Indeed, the Paleozoic portions of the polar wandering curves can be brought into close accord if North America and its curve were to be slid eastward about 30° toward Europe. This sort of information gave new life to the old notion of continental drift. More important, it has become significant in the recent modification of that earlier concept. The modification is known as **plate tectonics.**

The Basic Concept of Plate Tectonics

When considered at the general level, plate tectonics is a remarkably simple concept. The lithosphere, or outer shell of the earth (Fig. 8–22), is constructed of 7 huge slabs and about 20 smaller plates that are squeezed in between them. The larger plates (Fig. 8–23) are approximately 75 to 125 km thick. Movement of the plates causes them to converge, diverge, or slide past one another, and this results in frequent earthquakes along plate margins. When the locations of earthquakes are plotted on the world map, they clearly define the boundaries of tectonic plates (Fig. 8–24).

A tectonic plate containing a continent would have the configuration shown in Figure 8–25. Plates "float" upon a weak, partially molten region of the upper mantle called the **asthenosphere** (from the Greek *asthenos,* meaning "weak"). Geophysicists view the asthenosphere as a region of rock plasticity and flowage. Its presence was first detected by Beno Gutenberg on the basis of changes in seismic wave velocities. It should be noted that the boundary between the lithosphere and the asthenosphere does not coincide with the crust-mantle boundary.

Plate Boundaries and Sea Floor Spreading

Central to the idea of plate tectonics is the differential movement of lithospheric plates. For example, plates tend to move apart at **divergent plate boundaries,** which may manifest themselves as midoceanic ridges complete with tensional ("pull apart") geologic structures. Indeed, the mid-Atlantic ridge approximates the line of separation between the Eurasian and African plate on the one hand and the American plate on the other (Fig. 8–26). As is to be expected, such a rending of the crust is accompanied by earthquakes and enormous outpourings of volcanic materials that are piled high to produce the ridge itself. The void between the separating plates is also filled with this molten rock, which rises from below the lithosphere and solidifies in the fissure. Thus, new crust (i.e., new sea floor) is added to the **trailing edge** of each separating plate as it moves slowly away from the midoceanic ridge (Figs. 8–27 and 8–28). The process has been appropriately named **sea floor spreading.** Zones of divergence may originate beneath continents, rupturing the overlying land mass and producing rift features like the Red Sea and Gulf of Aden.

The axis of spreading is not a smoothly curving line. Rather, it is abruptly offset by numerous

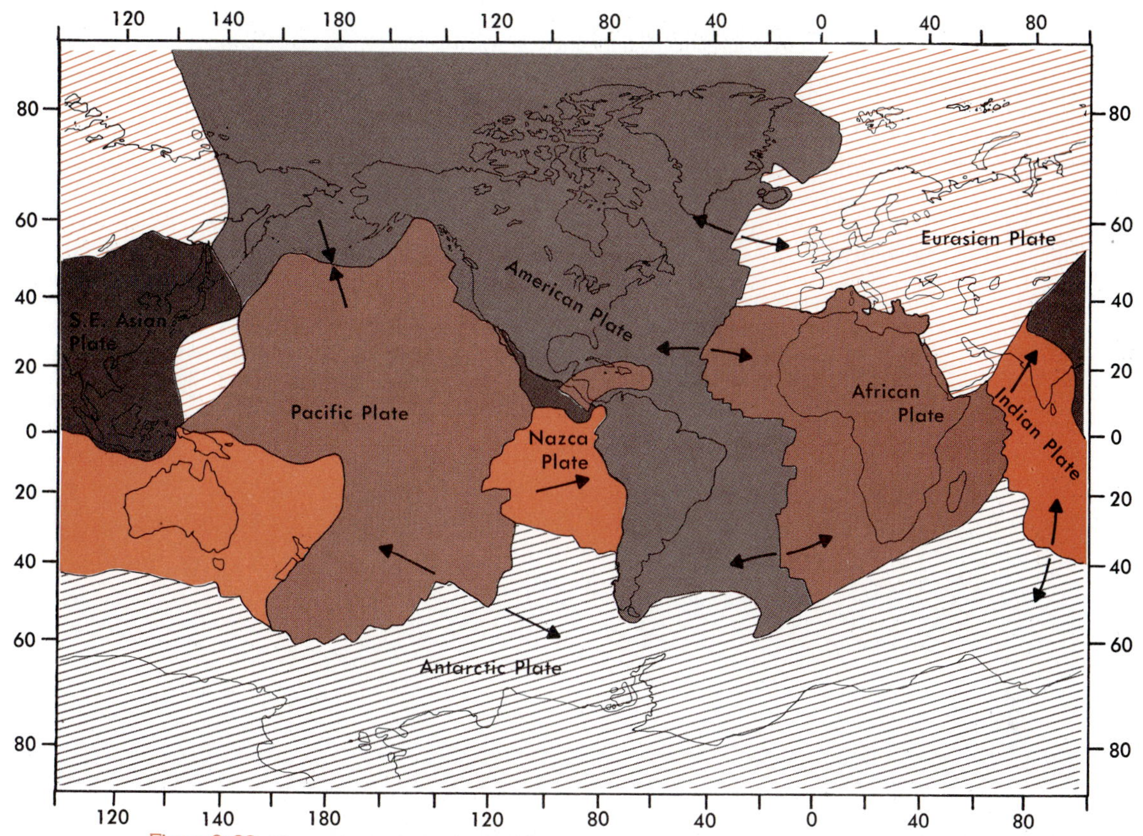

Figure 8–23 The major tectonic plates of the earth. Arrows indicate relative motion of plates as determined from magnetic data. (Modified from Dewey and Bird, 1970; Isacks et al., 1968; Morgan, 1968; and Vine, 1969.)

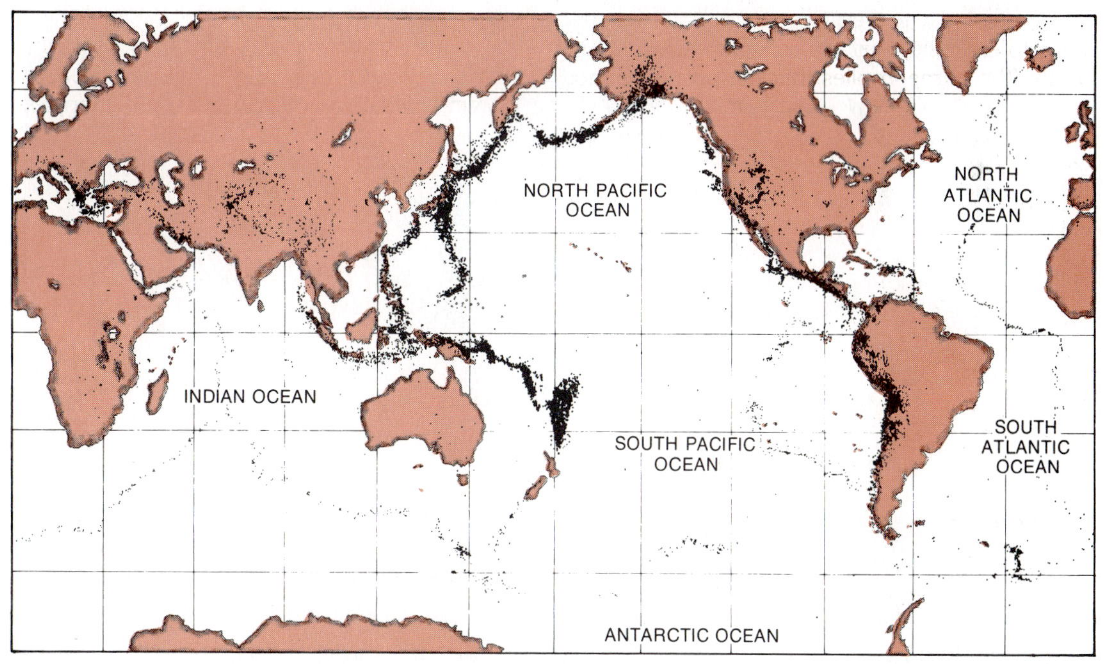

Figure 8–24 World distribution of earthquakes. Notice the major earthquake belt that encircles the Pacific and another that extends eastward from the Mediterranean toward the Himalayas and East Indies. The midoceanic ridges are also the site of many earthquakes. (From J. Turk and A. Turk. Physical Science. 2nd ed. Philadelphia, Saunders College Publishing, 1981.)

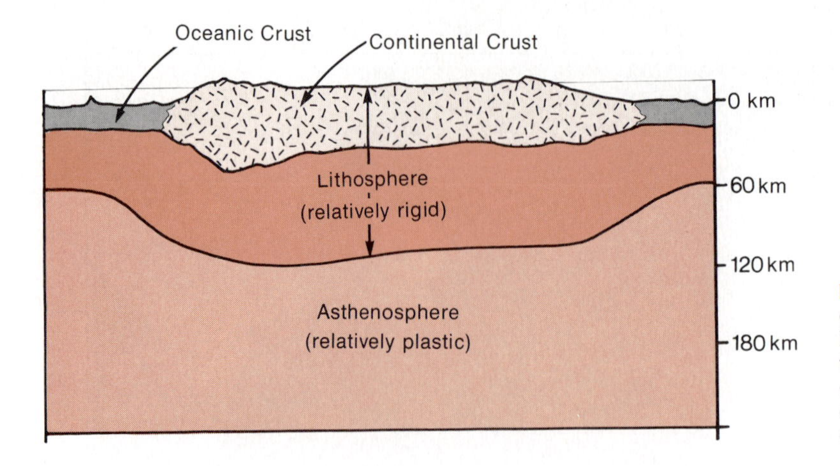

Figure 8–25 The outermost part of the earth consists of a strong, relatively rigid lithosphere, which overlies a weak, plastic asthenosphere. The lithosphere is capped by a thin crust beneath the oceans and a thicker continental crust elsewhere.

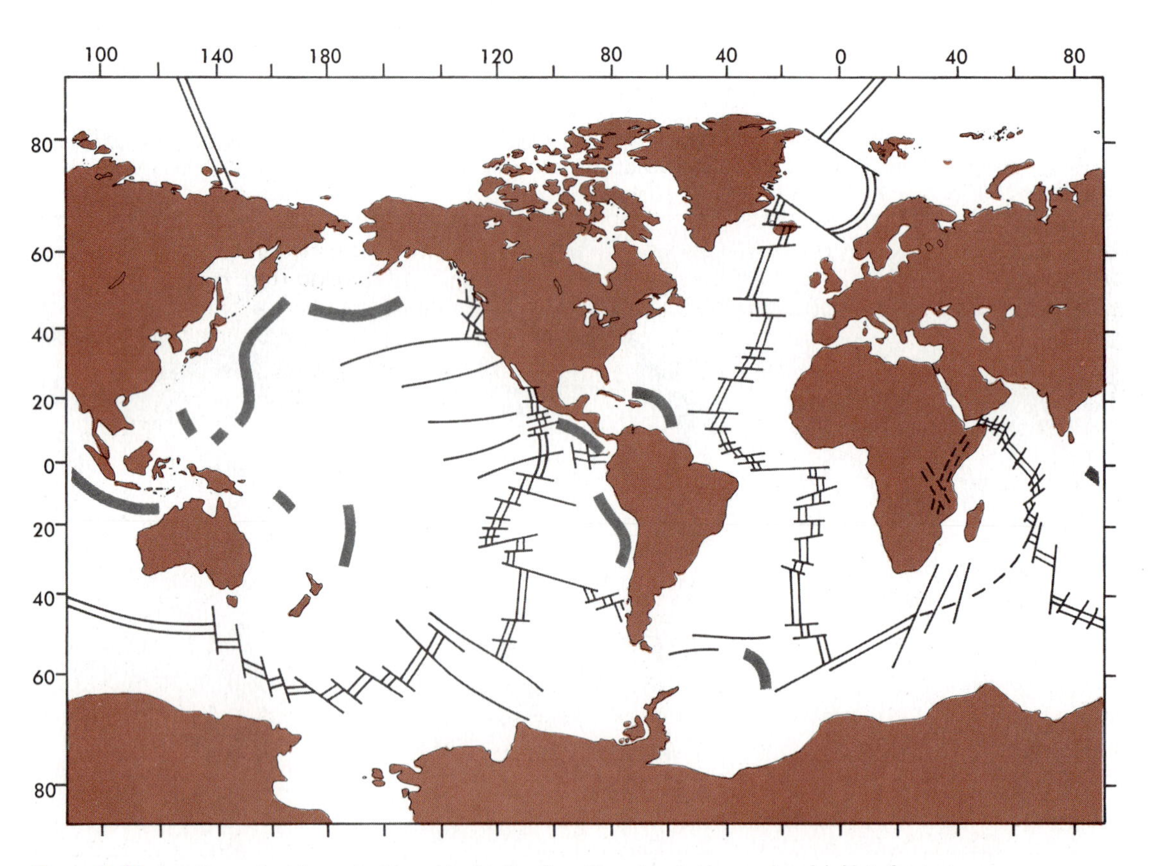

Figure 8–26 Locations of midoceanic ridges (double lines) and trenches (wide gray bands). Note fracture zones offsetting the ridges. Dashed lines in Africa represent the East African Rift Zone. (After Isacks et al. *J. Geophys. Res.*, 73:5855–5899, 1968.)

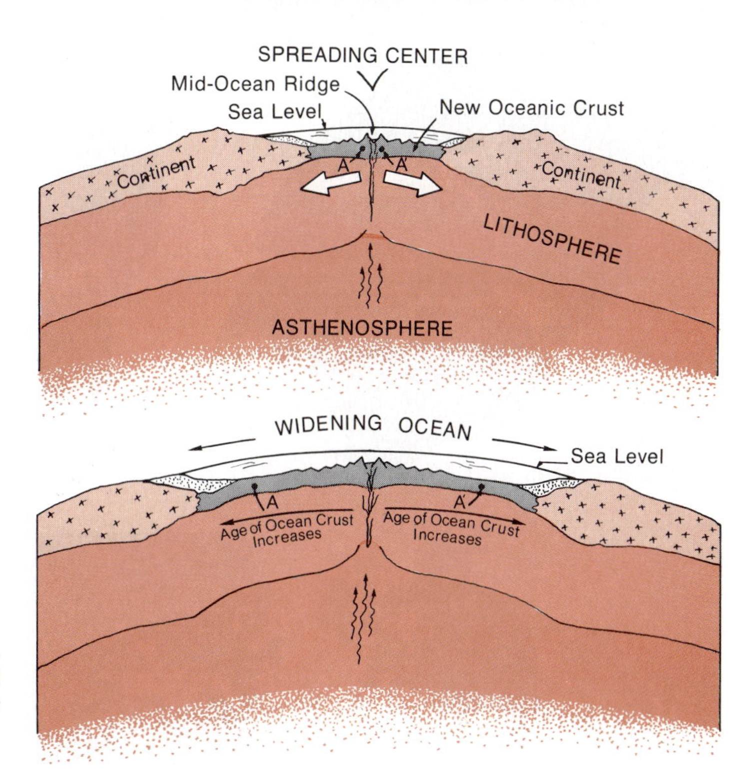

Figure 8–27 Sea floor spreading. As the rift widens, basaltic sea floor upwells to fill the space and becomes a new part of the trailing edge of the tectonic plate. A and A′ are reference points.

faults. These features are termed **transform faults** and are an expected consequence of horizontal spreading of the sea floor along the earth's curved surface. Transform faults take their name from the fact that the fault is "transformed" into something different at its two ends. The relative motions of transform faults are shown in Figure 8–29. The ridge acts as a spreading center that exists both to the north and to the south of the fault. The rate of relative movement on the opposite sides along fault segment X to X′ depends upon the rate of extrusion of new crust at the ridge. Because of the spreading of the sea floor outward from the ridge, the relative movement is opposite to that expected by ordinary fault movement. Thus, at first glance, the ridge-to-ridge transform fault (Fig. 8–29A) appears to be a left lateral fault, but the actual movement along segment X-X′ is really right lateral. Notice, further, that only the X-X′ is really right lateral and that only the X-X′ segment is active. Along this segment, seismic activity is particularly frequent. To the west of X and the east of X′ there is little or no relative movement except for sinking as crustal materials cool.

If plates are receding from one another at one boundary, they may be expected to collide or slide past other plates at other boundaries. Thus, in addition to the divergent plate boundaries that occur along midoceanic ridges, there are **convergent** and **shear boundaries.** Convergent plate boundaries develop when two plates move toward one another and collide. As one might guess, these convergent junctions are characterized by a high frequency of earthquakes. In addition, they are thought to be the zones along which folded mountain ranges or deep sea trenches may develop (Fig. 8–30). The structural configuration of the convergent boundary is likely to vary according to the rate of spreading and whether the leading edges of the plates are composed of oceanic or continental crust. Geophysicists speculate that when the plates collide, one slab may slip and plunge below the other, producing what is called a **subduction zone.** The sediments and other rocks of this plunging plate are pulled downward (subducted), melted at depth, and, much later, rise to become incorporated into the materials of the upper mantle and crust. In some instances, the silicate melts provide the lavas for chains of volcanoes.

An example of a shear plate boundary is the well-known San Andreas Fault in California. Along this great fault, the Pacific Plate moves laterally against the American Plate (Fig. 8–31). Shear plate boundaries are decidedly earthquake-prone but are less likely to develop intense igneous activity. They are the active segments of transform faults along which no new surface is formed or old surface consumed.

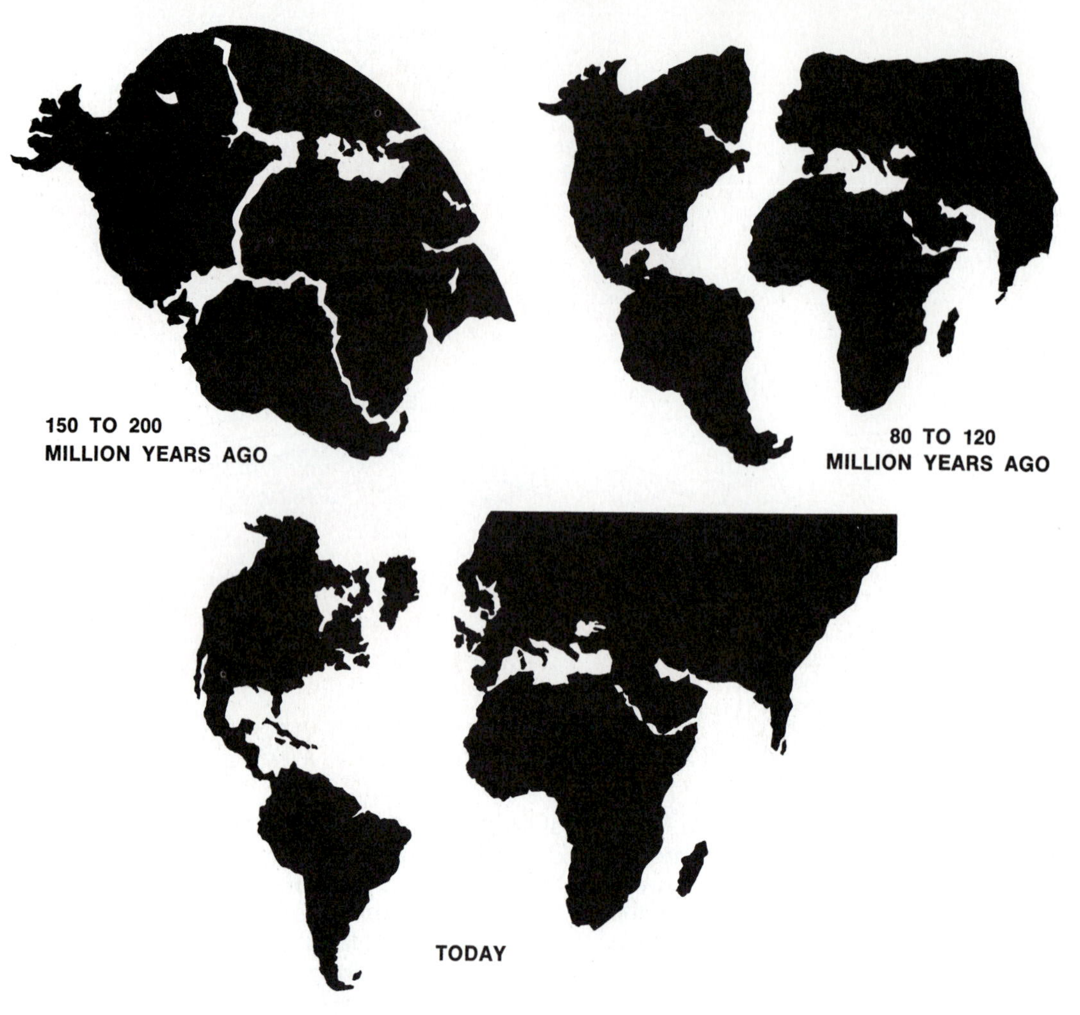

Figure 8–28 Three stages in the opening of the Atlantic Ocean. (From *Deep Sea Searches,* National Science Foundation.)

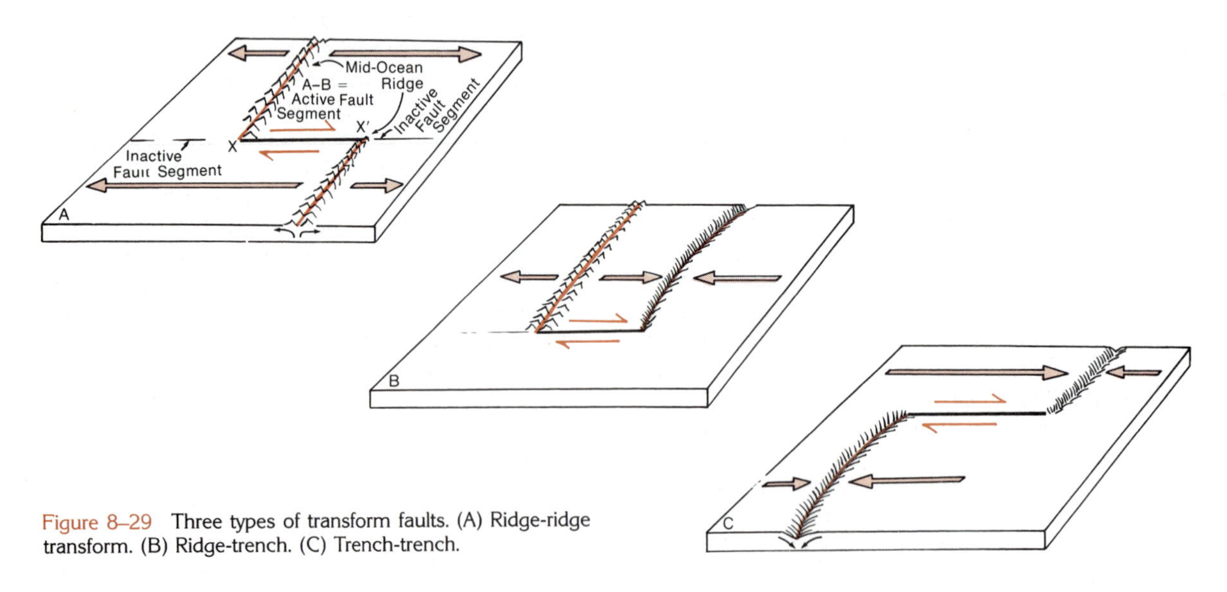

Figure 8–29 Three types of transform faults. (A) Ridge-ridge transform. (B) Ridge-trench. (C) Trench-trench.

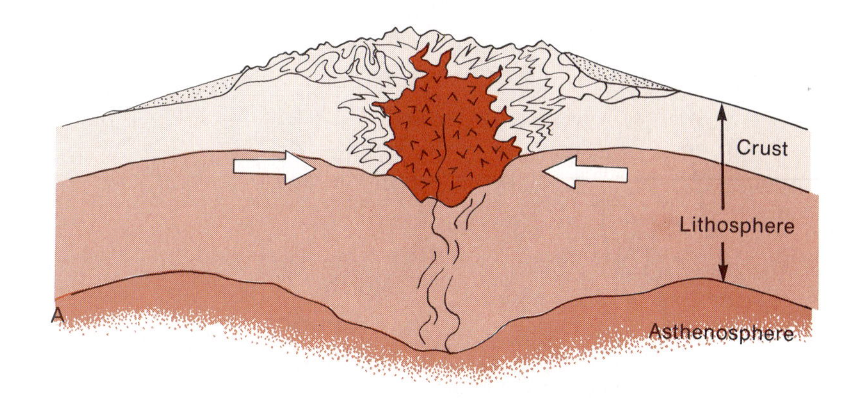

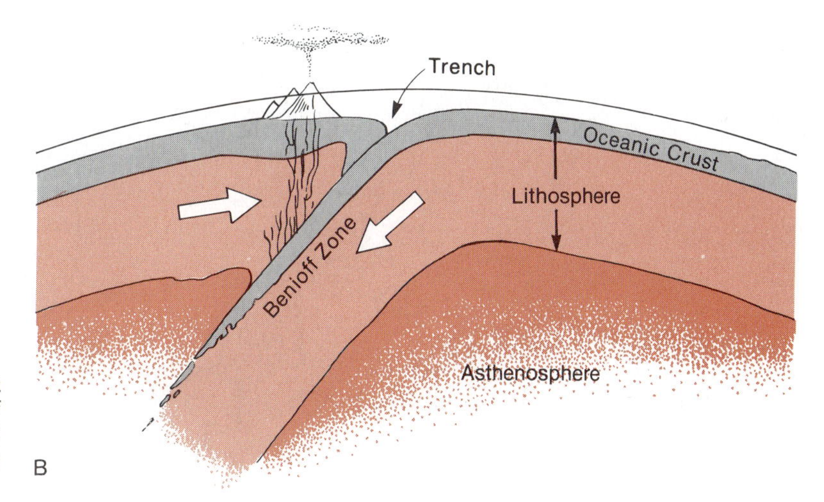

Figure 8–30 Two types of convergent plate boundaries. (A) Convergence of two plates, both bearing continents. (B) Convergence of two plates, both bearing oceanic crust.

Crustal Behavior at Plate Boundaries

We have noted that there are three basic kinds of plate boundaries: divergent, compressional, and shear. There are also three kinds of crustal behavior that may occur at colliding plate edges; however, the behavior may change from one region of a boundary to another. If the leading edge of a plate happens to be composed of continental crust, and it collides with a similarly continental opposing plate margin, the result is a folded mountain range in which are developed igneous rocks of granitic composition (see Figure 8–30A). In this kind of collision, subduction does not occur because the continental plates are too buoyant to be carried down into the asthenosphere. The Himalayan and related ranges were formed when the plates carrying Africa, Arabia, and India converged on the Eurasian Plate. The zone of convergence between the once separate continental masses is called a **suture zone.** The Appalachian and Ural Mountains may also have formed in such a smash-up of continental plate margins. It is likely, for example, that there was an earlier supercontinent than the one envisioned by Wegener. This earlier supercontinent, which can be termed *Pangaea I,* was probably present at least by the end of the Precambrian. *Pangaea I* evidently broke apart early in the Paleozoic, and as the segments began to move away from one another, they created an Appalachian Ocean and Uralian Ocean along the widening rifts. Much later the segments moved back toward one another and converged along a suture line to form the Appalachians and Urals. The recombined continent was the Pangaea of Alfred Wegener, or, as some would name it, *Pangaea II.* Research over the past decade has indicated that as many as six major continents were in existence during the Paleozoic prior to the consolidation that resulted in Wegener's *Pangaea II.* In addition, the matching of structures and stratigraphic sequences in the Late Precambrian suggests that continents have been breaking apart and moving about for at least the last 2.5 billion years.

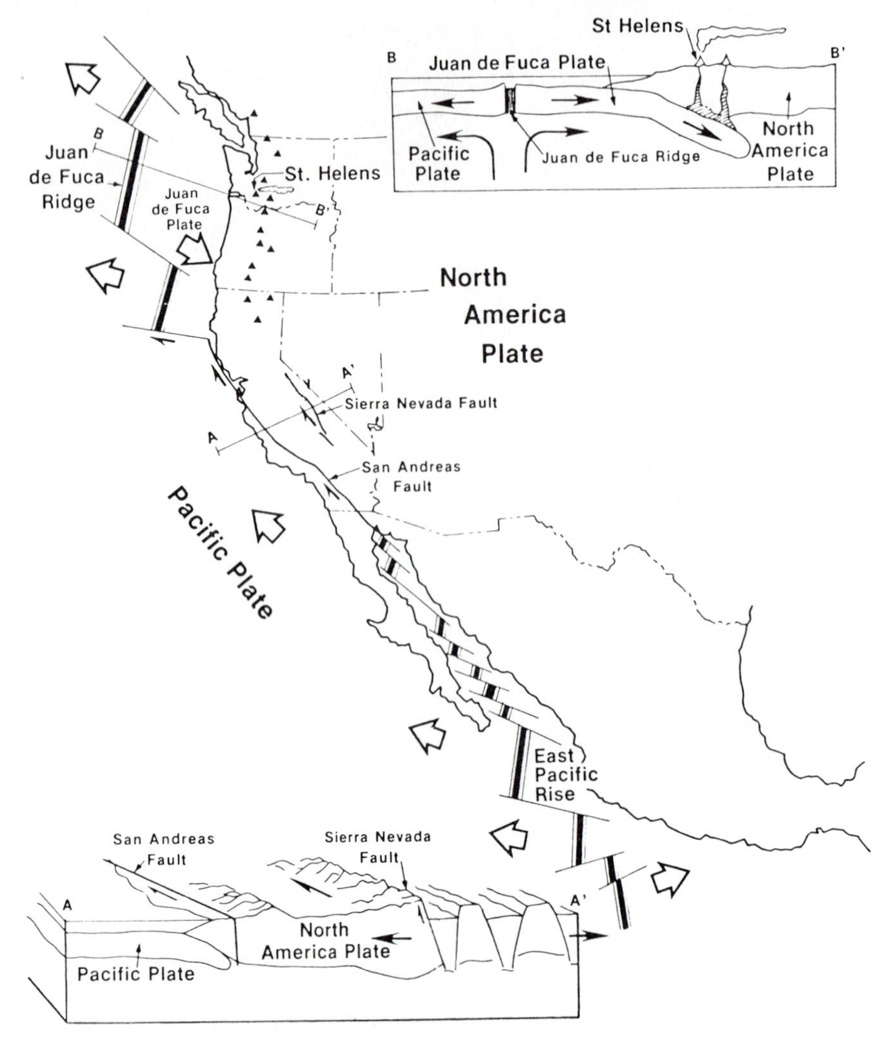

Figure 8–31 The juncture of the North American and Pacific Tectonic Plate. The heavy lines are spreading centers. Note the trace of the San Andreas Fault. To the north in Oregon and Washington, the small Juan de Fuca Plate plunges beneath the North American Continent to form the Cascades. (Courtesy of United States Geological Survey.)

The second kind of convergence situation at plate boundaries involves the meeting of two plates that both have oceanic crust at their converging margins (Fig. 8–30B). Although the rate of plate movement in an ocean-ocean convergence will affect the kinds of structures produced, it is likely that such locations will develop deep sea trenches with bordering volcanic arcs such as those of the southwestern Pacific.

Finally, there is the third possibility. It involves the collision of a continental (granitic) plate boundary with an oceanic (basaltic) one. The result of such a collision might be a deep sea trench located offshore from an associated range of mountains. Numerous volcanoes would be expected to develop, and the lavas pouring from their eruptions would probably be a compositional blend of granite and basalt. The Andes are such a mountain range, and the igneous rock *andesite,* named for its prevalence in the Andes, may represent such a silicate blend.

Regions of ocean-continent convergence (Fig. 8–32) are characterized by rather distinctive rock assemblages and geologic structure. As we have seen, the convergence of two lithospheric slabs results in subduction of the oceanic plate, whereas the more buoyant continental plate maintains its position at the surface but experiences intense deformation, metamorphism, and melting. A great mountain range begins to take form as a re-

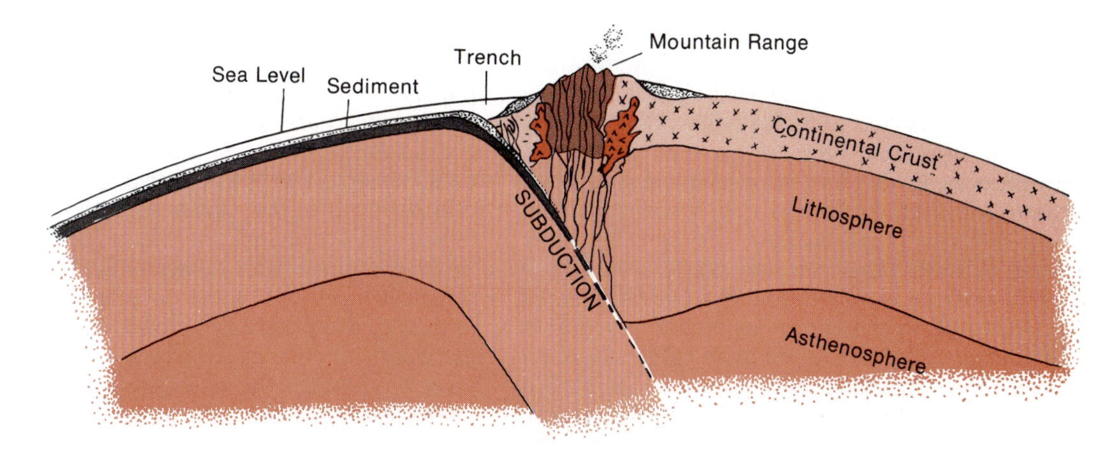

Figure 8–32 Collision of a plate bearing a continent with an oceanic plate. The leading edge of the overriding continental plate is crumpled, whereas the oceanic plate buckles downward, creating a deep sea trench. The situation generalized here is believed to be similar to that off the west coast of South America, where the Nazca Plate plunges beneath the South American Plate.

sult of all of this dynamic activity. At the same time, sediments and submarine volcanic rocks along the subduction zone are squeezed, sheared, and shoved into a gigantic, chaotic medley of complexly disturbed rocks called a **mélange.** Within the mélange one finds a distinctive assemblage of deep sea sediments containing microfossils, submarine lavas, serpentinized peridotite, and gabbro. Altogether, these rocks constitute an **ophiolite suite** (Fig. 8–33). The ophiolite suite rock masses are actually splinters of the oceanic plate that were scraped off the upper part of the descending plate and inserted into the crushed forward edge of the continent. Ophiolites are found in narrow zones within eugeosynclinal belts such as those described in the previous chapter. Indeed, the presence of ophiolites provides one means of identifying the subduction zone once present in old folded mountains. Another clue to the presence of ancient subduction zones is a distinctive kind of metamorphic rock containing blue amphiboles. These rocks are called **blue schists.** They form at high pressures but relatively low temperatures. This rather unusual combination of conditions is characteristic of subduction zones where the relatively cool oceanic plate plunges rapidly into deep zones of high pressure.

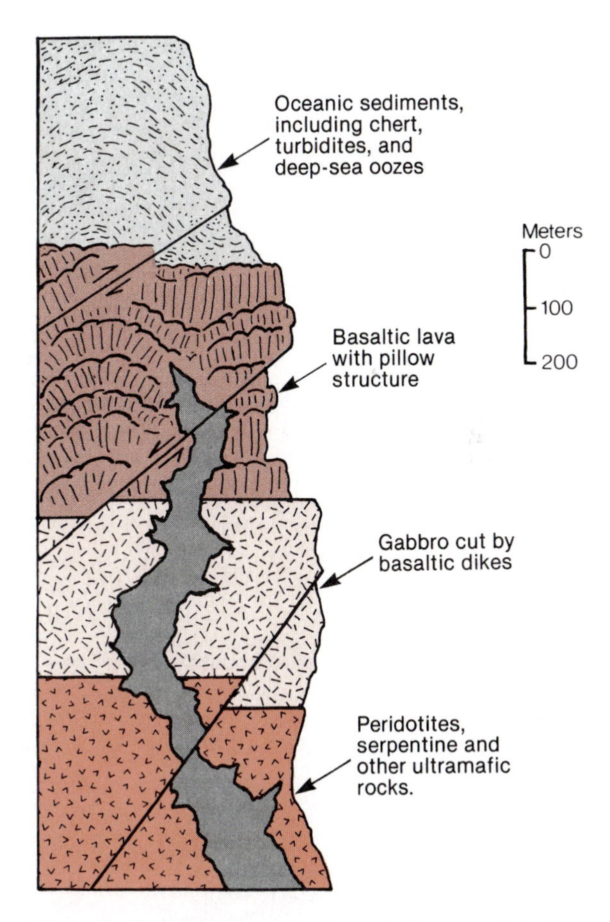

Figure 8–33 Idealized section of an ophiolite suite. Ophiolites are thought to be splinters of the ocean floor squeezed into the continental margin during plate convergence.

What Drives It All?

If we accept the fact that plates of lithosphere do move across the surface of the globe, then the next question to ask concerns how they are moved. The propelling mechanism is thought by some scientists to consist of large thermal convection cells that flow like currents of thick liquid and are provided with heat from the decay of radioactive minerals (Fig. 8–34A). The convection cells are believed to be located in the asthenosphere or, less likely, in the deeper parts of the mantle. Currents are thought to rise in response to heating from below and then to diverge and spread to either side. As they move laterally, the currents carry along with them the overlying slab of lithosphere and its surficial layers of sediment. Mantle material upwelling along the line of separation would join the trailing edges of the plates on either side. Ultimately, the convecting current would encounter a similar current coming from the opposite direction, and both viscous streams would descend into the deeper parts of the mantle to be reheated and moved toward the direction of an upwelling. Above the descending flow, one might expect to find subduction zones and deep sea trenches, whereas midoceanic ridges would mark the locations of ascending flows.

It is possible only to speculate about the size and number of convection cells; it is not improbable that they might have changed through the long course of earth history. Initially, there may have been only one large convecting system that swept together scumlike slabs of continental crust into a single land mass. Later systems may have been characterized by larger numbers of convection cells located closer to the base of the lithosphere. A concept such as this may account for an apparent increase in continental fragmentations in the Mesozoic compared with earlier eras. This idea was originally advanced by the geophysicist S. K. Runkorn, who postulated that growth of the core of the earth and consequent shrinkage of the mantle may have brought on the changes in convection cell characteristics. Later information, however, suggests that after the core formed early in the earth's history, it barely increased in size. This does not, of course, preclude changes in the convection cell patterns for other reasons.

As a mechanism for moving plates of lithosphere, thermal convection may be the best hypothesis thus far advanced. This does not mean that this mechanism should be regarded as an established fact. As yet there has not been an entirely satisfactory way to test the concept, although such

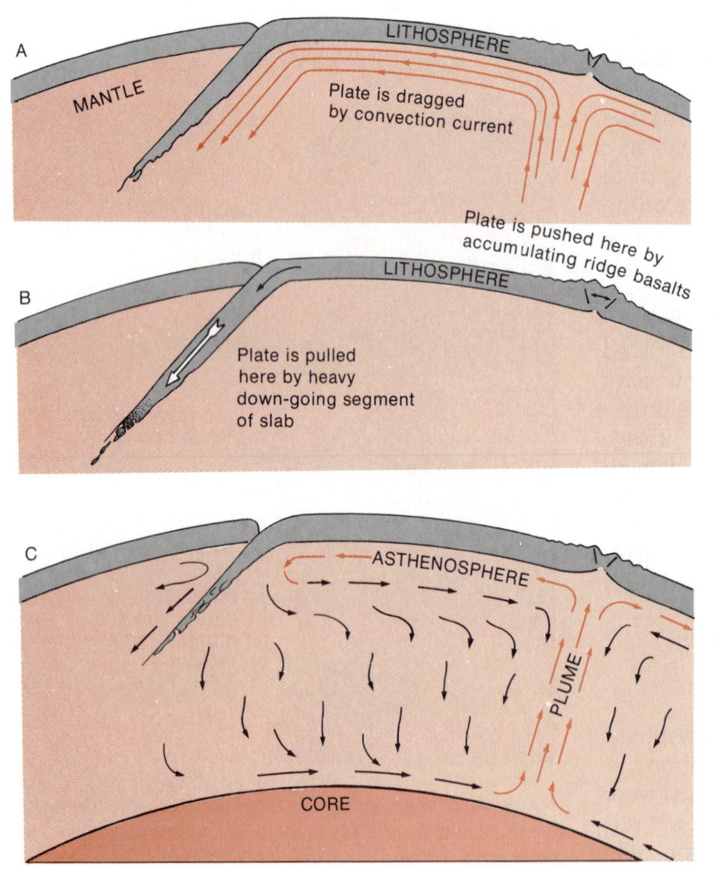

Figure 8–34 Three models that have been suggested as possible driving mechanisms for plate movements. (A) Plate dragged by roll-like convection movements in the mantle. (B) The "push-pull" model in which plates are pushed laterally at spreading centers, and pulled by the plunging cool and dense leading segment of the plate. (C) Thermal plume model in which upward movement is confined to thermal plumes that spread laterally and drag the lithosphere along.

currents may be physically plausible, and there is some tenuous evidence that they exist. For example, heat flows from the earth's interior at a greater rate along midoceanic ridges than from adjacent abyssal plains. Yet analyses of hundreds of measurements by geophysicists at Columbia University suggest that although heat flow along the Atlantic Ridge is 20 per cent greater than on the adjacent floors of the ocean basins, it is still not great enough to move the sea floor at the rates suggested by paleomagnetic studies. Another problem relates to the layering in the upper mantle that has been detected in seismologic studies. The mixing that accompanies convectional overturn would seem to preclude such layering.

The search for a mechanism to drive lithospheric plates has produced another convection-related model. In this so-called *thermal plume model* (Fig. 8–34C), mantle material does not circulate in great rolls but rather rises from near the core-mantle boundary in a manner suggestive of the shape and motion of a thundercloud. Proponents of this model suggest there may be as many as 20 of these thermal plumes, each a few hundred kilometers in diameter. According to the model, when a plume nears the lithosphere, it spreads laterally, doming surficial zones of the earth and moving them along in the directions of radial flow. The center of the Afar Triangle in Ethiopia (Fig. 8–35) has been suggested as the site of a plume that flowed upward and outward, carrying the Arabian, African, and Somali plates away from the center of the triangle. Geophysicists have suggested that other such triple-rift systems in the world have had a similar origin.

Faced with the uncertainties in all existing theories for plate-moving forces, one should keep an open mind about other, perhaps less popular, hypotheses. Among other mechanisms proposed are "slab-pull" and "ridge-push." Slab-pull is thought to operate at the subduction zone, where the subducting oceanic plate, being colder and denser than the surrounding mantle, sinks actively through the less dense mantle, *pulling* the rest of the slab along as it does so. The ridge-push mechanism, which may operate independently of or along with slab-pull, results from the fact that the spreading centers stand high on the ocean floor and have low-density roots. This causes them to spread out on either side of the midoceanic ridge and transmit this "push" to the tectonic plate. Other hypotheses suggest rafting of plates in response to stresses induced by differential rotation between the core and mantle, which might then be transmitted to the crust. In the years ahead, studies of rock behavior under high temperature and pressure, geophysical probings of the mantle, and careful field studies will reveal which of the proposed mechanisms can be most confidently accepted.

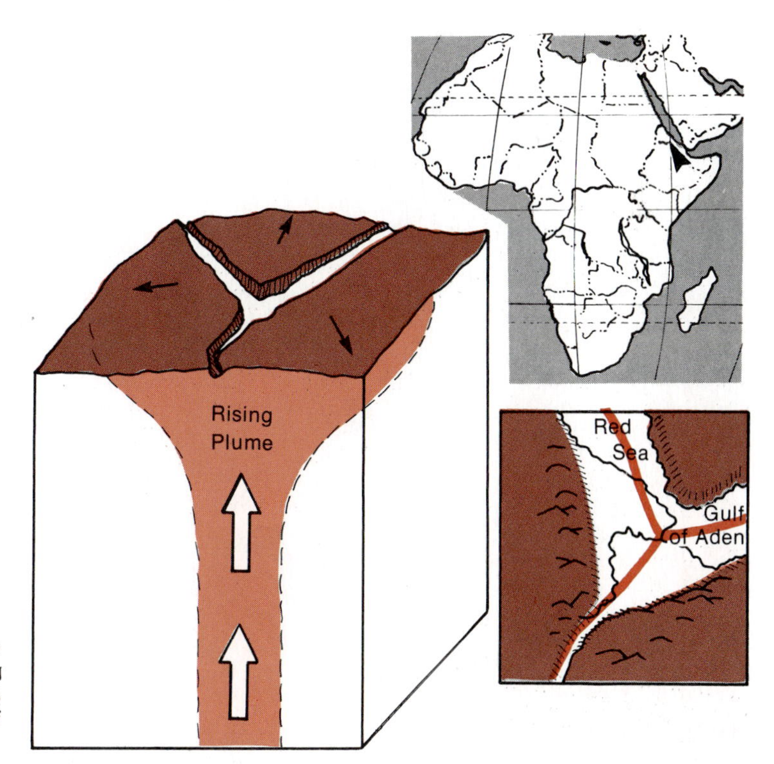

Figure 8–35 Rising plumes of hot mantle material may cause severe rifts, often forming 120° angles with one another. An example is the Afar Triangle, shown at the south end of the Red Sea on the small map of Africa.

Tests of the Theory of Plate Tectonics

We have examined the various lines of evidence supporting Wegener's notion of continental drift. Nearly all of these clues also support the newer concepts embodied in the theory of plate tectonics. The earlier clues were based mostly on evidence found on land. The new theory, with its keystone concept of sea floor spreading, was developed from evidence gleaned from the sea floor.

Slightly before the time when geophysicists like Harry Hess of Princeton University and Robert Dietz, formerly of the National Oceanographic and Atmospheric Agency, began formulating their ideas of sea floor spreading, other scientists were puzzling over findings related to paleomagnetism. The new data were obtained from sensitive magnetometers that were being carried back and forth across the oceans by research vessels. These instruments were able to detect not only the earth's main geomagnetic field but also local magnetic disturbances or magnetic anomalies frozen into the rocks along the sea floor. Maps were produced that exhibited linear anomaly bands of high and low field-magnetic intensities parallel to the west coast of North America (Fig. 8–36). Surveys of magnetic field strength on traverses across the Mid-Atlantic Ridge revealed a similar pattern of symmetrically distributed belts of anomalies (Fig. 8–37). Directly over the ridge, the earth's magnetic field was 1 per cent stronger than expected. Adjacent to this zone the field was somewhat weaker than would have been predicted, and the changes from weaker to stronger and back again often occurred in distances of only a few score miles.

In 1963, F. J. Vine, a 23-year-old research stu-

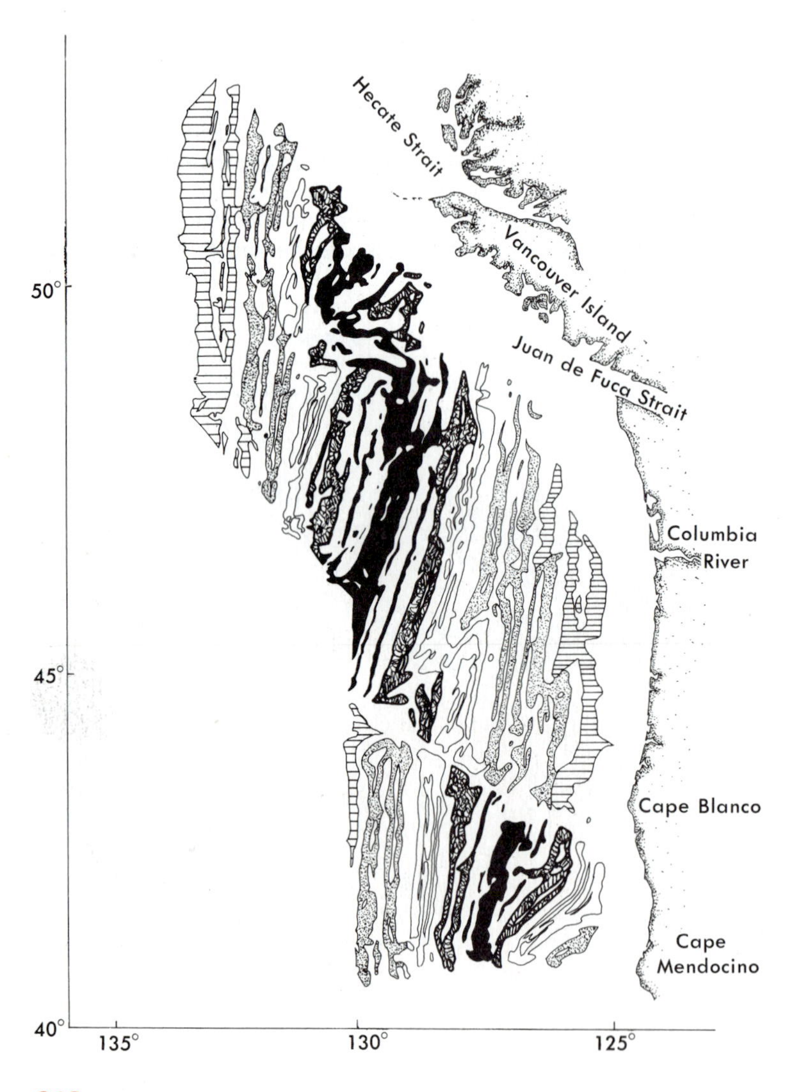

Figure 8–36 Magnetic field produced by the rocks on the floor of the northeast Pacific. Note the symmetry with respect to the ridges (the darkest shaded stripes). The ages of the rocks increase away from the ridges. The larger cross-hatched areas are about 8 to 10 million years old. (After A. D. Raff and R. G. Mason. Magnetic survey of the west coast of North America, 40° N to 52° N. latitude. *Bull. Geol. Soc. Am.,* 72:1267–1270, 1961.)

believe that the organelles in eukaryotic cells were once independent microorganisms that entered other cells and then established symbiotic relationships with the primary cell. For example, an anaerobic, heterotrophic prokaryote like a fermentative bacterium might have engulfed a respiratory prokaryote and thereby would have an internal consumer of the oxygen that might otherwise threaten its existence. A nonmotile organism might acquire mobility by forming a symbiotic association with a whiplike organism like a spirochete. The resulting cell would appear to have a flagellum for locomotion. Natural selection would clearly favor such advantageous symbiotic relationships.

THE FOSSIL RECORD FOR THE PRECAMBRIAN

Problems in Interpreting the Early Precambrian Fossil Record

Paleontologists seeking traces of the earth's earliest organisms are confronted with rather formidable problems. They must determine not only if a given microstructure is of biologic origin but also if it is truly contemporaneous with the enclosing sediment. The possibility of contamination by microorganisms coming into the rock postdepositionally must be recognized. In cases where chemical traces of life are detected, investigators must evaluate whether the compounds might have leaked into the rocks long after they were deposited or might have formed after deposition by more recent bacterial activity. There are also hazards associated with dating the presumably fossiliferous strata. Rather than directly dating the fossil-bearing formation, investigators often determine its age by correlation and superpositional relationships to rocks dated by radiometric methods.

Fossils Older than 2 Billion Years

The most ancient rocks that contain fossil microorganisms outcrop in South Africa and belong to the Precambrian Onverwacht Series. Onverwacht sediments are somewhat older than 3.2 billion years and are very likely the oldest little-altered sedimentary rocks on earth. Thus, the probability of finding significantly older fossils is remote. The microfossils of the Onverwacht consist of spheroidal and cup-shaped carbonaceous alga-like bodies. They not only are morphologically similar to some living forms of primitive algae but also are closely associated with filamentous structures that contain carbon compounds of biogenic origin.

Approximately 10,000 meters above the Onverwacht beds is another formation that has achieved fame because of its content of primitive fossils. Its name, somehow reminiscent of another story of genesis, is the **Fig Tree Formation.** The Fig Tree rocks consist of variously colored cherts, slates, ironstones, and tough sandstones. By means of radioactive isotopes, the formation has been dated as 3.1 billion years old. In 1967, samples of the Fig Tree Formation were studied by Harvard University paleobotanist, Elso S. Barghoorn, and his former graduate student, J. William Schopf. With the use of the electron microscope, these scientists were able to find a number of tiny, double-walled, rod-shaped structures (Fig. 9–23) that had a striking resemblance to modern bacteria. Barghoorn and Schopf named their find *Eobacterium isolatum*—the "isolated dawn bacteria." In their search for larger fossils, the scientists prepared thin sections of the chert and examined these with the optical microscope. The examination disclosed numerous spheroidal bodies very similar to certain blue-green algae. Because the samples were collected near the town of Barberton, Barghoorn and Schopf named the fossils *Archaeosphaeroides barbertonensis* (Fig. 9–24). The filaments of organic matter and hydrocarbons of organic derivation found in the Fig Tree Formation (and later in the older Onverwacht rocks) confirmed the existence of a primitive but vigorous flora of microscopic life in the Early Precambrian.

The presence of photosynthetic organisms among the Fig Tree fossils has not yet been posi-

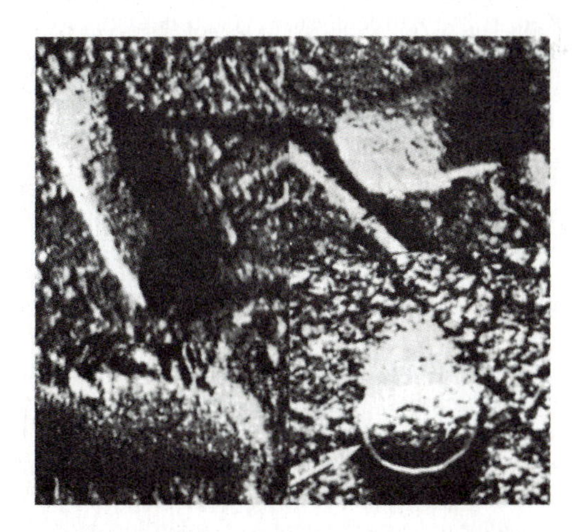

Figure 9–23 Electron micrographs of rod-shaped *Eobacterium isolatum,* about 0.6 micron in length. Specimen at lower right has been interpreted as a transverse section through a cell. (Courtesy of J. Wm. Schopf.)

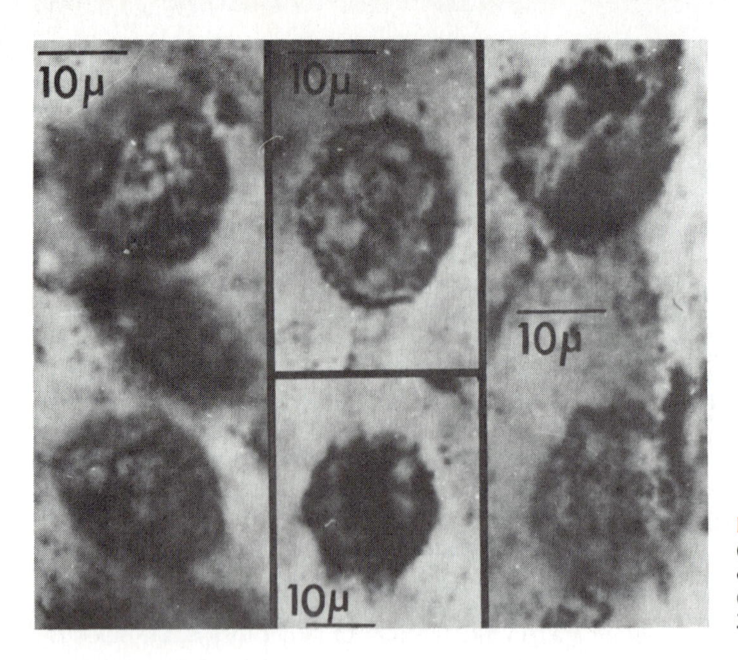

Figure 9–24 Spheroidal alga-like microfossils (*Archaeosphaeroides barbertonensis*) in black chert of the Early Precambrian Fig Tree Series. (The symbol µ means "microns.") These fossils are 3.1 billion years old. (Courtesy of J. Wm. Schopf.)

tively established. Like the Onverwacht rocks, the Fig Tree chert is black and rich in carbon. Biochemists have analyzed the carbonaceous materials extracted from these rocks and have detected organic compounds normally formed during the alteration of a particular component of chlorophyll called the porphyrin-magnesium complex. However, vanadium-porphyrin complexes also result from geochemical alteration of chlorophyll, and these have not yet been detected in the Fig Tree rocks. Two other organic compounds, phytane and pristane, are also regarded as breakdown products of chlorophyll. Both of these substances were detected in the extracts. However, this evidence has been judged inconclusive because some nonphotosynthetic microorganisms may also be capable of producing phytane and pristane.

Additional evidence for the existence of photosynthetic organisms in the Early Precambrian is provided by structures called stromatolites (Fig. 9–25). **Stromatolites** are distinctly laminated accumulations of calcium carbonate having rounded, cabbage-like, branching or frondose shapes. Today, the metabolic activities of certain marine colonial blue-green algae result in the formation of similar structures. The fine particles of calcium carbonate settle between the minute filaments of the matlike algal colonies and are temporarily bound within a film of gelatinous organic matter. Successive additional layers result in the laminations. The stromatolites are thought to have had a similar origin, particularly because of the discovery of concentrations of filamentous and spherical blue-green algae in the laminations of Precambrian stromatolites.

Near Bulawayo in southern Rhodesia, stromatolites have been found in limestone that is 2.7 billion years old. The carbon 12/carbon 13 ratio in the Bulawayo rocks suggests the occurrence of biologic fixation of carbon dioxide by photosynthetic organisms. Of equal importance is that it suggests biologically generated atmospheric oxygen was present about 2.7 billion years ago.

Although stromatolites are found in the Early Precambrian, they do not become common until Middle Precambrian. Late Precambrian stromatolites are sufficiently widespread and abundant to serve as guide fossils. They have formed extensive reeflike structures in Precambrian limestones. In the United States, the most notable of these stromatolitic structures occurs in rocks of the Belt series. They are exceptionally well exposed in Belt strata of Glacier National Park (Fig. 9–26). The photosynthetic activity of stromatolite microorganisms was probably important in causing the precipitation of the calcium carbonate that formed the thick layers of Belt limestone.

Modern stromatolites grow in the intertidal zone with their tops at the high-water mark. In this regard, it is interesting to note that some Late Precambrian stromatolites were approximately 6 meters in height. If these forms were also restricted to the intertidal zone, then Precambrian tides must have been considerably higher than they are today. Such high tides would indicate that the moon was closer to the earth during the Precambrian. More

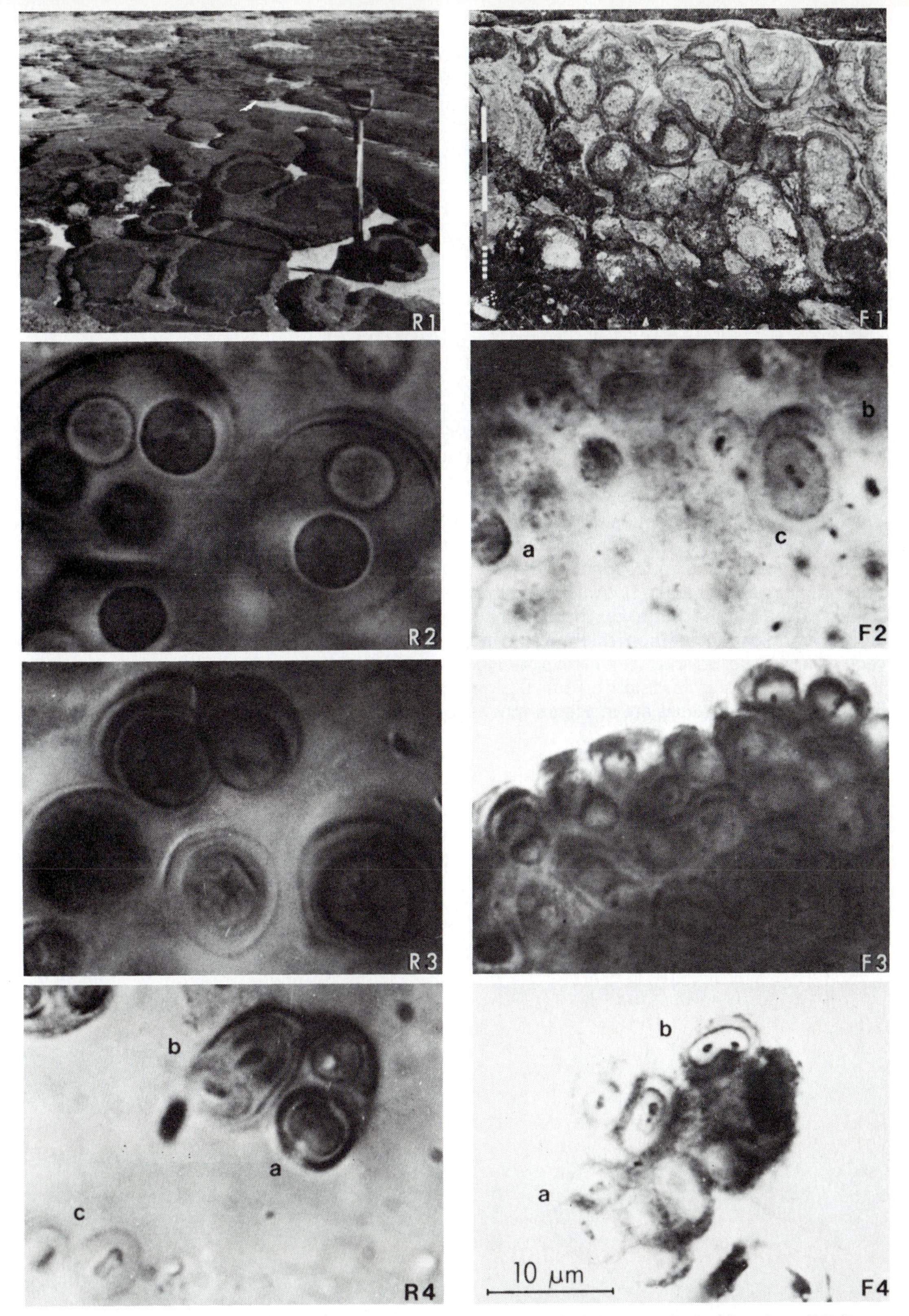

Figure 9–25 Comparison of present-day (left) and Precambrian (right) stromatolite-building algae. (R1) Recent stromatolite heads on wave-exposed shore, Shark Bay, Australia. (F1) Precambrian stomatolites preserved in chert, Belcher Islands, Canada. (R2, R3, R4) Cells of the living stromatolite algae *Entophysalis.* (F2, F3, F4) Cells of fossil Precambrian stromatolite algae of *Eoentophysalis* as they appear in thin sections of chert examined with the microscope. (From S. Golubic and H. J. Hoffman. *J. Paleontol.* 50(6), 1976; reprinted with permission of author and publisher.)

Figure 9–26 Stromatolites developed in carbonate rocks of the Precambrian Belt Series, Glacial National Park, Montana.

recent stromatolites are not as tall, suggesting the distance between the earth and the moon has been increasing, causing a corresponding decrease in tidal amplitude. In Chapter 1, it was noted that the length of the day on earth has been increasing. Conservation of angular momentum would indeed require a slowing down of the earth's rotation as the moon increased its distance from the planet. Thus, stromatolite studies are in accord with astronomic observations.

The Gunflint Flora

Extending eastward from Thunder Bay in the northern portion of Lake Superior are outcrops of a rock unit called the **Gunflint Chert** (Fig. 9–27). Radiometric age determinations indicate that the formation is approximately 1.9 billion years old. It contains a varied and abundant flora of the so-called thread bacteria and nostocalean blue-green algae. Unbranched filamentous forms, some of which are septate, have been given the name *Gunflintia* (Fig. 9–28). More finely septate forms, such as *Animikiea* (Fig. 9–29), are remarkably similar in appearance to such living algae as *Oscillatoria* and *Lyngbya.* Other Gunflint fossils, such as *Eoastrion* (the "dawn star"), resemble living iron- and magnesium-reducing bacteria. *Kaka-*

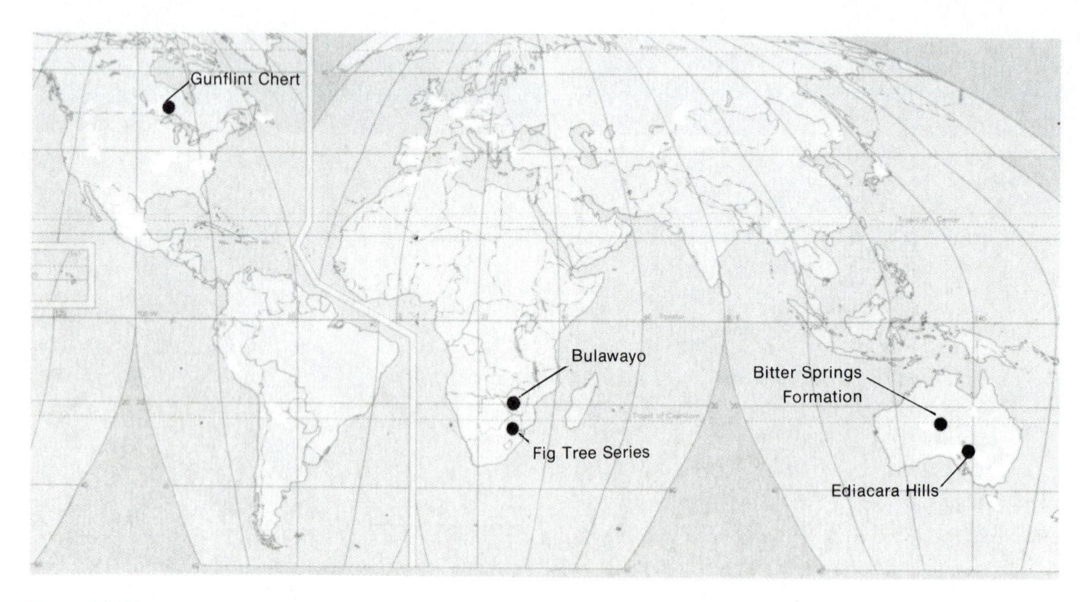

Figure 9–27 Location of some of the better known fossil-bearing Precambrian formations.

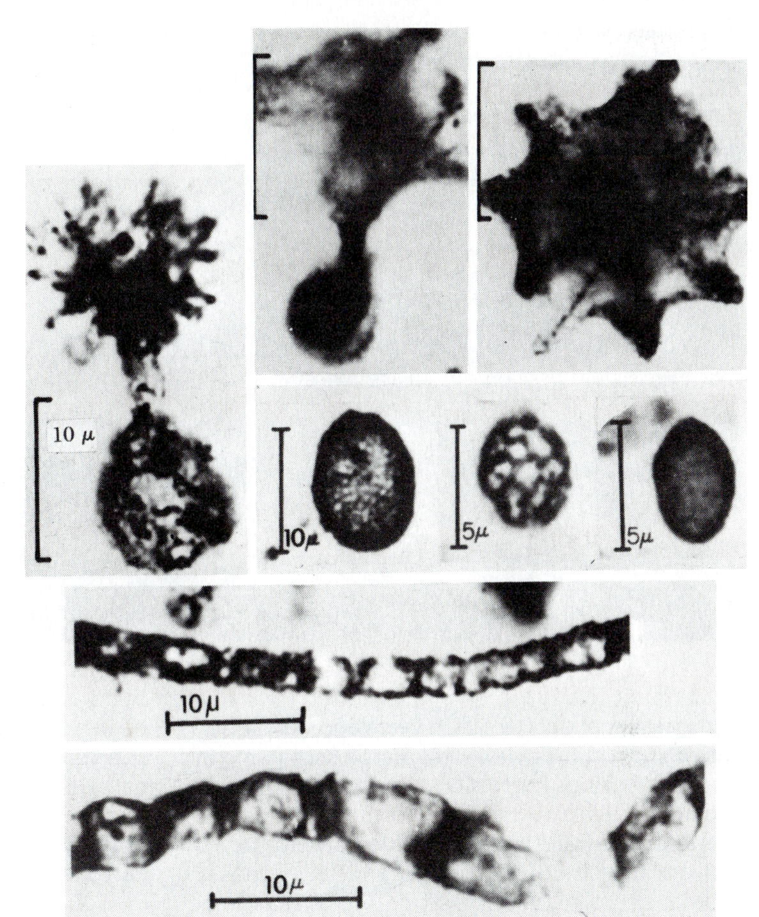

Figure 9–28 Fossil remains of microorganisms from the Gunflint Chert. The three specimens across the top with umbrella-like crowns are *Kakabekia umbellata*. The three subspherical fossils are species of *Huroniospora*. The filamentous microorganisms with cells separated by septa are species of *Gunflintia*. (Photographs provided by J. Wm. Schopf, courtesy of J. Wm. Schopf and Elso S. Barghoorn.)

bekia and *Eosphaera* (Figs. 9–29 and 9–30) are so different from any known microorganism that their classification is uncertain. That these and other Gunflint and stromatolitic organisms were actively producing oxygen and thereby altering the composition of the atmosphere does seem certain. Not only do they resemble living photosynthetic organisms but also their host rock contains phytane and pristane—organic compounds regarded as the breakdown products of chlorophyll.

One of the major events in biologic evolution was undoubtedly the origin of the eukaryotic cell, with its enclosed nucleus for the storage and transmission of genetic information. Although the potential for sexual reproduction provided by eukaryotes enormously increased the possibilities for evolutionary change, the fossil record for earliest eukaryotic organisms is ambiguous. This is not surprising, since the first eukaryotes probably were simple microscopic unicellular organisms that in the altered fossil state would be difficult to distinguish from prokaryotes. At the present time, one group of paleobotanists suggests that eukaryotes

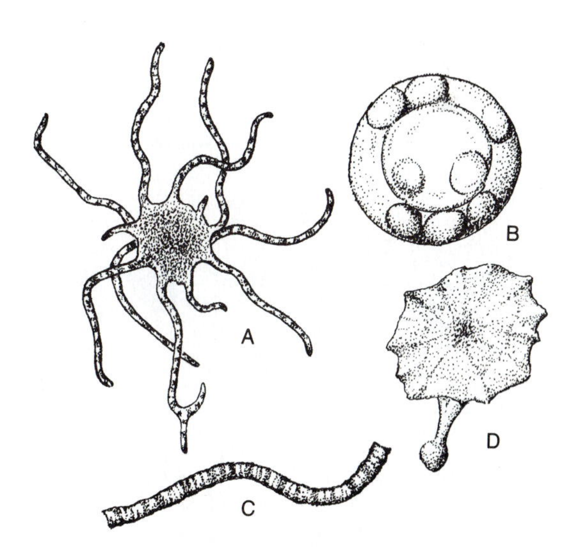

Figure 9–29 *Eoastrion* (A), *Eosphaera* (B), *Animikiea* (C), and *Kakabekia* (D) from the Gunflint Cherts. All specimens are drawn to the same scale. *Eosphaera* is about 30 microns in diameter.

Figure 9–30 The microfossil *Kakabekia* takes its name from Kakabeka Falls on the Kaministikwia River. Shales are overlain by layers of cherty rocks that form the resistant lip of the falls. (From E. G. Pye. Ontario Bureau of Mines, Circular No. 10, 1962; reprinted with permission.)

were present about 0.9 billion years ago. These scientists support their view with microfossils found in the Bitter Springs Formation of central Australia. In addition to a prokaryote assemblage of fossils, the cherts of the Bitter Springs yielded several forms that are the size of eukaryotes and have ghostly internal structures that might be the remnants of nuclei or organelles of eukaryote green algae (Figs. 9–31, 9–32, and 9–33). If these are indeed the structures they appear to be, then the passage from prokaryotic to eukaryotic life with its capacity for sexual reproduction and genetic variability had been accomplished at least as long ago as 0.9 billion years. However, the identification of some unicellular organisms in the Bitter Springs as eukaryotic is currently being challenged by those who doubt that the internal structures are indeed eukaryotic organelles.

Dawn Animals

As the search for Precambrian fossils continued, it was almost inevitable that some trace of the

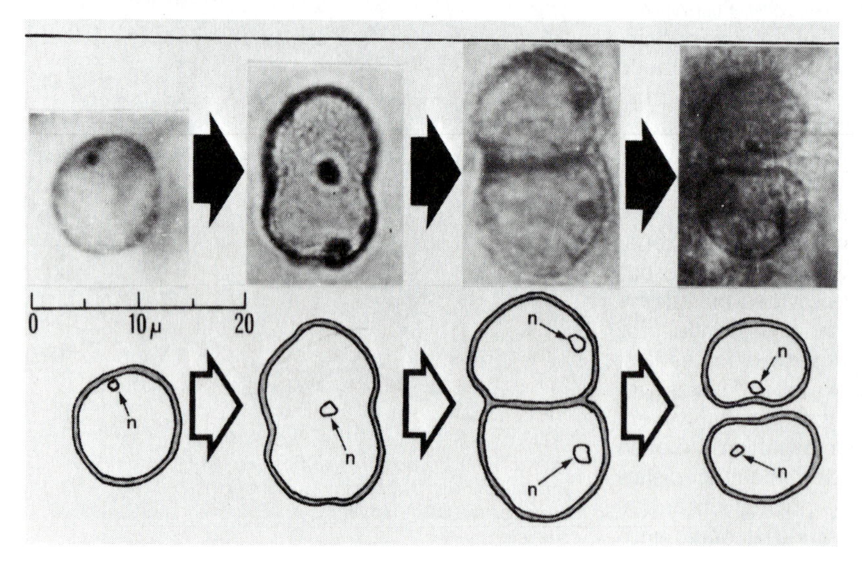

Figure 9–31 Sequence of microfossils from the Bitter Springs Formation of Australia arranged by J. W. Schopf in a sequence presumed to represent stages of cell division. It has been suggested that the microorganism (*Glenobotrydion*) may have been a species of green algae. The letter "n" indicates nuclear material. (Courtesy of J. Wm. Schopf, from photographs used during Proceedings of XXIV International Congress, Sec. 1, 1972.)

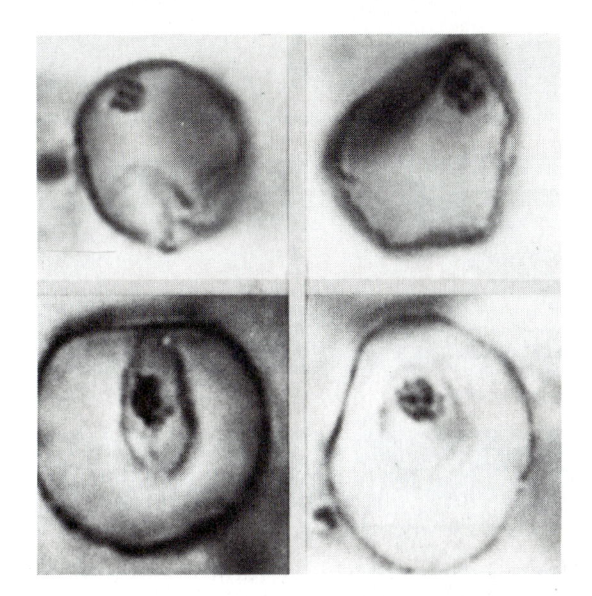

Figure 9–32 Optical photomicrographs showing unicellular fossils suggestive of eukaryotic organisms. Fossils are revealed in thin sections of black chert from the late Precambrian Bitter Springs Formation. Note vague internal (possibly nuclear) structures. (Courtesy of J. Wm. Schopf.)

more advanced forms of life would be discovered. Indeed, since the early 1960's, a fascinating assortment of advanced Precambrian fossils has been found in Australia, Siberia, and Africa. The fossils consist primarily of impressions in sedimentary rocks of the types of animals that can be loosely grouped under the category of **metazoans.** Metazoans are multicellular, possess more than one kind of cell, and have different kinds of cells organized into tissues and organs.

The best known fossils of Precambrian metazoans were found as well-preserved impressions in the consolidated sands of an ancient beach now called the Pound Quartzite. Exposures of the rock in the Ediacara Hills of south Australia (see Fig. 9–27) have yielded over 600 specimens of soft-bodied animals (Fig. 9–34). The collection includes several different kinds of jellyfish, soft corals resembling the modern "sea pen," polychaete worms, echinoderms, arthropod-like animals, and a number of unique forms whose precise biologic affinity is uncertain. There are also large numbers of tracks and burrows made by the members of this nearshore community. As nearly as can be determined, the "Ediacara fauna," as it is known, is around 650 to 700 million years old. Its importance lies in the glimpse it provides of the ancestors to the vast array of invertebrates that were to populate the seas in subsequent geologic time.

Another occurrence of Late Precambrian metazoan fossils that is rather similar in preservation to the Ediacara forms was discovered in the rocks of the Conception Group of Southeastern Newfoundland. The fossils at that site occur as hundreds of imprints on ripple-marked surfaces of graywackes. They include leaf-shaped animals (Fig. 9–35) that resemble sea pen corals, lobate forms that appear to be jellyfish, and other probable coelenterates with either branching or spindle shapes.

The sudden appearance of the diverse Ediacara and Conception Group fossils, and the apparent absence of any fossils or even traces of animal life in rocks older than 700 million years, suggests that metazoans expanded abruptly near the end of the Precambrian. According to Berkner and Marshall (1964), their rapid expansion may very well correlate with the accumulation of sufficient free oxygen to permit oxidative metabolism in organisms. On the other hand, the Ediacara life may have evolved more gradually from earlier small and naked forms that were essentially incapable of leaving a fossil record. Perhaps, as suggested by A. G. Fischer (1965), the ancestral metazoans lived in "oxygen oases" in which marine plants were concentrated. After the atmospheric oxygen content had reached about 1 per cent of the present level, these as-yet-undiscovered animals would have been free to leave their oases and spread widely in the seas. According to this view, the evolutionary development of metazoans may not have been abrupt, but their dispersal may have been rapid after suitable conditions became prevalent.

The Fig Tree, Gunflint, and Pound Formations represent only brief and isolated glimpses of the progress of biologic evolution during an almost incomprehensible span of more than 3 billion years. Unlike the fossil record, life is a continuous progression. Other exposures of Precambrian rocks will continue to be scrutinized in the years ahead to find the many missing chapters in the history of Precambrian life. Figure 9–36 depicts the current state of our knowledge.

ASPECTS OF PRECAMBRIAN CLIMATE AND ENVIRONMENT

Because so much of the Precambrian rock sequence is either altered or devoid of fossils, interpretations of climate and environment are more difficult than in the rocks of younger eras. This is especially true for the oldest Precambrian rocks. As noted previously, the earth's early atmosphere and hydrosphere were largely devoid of free oxygen. Then about 2 billion years ago, the atmospheric

Figure 9–33 Structurally and organically preserved Precambrian microorganisms in thin sections of chert from the Bitter Springs Formation, estimated at approximately 900 million years old. The fossils shown here are species of filamentous blue-green algae (cyanophytes). These fossil organisms are similar to those present in modern mat-building algal communities that produce stromatolites in warm, nearshore marine environments. (Courtesy of J. Wm. Schopf, University of California, Los Angeles.)

oxygen began to accumulate because of increased plant activity. One result of the oxygen buildup was the accumulation on land of considerable amounts of ferric iron oxide, which stained terrestrial sediments a rust-red color. Such sedimentary rocks are known as **red beds** and are considered a valid indication of the advent of an oxygenic environment. Of course, the oxygen level probably rose slowly and very likely did not approach 10 per cent of present atmospheric levels of free O_2 until the Cambrian Period.

In general, Precambrian rocks provide evidence for a wide range of climatic conditions, but there is no indication that these climates were especially unique in comparison with those of the Phanerozoic Eras. Thick limestones and dolostones with reeflike algal colonies were deposited along the Precambrian equator, where warm tropical conditions prevailed much as they do today. During the Early Proterozoic, the equator lay close to the northern border of North America. Precambrian evaporite deposits in eastern Canada and in Australia suggest that conditions were periodically rather arid during the Middle Proterozoic. In the

Figure 9–34 Three members of the Ediacara fauna.

(A) *Pseudohizostomites,* a wormlike form.
(B) *Tribrachidium,* a possible echinoderm.
(C) *Parvancorina,* a possible arthropod.

(Specimen A is an imprint in the Pound Quartzite; B and C are plaster molds made from original fossils. All specimens courtesy of M. F. Glaessner.)

Figure 9–35 Fossil metazoans from the Conception Group, Avalon Penninsula, Newfoundland. These early metazoans are similar to those of the Ediacaran fauna of Australia. The rocks in which they occur are thought to be very late Proterozoic in age. The form directly above the geology hammer is probably the impression of a jellyfish. The spindle-shaped form was made by a creature similar to a modern sea pen. (Photograph courtesy of S. B. Misra, Directorate of Geology and Mining, Rewa, India.)

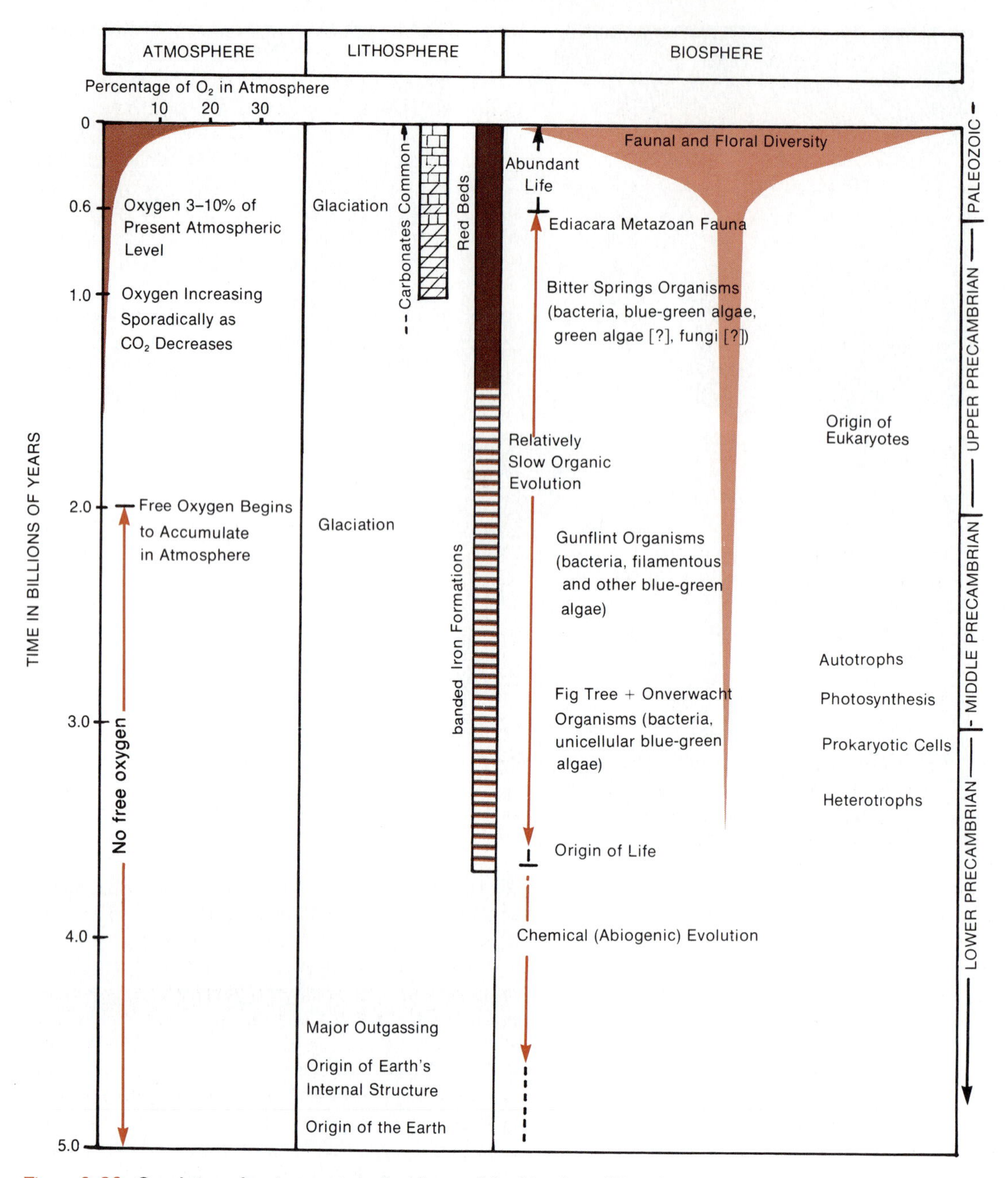

Figure 9–36 Correlation of major events in the history of the biosphere, lithosphere, and atmosphere.

middle and low latitudes, climates were more severe, as indicated by consolidated deposits of glacial debris (tillites) and glacially striated basement rocks. The best known of these poorly sorted, thick, boulder-like deposits is the Gowganda Tillite. Its widespread distribution over the southern portion of the Canadian Shield indicates the presence of continental, rather than mountain, glaciers. Indeed, geologic mapping suggests that the ice sheet was probably more than 1500 km in diameter and covered most of central Canada. The Gowganda Formation has been dated by radiometric methods as about 2.3 billion years old. Striations on the rock surfaces beneath the tillites suggest that the ice moved toward the Precambrian North Pole, which was located near lat. 22°N, long. 97°W at the time. These Precambrian glaciers ground their way northward into the Arctic Ocean from the Canadian and Baltic Shields.

The Gowganda tillites, although impressive, are not the only evidence of glaciation during the Precambrian. In southern Africa, the Chuos Tillite attains thicknesses of over 450 meters and has been recognized across an expanse of over 30,000 sq km. Tillites of Late Proterozoic age have been found on all continents except Australia, although they are not necessarily synchronous. The ample evidence of Precambrian glaciation around the world clearly indicates that the recent Pleistocene ice age was not at all a unique event in the earth's geologic history.

THE MINERAL WEALTH OF THE PRECAMBRIAN

Precambrian rocks contain a host of metal ores that have been of inestimable economic importance. Major sources of iron, nickel, gold, silver, chromium, and uranium are derived from Precambrian rocks. Iron is the most notable of these ores in terms of tonnages that have been mined. The world's largest group of iron ore localities is in the Canadian Shield. Of these, the most famous are the sedimentary ores of the Lake Superior region. The ore deposits were formed in local areas where the iron-bearing sedimentary rocks were altered and enriched by removal of nonmetallic constituents. Other major sources of Precambrian iron occur in Sweden, the Ukraine, South Africa, and South America.

About half of the world's gold is mined from Precambrian quartz conglomerates near Johannesburg, South Africa. A Precambrian gabbro intrusion at Sudbury, Ontario, provides 70 per cent of the world's nickel. Uranium of Precambrian age is mined in largest quantities in Ontario and South Africa. Enormous amounts of copper were once recovered from Proterozoic rocks along the Keweenaw Peninsula of Lake Superior; however, today the ores are almost exhausted. These abandoned mines provide mute testimony to the fact that every mineral deposit is exhaustible and irreplaceable.

Summary

The great interval in the history of the earth that began after the formation of the planet and lasted until the appearance of abundant fossils 600 million years ago is known as the Precambrian. Although terminology varies, the older, frequently deformed and metamorphosed Precambrian rocks are referred to as either Archaeozoic or Archean, whereas younger, overlying less altered portions of the rock record are given the name Proterozoic. Because the bulk of Precambrian rocks contain almost no fossils, radiometric dates are used to determine correlations. In many regions around the world, these dates record igneous activity accompanying a cycle of geologic events that begins with the depositon of thick sequences of geosynclinal sediments and is followed by compression of the geosynclinal materials into mountain ranges. Erosion of the intruded and granitized ranges, as well as stabilization of the mountain belt, provides a final episode in the cycle.

Precambrian rocks occur in many geologic settings. Splendid exposures are often visible in the deep valleys and canyons of mountains and plateaus. However, the most widespread and striking outcrops of Precambrian rocks occur across the long-denuded, upwarped, but now stable heartlands of the continents. Such regions are termed Precambrian Shields. They are composed of granitic intrusions and downfolded masses of altered sediments that are the stabilized record of numerous geologic cycles of deposition, orogeny, and erosion. In North America, the Canadian Shield forms the largest single region of Precambrian rocks in the world. The oldest portion trends southwesterly across the central portion of the shield from Labrador to Wyoming. Four successively younger belts lie adjacent to the most ancient terrain. Other important Precambrian regions are the Baltic Shield in Scandinavia, the Angaran Shield of Asia, three distinct South American shields, the Indian Shield, and the Australian Shield. Africa seems to be composed of several shield segments that were gradually welded together by orogenic activity occurring along their borders. Most shields contain sporadic but rich deposits of iron, gold, nickel, silver, copper, and uranium. Of these metal-containing ores, iron is especially characteristic of Late Precambrian rocks.

The Precambrian was the time when life originated on earth. There is, of course, no primary evidence of that event in Precambrian rocks. However, biochemists have shown experimentally that, in an environment largely devoid of free oxygen but containing carbon, hydrogen, nitrogen, and water in which inorganic salts are dissolved, complex organic molecules can be formed when the mixture is activated by ultraviolet light or lightning-like electrical discharges. These compounds and their derivatives are found in every living organism. They are the molecular basis of all organic structures, heritable information, and energy. A long episode of Precambrian chemical evolution eventually led to the formation of amino acids, sugars, and nucleic acids. These materials somehow acquired the organization and behavior of replicating living systems. The first living cells might have subsisted on the chemical energy they derived by consuming other cells and molecules. They were heterotrophs and were followed by cellular organisms (autotrophs) that could use the energy of the sun through photosynthesis. Free oxygen accumulated as a byproduct of photosynthesis, making possible the evolution of organisms that use oxygen to burn their food for energy.

The oldest currently known fossil evidence of life consists of spheroidal algal bodies, stromatolites, and filamentous structures from the 3.2-billion-year-old Onverwacht rocks of South Africa. Of only slightly younger age, the South African Fig Tree chert has yielded rod-shaped bodies resembling bacteria. A more diverse assemblage of ancient microorganisms has been found in samples of the Gunflint cherts of the Lake Superior region. The greater variety of organisms in the Gunflint is not unexpected, for the Gunflint Formations are roughly a billion years younger than the Onverwacht and Fig Tree. The most abundant of these microscopic fossils consist of septate, threadlike bodies presumed to be the remains of blue-green algae. Some of the filaments contain tiny "spores" resembling those in living algae and iron bacteria. Star-shaped and parachute-like forms also occur and are presumed to be prokaryotic organisms of unknown affinities.

Abundant, rather well-preserved flora of blue-green algae, colonial bacteria, probable fungi, and filamentous organisms are found in stromatolitic rocks of the Bitter Springs Formation of Australia. These 0.9-billion-year-old rocks also contain spheroidal organisms that may be the remains of eukaryotic green algae. Such an interpretation, however, is currently the object of vigorous debate among paleobotanists.

A final major discovery of Precambrian life is quite unlike the microorganism assemblages of the Gunflint and Bitter Springs Formations, because the fossils are of metazoans (multicellular animals). They consist largely of impressions in 0.7-billion-year-old sandstones of the Ediacara Hills of South Australia. Hundreds of specimens, mostly of coelenterates, annelids, and arthropods, have been collected at the Ediacara Hills locality.

There is little evidence upon which to base specific statements about the climate of the early Precambrian. Thick beds of carbonate rocks suggest deposition in warm or tropical seas, and the appearance of terrestrial sediments indicates that the conditions causing weathering and erosion were much the same as they are today. Indeed, from the time the atmosphere resembled its present composition, it is likely that the physical conditions of the environment were not radically different from those of subsequent geologic time. The most striking climatic feature of the Late Precambrian appears to have been episodes of continental glaciation that occurred on all of the large shield regions.

Questions for Review

1. What are Precambrian Shields? What events are thought to have occurred in the history of shields to account for their strongly deformed and eroded sedimentary, metamorphic, and igneous rocks?
2. Why are Precambrian rocks generally more difficult to correlate than more recent strata? What is the major method used in correlation?
3. What characteristics of the Canadian Shield favor the hypothesis that growth of continents is by marginal accretion?
4. Differentiate between the terms Phanerozoic and Cryptozoic. In the Canadian Shield, how may the Cryptozoic be further divided?
5. Account for the observation that rocks of granitic character are far more prevalent in shields than are rocks of basaltic composition.
6. What was the mixture that was used in the famous S. C. Miller experiment for artificial production of amino acids? Why was oxygen excluded?
7. In general, what is the inferred sequence of events that led to the origin of life?
8. What are metazoans? What is the age of the first occurrence of metazoans? How might the eventual accumulation of molecular oxygen in the atmosphere be correlated to the rather abrupt earliest appearance of metazoans?
9. What evidence indicates that oxygen was being produced by photosynthesis about 3 billion years ago?
10. Differentiate between a prokaryotic organism and a eukaryotic organism; between a heterotroph and an autotroph; between an anaerobic and aerobic organism.
11. Where, and in what type of rock, do the oldest

known fossils occur? What is the general nature of these remains?

12. What is the importance of amino acids in the molecular evolution and origin of life?
13. What is the environmental significance of the following Precambrian rock features: (a) banded iron formations, (b) stromatolitic limestones, (c) widespread tillites, (d) red beds, and (e) evaporites?
14. What is the explanation for evidence of continental glaciation in shield areas now located near the equator?
15. What important metal-bearing ores are extracted from Precambrian rocks? Why are Precambrian rocks largely devoid of mineral fuels like coal and oil?

Terms to Remember

amino acid
Angaran Shield
Animikean System
Animikiea
Aphebian Era
Archaeosphaeroides barbertonensis
Archaeozoic Era
Archean Era
autotroph
Baltic Shield
Canadian Shield
Cryptozoic
Eoastrion
Eobacterium isolatum
Eosphaera
eukaryote
fermentation
Gowgonda Formation
greenstone belts
Gunflint flora
Gunflintia
Hadrynian Era
Helikian Era
heterotroph
Huronian Succession
Kakabekia
Kenoran Orogeny
Mazatzal Orogeny
metazoan
nucleic acid
Phanerozoic
Precambrian
Precambrian shield
prokaryote
protein
Proterozoic Era
red beds
Rhiphaean rock
stromatolites

Supplemental Readings and References

Barghoorn, E. S. 1971. The oldest fossils. *Sci. Am.* 224(5):30–42.

Barghoorn, E. S., and Schopf, J. W. 1966. Micro-organisms three billion years old from the Precambrian of South Africa. *Science* 152(3723):758–763.

Cloud, P. E., Jr. 1976. Beginnings of biospheric evolution and their biogeochemical consequences. *Paleobiology* 2:351–387.

Engle, A. E. J., et al. 1968. Algal-like forms in the Onverwacht Series, South Africa: oldest recognized life-like forms on earth. *Science* 161:1005–1008.

Fox, S. W., and Dose, K. 1972. *Molecular Evolution and the Origin of Life.* San Francisco. W. H. Freeman & Co.

Glaessner, M. F. 1961. Precambrian animals. *Sci. Am.* 204(3):72–78.

Kummel, B. 1970. *History of the Earth* (2d ed). San Francisco, W. H. Freeman & Co.

Levin, H. L. 1975. *Life Through Time.* Dubuque, Wm. C. Brown & Co.

Margolis, L. 1981. *Symbiosis in Cell Evolution.* San Francisco, W. H. Freeman & Co.

Rankama, K. (ed.) 1963–68. *The Precambrian,* Vol. 1–4. New York, Interscience.

Reed, H. H., and Watson, J. 1975. *Introduction to Geology,* Vol. 2, Pt. 1. New York, John Wiley & Sons.

Rutten, M. G. 1971. *The Origin of Life by Natural Causes.* Amsterdam, Elsevier Books International.

Schopf, J. W. 1978. The Evolution of the Earliest Cells. *Sci. Am.* 239(3):110–138.

Niagara Falls, formed where the Niagara River flows from Lake Erie into Lake Ontario. The classic section of the American Silurian System is exposed along the walls of the gorge below the falls. (Photograph courtesy of Ontario Ministry of Industry and Tourism.)

10

The Antique World of the Early Paleozoic

> We need not be surprised if we learn from geology that the continents and oceans were not always placed where they are now, although the imagination may well be overpowered when it endeavors to contemplate the quantity of time required for such revolutions.
>
> Sir Charles Lyell, *The Student's Elements of Geology*, 1882.

PRELUDE

Anyone who has experienced a course in history is likely to appreciate the way historians group seemingly endless facts into tidy packages bearing such labels as "The Renaissance" or "The Hellenistic Period." With the help of such groupings, one is better able to retain an overall impression of the period. This is also a reason for dividing the history of the Paleozoic Era into "early" and "late" rather than providing a period by period account. The scheme is also convenient in that each of the two increments of time is of approximately equivalent duration. The Cambrian, Ordovician, and Silurian Periods constitute the Early Paleozoic, which lasted about 190 million years. In the subsequent 180 million years, the Devonian, Mississippian, Pennsylvanian, and Permian Periods ran their course. Each of the two parts of the Paleozoic was characterized by very general similarities in the historical sequence of events. For example, on most continents the Early Paleozoic began with gradual marine invasions of low-lying regions of the interior. These wide expanses of shallow epicontinental seas moderated climate and provided habitats for a multitude of marine organisms.

Subsequently, the continents began to stir. Either by uplift of the lands or by decline in sea level, the inland seas began an oscillatory regression back toward the major ocean basins. Increasing thicknesses of terrestrial sediments and volcanic rocks appeared in the stratigraphic sequence, and contorted strata attest to the growth of mountain ranges. It was a time of lands, just as the preceding episode was a time of seas. However, gradually the lands were reduced and the seas returned as we pass from the Early into the Late Paleozoic.

After the total physical history of the Paleozoic Era is examined, it becomes evident that long periods of quiet sedimentation were punctuated at intervals by severe change involving earth movements and mountain building. European geologists have recognized three such mountain-building, or orogenic, events. They refer to them as

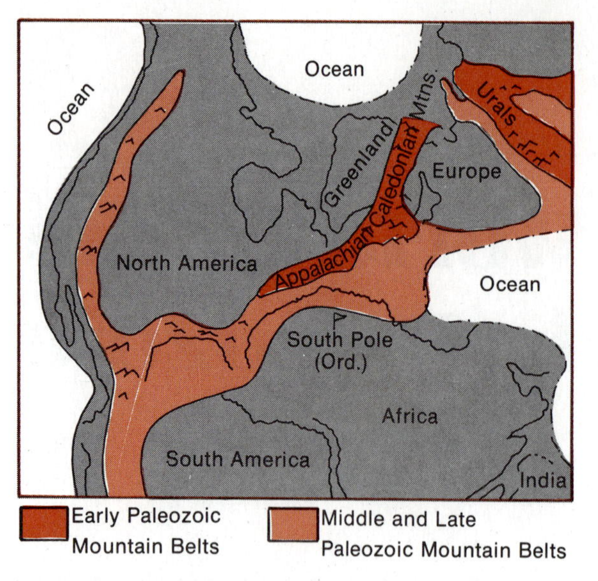

Figure 10–1 Pangaea II with mountain belts formed during the Paleozoic.

the **Caledonian, Hercynian,** and **Alpine** Orogenies. You will recall our discussion of how Pangaea I disaggregated. The return of the errant lands to form Pangaea II (Fig. 10–1) may very well be associated with the Caledonian and Hercynian Orogenies. The Alpine Orogeny records a similar but much later event.

LANDS, SEAS, AND MOBILE BELTS

Clues to Paleogeography

Because of the many processes that have ceaselessly worn away or altered the older parts of the geologic record, it is difficult to reconstruct details of Early Paleozoic geography. From paleomagnetic studies we have learned that most of North America during the Cambrian lay along the paleoequator, which was aligned nearly perpendicular to present parallels of latitude from north-central Mexico to Ellesmere Island in the Arctic (Fig. 10–2). The Ordovician equator extended slightly more northeastward, from what is now Baja California to Greenland and thence across the British Isles and Central Europe. These paleoequatorial positions are substantiated by paleoclimatic indicators. Salt and gypsum accumulations, for example, are found in Ordovician rocks of Arctic Canada at a paleolatitude of about 10° from the paleoequator. Great thicknesses of carbonate rocks bearing corals and other relatively warm water marine invertebrates are located within 30° of the paleomagnetically determined equator.

Both paleomagnetic evidence and paleontologic evidence suggest either the existence of a supercontinent at the very beginning of the Paleozoic or that the continents were gathered into a relatively compact assemblage. The land area that in a previous chapter we termed Pangaea I was situated primarily in the southern hemisphere. The Gondwana region was turned about 180° from its present orientation, whereas what is now North America and Europe lay somewhat east of South America. Very early in the Cambrian, or possibly late in the Precambrian, Pangaea I began to split. Before the Cambrian Period ended, the lines of separation had widened to admit the seas. These elongate tracts became great collecting troughs for the sediments that are now metamorphosed and folded in the highlands of northwest Europe, in the Urals, and in the Appalachians. By Ordovician time large continental fragments were drifting away from the parent mass.

Figure 8–47 illustrates in simplified diagrams the sequence of events as they are postulated by the earth scientists James Valentine and Eldridge Moores. These configurations show Pangaea I divided into four continents by Late Cambrian and Early Ordovician. Then, during Late Ordovician and Silurian time, Europe west of the Uralian trough pushed into North America, forming a chain of ancient mountains called the **Caledonides** and welding the two continental segments into a "Euramerica." The next major collision occurred about 300 million years ago when Euramerica joined Gondwanaland, forming the ancient Hercynian Mountains, which have since been largely eroded away and covered with younger strata over much of Europe. Finally, late in the Permian, Asia was joined to Euramerica. The sediments caught and crushed in the closing ocean formed the Urals. Pangaea II was born.

It must be remembered that these theories of continental splitting and drift, and our methods of locating the positions of continents long ago, are fraught with possibilities for error or misinterpretation. We are dealing with speculations based on limited data, and we should not be surprised if in a decade hence our views of Pangaea I and II are altered.

The Continental Framework

The Cambrian Period of the Paleozoic Era began approximately 570 million years ago. By that time most of the continental masses, even though clustered together, could be considered to have two parts that differed in their history and structure. These two major geologic divisions have been termed **cratons** and **mobile belts.**

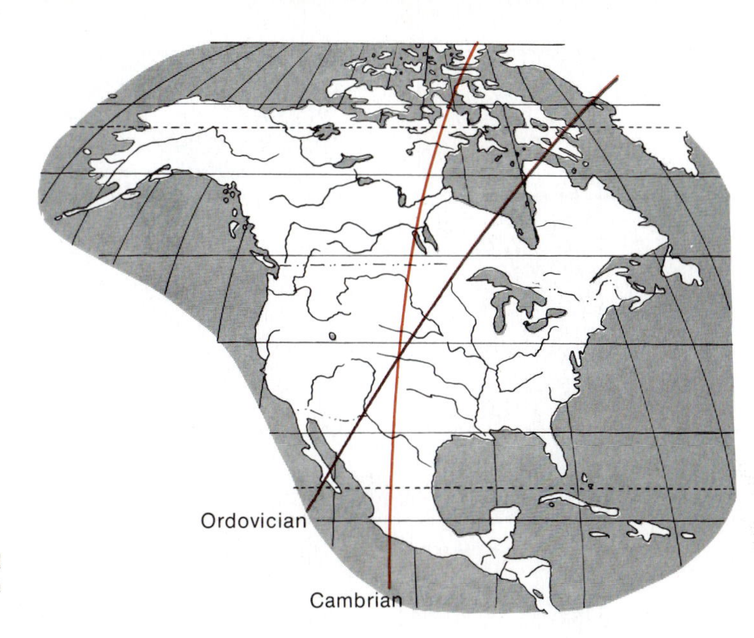

Figure 10–2 Inferred position of Cambrian and Ordovician paleoequator relative to North America.

Cratons

Cratons form the ancient central nuclei of continental masses. Cratons may include either or both of two kinds of surficial geology. As explained in Chapter 9, cratonic regions consisting of surficial Precambrian igneous and metamorphic rocks are termed **shields.** Most shield areas were exceptionally stable during the Paleozoic, in contrast to their Precambrian history, which included intense orogenic activity. The shield portion of the craton may be partially clothed in a veneer of relatively flat post-Precambrian sedimentary strata. This portion is called the **platform.** Like shields, platforms were generally stable throughout most of Phanerozoic geologic history. Platform strata consist of wave-washed sandstones and carbonates deposited in shallow seas that periodically flooded continental regions of low relief. The strata of platforms are not, however, entirely horizontal. Here and there they are gently warped into broad synclines, basins, domes, and arches (Fig. 10–3). The resulting tilt to the strata is so slight that it is usually expressed in feet per mile rather than degrees. In the course of geologic history, the arches and domes stood as low islands or barely awash submarine banks.

Domes and arches seem to have developed in response to vertically directed forces quite unlike those that formed the compressional folds of mountain belts. Whether domes or basins, structures of the platform can be recognized by their pattern of outcropping rocks. Erosional truncation of domes exposes older rocks near the center and younger rocks around the periphery. Sequences of strata over arches and domes tend to be thinner. Also, because these structures were periodically above sea level, they characteristically exhibit a greater frequency of erosional unconformities. Basins, on the other hand, were more persistently covered by inland seas, have fewer unconformities, and develop greater thickness of sedimentary rocks. In erosionally truncated basins, younger rocks are centrally located, and older strata occur successively farther toward the periphery (Fig. 10–4).

Mobile Belts

The North American craton is bounded on four sides by great elongate tracts that have been the sites of intense deformation, igneous activity, and earthquakes. Such tracts are called **mobile belts,** and one or more are present on all continents (Fig. 10–5). Most, like the North American Cordilleran Belt, are located along the margins of continents. It is also possible to have an intracratonic mobile belt, such as occurs along the Urals. Systems of deep sea trenches with associated volcanic island arcs are intraoceanic mobile belts. A **geosyncline** is a mobile belt that, in its early history, received a great accumulation of sediment. As described in Chapter 7, mountain ranges are related to geosynclines in that they are thought to have evolved through several stages of development, beginning with a period of sedimentation and ending with an episode of severe crustal deformation (Fig. 7–13).

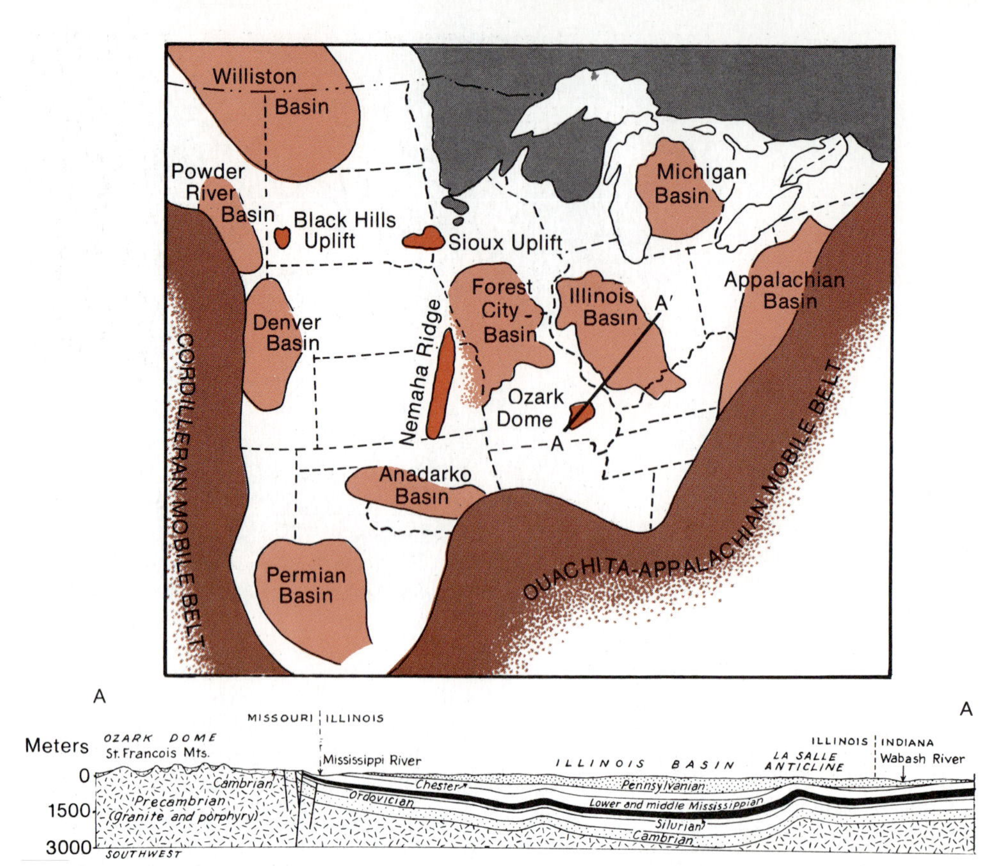

Figure 10–3 Map of the central platform of the United States showing major basins and domes. Structural section below the map crosses part of the Ozark Dome and the Illinois Basin.

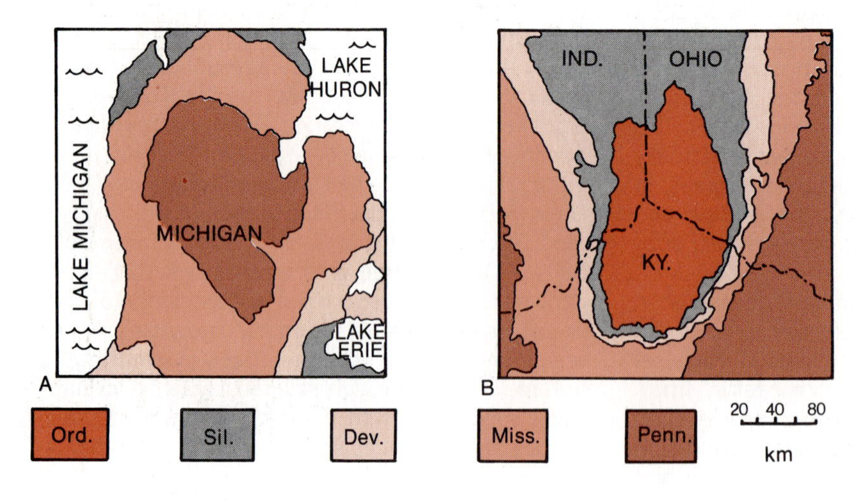

Figure 10–4 In an erosionally truncated basin like the Michigan Basin (A), youngest beds are centrally located. In a domelike structure such as the Cincinnati Arch (B), oldest beds are located in the center. (Compare with Figure 8–8D and E.)

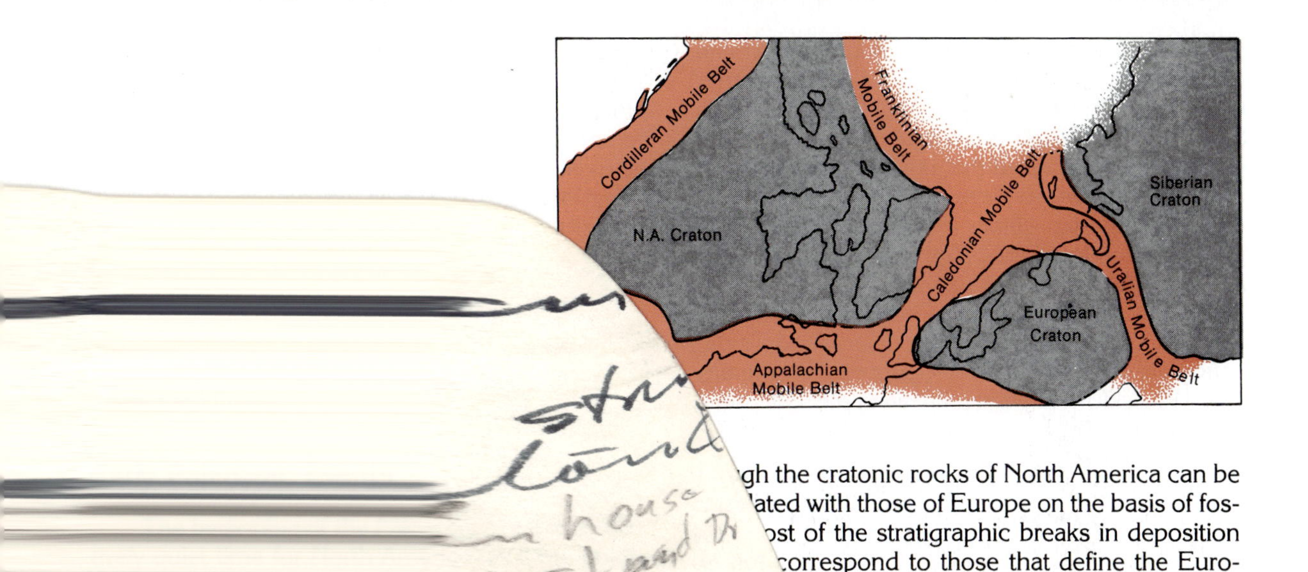

...However, oceans did exist and gradually spilled out of their basins onto low regions of the continents. Although a large part of the sedimentary record has been removed by erosion, enough remains to indicate that practically all of the Canadian Shield was inundated at times. Initially, the dominant deposits were sands and clays derived from the weathering and erosion of igneous and metamorphic rocks of the shield. Later, as these quartz- and clay-producing terrains were reduced, limestones and dolostones became increasingly more prevalent. These rocks contain abundant remains of lime-secreting marine organisms and indicate that shallow seas were common throughout much of the equatorial region of the earth during the Early Paleozoic. Indeed, the advances and retreats of these shallow seas were the most apparent events in the Early Paleozoic history of the continental interiors.

The North American Craton

It is convenient to discuss the geologic history of the continental mass that now constitutes North America in terms of its cratonic and geosynclinal regions. It is also of advantage to subdivide the Early Paleozoic of the craton on the basis of advances (transgressions) and retreats (regressions) of the inland or epeiric seas. Regressions, of course, exposed the old sea floors to erosion and produced extensive unconformities that are used as stratigraphic markers.

In Chapter 5 we discussed the largely European basis for defining the geologic periods. Al-

...gh the cratonic rocks of North America can be ...lated with those of Europe on the basis of fos- ...ost of the stratigraphic breaks in deposition ...correspond to those that define the Euro- ...od boundaries. Indeed, if the relative time ...d been worked out in North America in-

Table 10–1 RELATIONSHIP OF NORTH AMERICAN CRATONIC SEQUENCES TO THE STANDARD GEOLOGIC TIME SCALE

Geologic Time		Cratonic Sequences
Late Paleozoic	Permian	*Absaroka*
	Pennsylvanian	
		unconf.
	Mississippian	*Kaskaskia*
	Devonian	
		unconf.
Early Paleozoic	Silurian	*Tippecanoe*
	Ordovician	
		unconf.
	Cambrian	*Sauk*
		unconf.
	Precambrian	

stead of Europe, there would probably have been only two periods in the "Early Paleozoic" rather than three. The American stratigrapher Laurence Sloss has suggested a workable remedy for this disparity. He suggested that the Paleozoic rocks of the North American craton be divided into "sequences" of deposition. For the Paleozoic, he named these the **Sauk, Tippecanoe, Kaskaskia,** and **Absaroka** sequences. As shown in Table 10–1, the boundaries of these sequences are marked by widespread unconformities. The Sauk and Tippecanoe are the Early Paleozoic sequences, whereas the Kaskaskia and Absaroka are Late Paleozoic.

During the earliest years of the Sauk Sequence, the seas were largely confined to geosynclinal troughs, so that most of the craton was exposed and undergoing erosion. No doubt it was a bleak and barren scene, for vascular land plants had not yet evolved. Uninhibited by protective vegetative growth, erosional forces gullied and dissected the surface of the land. Over a span of at least 50 million years, the Precambrian crystalline rocks were deeply weathered and must have formed a thick, sandy "soil." Eventually, marine waters spilled out of the geosynclines and began a slow encroachment over the eroded and weathered surface of the central craton.

By Late Cambrian, seas extended across the southern half of the craton from Montana to New York. A vast apron of clean sand (Fig. 10–6) was spread across the sea floor for many miles behind the advancing shoreline (Figs. 10–7 and 10–8). This sandy aspect, or facies, of Cambrian deposition was replaced toward the south by carbonates. Here the waters were warm, clear, and largely uncontaminated by clays and silts from the distant shield. Marine algae flourished and contributed to the precipitation of calcium carbonate. Invertebrates, although present, did not contribute to the volume of sediment to the degree they did in later periods.

Along the Cordilleran Geosyncline, the earliest deposits are sands, which graded westward into finer clastics and carbonates. An excellent place to study this Sauk transgression is along the walls of the Grand Canyon of the Colorado River (Fig. 10–9). In this region one finds the Lower Cambrian Tapeats Sandstone as an initial strandline deposit above the old Precambrian surface. The Tapeats can be traced both laterally and upward into the Bright Angel Shale, which was deposited in a more offshore environment. Next is the Muav Limestone, which originated in a still more seaward environment. As the sea continued its eastward transgression, the early deposits of the Tapeats were covered by clays of the Bright Angel Formation, and the Bright Angel was in turn covered by limy deposits of the Muav Formation. Together these formations form a typical transgressive sequence, recognized by coarse deposits near the base of the section and increasingly finer (and more offshore) sediments near the top.

Cambrian rocks of the Grand Canyon region not only provide a glimpse of the areal variation in depositional environments as deduced from changing lithologic patterns but also illustrate that particular formations are not usually the same age everywhere they occur. Detailed mapping of the Tapeats, Bright Angel, and Muav units combined with careful tracing and correlation of trilobite occurrences indicates that deposition of the highest facies (Muav) had already begun in the west before

Figure 10–6 The Upper Cambrian Galesville Sandstone exposed in a quarry near Portage, Wisconsin. (Photograph courtesy of M. E. Ostrom, Director, Wisconsin Geological and Natural History Survey.)

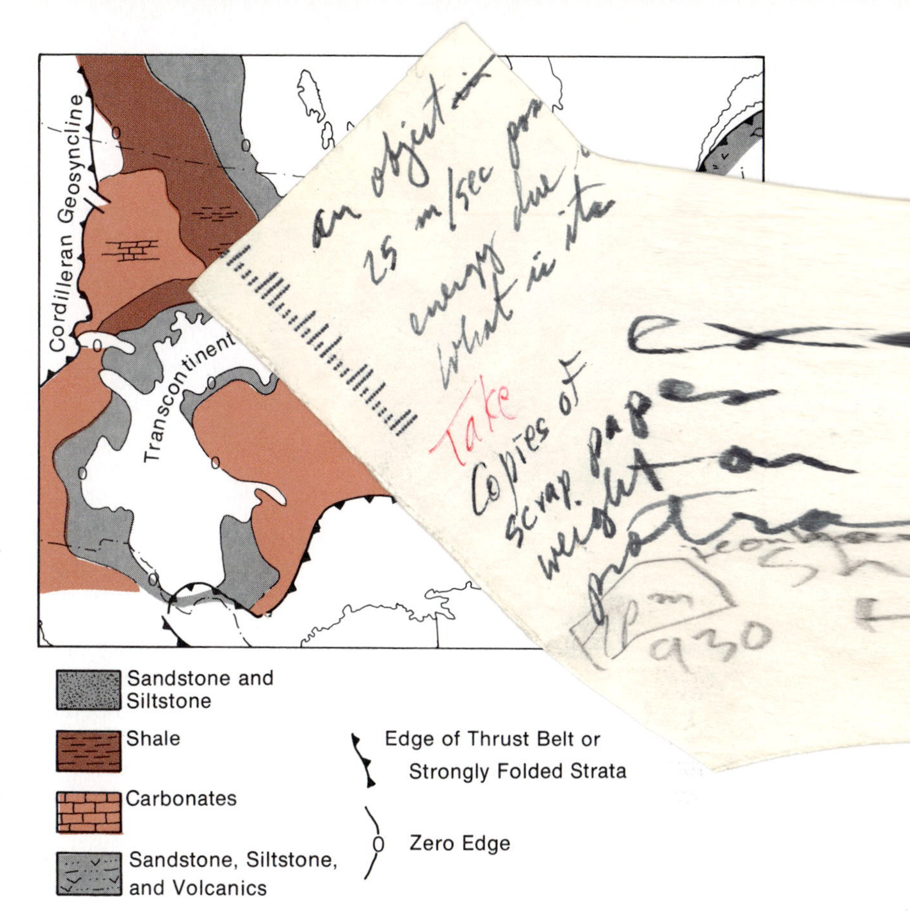

Figure 10–7 Upper Cambrian lithofacies map. (Simplified and adapted from *Stratigraphic Atlas of North America and Central America.* Shell Oil Company, Exploration Department.)

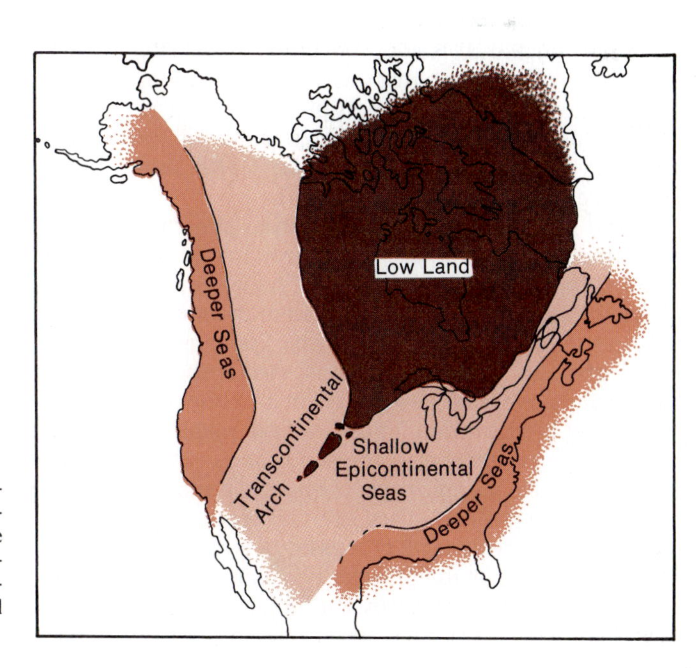

Figure 10–8 An interpretation of Late Cambrian paleogeography. During the Cambrian, the seas transgressed the craton until by late in the period marine waters covered most of the United States. The Canadian Shield remained as a lowland area, and large islands existed along the trend of the Transcontinental Arch.

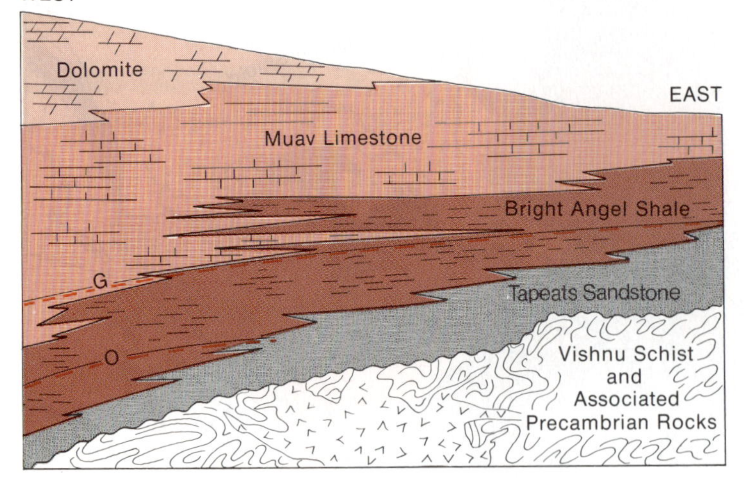

Figure 10–9 East-West section of Cambrian strata exposed along the Grand Canyon. Dashed line labeled "O" indicates the position of the high Lower Cambrian *Olenellus* trilobite zone. The dashed line labeled "G" indicates the location of the low Middle Cambrian *Glossopleura* trilobite zone. The section is approximately 220 km long, and the section along the western margin is about 600 meters thick. (Adapted from E. D. McKee, Cambrian stratigraphy of the Grand Canyon Region. Carnegie Institute, Washington, Publ. 563, 1945.)

deposition of the lowest facies (Tapeats) had stopped in the east. Correlations based on the guide fossils provided reliable evidence that the Bright Angel sediments were largely Lower Cambrian in age in California and mostly Middle Cambrian in age in the Grand Canyon National Park area. This example illustrates the **principle of temporal transgression,** which stipulates that sediments deposited by advancing or retreating seas are not necessarily of correlative geologic age throughout their areal extent.

The episode of carbonate deposition that was so characteristic of the southern craton continued into the Early Ordovician almost without interruption. Then, near the end of the Early Ordovician, the seas regressed, leaving behind a landscape underlain by limestones that was subjected to deep subaerial erosion. That erosion produced a widespread unconformity which geologists use as the boundary between the Sauk and the Tippecanoe sequences (Figs. 10–10 and 10–11).

The second major transgression to affect the platform occurred when the Tippecanoe Sea flooded the region vacated as the Sauk Sea regressed. Again the initial deposits were great blankets of clean quartz sands. Perhaps the most famous of these Tippecanoe sandstones is the St. Peter Sandstone (Fig. 10–12), which is nearly pure quartz and is thus prized for use in glass manufacturing. Such exceptionally pure sandstones cannot be developed in a single cycle of erosion, transportation, and deposition. They are the products of chemical and mechanical processes acting on still older sandstones. In the St. Peter Formation, waves and currents of the transgressing Tippecanoe Sea reworked Late Cambrian and Early Ordovician sandstones and spread the resulting blanket of clean sand over an area of nearly 7500 sq km. As described in Chapter 3, sandstones can be described by their texture and composition as either mature or immature. These textural and compositional traits are, of course, related to the processes that produced the sandstone. Sandstones with a high proportion of chemically unstable minerals and angular, poorly sorted grains are considered immature, whereas aggregates of well-rounded, well-sorted, highly stable minerals (like quartz) are considered mature. The St. Peter Sandstone is so pure as to be geologically unusual and perhaps should be termed an "ultramature sandstone."

Another noteworthy basal sandstone of the Tippecanoe Sequence is the Harding Sandstone of the Rocky Mountain Front ranges. Within this formation are found worn fragments of bone and bony scales that are the earliest fossil record of fishes in North America.

The sandstone depositional phase of the Tippecanoe was followed by the development of extensive limestones, often containing calcareous remains of brachiopods, bryozoans, mollusks, corals, and algae (Fig. 10–13). Some of these carbonates were chemical precipitates, many were fossil fragment limestones (bioclastic limestones), and some were great organic reefs. Frequently, the deposited lime underwent chemical substitution of some of the calcium by magnesium and in the process was converted to the carbonate rock called dolostone.

In the region east of the Mississippi River, the dolostones and limestones were gradually supplanted by shales. As we shall see, these clays are the peripheral sediments of the Queenston Delta,

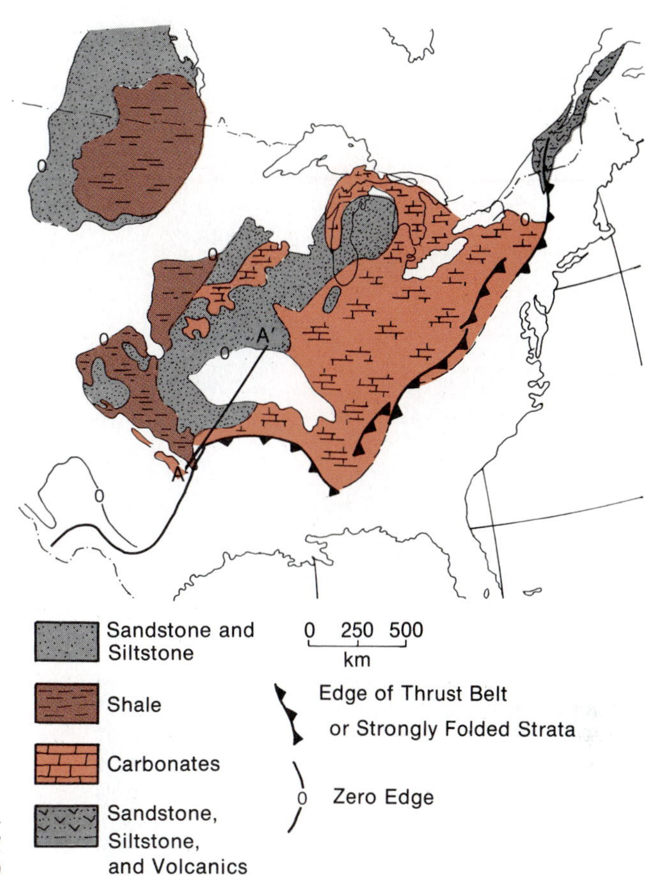

Figure 10–10 Lower Tippecanoe lithofacies map. (Simplified from *Stratigraphic Atlas of North and Central America*. Shell Oil Company, Exploration Department.)

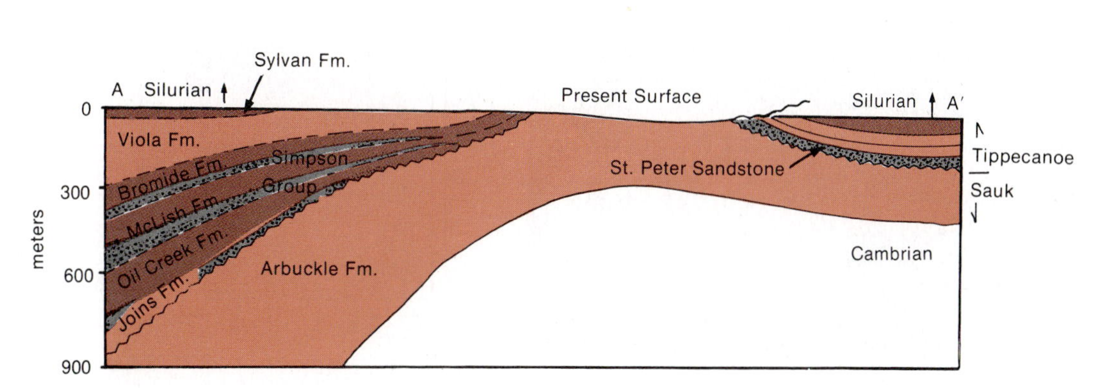

Figure 10–11 Cross-section along line A to A′ of Figure 10–10, showing that widespread unconformity separates Sauk and Tippecanoe sequences. (Adapted from *Stratigraphic Atlas of North and Central America*. Shell Oil Company, Exploration Department.)

Figure 10–12 Exposure of St. Peter Sandstone near Pacific, Missouri, Joachim dolomite overlies the massive sandstone and can be seen as well-stratified beds at the top right of the photograph. (See also Fig. 3–4.)

Figure 10–13 Slab of fossiliferous limestone (Kings Lake Formation of Decorah Subgroup). Among the fossils are brachiopods (a), trilobites (b), twiglike bryozoa (c), and gastropods (d). (Photograph courtesy of M. L. Shourd, Washington University.)

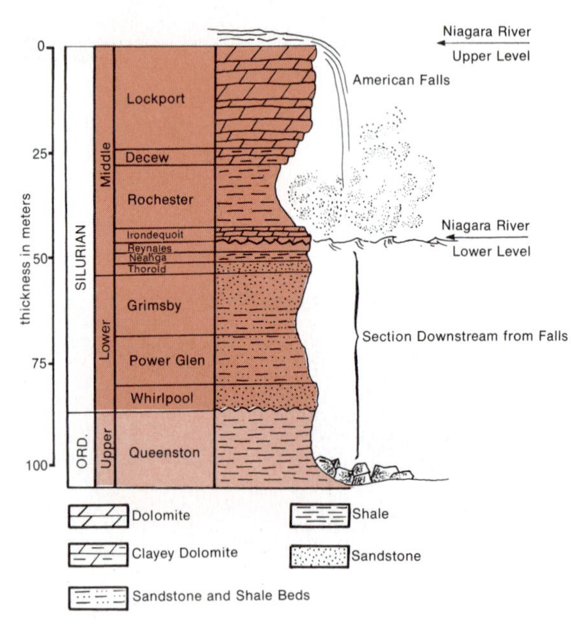

Figure 10–14 The stratigraphic section at Niagara Falls.

which lay far to the east. The geologic section exposed at Niagara Falls is a classic locality to examine these rocks (Figs. 10–14 and 10–15).

Near the close of the Tippecanoe Sequence, landlocked, reef-fringed basins developed in the region now occupied by the Great Lakes. The evaporation of these basins caused precipitation of salt and gypsum on a vast scale (Fig. 10–16). Salt from the Salina Group of the Michigan Basin is about 500 meters thick and is extensively mined. The Tippecanoe Seas were well into their regressive phase when these evaporites were being precipitated.

The Western Geosyncline

The great Cordilleran Geosyncline (Fig. 10–17), which covered a wide belt from Alaska down through western Canada and the United States, was the dominant feature of western North America during the Early Paleozoic. During much of this time, the Pacific Ocean extended inland all the way to the **transcontinental arch:** an elongate, ridgelike structure extending from Southern California to the Lake Superior region of Canada (Fig. 10–7).

The Cordilleran Geosyncline appears to have originated during the Proterozoic, when North America began to separate from a segment of the continent that extended westward beyond the present continental margin. The geosyncline formed along the passive trailing edge of the continent. One would expect to find evidence of rifting if such a continental separation occurred, and indeed such evidence does exist. The Belt Supergroup of Montana, Idaho, and British Columbia, the Uinta Series of Utah, and the Pahrump Series of California were all deposited in deep fault-controlled basins formed during an episode of rifting about 0.85 to 1.4 billion years ago.

One of the grandest sections of Cambrian rocks in the world outcrops in the Canadian Rockies of Alberta. The formations have been erosionally sculptured into magnificent mountain scenery. Lower Cambrian rocks include ripple-marked quartz sandstones, probably derived from the Canadian Shield. By Middle Cambrian time the seas had transgressed farther eastward, and shales and carbonates became more prevalent. An interesting section of these Middle Cambrian rocks is exposed along the walls of Kicking Horse Pass in Alberta. One of the units in this section, the Burgess Shale, has excited the keen interest of paleontologists around the world because it contains abundant remains of Middle Cambrian soft-bodied

Figure 10–15 American Falls at Niagara Falls. (Photograph courtesy of Ontario Ministry of Industry and Tourism.)

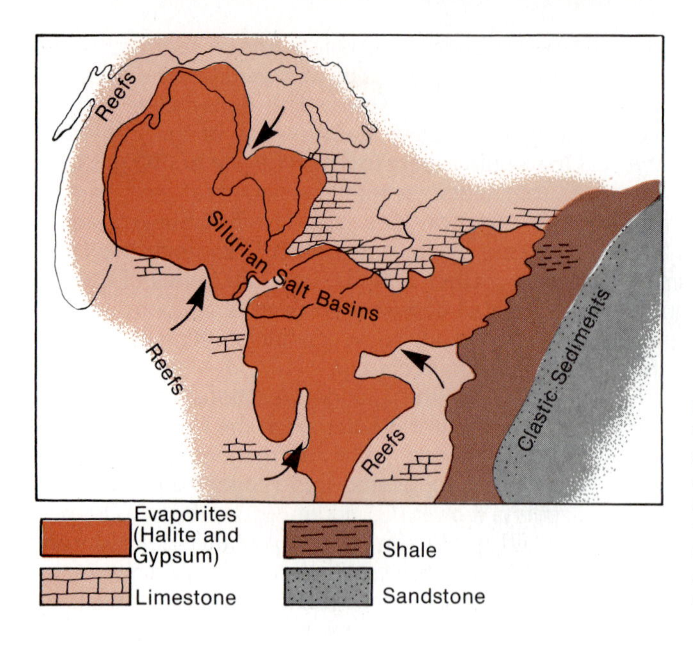

Figure 10–16 Development of evaporite basins during the Late Silurian. Areas of evaporite precipitation were surrounded by carbonate banks and reefs. Tongues of carbonate rocks extending into basins (shown by *arrows*) may have been locations for influx of normal sea water. Such influx would have been necessary in order to replace water lost by evaporation. (Adapted from H. L. Alling and L. I. Briggs. *Bull. Am. Assoc. Petr. Geol.,* 45:515–547, 1961.)

animals, most of which have never been discovered elsewhere (Fig. 10–18).

By viewing the patterns of sedimentation in the Cordilleran region, it is apparent that the geosyncline can be naturally divided into an eastern miogeosynclinal belt, in which continental shelf sediments predominate, and an adjacent eugeosyncline, which received great thicknesses of continental rise siliceous shales, cherts, graywackes, and volcanics. The stratigraphic sequence of the miogeosyncline resembles carbonate and quartz sandstone associations of the platform. The thickness of sediment, however, was much greater. Because of the large numbers of calcareous invertebrates fossilized in the miogeosynclinal limestones, they are often referred to as the "shelly facies." Eugeosynclinal deposits, on the other hand, often include dark shales containing carbonized

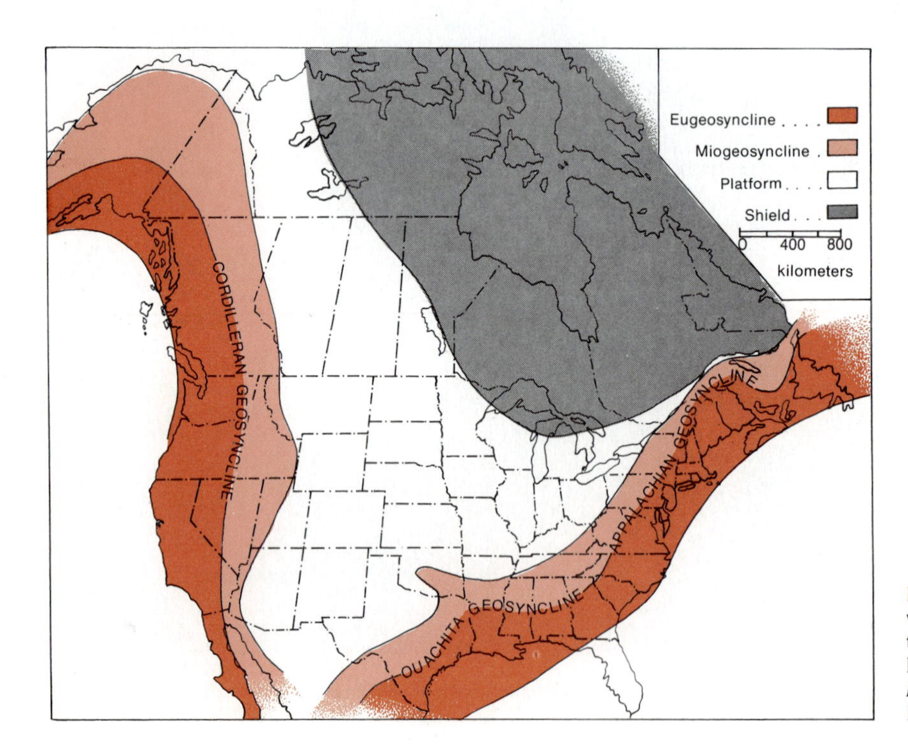

Figure 10–17 Tectonic framework of North America during the Early Paleozoic (After Eardley. *Structural Geology of North America.* New York, Harper Bros., 1951.)

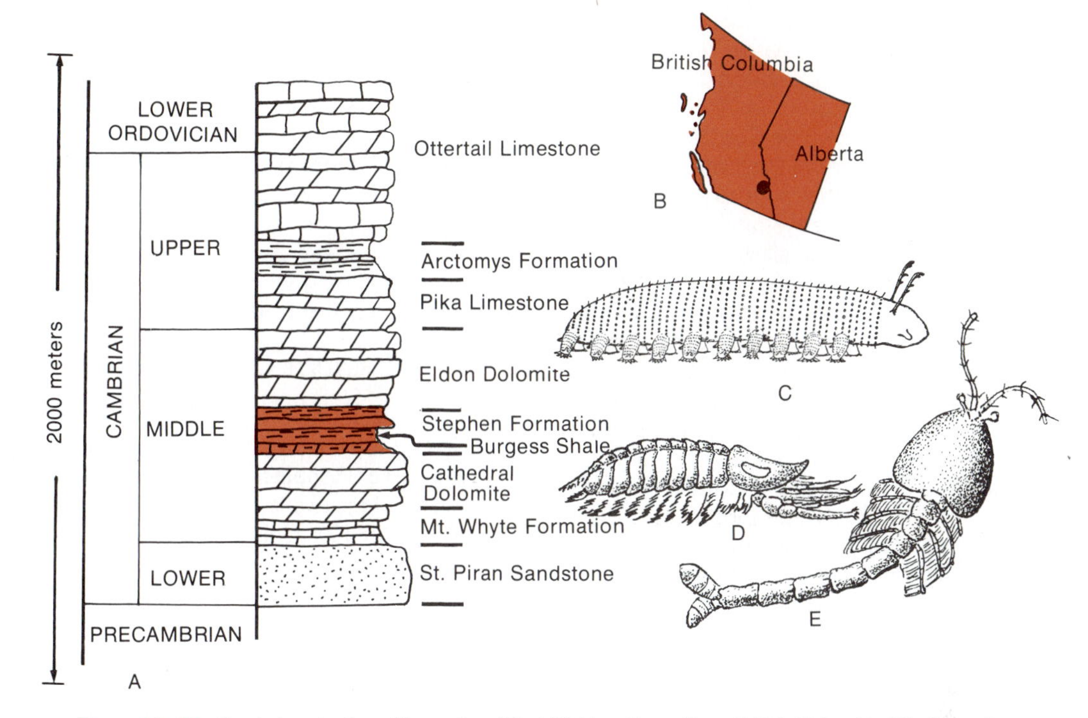

Figure 10–18 Cambrian stratigraphic section (A) at Kicking Horse Pass, British Columbia (B). According to a now famous story, this is where a slab of fossiliferous shale was kicked over by a pack horse so as to catch the eye of paleontologist Charles Walcott, who followed the trail of rock debris up the side of the mountain to discover the Burgess Shale Beds with their rich fauna of Middle Cambrian fossils. Among these were *Aysheaia* (C), an "onycophoran" believed to be intermediate in evolutionary position between segmented worms and arthropods. Among the thousands of specimens were trilobite-like arthropods such as *Leanchoilia superlata* (D) and *Waptia fieldensis* (E).

Figure 10–19 Slab of rock containing hundreds of graptolites, mostly species of *Monograptus* (Silurian).

remains of colonial organisms called **graptolites** (Fig. 10–19). For this reason the eugeosynclinal sediments have been dubbed the "graptolite facies."

The Ordovician of the Cordilleran Geosyncline is represented by several thousand meters of fossiliferous carbonates that were deposited along the miogeosyncline. These strata contrast markedly with the dark graptolitic siltstones and graywackes of the eugeosyncline to the west (Fig. 10–20). The eugeosynclinal facies are best exposed in great thrust sheets in central Nevada, where the section measures over 5000 meters in thickness. Similar eugeosynclinal rocks outcrop in central Idaho, in eastern Washington, and in the central Sierra Nevada Mountains of California.

Silurian deposits of the Cordilleran are generally similar to those of the Ordovician. Eugeosynclinal facies can be recognized in southeastern Alaska, California, and Nevada. Nevada also contains exposures of Silurian miogeosynclinal carbonates.

The Northern Geosyncline

The geosyncline at the top of North America has been named the Franklinian Geosyncline. It extended over a distance of 2250 km from northernmost Canada to northeastern Greenland (Fig. 10–21). Cambrian rocks are not exposed in this frigid region, but Ordovician formations of the shelly facies are widespread. Ordovician limestone reefs and evaporites are also prevalent. On the Ellesmere and Axel Heiberg Islands, thick sequences of graywackes, black shales, conglomerates, and volcanics indicate the existence of a well-developed eugeosyncline.

The Silurian section in this region is one of the thickest in the world. Here, too, it contains extensive reef deposits and thick accumulations of evaporites. Such sedimentary rocks imply former tropical conditions, an interpretation strengthened by paleomagnetic data indicating that the Franklinian Geosyncline was located about 15° from the equator during the Early Paleozoic.

The Eastern Geosyncline

Bordering the interior platform of North America on the east from Newfoundland to Georgia was the great Appalachian Geosyncline. Its southwestward extension is called the Ouachita Geosyncline (Fig. 10–17). Unlike the Cordilleran Geosyncline, which was a relatively passive plate margin during most of the Early Paleozoic, the Appalachian belt was subjected to frequent episodes of intense orogenic activity resulting from plate collisions. Sediment was poured into this great arcuate trough from both the craton on the one side and lands and volcanic islands that existed just beyond the present continental boundaries on the other. Once again, Cambrian sands derived from shield source areas are the initial deposits of the miogeosyncline. These sandy deposits are followed by relatively shallow water Middle Cambrian and Ordovician carbonates that have great thicknesses (in excess of 3000 meters in the Ouachita part of the geosyncline). Eastward of the miogeosyncline, along what

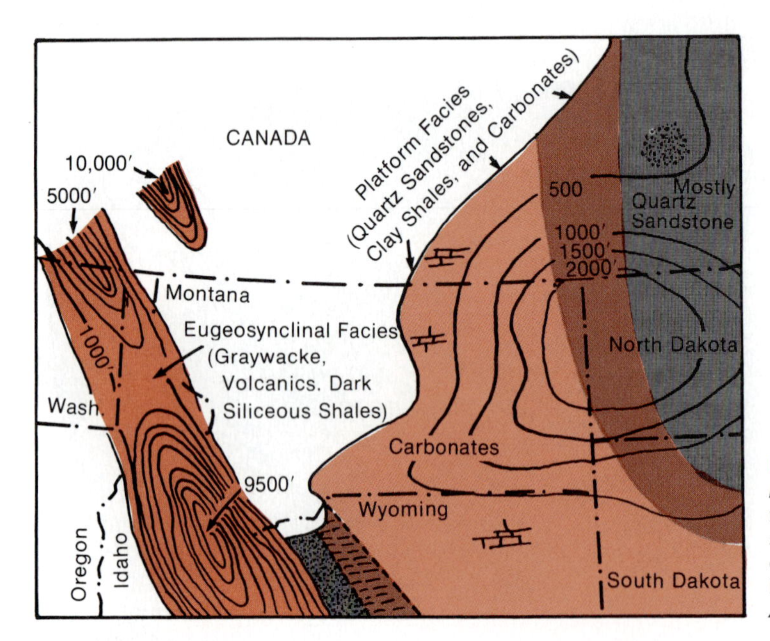

Figure 10–20 Ordovician sedimentation in the Montana region. Isopach lines indicate thickness of the Ordovician time-rock unit. Eugeosynclinal sediments reach 9500 ft (2900 meters) in Idaho and 10,000 ft (3080 meters) in British Columbia. (After L. L. Sloss. *Bull. Am. Assoc. Petr. Geol.*, 34:423–451, 1950.)

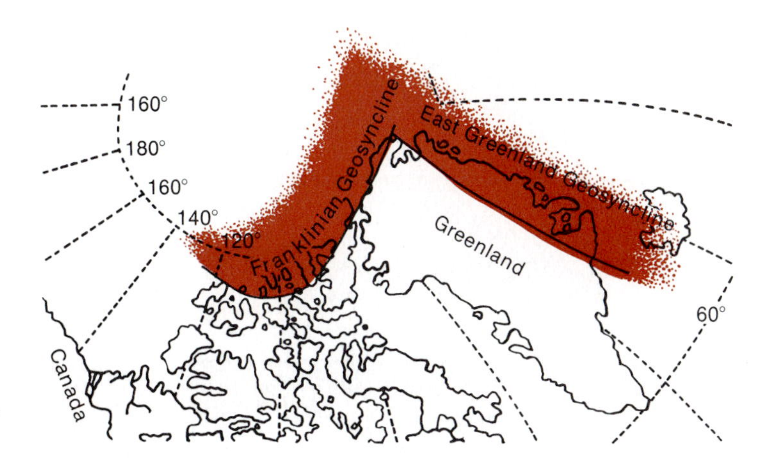

Figure 10–21 Location of the Franklinian Geosyncline.

is now the Atlantic coastal region, lay the more active eugeosyncline. Outcrops in the New England states and Newfoundland consist of siliceous shales, volcanics, and graywackes deposited within that eugeosynclinal belt (Fig. 10–22).

From our earlier discussion of the breakup of Pangaea I, you may recall that, in theory at least, the Cambrian may have been the time of opening of a proto-Atlantic. That oceanic tract may then have begun to close again during the Ordovician and Silurian. The closure was not continuous along the entire tract, but different parts came together at somewhat different times and with varying degrees of severity. The closures resulted in the Paleozoic orogenic events of eastern North America, northwestern Europe, and the northwestern tip of Africa.

The coming together of the broken continental margins was not without a depositional record. With increasing frequency one finds ever thicker and coarser clastics that signify orogenic activity in the eugeosyncline during the Early Paleozoic (Fig. 10–23). The clastics covered most of the miogeosyncline; they are coarsest on the eastern margin, grade westward to finer and more clayey materials, and ultimately merge into carbonate strata along the edge of the platform. Long ago, geologists viewing this great wedge of sediment that became thicker, coarser, and more volcanic toward the Atlantic speculated that some now vanished highland area must have supplied the detritus.

Pulses of orogenic activity became relatively frequent during the Early Ordovician. This preliminary unrest was followed by several more intense deformational events in Middle and Late Ordovician that comprise the **Taconic Orogeny.** The Taconic Orogeny was caused by the collision of another tectonic plate, probably part of ancestral western Europe, with the eastern coast of North America. Later orogenic episodes resulted from further compression between the opposing plates. The effects of the Taconic Orogeny are most apparent in the northern part of the Appalachian belt. Caught in the vise between closing lithospheric plates, geosynclinal sediments were crushed, metamorphosed, and thrust northwestward along a great fault. In this fault, named **Logan's Thrust** (Fig. 10–24), eugeosynclinal sediments have been shoved over and across some 48 km of miogeosynclinal and shield rocks. Today we are able to view the remnants of this activity in the Taconic Mountains of New York. Ash beds, now weathered to a yellow clay called *bentonite,* attest to the violence of volcanism. Masses of granite now exposed in the Piedmont help to record the great pressures and heat to which the geosynclinal sediments were subjected. But even if the igneous rocks were never found, geologists would know mountains had formed, for the great apron of sandstones and shales that outcrop across Pennsylvania, Ohio, New York, and West Virginia must have had their source in the rising Taconic ranges.

From a feather edge near Cincinnati, this barren wedge of rust-red terrestrial clastics called the **Queenston Delta** becomes increasingly thicker and coarser toward the ancient source area to the east (Figs. 10–25 and 10–26). It has been estimated that over 600,000 cu km of rock were eroded to produce the enormous volume of sediment in the Queenston Delta. It is not unlikely that the Taconic ranges exceeded elevations of 4000 meters.

During the Silurian, the locus of most intense orogenic activity seems to have shifted northeastward into the Caledonian belt. Meanwhile, erosion of the Taconic Highlands continued. Early Silurian beds are often coarsely clastic, as evidenced by the

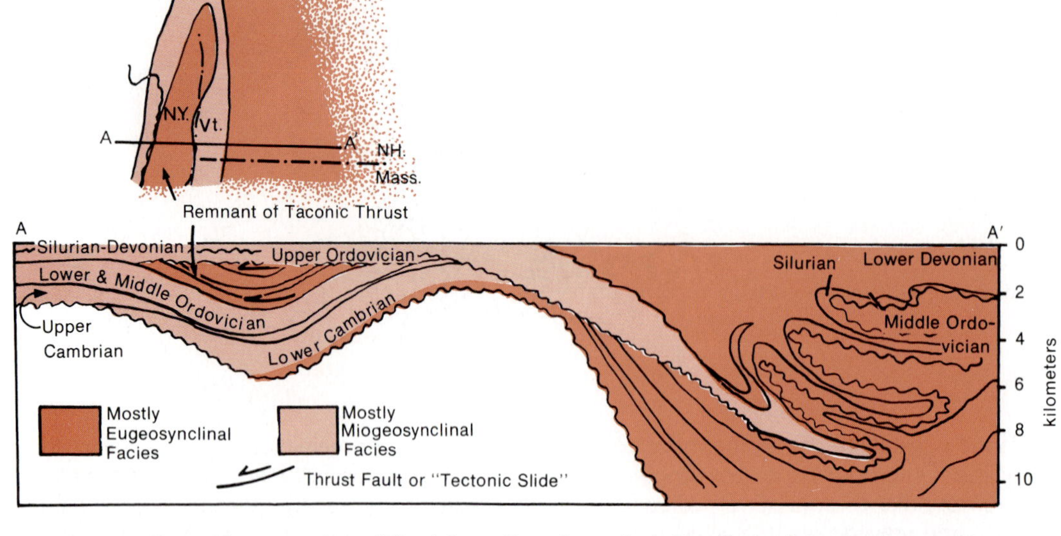

Figure 10–22 Restored cross-sections of Cambrian to Devonian rocks in New England showing the transition from eugeosynclinal to miogeosynclinal zones. Eugeosynclinal deposits are nearly three times as thick as miogeosynclinal deposits of the same age. (Simplified from W. M. Cady. *In* Zen, White, Hadley, and Thompson (eds.). *Studies of Appalachian Geology: Northern and Maritime.* New York, Interscience Publishers, Division of John Wiley & Sons, Inc., Chapter 11, p. 155, 1968.)

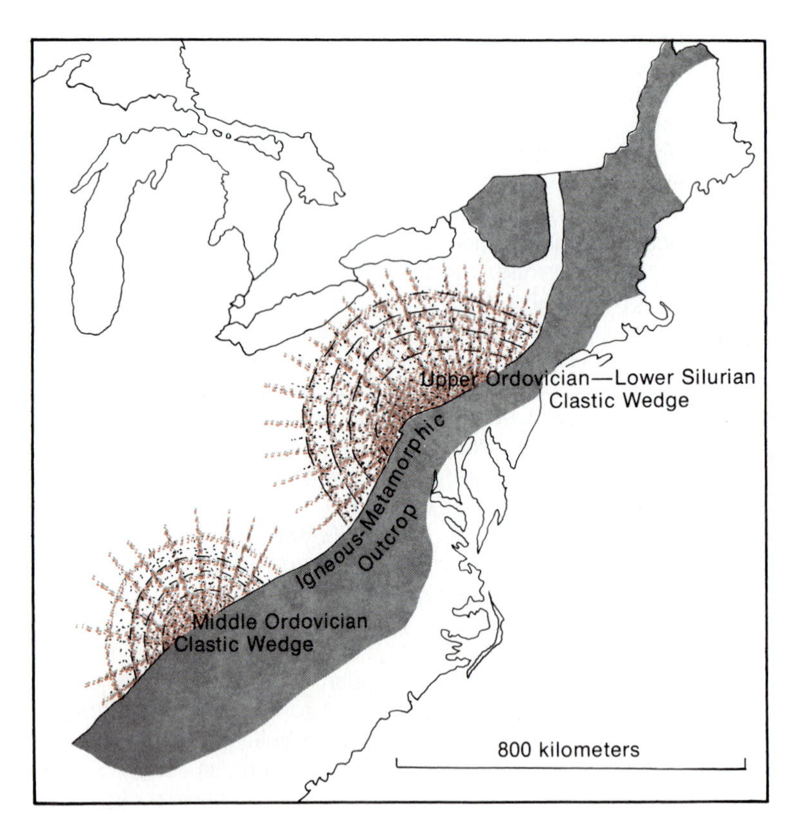

Figure 10–23 Great wedges of clastic sediments spread westward as a result of erosion of mountain belts developed during the Early Paleozoic. (After J. B. Hadley. *In Tectonics of the Southern Appalachians* V.P.I., Dept. Geol. Sci., Memoir. 1, 1964.)

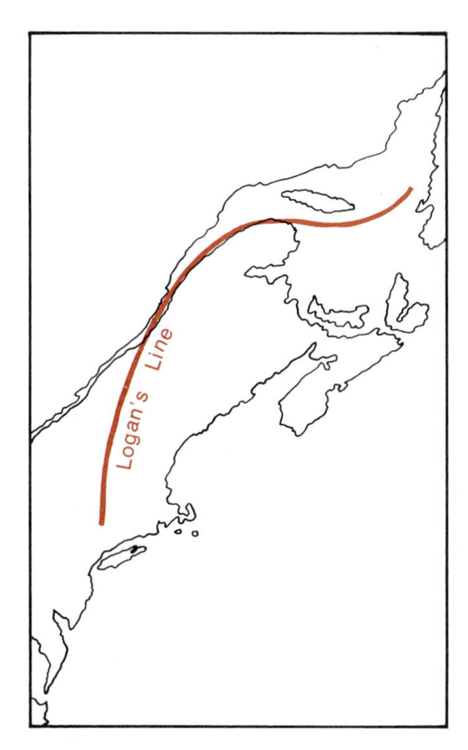

Figure 10–24 "Logan's Line." As a result of folding and thrust-faulting, the New England region of North America was compressed by several hundred kilometers. Cambrian and Ordovician sediments originally deposited near the present coastline are found along the Hudson River and Lake Champlain as remnants of great thrust sheets. The boundary between the thrust sheets and the relatively less disturbed strata is named Logan's Line after Sir Joshua Logan, a pioneering Canadian geologist who first described its significance.

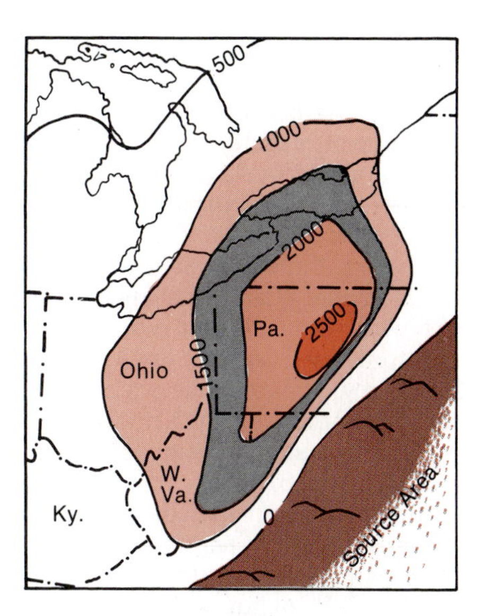

Figure 10–25 Isopach map illustrating the regional variation in thickness of Upper Ordovician sedimentary rocks in Pennsylvania and adjoining states. (After M. Kay. North American Geosynclines. *Geol. Soc. Am. Mem.* 48, 1951.)

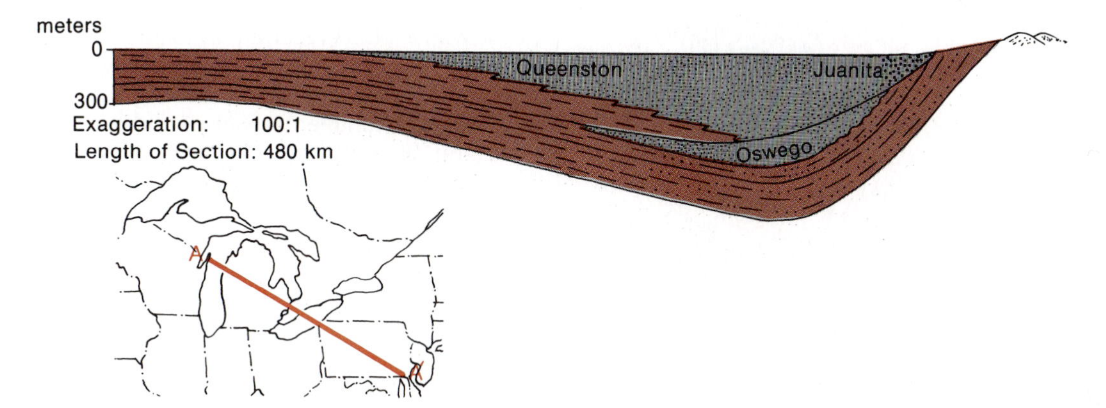

Figure 10–26 Restored section of Upper Ordovician rocks from the Atlantic to a point northeast of Lake Michigan. The interpretation given to the cross-section is that there was a rising highland area to the east that supplied clastic sediments to the geosyncline until it filled. Continued sedimentation forced the retreat of the sea westward and extended a clastic wedge toward the west (Queenston and Juanita Strata) of terrigenous red beds. (After M. Kay. *Bull. Geol. Soc. Am.*, 1942.)

Shawangunk Conglomerate of New York (Fig. 10–27). These pebbly strata give way upward and laterally to sandstones such as those found in the Cataract Group of Niagara Falls. Silurian iron-bearing sedimentary deposits accumulated in the southern part of the geosyncline. The greatest development of this sedimentary iron ore is in central Alabama, where there are also coal deposits of Pennsylvanian age. The coal is used to manufacture coke, which is needed in the process of smelting. Limestone, used as blast furnace flux, is also available nearby. The fortunate occurrence of iron ore, coal, and limestone in the same area accounts for the important steel industry of Birmingham, Alabama (Fig. 10–28).

Extending across the southern margin of the North American craton, and continuous with the Appalachian Geosyncline, is the Ouachita Geosyncline (Fig. 10–17). Although this segment of the mobile belt is over 1500 km long, only about 300 km of its folded strata are exposed. Additional information about the distribution of Paleozoic rocks in this region is derived mostly from oil well drilling operations.

Altogether, nearly 10,000 meters of Paleozoic sediment fill the Ouachita Geosyncline. However, of this thickness, only about 1600 meters were deposited in the Early Silurian. This indicates more rapid subsidence and deposition in the trough during the Late Paleozoic. Unfossiliferous metamorphosed graywackes, quartzites, and cherty beds of probable Early Cambrian age are the oldest Paleozoic sediments known from the Ouachita Geosyncline and indicate eugeosynclinal deposition. The next oldest rocks are Middle Cambrian igneous bodies that yield dates of 525 to 535 million years. These are overlain unconformably by basal sandstones and fossiliferous limestones of platform facies. Ordovician and Silurian deposits in the Ouachita region can be differentiated into distinctly eugeosynclinal (graptolitic shales) and miogeosynclinal facies. The eugeosyncline lay to the south, whereas the miogeosyncline extended along the edge of the craton. Northward from the miogeosyncline, the Early Paleozoic section thins as it becomes part of the interior platform. Paleozoic rocks of the Ouachita Geosyncline are noted for their unusually siliceous and cherty character. Very likely, the abundant silica was derived from the submarine weathering of volcanic ash ejected by volcanoes that lay to the south of the Ouachita trough.

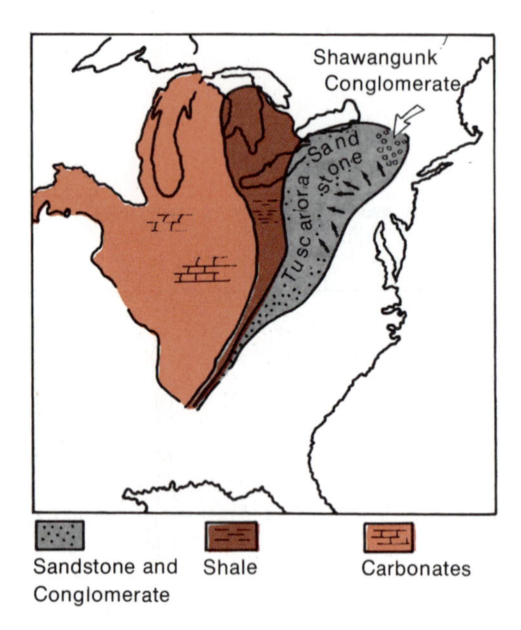

Figure 10–27 Lower Silurian lithofacies. (Simplified from *Stratigraphic Atlas of North and Central America.* Shell Oil Company, Exploration Department.)

The Caledonides

Scotland's ancient name, still used in poetry, is **Caledonia.** Eduard Suess used the name for the Scottish remnants of an Early Paleozoic mountain range, the **Caledonides.** It is also the basis for the name of the Caledonian Geosyncline, which extended along the northwest border of Europe (see Fig. 10–5).

The Caledonian and Appalachian Geosynclines have had a generally similar history. This is not surprising, for both are really part of one greater Appalachian-Caledonian system. The geosyncline evolved as a result of a cycle of ocean expansion and contraction such as is schematically illustrated in Figure 10–29. The cycle began with a Late Precambrian to Middle Ordovician episode of spreading, as the Proto-Atlantic widened to admit new oceanic crust along a spreading center. On the continental shelves and rises of the separating blocks, thick lenses of sediment accumulated. This depositional phase is marked by the development of two distinct facies. The graptolite facies consists of more than 6000 meters of volcanics, graywackes, and shales. Graptolites are prevalent in this facies and are used for subdividing the Ordovician and Silurian into biostratigraphic zones. A shelly facies with clean sandstones and fossiliferous limestones help us to identify the miogeosynclinal or continental shelf facies. Here and there on the margins of the Caledonian Miogeosyncline are found freshwater deposits of Late Silurian age that are noteworthy for fossil remains of early fishes and strange arthropods called **eurypterids** (Fig. 10–30).

Figure 10–28 The Red Mountain Expressway road cut near Birmingham, Alabama. Iron ore and underlying limestone are exposed in the cut. Coal to fuel Alabama's iron industry outcrops nearby. (From P. E. LeMoreaux and T. A. Simpson. Birmingham's Red Mountain Cut. *Geotimes,* pp. 10–11, 1970.)

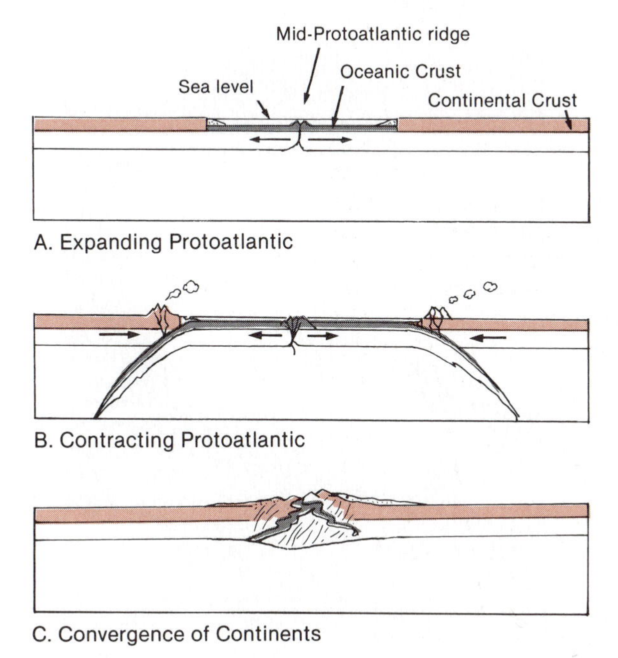

Figure 10–29 The cycle of oceanic expansion and contraction that controlled the depositional and deformational history of the Appalachian and Caledonian Geosynclines. (After J. F. Dewey. *Nature,* 272:124–129, 1969.)

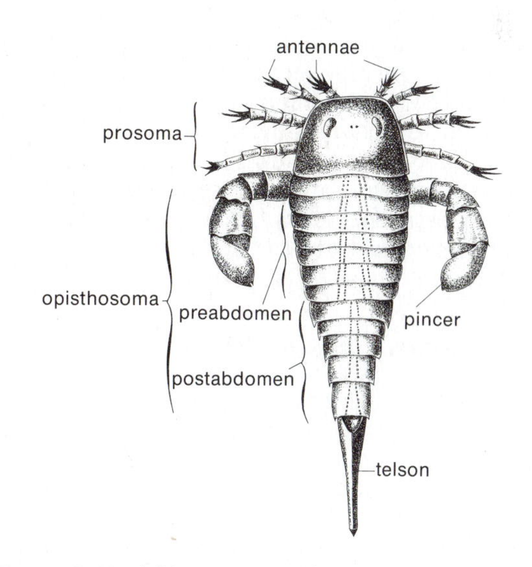

Figure 10–30 A Silurian eurypterid.

The closure of the Proto-Atlantic and crumpling of the Caledonian Geosyncline began in Middle Ordovician time, when subduction zones developed along the margins of the formerly separating continents. This event is recognized from the volcanic rocks that occur in the Canadian Maritime Provinces, northwestern England, northeastern Greenland, and Norway. Little by little, the Proto-Atlantic closed until the opposing continental margins converged in a culminating mountain-building event termed the **Caledonian Orogeny.** The orogeny reached its climax in Late Silurian to earliest Devonian time. It was most intense in Norway, where Precambrian and Lower Paleozoic sedimentary rocks are drastically metamorphosed and thrust-faulted. Southwestward, in the British Isles, the effects were not as severe, although mountainous terrains also developed there.

South of the unruly Caledonian belt there is evidence of a land mass that during Late Silurian and Devonian received sandy detritus from the growing Caledonides. This region, on which was deposited clays and sands of the Old Red Sandstone Formation, is often referred to as the "Old Red Continent." It was a thickened wedge of nonmarine clastics not unlike the Queenston Delta (Fig. 10–31).

Rocks of the Caledonian Geosyncline are well exposed not only in the British Isles but also across northeastern Greenland and Spitzbergen. The geologic succession in the Caledonides of East Greenland begins with a thick (10,000 meter) sequence of Precambrian rocks. The uppermost unit in the Precambrian sequence contains tillites, indicating an episode of glaciation just prior to the beginning of the Paleozoic. In eastern Greenland, Cambrian deposition began with sandstones followed by shales, dolostones, and limestones. Carbonates continued to be prevalent in the Early Ordovician, but thereafter sedimentational patterns began to change in response to the closing of the Proto-Atlantic as it was manifested in the Caledonian Orogeny. Erosion of the metamorphosed and folded geosynclinal rocks led to the accumulation of the continental Old Red Sandstone facies in basins adjacent to the growing mountains.

The Uralian Geosyncline

The Uralian Geosyncline extended southward from the Russian Arctic along the course of the present Ural Mountains and eastward to central Mongolia. A eugeosynclinal belt developed here also and was bordered on either side by miogeosynclines. However, the eastern miogeosyncline is poorly known because of a thick cover

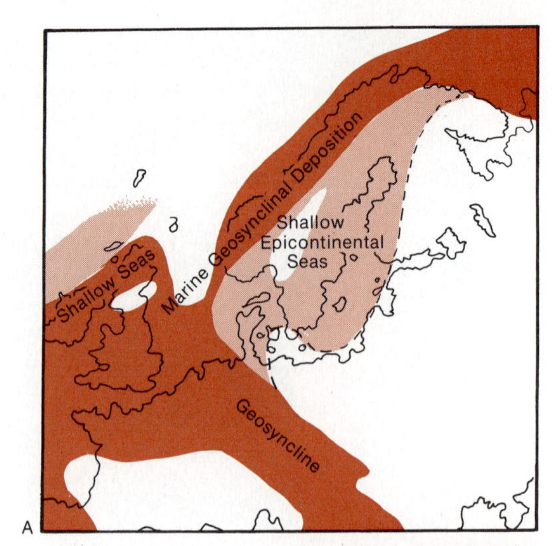

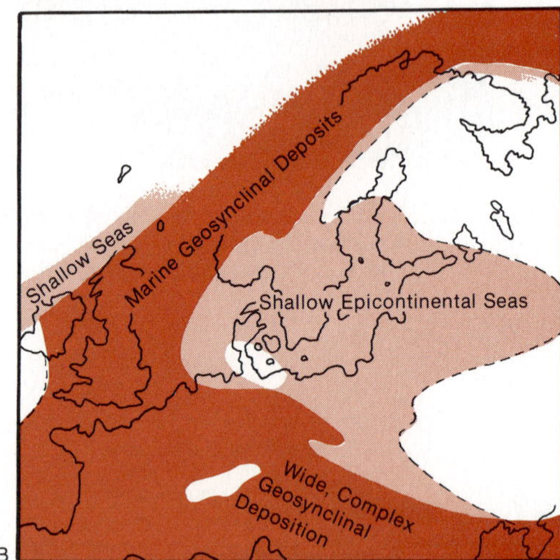

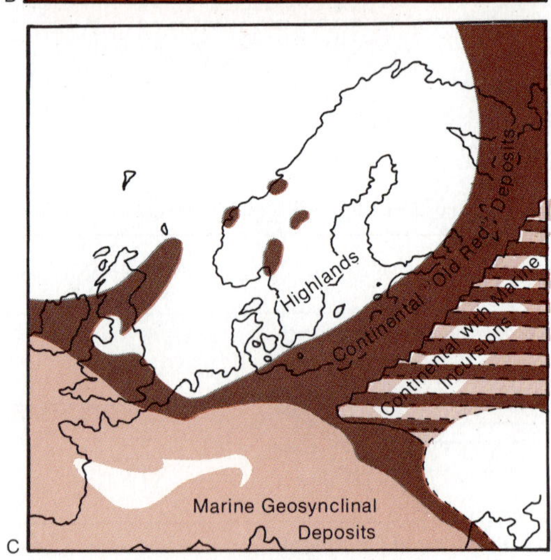

Figure 10–31 Paleogeographic maps of Europe during the Cambrian (A), Ordovician-Silurian (B), and Devonian (C). (After R. Brinkman. *Geologic Evolution of Europe.* Stuttgart, Enke, New York, Hafner Publishing Co., 1969.)

of Cenozoic strata. This is not so for the western miogeosyncline, where carbonates and sandstones grade eastward into thick sequences of shale, coarse arkosic sandstones, and volcanics (Fig. 10–32).

Although the Uralian orogenic climax was not to come until the Late Paleozoic, folded strata, unconformities, and mafic intrusions attest to plate convergence and subduction in the region. The history of the geosyncline seems to have followed an ocean expansion and contraction cycle in some ways similar to that of the Caledonian Geosyncline. Paleomagnetic studies indicate that the Russian and Siberian platforms (Fig. 10–33) were separated by a wide oceanic tract during the Early Paleozoic. During the Silurian and Devonian, subduction zones developed between the two opposing continental masses. As the oceanic crust was being subducted, the intervening ocean narrowed, and the continents began to converge upon one another. By the end of the Paleozoic, and possibly continuing into the Triassic, the geosyncline experienced its orogenic climax as the two continental blocks collided.

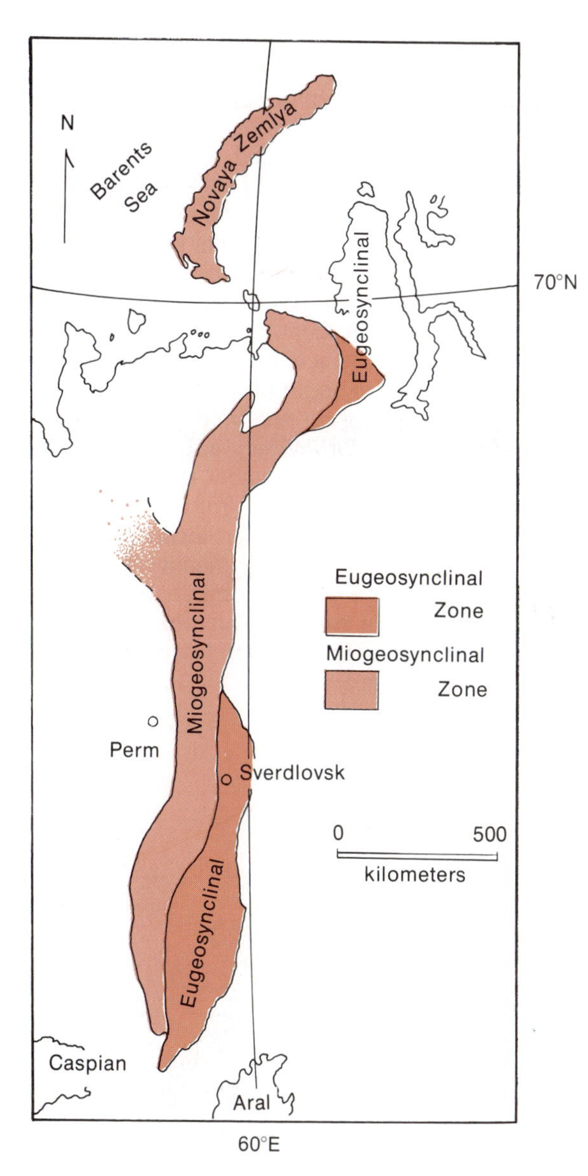

Figure 10–32 The Uralian Geosyncline. (Adapted from H. H. Read and J. Watson. *Introduction to Geology.* New York, Halsted Press, John Wiley & Sons, Inc., Vol. 2, Part II, p. 93, 1975.)

Asia Begins to Take Form

Asia east of the Urals is a very ancient and complex terrain formed from the convergence and accretion of several crustal blocks. The sutures along which the continents collided are marked by ophiolite belts, narrow zones of high-pressure metamorphism, and volcanic rocks that characterize crustal tracts above subduction zones that descend at continental margins. Some geologists believe as many as nine separate crustal blocks or microcontinents may have combined to form Asia. In the Early Paleozoic, faunal and paleomagnetic evidence suggests that most of these blocks were separated by considerable widths of oceanic crust. By the Middle Ordovician, narrowing of these oceans and convergence of the continental crustal blocks was in progress, although major episodes of collisional orogenesis did not occur in most regions until late in the Paleozoic and in the Triassic and Jurassic.

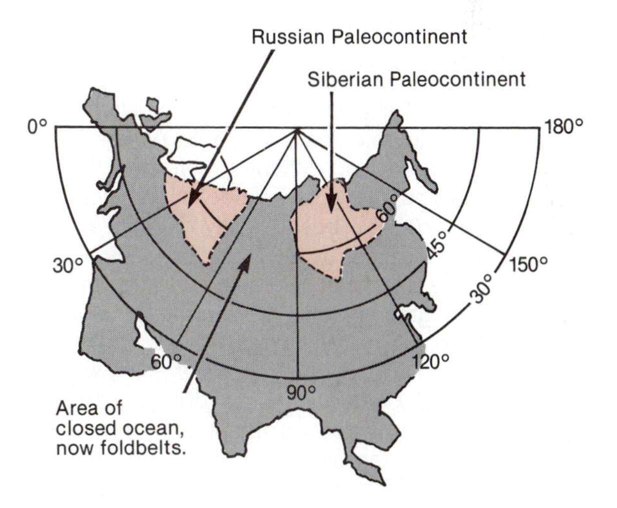

Figure 10–33 Present positions of the Russian and Siberian continental blocks relative to the Uralian fold belts.

Way Down South

Unlike the stratigraphic record of the Early Paleozoic for North America and Europe, the record for the southern hemisphere land masses is relatively poor. This may be the result of the continents being prevailingly above sea level, which would favor erosion rather than deposition. However, there is a fairly complete record of sedimentation along the north-south–trending Tasman Geosyncline of Australia (Figs. 9–19 and 10–34). Sedimentation there began just before the Paleozoic in a time period sometimes designated the "Eocambrian." These uppermost Precambrian rocks are mostly sandstones and grade into overlying Lower Cambrian formations without a significant erosional unconformity. The relatively quiet period represented by these sediments was followed by a full-scale orogeny in the Late Cambrian. Thick graywackes and volcanics record the orogenic disturbances. During the Ordovician, a broad shelf sea spread across central Australia and received deposits of well-sorted sandstones and limestones.

The African record of the Early Paleozoic is far more obscure than that of Australia. The most complete sequences of strata are located north of the equator, where one finds Cambrian through Silurian rocks that were laid down in shallow embayments of a seaway that was the precursor of the Tethys Geosyncline. Africa south of the equator was persistently above sea level and was being eroded during most of the Early Paleozoic. Strata of this age are indeed uncommon.

One feature in the geologic record of the northern part of Africa is rather astonishing: There is remarkable evidence of an Ordovician episode of glaciation in the Sahara. One can only surmise that this part of Africa was close to the Ordovician South Pole (as indicated by paleomagnetic data) or was at sufficient altitude to sustain glaciation.

Lower Paleozoic rocks in South America are also poorly known and difficult to interpret. Those strata deposited along the Andean Geosyncline (see Fig. 9–16) have been greatly deformed, and frequently obliterated, by Mesozoic orogenic activity. Elsewhere, the Paleozoic sequence lies buried beneath the great blanket of sediment resulting from the erosion of the Andes. We do, however, have scattered bits of the record. We know that Cambrian and Ordovician seas did occupy the western margin of the continent and that sediments of both graptolitic and shelly facies were deposited there. Platform deposits were laid down in a shallow basin in what is now the Amazon region. In Silurian time, an elongate seaway covered Ecuador and Peru and then swung eastward across Argentina in apparent response to uplift in the southern Andean mobile belt. Finally, we note a short period of glaciation that left its traces in Ordovician rocks of western Argentina.

Early Paleozoic Global Geography

We are so accustomed to the shapes and locations of today's continents that earth's Early Paleozoic geography seems alien to us. Geologists have attempted to reconstruct the positions of continents during this ancient time by combining all relevant paleomagnetic, paleoclimatologic, biogeographic, and tectonic information. This synthesis of data has resulted in global paleogeographic maps such as those provided in Figure 10–35. These maps indicate that by Late Cambrian time the major land areas were positioned at low latitudes as rather isolated continents in a well-interconnected world ocean. The main continental blocks at this time were Kakzakhstania (a piece of crust now located in Central Asia north of Iran), Siberia, Laurentia, Baltica, and Gondwana. There was no land north or south of 60° latitude, and the poles were therefore in the middle of oceanic areas. During the Ordovician and Silurian, the most dramatic change in global geography was the movement of Gondwana from its equatorial position southward to a polar location. Baltica (northern Europe and Russia), meanwhile, gradually circled and set a collision course toward eastern Laurentia. The actual collision took place in the Devonian as part of the Acadian Orogeny and the assembly of Pangaea II.

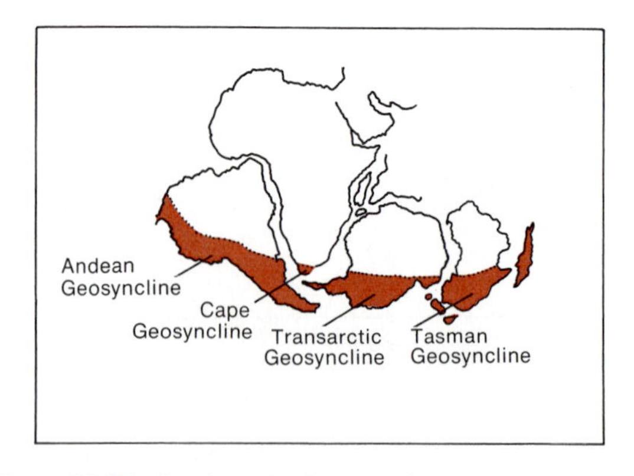

Figure 10–34 Gondwanaland geosynclines.

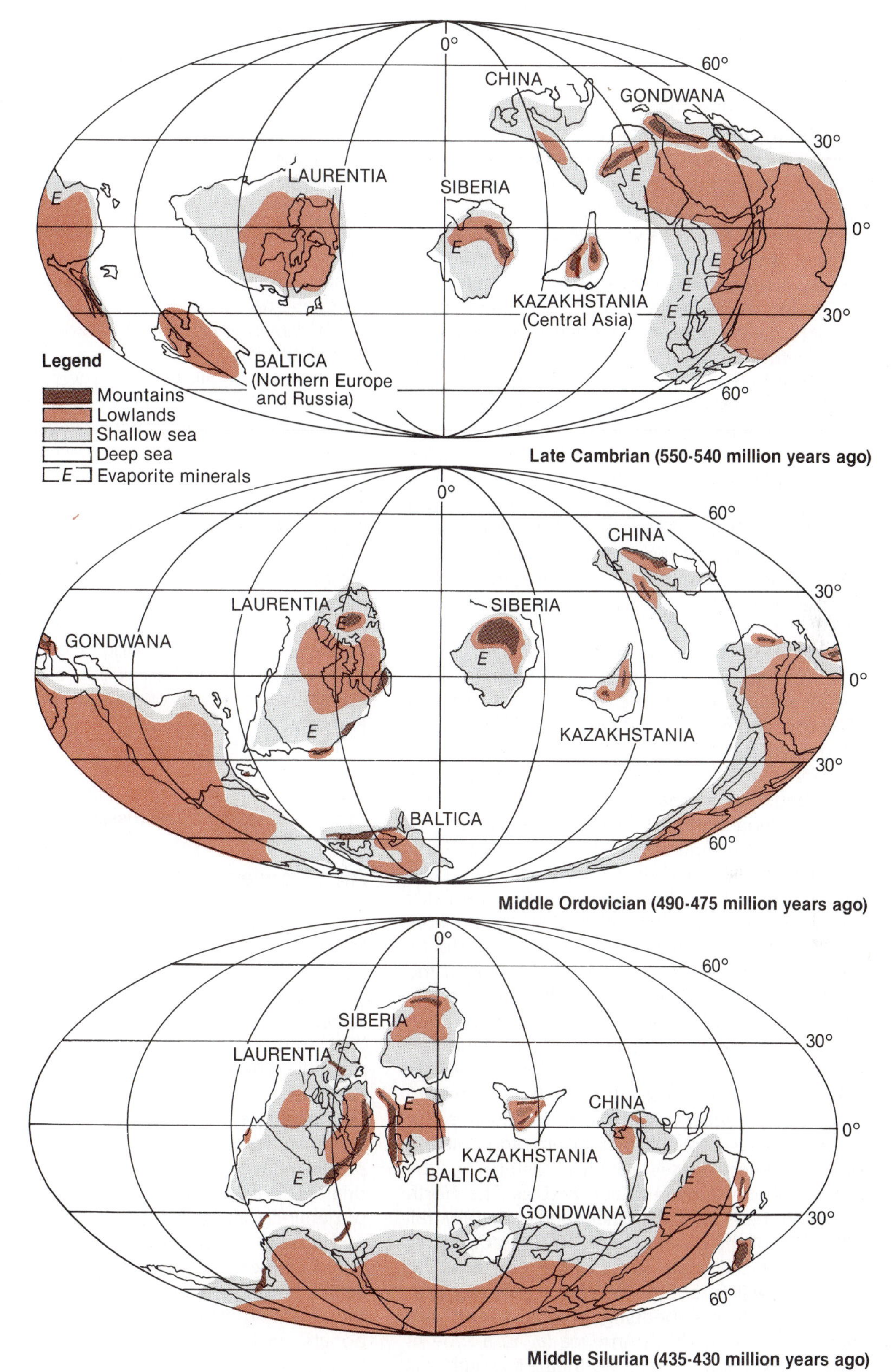

Figure 10–35 Global paleogeography of the Early Paleozoic. (From C. R. Scotese, R. K. Bambach, C. Barton, R. Van Der Voo, and A. M. Ziegler. *J. Geol.,* 87(3):217–277, 1979.)

ASPECTS OF EARLY PALEOZOIC CLIMATES

During the Early Paleozoic, the earth had a north and south pole, was encircled by an equator, and must have been characterized by latitudinal and topographic variations in climate just as such variations exist today. As we have seen, during episodes of great marine transgressions and low-lying terrains, climates were probably somewhat milder; and during those periods when continents stood high and great ranges diverted atmospheric circulation patterns, climates were more diverse and extreme, much as is the case today. However, there were also factors that would have made those climates of long ago rather unique. During the Early Paleozoic, the earth turned faster and the days were shorter. Tidal effects were stronger, and there was no green covering of vegetation to better absorb the sun's radiation. There may even have been changes in solar radiation or atmospheric composition that left no unambiguous clues. The total climatic effects of these possibilities can only be surmised.

At the very beginning of the Cambrian, the climate was probably somewhat cooler than the average for the entire Early Paleozoic. The earth was still recovering from a great ice age that had culminated near the close of the Precambrian. Soon, however, climates became warmer. Epeiric seas slowly began to encroach upon the continental interiors. Paleogeographic reconstructions such as those shown in Figure 10–35 suggest that North America, Europe, and even Antarctica may have lain astride or near the equator during the Early Paleozoic. These geophysical data are compatible with the depositional record of thick limestones and extensive reefs, which today form only in the warmer regions of the oceans. However, some evaporites and fossiliferous limestones have been found in rather high paleolatitudes, indicating that worldwide climates may have been somewhat warmer than average for a time. In the Ordovician stratigraphic sequences of Northern Canada, for example, thick deposits of gypsum and salt are found at paleolatitudes (latitudes determined by paleomagnetic studies) of about 10°S. The North American reefs of Ordovician age generally fall within 30° of the paleoequator, whereas the richly fossiliferous carbonate rocks of the craton would have been within about 40° of the paleoequator. The study of cross-bedding in sandstones suggests that eastern North America lay in a zone of trade winds that blew from northeast to southwest with respect to the present directional grid. These winds thus moved in a direction opposite to those prevalent in eastern North America today.

That there were severe as well as equable climates in the Early Paleozoic is indicated by extensive Late Ordovician glacial deposits in the region of the present Sahara Desert. These deposits are sufficiently widespread that they must have been emplaced by continental glaciers and certainly must have been accompanied by a lowering of mean annual temperatures at middle and high latitudes. Study of the distribution of the glacial deposits, directions of glacial striations, and paleomagnetic data indicate that in Ordovician time the north African part of Gondwana was positioned astride the North Pole.

During the Silurian, there were latitudinal variations in climate somewhat similar to those of today. Glacial deposits of this age are found only in higher locations, usually in excess of 65° from the paleoequator. Coral reefs, evaporites, and desert sandstones are found within 40° of the Silurian equator. Certainly there were regions of marked aridity during the Silurian. Epeiric seas became relatively less extensive as the Silurian drew to a close. We can speculate that the "Proto-Atlantic," formed by the division of that hypothetic continent Pangaea I, had already begun closing and that the Caledonian crunch was about to take place. The curtain was ready to rise on the world of the Late Paleozoic.

LIFE OF THE EARLY PALEOZOIC

An Overview

In the previous chapter we noted that fossil indications of life during most of the time that preceded the Paleozoic consisted largely of algae, fungi, and bacteria. Unquestioned animal remains appear only in uppermost Precambrian strata—"Eocambrian" strata—like those of Ediacara Hills, Australia. The Ediacara fauna includes some rather complex animals. They may have been the products of a previous lengthy period of evolution for which we have scant fossil evidence, or they may possibly have evolved rapidly just before the Cambrian. The Eocambrian, as mentioned, was a time of generally emergent lands, worldwide glaciation, and limited epeiric seas. It was a period that favored neither fossilization nor development of diversity among marine invertebrates. It is likely that the ability of invertebrates to secrete calcium carbonate shells was limited.

In passing from the Late Precambrian to the Early Cambrian, the record gradually improves. Here and there around the shallow margins of the continents, isolated groups of invertebrates had begun to establish themselves. Soon the great inland seas began to creep across the cratons, providing as they did so a multitude of opportunities for the diversification and expansion of the phyla.

Certainly the improvement in the fossil record following the Eocambrian can be partly attributed to the spread of shell-building abilities. During the Cambrian, shell-building brachiopods and trilobites were abundant, but the most rapid expansion and diversification of shell-bearing groups began later, in Early Ordovician time. Many theories have been proposed to account for this expansion of shell-builders, including some that suggest pre-Ordovician seas were chemically unsuitable for the secretion of calcium carbonate by organisms. However, Cambrian trilobites and brachiopods apparently had no difficulty in manufacturing exoskeletons of calcium carbonate. In modern shell-bearing invertebrates, the shell serves as protection and provides support for soft tissues. It seems reasonable that these were also the functions served by the shells of Early Paleozoic invertebrates. Because there is little fossil evidence of numerous predatory animals until about Late Cambrian, support may have initially been the more important function. However, with the advent of abundant and vigorous carnivores (such as the predatory cephalopods) during the Late Cambrian, active predation may have provided the selective pressures that would have favored rapid evolutionary expansion of protective exoskeletons. The result was the multitude of shellfish that have populated the oceans down to the present day.

Figure 10–36 Cambrian stromatolites exposed along the valley of the Black River in Missouri.

Figure 10–37 The algal fossil *Receptaculites.* (Photograph courtesy of Wards Natural Science Establishment, Rochester, N.Y.)

Plants

The history of plants has an obscure beginning among the bacteria and algae of the Precambrian. Indeed, the Precambrian sedimentary iron ores are presumed to have been produced by bacterial activity. However, as has been noted, fossils of such tiny, soft-bodied creatures are very rare.

An exception to the poor fossil record of earliest plants is provided by the stromatolites (Fig. 10–36), which are found frequently in Late Precambrian rocks and continue to be found in limestones of all younger ages. Laminations are the most apparent feature of stromatolites. They may vary in form from massive and incrusting to branching or finger-like. Present-day colonies of blue-green algae cause the construction of similar masses. As a result of the photosynthetic activity of the algal mat, calcium carbonate is precipitated in the surrounding sea water, settles onto the algal colony, and adheres to the gelatinous film. Successive additional layers result in the laminations. Stromatolitic reefs were widespread during the Cambrian but were more restricted during the Ordovician, perhaps because of the rise of grazing marine invertebrates.

The green algae also produced some interesting fossils during the Early Paleozoic. A lime-secreting family, the Dasycladaceae, includes fossils known as **receptaculids** (Fig. 10–37). When first encountering one of these fossils, one is reminded of the appearance of the seed-bearing central area of a large sunflower; this has caused fossil collectors to designate them "sunflower corals." They are, however, neither sunflowers nor corals. Because the fossils contained spicule-like rods and circulatory passages, receptaculids were for many years classified as ancient varieties of sponges. Although most frequently found in Ordovician rocks, the receptaculids also occur sparsely in Silurian and Devonian strata.

The Late Precambrian and Early Paleozoic seas harbored a variety of unicellular, solitary, or colonial phytoplankton. Life in the oceans then, as today, would have been dependent upon those plants, which are known primarily from the chemically resistant coverings of their spores.

But what of life on land? The appearance of our landscapes today owes much to a verdant covering of forests and grasslands. At the beginning of the Paleozoic, however, there were no forests or grassy meadows. Most living organisms were confined to the seas. It is very likely that some species of algae invaded freshwater streams and lakes, and some may even have evolved means of moisture control independent of their environment, moved onto damp areas of the land, and begun the evolutionary progression toward more complex land plants.

Most land plants are *vascular*. Because they are not continuously immersed in water, they require systems of tiny tubes to transport fluids and nutrients. The tissues of land plants are tougher and more resistant to decay than those of the simple, nonvascular plants, and thus preservation is enhanced. The most reliable clues to the early presence of vascular plants are microscopic fossil spores discovered in mid-Ordovian rocks from Libya. The basic configuration of these microfossils corresponds to that in simpler types of spores known from Lower Devonian land plants of the Family Psilophytaceae. Actual remains of the parent plants of these so-called **psilophytes** have been found in Upper Silurian strata.

The psilophytes (Fig. 10–38) were small plants characterized by a horizontal stalk that grew just beneath the ground in moist soil and from which grew short, vertical "stems" bearing smaller branches and spore sacs. Although they had neither true leaves nor roots, psilophytes were more advanced than the algae and mosses in that they did have simple vascular tissue and even delicate strands of wood cells. The vascular system permitted part of the plant to exist underground, where there was water but no light, and another part to grow where the water supply was uncertain but

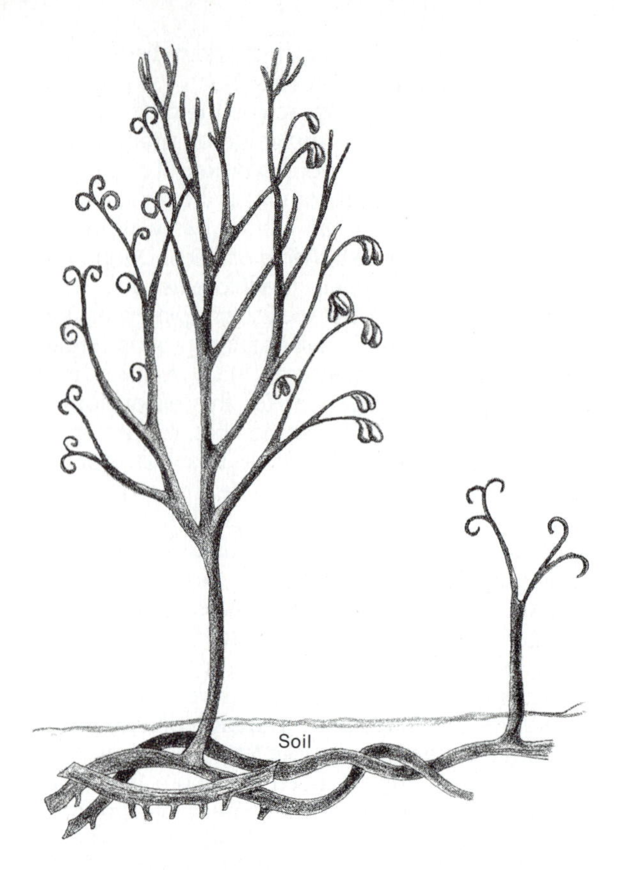

Figure 10–38 A psilophyte from the Lower Devonian of Gaspé.

sunlight for photosynthesis was insured. The psilophytes paved the way for the evolution of large, upright trees that covered large regions of the continents during the Late Paleozoic.

Invertebrate Animals of the Early Paleozoic

People are accustomed to thinking of the ocean as the birthplace of life. Thus, it is no surprise to learn that the earth's earliest animals lived in the sea. The only fossils ever found of animals that lived during the Cambrian and Ordovician have been those of ocean dwellers. As is the case today, these animals varied in their choices of habitat as well as in the kinds of food they required. In the geologic past, as in the present, there were basically three modes of life. Organisms that could live on the sea floor or burrow in sea floor sediment are termed **benthic,** those that floated are termed **planktonic,** and those that swam are termed **nektonic.** In addition, many animals in the sea spend a phase of their lives in one of these modes and then change to accommodate themselves to another. In fact, the mobile larvae of a multitude of otherwise sedentary (stationary) "critters" are dispersed in this way. In the fossil record of the Early Paleozoic, the bottom-dwellers are abundantly represented and often splendidly preserved. At least part of the fossil abundance of benthic organisms can be attributed to the fact that these bottom-dwellers characteristically develop shells or mineralized carapaces.

The fossil assemblages of the Early Paleozoic are clearly dominated by shell-bearing organisms; however, there must also have lived many soft-bodied creatures that were not fossilized. The Late Precambrian Ediacara Hills fauna suggest this was true, but more convincing proof derives from the discovery of the Middle Cambrian Burgess Shale fauna mentioned in the previous section. This unusual find was made by Charles D. Walcott in 1910 while he was on an expedition in the Canadian Rockies. Near Field, British Columbia, while ascending the southwestern slope of Mt. Wapta, Walcott happened to notice some slabs of black shale that contained shiny, jetlike impressions of a variety of arthropods, worms, jellyfish, sea cucumbers, and sponges (see Fig. 10–18). The organisms are preserved as flattened, carbonaceous films with such perfection that fine, hairlike appendages, intestinal organs, and delicate dermal patterns are revealed. Walcott described 130 species from this locality. The Burgess Shale fauna is an excellent example of Early Paleozoic marine biota. It also reminds us of the bias of the fossil record, which only occasionally provides fossils of soft-bodied animals.

The diversification of marine invertebrates during the Early Paleozoic appears to have proceeded in two stages. An initial "experimental" evolutionary radiation during the Cambrian established most of the invertebrate phyla with preservable hard parts (the bryozoa do not appear until Ordovician). Many taxonomic groups within this original radiation did not expand widely, and many became extinct. Those that were best adapted survived and formed the ancestral stock for a grander radiation that took place during the Ordovician. Indeed, most of the classes of marine invertebrates with representatives in modern seas did not appear until Late Ordovician. Nearly 170 million years of evolution was necessary for this diversification.

The buildup to a complete marine fauna involved the expansion of life into a multitude of habitats and lifestyles. The many factors involved in evolutionary adaptation produced **epifaunal** animals that lived on the surface of the sea floor, as well as **infaunal** creatures that burrowed beneath the surface (Fig. 10–39). There were borers as well as burrowers, attached forms and mobile crawlers, swimmers, and floaters. Some were *filter-feeders,* which strained tiny bits of organic matter or micro-organisms from the water; others were *sediment-*

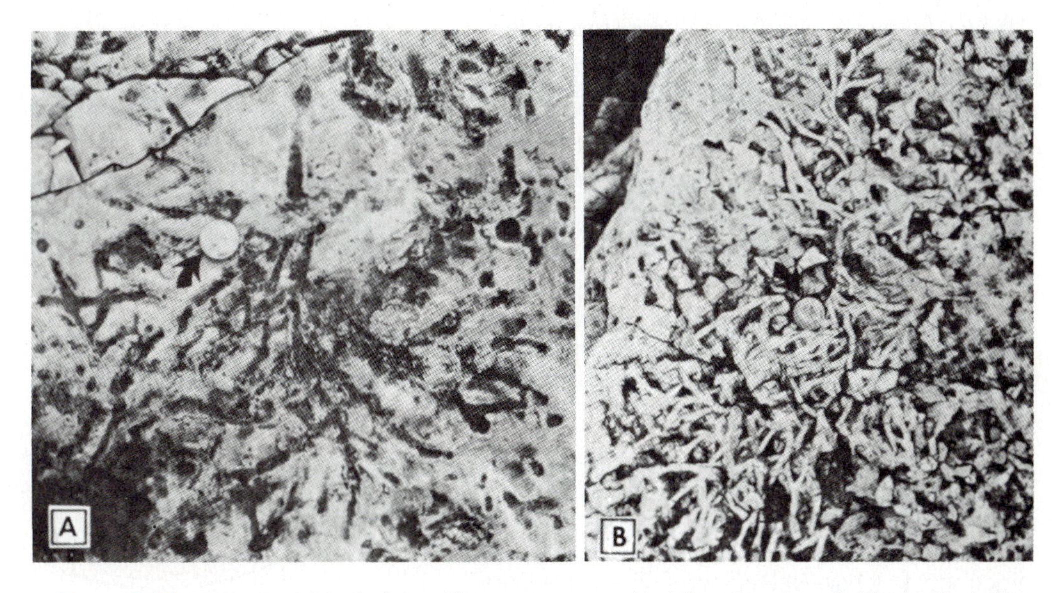

Figure 10–39 "Trace fossils" in the form of burrows constructed by infaunal creatures that lived in the bottom sediment of Ordovician seas. The fossils have been interpreted as the trace fossil *Chondrites.* The host rock is the Plattin Limestone of Middle Ordovician Age. In A, the filling of the burrow has been removed by weathering processes, whereas in B the filling remains so that the burrows appear twiglike. (From M. L. Shourd and H. L. Levin. *J. Paleo.* 50(2):263, 1976.)

feeders, which passed the mud of the sea floor through their digestive tracts in order to extract the nutrients within. There were animals that grazed upon algae that covered parts of the ocean bottom, and carnivores that consumed these grazers. A host of scavengers processed organic debris and aided in keeping the seas suitable for life. Some classes of invertebrates maintained a single mode of life, whereas groups within other classes adapted to several different life styles. Snails, for example, included scavengers, herbivores, and carnivores. With the passage of geologic time, many evolutionary changes and extinctions were to occur among the invertebrates, but these changes tended to occur *within* the taxonomic classes that had been estabished during the Early Paleozoic.

The Unicellular Animals

The postulated biochemical steps leading to the first unicellular animals were discussed in the previous chapter. The first unicellular organisms were, of course, plants, but among modern unicells are creatures that defy classification as either animal or plant. Fortunately, there are also some single-celled forms that can with confidence be identified as animals. Two groups of creatures in the last category that are also capable of building shells or supportive structures are the **foraminifers** and **radiolarians.**

Foraminifers build their tiny shells by adding chambers singly, in row, in coils, or in spirals (Fig. 10–40). Some species construct the shell, or test, of tiny particles of silt; others secrete tests composed of calcium carbonate. The test is characteristically provided with holes through which extend "tentacles" of protoplasm for feeding. It is from these holes, or foramina, in the test that foraminifers take their name. Foraminifers range from the Cambrian to the present, but they are rare and poorly preserved in Lower Paleozoic strata. During these early stages of their history they were mostly simple, saclike, tubular, or loosely coiled benthic creatures with tests composed of sedimentary particles. In modern oceans the tests of calcareous foraminifers rain down continuously on the sea floor, much like an unending limy snow. The accumulated debris forms, in part, a deep sea sediment called globigerina ooze because of the prevalence of the remains of species of *Globigerina*.

Radiolarians are also single-celled planktonic organisms that have been present on earth at least since the Paleozoic began. Like the foraminifers, they have threadlike pseudopodia that project from an ornate lattice-like skeleton of opaline silica or a proteinaceous substance. In some regions of the oceans today, radiolarian skeletons accumulate to form deposits of radiolarian or siliceous ooze. Although radiolarians do occur in the Early Paleozoic (Fig. 10–41), they are rare and not yet useful for stratigraphic correlation. They are more abundant during the Mesozoic and Cenozoic and have been shown to be particularly useful in correlating Pleistocene deep sea deposits. Of special interest is the observation that Pleistocene radiolarian extinctions appear to be associated with dates of reversal in the earth's magnetic field.

The Cup Animals

Archaeocyatha means "ancient cups." It is an appropriate name for a somewhat enigmatic group of Cambrian organisms that constructed conical or vase-shaped skeletons out of calcium carbonate (Fig. 10–42). Archaeocyathids hold two paleontologic records. They are the earliest abundant reef-building animals on earth, and they are members of a major phylum that suffered extinction. Although quite abundant during the Early Cambrian, they had entirely died out before the end of that period. Gregarious in habit, they carpeted the floors of the warm inland Early Paleozoic seas. Low reefs formed primarily of archaeocyathids can be studied today in North America, Siberia, Antarctica, and Australia. The Australian archaeocyathid reefs are often over 60 meters thick and extend horizontally in narrow bands for over 200 km.

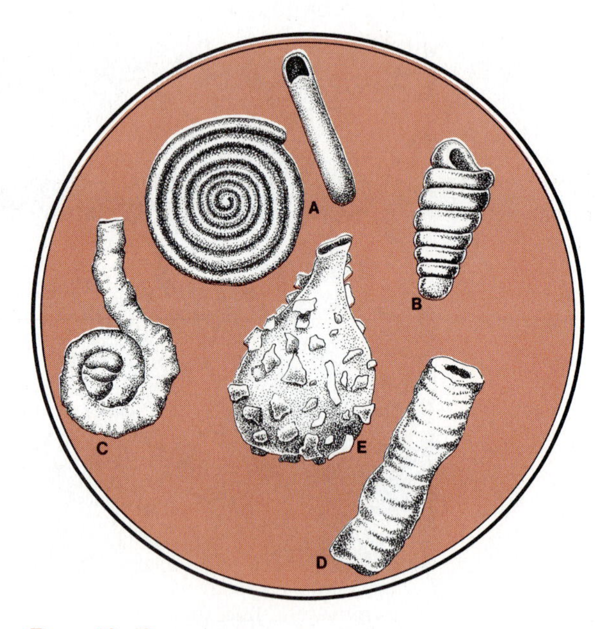

Figure 10–40 Foraminifers that are found in Early Paleozoic strata. (A) *Ammodiscus.* (B) *Turritellella.* (C) *Lituotuba.* (D) *Bathysiphon.* (E) *Lagenammina.* (All have a Silurian-Recent geologic range; all are enlarged about 80 times.)

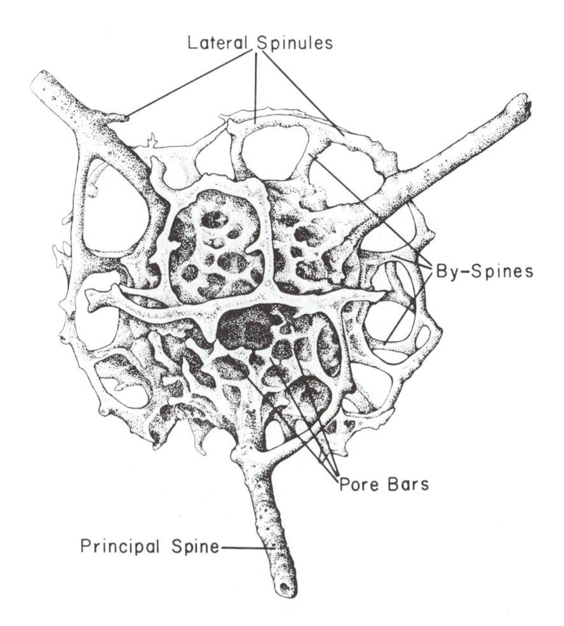

Figure 10–41 A Late Ordovician radiolarian from the Hanson Creek Formation of Nevada. It is unusual to find well-preserved radiolarians in rocks of the Early Paleozoic. (From J. B. Dunham and M. A. Murphy. *J. Paleo.,* 50(5):883, 1976.)

The Pore-Bearers

Among the many stationary animals to colonize the Early Paleozoic sea floor were the sponges. Sponges are members of the Phylum **Porifera.** They appear to have evolved from colonial flagellated unicellular creatures and thus provide insight into how the transition from unicells to multicellular animals may have occurred.

Sponges are a relatively conservative branch of invertebrates that have a long history. Cambrian representatives of all but one modern class of Porifera are known as fossils. Some fossil sponges, like *Protospongia* from the Cambrian and *Astraeospongium, Microspongia,* and *Astylospongia* from the Silurian, are well known to geologists as index fossils (Fig. 10–43).

Sponges have always been predominantly marine creatures, although a few modern species live in fresh water. Spicules formed in the walls and cell layers of sponges are distinguishing characteristics of the phylum and provide both protection and support. The spicules may be composed of silica, calcium carbonate, or a proteinaceous material termed **spongin.** Naturally, the mineralized spicules are more commonly preserved and are frequently found by geologists examining rocks that have been disaggregated for study. Spicules are also important in the classification of sponges. For example, the *Desmospongea* consist entirely or in part of spongin, which may be reinforced with siliceous spicules. *Hyalospongea* develop siliceous spicules of distinctive shape, and *Calcispongea* are characterized by spicules made of calcium carbonate.

Although sponges vary greatly in size and shape, their basic structure (Fig. 10–44) is that of a highly perforated vase modified by folds and canals. The body is attached to the sea floor at the base, and there is an excurrent opening, or **osculum,** at the top. The wall consists of two layers of cells. Facing the internal space is a layer of collar cells (choanocytes), and on the outside is a protective wall of flat cells that somewhat resemble the bricks of a worn masonry pavement. Between these two layers one finds a gelatinous substance called **mesenchyme.** Here amoeboid cells go about the work of secreting the spicules. Sponges lack true organs. Water currents moving through the sponge are created by the beat of **flagellae.** These currents bring in suspended food particles, which are ingested by the collar cells. In a simple

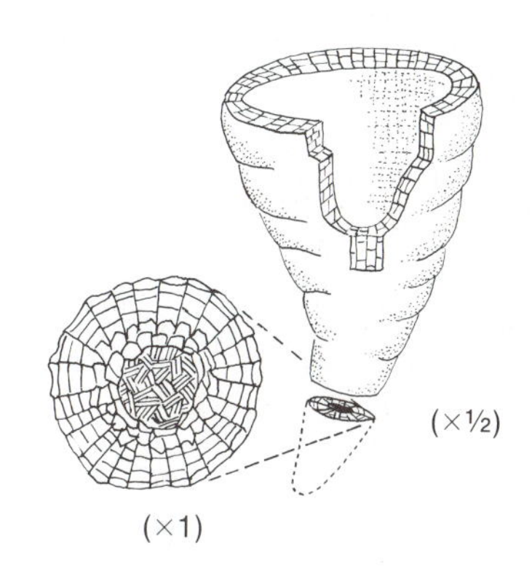

Figure 10–42 An archaeocyathid.

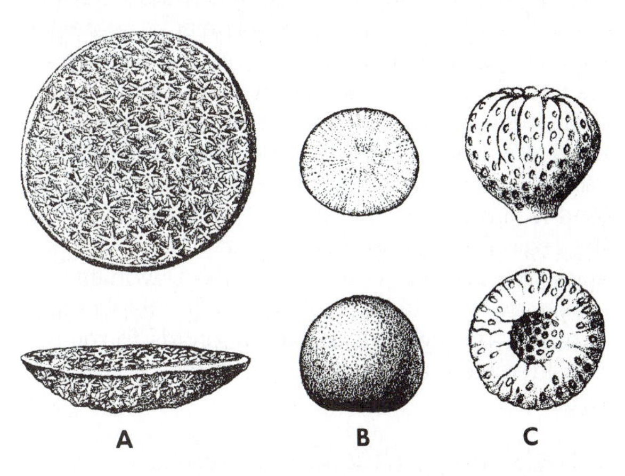

Figure 10–43 Early Paleozoic (Silurian) sponges. (A) *Astraeospongea* (it takes its name from the starlike spicules). (B) *Microspongea.* (C) *Astylospongea.*

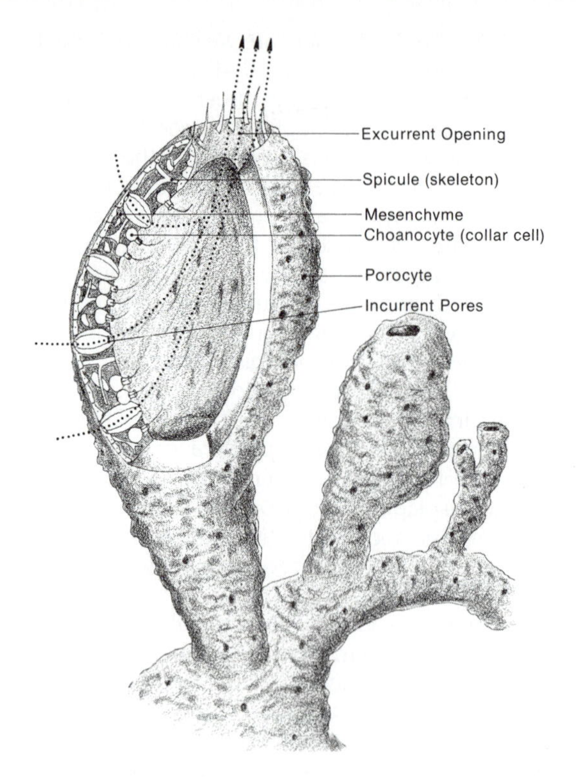

Figure 10–44 Schematic diagram of a sponge having the simplest type of canal system. The path of water currents is indicated by *arrows*.

sponge, water enters through the pores, flows across sheets of choanocytes in the central cavity, and passes out through the osculum.

One group of Early Paleozoic sponges that are particularly interesting because of their reef-building capabilities is the **stromatoporoids** (Fig. 10–45). These organisms constructed fibrous, calcareous skeletons of pillars and thin laminae that can nearly always be found in reef-associated carbonate rocks of the Silurian and Devonian. Apparently, stromatoporoids grew profusely and in close association with corals, brachiopods, and other invertebrate reef-dwellers. Silurian coral-stromatoporoid reefs are well known to geologists primarily because of extensive study of reef exposures on Gotland Island off the Baltic coast of Sweden and in the region around and southwest of the Michigan Basin. Reef development during the Devonian was similar to that in the Silurian, with blanket-like, massive, and cylindric stromatoporoids forming a considerable part of the total mass of the reef structures.

Corals and Other Coelenterates

Sea anemones, sea fans, jellyfish, the tiny *Hydra,* and the reef-forming corals are all repre-

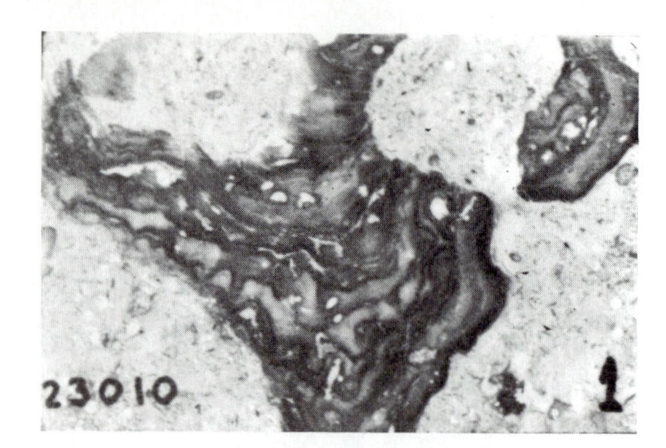

Figure 10–45 Stromatoporoid in polished slab of limestone.

sentatives of the Phylum **Coelenterata,** which is known for the great diversity and beauty of its members (Fig. 10–46). The coelenterate body wall is composed of an outer layer of cells, the **ectoderm,** an inner layer, the **endoderm,** and a thin, noncellular intermediate layer, the **mesoglea.** In the endoderm are found primitive sensory cells, gland cells that secrete digestive enzymes, flagellated cells, and nutritive cells to absorb nutrients. A distinctive feature of many coelenterates is the presence of stinging cells, which, when activated, can inject a paralyzing poison.

Body form in coelenterates may be either **polyp** or **medusoid** (Fig. 10–47). The medusoid form is seen in the jellyfish, which somewhat resembles an umbrella in shape. Jellyfish have a

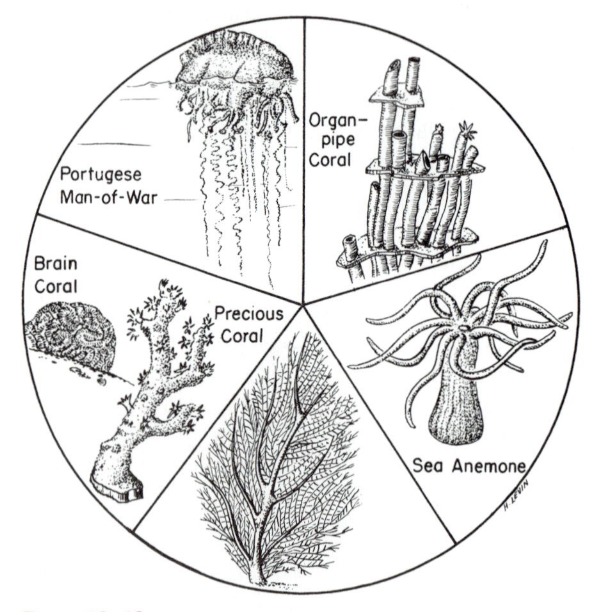

Figure 10–46 Diversity among the coelenterates. (From H. L. Levin. *Life Through Time.* Dubuque, William C. Brown Co., 1975.)

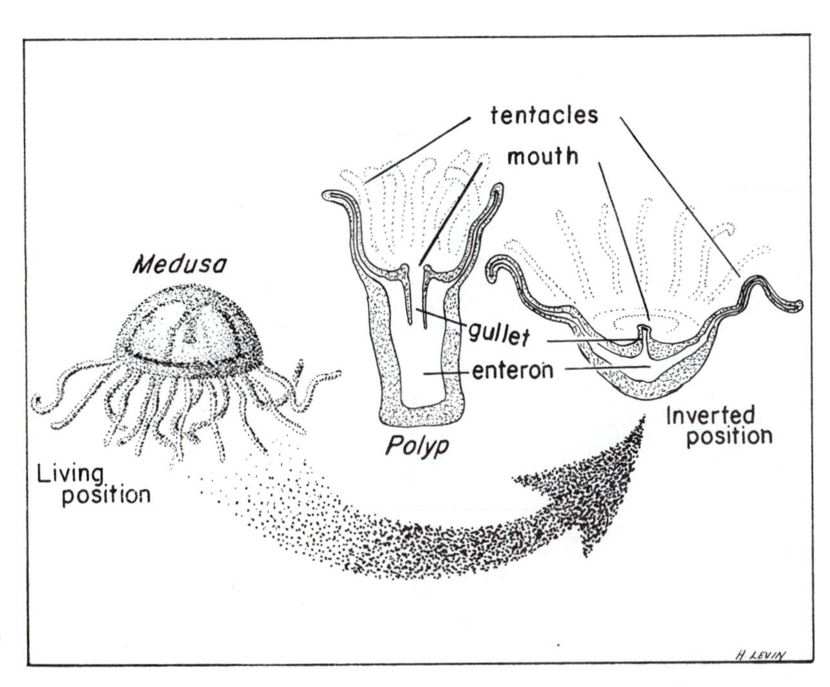

Figure 10–47 Comparison of polyp and medusa form in coelenterates.

concave undersurface that contains a centrally located mouth; for this reason, it is designated the oral surface. In jellyfish such as the living *Aurelia,* which is common along the eastern shore of the United States, the mouth is surrounded by four oral arms. All jellyfish have tentacles, and in *Aurelia* these are located around the margin of the umbrella. Most jellyfish swim by rhythmic contractions of the umbrella.

In the polyp form, as exemplified by *Hydra,* there is a circle of tentacles above the saclike body. Corals and sea anemones are among the frequently encountered coelenterates with this polyp form. In stony corals (Class Anthozoa), the polyp secretes a calcareous cup, in which it lives. The animal may live alone or may combine with other individuals to form large colonies. The cup, or **theca,** may be divided by vertical plates called **septa,** which serve to separate layers of tissue and provide support. As the animal grows, it also secretes horizontal plates termed **tabulae.** Corals are identified according to the nature of their septa, tabulae, and other skeletal features. For example, after the development on an initial embryonic set of six protosepta, the Paleozoic **rugose corals,** or **Rugosa,** insert new septa at only four locations during growth. In other Paleozoic corals, septa are absent or poorly developed, so that tabulae are the most important features in classification. These are the **Tabulata.** Tabulates include many interesting colonial forms, such as the honeycomb and chain corals. The rugose corals and tabulate corals became extinct at the end of the Paleozoic and were followed in the Mesozoic by anthozoans of the Order *Scleractinia.* In all scleractinian corals, septa are inserted between the mesenteries in multiples of six.

The fossil record for coelenterates begins with the discovery of fossil jellyfish impressions in the Late Precambrian. The phylum is still poorly represented in Cambrian rocks. In the succeeding Ordovician the record improves dramatically, for the lime-secreting anthozoans begin to expand and diversify. The first stony corals were the tabulates, recognized by their simple, often clustered or aligned tubes divided horizontally by transverse tabula. During the Silurian, *Halysites,* the chain coral, and *Favosites,* the honeycomb coral (Fig. 10–48), contributed their skeletons to extensive reef developments. The reefs in turn provided a variety of habitats for use by other marine invertebrates. Simple representatives of the horn-shaped rugose corals (Fig. 10–49) appeared in the Ordovician only slightly later than the tabulates. The tabulates, however, were the dominant corals of Silurian reefs. By Devonian time, tabulates began to be overshadowed by numerous and diverse members of the expanding rugose corals. These rugose corals joined with the declining tabulates and the ubiquitous stromatoporoids to form the reefs that were common in the epicontinental seas that covered the craton. The reefs varied in size. Some were only a few meters thick, whereas others attained thicknesses of nearly 300 meters. Many of the reefs in

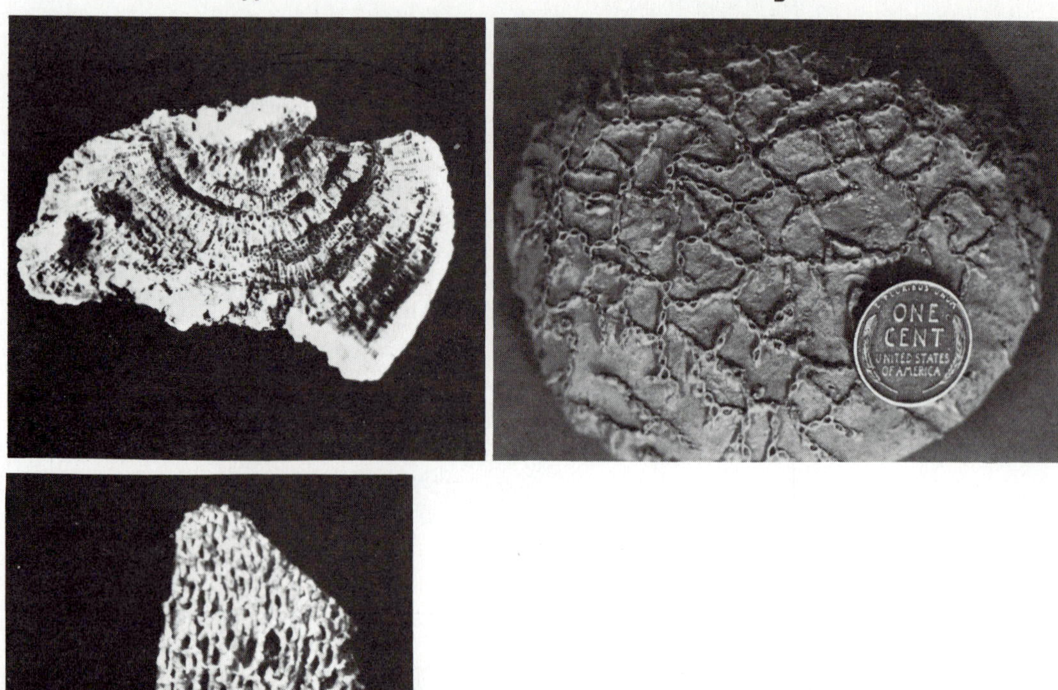

Figure 10–48 Tabulate corals. (A) *Favosites,* the "honeycomb coral." (B) *Halysites,* the "chain coral." (C) *Syringopora,* the "fossil organ pipe coral." (Specimen A, vertical dimension of 7 cm; specimen C, height of 10 cm.)

Illinois and Ohio are covered by younger strata and were discovered during oil explorations. Because of their porous nature, buried reefs provide excellent structures for the accumulation of petroleum. Corals have also been remarkably useful in local and regional stratigraphic correlations. Forms such as *Tetradium* were common Ordovician tabulates, whereas *Favosites* and *Heliolites* populated Silurian seas. Among Early Paleozoic rugosa, *Favistella* (Fig. 10–50) and *Streptelasma* are used widely in correlation.

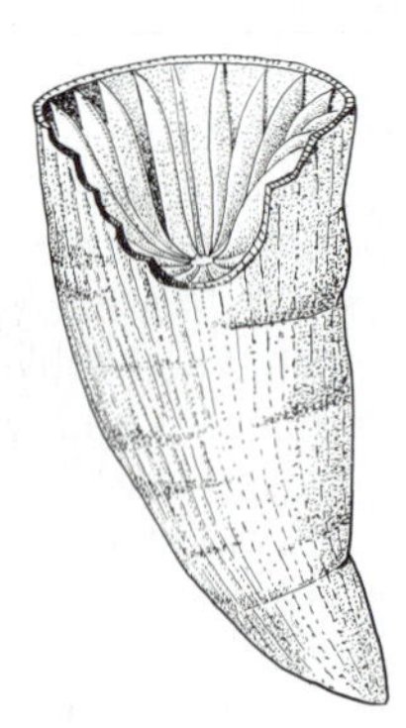

Figure 10–49 An Ordovician rugose or "horn coral." (Height, 12 cm.)

Moss Animals

Bryozoans are minute, bilaterally symmetric animals that grow in colonies that frequently appear twiglike when viewed without the aid of a magnifier. The individuals, called **zooids,** are housed in a capsule, or **zooecium** (Fig. 10–51), that is often preserved as a result of its being calcified. The zooecia appear as pinpoint depressions on the outside of the colony, or **zoarium.** The zooid has a complete, U-shaped digestive tract with mouth surrounded by a tentacled feeding organ called the **lophophore.**

Today, there are more than 4000 living species, and nearly four times that number are known as fossils. Their earliest reliably reported occurrence is from Lower Ordovician strata of the Baltic, but they did not become abundant until Middle Ordovician and Silurian. In rocks of these ages,

Figure 10–50 The colonial rugose coral *Favistella,* one of the commonest Ordovician corals.

M, L&F also calls it a rugose

their remains sometimes make up much of the bulk of entire formations. Like the corals, bryozoa contributed to the framework of reefs. One of the commonest of all the Early Paleozoic bryozoans was *Fistulipora,* different species of which develop massive, incrusting, or arborescent colonies. The genus *Hallopora* (Fig. 10–52) is characterized by abundant nodes called monticules. Star-shaped patterns on the surface of *Constellaria* provided the inspiration for its generic name.

Like corals, bryozoa are useful in stratigraphic correlation. Biostratigraphic zones based on overlapping ranges of fossil bryozoans have made possible correlation of Ordovician strata from Oklahoma across the central United States to Ontario and from North America to the Baltic area of Europe.

Brachiopods

Second only to the trilobites, brachiopods are the most abundant, diverse, and useful fossils in Paleozoic rocks. They are characterized by a pair of enclosing valves, which together constitute the shell of the animal (Fig. 10–53). They resemble clams in this regard, but, in symmetry of the valves and soft part anatomy, brachiopods are quite different from pelecypods. Brachiopod valves are almost always symmetric on either side of the midline, and the two valves differ from each other in size and shape. The valves of a clam are right and left, whereas those of a brachiopod are dorsal and ventral. The valves may be variously ornamented with radial ridges or grooves, spines, nodes, and growth lines. Calcium carbonate is the usual hard tissue of brachiopods, although the valves of some families are composed of mixtures of chitin and calcium phosphate. This is particularly true of the group called **inarticulates.**

In the **articulate** brachiopods, the valves are hinged along the posterior margin and are prevented from slipping sideways by teeth and sockets. The less common inarticulates lack this definite hinge, are held together by muscles, and characteristically have simple, spoon-shaped or circular valves. Although brachiopod larvae swim about freely, the adults are frequently anchored or cemented to objects on the sea floor by a fleshy stalk (**pedicle**) or by spines. Some simply rest on the sea floor. One of the more conspicuous soft organs of brachiopods is the **lophophore,** a struc-

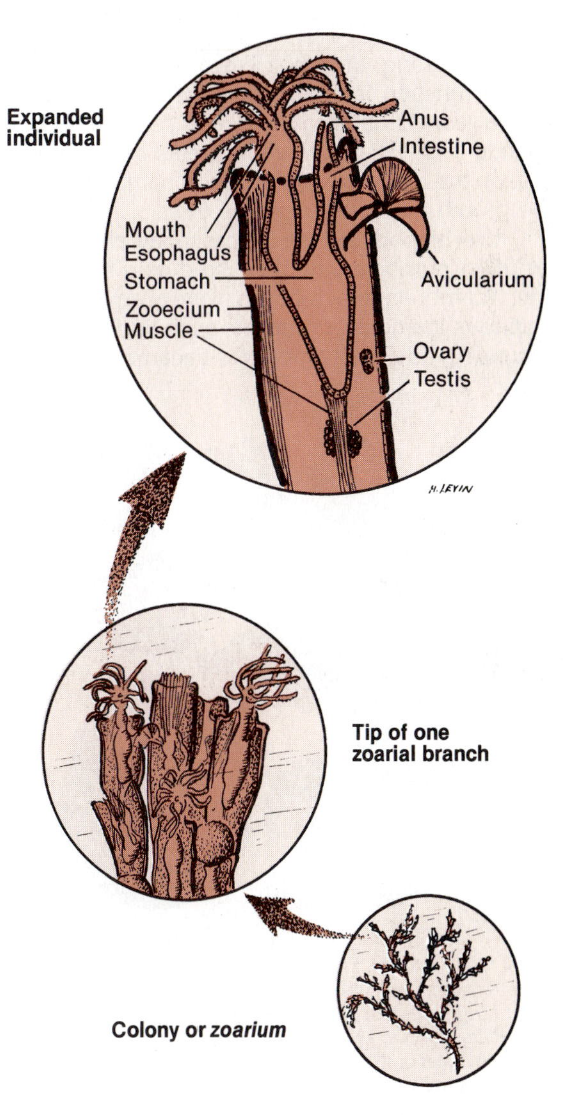

Figure 10–51 Relation of bryozoan individuals to the zoarium. (From H. L. Levin. *Life Through Time,* Dubuque, William C. Brown Co., 1975.)

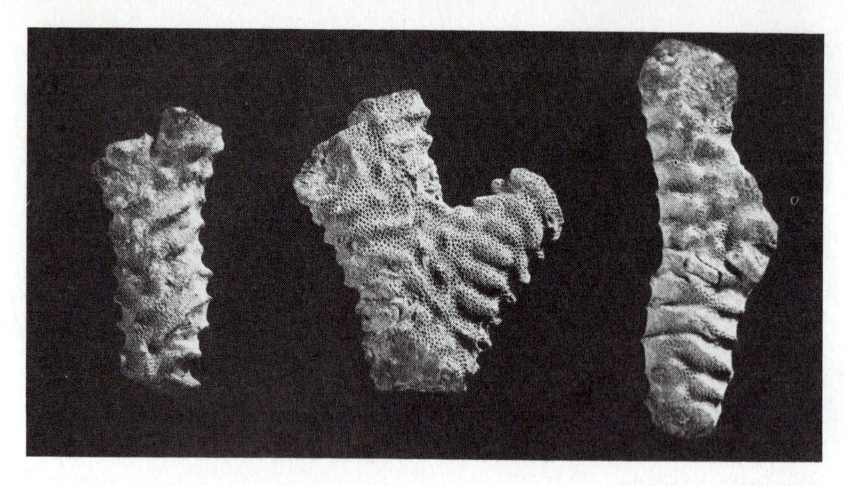

Figure 10–52 The "branching twig" bryozoan *Hallopora,* from Ordovician strata near Cincinnati. (Width of field, 8 cm.)

ture consisting of two ciliated coiled tentacles whose function is to circulate the water between the valves, distribute oxygen, and remove carbon dioxide (Fig. 10–54). Water currents generated by **cilia** on the lophophore move food particles toward the mouth and short digestive tract.

Brachiopods still live in seas today, although in far fewer numbers than they did during the Paleozoic. Cambrian brachiopods were frequently the chitinous inarticulate varieties, such as A and B in Figure 10–53. The inarticulates became somewhat more diverse during the Ordovician and then declined. A few species of only three families remain today.

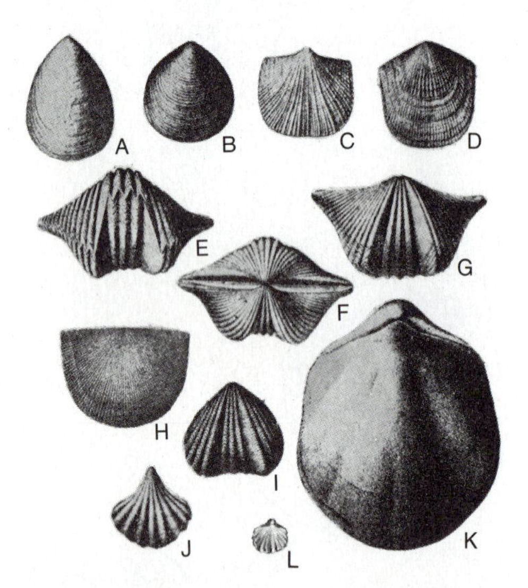

Figure 10–53 Some Early Paleozoic brachiopods. A–D, brachiopods common in the Cambrian. Pictured here are species of (A) *Lingulella.* (B) *Dicellomus.* (C) *Eoorthis.* (D) *Billingsella. Platystrophia,* which was abundant in the Ordovician, is shown in anterior (E), posterior (F), and ventral (G) views. *Strophomena* (H) and *Lepidocyclus* (I) are also common in the Ordovician. *Rhynchotreta* (J) and *Pentamerous* (K) are frequently found Silurian brachiopods. *Zygospira* (L) is an Ordovician spiriferid. A and B are inarticulates; C–L are articulate brachiopods. (Magnification of specimens ½ ×.)

The articulates also first appeared in the Cambrian but became truly abundant during the succeeding Ordovician Period, when there was a great expansion of all sorts of shelled invertebrates. Across the floors of Early Paleozoic epeiric seas, large and small aggregates of these filter-feeders could be found. Today, their skeletons compose much of the volume of thick formations and provide the stratigrapher with essential markers for correlation. Among the articulate brachiopods that were particularly abundant during the first three periods of the Paleozoic were the orthids (Fig. 10–53E, F, G), strophomenids (Fig. 10–53H), pentamerids (Fig. 10–53K), rhynchonellids (Fig. 10–53I), and spiriferids (Fig. 10–53L).

The Mollusks

A stroll along almost any seashore will provide evidence that mollusks are today's most familiar marine invertebrates. The shells of most members of this phylum have been readily preserved and provide enjoyment to collectors and conchologists today. Such well-known animals as snails, clams, chitons, tooth shells, and squid are included within the **Phylum Mollusca** (Fig. 10–55). Although most mollusks possess shells, some, like the slugs and octopods, do not. The various member classes of the Mollusca may differ considerably in external appearance, yet they have fundamental similarities in their internal structure. There is a muscular portion of the body called the **foot** that functions primarily in locomotion. In cephalopod forms, the foot is modified into tentacles. A fleshy fold called the **mantle** has the function of secreting the shell. In aquatic mollusks, respiration is accomplished by

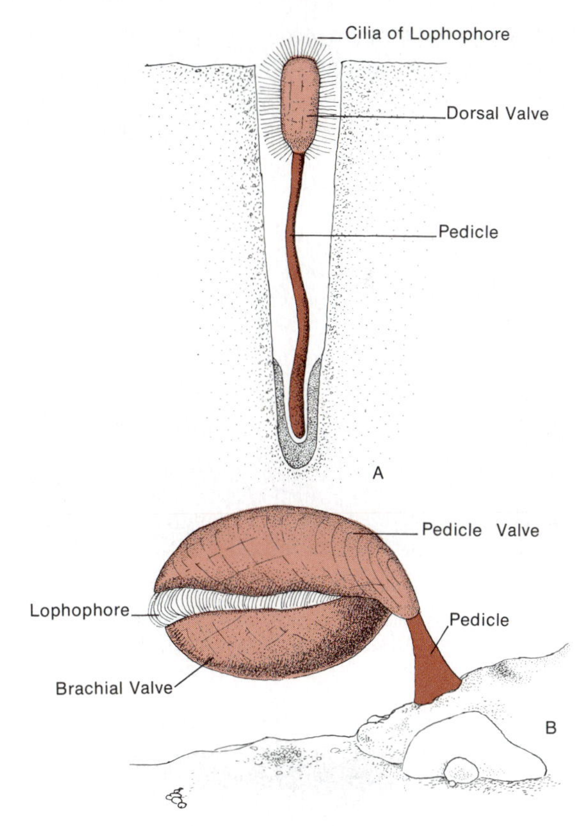

Figure 10–54 Living positions of two modern brachiopods. The inarticulate brachiopod *Lingula* excavates a tube in bottom sediment and lives within it. The pedicle secretes a mucus at the end that glues the animal to the tube. The life position of a living articulate is shown in B, with the lophophore barely visible through the gape in the valves.

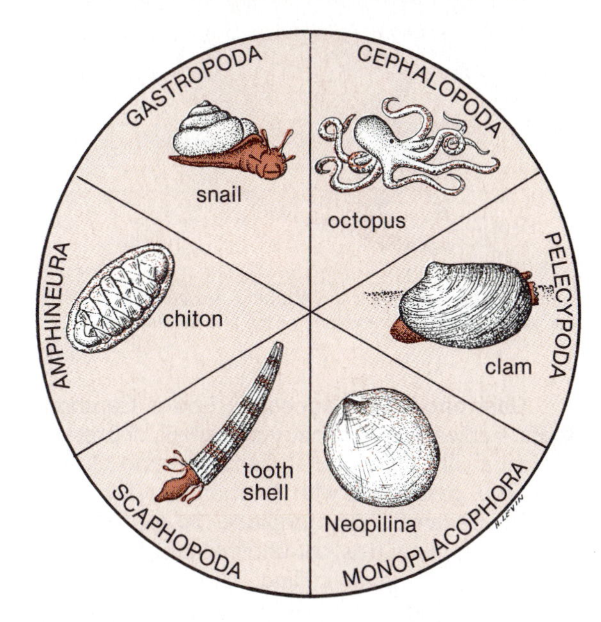

Figure 10–55 Diversity among the members of the Phylum Mollusca. (H. L. Levin. *Life Through Time,* Dubuque, William C. Brown Co., 1975.)

means of gills. Well-developed organs for digestion, sensation, and circulation attest to the advanced stage of evolution that mollusks have attained.

As presently classified, the Phylum Mollusca consists of eight classes. Three of these can be collectively termed **placophorans.** They are relatively primitive mollusks with multiple paired gills. The placophorans are divided into the *Monoplacophora,* which have simple cap-shaped shells; the *Polyplacophora,* represented by chitons, which possess eight overlapping shell plates; and the *Aplacophora,* which lack mineralized plates. The remaining classes of Mollusca include *Scaphopoda, Hyolitha* (small, conical mollusks with lids, or opercula), and the more familiar *Gastropoda, Pelecypoda,* and *Cephalopoda.*

Fossil monoplacophorans are known to occur in rocks as old as Lower Cambrian. This group persists today in the form of small, limpet-like mollusks named *Neopilina* (Fig. 10–56).

Of great interest to biologists is the fact that the "living fossil" *Neopilina* displays a segmental arrangement of gills, shell muscles, and other organs. This evidence of segmentation provides some support for the theory that the mollusks branched off the family tree at some point below the branch represented by segmented annelid worms and that the unsegmented condition in mollusks is not primitive but rather is a secondary development.

The **Pelecypoda** (sometimes termed the Bivalvia) are a class of mollusks that include clams, oysters, and mussels. Pelecypods are generally characterized by layered gills, a muscular foot (Fig. 10–57), bilobed mantle, and absence of a definite "head." Although members of the class (Fig. 10–58) made their first appearance in Cambrian time, they did not become notably diverse or numerous until the Mesozoic and thus are infrequently used as Paleozoic index fossils.

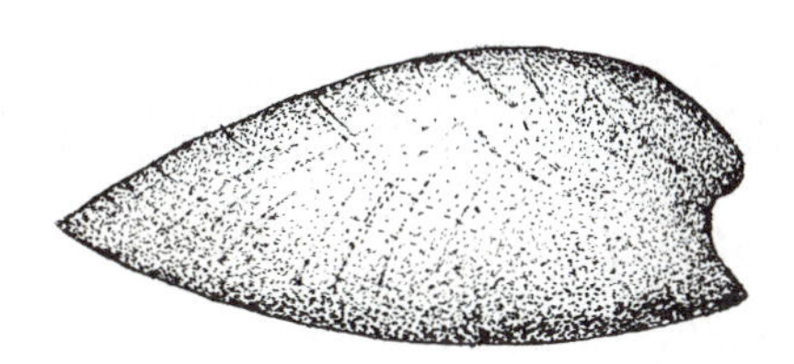

Figure 10–56 The "living fossil" *Neopilina ewingi.* (Magnification ½ ×.)

Figure 10–57 Modern clam showing muscular foot, siphon (for water circulation), and one of the two valves.

Figure 10–58 The Ordovician Pelecypod *Cuneamya*, from the Trenton Limestone of New York. a, Left lateral; b, anterior; c, ventral. (From B. Runnegar. *J. Paleo.*, 48(5):904–940, 1974.)

Gastropods first appear in Lower Cambrian strata. Earliest forms constructed small, depressed, conical shells. During later Cambrian and Ordovician time, gastropods with the more familiar coiled conchs became commonplace. It appears that coiling in a plane was not unusual initially but was supplanted by spiral coiling. Gastropods of the Ordovician (Fig. 10–59) and Silurian had shapes similar to those of living species. One presumes that their soft parts were also similar, with distinct head, mouth, eyes, tentacles, and a vertically flattened foot to provide for gliding movement (Fig. 10–60). Certain species of gastropods had relatively short geologic ranges and are therefore useful as guide fossils.

The **cephalopods** may very well be the most complex of all the mollusks. Today, this marine group is represented by the squid, cuttlefish, octopods, and the lovely chambered nautilus. The nautilus in particular provides us with important information about the soft anatomy and habits of a vast array of fossils known only by their preserved conchs. In the genus *Nautilus* (Fig. 10–61), one finds a bilaterally symmetrical body; a prominent head with paired, image-forming eyes; and tentacles developed on the forward portion of the foot. Water is forcefully ejected through the tubular "funnel" to provide swift, jet-propelled movement.

At first glance the conchs of cephalopods resemble snail shells, but the resemblance is only superficial. Although we have mentioned an exception to this rule, the gastropod shell generally coils in a spiral, whereas the cephalopod conch characteristically coils in a plane. More importantly, the planispirally coiled conch is divided into a series of chambers by transverse partitions, or septa. The bulk of the soft organs reside in the final chamber. Where the septa join the inner wall of the conch, **suture** lines are formed. These lines are enor-

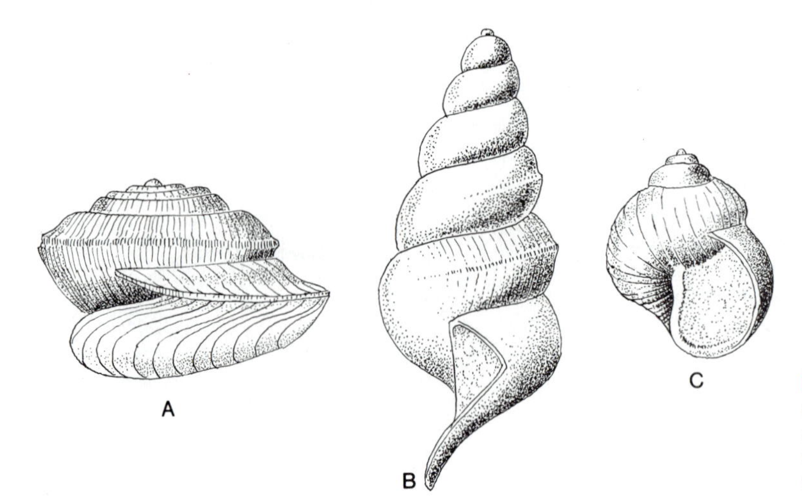

Figure 10–59 Three wide-spread Ordovician marine gastropods. (A) *Eotomaria supracingulata.* (B) *Hormotoma trentonensis.* (C) *Cyclonema limatum.* (Magnification 1 ×.)

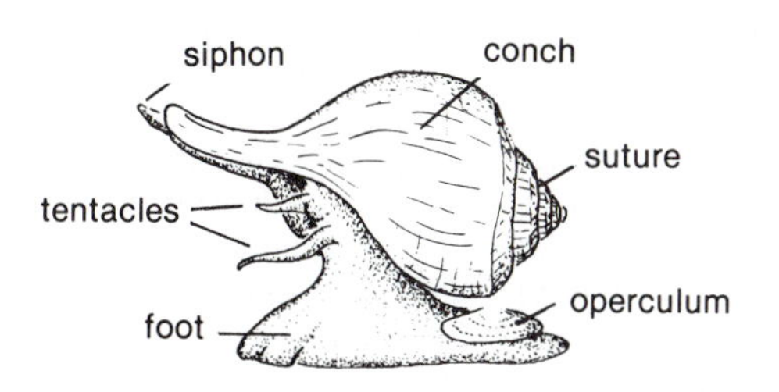

Figure 10–60 Externally visible features of a marine gastropod. (From H. L. Levin. *Life Through Time,* Dubuque, William C. Brown Co., 1975.)

mously useful in the identification and classification of cephalopods. For example, cephalopods placed in the **Nautiloidea** have straight or gently undulating sutures, whereas the **Ammonoidea** have complicated wriggly sutures (Fig. 10–62). Ammonoids became widespread during the Mesozoic; for this reason we will delay our discussion of them until a later, more appropriate, chapter.

The oldest fossils to be classified as cephalopods were small, conical conchs discovered in Early and Middle Cambrian rocks of Europe. The class gradually increased in number and diversity

Figure 10–61 *Nautilus,* a modern nautiloid cephalopod. (A) Conch of a nautiloid sawed in half to show large living chamber, septa, and septal necks through which the siphuncle passed. (B) Living animal photographed at a depth of about 200 meters. (Photograph courtesy of W. Bruce Saunders.)

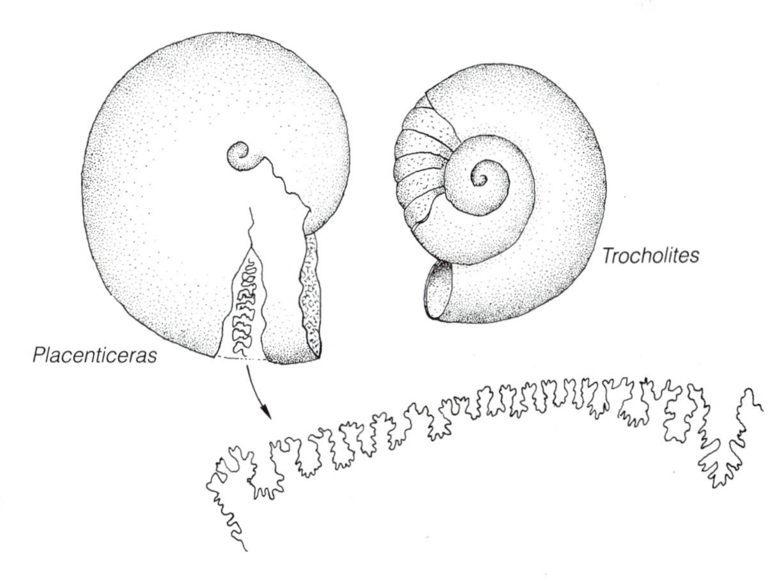

Figure 10–62 Nautiloid cephalopods are characterized by relatively straight sutures, as shown in *Trocholites* from the Ordovician of New York. Ammonoids have complexly wrinkled sutures, as exemplified in *Placenticeras* from the Cretaceous.

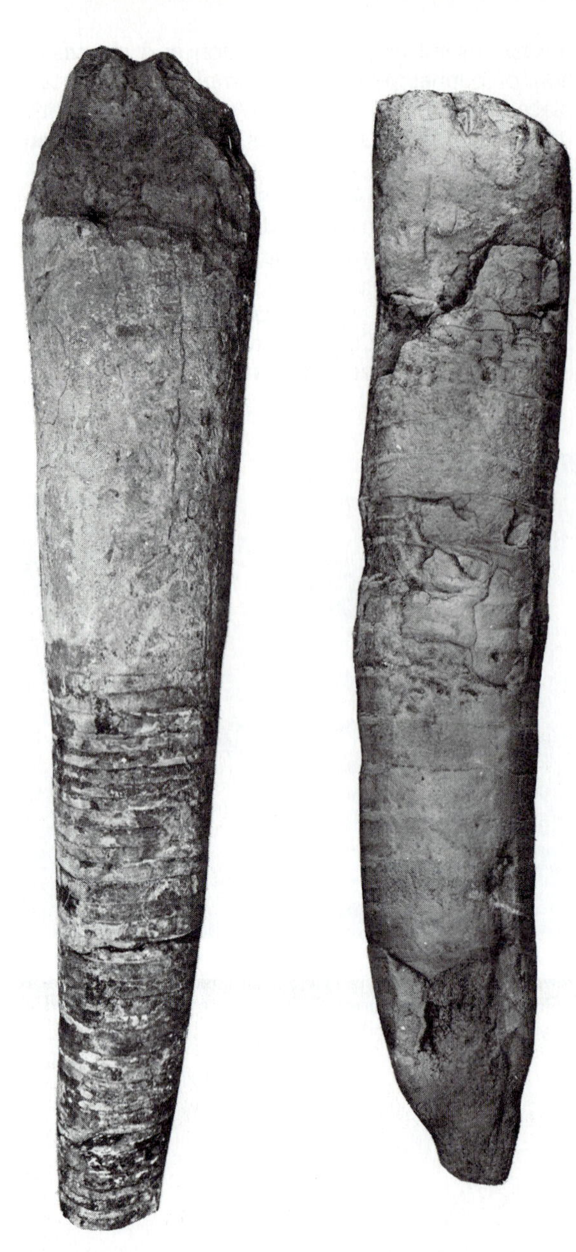

Figure 10–63 Two specimens of straight-conch cephalopods from the Saluda Formation of Indiana. The specimens are Upper Ordovician in age. (Photograph courtesy of Rousseau H. Flower.)

and became ubiquitous inhabitants of Ordovician and Silurian seas. Indeed, by Ordovician time, a great variety of conch forms—from straight to tightly coiled (Figs. 10–63 and 10–64)—had developed. Some of the elongate forms exceeded 4 meters in length. The first signs of a decline in nautiloid populations can be detected in Silurian strata. After Silurian time the group continued to dwindle until today only a single genus, *Nautilus,* survives.

Arthropods

The arthropod phylum is enormous. It includes such living animals as lobsters, spiders, insects, and a host of other "critters" that possess chitinous exterior skeletons, segmented bodies, paired and jointed appendages, and highly developed nervous system and sensory organs. Members of the Arthropoda that have left a particularly significant fossil record are the **trilobites, ostracodes,** and **eurypterids.**

Trilobites (Figs. 10–65 and 10–66) were swimming or crawling arthropods that take their name from division of the dorsal surface into three longitudinal segments, or *lobes.* There are, for example, a central axial lobe and two lateral (pleural) lobes (Fig. 10–67). There was also a transverse differentiation of the shield into an anterior **cephalon,** a segmented **thorax,** and a posterior **pygidium.** The skeleton was composed of chitin strengthened by calcium carbonate in parts not requiring flexibility. As in many other arthropods, growth was accomplished by molting. Although many trilobites were sightless, the majority had either single-lens eyes or compound eyes composed of a large number of discrete visual bodies. The earliest trilobites lacked a pygidium and had a large number of thoracic segments. These characteristics have led paleontologists to speculate that trilobites may have evolved from annelid worms sometime during the Precambrian.

If one considers the entire fossil record for only the Cambrian, the trilobites are clearly the most abundant and diverse. More than 600 genera are known from the Cambrian. Their first appearance, marked by remains of the genus *Olenellus,* is used to identify the initial strata of the Paleozoic.

The earliest trilobites were apparently bottom-dwelling, crawling scavengers and mud processers. Some preferred limy bottoms of cratonic and miogeosynclinal areas, and others, like *Paradoxides* (Fig. 10–67) of the Middle Cambrian, inhabited the muddy or silty floors of eugeosynclinal tracts. As a result of such environmental preferences, it has been possible to delineate trilobite faunal provinces in North America and Europe that suggest these two continents were once close together. (For example, the Lower Cambrian *Olenellus* assemblage is found in exposures in Pennsylvania, New York, Vermont, Newfoundland, Greenland, and Scotland.)

The optimum time for trilobites was reached in the Late Cambrian. After that they began to decline, perhaps in response to predation from cephalopods and fishes; however, they remained fairly abundant throughout the Ordovician and Silurian.

Figure 10–64 Reconstruction of an Ordovician beach on which are stranded large cephalopods with conical conchs, coiled cephalopods, and trilobites. (Courtesy of the Field Museum of Natural History, Chicago.)

Except for a temporary increase in diversity during the Devonian, they continued to wane until they were overtaken by extinction near the close of the Paleozoic. Nevertheless, they were not at all biologic failures, for they had been important animals in the oceans for over 300 million years.

Arthropod companions to the trilobites in the Paleozoic seas were the small, bean-shaped ostracods (Fig. 10–68). At first glance, ostracods appear so different from trilobites as to cause one to question their classification within the same taxonomic phylum. The ostracods have a bivalved shell vaguely suggestive of some sort of tiny clam. However, this bivalved carapace encloses a segmented

Figure 10–65 A slab of rock containing specimens of *Ellipsocephalus,* a Cambrian trilobite. (Courtesy of Wards Natural Science Establishment, Inc., Rochester, N.Y.)

Figure 10–66 *Flexicalymene meeki,* an Ordovician trilobite frequently found enrolled. (Photograph courtesy of Wards Natural Science Establishment, Inc., Rochester, N.Y.)

body from which extend seven pairs of jointed appendages. Adult animals are about ½ to 4 mm in length. The valves are composed of both chitin and calcium carbonate and are hinged along the dorsal margin.

Ostracods first appeared early in the Ordovician, and some limestones of this age are almost completely composed of their discarded carapaces. *Leperditia, Euprimitia,* and *Ceratopsis* are common Ordovician genera. Ostracods continue in relative abundance to the present day. They occur in both marine and fresh water sediments. Because of their small size, they are brought to the surface in wells drilled for oil and are used by exploration geologists in correlating the strata of oil fields.

Eurypterids (see Fig. 10–30) are a group of arthropods that, because of their rarity, are less useful in stratigraphic studies than are either trilobites or ostracods. Nevertheless, they are among the most impressive of Early Paleozoic marine invertebrates. Although many were of modest size, some were nearly 3 meters long and, had they survived, would be suitable subjects for a Hollywood monster film. On their scorpion-like bodies were five pairs of appendages and a fearful-looking pair of pincers. They were also equipped with a venomous stinger. Eurypterids ranged across portions of the sea floor and brackish estuaries from Ordovician until Permian time but were especially abundant during the Silurian and Devonian.

Spiny-Skinned Animals

Just as modern seas abound with starfish, sea urchins, and sea lilies, so also were the oceans of the Early Paleozoic populated with members of the Phylum Echinodermata (Fig. 10–69). Echinoderms are animals with pentamerous symmetry that masks an underlying primitive bilateral symmetry. They are well named "the spiny skinned," for spines are indeed present in many species. A unique characteristic of the phylum is the presence of a system of vessels—the water vascular system—which functions in respiration and locomotion (Fig. 10–70). Members of the phylum are exclusively marine, typically bottom-dwelling, and either attached to the sea floor or able to move about slowly.

Of considerable interest is the evolutionary relationship between echinoderms and vertebrates. There are so many allied embryologic parallelisms among primitive chordates and echinoderms that some zoologists speculate that both may have arisen from similar ancestral forms. For example, the larvae of echinoderms closely resemble those of the living protochordates called acorn worms. Also, in embryologic development, the mesoder-

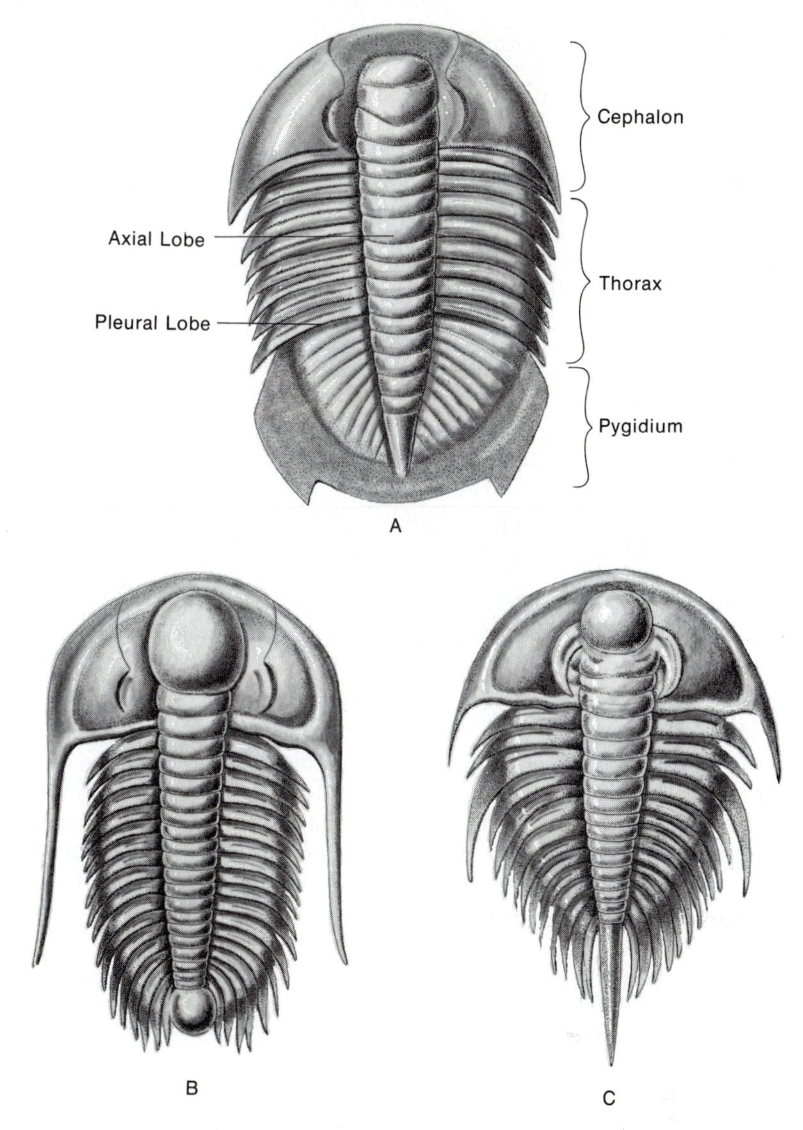

Figure 10–67 Three well-known Cambrian trilobites. (A) *Dikelocephalus minnesotensis* (Upper Cambrian), (B) *Paradoxides harlani* (Middle Cambrian), (C) *Olenellus thompsoni* (Lower Cambrian).

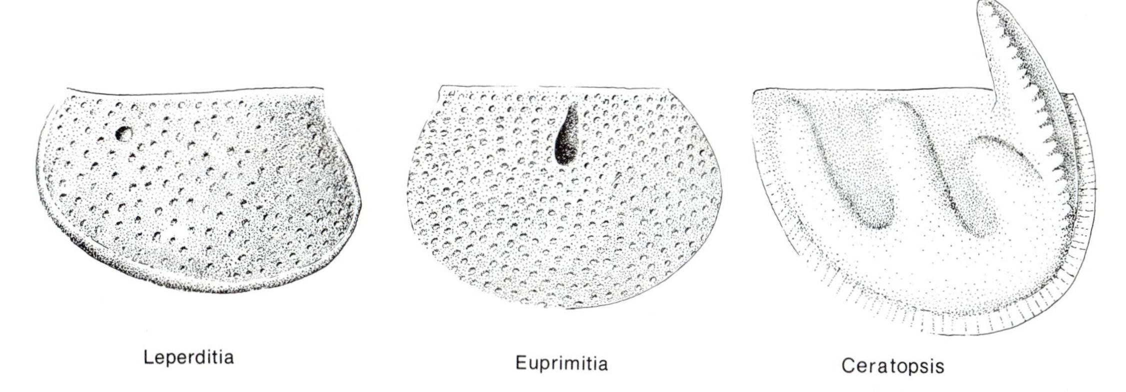

Figure 10–68 Three common Ordovician ostracods.

ity and oxygen-carrying pigments in the blood.

Among the many classes of the Phylum Echinodermata, the Asteroidea (starfish), Ophiuroidea (brittle stars), Edrioasteroidea, Crinoidea (sea lilies), Blastoidea, and Cystoidea are the most abundant and useful in geologic studies. The last four of these classes constitute echinoderms that were sedentary and attached. They were particularly characteristic of the Paleozoic Era, whereas the other unattached and more mobile forms were more frequent in later eras. For this reason, we will discuss the stemless echinoderms in a later chapter dealing with Mesozoic life.

The oldest probable echinoderm thus far discovered may be *Tribrachidium* (see Fig. 9–34B) of the Late Precambrian Ediacara Hills fauna. This peculiar echinoderm appears to be related to a group called **edrioasteroids,** which are among the oldest members of the phylum and considered by many paleontologists to be ancestral to starfish and sea urchins. Edrioasters like *Edrioaster bigsbyi* (Fig. 10–71) had developed several general echinoderm characteristics, such as calcareous plates, a central mouth, radiating food grooves (ambulacra), and a water vascular system.

The stemmed or stalked echinoderms first occur in Middle Cambrian strata but do not become abundant until Ordovician and Silurian time. Stalked forms called **cystoids** (Fig. 10–72) are the most primitive among this group. The striking pen-

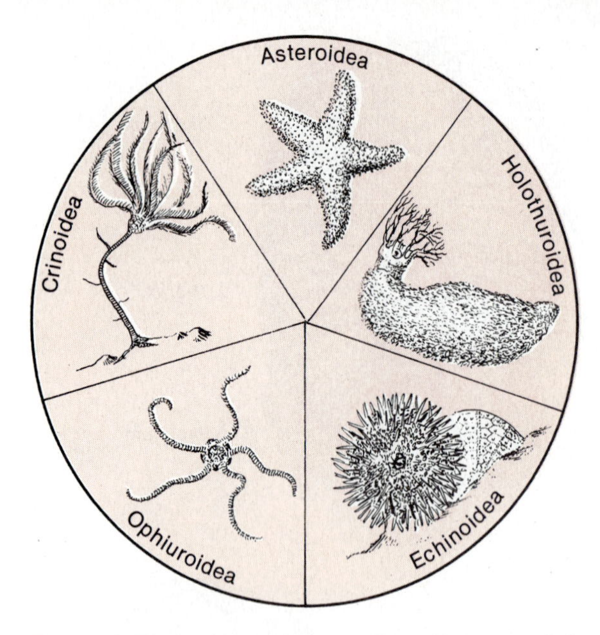

Figure 10–69 Living echinoderms. (From H. L. Levin. *Life Through Time.* Dubuque, William C. Brown, 1975.)

mal layer of cells, as well as certain other elements of the body, arises in the same way. Biochemistry has provided additional evidence for a relationship between echinoderms and chordates by revealing chemical similarities associated with muscle activ-

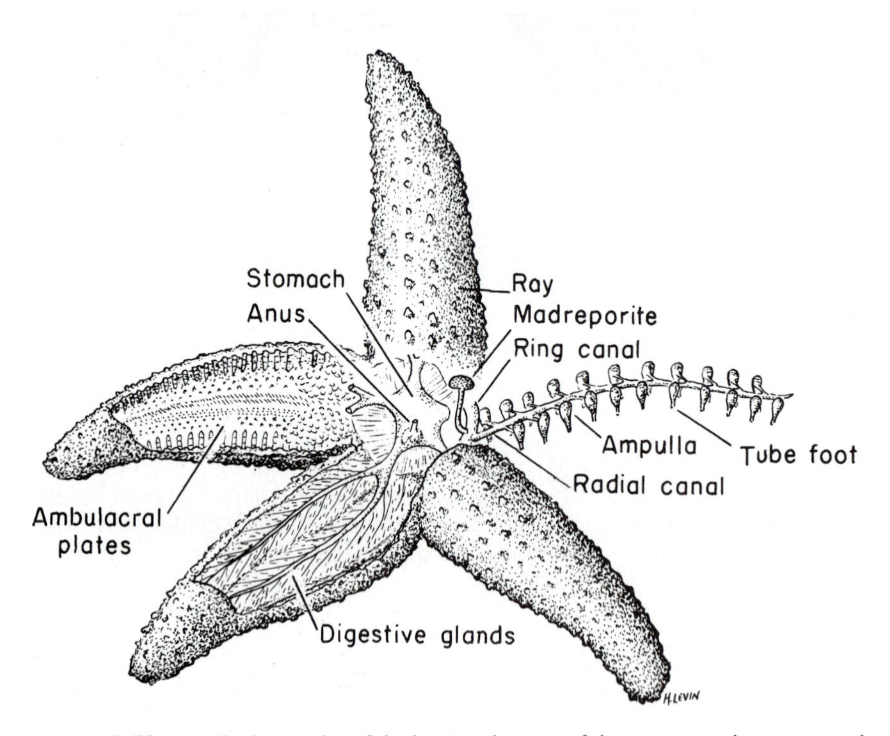

Figure 10–70 Partially dissected starfish showing elements of the water vascular system and other organs.

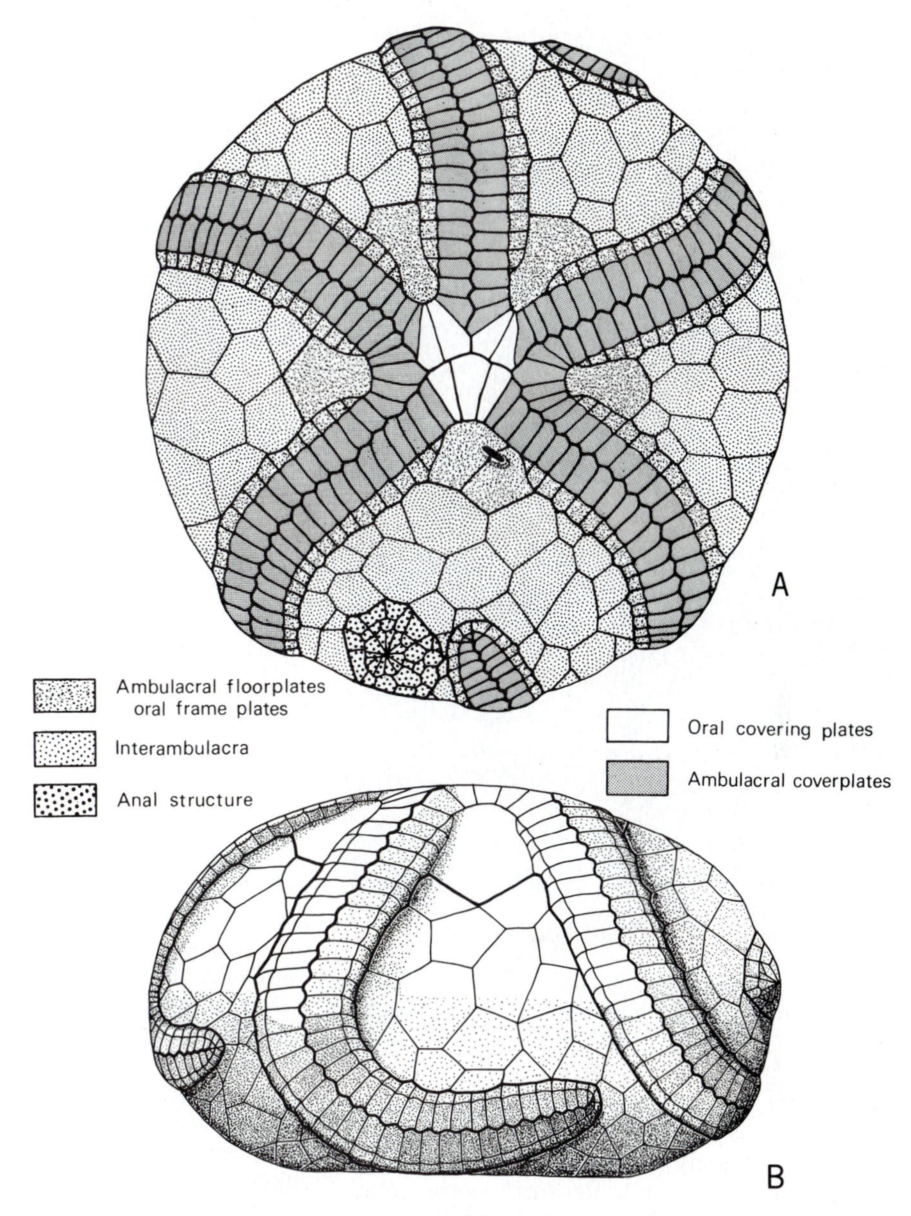

Figure 10–71 *Edrioaster bigsbyi,* a Middle Ordovician edrioasteroid. Specimen is 45 mm in diameter. (A) Oral surface. (B) Lateral view of the globoid fossil. (From B. M. Bell. *J. Paleo.,* 51(3):620, 1977.)

tamerous symmetry that is evident in most echinoderms is often less well developed in cystoids. Beginning students of paleontology often recognize cystoids by the characteristic pores that occur on the plates of the calyx and that compose part of the water vascular system. Although cystoids range from Cambrian into Late Devonian, they are chiefly found in Ordovician and Silurian rocks.

Unlike some of the cystoids, stalked echinoderms known as **blastoids** (Fig. 10–73) have a beautifully symmetric arrangement of plates. The ambulacra are prominent, bear slender branches or brachioles along their margins, and have a well-developed and unique water transport system. The blastoids first appear in Silurian time, expand in the Mississippian and decline to extinction in the Permian.

Crinoids, like most cystoids and blastoids, are composed of three main parts: the *calyx,* which contains the vital organs; the *arms;* and the *stem* with its rootlike *holdfast* (Fig. 10–74). The arms bear ciliated food grooves and, like the brachioles

Figure 10–72 A well-preserved specimen of the Silurian cystoid *Caryocrinites ornatus* from the Lockport Shale of New York. (From J. Sprinkle. *J. Paleo.,* 49(6):1062–1073, 1975.)

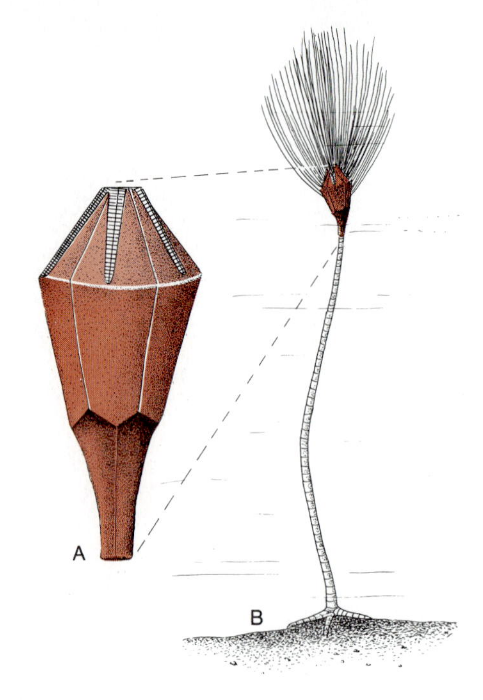

Figure 10–73 The blastoid *Troostocrinus reinwardti* from the Silurian of Kentucky. (A) Calyx. (B) Reconstruction of entire animal.

of blastoids, serve to move food particles toward the mouth. Crinoids are found in rocks that range in age from Ordovician to Recent. In some areas, Ordovician and Silurian rocks (and later Mississippian and Pennsylvanian) have such great quantities of the disaggregated plates of crinoids that they are formally named crinoidal limestones. One can visualize the crinoid "gardens" that occurred sporadically across the floors of Early Paleozoic epeiric seas. However, even these impressive communities would be dwarfed by those that developed later during the Carboniferous Period.

Graptolites

Just as the Early Paleozoic sea floors bustled with the activities of a myriad of invertebrates, so also did the overlying waters. However, many of these planktonic creatures lacked skeletons and so are poorly known. One group, the graptolites (Fig. 10–75), had preservable corneous skeletons that housed colonies of tiny individual animals.

The graptolites made their appearance at the very end of the Cambrian but were rare until the Ordovician, when they became quite abundant and diverse. Unfortunately for paleontologists, they did not survive beyond the Mississippian, and the nature and functions of their soft organs can only be surmised. The structure of the exoskeleton is relatively simple. The tiny cups, or thecae, are arranged along a branch, or *stipe.* The stipes may be solitary or formed into a system of two or more branches. The entire colony is referred to as the *rhabdosome* and is supported or attached at one end by a thin filament—the *nema.* Where the lower end of the nema reaches the base of a *stipe,* there is a conical *sicula,* which may have served as the theca for the first individual. From the sicula subsequent members of the colony add their thecae by budding. Although most graptolites were planktonic creatures and often developed flotational structures, some were apparently attached to objects by the threadlike nema.

Graptolites occur as streaks of flattened, carbonaceous matter in fine-grained rocks. Recently, however, uncompressed graptolites have been studied that reveal an unexpected relationship to

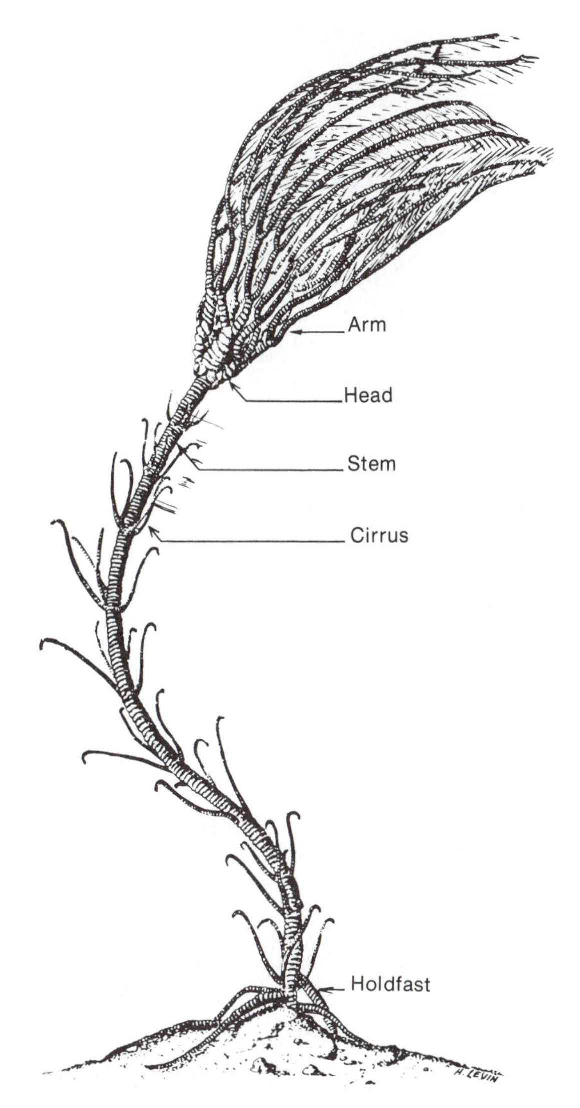

Figure 10–74 Crinoid in living position on sea floor.

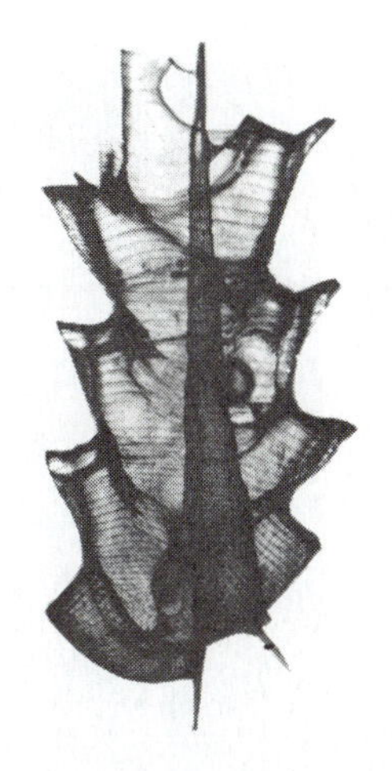

Figure 10–75 Part of a stipe of the Ordovician graptolite *Orthograptus quadrimucronatus* (×15). The specimen has been bleached for better visualization of internal features. The cuplike thecae are clearly visible on either side. One can also see the inverted cone of the sicula. The nema would rise from the pointed end of the sicula. (See the text for more complete descriptions of these features.) (From S. R. Herr. *J. Paleo.,* 45(4):628–632, 1971.)

primitive chordates. For example, both graptolites and the living protochordates known as pterobranchs secreted tiny enclosed tubes, and in both the unique structure of the enclosing wall of the sheath was so similar that a relationship is virtually certain.

During the Early Paleozoic, graptolites were at times so abundant that they have been visualized as forming sargasso sea–like floating masses. They were apparently carried about by ocean currents and thus achieved a worldwide distribution, which has enhanced their use as index fossils. There is, however, a limitation, in that they are seldom found preserved except in fine-grained, usually carbonaceous, shaly sediments. Perhaps these occurrences came about as masses of graptolite colonies floated into areas of toxic conditions, died, and settled to the sea floor. It is also possible that such accumulations resulted from fortuitously favorable preservation.

Conodonts

One group of well-known but enigmatic fossils may have some affinity to a prechordate animal. However, their true biologic relationships are still being hotly debated, and many paleontologists are convinced that they are really food-filtering parts of some ancient invertebrate animal. These problematic fossils are called conodonts. Some are platelike or cone-shaped (Fig. 10–76), and all are composed of calcium phosphate.

At the base of conodont fossils, one finds cavities or attachment scars where they were presumably attached to a supportive part of the parent organism. Conodonts occur mostly as single disarticulated fossils, but occasionally, paired groupings are discovered that are suggestive of an animal's feeding mechanism. Conodonts are excellent guide fossils. They occur in a variety of marine sedimentary rocks and range from latest Precambrian into the Triassic.

Vertebrate Animals of the Early Paleozoic

A momentous biologic occurrence took place in the Early Paleozoic. It came gradually and in a

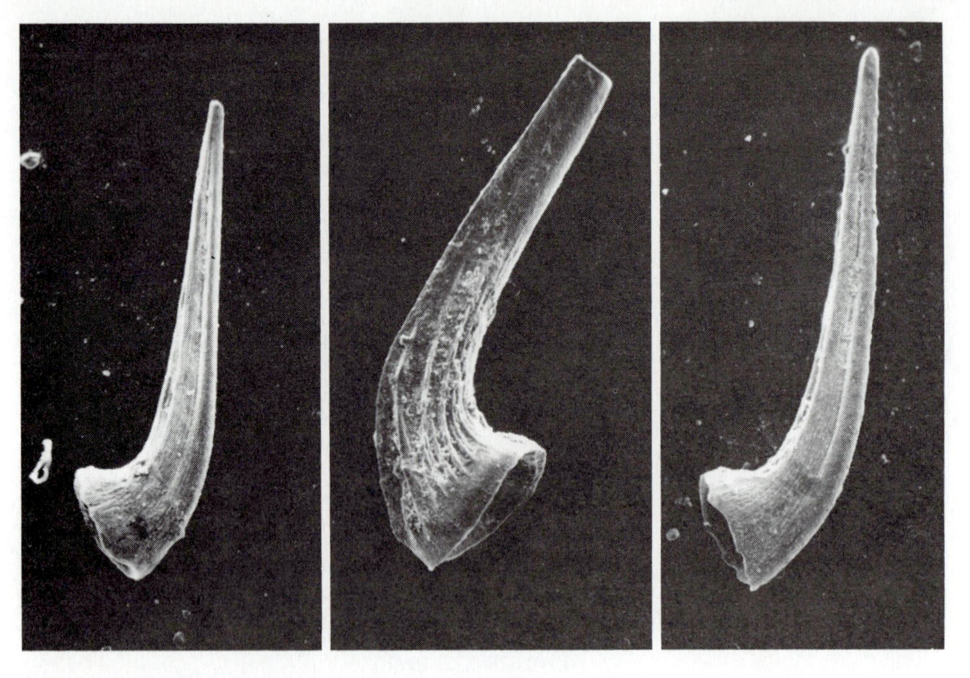

Figure 10–76 Scanning electron micrographs of Ordovician conodonts. These simple, recurved, cone-shaped forms have a deeply excavated base resembling a pulp cavity. (Courtesy of D. Hearns.)

manner that left no fossilized traces. The event, of course, was the birth of the chordate line. A chordate is an animal that has, at least at some stages in its life history, some kind of stiff, elongate supporting structure, a dorsal hollow central nervous system, gill slits, and blood that circulates forward in a main ventral vessel and backward in the dorsal. The supportive structure in primitive chordates is called a notochord and has been studied by generations of biology students in such animals as the lancelet *Branchiostoma* (Fig. 10–77). In the taxonomic hierarchy, vertebrates are simply those chordates in which the notochord is supplemented or replaced by a series of cartilaginous or bony vertebrae.

As we have mentioned previously, the ancestors of the vertebrate lineage may lie somewhere among the echinoderms. The theoretic evolutionary progression that was to lead to vertebrates may have begun with sedentary, filter-feeding animals that had exposed cilia located along their arms, somewhat in the fashion of crinoids. From such a beginning there may have evolved filter-feeders with cilia brought inside the body in the form of gills. In a subsequent stage, the organisms may have become free-swimming, gilled animals, in superficial appearance not unlike small, simple fishes or frog tadpoles. With the advent of vertebrae, the evolutionary progression toward true vertebrates would have been achieved. Presumably, we can call those earliest vertebrates fishes.

The oldest fossils that are considered to be the possible remains of fishes are found in strata of Ordovician age. They consist of tiny mouth parts bearing still smaller, pointed teeth. Scales and scraps of bony armor are found in Ordovician rocks in both North America and Russia.

Recently, small fish scales and plates were recovered from Middle Ordovician Viola Limestone exposures in the Arbuckle Mountains of Oklahoma. The Viola also contains a rich assemblage of marine invertebrates, thus providing evidence that the earliest vertebrates now known lived in the sea rather than in freshwater bodies.

The vertebrates that we loosely call fishes are actually divided into at least five distinct taxonomic classes (Fig. 10–78). There are the jawless fishes, or *Agnatha;* two groups of archaic jawed fishes, the *Acanthodii* and *Placodermi;* the cartilaginous fishes, or *Chondrichthyes;* and the familiar *Osteichthyes* with their highly developed bony skeletons. It is the first three of these categories—namely, the Agnatha, Acanthodii and Placodermi—that frequented freshwater and salt water bodies of the Early Paleozoic.

The agnaths still live today in the guise of the inelegant hagfish and lamprey. However, these specialized survivors are quite unlike the jawless

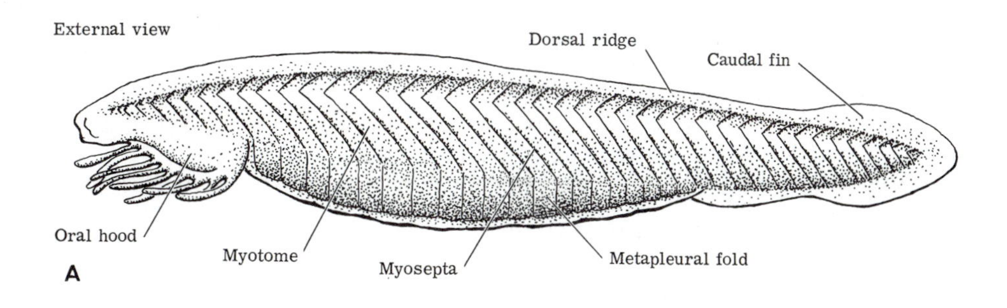

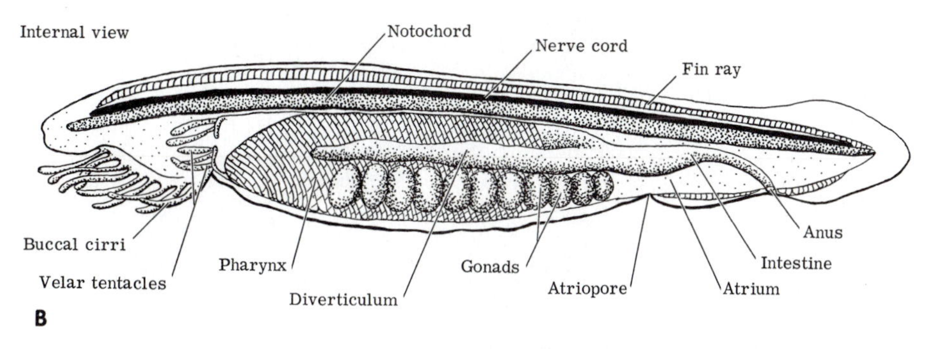

Figure 10–77 Structure of the lancelet *Branchiostoma* (formerly called *Amphioxus*). (From E. L. Cockrum and W. J. McCauley. *Zoology*. Philadelphia, Saunders College Publishing, 1965.)

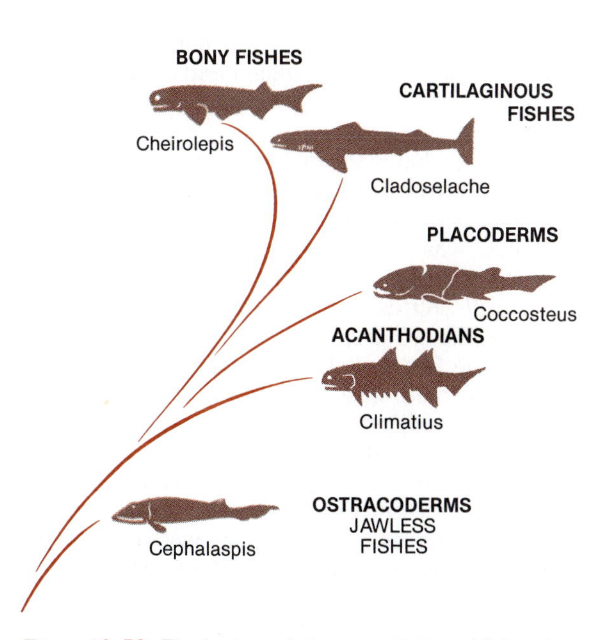

Figure 10–78 The basic evolutionary radiation of fishes that began during the Early Paleozoic and produced the sharks and bony fishes during the Devonian. (From E. H. Colbert. *Evolution of the Vertebrates.* John Wiley & Sons, 1969. With permission of the author, artist Lois Darling, and publisher.)

fishes of the Ordovician and Silurian. Early Paleozoic agnaths are collectively termed **ostracoderms.** The name means "shell skin" and refers to the bony exterior that was a distinctive trait of many of these fishes. Ostracoderms composed a rather diverse group that included the unarmored forms *Thelodus* (Fig. 10–79) and *Jamoytius,* as well as such armored creatures as *Pteraspis.* The purpose of the bony armor that is the hallmark of many of the ostracoderms is still being debated. A widely held view is that the armor provided protection, although the fossil record shows few creatures of the same age and habitat that could have preyed on the ostracoderms. Nevertheless, it does not seem unreasonable that some of the smaller ostracoderms might have provided an occasional meal for a large cephalopod or eurypterid. Another theory stipulates that the dermal armor was not primarily for protection but rather was a device for storing seasonally available phosphorus. According to this idea, phosphates could be accumulated as calcium phosphate in the dermal layers during times of greater availability and utilized during periods when supply was deficient. This cache of phosphorus may have been vital to early vertebrates in order for them to maintain a suitable level of muscular activity.

Hemicyclaspis (Fig. 10–80) is one of the most widely known of the ostracoderms. It is rec-

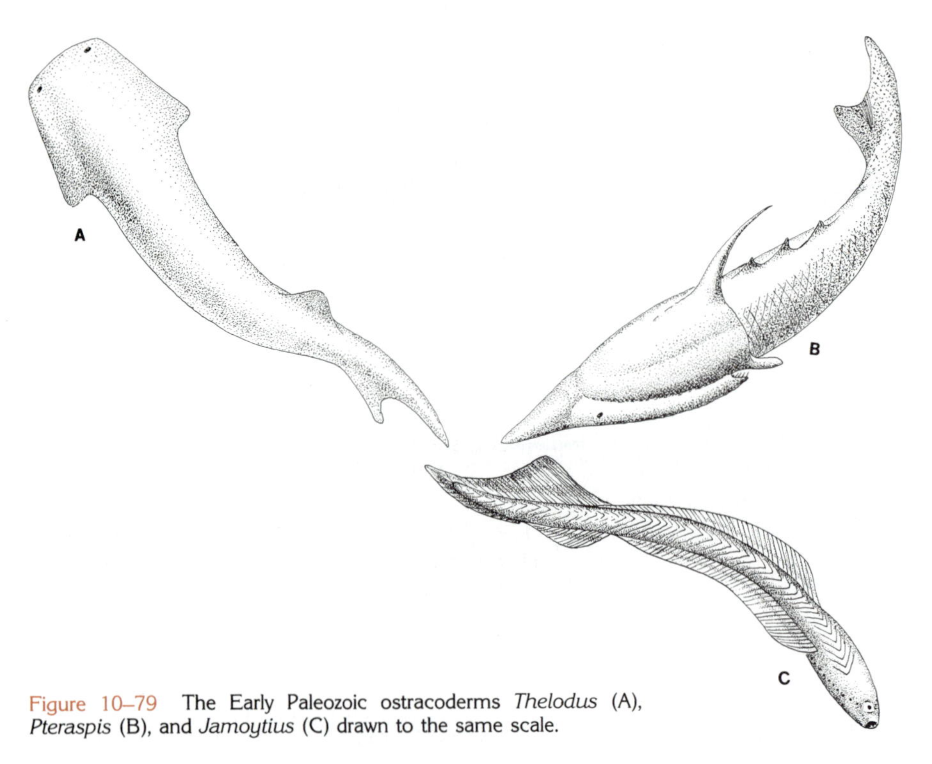

Figure 10–79 The Early Paleozoic ostracoderms *Thelodus* (A), *Pteraspis* (B), and *Jamoytius* (C) drawn to the same scale.

ognized by its large, semicircular head shield, on the top of which can be found four openings: two for the heaven-directed eyes, a small pineal opening, and a single nostril. The pineal opening may have housed a light-sensitive third eye. The mouth was located ventrally in a position suitable for taking in food from the surface layer of soft sediment. Along the margin of the head shield were depressed areas that may have had a sensory function.

Internally, the head contained a regular segmental arrangement of nerves, blood vessels, and gill pouches. These and other features of the soft internal anatomy have been discerned by careful dissection of exceptionally well-preserved specimens.

The ostracoderms continued into the Devonian but did not survive beyond that period. For the most part they were small, rather sluggish creatures restricted to mud-straining or filter-feeding modes of life. They were to be gradually replaced by fishes that had developed bone-supported, movable jaws. The evolution of the jaw was no small accomplishment, for it enormously expanded the adaptive range of the vertebrates. Fishes with jaws were able to bite and grasp. These new abilities led to more varied and active ways of life and to new sources of food not available to the agnathans.

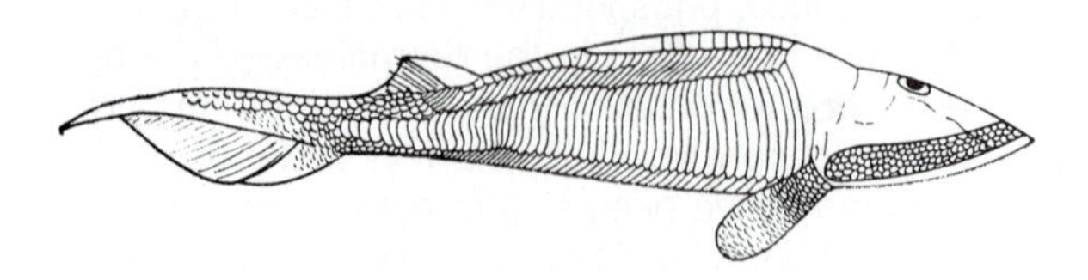

Figure 10–80 The ostracoderm *Hemicyclaspis*. The creature was about 15 cm long. (After A. E. Stensiö. Cephalaspids of Great Britain. *Br. Mus.*, p. 3, Fig. 15, 1932.)

The evolutionary development of the vertebrate jaw involved a remarkable transformation in which an older structure was modified to perform a role that was entirely different from its original function. Those older structures were supports called *gill arches* that were located between the *gill slits* (Fig. 10–81). As the mouth extended posteriorly, the first two of these arches were sequentially eliminated. The third set of gill arches was gradually modified into jaws. Classic anatomic studies have clearly shown the similarities in jaw architecture and neurology to elements of the gill arches and the nerves that serve these supports. In the earliest fishes, the upper jaw was attached to the skull by ligaments only. A full gill slit occurred immediately behind the jaw. In more advanced fishes, the upper jaw was anchored to the skull by the upper part of the next gill arch, the **hyomandibular.** The pair of gill slits that had once existed behind the jaw was squeezed into a smaller space and ultimately developed into the **spiracle**

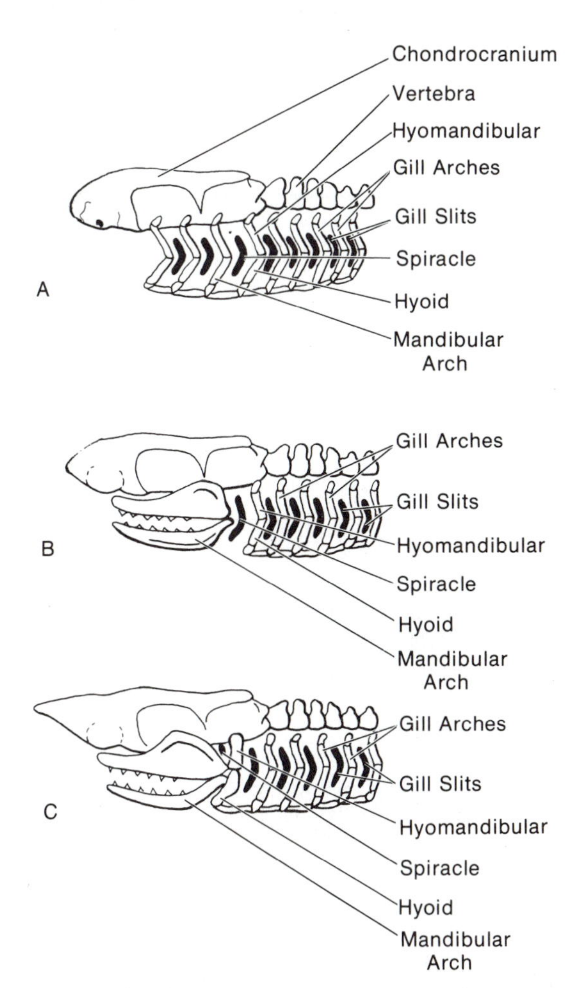

Figure 10–81 Origin of jaws. (A) Gill arches as they might exist in a primitive jawless fish. (B) Early jawed fish with a complete gill slit behind jaws. (C) A jawed fish (like a shark) with the first gill slit reduced to a spiracle. (Modified from E. L. Cockrum and W. J. McCauley. *Zoology.* Philadelphia, Saunders College Publishing, 1965.)

Figure 10–82 The Early Devonian acanthodian fish *Climatius.* (After A. S. Romer. *Vertebrate Paleontology.* Chicago, University of Chicago Press, Fig. 25, p. 41, 1945.)

openings found in members of the shark family today. The spiracle was a structure destined to become part of the ear apparatus in higher vertebrates.

The oldest fossil remains of jawed fishes are found in nonmarine rocks of the Late Silurian. These fishes, called *acanthodians* (Fig. 10–82), became most numerous during the Devonian and then declined to extinction in the Permian. Acanthodians were "archaic" jawed fishes and were quite distinct from the great orders of modern fishes. Another group of archaic fishes includes the *placoderms,* or "plate-skinned" fishes. They too arose in the Late Silurian, expanded rapidly during the Devonian, and then in the latter part of that period began to decline and be replaced by the ascending sharks and bony fishes.

There was considerable variety among the placoderms. The most formidable of these "plate-skinned" fishes was a carnivorous group called *arthrodires. Dunkleosteus* (Fig. 10–83) was a Devonian arthrodire whose length exceeded 9 meters and whose huge jaws could be opened exceptionally wide to engulf even the largest of available prey. Other placoderms, called *antiarchs* (Fig. 10–84), had the heavily armored form and mud-grubbing habits of their predecessors, the ostracoderms. The placoderms known as *rhenanids* anticipated the form and habits of living rays and skates (Fig. 10–85).

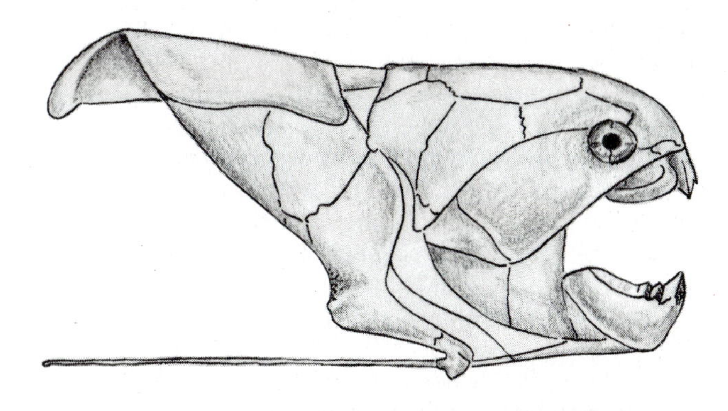

Figure 10–83 *Dunkleosteus,* a highly predaceous Upper Devonian placoderm. (From A. Heintz. *Acta Zool.,* 12:225–239, 1931.)

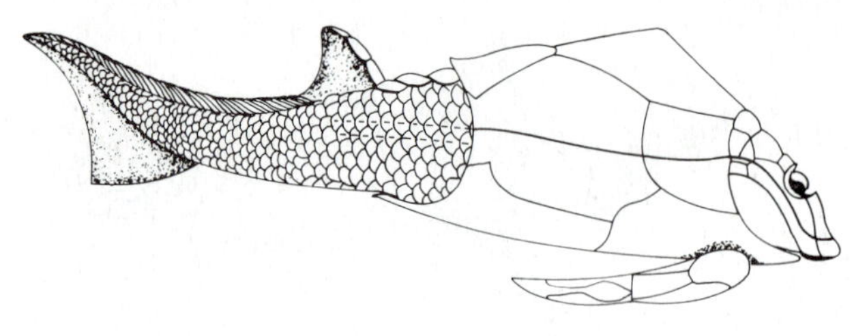

Figure 10–84 The Devonian antiarch fish *Pterichthyodes*. (From A. S. Romer. *Vertebrate Paleontology*. Chicago, University of Chicago Press, Fig. 38, p. 54, 1945.)

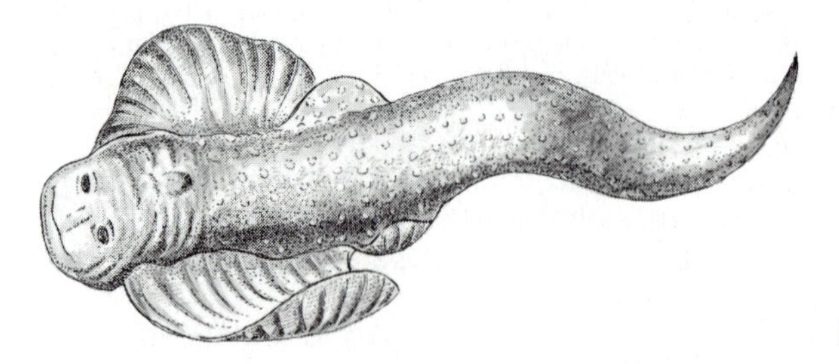

Figure 10–85 The rhenanid skatelike placoderm *Gemuendina*.

Among the Late Silurian acanthodians and placoderms were the ancestors of the bony and cartilaginous fishes that were to dominate the marine realm in succeeding periods. The precise ancestor has not yet been discovered but will be keenly sought in the years ahead.

Summary

It is generally agreed that the Paleozoic Era began with the appearance of abundant marine life. It also began at a time when the continents were separating from a presumed Precambrian cluster, so that there were oceanic tracts between the land masses. The borders of the continents and the geosynclines were the first to experience submergence. Throughout the Early Paleozoic most of the major geosynclines were submerged by shallow seas and undergoing subsidence. Great thicknesses of clastic sediments and volcanics were deposited in the eugeosynclines, and thick carbonates were deposited in the miogeosyncline belts. Shield areas were also inundated and became the sites of major cycles of transgression and regression that left behind widespread dolostones, clean sandstones, and limestones. During the Ordovician Period, the Appalachian Geosyncline was subjected to a mountain-building disturbance called the Taconic Orogeny. Periodic crustal disturbances associated with this orogeny raised elongate ridges in the geosyncline and caused the deposition of a great apron of sediment—the Queenston Delta—on the miogeosyncline and part of the eastern craton. Unrest during the Silurian Period occurred chiefly in northwestern Europe, where the Caledonian geosyncline trough was compressed and subjected to granitic intrusions in the course of the Caledonian Orogeny. These disturbances may have been caused by tectonic plate convergence associated with the construction of the supercontinent Pangaea II.

Invertebrate life of the Paleozoic expanded rapidly after the beginning of the Cambrian Period, with trilobites and brachiopods being initially the most abundant fossilized animals. However, all major invertebrate phyla had become common by Ordovician time. Land vegetation appeared in the Silurian with the advent of a group of primitive vascular land plants called psilopsids. The first uncontested remains of fishes occur in rocks of Ordovician age. They are representatives of a group known as ostracoderms. Archaic jawed fishes appear late in the Silurian, but in general, fishes did not begin to dominate the marine realm until the Late Paleozoic.

Climates were relatively cool at the beginning of the Cambrian but warmed steadily. Mild conditions prevailed over most of the northern continents during the Early Paleozoic interval. In Late Ordovician there was an episode of cooling in the Southern Hemisphere, as indicated by evidence of glaciation in northwestern Africa.

Questions for Review

1. What are the names and locations of the major orogenic events of the Early Paleozoic, and when did they occur? What large-scale sedimentary features resulted from the erosion of the mountains developed by the orogenic activity?
2. Domes and basins are characteristic structural fea-

tures of the platform region of North America and other continents. Define each and explain how they differ in thickness of strata and unconformities. How might one distinguish between these two structures on a geologic map?

3. Geologists have suggested that erosion by wind and water proceeded more vigorously during the Cambrian and Ordovician than today. What might have been the reason?
4. What must have been occurring to the sea floor along the eastern side of North America during the Early Paleozoic to account for the accumulation of over 7000 meters of shallow-water sediment? How do geologists know that the sediment was deposited in shallow water?
5. What are the relationships between the solubility of calcium carbonate, the temperature, and the carbon dioxide content of sea water? What do these relationships tell us about the environment of deposition of the Early Paleozoic limestones?
6. Prepare definitions for the following organisms that were prevalent in Early Paleozoic seas.

   a. stromatolites
   b. archaeocyathids
   c. graptolites
   d. eurypterids
   e. ostracoderms
   f. acanthodians
   g. placoderms
   h. trilobites
   i. psilopsids

7. How might climatic conditions just prior to the beginning of the Paleozoic explain the great improvement in the fossil record of the Paleozoic compared to the Precambrian?
8. During what geologic period and under what climatic conditions did extensive evaporites form in the Early Paleozoic?
9. Geosynclines can be considered as having two parts: eugeosynclines and miogeosynclines. How do these two parallel tracts differ in terms of thickness of section, nature of sediment, and location?
10. Where is the Transcontinental Arch located? How might you convince a sceptic that it really existed during the Early Paleozoic?
11. Why is it logical to consider the Cambrian, Ordovician, and Silurian Periods together as "Early Paleozoic"? What two North American cratonic sequences are represented in the Early Paleozoic? How do these sequences relate to major marine transgressions and regressions?
12. What is the difference between the terms chordate and vertebrate? What phylum of invertebrates has been considered ancestral to the earliest chordates or protochordates? Why?
13. What are conodonts? Some paleontologists regard conodonts as the earliest evidence of chordates. On what do they base their arguments?
14. According to the concept of plate tectonics, what may have been the reason for the development of the Appalachian Geosyncline?

Terms to Remember

Absaroka Sequence
basin
blastoid
brachiopod
bryozoan
cephalopod
coelenterate
conodont
craton
cystoid
dome
edrioasteroids
epifaunal
eurypterids
foraminifer
graptolite
infaunal
Kaskaskia Sequence
Logan's Thrust
lophophore
medusoid coelenterate
mobile belt
Mollusca
nektonic
ostracoderms
ostracods
placoderms
placophoran
planktonic
polyp
Porifera
Principle of Temporal Transgression
Queenston Delta
radiolarian
receptaculid
rugose coral
Sauk Sequence
septa
stromatoporoid
tabulate coral
Taconic Orogeny
theca
Transcontinental Arch
trilobite
zoarium
zooecium
zooid

Supplemental Readings and References

Bambach, R. K., Scotese, C. R., and Ziegler, A. M. 1980. Before Pangaea: The Geographies of the Paleozoic World. *Amer. Scientist* 68:26–38.

Dott, R. H., and Batten, R. L. 1981. *Evolution of the Earth,* 3d ed. New York, McGraw-Hill Book Co.

Fairbridge, R. W. 1970. Ice age in the Sahara. *Geotimes* 15 (6):18.

Hamilton, W. 1970. The Uralides and the Motion of the Russian and Siberian Platforms. *Bull. Geol. Soc. Amer.* 81:2553–2576.

Kozlowski, R. 1966. On the structure and relationships of graptolites. *Paleo.* 40 (3):501–509.

Kummel, B. 1970. *History of the Earth,* 2d ed. San Francisco, W. H. Freeman Co.

Sloss, S. L. 1963. Sequences in the cratonic interior of North America. *Geol. Soc. Amer. Bull.* 74:93–111.

Stearn, C. W., Carroll, R. L., and Clark, T. H. 1979. The Geological Evolution of North America, 3d ed. New York, Ronald Press.

Valentine, J. W., and Moores, E. M. 1972. Global tectonics and the fossil record. *J. Geol.* 80 (2):167–184.

Restoration of a Pennsylvanian sea floor.

11

The Late Paleozoic

Amid all the revolutions of the globe, the economy of nature has been uniform, and her laws are the only thing that have resisted the general movement. The rivers and the rocks, the seas and the continents have changed in all their parts; but the laws which direct those changes and the rules to which they are subject have remained invariably the same.

John Playfair, *Illustrations of the Huttonian Theory of the Earth*, 1802

PRELUDE

The Devonian, Mississippian, Pennsylvanian, and Permian Periods compose the Late Paleozoic. This was a time when most of the separate land masses of earlier periods were assembled into the great supercontinent of Pangaea II. The process of gathering the errant continents had actually begun earlier. In the Silurian, for example, the ocean tract between North America (Laurentia) and northern Europe (Baltica) gradually narrowed as the two continents came together. The Silurian Caledonian Orogeny was the result of the collision that continued into the Devonian of eastern North America as the **Acadian Orogeny.** North America was now sutured to Europe, and the combined land mass has been named Laurussia (Fig. 11–1).

After the Devonian, the process of assembling Pangaea II continued as the plate-bearing Gondwana made its way northward at the expense of the intervening oceanic area. By late in the Carboniferous, Gondwana had come into contact with Laurussia, generating as it did so the Hercynian Orogeny of central Europe and the Allegheny Orogeny of eastern North America. It seems startling to us that mountains in the southeastern United States originated as a result of a collision with what is now northwestern Africa, but this is indeed what occurred. The clustering to form Pangaea II was nearly complete by Late Permian. Great mountain ranges marked the suture zones of the once separated paleocontinents, and an enormous ocean (Panthalassa) spanned the globe across nearly 300° of longitude.

The Late Paleozoic was a time of diverse sedimentation, progress in organic evolution, and diverse climatic conditions. During this time there was widespread colonization of the land by both large land plants and vertebrates. Amphibians and reptiles, along with spore-bearing trees and seed ferns were especially evident in the landscapes of the Late Paleozoic. Near the end of the era, conifers and reptiles tolerant of dryer, cooler climates become abundant. Marine invertebrates thrived in shallow epicontinental seas that were especially

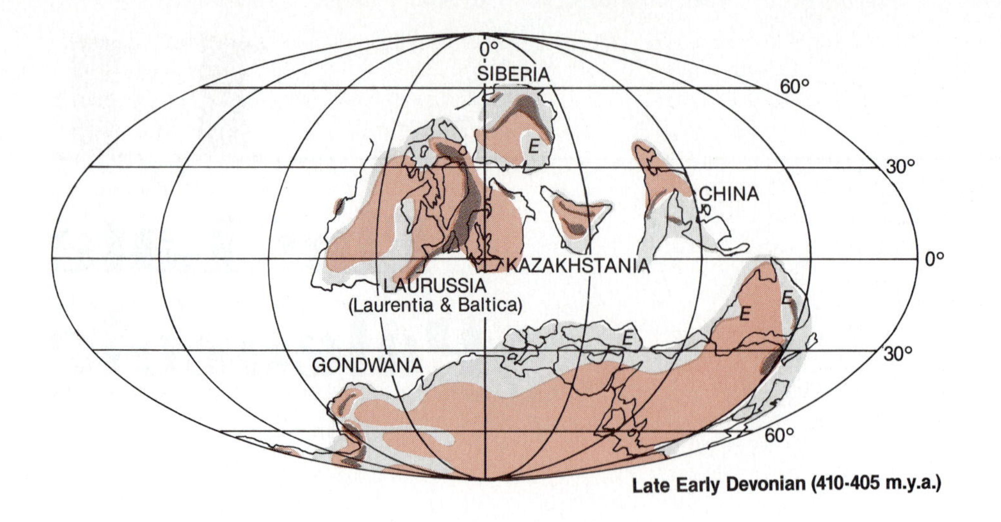

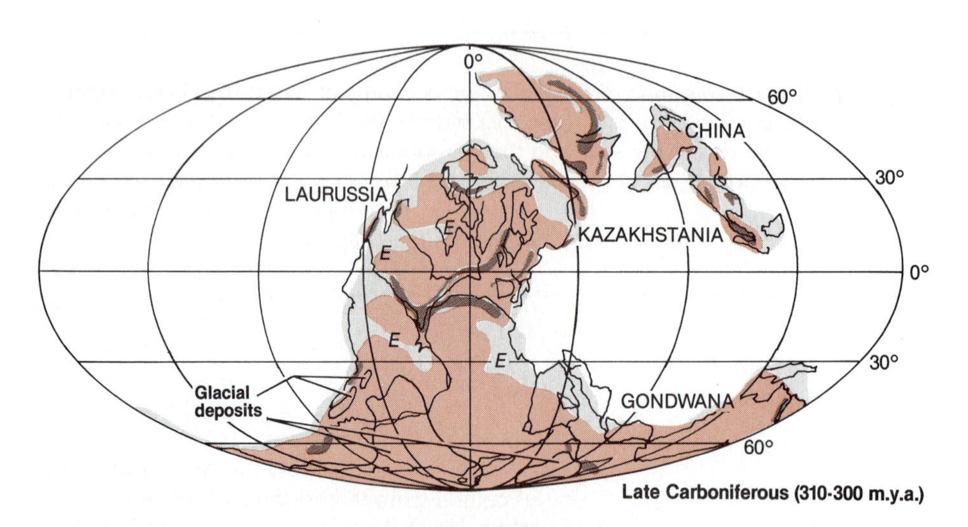

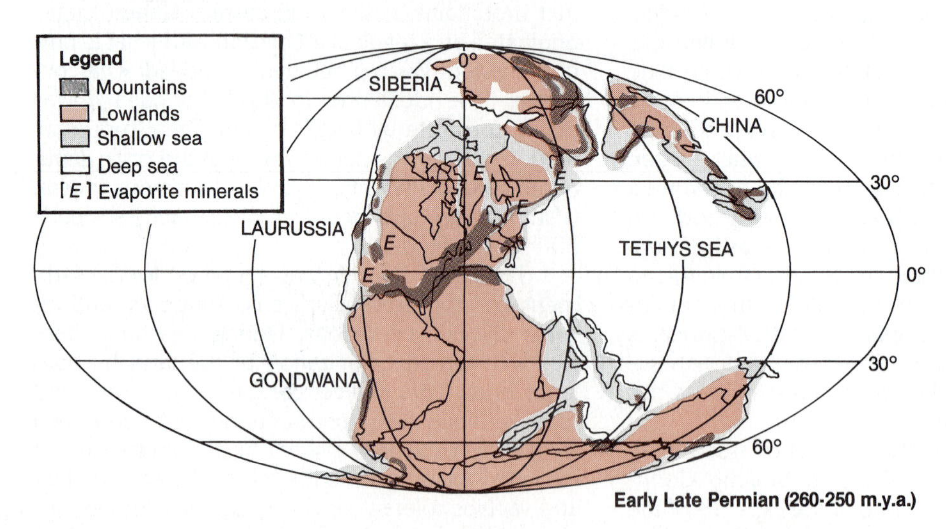

Figure 11–1 Global paleogeography of the Late Paleozoic. (From C. R. Scotese, R. K. Bambach, C. Barton, R. Van Der Voo, and A. M. Ziegler. *J. Geol.,* 87(3):217–277, 1979.)

extensive in North America during the Mississippian. In the central interior of the United States limestones deposited in these inland seas are over 700 meters thick and are among the most extensive sheets of limestone in the world. Adjacent to the orogenic belts and in regions where the epicontinental seas were drained as a result of uplift of the land or lowering of sea level, sediment-laden streams deposited their loads of sand and mud. These nonmarine sediments contain the fossil plants and animals that aid in inferring climatic conditions on the continents. The accumulation of abundant vegetation in swampy areas provided the raw materials from which the famous coal deposits of the Pennsylvania were formed. Now consolidated sands of desert dunes and layers of evaporites attest to arid conditions at many localities around the globe (Fig. 11–2). Tillites and glacial striations on bedrock provide evidence of glaciation. This is not unexpected, for the more southerly parts of Gondwana were located near the poles. Evidence of Late Carboniferous or Permian ice sheets can be found on the now separated continents of South America, Africa, Australia, Antarctica, and India. Decades before geologists had an understanding of plate tectonics, the ancient tillite deposits on these continents were cited as evidence for continental drift.

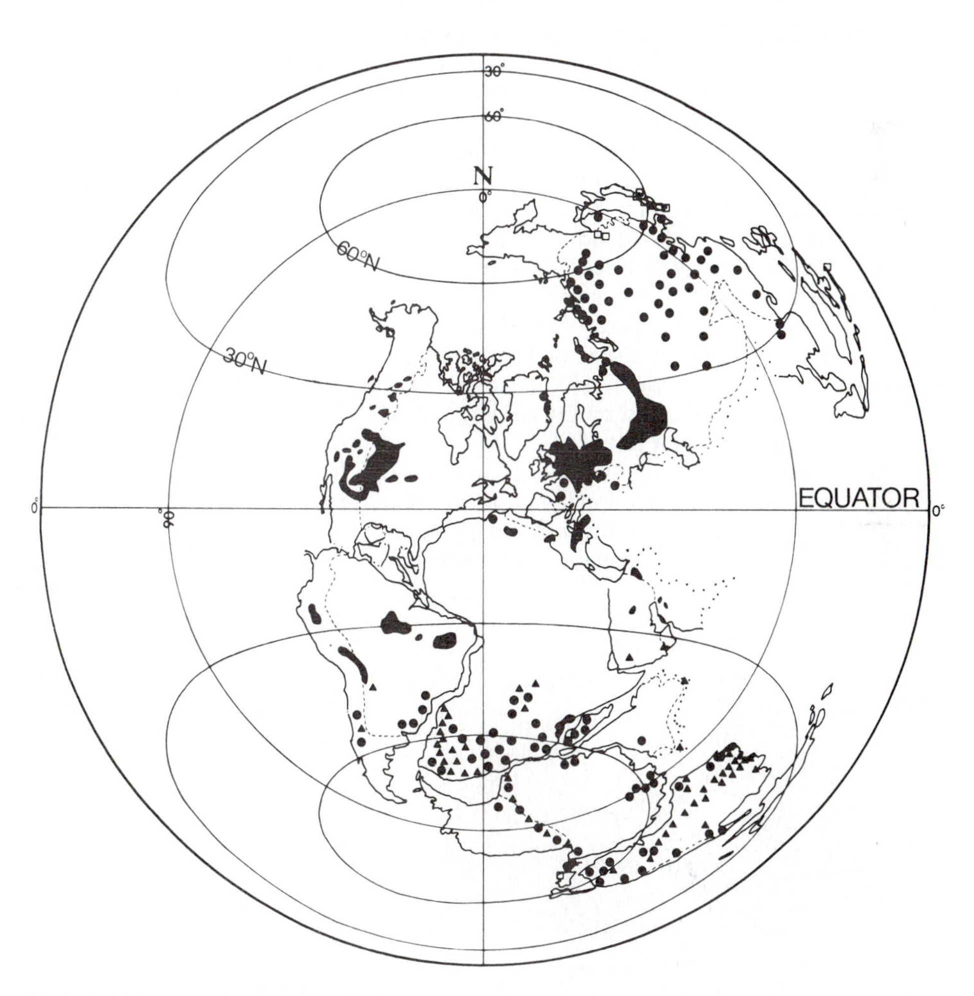

Figure 11–2 World geography about 250 million years ago during the Permian. The distribution of glacial tillites is represented by filled triangles. Coal is represented by filled circles and evaporites by the irregularly shaped filled areas. (From G. E. Drewey, T. S. Ramsay, and A. G. Smith. *J. Geol.*, 82(5):539, 1974.)

North America

During the Late Paleozoic, the North American portion of Pangaea II was nearly encircled by geosynclines. To the northeast lay the Northern Appalachian belt, which continued southward as the Southern Appalachian and Ouachita Geosynclines. The western edge of the continent was occupied by the western or Cordilleran Geosyncline. The encirclement was completed along the northern periphery of the continent by the Franklinian Geosyncline, which extended eastward for over 2000 km. At the center of the ring of geosynclines lay the stable cratonic region (Fig. 11–3).

Sedimentation on the Craton

THE KASKASKIA SEQUENCE. The final event of the Early Paleozoic on the North American craton had been the withdrawal of the inland seas. The Early Paleozoic formations were subjected to erosion, which exposed older and deeper formations in those locations that had experienced uparching. The gradual flooding of the old Early Paleozoic surface was the first event of Late Paleozoic cratonic history. This inland water body has been termed the **Kaskaskia Sea.** Except for minor regressions, it persisted over extensive areas of the craton well into the Mississippian Period, at which time the seas withdrew. Erosion of the layers of sediment deposited on the Kaskaskia Sea floor began, only to be interrupted by the advance of a second sea, the **Absaroka Sea.**

Along the eastern side of the North American craton, Kaskaskia sedimentation consists initially of clean quartz sands. The most famous of these sandy formations is the Oriskany Sandstone of New York and Pennsylvania (Fig. 11–4). Because of its purity, Oriskany sandstone is extensively used in making glass.

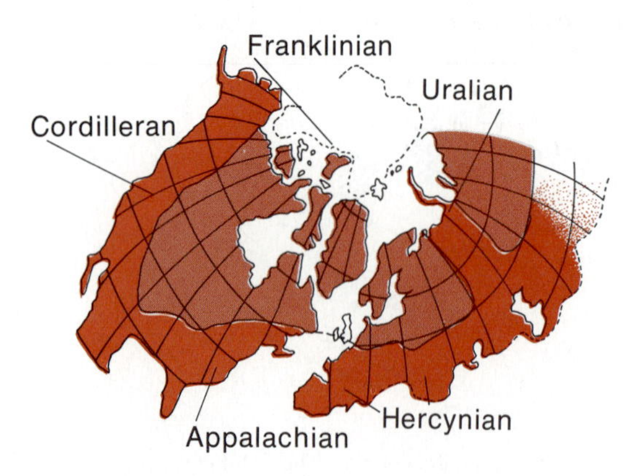

Figure 11–3 Location of Laurasian mobile belts (orange) and cratons (brown).

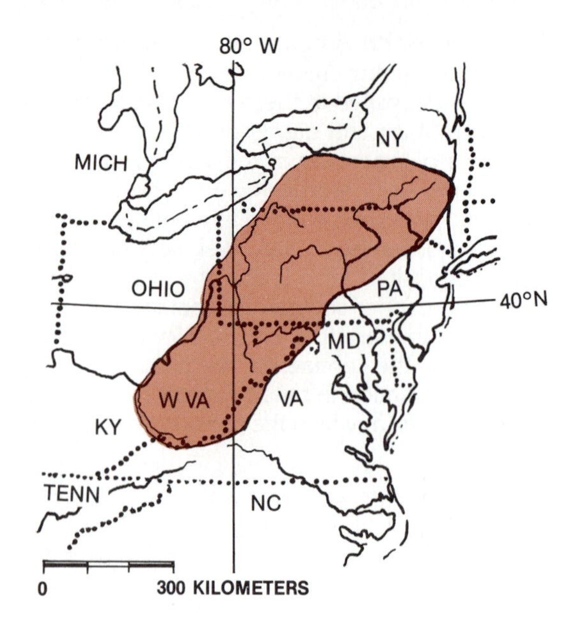

Figure 11–4 Areal extent of the Oriskany Sandstone. Surface exposures occur along the northern and eastern borders of the shaded area.

Although quartz is by far the dominant mineral in the Oriskany (Fig. 11–5), the formation also contains a very small percentage of other silicate minerals, which, like quartz, are resistant to weathering and erosion but are heavier than quartz. These so-called heavy minerals can be used to determine the kinds of parent rocks and source areas from which clastic sediments were derived. After study of the heavy minerals from scattered exposures of Oriskany sandstones, it became apparent that there were two distinctly different assemblages of heavy minerals. In the area south of New York, the formation contained exceedingly small amounts of only the most stable heavy minerals: tourmaline, zircon, and rutile. The grains in these sandstones were exceptionally well rounded and worn, and less stable heavy minerals were lacking. This part of the Oriskany was considered to have been derived from older clastic sedimentary units to the east and north. In contrast, the Oriskany of the New York State area contains a heavy mineral assemblage of relatively unstable pyroxene, amphibole, biotite, and garnet. Garnet is ordinarily derived from metamorphic source rocks, whereas the others are regular components of a variety of igneous and metamorphic rocks. The grains are unaltered. One may infer, therefore, that the sands composing the Oriskany of the New York area may

Figure 11–5 Thin section of the Oriskany Sandstone made from an exposure near Union Springs, New York. Note secondary outgrowths of silica around originally well-rounded quartz grains.

have had an igneous or metamorphic source area and were not recycled from older sandstones. The Oriskany is an example of one method utilized by geologists to infer the history of a rock unit.

The Oriskany Sandstone is the initial deposit of a transgressing sea, as is indicated by the way the sands overlap older strata. As this marine advance continued, limy sediments were deposited over the Oriskany, and corals began to build reef structures. In areas where water circulation was restricted, salt and gypsum were deposited. During the Middle and Late Devonian of the region east of the Mississippi Valley, carbonate sedimentation gave way to shales. The change to clastic deposition was a consequence of mountain building associated with the Acadian Orogeny in the Northern Appalachians. Highlands developed during this orogenesis were rapidly eroded, and clastics were transported westward as a vast apron of sediments that geologists refer to as the **Catskill Delta.** This great wedge of sediment is coarser and thicker near the source areas and merges into a thinner but extensive sequence of black shales in the east-central region of the United States. The Catskill Delta was located in the equatorial zone. Hence it is not surprising that the sediments are deeply weathered and stained red by iron oxides. In these tropical stream and lake deposits, the first abundant remains of truly large land plants are found.

In the far west part of the craton, Middle and Late Devonian rocks are largely limestones, although there are shales as well. In a depressed area known as the Williston Basin (from South Dakota and Montana northward into Canada), extensive reefs were developed. Arid conditions and restricted circulation in reef-enclosed basins resulted in the deposition of impressive thicknesses of gypsum and salt. Some of Canada's largest oil fields derive their product from permeable reefs into which oil migrated and became trapped by enclosing layers of impermeable shale.

Late Devonian and early Mississippian seas deposited a remarkably uniform blanket of black shales that extended from the eastern geosyncline to the Mississippi Valley. Although there are many local names for this sediment, the most widely used is **Chattanooga Shale.** The clayey sediment had as its source the continuing erosion of highlands to the east that had periodically uplifted during the Late Devonian and Mississippian. In places, the Chattanooga consists largely of fine-grained calcite and compressed carbonized plant material. Gradually, as the highland source areas were reduced, the quantity of clastic materials decreased, and carbonates became the most abundant and widespread deposit (Figs. 11–6 and 11–7). Cherty limestones, limestones composed entirely of the remains of billions of crinoids and wave-worn debris of other invertebrates, and limestones formed from myriads of tiny calcium carbonate spheres (oölites) formed massive beds and covered the entire central and western regions of the craton (Figs. 11–8 and 11–9). Oölites are spherical grains formed by the precipitation of calcium carbonate around a nucleus. Oölites are known to occur today in shallow coastal areas with high turbulence, such as breaker zones along shorelines. The seaway in which these carbonates were deposited was the most extensive North America had experienced since Ordovician time and was the last of the great

Figure 11–6 The Mississippian St. Louis Limestone exposure in a road cut near Red Rock Park, Illinois. (Photograph courtesy of J. C. Brice.)

Paleozoic floodings of the North America craton.

In Late Mississippian time, as the Kaskaskia Seas began their final withdrawal, large quantities of clastics and thin, laterally nonpersistent limestones were deposited. These rocks are reservoirs for petroleum in Illinois and therefore have been studied extensively. Detailed maps of thicknesses of some of the sandstone units show that they are thickest along branching, sinuous trends that suggest old stream valleys developed on the former sea floor. Studies of grain size, cross-bedding (Fig. 11–10), and current-produced sedimentary structures strongly indicate that the clastics were derived from the northern Appalachians and were transported southwestward across the central interior.

THE ABSAROKA SEQUENCE. When the Kaskaskia Seas had finally left the craton, at the end of Mississippian time, the exposed terrain was subjected to erosion that resulted in one of the most widespread regional unconformities in the world. Not only was erosion areally extensive, but also over arches and domes it beveled away entire systems of older rocks. The unconformity provides a criterion for separating those strata equivalent to the Carboniferous of Europe into the Mississippian and Pennsylvanian systems. It is appropriate to use these two names in eastern and central North America, not only because of the unconformity but also because the rocks above the erosional hiatus differ markedly from those below. Indeed, the overlying Pennsylvanian strata are a consequence of quite different tectonic circumstances.

It was not until near the beginning of Middle

Figure 11–7 The St. Genevieve Limestone of the Kaskaskia (Mississippian) sequence in western Illinois. Notice the well-developed ripple marks on bedding surfaces. (Photograph courtesy of J. C. Brice.)

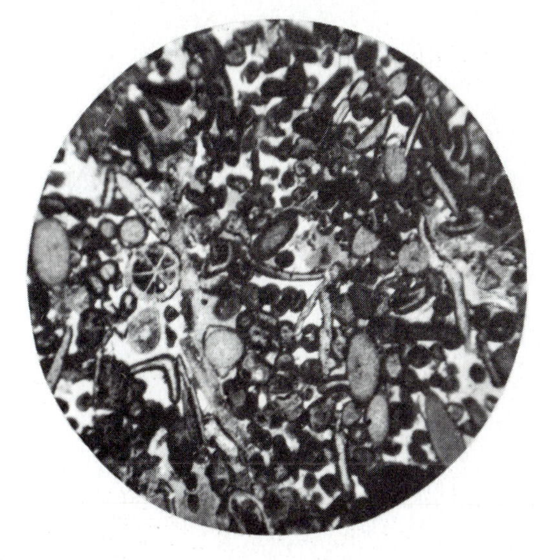

Figure 11–8 Photomicrograph of the Salem Limestone (Mississippian), which consists of assorted fossil debris and oölites cemented with clear calcite. (Photograph courtesy of J. C. Brice.)

Figure 11–9 Modern oölites from the Bahama Banks. (Photograph courtesy of J. C. Brice.)

Pennsylvanian time that the seas were able to encroach onto the long exposed surface of the craton. The deposits of this seaway are those of the Absaroka Sequence. In general, the Pennsylvanian section of rocks near the eastern geosyncline are thicker, and virtually all are continental sandstones, shales, and coal beds. This eastern section of Pennsylvanian rocks gradually thins away from the Appalachian belt and changes from predominantly terrestrial to about half marine rocks and half nonmarine rocks. Still farther west, the Pennsylvanian outcrops are largely marine limestones, sandstones and shales.

One of the most notable aspects of Pennsylvanian sedimentation in the middle and eastern states is the repetitive alternation of marine and nonmarine strata. A group of strata deposited during one alternation of such a cycle is called a **cyclothem.** A typical cyclothem in the Pennsylvanian of Illinois frequently contains 10 units (Fig. 11–11). Units one through five are continental deposits, the uppermost of which is a coal bed (Fig. 11–12). The strata deposited above the coal bed represent an advance of the sea over an old forested area. In Missouri and Kansas, at least 50 changes of this kind are recognized within a section about 750 meters thick. Some extend across thousands of kilometers. It is apparent that the advances and retreats of the seas were frequent and widespread. What caused these oscillations? One explanation involves periodic regional subsidence of the land to a level slightly below sea level so that marginal seas could spill onto the level swampy lowlands. A short time later, subsidence might cease and sediments be built up above sea level to extend the shoreline seaward and re-establish continental conditions. Alternatively, the re-establishment of dry lands may have resulted from temporary regional uplifts. Finally, worldwide or eustatic changes in sea level, possibly caused by periodic glacial advances in Gondwanaland, might also produce marine invasions and regressions, particularly along low-lying coastal plains. Perhaps the cyclothems were caused by some combination of these conditions. Whatever their cause, it is evident

Figure 11–10 Cross-bedding in sandstones of the Upper Mississippian Tar Springs Formation, southern Illinois. (Photograph courtesy of J. C. Brice.)

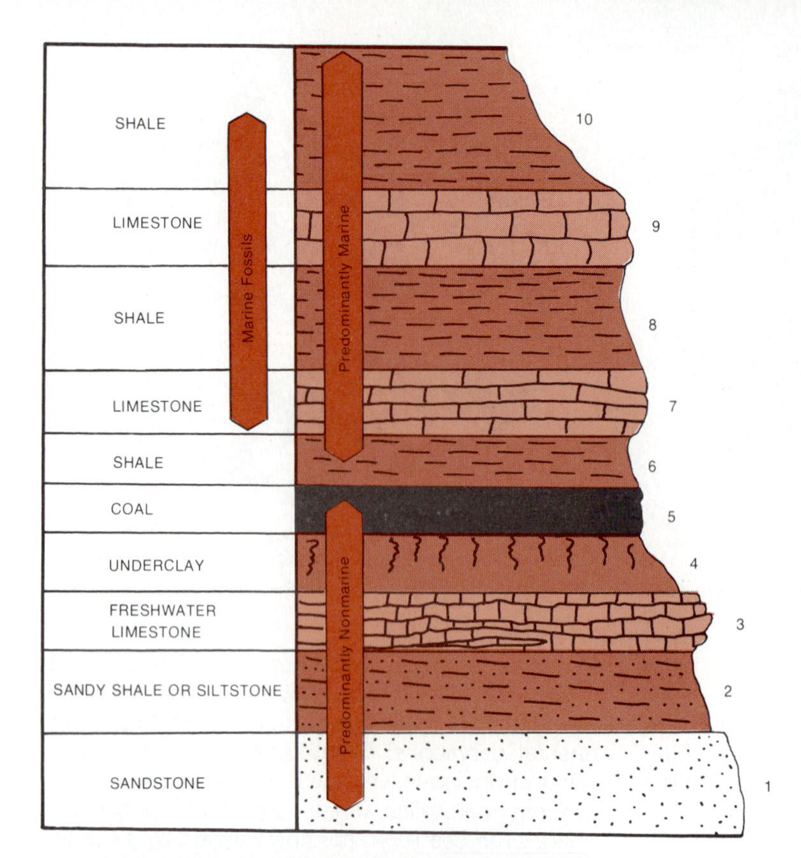

Figure 11–11 A coal-bearing cyclothem showing the ideal sequence of layers. Many cyclothems do not contain all 10 units, as in this illustration at an ideal sequence. Some units may not have been deposited because the changes from marine to nonmarine conditions were abrupt, and units may also have been removed by erosion following marine regressions.

that they represent short-term oscillations superimposed on the long-term regional uplift associated with the development of the Appalachian and Ouachita Mountains.

Ordinarily, cratonic areas of continents are characterized by stability. The southwestern part of the North American craton, however, provides an exception to this general rule, for during the Pennsylvanian this was a region of mountain building. The resulting highlands are generally termed the Colorado Mountains (also called the ancestral Rockies) and the Oklahoma Mountains. These mountains and related uplifts appear to have resulted from movements of crustal blocks along large, nearly vertical, reverse faults. The Colorado Mountains included a range that extended north-south across central Colorado (the Front Range–Pedernal Uplifts) and a segment curving from Colorado into eastern Utah (the Uncompahgre Uplift).

Figure 11–12 Part of an Illinois cyclothem. The lowermost layer is the coal seam (cyclothem bed 5), followed upward by shale (bed 6) near geologist's hand, limestone (bed 7), shale (bed 8), another limestone (bed 9), and the upper shale (bed 10). Part of another sequence caps the exposure. This cyclothem is part of the Carbondale Formation. (Photograph courtesy of D. L. Reinertsen and the Illinois Geological Survey.)

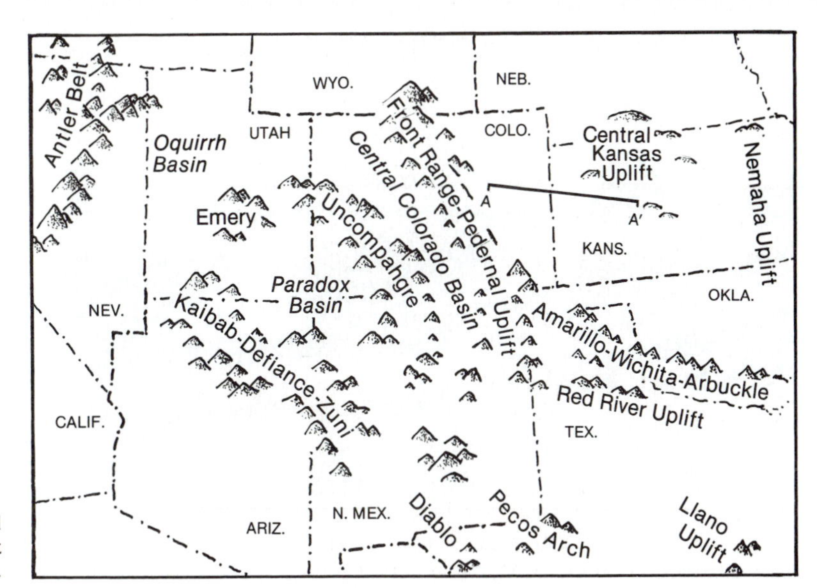

Figure 11–13 Location of the principal highland areas of the southwestern part of the craton during Pennsylvanian time.

The Central Colorado Basin lay between these highland areas (Fig. 11–13). A separate range, the Zuni–Fort Defiance Uplift, extended across northeastern Arizona on the southwestern perimeter of the Paradox Basin. To the east of the Colorado Mountains lay the southeastward-trending Oklahoma Mountains. Eroded stumps of this once rugged range form today's greatly reduced Arbuckle and Wichita Mountains. Remains of the Amarillo Mountains are now buried beneath younger rocks and are known principally from exploratory drilling for oil.

Judging from the tremendous volume of sediments eroded from the Colorado Mountains, it is likely that they attained heights in excess of 1000 meters. They also were probably subjected to repeated episodes of uplift, continuing in some places into the Permian. Erosion of these highlands eventually exposed their Precambrian igneous and metamorphic cores. As erosion and weathering continued, great wedge-shaped deposits of red arkosic sediments were spread onto adjacent and intervening basins (Fig. 11–14). A small part of this massive accumulation of clastic sediment is dramatically exposed in the Red Rocks Amphitheatre just a few miles southwest of Denver and the "flat irons" near Boulder, Colorado (Fig. 11–15).

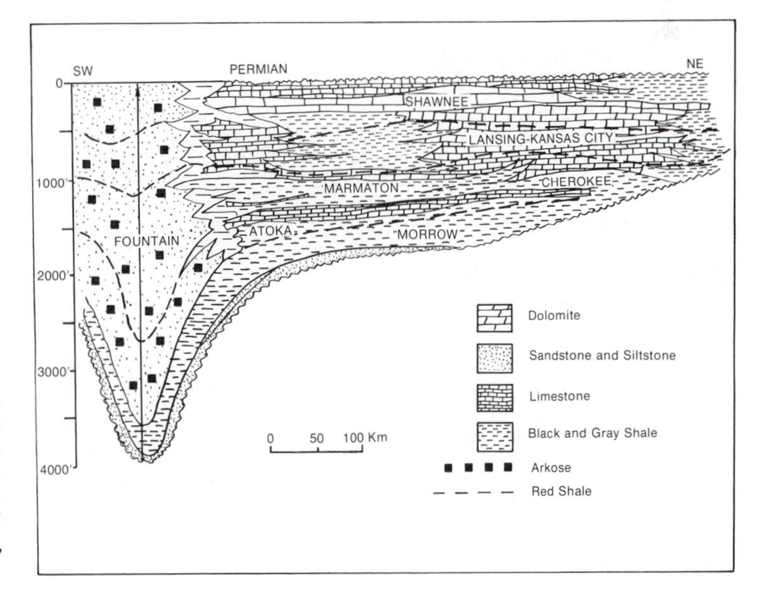

Figure 11–14 Pennsylvanian cross-section across eastern Colorado and western Kansas showing great accumulation of coarse arkosic sandstones east of the Colorado Mountains. Line of section is along A–A′ in Figure 11–13. (From *Stratigraphic Atlas of North and Central America.* Shell Oil Company, Exploration Department.)

Figure 11–15 Coarse arkosic red beds of the Pennsylvanian Fountain Formation exposed around the famous Red Rocks Amphitheatre a few miles southwest of Denver. The strata are tilted eastward by a post-Cretaceous orogeny.

Geologists have long puzzled over this episode of deformation of the craton, but recently plate tectonics has provided a hypothesis for this unusual event. It seems likely that the collision of Gondwana with North America along the site of the Ouachita Geosyncline generated stress in the bordering area of the craton to the north. Crustal adjustments to relieve these stresses resulted in the

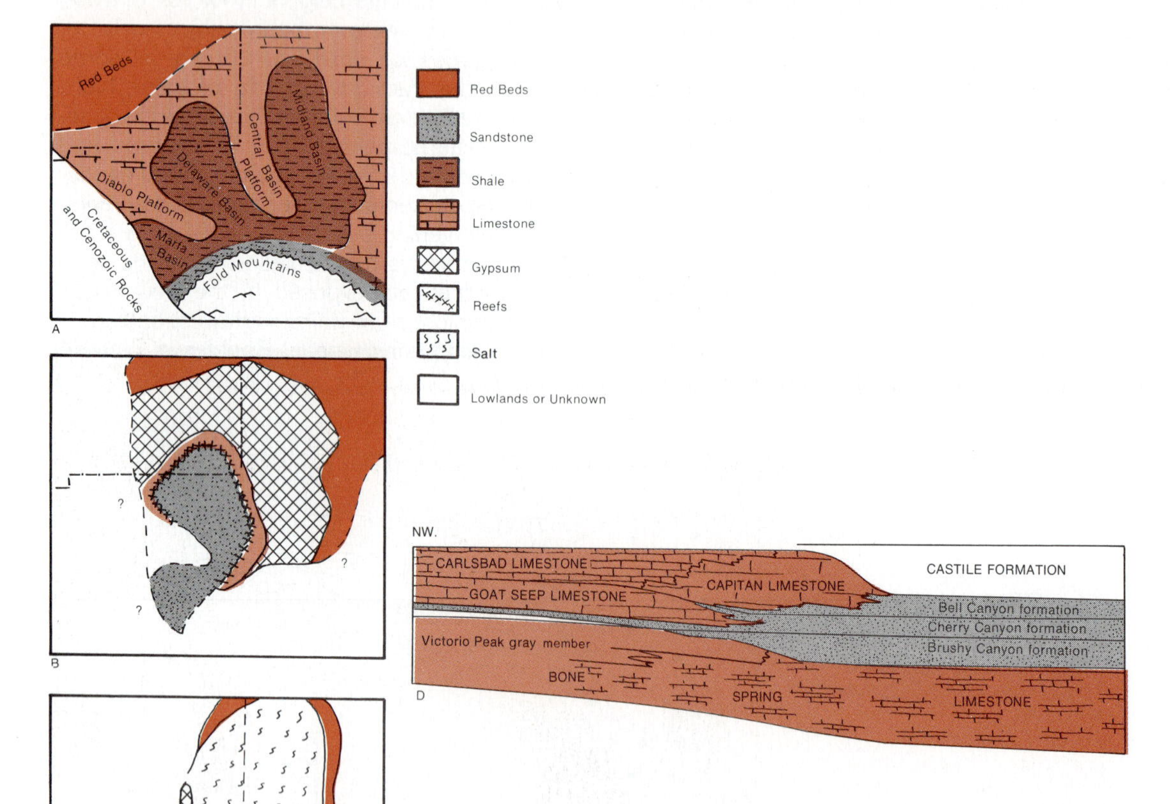

Figure 11–16 The Permian of North America can be divided into four stages. The oldest is the Wolfcampian, which is followed by Leonardian, Guadalupian and Ochoan. A, B, and C show the paleogeography and nature of sedimentation in western Texas during the Wolfcampian (A), Guadalupian (B), and Ochoan (C). A simplified cross section of Leonardian and Guadalupian sediments of the Guadalupe Mountains (D) indicates the relationship of the reef to other facies. (After many sources, but primarily P. B. King. U.S. Geol. Survey, Professional Paper 215, 1948.)

deformation that produced the highlands and associated basins.

The basins that lay adjacent to the Colorado Mountains are important in working out the geologic history of the southwestern craton because of the excellent stratigraphic record they contain. The Paradox Basin is especially interesting. It was inundated by the Absaroka Sea in Early Pennsylvanian time. The initial Pennsylvanian deposits, largely shales, were deposited over a karst topography developed on Mississippian limestones. By Middle Pennsylvanian time, the western access of the Absaroka Sea to the Paradox Basin had become restricted, and thick beds of salt, gypsum, and anhydrite were deposited. Fossiliferous and oölitic limestones developed around the periphery of the basin, and patch reefs grew abundantly along the western side. The association of porous reefs and lagoonal deposits resulted in several sites suitable for the later entrapment of petroleum.

Near the end of the Pennsylvanian, the Paradox Basin was filled to above sea level by arkosic sediments shed from the recently uplifted Uncompahgre highlands. Also at this time, the Absaroka Sea, which had begun its transgression at the beginning of the Pennsylvanian, began a slow and irregular regression near the end of the same period. The withdrawal was still incomplete in Early Permian time, so that marine sediments continued to be deposited in a rather narrow zone from Nebraska through western Texas. Fossiliferous limestones characterized these inland seas, although near highlands in Colorado, Texas, and Oklahoma, coarse clastics accumulated to sufficient thicknesses to bury surrounding uplands. The deposits along what was the eastern edge of the seaway have been eroded away, but at several places along the western side, one can observe the change from richly fossiliferous beds below to barren shales, red beds, and evaporites above. The thick and extensive salt beds of Kansas provide testimony to the gradual restriction and evaporation of Permian seas in the central United States. The "last stand" for Permian marine conditions occurred in the western part of Texas and southeastern New Mexico, where a remarkable sequence of interrelated lagoon, reef, and open-basin sediments were deposited (Fig. 11–16). In this region, several irregularly subsiding basins developed between shallowly submerged platforms. Dark-colored limestones, shales, and sandstones were deposited in the deep basins, whereas massive reefs formed along the basin edges. Behind the reefs, in what must have been the shallow waters of extensive lagoons, the deposits are thin limestones, evaporites, and red beds. Late in the Permian, the

Figure 11–17 Example of a facies change in rocks at the south end of the Guadalupe Mountains about 90 mi south of El Paso, Texas. The prominent cliff is called El Capitan. It is composed of a Permian limestone reef deposit, rich in corals and associated marine invertebrate fossils. Behind the reef there once existed lagoons with abnormally high salinity, as suggested by evaporites, dolostones, and the virtually unfossiliferous limestones. These lagoonal beds are exposed along the ridge on the right side of the photograph. Thus a massive reef limestone facies lies adjacent to a lagoonal facies. To the south of the reef lay a marine basin in which normal marine sediments were deposited. (Photograph courtesy of the National Park Service.)

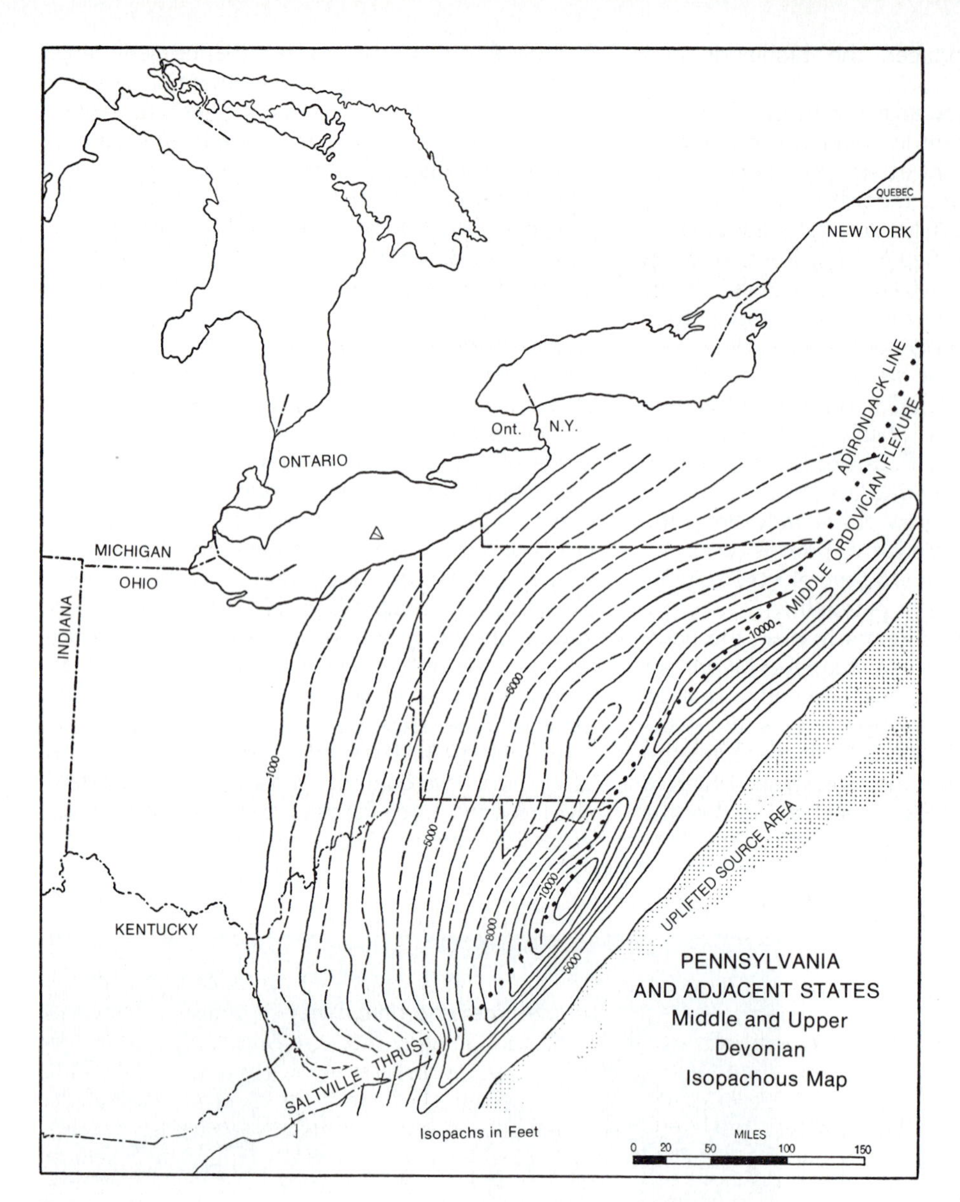

Figure 11–18 Isopach (thickness) map of Middle and Upper Devonian strata in West Virginia, Pennsylvania, and parts of adjacent states. (After M. Kay. North American geosynclines. *Geol. Soc. Am.*, Memoir 48, 1951.)

connections of these basins to the south became so severely restricted that the waters gradually evaporated, leaving behind great thicknesses of gypsum and salt.

Much paleoenvironmental information has been gleaned from a study of the western Texas Permian rocks. From the lack of medium- and coarse-grained clastics, one may assume that surrounding regions were low-lying. The gypsum and salt suggest a warm, dry climate in which basins were periodically replenished with sea water and experienced relatively rapid evaporation. Careful mapping of the rock units has helped to establish an estimated depth of about 500 meters for the deeper basins and only a few meters for the intrabasinal platforms. The basin deposits are dark in color and rich in organic carbon as a likely consequence of accumulation under stagnant, oxygen-poor conditions. Perhaps upwelling of these deeper waters may have provided a bonanza of nutrients on which phytoplankton and reef-forming algae thrived. Along with the algae, the reefs contain the skeletal remains of over 250 species of marine invertebrates. Today, because of their relatively

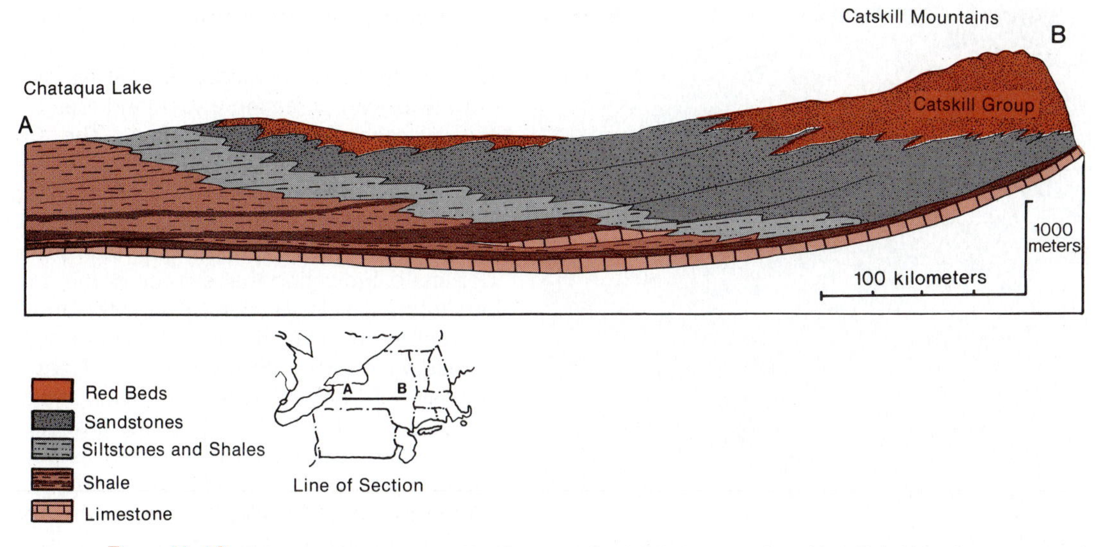

Figure 11–19 East-west section across the Devonian Catskill Delta in southern New York. Note that continental red beds interdigitate with nearshore marine sandstones, and these in turn grade toward the west into offshore siltstones and shales. Continental deposits prograded upon the sea, pushing the shoreline progressively westward. (Based on several classic studies by G. H. Chadwick and G. A. Cooper completed between 1924 and 1942.)

greater resistance to erosion, these ancient reefs form the steep El Capitan promontory in the Guadalupe Mountains of Texas and New Mexico (Fig. 11–17). Here one can examine the forereef composed of broken reef debris that formed a sort of talus deposit caused by the pounding of waves along the southeast side.

The Eastern Geosynclines

During the latter half of the Paleozoic Era, the great Appalachian and Ouachita Geosynclines experienced their culminating and most intense episodes of orogenesis. The crumpling of these geosynclinal tracts was a consequence of the reassembly of the formerly separated continents into Pangaea II. The northern part of the Appalachian-Caledonian belt had taken the shock of a collision with Europe during the Silurian and Devonian. That collision was the cause of the Caledonian Orogeny in Europe and the Acadian Orogeny in northeastern North America. As indicated in Figure 11–1, the southern half of the Appalachian Geosyncline and the Ouachita Geosyncline as well were crushed into mountainous tracts when they were struck by the northwestern bulge of Africa during the Late Carboniferous. This great encounter between Laurussia and Gondwana was the cause of the Allegheny Orogeny. As noted earlier, the collision from the south not only raised the old geosynclinal tracts but also transmitted stresses into the interior, causing deep-seated deformations such as those that produced the Colorado and Oklahoma Mountains.

The effects of the Devonian deformation, called the Acadian Orogeny, are clearly seen in a belt from Newfoundland into West Virginia. Here one finds thick, folded sequences of eugeosynclinal clastics interspersed with rhyolitic volcanic rocks and granitic intrusions. The intensity of the compression that affected these rocks is reflected in their metamorphic minerals, which indicate that mineralization occurred at temperatures exceeding 500°C and pressures equivalent to burial under 15,000 meters of rock. The overall result of the Acadian Orogeny was to demolish forever the Appalachian eugeosyncline as a marine depositional trough and establish in its place mountainous areas in which erosion was the prevalent geologic process. Here and there in isolated basins among the mountains, Devonian nonmarine sediments were deposited. However, the greatest volume of erosional detritus was spread outward from the highlands as a great wedge of terrigenous sediment called the Catskill Delta (Figs. 11–18 and 11–19). The pattern of sedimentation seen in the Taconic Orogeny with its Queenston Delta was repeated, except that the Catskill deposits came at a time when land plants were abundant and were able to provide a green mantle for the alluvial plains and hills. The Catskill sediments are dominated by sandstones, and shales in which the contained iron is deeply oxidized. The oxide takes the mineral form of hematite and causes the reddish

Figure 11–20 Conglomeratic facies of the Old Red Sandstone Formation, Somerset, England. (Photograph courtesy of Institute of Geological Sciences, London.)

or brownish coloration characteristic of Catskill "red beds." Because they are so deeply oxidized, red beds imply subaerial deposition. However, this inference must be confirmed by other evidence of continental deposition, because the color may be secondarily derived from reddish sediments in the source area. In the case of the Catskill red beds, there are numerous traces of rootlets and shrublike vegetation. Toward the west, the red beds give way to gray sandstones and shales that contain fossil tree stumps. Here, the grayish coloration results from the ferrous variety of iron oxide and may imply deposition in frequently swampy or marshy environments where there is a relative deficiency of oxygen.

On the European side of the adjoined continents, the blanket of coarse debris (Fig. 11–20) derived from the Caledonian highlands spread southwestward, creating a vast land area named—for its most famous formation—**The Old Red Sandstone.**

Mississippian strata crop out in the Appalachian region from Pennsylvania to Alabama (Fig. 11–21). Nonmarine shales and sandstones predominate, indicating that erosion of the mountainous tracts developed during the Acadian Orogeny was still in progress. Some of the finer clastics were spread westward onto the craton to form the vast deposits of Early Mississippian black shales mentioned earlier. Particularly coarse sediments, including conglomerates, were deposited as part of the **Pocono Group.** Pocono sandstones (Fig. 11–22) form some of the resistant ridges of the Appalachian Mountains in Pennsylvania. Westward, the Pocono section thins and changes imperceptibly into marine siltstones and shales. Evidently, the depositional framework consisted of a great complex of alluvial plains that sloped westward and merged into deltas that were being built outward into the epicontinental seas. The plains and deltas, standing only slightly above sea level, were backed by the rising mountains of the Appalachian fold belt. A large part of the coarser clastics had settled before reaching the southern portions of the geosyncline, and marine limestones are the most prevalent rocks.

Pennsylvanian rocks of the Appalachians are characterized by cross-bedded sandstones and gray shales that were deposited by rivers or within lakes and swamps. Coal seams are, of course, prevalent in the Pennsylvanian System of the east-

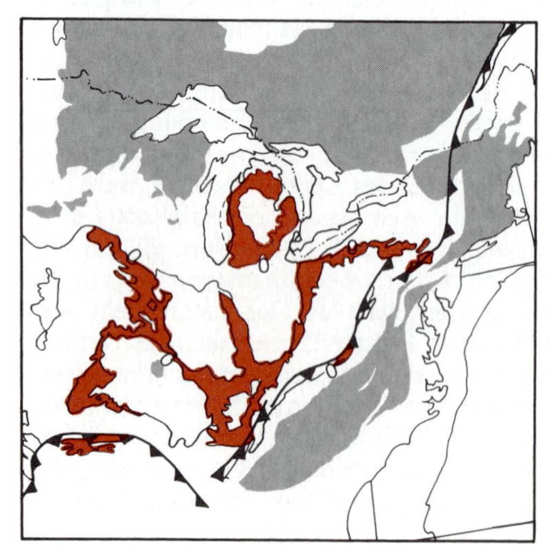

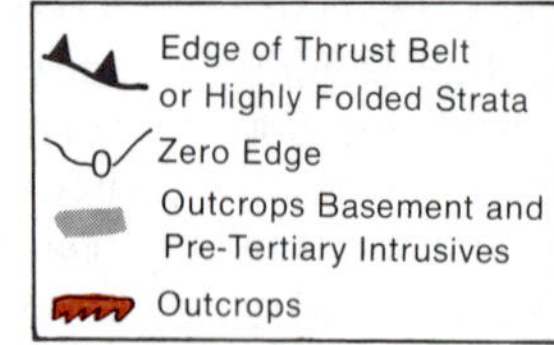

Figure 11–21 Areas of Mississippian outcrops. (From *Stratigraphic Atlas of North and Central America.* Shell Oil Company, Exploration Department.)

Figure 11–22 Yellowish-gray, olive-brown, and yellowish-brown sandstones and siltstones of the Pocono Formation exposed at Mile 103 of I-81, where the road is cut through Second Mountain in Pennsylvania. (Photograph courtesy of the Commonwealth of Pennsylvania, Bureau of Topographic and Geologic Survey.)

ern United States (Fig. 11–23) and reflect the luxuriant growths of mangrove-like forests that clothed the lands. This was an ideal environment for coal formation. Vegetation that accumulated in the poorly drained swampy areas was frequently inundated and killed off. Immersed in water or covered with muck, the dead plant material was protected from rapid oxidation; however, it was attacked by anaerobic bacteria. These organisms broke down the plant tissues, extracted the oxygen, and released hydrogen. What remained was a fibrous sludge with a high content of carbon. Later, such peatlike layers were covered with additional sediments—usually siltstones and shales—and then compressed and slowly converted to coal.

Figure 11–23 Anthracite bed in the Brooks Mine, Scranton, Pennsylvania. Anthracite beds (often called "veins") are included in the Llewellyn Formation. (Photograph courtesy of the Commonwealth of Pennsylvania Bureau of Topographic and Geologic Survey.)

The culminating deformational event in the southern Appalachians has been termed the **Allegheny Orogeny.** This great episode of mountain building probably began during Pennsylvanian time and continued throughout the Permian and into the Triassic. It affected a belt that extended for over 1600 km from southern New York to central Alabama. The results of the orogeny were profound and included Permian compression of geosynclinal strata as well as the bordering tract of the craton. The great folds now visible in the Valley and Ridge Province were developed during this orogeny (Fig. 11–24). Less visible at the surface but no less impressive are enormous thrust faults formed along the east side of the southern Appalachians. The folds are asymmetrically overturned toward the northwest, and the fault surfaces are inclined southeastward, suggesting that the entire miogeosyncline was moved forceably against the central craton. Not unexpectedly, erosion of the rising mountains produced another great clastic blanket of nonmarine sediments. These mostly continental reddish sandstones and gritty shales compose the **Dunkard** (Fig. 11–25) and **Monongahela** series of Permo-Pennsylvanian age.

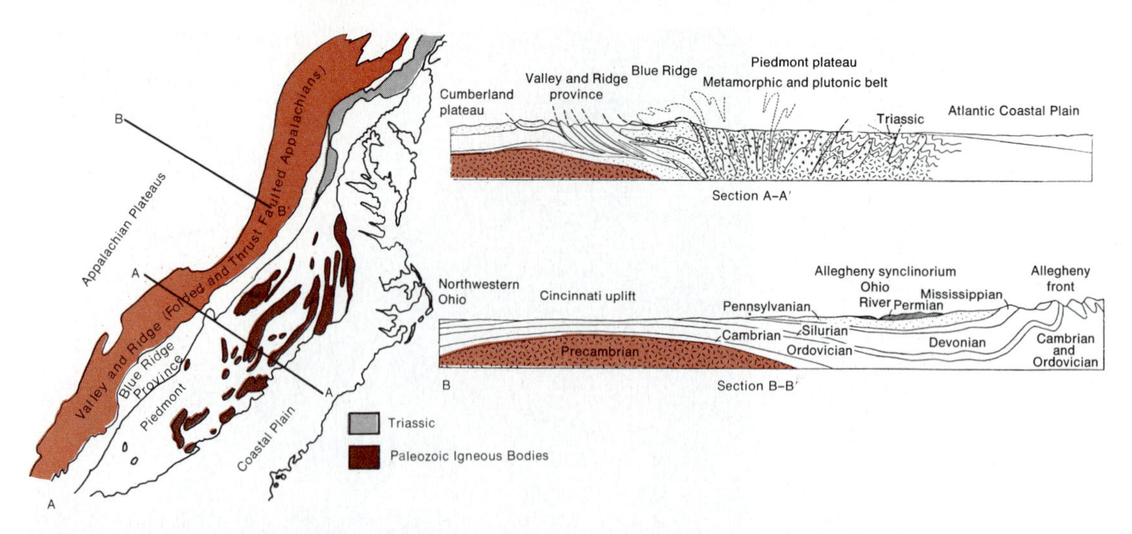

Figure 11–24 (A) Map of physiographic provinces of the Appalachian region. (B) Two cross-sections showing the structural relations of the Appalachian Mountains. (A from *Tectonic Map of the United States.* Geological Society of America, 1944; B from P. B. King. *Bull. Am. Assoc. Petr. Geol.,* 1950.)

The Ouachita Deformation

As suggested earlier, the deformation of the Ouachita Geosyncline is believed to have been caused by collision of the African edge of Gondwanaland with the southeastern margin of the American craton. However, deformation began rather late, and from Early Devonian until Late Mississippian, the region was undergoing slow sedimentation interrupted by only minor disturbances. Carbonates predominated in the more northerly miogeosynclinal zone, whereas cherty rocks

Figure 11–25 The Waynesburg Sandstone of the Dunkard Series exposed along the Ohio River in Meigs County, Ohio. (Photograph courtesy of the Ohio Division of Geological Survey.)

Figure 11–26 (A) Tightly folded beds of Arkansas novaculite. (B) Close-up view of complex, small folds in the novaculite beds. (Photograph courtesy of C. G. Stone and the Arkansas Geological Commission.)

known as **novaculites** accumulated in the eugeosyncline (Fig. 11–26). The pace of sedimentation increased dramatically toward the end of the Mississippian, when a thickness of over 8000 meters of graywackes and shales was spread into the geosyncline. The flood of clastic deposition continued into the Pennsylvanian, forming a great wedge of sediment that thickened and became coarser

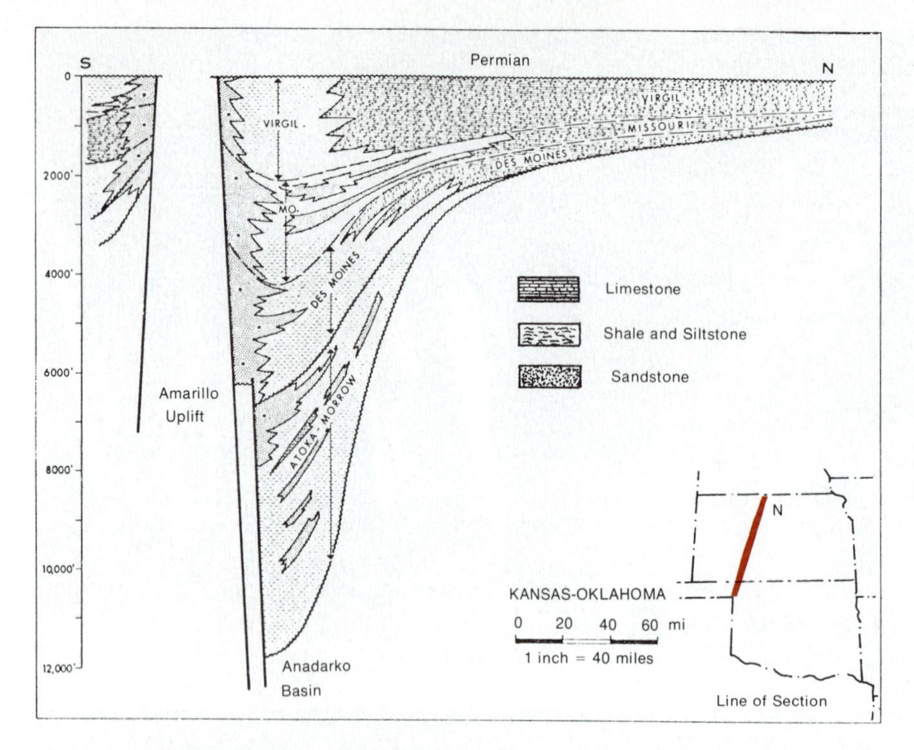

Figure 11–27 Geologic section of Pennsylvanian rocks across Oklahoma and Texas, showing thick wedge of sediment shed from mountain ranges to the south. (From *Stratigraphic Atlas of North and Central America.* Shell Oil Company, Exploration Department.)

toward the south, where mountain ranges were being formed (Fig. 11–27). Radiometric dating of now deeply buried basement rocks from the Gulf Coastal states indicates that these rocks were metamorphosed in the Late Paleozoic and were the likely source for the Pennsylvanian clastics spread into the geosyncline. These coarse sediments document several pulses of orogenesis that were ultimately to produce mountains along the entire southern border of North America. In the faulting that accompanied the intense folding, eugeosynclinal rocks were thrust northward onto the miogeosyncline. By Permian time, stability returned, and strata of this final Paleozoic period are relatively undisturbed.

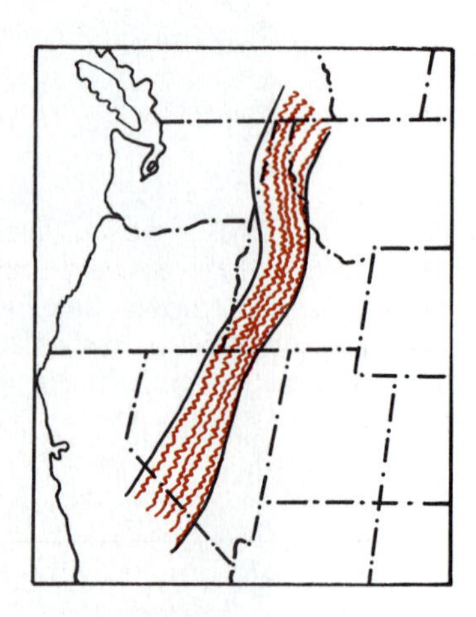

Figure 11–28 Location of the Antler Orogenic Belt.

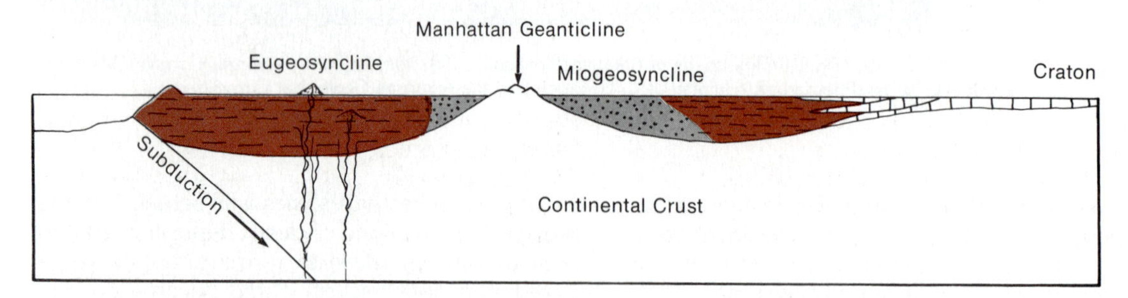

Figure 11–29 Interpretive cross-section of the Cordilleran Geosyncline during Mississippian time.

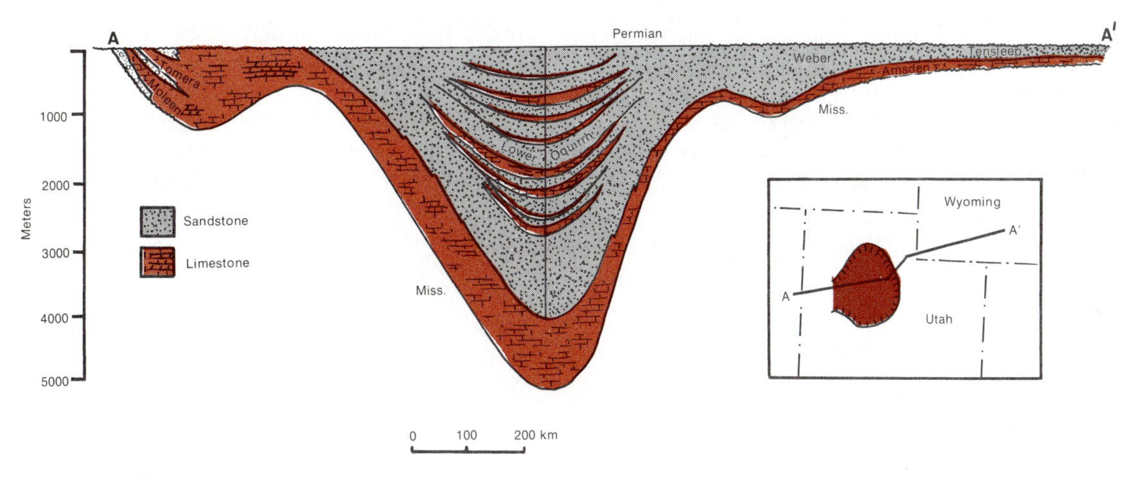

Figure 11–30 Section across the Oquirrh Basin. (Section courtesy of Shell Oil Company; from *Stratigraphic Atlas of North and Central America.* Shell Oil Company, Exploration Department.)

Since the time of active mountain building, erosion has leveled most of the highlands, leaving only the Ouachita Mountains of Arkansas and Oklahoma and the Marathon Mountains of southwest Texas as remnants of once lofty ranges. Although the Ouachita Geosyncline has been traced over a distance of nearly 2000 km, only about 400 km are exposed. Thus, the actual configuration of the geosyncline has been determined by the examination of millions of well samples and other data obtained during drilling activities associated with petroleum exploration.

The Late Paleozoic history of the western or Cordilleran Geosyncline was almost as lively as that of the Appalachian. In the far west, patches of highly deformed and intruded Late Paleozoic cherts and volcanics and thick sections of graywacke attest to deformation that is thought to have been caused by the collision of an eastward-moving island arc against the western border of North America. As a result of this encounter, an elongate area of uplift and instability known as the Antler Orogenic Belt (Fig. 11–28) developed along the boundary between the eugeosyncline and miogeosyncline. Tectonic activity along this belt resulted in highlands that provided source materials for over 1000 meters of conglomerates and sandstones spread across Nevada and Idaho. The Antler Orogeny was accompanied by thrust-faulting on a colossal scale, with eugeosynclinal sequences moved as much as 80 km eastward over miogeosynclinal sediments of the former continental shelf.

The Antler Orogeny, which had begun in the Late Devonian, continued actively into the Mississippian and Pennsylvanian. An upland tract known as the Manhattan Geanticline shed sediments into both the miogeosyncline on the east and the eugeosyncline on the west (Fig. 11–29). Especially thick deposits of Pennsylvanian and Permian miogeosynclinal sediments accumulated in the area now occupied by the Wasatch and Oquirrh Mountains (Fig. 11–30). In the latter area, the Oquirrh Formation is over 9000 meters thick.

Mississippian and Pennsylvanian eugeosynclinal deposits west of the Manhattan Geanticline include a great volume of coarse clastics and volcanics. Over 2000 meters of sandstones, shales, lavas, and ash beds are found in the Klamath Mountains of northern California. Volcanic rocks in western Idaho and British Columbia attest to a continuation of vigorous volcanism from the Carboniferous through the Permian. Crustal deformation along the west side of the Manhattan Geanticline is indicated by areally extensive angular unconformities between Permian and Triassic sequences. These Permo-Triassic disturbances of the Cordilleran Belt have been named the **Cassiar Orogeny** in British Columbia and the **Sonoma Orogeny** in the southwestern United States. Like the earlier Antler Orogeny, the Sonoma event was probably caused by the collision of an eastward-moving island arc against the North American continental margin in west-central Nevada. Oceanic rocks and remnants of the arc were thrust onto the edge of the continent and became part of North America.

Permian miogeosynclinal conditions were quieter than those in the eugeosyncline to the west. Quartz sandstones and limestones are the predominant deposits. One of the better known of these Permian miogeosynclinal rock units is the **Phosphoria Formation,** which was deposited in a shallow sea that covered Wyoming and eastern Idaho. As indicated by its name, the formation includes phosphatic limestone and phosphorite deposits. The **phosphorite,** a dark gray, concretionary variety of calcium phosphate, is mined for the manufacture of fertilizers and other chemical products. The unusual concentration of phosphates may have resulted from upwelling of phosphorus-rich sea water from deeper parts of the basin. Once within the area of phosphoria deposition, microorganisms may have caused the precipitation of the phosphate salts.

The Northern Geosyncline

The Late Paleozoic record of the Northern or Franklinian Geosyncline is initiated by Devonian shales and limestones. Coral reefs are abundantly preserved in the Devonian System. The existence of these reefs is not surprising, for the region lay at about 15° north latitude during the Devonian and temperatures were much warmer than today. Following the deposition of the Devonian marine rocks, a highland area was developed along the northern border of the geosyncline. Coal-bearing deltaic strata were deposited in the adjacent lowland. These and older deposits were folded in Early Mississippian time, then eroded, and ultimately covered by a series of limestones and evaporites.

Europe During the Late Paleozoic

During the Late Paleozoic, Europe was bordered by the Uralian Geosyncline on the east and the Hercynian on the south. Along the northern margin of Europe, the Caledonian Orogeny had created the vast land area of the Old Red Continent. Uplands provided a source for clastic sediments, which were swept out into numerous basins to accumulate to thicknesses exceeding 10,000 meters. Judging from the many interlayers of ash and lava, volcanic activity was frequent on the Old Red Continent. Fossil fishes and sedimentary evidence indicate that Europe's climate was tropical and possibly semiarid at the time.

South of the Caledonian Mountains, the continental deposits gradually thin and grade into marine shales and limestones of the Hercynian Geosyncline (see Fig. 11–3). For a time, quiet prevailed in the Hercynian Geosyncline. However, in Late Devonian and Early Mississippian, the belt was intensely folded, metamorphosed, and intruded by granites as ancestral Europe and Gondwanaland collided. The event has been named the **Hercynian Orogeny,** and its result was a great range of mountains across southern Europe. For the most part, the eroded stumps of these ranges are now buried, but here and there patches of the covering younger rocks have been eroded away, revealing the intensely folded, faulted, and intruded older rocks that lie beneath.

From the Hercynian uplands, gravel, sand, and mud were carried down into basins and coastal environments. The clastic deposits were quickly clothed in dense tropical forests. Burial and slow alteration of vegetative debris from these forests provided the material for coal formation in the great European coal basins. Plant fossils found in these coal seams are of the tropical kind and differ from the more temperate Gondwana flora of the Southern Hemisphere.

Although the greatest amount of Hercynian deformation occurred toward the end of the Early Carboniferous (Mississippian) in Europe, spasms of unrest continued in both the Late Carboniferous and the Permian. These later episodes of folding correlate with similar deformations in the Southern Appalachian and Ouachita orogenic belts of North America.

No less important than the Hercynian Geosyncline in Europe was the lengthy Uralian Geosyncline, which extended along a belt now occupied by the Ural Mountains. The geosyncline was already in existence at the beginning of the Paleozoic Era. Interestingly, there are discrepancies between the polar wandering curves for Paleozoic rocks from the cratons on either side of the Urals. The discrepancies indicate that the Russian and Siberian platforms were widely separated during Early Paleozoic time and that they began to converge in the middle part of the era and ultimately collided by the end of the Paleozoic (Fig. 11–1). Indeed, the eugeosynclinal tract of the Urals consists largely of oceanic material scraped off against the edges of the converging plates. The collision resulted in the formation of a great mountain system along the entire Uralian orogenic belt and unified ancestral Siberia with eastern Europe.

In central Europe and parts of Russia, there were temporary Late Permian marine incursions that precipitated evaporites. The famous German potassium salts are a product of one of these inland seas, which has been named the **Zechstein Sea** (Fig. 11–31). Apparently, aridity was as characteristic of western Europe during the Permian as it was of Texas and New Mexico.

Gondwana During the Late Paleozoic

During the Late Paleozoic, the great land mass of Gondwana remained fairly intact. It moved across the south pole and more fully entered the

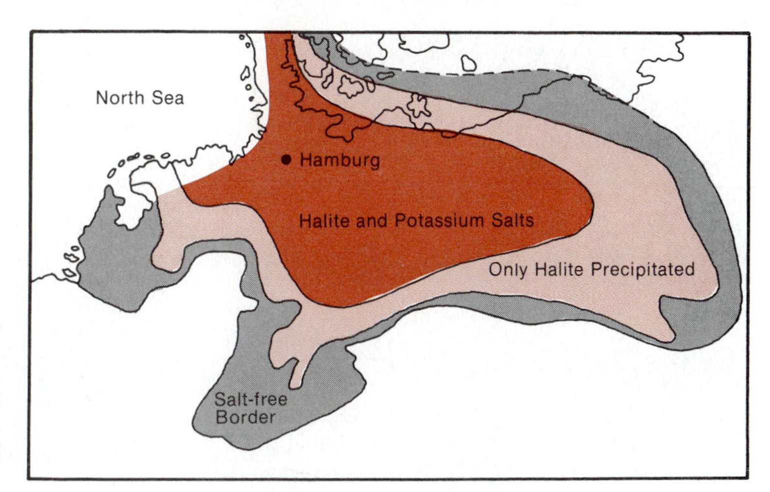

Figure 11–31 Outcrop area of sedimentary rocks deposited in the Zechstein Sea. (After R. Brinkman. *Geological Evolution of Europe.* 1969. Fig. 20, p. 65. Translation from German by J. E. Sanders, Ferdinard Enke. Verlag, Stuttgart, and Hafner Publishing Co., New York.)

side of the earth on which Laurussia was located. In its northward migration, Gondwana caused the closure of the ocean that separated it from Laurussia, causing the Hercynian Orogeny of Europe and the Allegheny Orogeny of North America. Orogenic activity associated with subduction zones was also evident in the Late Paleozoic history of Gondwana, particularly along the Andean Geosyncline of South America and the Tasman Geosyncline of eastern Australia.

The most dramatic paleoclimatologic event of Gondwana's Late Paleozoic history was the growth of vast continental glaciers (see Fig. 11–1). Extensive layers of tillite and the scour marks of glaciers (Fig. 11–32) have been found at hundreds of locations in South America, South Africa, Antarctica, and India. There are indications of at least four and possibly more glacial advances, suggesting a pattern of cyclic glaciation not unlike that experienced by North America and Europe less than 100,000 years ago. The orientation of striations chiseled into bedrock by the moving ice suggests the glaciers moved northward from centers of accumulation in southwestern Africa and eastern Antarctica. During the warmer interglacial stages and in outlying less frigid areas, *Glossopteris* and other plants tolerant of the cool, damp climates grew in profusion and provided the materials for thick seams of coal.

In time the ice receded, and Gondwana's vast cratonic areas became sites for deposition of Permian nonmarine red beds and shales. As we shall

Figure 11–32 Glacial pavement, Hallet Cove, South Australia, formed in very early Permian time on upper Precambrian quartzite. Ice flowed to the right parallel to the pencil; note striations, downflow smoothing, and crescentic chatters facing upflow. (Photograph courtesy of Warren Hamilton, U.S. Geological Survey.)

Figure 11–33 Restoration of a Middle Devonian forest in the eastern area of the U.S. (A) An early lycopod, *Protolepidodendron.* (B) *Calamophyton,* an early form of the horsetail rush. (C) Early tree fern, *Eospermatopteris.* (Courtesy of Field Museum of Natural History; painting by C. R. Knight.)

see in the next section, some of these sediments contain the fossil remains of the ancestors of the earth's first mammals.

LIFE OF THE LATE PALEOZOIC

An Overview

One of the lessons learned from the fossil record is that once a major taxonomic category of organisms has evolved and proliferated, that phylum tends to persist. Of course, members of the phylum diversify according to evolutionary factors that act upon them. For this reason, life of the Late Paleozoic represents a continuation of the evolutionary trends initiated in the Early Paleozoic. There are some important innovations, however. During the Early Paleozoic, most life appears to have been marine, whereas early in the Late Paleozoic, plants and animals had begun to proliferate upon the continents. Great forests appeared and changed the appearance of the landscape. From the fishes, which made their debut in the Early Paleozoic, amphibians and reptiles evolved. For the first time ever, vertebrates walked across the lands, and insects began to diversify and populate the continents. In the sea, continuations rather than new appearances were the rule. For example, trilobites persisted into the Permian, although their abun-

Figure 11–34 Fossilized bark of the Carboniferous tree *Lepidodendron.* (Photograph courtesy of Wards Natural Science Establishment, Inc., Rochester, N.Y.)

dance and diversity were markedly decreased. Corals and bryozoans expanded and replaced stromatoporoids as the principal reef-forming organisms. Graptolites and other once flourishing groups suffered extinction.

Plants

The first unquestioned occurrences of vascular land plants are in late Ordovician rocks. These early invaders of the terrestrial environment were the **psilopsids** (see Fig. 10–38) described in the previous chapter. The psilopsids preceded in evolution a variety of higher plants, many of which were lofty, well-rooted leafy trees that grew in Devonian forests (Fig. 11–33). One of these forests has left an ample fossil record in the vicinity of Gilboa, New York. The Gilboa Forest contained trees over 7 meters tall. However, impressive as they were, the Devonian forests were dwarfed by their Carboniferous descendants.

When one surveys the entire history of vascular plants, three major advances become apparent. Each involved the development of increasingly more effective reproductive systems. The first advance led to seedless, spore-bearing plants, such as those that were ubiquitous in the great coal-forming swamps of the Carboniferous. The second saw the evolution of seed-producing, pollinating, but nonflowering plants ("gymnosperms"). This also was a Late Paleozoic event. The evolution of plants with both seeds and flowers ("angiosperms") came late in the era that followed the Paleozoic.

Among the moisture-loving plants of the Carboniferous were the so-called scale trees, or **lycopsids.** Today, they are represented by their smaller survivors, the club mosses, of which the ground pine *Lycopodium* is a member. Small size was not particularly characteristic of the Late Paleozoic lycopsids. The forked branches of *Lepidodendron* (Fig. 11–34) reached 30 meters into the sky. The elongate leaves of the scale trees emerged directly from the trunks and branches. After being released, they left a regular pattern of leaf scars. In *Lepidodendron,* the scars are arranged in diagonal spirals, whereas species of *Sigillaria* have leaf scars in vertical rows.

Another dominant group of plants that grew side by side with the Carboniferous lycopsids were the **sphenopsids.** Living sphenopsids include the scouring rushes and horsetails. Fossil sphenopsids, such as *Calamites* (Fig. 11–35) and *Annularia* (Fig. 11–36), possessed slender, unbranching, longitudinally ribbed stems with a thick core of pith and rings of leaves at each transverse joint. At the top, a cone bore the spores that would be scattered in the wind.

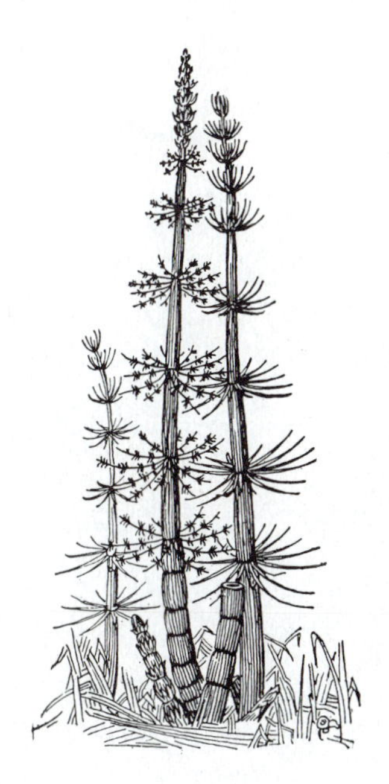

Figure 11–35 *Calamites.*

True ferns were also present in the coal forests. Many were tall enough to be classified as trees. Like the lycopsids and sphenopsids, they reproduced by means of spores carried in regular patterns on the undersides of the leaves.

Seed plants also made their debut during the Late Paleozoic (Fig. 11–37). They probably arose from Devonian fernlike plants. These plants, appropriately dubbed "seed ferns," had fernlike leaves but unlike true ferns reproduced by means of seeds.

One of the most widely known seed fern groups was *Glossopteris,* which was widespread in the southern hemisphere during the Carboniferous and Permian (see Fig. 8–16). Most species of *Glossopteris* were plants with thick, tongue-shaped leaves. Because of certain anatomic traits, and of their association with glacial deposits, *Glossopteris* and associated plants are thought to have been adapted to cool climates.

Fossils of seed plants are present in the northern hemisphere also. Cordaites (primitive members of the conifer lineage) were abundant, and some towered 50 meters (Fig. 11–38). Their branching limbs were crowned with clusters of large, straplike leaves. These and somewhat more modern cone-bearing plants spread widely during the Permian, perhaps as a consequence of dryer climatic conditions. The first ginkgoes made their appearance during the Permian. Today, only a sin-

Figure 11–36 *Annularia,* a common carboniferous sphenopsid. Specimen from Pennsylvanian strata near St. Louis.

gle species, *Ginkgo biloba,* remains as a survivor of this once flourishing group.

Continental Invertebrates

Because of the greater hazards of postmortem destruction of continental invertebrates, their fossil record is not as complete as that for marine invertebrates. Nevertheless, logic dictates that many invertebrate animals populated the soil, rivers, and lakes of the Late Paleozoic. The oldest known insects are wingless species found in Devonian rocks. Carboniferous strata contain a more complete but still inadequate insect record. Included were giant dragonflies with wingspans of over 70 cm. Cockroaches (Fig. 11–39) that reached lengths of 10 cm creeped about among the rotting vegetation. Eurypterids, although not common, persisted throughout the Late Paleozoic. Several species of Devonian spiders have been recovered from Devonian and Carboniferous strata in North America, Germany, and Great Britain. In Nova Scotia, an interesting collection of land snails has been collected from hollows within the preserved stumps of trees. Scorpions and centipedes

Figure 11–37 *Alethopteris,* a seed fern of the Pennsylvanian.

Figure 11–38 *Cordaites.*

are found in abundance in a concretionary Pennsylvanian siltstone exposed along Mazon Creek in northern Illinois.

Of all the land invertebrates, it is apparent that the arthropods and gastropods have been the most persistently successful. Many paleontologists believe this may be because of attributes already evolved by their aquatic ancestors. As protection against desiccation, arthropods had evolved relatively impervious exoskeletons. Snails derived similar benefits from their shells. Both groups included very active animals with sufficient mobility to seek out food aggressively.

Marine Invertebrates

Most of the familiar groups of Early Paleozoic marine invertebrates continued into the Late Paleozoic, although some groups diminished in number and variety, whereas others expanded and diversified. The foraminifers, which had a modest start in the Early Paleozoic, experienced their first major expansion in the Carboniferous. One reflection of their numerical increase is seen in the Mississippian Salem Limestone, which locally consists almost entirely of specimens of *Endothyra* (Fig. 11–40). The **fusulinids**—spindle-shaped descendants of endothyrids—proliferated during the Pennsylvanian and Permian (Fig. 11–41). As a group, fusulinids looked superficially similar, but individual species evolved complex and distinctive internal structures that permit their use as index fossils in many parts of the world.

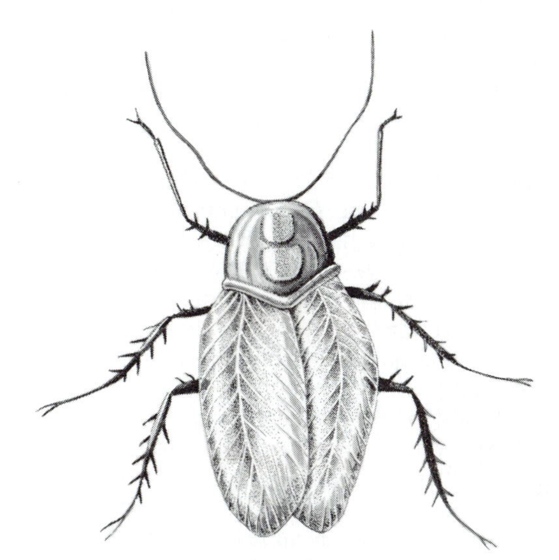

Figure 11–39 Reconstruction of a primitive Pennsylvanian cockroach.

Figure 11–40 *Endothyra,* Mississippian foraminifers from the Salem Limestone of Missouri.

Figure 11–41 Slab of limestone composed predominantly of fusulinids. The rock is Pennsylvanian in age and is part of the Modesto Formation in Peoria County, Illinois. (Photograph courtesy of the Illinois State Geological Survey.)

The most striking innovation among **sponges** was the expansion of siliceous forms during the Devonian. Localities in New York yield great numbers of fossilized siliceous sponges (Fig. 11–42), many of which were similar to the modern Venus flower basket.

Rugose corals continued to diversify and increase during the Late Paleozoic (Figs. 11–43 and 11–44). In general, the rugose group included both solitary "horn corals" and large compound colonies composed of hundreds of closely packed individuals. Although the earlier tabulate species continued for a time, they had nearly vanished after the Devonian. Following their Devonian and Carboniferous abundances, the rugosans also began to decline and apparently suffered complete extinction by the end of the Paleozoic.

Among other stationary animals that colonized the Late Paleozoic sea floors were the *Bryozoa*. The varieties that constructed lacy, delicate, fanlike colonies (Fig. 11–45) were especially abundant in the shallow tropical waters of the Mississippian. Associated with these so-called **fenestellid** colonies were the bizarre corkscrew bryozoans known by the generic name *Archimedes* (Fig. 11–46).

Brachiopods were present by the millions throughout the Paleozoic, but they may have reached their zenith in number and variety during the Devonian. The spiriferid brachiopods, so named because they possessed internal helicoid spirals to support the lophophores, were especially abundant during the Devonian (Figs. 11–47 and 11–48). During the Carboniferous and Permian, the productid brachiopods (Fig. 11–49), which had distinctly spinose and inflated convex ventral valves, became the most abundant group. Indeed, among invertebrate paleontologists, the Late Paleozoic is frequently proclaimed the "Age of Productids."

Although the greatest advances among the **mollusk** clan came in eras that followed the Paleozoic, they were nevertheless persistently present throughout the Late Paleozoic. **Pelecypods** flourished in the frequently sandy and muddy sea bottoms of the time. Air-breathing **gastropods** made their first appearance, whereas the older, aquatic group persisted. However, the single most impor-

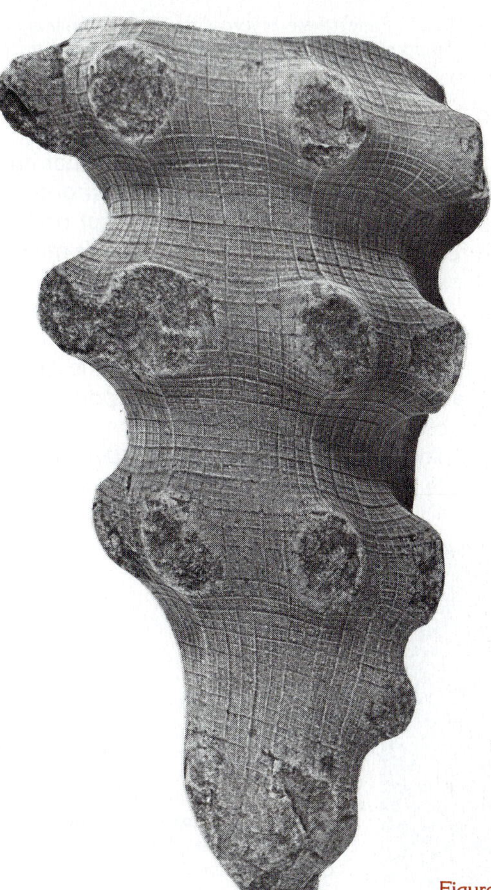

Figure 11–43 The colonial rugose coral *Acrophyllum,* from the St. Louis Limestone in Illinois. (Photograph courtesy of Wards Natural Science Establishment, Rochester, N.Y.)

Figure 11–42 The Devonian siliceous sponge *Hydnoceras.* (Photograph courtesy of J. Keith Rigby, Brigham Young University.)

Figure 11–44 Restoration of part of the Devonian Sea floor occupied largely by colonial rugose corals and occasional horn corals, cephalopods, crinoids, straight nautiloid cephalopods, and trilobites. (Courtesy of the Smithsonian Institution.)

Figure 11–45 Fenestellate bryozoans from the Mississippian Chester Group of Kentucky. (Photograph courtesy of Frank K. McKinney.)

tant molluscan event was the debut of that important cephalopod group known as **ammonoids.** In this group, sutures were developed that no longer had relatively straight edges like those of nautiloids but rather were bent into lobes and saddles (Fig. 11–50). The ammonoids known as **goniatites** persisted in marine deposits throughout the Late Paleozoic and gave rise to the **ceratites** and **ammonites,** which became important in the Mesozoic.

Trilobites were generally on the decline during the Late Paleozoic, but, locally, particular families

Figure 11–46 Shown at left are three of the screw-shaped axes of the bryozoan *Archimedes.* At the right is a slightly compressed specimen of *Archimedes* with the edge of the continuous spiral of radiating branches exposed. Specimens are from the Mississippian Bangor Limestone of Alabama. Height of all specimens is about 9 cm. (Photograph courtesy of Frank K. McKinney.)

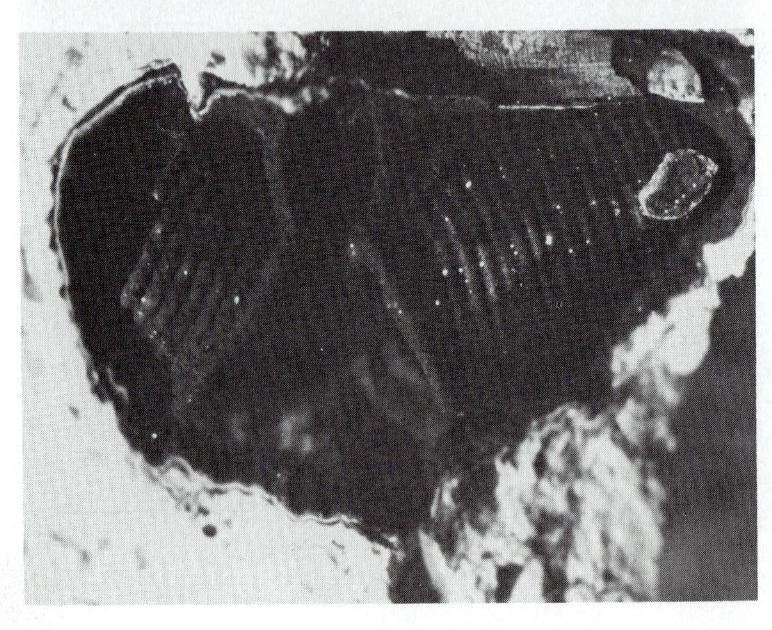

Figure 11–47 A spiriferid brachiopod in which by an accident of preservation one valve has been dissolved away to reveal the internal spiral structure.

Figure 11–48 The spiriferid brachiopods *Mucrospirifer* from the Devonian of Ontario. (Photograph courtesy of Wards Natural Science Establishment, Rochester, N.Y.)

thrived. The group known as proparians (Fig. 11–51) was particularly abundant during the Devonian. The largest trilobites ever found lived during the Devonian, one giant reaching a length of over 70 centimeters. Decline in trilobite populations and diversity became severe after the Devonian. By Late Permian, this great group of arthropods, which had persisted over a span of 350 million years, became extinct.

Members of the Phylum Echinodermata, including crinoids, blastoids, starfish, and echinoids, were abundantly represented among the shallow marine invertebrate faunas of the Late Paleozoic. The crinoids and blastoids were particularly profuse during the Mississippian (Figs. 11–52 and 11–53). Indeed, some Mississippian limestones are composed almost entirely of crinoidal skeletal debris (Fig. 11–54). The period has been nicknamed the "Age of Crinoids." Although one group of crinoids survived the hard times at the end of the Paleozoic and gave rise to those still living today, most died out before the end of the Permian. The

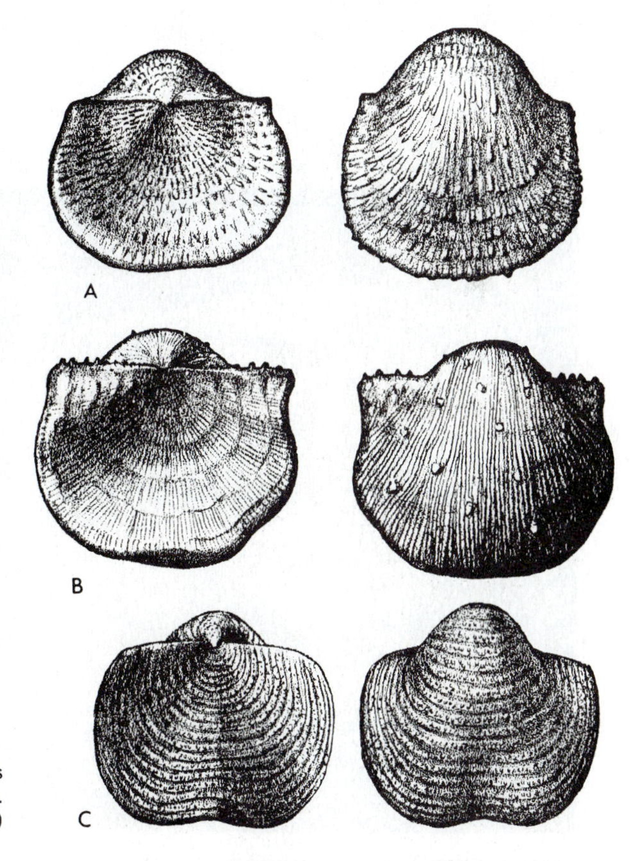

Figure 11–49 The Pennsylvanian productid brachiopods *Juresania* (A), *Linoproductus* (B), and *Echinoconchus* (C). (From Guidebook series 8, Illinois Geological Survey, 1970.)

Figure 11–50 The Devonian goniatite cephalopod *Tornoceras* from the Hamilton Group of western New York.

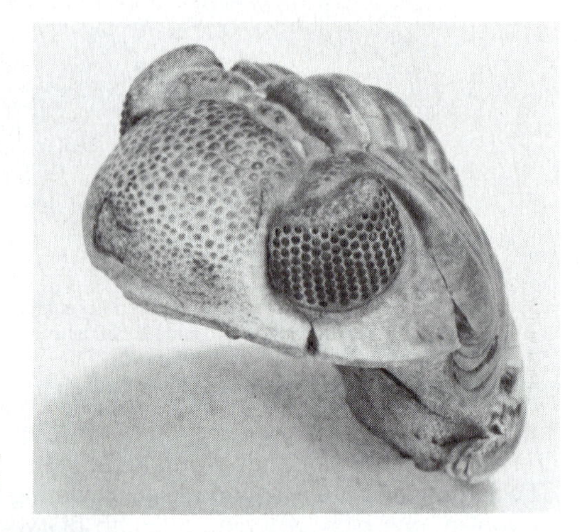

Figure 11–51 The Devonian trilobite *Phacops rana* from Devonian strata in Ohio. (Photograph courtesy of Wards Natural Science Establishment, Rochester, N.Y.)

Figure 11–52 Limestone slab containing exceptionally well-preserved Mississippian crinoids. (Photograph courtesy of Wards Natural Science Establishment, Rochester, N.Y.)

Figure 11–53 Crinoid "garden" living upon the Mississippian sea floor. (Courtesy of the Smithsonian Institution.)

blastoids (Figs. 11–55 and 11–56) became totally extinct before the end of the Paleozoic. During the post-Paleozoic eras, free-living echinoderms, such as starfish and echinoids, were the most ubiquitous representatives of the phylum.

The Vertebrates

The Late Paleozoic was a particularly eventful time span for the vertebrates. This was the interval during which animals evolved the limbs and other

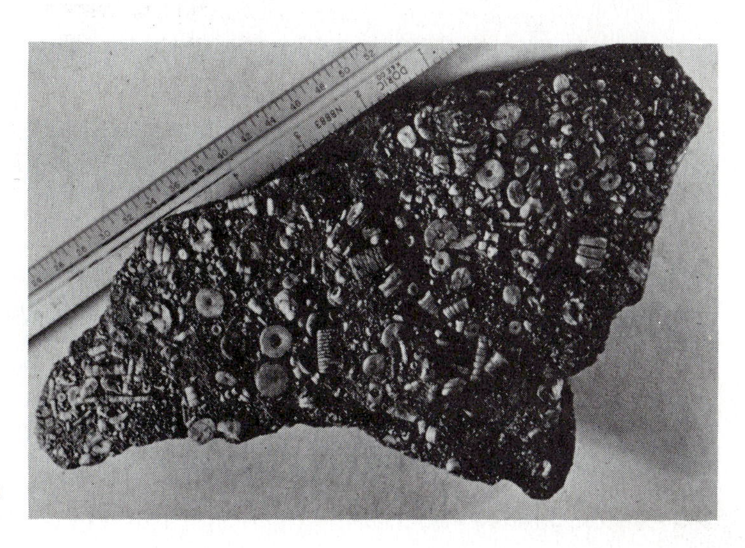

Figure 11–54 Crinoidal limestone, Burlington Formation (Mississippian).

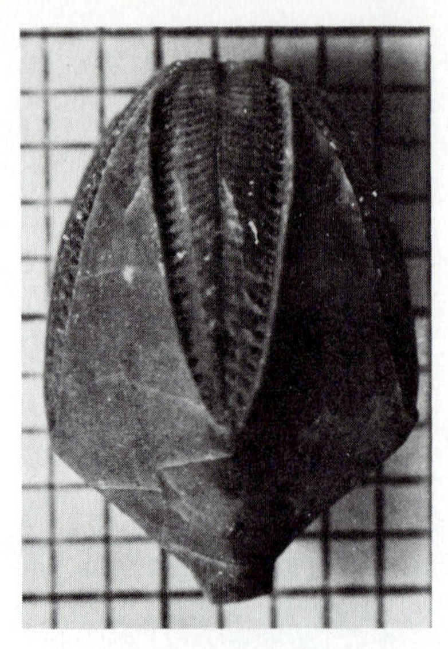

Figure 11–55 The Upper Mississippian blastoid *Pentremites* from the Paint Creek Formation of Illinois. (Magnification 4×.)

Figure 11–56 The blastoid *Calycoblastus tricavatus* from the Permian of Timor. This rare fossil is one of the largest blastoids ever discovered. The specimen shown here is 50 mm in length. (From D. B. Macurda. *J. Paleontol,* 46(1):97, 1972.)

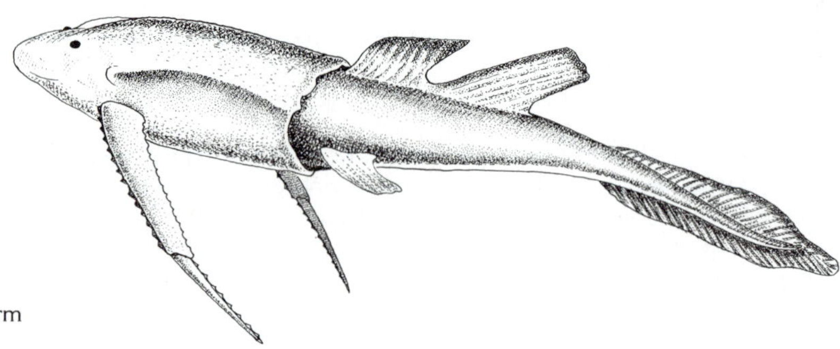

Figure 11–57 The Devonian placoderm *Bothriolepis.*

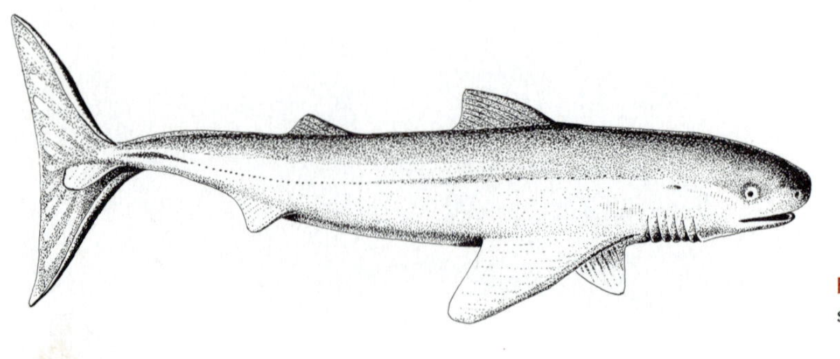

Figure 11–58 *Cladoselache,* a Devonian shark.

features that permitted them to invade the land. It was the time that two great classes of vertebrates arose, the amphibians and the reptiles.

Fishes

In the previous chapter, the evolutionary progression from the jawless ostracoderms to the first fishes with movable lower jaws was described. Those earliest jawed fishes, the acanthodians (see Fig. 10–82), first appear in nonmarine rocks of the Late Silurian. They increased in numbers during the Devonian and then slowly declined toward extinction in the Permian. Placoderms (see Fig. 10–83), another group of archaic jawed fishes, had a similar duration on earth. Among the placoderms were the savage predators *Dinichthyes.* These placoderms, some of which were over 9 meters in length, had huge jaws lined with scissors-sharp bone. Other placoderms were less formidable. *Bothriolepis* (Fig. 11–57), for example, seems to have been well adapted for gleaning food from the floors of lakes or streams, in somewhat the same manner as some ostracoderms had done in an earlier age.

During the Devonian there was a veritable piscine explosion as fishes began their dominance of the oceans as well as streams and lakes. Two important categories of fishes, the cartilaginous chondrichthyans and the bony osteichthyans, made their debut in the Devonian. Today the cartilaginous fishes are represented by sharks, rays, and skates. Among the better known of the Late Paleozoic sharks were species of *Cladoselache* (Fig. 11–58). Remains of this shark are frequently encountered in the Devonian shales that outcrop on the south shore of Lake Erie. During the Late Carboniferous, a group represented by *Xenacanthus* (Fig. 11–59) managed to penetrate the freshwater environment. A third group of Paleozoic cartilaginous fishes were the bradyodonts. These had flattened bodies like modern rays and blunt, rounded teeth for crushing shellfish. Apparently, modern sharks arose from cladoselachian ancestors but retained their archaic traits until the Jurassic.

Because of the role of bony fishes in the evolution of tetrapods (four-legged animals) and because they are the most numerous, varied, and successful of all aquatic vertebrates, their evolution is of particular importance. Bony fishes may be divided into two categories, namely the familiar "ray-fin," or **Actinopterygii,** and the "lobe-fins," or **Choanichthyes.**

As implied by their name, ray-fin fishes lack a muscular base to their paired fins, which are thin structures supported by radiating bony rays. Unlike the Choanichthyes, they do not possess paired nasal passages that open into the throat. The ray-fins began their evolution in Devonian lakes and streams and quickly expanded into the marine realm. They became the dominant fishes of the modern world. The more primitive Devonian bony fishes are well represented by the genus *Cheiro-*

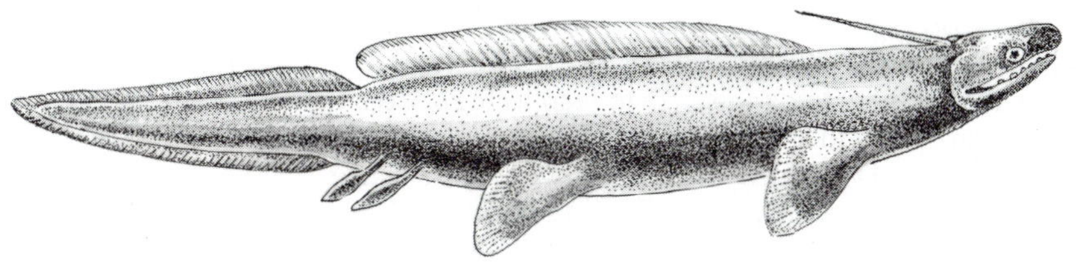

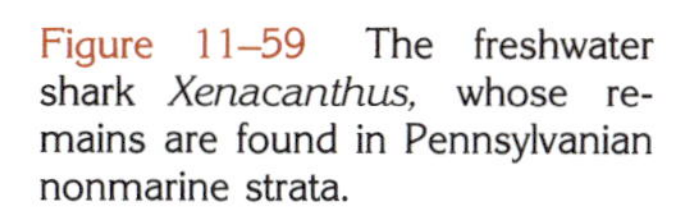

Figure 11–59 The freshwater shark *Xenacanthus,* whose remains are found in Pennsylvanian nonmarine strata.

Figure 11–60 The ancestral bony fish *Cheirolepis* from the Devonian.

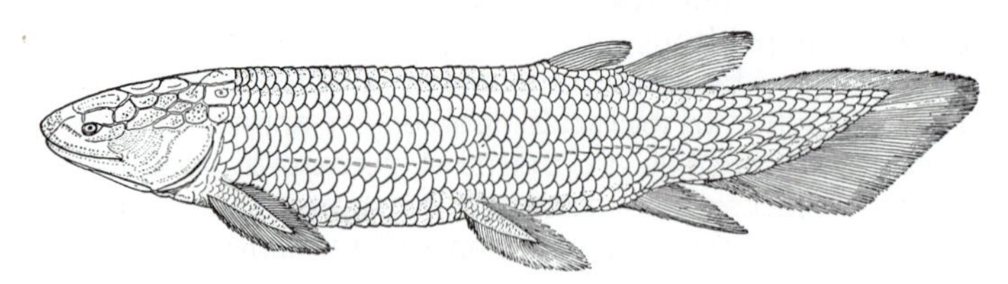

Figure 11–61 *Dipterus,* a Devonian lungfish.

lepis (Fig. 11–60). From such fishes as these evolved the more advanced bony fishes during the Mesozoic and Cenozoic.

The second category of bony fishes, the Choanichthyes, take their name from the Greek word *choana* meaning "internal nostril." Choanichthyids have a pair of openings in the roof of the mouth that lead to clearly visible external nostrils. Such fish were able to rise to the surface and take in air through their nostrils. Choanichthyans also possessed sturdy muscular fins and lungs. Lungs and fins do not seem to go together in modern fishes, but in Late Paleozoic fishes the combination was not uncommon. Studies of living choanichthyans indicate that lungs probably began their evolution as saclike bodies developed on the ventral side of the esophagus and then became enlarged and improved for the extraction of oxygen. (In modern fishes, the lung has been converted to a swim bladder, which aids in hydrostatic balance.)

Two major groups of lungfishes lived during the Devonian. They are designated the **Dipnoi** and **Crossopterygii.** The Dipnoi, represented in the Devonian by *Dipterus* (Fig. 11–61), were not on the evolutionary track that was to lead to tetrapods. They are, nevertheless, an interesting group, which includes the living freshwater lungfish of Australia, Africa, and South America. Their restricted presence south of the equator suggests that Gondwanaland was the probable center of dispersal for dipnoans. Dipnoi means "double breather." The name was suggested by the observation that living species are able to breathe by means of lungs during dry seasons. At such difficult times, they burrow into the mud before the water is gone. When the lake or stream is dry, they survive by using their accessory lungs, and when the waters return they switch to gill respiration.

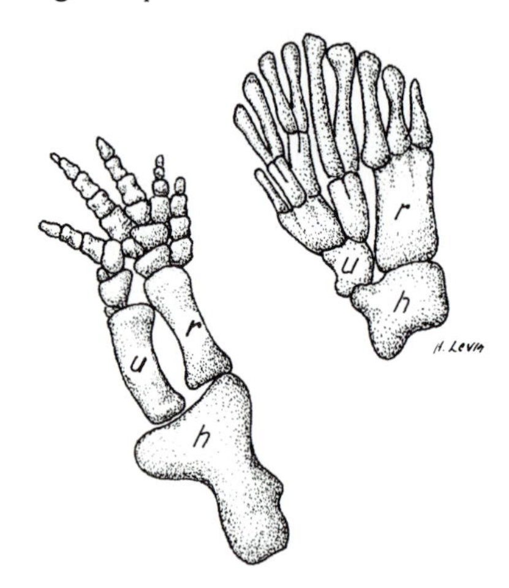

Figure 11–62 Comparison of the limb bones of a crossopterygian fish (upper right) and an early amphibian. (From H. L. Levin. *Life Through Time.* Dubuque, William C. Brown, 1975.)

Because of the arrangement of bones in their muscular fins (Fig. 11–62), the pattern of skull elements (Fig. 11–63), and the structure of their teeth (Fig. 11–64), the fossil Crossopterygii are considered the ancestors of the amphibians. A rather advanced fish that exemplifies Devonian crossopterygians is *Eusthenopteron* (Fig. 11–65). In this genus, the paired fins were short and muscular. Internally, a single basal limb bone, the humerus (or femur for the pelvic structure), articulated with the girdle bones and was followed by two bones, the ulna and radius, for the pectoral fins, and the tibia and fibula for the posterior fins.

Of course, the robust skeleton and sturdy limbs of the crossopterygians did not evolve because fishes had miraculously taken on a desire for life on land. These were adaptations for moving to water and remaining in water during periods of drought. The air-gulping fishes found land a hostile environment and occasionally dragged themselves onto the land only as a means of reaching another pond or fresher body of water.

During the Devonian, two distinct branches of crossopterygians had evolved. One of these led ultimately to amphibians, and the other led to salt water fishes called coelacanths. Coelacanths were thought to have undergone extinction during the late Cretaceous, but their survival down to the present day has been documented by several catches of the coelacanth *Latimeria* (Fig. 11–66) near Madagascar.

Amphibians

It required tens of millions of generations to convert the crossopterygian fishes into animals that could live comfortably on solid ground (Fig. 11–67). Even so, the conversion was not complete, for amphibians continued to return to water to lay their fishlike naked eggs. From these eggs came fishlike larvae, which, like fish, used gills for respiration.

A number of changes accompanied the shift to land dwelling. A three-chambered heart devel-

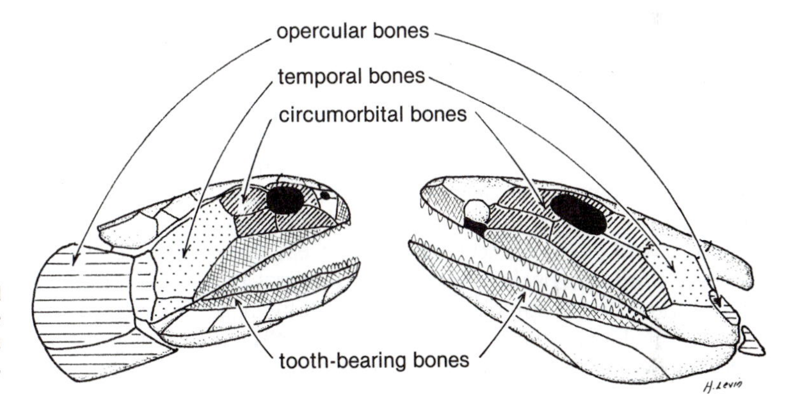

Figure 11–63 Comparison of skulls and lower jaws of a crossopterygian (left) and the Devonian amphibian *Ichthyostega*. (From H. L. Levin. *Life Through Time*. Dubuque, William C. Brown, 1975.)

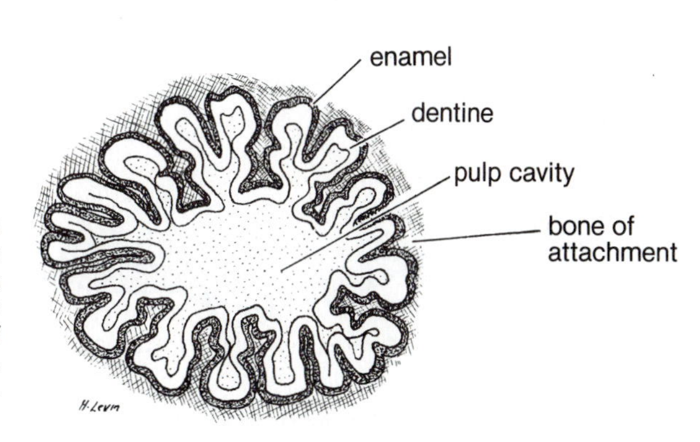

Figure 11–64 Cross section of a crossopterygian tooth clearly exhibits the distinctive pattern of infolded enamel. This same characteristic is a characteristic of the teeth of the amphibian descendents of crossopterygian fishes. (From H. L. Levin. *Life Through Time*. Dubuque, William C. Brown, 1975.)

Figure 11–65 *Eusthenopteron,* a Devonian crossopterygian fish that had evolved in the direction of early amphibians.

oped to route the blood more efficiently to and from the now more efficient lungs. The limb and girdle bones were modified to overcome the constant drag of gravity and better hold the body above the ground. The spinal column, a simple structure in fishes, was transformed into a sturdy but flexible bridge of interlocking elements. In order to improve hearing in gaseous rather than fluid surroundings, the old hyomandibular bone, used in fishes to prop the braincase and upper jaw together, was pressed into service as an ear ossicle—the **stapes.** The fish spiracle (a vestigial gill slit) became the amphibian eustachian tube and middle ear. To complete the auditory apparatus, a tympanic membrane ("ear drum") was developed across a prominent notch in the rear part of the amphibian skull.

The fossil record for amphibians begins with a group called **ichthyostegids** (Fig. 11–68). As suggested by their name, these creatures retained many features of their piscine ancestors. The amphibians that followed the ichthyostegids fall mostly within a group collectively termed the **labyrinthodonts.** The labyrinthic wrinkling and folding of tooth enamel provided the inspiration for the labyrinthodont name (see Fig. 11–64). During the Carboniferous, large numbers of labyrinthodonts wallowed in swamps and streams, eating insects, fish, and one another. A labyrinthodont that exemplifies the culmination of their lineage is *Eryops* (Fig. 11–69). Eryops was a bulky, inelegant creature with the flattish skull so typical of labyrinthodonts and bony nodules in the skin for protection against some of its more vicious contemporaries. The labyrinthodonts declined during the Permian, and only a relatively few survived into the Triassic.

Figure 11–66 *Latimeria,* a surviving coelacanth living in the ocean near Madagascar and southeast Africa. (From H. L. Levin. *Life Through Time.* Dubuque, William C. Brown, 1975.)

Reptiles

The evolutionary advance from fish to amphibians was no trivial biological achievement, yet the Late Paleozoic was the time of another equally significant event. Among the evolving land vertebrates were some that had developed a way to reproduce without returning to the water. They accomplished this extraordinary feat by evolving enclosed eggs in which the embryonic animal was allowed to pass through larval and other developmental stages before being hatched in an essentially adult form. This enclosed egg liberated the

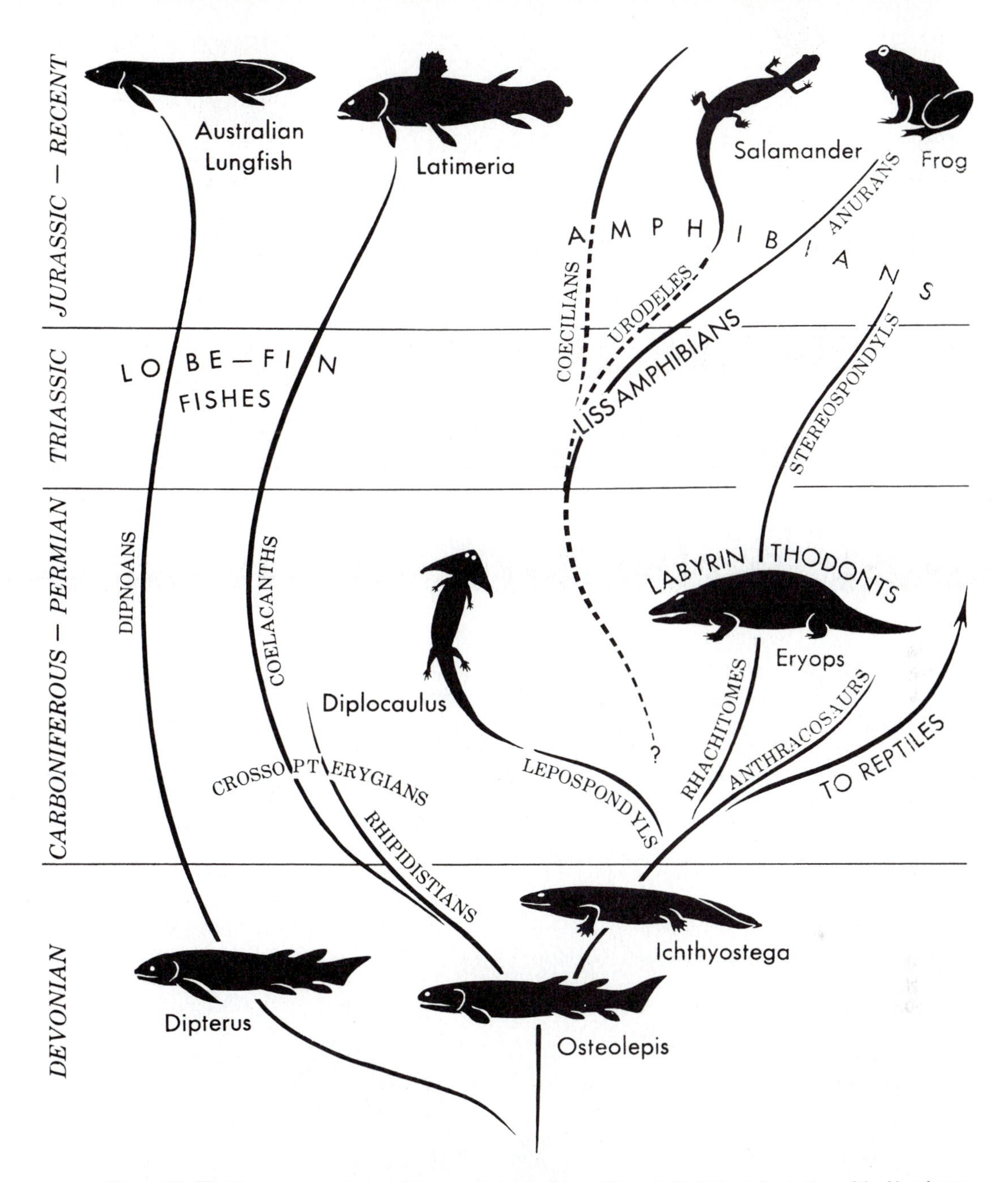

Figure 11–67 The evolution of amphibians and lobe-fin fishes. (From E. H. Colbert. *Evolution of the Vertebrates.* New York, John Wiley & Sons, 1969. With permisson of the author, artist Lois Darling, and Publisher.)

tetrapods from their reliance on water bodies and has been hailed as a major milestone in the history of vertebrates. The animals that first evolved the so-called amniotic egg were amphibians. Evolutionary processes had provided them with a method for protecting developing young from predation and desiccation.

The most ancient remains of reptiles are Late Carboniferous in age and poor in preservation. The fossils are representatives of a group of transitional reptiles called **cotylosaurs.** The preservation of cotylosaurs is somewhat better in Permian sediments of Texas, where the representative form *Seymouria* (Fig. 11–70) was discovered. From cotylosaurian stock, more advanced, large and small reptiles, carnivores, and herbivores evolved

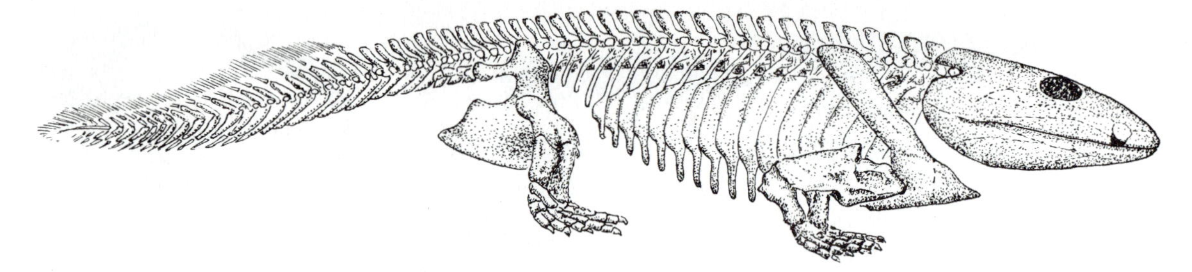

Figure 11–68 The skeleton of *Ichthyostega* still retains the fishlike form of its crossopterygian ancestors. (From H. L. Levin. *Life Through Time.* Dubuque, William C. Brown Co., 1975.)

Figure 11–69 The large Permian amphibian *Eryops.* (Length is about 2 meters.)

Figure 11–70 The primitive reptile *Seymouria,* sometimes called the "stem reptile" because it possessed skeletal characteristics transitional between amphibians and reptiles.

Figure 11–71 Permian reptiles. The sailback reptiles with the larger skulls and teeth are *Dimetrodon. Edaphosaurus,* a herbivorous form, is just right of center. The smaller lizard-like reptiles are *Casea.* (Courtesy of Chicago Field Museum of Natural History; painting by C. R. Knight.)

and came to dominate Permian landscapes. The most spectacular of these reptiles were the **pelycosaurs,** several species of which sported erect "sails" supported by rodlike extensions of the vertebrae. The pelycosaurs were a varied group. Some, like *Edaphosaurus,* were plant eaters, whereas others, as indicated by their great jaws and sharp teeth, ate flesh. *Dimetrodon* (Figs. 11–71 and 11–72) was one such predator. There have been many attempts to explain the function of the pelycosaurian sail. The most reasonable explanation is that the sail-fin acted in temperature regulation by serving sometimes as a collector of solar heat and at other times as a radiator giving off surplus heat. Because pelycosaurs are considered the group from which mammal-like reptiles arose, it is especially interesting that they may have been attempting some sort of body temperature control.

To humans, the mammal-like reptiles are among the most fascinating of fossil vertebrates. These creatures, collectively known as **therapsids,** were widely dispersed during Permian and Triassic time. There is a fabulous record of these reptiles and their contemporaries in the Karoo Basin of South Africa and more modest discoveries in Russia, South America, Australia, India, and Antarctica. Therapsids were predominantly small to moderate-sized animals that displayed at least the beginnings of several mammalian skeletal traits. There were fewer bones in the skull than in reptile skulls generally, and there was a mammal-like enlargement of the lower jaw bone (dentary) at the expense of more posterior elements of the jaw. A double ball and socket articulation had evolved between the skull and neck. Teeth showed a primitive but distinct differentiation into incisors, canines, and cheek teeth. The limbs were swung more directly into vertical alignment beneath the body, and the ribs were reduced in the neck and lumbar region for greater overall flexibility. These features are well developed in the Permian therapsid *Cynognathus* (Fig. 11–73). The therapsids continued from the Permian through the Triassic but became extinct early in the Jurassic. However, before they died out completely, they gave origin to the early mammals.

Extinctions

The fossil record indicates that there has been a persistent increase in the numbers of different species through geologic time. It also shows that this trend has been upset or even reversed at times as a result of environmental adversity. Extinctions of large numbers of previously successful animals and plants may have resulted from their inability to adapt to changing conditions. Early geologists were able to identify such times of biologic crisis and used extinctions to define further the termination of the Paleozoic and Mesozoic Eras.

As discussed earlier in this chapter, there were drastic geographic changes as the Paleozoic drew to a close. Pangaea was experiencing mountain building and regional warping, there was violent volcanism in several regions of the earth, and the seas had withdrawn from the extensive tracts they once covered. Climates became more severe, and

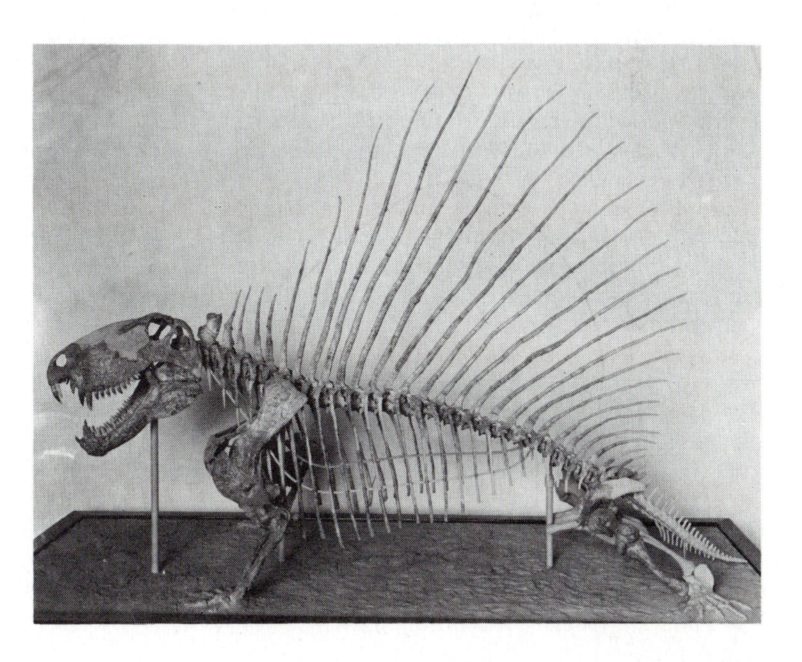

Figure 11–72 Mounted skeleton of the Permian "sail-reptile" *Dimetrodon gigas.* (The tail was actually somewhat longer.) (Courtesy of the Smithsonian Institution.)

Figure 11–73 Mammal-like reptiles. The scene depicts three carnivorous forms *(Cynognathus)* about to attack a plant-eating reptile *(Kannemeyeria)*. (Courtesy of Chicago Field Museum of Natural History; painting by C. R. Knight.)

a protracted glacial age occurred in the south. There was evidence of marked seasonal changes and aridity in many areas of the globe. As one or another group of animals gradually diminished in response to these changes, they contributed to the demise of others farther along in the food chain. Then, as now, the balance in the biologic environment could be easily upset, causing far-reaching waves of extinction.

Among the invertebrate groups that became extinct near the end of the Paleozoic were the trilobites, eurypterids, rugose and tabulate corals, productid brachiopods, many families of bryozoans, and blastoids. Entire families of other invertebrate taxa were lost as well. The archaic jawed fishes (acanthodians) became extinct during the Permian, whereas amphibians, cotylosaurs, and several groups of plants declined conspicuously. The misfortunes of the invertebrates may have been associated with the drastic reduction of the inland seas, whereas the amphibians may have suffered in the competition with evolving reptiles. Among those reptiles were a few hardy groups that were able to meet the new challenges and become the founders of the reptilian communities that were to dominate the Mesozoic terrestrial scene.

CLIMATES OF THE LATE PALEOZOIC

The main climatic zones of the Late Paleozoic tended to parallel latitudinal lines just as they do today. Of course, the continents were located differently, so that the south pole was in South Africa and the north pole was over open ocean. The Late Paleozoic equator trended northeastward across Canada and southward across Europe (Fig. 11–74). Coal beds, evaporites, coral reefs, and dune-deposited red beds developed within 40° of the paleoequator. Today, we cannot help but be startled to find the fossils of tropical amphibians and plants at locations within the arctic circle. As already noted, cooler climates prevailed in Gondwanaland because of its Late Paleozoic proximity to the south pole, and by Permian time the southern continents were in the grip of a major ice age. To the north, the apparent aridity of large regions

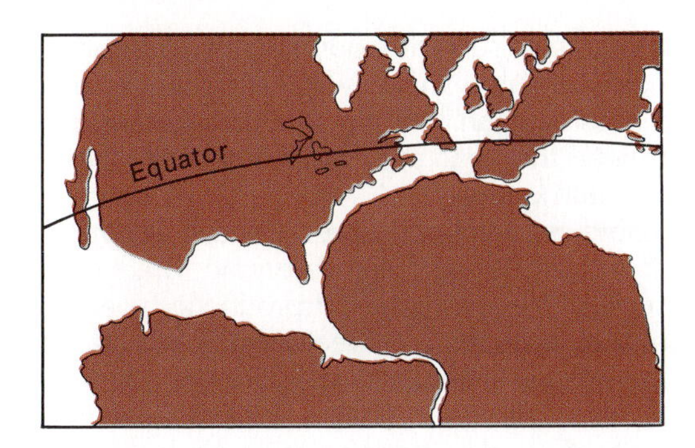

Figure 11–74 Approximate relationship of continents to the equator during the Carboniferous.

was influential in the evolution and extinction of terrestrial animals and plants.

MINERAL PRODUCTS OF THE LATE PALEOZOIC

Although a great variety of mineral deposits were formed during the Late Paleozoic, the fossil fuels (coal, oil, and gas) are particularly significant. Coal occurs in all post-Devonian systems. In northern hemisphere continents, it is particularly characteristic of the Late Carboniferous. Thick deposits of Pennsylvanian coal occur in the Appalachians, the Illinois Basin, and the industrial heartland of Europe. Single sequences of Pennsylvanian strata may include several coal beds, as in West Virginia, where 117 different layers have been named. In the so-called anthracite district of western Pennsylvania, orogenic compression has partially metamorphosed coal into an exceptionally high-carbon, low-volatile variety that is prized for its industrial uses. Permian coal seams are found in China, Russia, India, South Africa, and Australia. The coal industry in North America, which had deteriorated for the last few decades, is currently experiencing an upsurge as a consequence of decreasing supplies of petroleum and natural gas.

Commercial quantities of oil and gas are frequently found in Late Paleozoic strata. Devonian reefs within the Williston Basin of Alberta and Montana have been exceptionally productive reservoir rocks for petroleum. Devonian petroleum has also been produced in the Appalachians. Indeed, in 1859, the first American oil well was dug into a Devonian sandstone. (Oil was struck at a depth of only 20 meters.) Carboniferous formations of the Rocky Mountains, midcontinent, and Appalachians also contain oil reservoirs. However, wells drilled into reefs and sandstones of the Permian Basin of western Texas have yielded the greatest amounts of oil from Late Paleozoic formations of the western United States. Oil trapped in Late Paleozoic strata beneath the North Sea is now being actively utilized to supply the needs of Europe.

Arid, warm climatic conditions, particularly in northern continents during the Late Paleozoic, provided a suitable environment for deposition of sodium and potassium salts. In Late Permian time, enormous amounts of phosphates were also deposited as part of the Phosphoria Formation, which is exposed in Montana, Idaho, Utah, and Wyoming. The Phosphoria shales are extensively quarried for phosphate, which is sold as an important plant food.

Mountain building, with its attendant intrusive and volcanic igneous activity, is nearly always accompanied by the emplacement of metallic ores. The Hercynian and Allegheny Orogenies of the Late Paleozoic generated ores of tin, copper, silver, gold, zinc, lead, and platinum. Deposits of all the precious metals, as well as copper, zinc, and lead, are found in the Urals of Russia and in China, Japan, Burma, and Malaya. Tin, tungsten, bismuth, and gold are mined in Australia and New Zealand. It is self-evident that Late Paleozoic rock sequences, with their stores of both metallic and nonmetallic economic minerals and their content of fossil fuels, are vitally important to the welfare of modern civilizations.

Summary

The continental interior of North America experienced five major marine transgressions, in the course of which the Kaskaskia and Absaroka sequences of sedimentary rocks were deposited. The Kaskaskia rocks were predominantly fossiliferous limestones, with sandstones and shales increasing in volume toward the eastern and southern geosynclinal belts. As the Kaskaskia Seas withdrew, deltaic and fluvial deposits spread across the old sea floor. The regression resulted in a great regional unconformity that marks the boundary between Mississippian and Pennsylvanian systems in North America. The most distinctive feature of the Pennsylvanian sediments is their cyclic nature. In the midcontinent, they consist of alternate marine and nonmarine groupings. To the east, the Pennsylvanian sediments were largely terrestrial, whereas marine deposition prevailed in the western part of the craton. Near the end of the Late Paleozoic, the epicontinental seas gradually withdrew, and evaporite and red bed sequences developed in the Permian basins of New Mexico and western Texas.

The unrest that marked the Appalachian-Caledonian geosynclinal belt during the Early Paleozoic continued into the Devonian with the Acadian Orogeny and reached its climax in the great Allegheny Orogeny. Those events are thought to have been manifestations of the closing of the Proto-Atlantic Ocean as Pangaea II was assembled. The compressive forces of continental collision raised great mountain ranges along the eastern and southern margins of North America.

In the west, crustal disturbances during the Mississippian uplifted a long and narrow range of mountains from Arizona to Idaho. Thick deposits of sand and gravel attest to the former existence of these ranges. There was further activity during the Pennsylvanian when uplands still visible were formed in Texas, Arkansas, and Oklahoma. A chain of mountains (the "Colorado Mountains") also developed across Colorado. These Colorado Mountains were the source of sandstones that today form the famous "red rocks" of the Garden of the Gods along the Front Range of the Rockies.

The Old Red Continent was the dominant feature of Europe at the beginning of the Late Paleozoic. For a time, relative stability prevailed in Europe, and then much of the continent was thrown into an episode of folding, volcanism, and intrusion as it collided with Gondwana. These disturbances are included in the Hercynian Orogeny. The collision of Gondwana with Laurussia was not only the cause of the Hercynian Orogeny in central Europe but also was the cause of the Allegheny Orogeny along the eastern and southern margin of North America.

Gondwanaland was ringed by geosynclines during the Late Paleozoic, and the Andean and Tasman segments experienced Late Paleozoic orogenesis. The central regions were occasionally covered with inland seas in which were deposited fossiliferous limestones and shales. During the Permian, when the northern land masses were relatively warm and sometimes arid, Gondwanaland appears to have been cooler. The south pole for the Permian was located in what is now South Africa, and all of the now separated Gondwanaland continents exhibit the scars of widespread glaciations.

The marine faunas of the Late Paleozoic included representatives of most of the phyla that had existed previously. Some lesser taxa within phyla had become extinct, and a few new groups appeared. The evolution of ammonoid cephalopods was one such new appearance. Blastoids and crinoids reached their evolutionary zenith during the Mississippian. During the Pennsylvanian and Permian, the tiny protozoan fusulinids became both numerous and diverse. Trilobites and nautiloid cephalopods experienced a decline after their heyday in the Early Paleozoic. Brachiopods, especially the spiriferid and productid types, were common, whereas great reefs of rugose corals and bryozoa expanded across the floors of clear inland seas. Many of these marine invertebrates were profoundly affected by a reduction in the extent of epicontinental seas that took place near the end of the Paleozoic. A wave of extinction swept across the marine environment. All of the rugose corals, two orders of bryozoans, many taxa of attached echinoderms, and entire families of other invertebrates terminated their evolutionary history.

At the beginning of the Late Paleozoic, the lands were nearly barren of higher forms of vascular plants. Only the lowly psilophytes attempted to green the surface of the land. However, plants expanded rapidly during the Devonian and subsequent Paleozoic periods. Lycopsid scale trees, sphenopsids, and true ferns grew in profusion. Seed ferns began their colonization and were joined in Permian time by the conifers and ginkgoes. Other seed plants in more temperate southern hemisphere lands developed into a flora dominated by *Glossopteris.*

The most interesting story of all, however, is that revealed by the study of fossil vertebrates. The Late Paleozoic witnessed the rise of the higher fishes—the chondrichthyans and osteichthyans. From the latter arose a special group of bony fishes called the crossopterygians. These fishes could breathe air by means of accessory lungs and possessed muscular fins that provided short-distance overland locomotion. The bones within the fins of crossopterygians resembled those of primitive amphibians. Other skeletal traits clearly suggest that these Devonian fishes were the ancestors of the first amphibians—the ichthyostegids. From a start provided by ichthyostegids, amphibians called labyrinthodonts underwent a successful adaptive radiation that lasted until the close of the Triassic. Long before their demise, however, they provided a lineage from which evolved the reptiles.

The transition from amphibian to reptile entailed a significant breakthrough in evolution. It involved the development of the amniotic egg, a biologic device that liberated land vertebrates from the need to return to water bodies in order to reproduce. The first known reptiles are Pennsylvanian in age. The Late Paleozoic reptiles underwent an elaborate evolutionary radiation that initiated several of the major taxa. Permian reptiles included the pelycosaurs (represented by sail-back reptiles) and the enormously important therapsids (the mammal-like reptiles).

The Late Paleozoic was indeed an eventful span of geologic time. It witnessed the appearance of two new classes of vertebrates, the formation of great mountain ranges, the emplacement of important ore deposits, the formation of extensive layers of coal, and a major age of glaciation. The tempo of events, however, was to continue unabated into the Mesozoic Era.

Questions for Review

1. How may the orogenies that produced the southern Appalachian and Ouachita Mountains be related to modern concepts of plate tectonics?
2. What relationship might the Allegheny Orogeny have had to the Hercynian Orogeny of Europe?
3. Why is it logical to divide the Late Paleozoic of the North American craton into two sedimentary sequences—the Kaskaskia and Absaroka?
4. Where was the Old Red Continent located? When did it come into existence? What sort of sediments were deposited on it?
5. What is the paleoenvironmental significance of the extensive red sandstones, salt, and gypsum deposits of the Permian? Where are the more extensive deposits of Permian evaporites located?
6. What is the geologic evidence for the occurrence of a geographically extensive episode of continental glaciation in the southern hemisphere near the end of the Paleozoic?
7. What is a cyclothem? What may have been the cause of this type of cyclic deposition during the Pennsylvanian?
8. In terms of location, climatic preference, and reproduction, how did *Lepidodendron* differ from *Glossopteris*?
9. When and where were the Colorado Mountains ("ancestral Rockies") developed?
10. What is the evolutionary importance of crossopterygians and ichthyostegids in the history of vertebrates?
11. Geologists have suggested that erosion on land by water and wind probably proceeded more rapidly during the Early Paleozoic than during the remainder of the era. Can you suggest why this might be a valid inference?
12. What geologic interpretation is given to great "clastic wedges" such as the Catskill Delta and the Pocono Group?
13. From what previous anatomic structure was the vertebrate jaw derived?
14. List three anatomic traits of therapsids that indicate they were on the main line of evolution toward mammals.
15. Name three metallic and three nonmetallic mineral resources that are extracted from Late Paleozoic rocks.

Terms to Remember

Acadian Orogeny
Actinopterygii
Allegheny Orogeny
ammonite
ammonoid
amphibolite
brachiopod
Catskill Delta
ceratite
Chattanooga Shale
Colorado Mountains
Crossopterygii
cyclothem
Dipnoi
Dunkard Series
fenestellid bryozoan
fusulinid
goniatite
Hercynian Orogeny
ichthyostegid
Kaskaskia Sea
labyrinthodonts
lycopsid
Monongahela Series
mollusk
novaculite
Oklahoma Mountains
Old Red Sandstone
pelecypod
pelycosaur
Phosphoria Formation
Pocono Group
psilopod
sphenopsid
stapes
therapsid
trilobite
Zechstein Sea

Supplemental Readings and References

Bird, J. M., and Dewey, J. F. 1970. Lithosphere plate–continental margin tectonics and the evolution of the Appalachian orogeny. *Bull. Geol. Soc. Amer.* 81:1031–1060.

Hamilton, W., and Krinsley, D. 1967. Upper Paleozoic glacial deposits of South Africa and Southern Australia. *Bull. Geol. Soc. Amer.* 82:1581.

Hatcher, R. D., Jr. 1972. Developmental model for the southern Appalachians. *Bull. Geol. Soc. Amer.* 83:2735–2760.

Kay, M. (ed.) 1969. North Atlantic—Geology and Continental Drift. *Amer. Assoc. Pet. Geol.,* Memoir 12.

Levin, H. L. 1976. *Life Through Time.* Dubuque, William C. Brown.

Pelletier, B. R. 1958. Pocono paleocurrents in Pennsylvania and Maryland. *Bull. Geol. Soc. Amer.* 69:1033.

Reed, H. H., and Watson, J. 1975. *Introduction to Geology,* Vol. 2, Part II, *Later Stages of Earth History.* New York. Wiley Interscience.

Rodgers, J. 1970. *The Tectonics of the Appalachians.* New York, Wiley Interscience.

Stahl, B. J. 1974. *Vertebrate History: Problems in Evolution.* New York, McGraw-Hill.

United States Geological Survey, 1979. *The Mississippian and Pennsylvanian (Carboniferous) Systems in the United States.* Professional Paper 1110.

The Painted Desert viewed from Tawa Point, Arizona. Most of the landscape is eroded from the Triassic Chinle Formation. (Photograph courtesy of the United States National Park Service.)

12

The Mesozoic Era

> The earth, from the time of the chalk to the present day, has been the theater of a series of changes as vast in their amount as they were slow in their progress. The area on which we stand has been first sea and then land for at least four alternations and has remained in each of these conditions for a period of great length.
>
> Thomas Huxley, 1868, *On a Piece of Chalk*

PRELUDE

The extinction of Paleozoic animals and plants that occurred at the close of the Permian Period did not go unnoticed by the founders of geology. The fossil evidence for this time of crisis provided a natural boundary that was used to mark the end of the Paleozoic System. Overlying younger strata contained different and generally more progressive organisms, yet not as modern as those that live today. Thus, it seemed appropriate to formulate a middle chapter in the history of life. They named that chapter the Mesozoic Era. In its turn, the Mesozoic also ended in a biologic crisis that is used to mark its separation from the youngest era of all—the Cenozoic.

The Mesozoic Era lasted an estimated 160 million years, ending approximately 65 million years ago. The three periods of the Mesozoic are unequal in duration. The Triassic lasted about 30 million years, the Jurassic 55 million years, and the Cretaceous about 75 million years.

During the lengthy span of the Mesozoic, the earth witnessed many changes. New families of plants and animals evolved and experienced often spectacular radiations. It was the era in which two new vertebrate classes, the birds and the mammals, first appeared. It was also a time in which the supercontinent Pangaea II, formed in the previous era by the joining of errant ancestral continents, was gradually dismembered. As the continents moved slowly apart, the oceanic rifts between them deepened and broadened. A process of fragmentation and drift had begun that would ultimately lead to the present physical geography of the planet.

THE DISMEMBERMENT OF PANGAEA II

Any discussion of the history of the Mesozoic Era must have Pangaea II as its starting point (see Fig. 11–2). The same forces that had drawn the plate segments together to form this great supercontinent were still in operation at the beginning of the Mesozoic; however, they now moved in different directions. In general, the dismemberment of Pangaea II seems to have occurred in four stages.

The first stage began in the Triassic, with rifting and volcanism along tensional fault systems. Events such as these often result when a region of the lithosphere is subjected to tensional stresses. In this instance, the tension very likely was associated with the separation of North America and Gondwana. As the rifting progressed, Mexico was decoupled from South America, and the eastern border of North America parted from the Moroccan bulge of Africa (Fig. 12–1). Oceanic basalts were added to the sea floor of the newly formed and gradually widening Atlantic Ocean. The largely tensional geologic structures, as well as the volcanic and clastic sedimentary rocks in the Triassic System of both the eastern United States and Morocco exhibit many striking similarities. These now widely separated regions were on opposite sides of the axis of spreading and were affected by similar forces and events. Temporarily, at least, Laurussia north of the Maritime Provinces remained intact. In addition, South America seems to have maintained its hold on Africa.

The second stage saw the beginning of the breakup of Gondwanaland. A great rift separated Antarctica from the southern ends of South America and Africa. This rift developed a branch that extended eastward from South Africa along what is now the eastern side of India. Great volumes of basaltic lavas poured from volcanoes located along these great rifts while the separated Gondwanaland segments began to move slowly northward, turning gently counterclockwise as they did so.

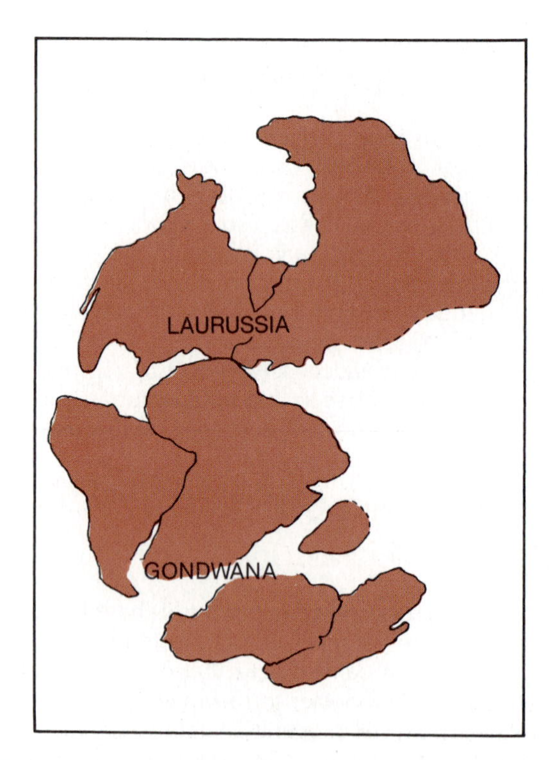

Figure 12–1 Conceptual paleogeographic map of the Triassic world about 180 million years ago, when the breakup of Pangaea II was beginning.

In stage three of the breakup of Pangaea II, the Atlantic rift began to extend itself northward, and clockwise rotation of Eurasia tended to close the eastern end of the Tethys Sea, a forerunner of the Mediterranean. By the end of Jurassic time, an incipient breach began to split South America away from Africa. The cleft apparently worked its way up from the south, creating a long seaway somewhat reminiscent of the present Red Sea. Australia and Antarctica remained intact, but India had moved well along on its long voyage to Laurasia. By Late Cretaceous, some 70 million years ago, South America had completely separated from Africa, and Greenland began to separate from Europe. However, northeasternmost North America still clung to Greenland and northern Europe.

The fourth stage in the dismemberment of Pangaea II was not a Mesozoic event. It occurred early in the Cenozoic, during which time the North Atlantic rift slowly penetrated northward until eventually the two great Laurasian continents were completely separated. Also during this final era of earth history, Antarctica and Australia parted company.

THE MESOZOIC HISTORY OF NORTH AMERICA

To the East and South

At the beginning of the Mesozoic, conditions in the eastern part of North America were much the same as they had been near the end of the Paleozoic. The rugged Appalachian ranges that had been raised during the Allegheny Orogeny were undergoing vigorous erosion. The coarse clastics derived from the uplands filled intermontane basins and other low areas during the Early and Middle Triassic. Then, during the Late Triassic, North America began to move away from Gondwanaland. As a possible result of the crustal stretching that accompanied the separation, as well as of gravitation adjustments associated with erosion of the mountains, a series of fault-bounded troughs developed along the eastern coast from Nova Scotia to North Carolina (Fig. 12–2). The downfaulted blocks received great quantities of poorly sorted, arkosic, red sandstones and shales;

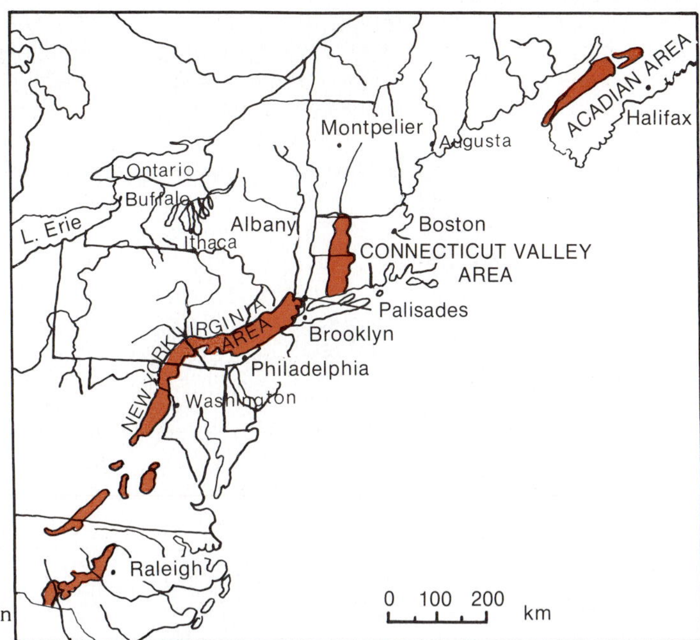

Figure 12–2 Outcrop areas of Triassic rocks in eastern North America.

these rocks are known as the **Newark Series.** The poor sorting of the coarser Newark clastics and their high content of relatively unweathered feldspar grains indicate transportation and deposition by streams flowing swiftly down from the granite highlands that bordered the fault basins (Fig. 12–3). Here and there, drainage in the basins became impounded, and lakes formed. The lake deposits contain the remains of freshwater crustaceans and fish. Ripple marks, mud cracks, raindrop impressions, and even the footprints of early members of the dinosaur line are frequently found on the lithified surfaces of Newark sediments. Lavas, steaming up along steeply dipping fault surfaces, flowed out upon the recently deposited sediment, and volcanoes spewed ash onto the surrounding hills and plains.

Three particularly extensive lava flows and an imposing sill are included within the Newark Group in the New Jersey-New York area. The exposed edge of a vertically jointed sill forms the Palisades of the Hudson River (Fig. 12–4). Radiometric dates obtained for the Palisades basalt indicate that it solidified about 200 million years ago. Although most of the Newark beds are Triassic, recent spore and pollen studies suggest that Newark sediments in some areas accumulated during the early Jurassic as well.

By Late Triassic, the fault block mountains that had been produced as a result of rifting of Laurussia had been severely reduced by erosion. The topography was further reduced during the Jurassic and Early Cretaceous, until only a broad, low-lying erosional surface remained. That old surface is called the Fall Line Surface because its profile can best be seen along the Fall Line, which separates more resistant rocks of the Piedmont from the softer sediments of the Atlantic Coastal Plain. Beneath the Fall Line Surface are Triassic faulted sediments and older crystalline rocks, and above it lie the more or less flat-lying Cretaceous sediments of the Atlantic Coastal Plain.

South of the old Appalachian-Ouachita geosynclinal belt, a new depositional province, the Gulf of Mexico, began to take form. The initial deposits were mostly evaporites and do not lend themselves to radiometric dating. However, analyses of pollen from a sample of salt taken from this sequence suggest an age of Late Triassic. Evaporites continued to be deposited well into the Jurassic, indicating aridity in the Gulf region. This interpretation is strengthened by paleomagnetic data that place the Gulf of Mexico between latitude 20° and 30° during the Jurassic. At times, the entire Gulf of Mexico must have served as a great evaporating pan, concentrating the waters of the Atlantic Ocean and precipitating salt and gypsum to thickness exceeding 1000 meters. Many geologists believe that Jurassic evaporite beds are the source of the salt domes of the Gulf Coast. Salt domes are economically important structures associated with the entrapment of petroleum. When compressed by a heavy load of overlying strata, salt tends to flow plastically. As it moves upward in the direction of

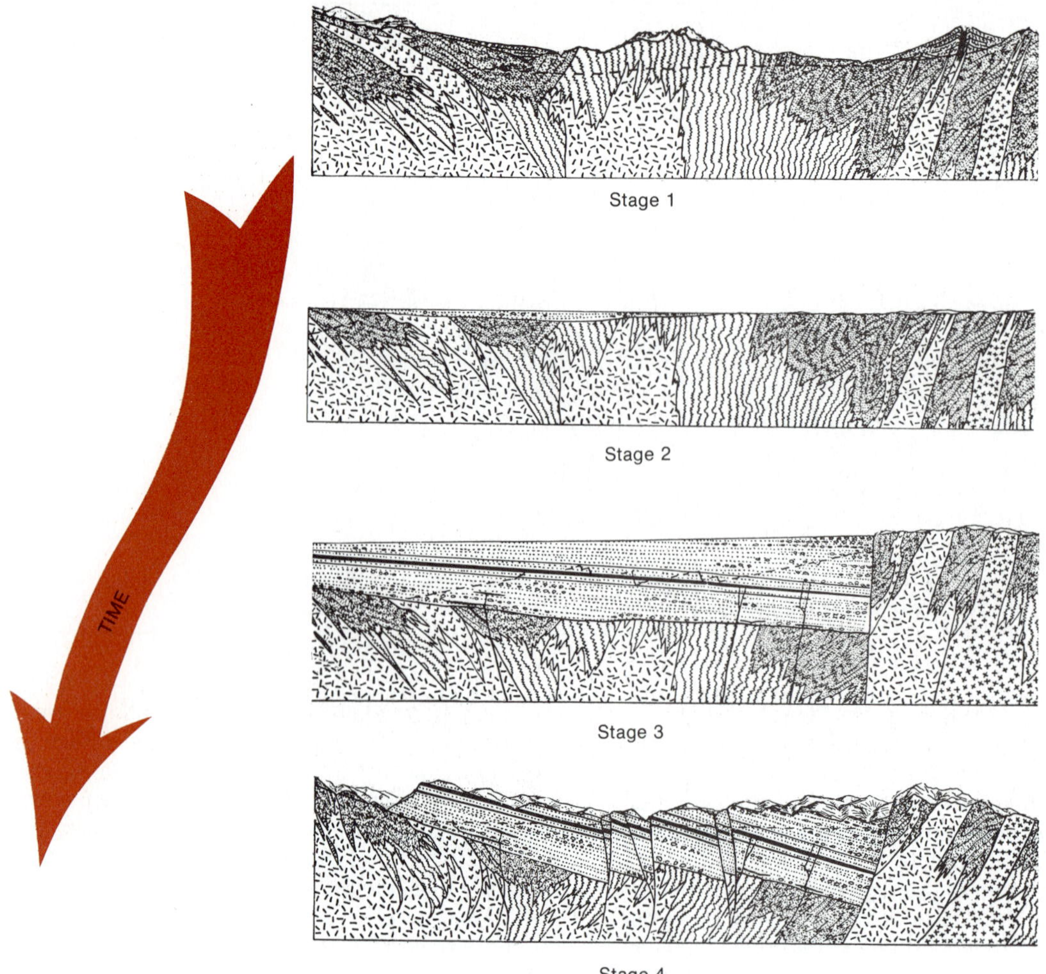

Figure 12–3 Four stages in the Triassic history of the Connecticut Valley. (Stage 1) Erosion of the complex structures developed during the Allegheny Orogeny. (Stage 2) Mountains have been eroded to a low plain, and Triassic sedimentation has begun. (Stage 3) Newark sediments and basaltic sills, flows, and dikes accumulate in troughs resulting from faulting. (Stage 4) In Late Triassic the area is broken into a complex of normal faults as part of the Palisades Orogeny. (Modified from the classic 1915 paper by J. Barrell titled Central Connecticut in the Geologic Past, *Conn. Geol. Nat. Hist. Bull.,* No. 23.)

lesser pressures, the salt bends and faults overlying strata. The permeable, often faulted beds that slope downward away from the salt, or are arched over the dome, afford excellent sites for petroleum accumulation (see Fig. 2–39).

Evaporative conditions abated later in the Jurassic, and several hundred meters of normal marine limestones, limy muds, shales, and sandstones accumulated in the alternately transgressing and regressing seas of the Gulf embayment (Fig. 12–5). These rocks and the evaporites beneath them are not exposed at the surface. They are deeply buried beneath a thick cover of Cretaceous and Cenozoic sediments. Were it not for the drilling activities of oil companies, the nature of these rocks could only be inferred from geophysical data.

The Cretaceous was a time of great marine inundations of land masses, and the Atlantic and Gulf coastal regions received their full share of flooding. Early in Cretaceous time, the Atlantic Coastal Plain, which had been experiencing erosion since the beginning of the Mesozoic, began to subside. At about the same time, the Appalachian belt that lay to the west was elevated. On the subsiding coastal plains, alternate layers of marine and deltaic terrestrial deposits accumulated and were gradually built into a great wedge of sediments that thickened seaward (Fig. 12–6). Today, the thinner,

Figure 12–4 Palisades of the Hudson River. (Photograph courtesy of Palisades Interstate Park Commission.)

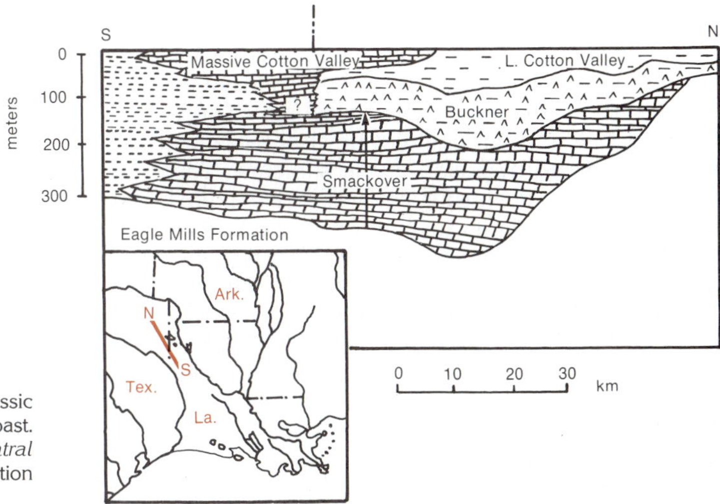

Figure 12–5 Cross-section of Early Upper Jurassic strata in the subsurface of the northern Gulf Coast. (From *Stratigraphic Atlas of North and Central America.* Shell Oil Company, Exploration Department.)

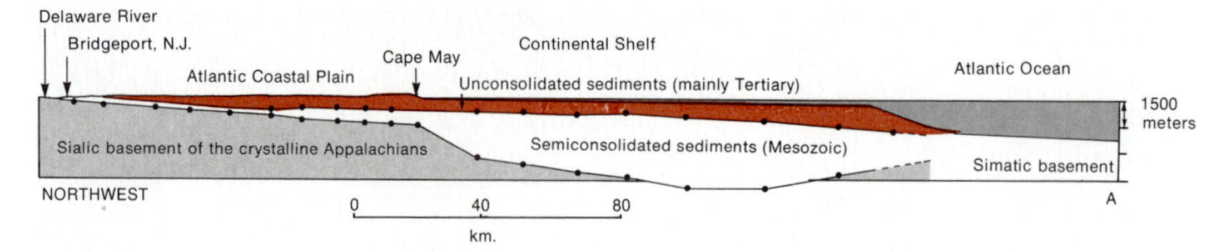

Figure 12–6 Section across the Atlantic Coastal Plain from the vicinity of the Fall Line to the edge of the continental shelf in east-central North Carolina. Data based on deep drilling. (From P. B. King. *The Tectonics of Middle North America.* Princeton, Princeton University Press, 1951, p. 165.)

eastern border of the wedge is narrowly exposed in New Jersey, Maryland, Virginia, and the Carolinas. But the greatest volume of Cretaceous sediments was deposited farther east, along the present continental shelf.

To the south, the Florida region was a shallow submarine bank during the Cretaceous. The *oldest* strata consist of limestones, but later in the period clastics were bought into the area by streams flowing from the southern Appalachians. Gradually, these source areas were worn down, and carbonate deposition prevailed.

The Cretaceous is noteworthy as a period of the Mesozoic in which carbonate reefs were particularly extensive. Among the invertebrates that contributed their skeletal substance to these reefs, a group of pelecypods called **rudists** were immensely important. In many Cretaceous reefs, the shells of these creatures (Fig. 12–7) form the basic framework of the reef. Because of their high porosity and permeability, rudistoid reefs serve as the reservoir formation for some of the world's greatest Cretaceous oil accumulations.

Following the withdrawal of Jurassic seas, the region west of Florida consisted of low-lying coastal plains. However, early in the Cretaceous, these relatively level lands were invaded by marine waters moving northward from the ancestral Gulf of Mexico. Nearshore deposits such as sandstones are overlain by the finer sediments that are characteristic of deeper water and provide clear evidence of the northward migration of the shoreline. The advance of the Cretaceous sea was not uniform. An extensive regression occurred near the end of the first half of the period; however, flooding resumed in Late Cretaceous time, when a wide seaway occupied a tract from the Gulf of Mexico to the Arctic Ocean (Fig. 12–8). Among the Upper Cretaceous formations deposited in this inland sea, chalk was particularly prevalent. Chalk is a white, fine-grained, soft variety of limestone that is composed largely of the microscopic calcareous platelets (called coccoliths) of golden brown algae. It is common among beds of Cretaceous age in many other parts of the world (Fig. 12–9). Indeed, the Cretaceous takes its name from *creta,* the Latin word for chalk.

To the West

As the eastern border of North America was experiencing the tensional effects of separation from Europe and Africa, compressional forces prevailed in the west. While the newly formed Atlantic Ocean was widening, North America moved westward, overriding the Pacific Plate. Thus deformation of western North America does have a relation-

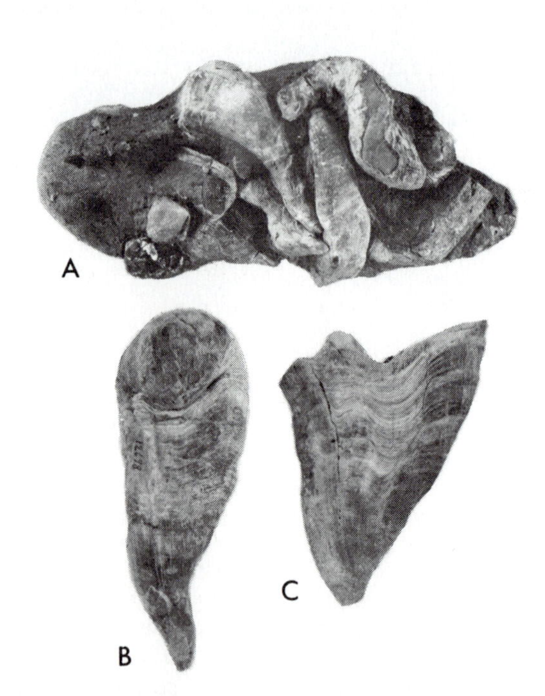

Figure 12–7 The Late Cretaceous rudist pelecypod *Coralliochama orcutti* from Baja, California. (A) Cluster of specimens from the reef deposit. (B and C) Two single specimens. (From L. Marincovich, Jr. *J. Paleontol.,* 49(1):212–223, 1975. Used with author's permission.)

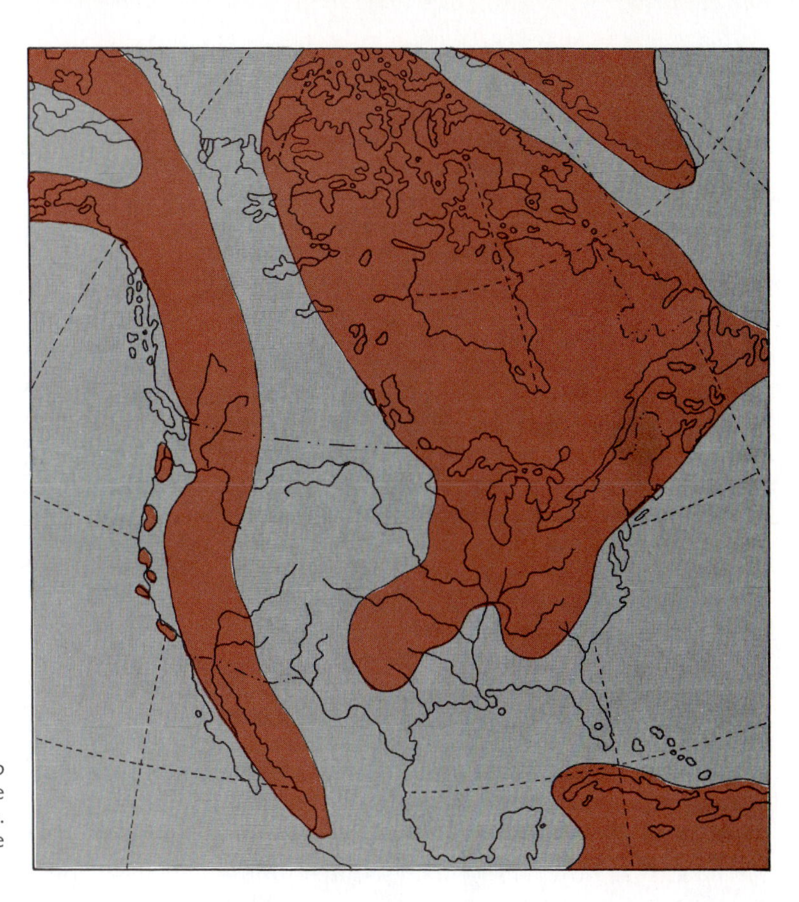

Figure 12–8 Paleogeographic map showing area of marine coverage during the maximum Late Cretaceous submergence. (Land areas are shown in brown; marine areas are shown in gray.)

ship to events in the east. It has been shown, for example, that the pace of tectonic activity in the North American Cordillera was most intense during the time when sea floor spreading was most rapid in the Atlantic.

During the Mesozoic, a steeply dipping subduction zone developed along the western margin of North America. The advancing Pacific Plate carried not only oceanic basalts and sea floor sediments to the subduction zone but microcontinents as well. The microcontinents have been incorporated into the Cordillera as allochthonous terrains. These terrains are distinctively coherent in age, lithology, and fossil content and are thought to be fragments of a once larger continent that existed somewhere in the Pacific Ocean Basin. Thus, the western border of North America grew, not only by accretion of geosynclinal materials but also by the incorporation of large chunks of continental crust formed elsewhere and conveyed to our shores by sea floor spreading. At times, some of these chunks were too much for the subduction zone to swallow. They choked the subduction zone and caused shifts in its orientation.

The Cordilleran Geosyncline as it existed during the Mesozoic can be divided into a western eugeosynclinal belt containing thick volcanic and siliceous deposits and a wide eastern miogeosynclinal tract adjacent to the more stable interior of the continent. Nearly 800 meters of clastic sediments and volcanics exposed in southwestern Nevada and southeastern California attest to the instability of the eugeosyncline. Geologists speculate that the belt may well have resembled the Indonesian Island Arc of today.

Triassic

The initial orogenic event of the Cordilleran region is difficult to date precisely, although evidence for deformation at or near the Permian-Triassic boundary can be found from Alaska to Nevada. In the United States, the event has been named the **Sonoma Orogeny.** Its effects can be studied in west-central Nevada. Like the Antler Orogeny, which preceded it, the Sonoma Orogeny was caused by the collision of an island arc system with the southwestern margin of North America. It was quite a smashup, as oceanic and island arc rocks were thrust scores of kilometers eastward

A

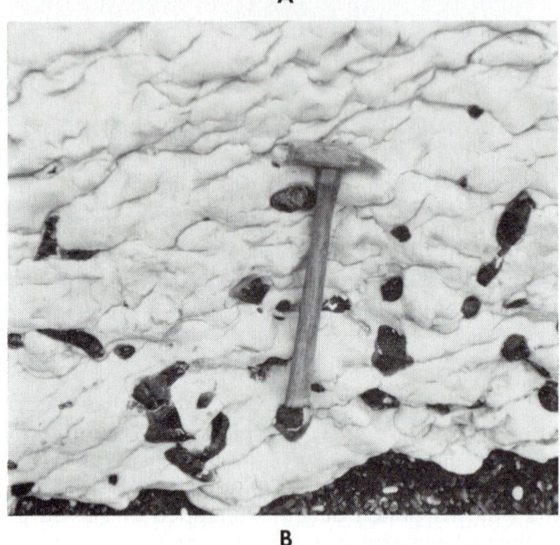

B

Figure 12–9 (A) Cliffs composed of chalk near Dorset, England. (B) Close-up of chalk, showing dark nodules of flint.

against the continental margin. The island arc itself became permanently welded to the western edge of the continent, thereby increasing its girth by 200 to 300 km.

The Triassic rocks of the far western part of the Cordilleran Geosyncline include great thicknesses of volcanics and graywackes presumably derived from the island arc. There is a question, however, of whether or not these rocks were deposited where they are now found or whether they are part of a displaced terrain, called the **Sonoma Terrain,** that actually originated in some unknown part of the Pacific.

Early Triassic sediments of the miogeosyncline consist of sandstones and limestones of mostly shallow marine origin. The thickest section of these marine strata occurs in southeastern Idaho, where nearly 1000 meters of Lower Triassic sediments accumulated. Eastward from the miogeosyncline, these marine beds interfinger with continental red beds. Although Triassic seas remained in Canada during the Middle Triassic, in the United States they regressed westward, leaving vast areas of the former sea floor subject to erosion. Upper Triassic formations rest directly above the unconformity resulting from this erosion. The Upper Triassic series consists mostly of continental deposits transported by rivers flowing westward across an immense alluvial plain. There were also upland source areas in western and southern Nevada, Arizona, and California. For example, the lowermost strata consist of the sandy **Moenkopi Formation,** followed by the pebbly **Shinarump Conglomerate** (Figs. 12–10 and 12–11), derived from uplifted neighboring areas in Arizona, western

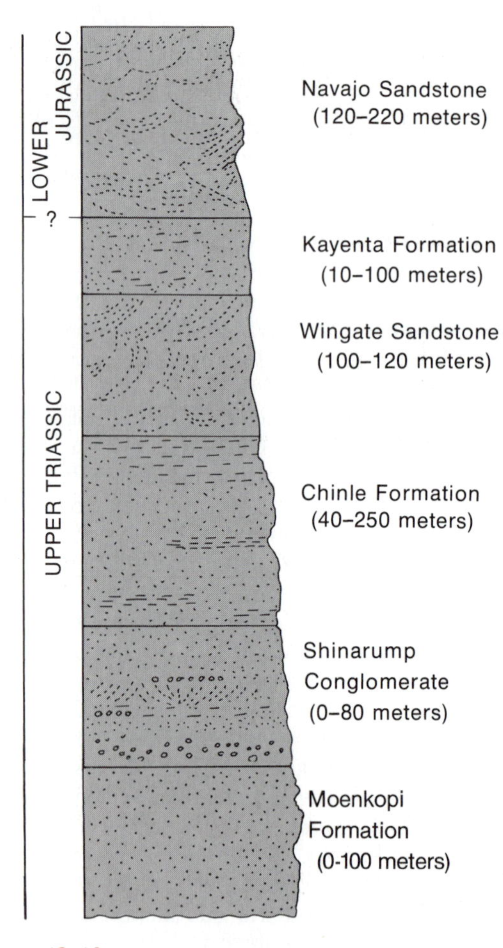

Figure 12–10 Generalized geologic section of Upper Triassic and Lower Jurassic sedimentary rocks of central Utah.

Figure 12–11 Shinarump Formation exposed above the adits of the Colt Mesa Copper Mine, Utah. (Photograph courtesy of the Utah Geological and Mining Survey.)

Colorado, and Idaho. Above the Shinarump lie the vividly colored shales, silts, and sandstones of the **Chinle** and **Kayenta** formations (Fig. 12–12). The sediments of these two formations were deposited in stream valleys and lakes. They alternate with sand dune accumulations of the **Navajo** (Fig. 12–13) and **Wingate** formations. The formations are beautifully exposed in the walls of Zion Canyon in southern Utah. They display sweeping cross-bedding, such as is formed in sandy deposits transported by winds. (However, such so-called "festoon cross-bedding" is not always caused by winds, for submarine dunes may be very similar in form.)

The Painted Desert of Arizona is developed mostly in Chinle rocks. The formation is known throughout the world for the petrified logs of conifers it contains. Each year, thousands of tourists examine these logs, now turned to colorful agate, in the Petrified Forest National Monument (Fig. 12–14). Apparently, during times of flood, the trees were left on sand bars or trapped in log jams and covered by sediment. Percolating solutions of underground water subsequently replaced the wood with silica.

Jurassic to Early Tertiary Tectonics

Because orogenic events in the Cordillera of the western United States overlapped one another in time of occurrence and location, it is difficult to apply names to them and list them in chronologic sequence. Perhaps at the risk of oversimplification, however, it is possible to examine the general cause of the major episodes of orogeny and describe the style of deformation. The fundamental cause of Cordilleran orogenic activity during the Mesozoic was the continuing eastward underthrusting of oceanic lithosphere beneath the continental crust of the North American Plate. That underthrusting varied in rate, in inclination, and, to a small degree, in direction. It resulted in eastward-shifting phases of deformation, which initially affected the eugeosyncline on the far west, then the miogeosyncline on the east, and finally the margin of the craton itself.

The deformational and magmatic activity associated with the eugeosynclinal tract is sometimes termed the **Nevadan Orogeny.** Beginning in the Triassic, and increasing in tempo during the

Figure 12–12 A typical scene in the Orange Cliffs area of Utah. The high cliffs along the horizon are Wingate Formation. They are capped by beds of Kayenta visible near the top center. The slope immediately below the cliffs is eroded in the Chinle Formation. The ridge in the foreground is capped by the Lower Triassic Moenkopi Formation. (Photograph courtesy of the Utah Geological and Mining Survey.)

Figure 12–13 Panoramic view of Triassic and Jurassic formations in Zion National Park, Utah. The top of the prominent peak on the left is the Carmel Formation (Middle Jurassic). The vertical cliffs are Navajo Sandstone (Lower Jurassic). The Chinle underlies the first slope. The ledge in the far left is Shinarump, followed by the Moenkopi Formation. (Photograph courtesy of the United States National Park Service; Zion National Park.)

Figure 12–14 Petrified Forest National Park, Arizona. All of the logs pictured here are *Araucarioxylon*, which is by far the most common in the park. The ground is littered with petrified wood chips and chert pebbles eroded from Triassic sandstones. (Photograph courtesy of the United States National Park Service.)

Jurassic and Cretaceous, sea floor sediments, graywackes, slates, cherts, and volcanics that had been swept into the subduction zone were severely folded, faulted, and metamorphosed. The intensely crumpled and altered rock sequences that were affected by compression between the converging plates are appropriately termed mélange, which means a jumble. The Franciscan Fold Belt of California (Fig. 12–15) provides a good example of a mélange. In addition to the deformation of the sedimentary rock sequences, enormous volumes of granodioritic rock were generated above the subduction zone and intruded overlying rocks repeatedly during the Jurassic and the Middle and Late Cretaceous. The Sierra Nevada, Idaho, and Coast Range Batholiths are impressive reminders of the vast scale of this Nevadan magmatic activity (Fig. 12–16).

Somewhat before the Sierra Nevada Batholith was emplaced in the Cretaceous, another series of deformations had begun east of the eugeosyncline. This second phase of the tectonic development of the Cordillera primarily affected miogeosynclinal rocks over a time span from Middle Jurassic to earliest Cenozoic. This new orogeny has been named the **Sevier,** and its style of deformation was distinctive. In response to compression transmitted eastward from the subduction zone, miogeosynclinal strata were sheared from underlying Precambrian rocks and broken along parallel planes of weakness to form multiple, imbricated, low-angle thrust faults (Fig. 12–17). The French word *décollement* (unsticking) has been used to describe this kind of structure, in which older rocks are thrust upon younger in multiple, nearly parallel slabs of crust. It has been estimated that the compressional structures produced during the Sevier Orogeny resulted in over 100 km of crustal shortening in the Nevada-Utah region. In addition to the major thrust faults, several large folds are known, as well as intricately folded strata that were deformed within the major thrust units.

Although the term Sevier is sometimes reserved for deformational episodes in the Nevada-Utah region, a similar style of deformation occurred to the north in Montana, British Columbia, and Alberta. Each of the major ranges in this region is a fault block composed of Paleozoic mioge-

Figure 12–15 An exposure of Franciscan rocks in San Francisco, California. (Photograph courtesy of J. C. Brice.)

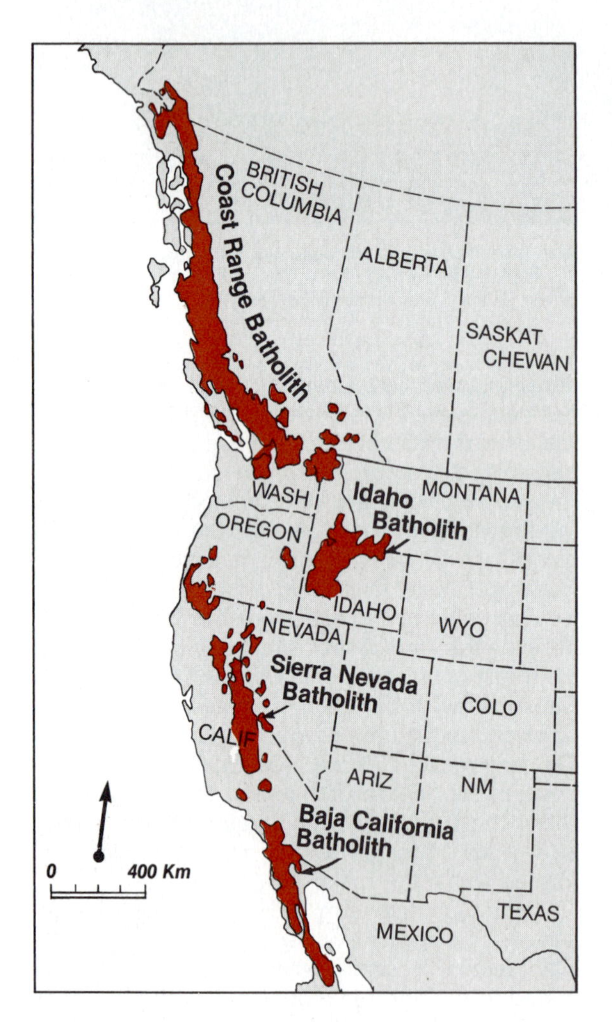

Figure 12–16 Mesozoic batholiths in west-central North America.

osynclinal strata that have been thrust toward the east along westwardly dipping fault surfaces. The most famous of these faults is the Lewis Thrust (Fig. 12–18), along which Proterozoic rocks of the Belt Group were carried 65 km toward the east.

Magmatic activity along the far western edge of the North American Plate had diminished near the end of the Cretaceous, and by Early Tertiary much of the major thrusting of the miogeosynclinal region had subsided. Once again, deformational events shifted eastward to the cratonic region where now are located the Rocky Mountains of New Mexico, Colorado, and Wyoming. These more eastwardly disturbances are sometimes called the Laramide Orogeny. Although thrust faults do occur in the region of Laramide activity, the more characteristic kind of structures are broadly arched domes, basins, and anticlines. Strata composing the domes and anticlines are draped over central masses of Precambrian igneous and metamorphic rocks. In several instances, erosion has stripped away the cover of strata, exposing the central crystalline mass. Resistant layers of inclined beds that surround the central cores stand as mountains today. Many of these uparched areas appear to have been produced by movements along underlying faults in the basement rocks. It has been suggested that these controlling structures were in turn developed by drag when the eastward-moving subducted oceanic lithosphere scraped along the sole of the cratonic margin of North America.

Most of the structures of the present Rocky Mountains are the result of the Laramide phases of orogeny. The landscape we see today, however, is the result of repeated episodes of Cenozoic erosion

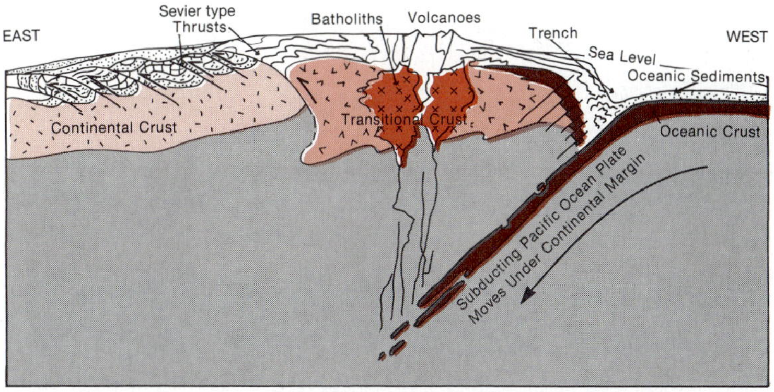

Figure 12–17 An advanced stage in the evolution of the North American Cordillera with structures developing as a consequence of underthrusting of the continent by the Pacific oceanic plate. (The diagram is simplified from J. F. Dewey and J. M. Bird. Mountain belts and the new global tectonics. *J. Geophy. Res.,* 75(14):2638, 1970.)

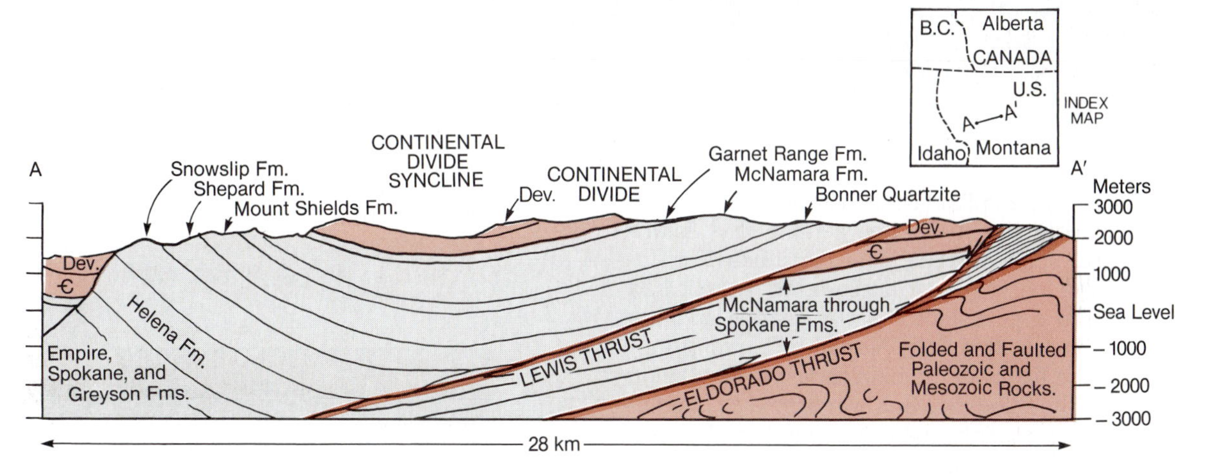

Figure 12–18 Northeast-southwest cross-section showing the Lewis and Eldorado thrust faults at a location about 66 km south of Glacier National Park. Formations shown in gray are Proterozoic in age. Orange color indicates Paleozoic and Mesozoic rocks. (After M. R. Mudge, and R. L. Earhart. *The Lewis Thrust Fault and related structures in the Disturbed Belt, Northwestern Montana.* U.S. Geol. Survey Professional Paper No. 1174, 1980.)

and uplift. Erosion, acting on the geologic structures already present, sculptured the final scenic design.

Jurassic Sedimentation

During the Jurassic, when the mélange was being formed in the eugeosyncline, the miogeosyncline was the site of a less dramatic kind of sedimentation. Early Jurassic deposits consist of clean sandstones, such as the **Navajo Sandstone.** Large-scale cross-bedding is well developed in the Navajo (Fig. 12–19). However, thin beds of fossiliferous limestone and evaporites occur locally and indicate that at least parts of the formation were deposited in water. Indeed, recent interpretations suggest that much of the cross-bedding in the Navajo resulted from water currents that formed sand dunes on the floors of shallow marine areas (Fig. 12–20). The Navajo and associated sand bodies were probably deposited in a nearshore environment, and it is likely that some of the deposits were part of a coastal dune environment. They do not seem to have been laid down on the floors of a vast interior desert, as has been often postulated. Judging from studies of cross-bedding orientations, the source area for these clean sandstones was probably the Montana-Alberta craton. Somewhat older but similar quartz sandstones were recycled and spread southward into Wyoming and Utah, and then westward across Nevada.

Marine conditions became more widespread in the Middle Jurassic, when the entire west-central part of the continent was flooded by a wide seaway that extended well into central Utah (Fig. 12–21). This great embayment has been dubbed the **Sundance Sea.** Its deposits, largely derived from the Mesocordilleran Highlands, are spread widely across the Northern Great Plains, where they make up the Sundance Formation. The **Mesocordilleran**

Figure 12–19 Broad, sweeping cross-bedding in the Navajo Sandstone, Circle Cliffs Area, Garfield County, Utah. (Photograph courtesy of the Utah Geological and Mining Survey.)

Highlands were upland tracts trending north-south roughly along the border between the miogeosyncline and eugeosyncline. The highlands first appeared late in the Triassic and by Late Jurassic had grown into a persistent chain of islands and uplands that stretched from Alaska to Central America. On the flanks of the highlands, the fluctuating nature of the Sundance Sea is indicated by interfingering of its marine strata with land-laid deposits. Apparently, the sea was never able to connect with contemporary embayments that spread northward from the ancestral Gulf of Mexico. Gradually, near

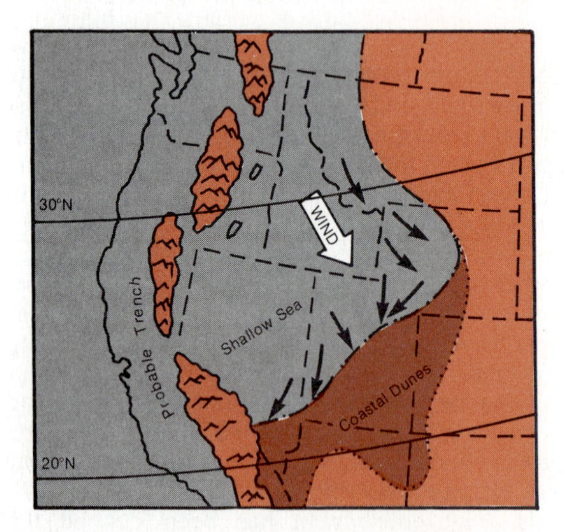

Figure 12–20 Paleogeographic map for the Early Jurassic of the western United States, showing general extent of sea and land as well as paleolatitudes. (From K. O. Stanley, W. M. Jordan, and R. H. Dott. *Bull. Am. Assoc. Petrol. Geol.,* 55(1):13, 1971).

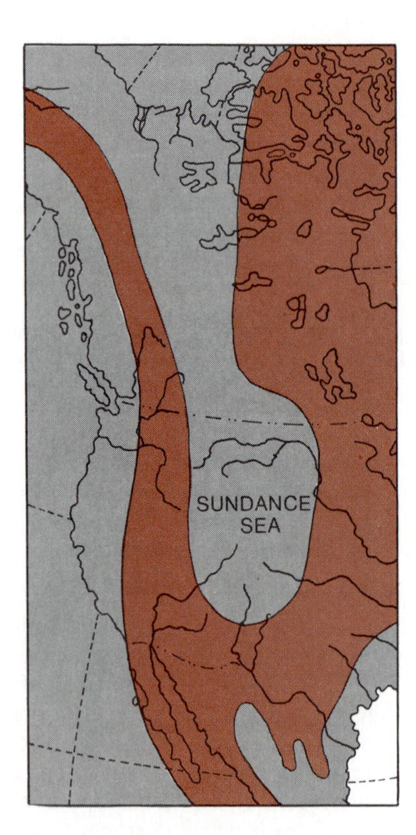

Figure 12–21 Region in western North America inundated by the Middle Jurassic Sundance Sea. (Land areas are shown in brown; marine areas, in gray.)

the end of the Jurassic, the seas withdrew, leaving a vast, swampy plain across which meandering rivers built wide floodplains of mud, gravel, and silt. These deposits compose the **Morrison Formation** (Fig. 12–22), which extends across millions of square kilometers of the American West. Enclosed within these floodplain deposits are the bones of more than 70 species of dinosaurs, including some of the largest land animals ever to have existed.

Cretaceous Sedimentation

During the Cretaceous, the Pacific border region of North America was a land of lofty mountains formed during the orogenesis that had begun in the Jurassic. Erosion of these ranges brought sediments downward into adjoining rapidly subsiding basins, many of which were open to the Pacific Ocean. In some places, over 15,000 meters of volcanics and clastics accumulated. Cycles of folding and intrusion occurred repeatedly, with rather exceptional deformations taking place during Middle and then Late Cretaceous time.

Eastward of the eugeosyncline, the Cretaceous began with the advance of marine waters both northward from the ancestral Gulf of Mexico and southward from arctic Canada. These Early Cretaceous invasions did not meet, so that an area of dry land existed in Utah and Colorado. Their advance was reversed by a general regression that resulted in the unconformity used in separating the Cretaceous into an early and late division. The flooding that followed in the Late Cretaceous was the greatest of the entire Mesozoic Era. This time, the embayment from the north joined with the southern seaway and effectively separated North America into two large islands.

Sedimentation in Cretaceous epicontinental seas was largely controlled by local conditions. Along the Gulf Coast, limestones and marls (clayey limestones) accumulated (Fig. 12–23). North of Texas, however, the great bulk of sediments consisted of clastics supplied by streams flowing off the western highlands. Typically, these deposits are sandstones that gradually thin and grade eastward into shales. Sandstones of the **Dakota Group** are particularly well known (Fig. 12–24). The dark brown sandstone layers of the Dakota are exposed in many places along the eastern front of the Rocky Mountains, where the inclined beds form prominent ridges called hogbacks (Fig. 12–25). In parts of the Great Plains, the Dakota sandstones are an important source of underground water.

The carbonate formations of the miogeosyn-

Figure 12–22 Morrison Formation exposed along the east flank of Circle Cliffs, Garfield County, Utah. (Photograph courtesy of the Utah Geological and Mining Survey.)

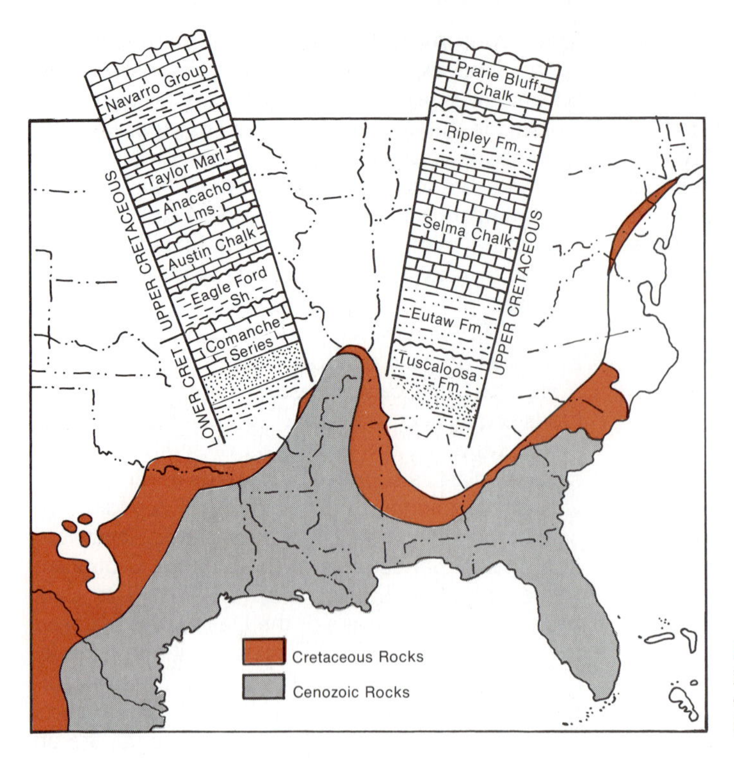

Figure 12–23 Area of outcrop of Cretaceous rocks in the Atlantic and Gulf Coastal Plains, a generalized columnar section of the Texas (left) and Alabama (right) Cretaceous.

Figure 12–24 Massive concretions that have weathered out of the Dakota Sandstone at "Rock City," Ottawa County, Kansas. (Photograph courtesy of the Kansas Geological Survey; photograph by J. Enyeart.)

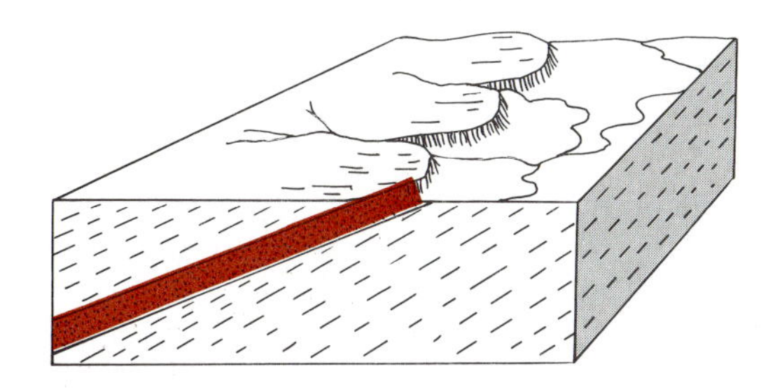

Figure 12–25 A hogback ridge.

cline include soft, clayey limestones and chalky shales of the **Niobrara Formation** (Fig. 12–26). The Niobrara has yielded the remains of a variety of marine creatures, including enormous numbers of oysters, a giant Cretaceous diving bird (*Hesperornis* [Fig. 12–27]), marine reptiles, and the large flying reptile *Pteranodon.* Toward the end of the Cretaceous, the seas that supported these creatures began a slow withdrawal. This regression of the Cretaceous epicontinental sea from central

Figure 12–26 An exposure of Niobrara chalk in Kansas sculptured by processes of erosion into a pinnacle. (Photograph courtesy of the Kansas Geological Survey; photograph by Jim Enyeart.)

North America was contemporaneous with the Laramide deformation that produced the Rocky Mountains. Coal-bearing deltaic and other continental sediments initiated a new phase of sedimentation on the old sea floor.

To the North

Because of the problems inherent in attempting geologic work in frigid polar climates, data on the geology of the northern geosyncline have been limited. However, within the last decade, the exploitation of petroleum reservoirs in Alaska has contributed new information about some areas along the northern margins of the continent.

For the most part, Mesozoic sediments accumulated in two basins along the northern geosynclinal belt (Fig. 12–28): the Brooks Basin of northern Alaska and the Sverdrup Basin farther east. The Mesozoic record for the Brooks Basin begins with Triassic clastics and limestones, some of which are fossiliferous. Jurassic and Cretaceous beds consist of volcanic rocks, sandstones, and shales.

Mesozoic formations of the Sverdrup Basin include both marine and nonmarine sandstones and shales. Some of the formations contain petroleum. Sandstones and siltstones predominate within the Triassic system. Similar clastics, interspersed with coal beds, were deposited during the Jurassic. Volcanic rocks, continental clastics, red beds, and additional coal beds characterize the Cretaceous rock record.

EURASIA AND THE TETHYS GEOSYNCLINE

Extending eastward from Gibraltar and across southern Europe and Asia to the Pacific is the great Alpine-Himalayan mountain belt. Most of the rocks composing the ranges within the belt were laid down in an important geosynclinal trough known as the **Tethys.** Long after their deposition, these rocks were deformed into mountain ranges as northward-moving segments of Gondwanaland collided with Eurasia. Here and there, however, the waters remained and persist today as the Mediterranean, Caspian, and Black Seas.

The depositional history of the Tethys Geosyncline can be traced well back into Early Paleozoic time. There is even some evidence of crustal mobility in parts of the belt as long ago as Late Precambrian. However, just as most of the depositional and tectonic activity for the Appalachian Geosyncline transpired within the Paleozoic, for the

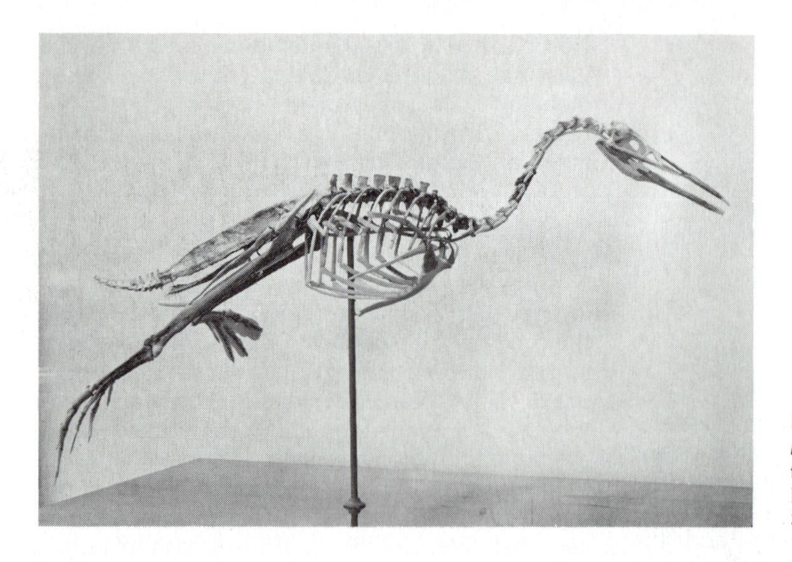

Figure 12–27 Skeleton of *Hesperornis regalis,* an extinct, aquatic, toothed bird from the Niobrara Formation of Logan County, Kansas. (Photograph courtesy of the Smithsonian Institution.)

Tethys, the Mesozoic and Cenozoic are the most eventful eras.

During the Triassic Period, the Tethys (Fig. 12–29) was the site of limestone deposition. In various locations, marine invertebrates flourished. Pelecypods, crinoids, reef corals, and algae are well represented in the fossil record. Of particular importance are the abundant Triassic ammonoid cephalopods found in the Alpine region, because these mollusks are the principal fossils used in Triassic correlation. North of the Tethys, a highland area known as the Vindelician Arch developed and separated the geosynclinal depositional environment from the quite different sedimentation of north-central Europe. In that northern region, the Triassic record begins with reddish, nonmarine clastic sediments that resemble those deposited under arid conditions during the preceding Permian Period. These largely fluvial deposits are overlain by shoreline sands, marls, evaporites, and limestones deposited during a temporary marine invasion of central Europe. The sea did not remain long. Nor was it able to reach as far north as Great Britain, where a nonmarine sequence known as the New Red Sandstone was being deposited. As the sea withdrew, continental sedimentation much like that of the earliest Triassic resumed.

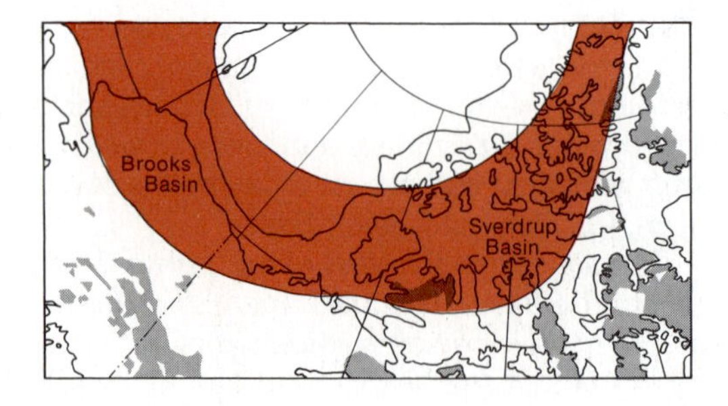

Figure 12–28 Location of the Brooks and Sverdrup depositional basins along the northern geosyncline. (Shaded areas are outcrops of pre-Cenozoic igneous rocks.)

Marine conditions were far more widespread during the Jurassic than they had been during the Triassic. Shallow seas spilled out of both the Tethys and the Atlantic and spread across Europe, leaving a rich sedimentary record in the basins that lay between the old Hercynian uplands. Although the predominant Jurassic sediment is limestone, there is often a gradation to fine clastics adjacent to the highlands. Eventually, marine conditions extended from the Tethys across Russia and into the Arctic Ocean. The invasion was short-lived, however, and the shallow seas gradually drained from the continent by the end of the Jurassic. Marine conditions persisted in the Tethys, and a complicated pattern of facies changes developed in response to block-faulting along the floor of the geosyncline.

The Cretaceous is the most eventful geologic period of European Mesozoic history. During this time, northward-moving Africa began to close the gap that was the Tethys Sea. The Alpine Geosyncline was subjected to an episode of powerful compressional folding. As evidence of this event, a zone of **ophiolites** can be traced along the southern margin of Europe. Ophiolites are greenish igneous rocks believed to have once been part of the crust of the ocean that existed between Gondwanaland and Laurasia. The ophiolites are found in association with radiolarian cherts and lithified deep sea oozes, which were very likely also part of that sea floor. The movement of the oceanic crust beneath Eurasia as the African Plate approached created the ophiolite zone as well as the deformation and

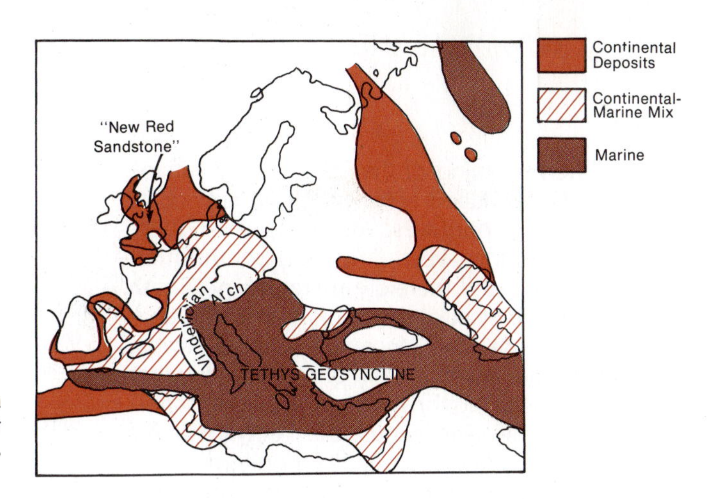

Figure 12–29 Paleogeographic map of the Triassic in Europe. (Modified from R. Brinkmann. *Geologic Evolution of Europe.* New York, Hafner Publishing Co., 1969.)

volcanism that characterized the southern margin of Eurasia during the Cretaceous.

The Late Cretaceous is noteworthy not only for its mountain-building events but also as a time of extensive marine transgressions. More of Europe was inundated during the Late Cretaceous than at any other time since the Cambrian (Fig. 12–30). A great embayment from the Tethys worked its way northward until it ultimately joined with a similar southward encroachment from the region of the North Atlantic. One result of the linking of the two seaways was a rapid interchange of marine organisms from the two formerly separated regions. In the Tethys-Alpine region, the block-faulting that had begun during Late Jurassic continued until about Middle Cretaceous. At that time, folding became the predominant structural style. The axes of the anticlines tended to parallel the east-west trend of the Tethys. Most of these great welts remained submerged, but the tops of some of the folds rose above the level of the waves, forming elongate islands along the north side of the narrowing Tethys seaway. Erosion of the anticlinal islands produced clastic debris that was periodically swept downward by turbidity currents into the adjoining basins, to be deposited as a thick sequence of interbedded sandstones and shales. These sediments and associated volcanic rocks foretell the coming of tumultuous events during the Cenozoic.

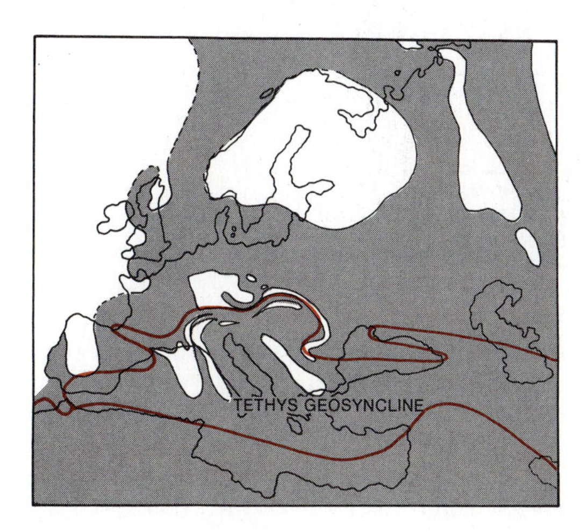

Figure 12–30 Areas covered by seas during the Cretaceous. (Simplified from R. Brinkmann. *Geologic Evolution of Europe.* New York, Hafner Publishing Co., 1969.)

An event that determined the geographic configuration of the northern coastlines of Spain and France occurred in Europe between Late Jurassic and Late Cretaceous time. As indicated by paleomagnetic data, there was no Bay of Biscay separating eastern France from Spain prior to the Late Jurassic. The Bay of Biscay opened as the Iberian Penninsula (Spain and Portugal) rotated about 35° from its former line of contact with France. The upheaval of the Pyrenees was another consequence of this rotation. The dating of the rotational event is confirmed by faults of Late Jurassic age on the northern border of the Bay of Biscay and by the fact that the oldest sediments in the bay are Late Cretaceous in age.

East of the Alpine region, the Tethys extended as a great loop past Burma and southeastward into Indonesia. The sediments of the Triassic and Jurassic of the eastern Tethys consisted of shales, fossiliferous limestones, and dolostones. During the Cretaceous there was volcanic activity throughout what is now the Himalayan region. Some areas seem to have experienced marked subsidence, and clastic sediments resembling those moved by submarine landslides or turbidity currents merged with volcanic tuffs, breccias, and radiolarian cherts.

THE GONDWANALAND CONTINENTS

Africa

By the beginning of the Triassic, separations had already developed between the eastern coast of America and northwest Africa. However, there was as yet no rift between Africa and South America (Fig. 12–1), and therefore there was no South Atlantic Ocean. Most of the portion of Gondwanaland that we now call Africa was relatively stable throughout the Mesozoic, with only relatively minor transgressions along the northern and eastern borders. During the Cretaceous, the Tethyan waters spread across what is today the northern tier of countries from Algeria to Egypt. Fine muds and carbonates were the most prevalent deposits.

Although the Mesozoic rock record in the more interior parts of Africa is often poor, some regions do yield particularly interesting geologic information. One such region is the Karoo Basin at the southern end of Africa (Fig. 12–31). The Karoo was a continental basin that was formed late in the Carboniferous and received swamp, lake, and river deposits until late in the Triassic Period. Rocks of the Karoo are known to paleontologists around the world for their wealth of fossilized mammal-like reptiles. The fauna included large and small herbivores, diverse carnivores, and a range of insectivores and omnivores. As the Triassic drew to a close, the reddish silts and sands of the Karoo sequence were covered by flow after flow of low-viscosity lava, which gushed onto the surface from fissures and, less commonly, from volcanoes in the southeast. The old Karoo landscape was buried beneath more than 1000 meters of basalt. Geologists speculate that these great outpourings of lava were associated with the pulling away of the Gondwanaland segments that once adjoined South Africa. Such fragmentation would very likely have caused severe fracturing of the foundations of the continents and provided multiple avenues for the upwelling of the molten rock. The extrusions continued well into the Jurassic Period. Contemporaneous lava floods and volcanism also occurred on the separating land masses of South America, Australia, and, somewhat later, Antarctica.

Africa appears to have been more at peace with itself throughout the remainder of the Jurassic and the Cretaceous. Marginal seas extended along the eastern edges of the continent. Periodic advance and retreat of these seas brought an alternation of marine and nonmarine beds.

Australia and New Zealand

During the Triassic, Australia was still attached to Antarctica but was pulling northward. In this earliest Mesozoic period, both Australia and Antarctica were predominantly emergent. Terrestrial sandstones and shales resembling those of the Permian were deposited. Except for brief marginal embayments, deposition of lake and stream sediments continued to dominate until the middle of Early Cretaceous. At that time, a marine transgression brought sandstones and chalk beds into the interior.

During the Mesozoic, New Zealand was geologically far more restless than Australia. It lay astride the Pacific mobile belt and thus experi-

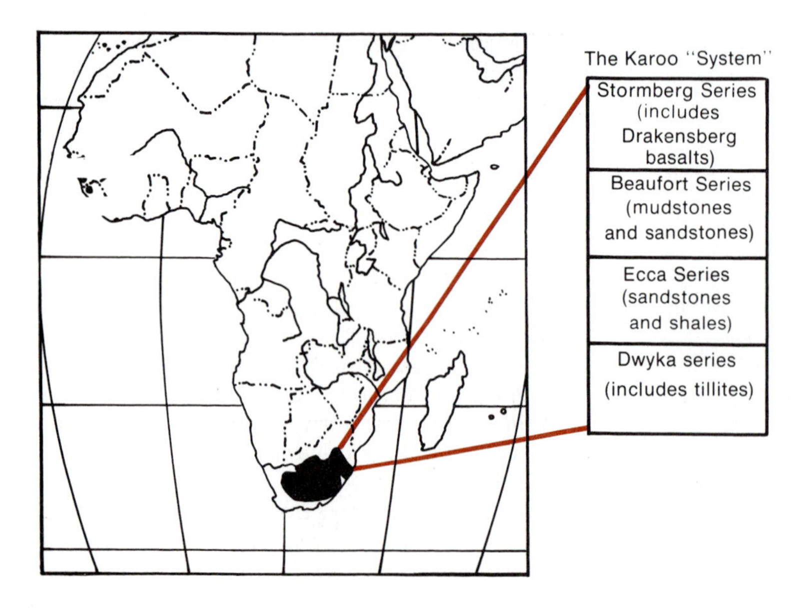

Figure 12–31 Location of the Karoo Basin of South Africa and the four principal units of the Karoo "System."

enced deposition of enormous thickness of Mesozoic geosynclinal sediments and volcanics. Near the end of the Jurassic, a series of orogenic events occurred as a result of the collision and subduction of the Pacific Plate beneath the island continent. In the course of the orogeny, mountains were erected, and deep, unstable trenches developed. The activity continued into the Cenozoic and, intermittently, even to the present day.

India

During the Mesozoic, India moved steadily northward on its remarkable voyage to Laurasia. The passage was for the most part quiet, but it was to have a tumultuous ending. The land mass experienced only relatively minor marginal incursions of the sea. Across the interior, erosion of upland areas spread terrestrial clastic sediments into the lowlands and plains. Some of these continental formations have yielded a wealth of dinosaur bones and superb fossils of plants. By Cretaceous time, India was nearing the Tethyan Geosyncline. Here and there, the floor of the Tethyan trough was buckled into elongate ridges that were early indications of the coming tectonic storm. While these ridges were developing, the northwestern half of India was flooded with immense quantities of low-viscosity basaltic lava. These now solidified lavas are known as the **Deccan Traps;** they are also termed "flood basalts." Radiometric dates obtained from the basalts indicate that the outpourings continued from Cretaceous time well into the Early Cenozoic. It is likely that these basalt floods record the passage of India across a fixed "hot spot" in the mantle.

South America

Much like Africa, South America stood well above sea level at the beginning of the Mesozoic. Along the western margin of the continent, a lengthy geosyncline was already in existence during the Triassic. Not unexpectedly, it was differentiated into a eugeosyncline on the Pacific side and a miogeosyncline along the edge that bordered the craton. Today the boundary between these two tracts would lie along the central ranges of the Andes. In some areas a geanticline similar to the Mesocordilleran Highlands of North America lay between the eugeosynclinal and miogeosynclinal tracts. Graywackes, conglomerates, siliceous sediments, and lavas accumulated in the eugeosynclinal belt, whereas carbonates and shales predominated in the neighboring miogeosyncline. In addition to these marine tracts, broad basins in the interior accumulated continental deposits. Particularly remarkable are the eolian and fluvial sands and silts that spread across the southeastern region of the continent during the Triassic. The beds contain a rich fauna of Triassic vertebrates. Lava flows, chronologically equivalent to those of the Karoo rocks in Africa, are also prevalent in the upper part of this stratigraphic sequence.

By Jurassic time, the initial narrow split between South America and Africa had widened into a configuration resembling the present-day Red Sea. New sea floor formed along the spreading center of the developing South Atlantic. The western side of South America was at the leading edge of the westward-moving tectonic plate and was being underthrust by an opposing oceanic plate, much as was the case in the far west of the United States.

Near the end of the Jurassic, the Andean belt experienced deformation and volcanism. This activity appears to have begun in the south and then sporadically shifted with time toward the north. The climax of orogenic deformation occurred in Late Cretaceous and Early Tertiary and was accompanied by regional uplift that led to widespread withdrawal of the seas. Here and there, evaporites were deposited in isolated basins that for a time retained marine waters.

The frequency and intensity of deformational and volcanic activity along the Andean Geosyncline increased during the Cretaceous. The subduction zone (created by the movement of the Pacific Plate against South America) received both Pacific floor sediments and older, crustal materials from the continents. These materials were melted at great depths. The melts in turn worked their way back toward the surface as great intrusions and outpourings of andesitic basalt. Deformation and igneous activity continued well into the Early Cenozoic and formed many of the structures now seen in the towering peaks of the Andes.

Antarctica

Antarctica was predominantly emergent throughout the Mesozoic. Eastern areas of the now frigid land mass were sites of continental deposition. Beds of volcanic ash and lava flows occur frequently between these lake and stream deposits.

The nature of sedimentation was markedly different in western Antarctica. Outcrops along the Antarctic Peninsula consist of volcanic and clastic formations that are very similar to those of equivalent age in the Andes. Very likely, the strata of the Antarctic Peninsula represent part of a continuous mobile belt that extended up through the Scotia Arc and along the western margin of South America.

ECONOMIC RESOURCES

Uranium Ores

Rocks of the Mesozoic System, like those of earlier eras, contain a wealth of important mineral resources. Notable among such resources are the nuclear fuels. In the United States, uranium ores are derived chiefly from continental Triassic and Jurassic rocks of New Mexico, Colorado, Utah, Wyoming, and Texas. The chief ore mineral, known as **carnotite,** was deposited in the form of uranium salts within the pore spaces of fluvial sandstones (Fig. 12–32). Apparently, the abundance of organic material in the sediment enhanced the precipitation of the uranium salts. In some instances, petrified logs have provided amazing concentrations of not only uranium but also vanadium and radium. These are striking exceptions, however, and the radioactive ores are mostly of very low concentration.

There is at present concern that the richer sources of fissionable uranium 235 in this country are likely to be exhausted by about the same time as our petroleum reserves are depleted. Among the alternate sources of nuclear energy now being studied are uranium 238 and thorium 232, which can be converted to fissionable isotopes by a process termed **breeding.** This will permit the use of the more abundant low-grade uranium ores. Of course, breeder reactors are not yet perfected and may not prove sufficiently safe or as enduring as other possible energy sources.

Until either the breeder reactor or devices to utilize solar energy are perfected, the nations of the world will continue to rely heavily on fossil fuels. Mesozoic rocks are hosts for these critical re-

Figure 12–32 Opening to small uranium mine in Shinarump Sandstone (massive bed). Slope debris covers Moenkopi Formation. (Photograph courtesy of the Utah Geological and Mining Survey.)

Figure 12–33 Coal beds in the Mesaverde Formation (Upper Cretaceous) exposed in a road cut on Highway 6-50 at Castlegate, Utah. (Photograph courtesy of J. C. Brice.)

sources also. The Jurassic, for example, is an important coal-producing system. The larger foreign mines are located in Siberia, China, Australia, Tasmania, and Spitzbergen. In North America, thick seams of Jurassic coal occur in British Columbia and Alberta. Cretaceous coal underlies more than 300,000 sq km of the Rocky Mountain region (Fig. 12–33). Much of this coal is now being vigorously mined, particularly because of its relatively low content of environmentally offensive sulfur.

Fossil Fuels

Mesozoic rocks in certain favorable areas around the world supply large quantities of oil and gas to energy-hungry industrial nations. The oil provinces of the Middle East and North Africa probably contain more oil than the combined reserves of all other countries. Middle East petroleum comes primarily from thick sections of Jurassic and Cretaceous sediments that had accumulated in the Tethys seaway. Other areas of petroleum production from Mesozoic rocks occur in the Rocky Mountains, Alaska (Fig. 12–34), Arctic Canada, the Gulf Coastal states, western Venezuela, southeast Asia, beneath the North Sea, and beneath the eastern offshore area of Australia. Notwithstanding the size of some recent discoveries in rather forbidding parts of the world, it appears likely that the present rates of oil and gas consumption will cause exhaustion of the world's resources in less than a century. Therefore, oil and gas must be replaced by other energy sources in the near future, especially if we are to conserve these valuable materials for the chemical industry.

Figure 12–34 Drilling for oil in a frigid offshore area of Alaska. (Photograph courtesy of the American Petroleum Institute Photo Library.)

Ores

Metalliferous deposits were formed widely throughout the active orogenic belts of the Mesozoic world. A variety of metals now found in the Rocky Mountains, the Pacific states, and British Columbia were emplaced during batholithic intrusions that accompanied Jurassic and Cretaceous mountain building. Among these ore deposits were the gold-bearing quartz veins known as the "Mother Lode." The California gold rush of 1849 was a consequence of the discovery of gold-bearing gravels eroded from the Mother Lode. The copper, silver, and zinc veins of Butte, Montana, and Coeur d' Alene mining districts were emplaced as a result of Cretaceous igneous activity. One belt of porphyritic copper-bearing rock extends from Denver to the Four Corners area. The zone has become known as the "porphyry copper belt." The intrusions that produced the belt range in age from Cretaceous to Early Cenozoic. Triassic rocks yield copper in Germany and Russia. Not all the ore deposits resulted from igneous activity. In England and Alsace-Lorraine, there are important Jurassic iron ores that are sedimentary in origin.

Nonmetallic deposits of the Mesozoic include sulfur and salt. Both are produced from the salt domes mentioned earlier. Diamonds are also obtained from Mesozoic igneous rocks. Siberia's diamond-bearing intrusions are believed to have penetrated the upper crust during the Triassic and Jurassic. Similar "diamond pipes" in Africa are probably of Cretaceous age.

Plate Tectonics and Ore Deposits

In recent years, geologists have begun to see numerous correlations between patterns of mineral distribution and locations of present and former tectonic plate boundaries. Plate boundaries are likely sites for movements of hot aqueous fluids that bear important metals in solutions. With changes in temperature or pressure, or on contact with reactive rocks, such hydrothermal solutions will precipitate their dissolved wealth. The process may operate at both convergent and divergent plate boundaries. Both the Cordilleran and Andean mobile belts are examples of convergent boundaries. An example of hydrothermal mineralization at a divergent boundary is provided by the Red Sea, which has "pools" of very hot and exceptionally salty water along its bottom. The pools are rich in iron, manganese, zinc, and copper. The brines apparently have percolated upward through the young oceanic crust, dissolving the metals en route. Metals brought to the surface in this way in the past may have been conveyed in thin layers across immense tracts of the ocean floor by sea floor spreading. Sediments containing metallic ions extracted from the sea water itself may have enriched the accumulation. Ultimately, sea floor spreading would move the sediments and their contained metals into collision with another plate, providing an opportunity for their inclusion as ore bodies within the deformed belt.

Of course, the formation of ores in orogenic belts is a far more complicated process than this conceptual model suggests. Every ore body requires very special local conditions and events for its emplacement. Many ore deposits have no relationship to plate boundaries. However, in the case of the metallic deposits of North America and South America, a relationship is possible, for both of these regions form overriding convergent boundaries with the plates of the Pacific.

Summary

Within the 162 million years of the Mesozoic Era, the pattern of lands and seas was extensively altered as large segments of Pangaea II were separated. The breakup appears to have begun during the Triassic, with the splitting off of North America from Gondwanaland. By Late Jurassic, rifts had developed between all of the Gondwanaland segments except Australia and Antarctica. These two continents, like northwestern North America, Greenland, and northwestern Europe, retained a hold on one another until early in the Cenozoic.

At the beginning of the Mesozoic, most of North America was emergent. In the east, crustal tension was expressed by the development of large-scale block-faulting and volcanism. In downfaulted valleys, arkosic sandstones, conglomerates, and lacustrine shales accumulated. While this was occurring, the ancestral Gulf of Mexico began to take form far to the south. During the Jurassic and Cretaceous, thick sequences of limestones, evaporites, sandstones, and calcareous shales accumulated along what are now the Gulf and Atlantic coastal states. Continental deposits, particularly red beds, characterize the Triassic of the Rocky Mountain regions of North America. Marine conditions persisted along the Pacific eugeosynclinal belt, where volcanic rocks and graywackes accumulated to great thicknesses. During the Jurassic, the eugeosynclinal belt was strongly deformed as western North America began to override the Pacific Plate. This activity, however, did not prevent marine transgressions during parts of the Jurassic and Cretaceous Periods. The surge of deformation that had begun along the Pacific Coast moved progressively eastward, until by late Cretaceous, inland seas were displaced and majestic ranges stood in their place. Crustal unrest continued to affect the western states even during the early epochs of the Cenozoic Era.

The focus on Eurasian Mesozoic events was the great east-west trending Tethys mobile belt. The initial Mesozoic deposits of the western Tethys were predominantly carbonates. Block-faulting developed along parts of the belt during the Jurassic. This middle part of the Mesozoic witnessed marine invasions that completely spanned the continent from the Tethys to the Arctic. During the Cretaceous, the Tethys was affected by powerful compressional forces as the northward-moving oceanic plate bearing peninsular India moved toward and under the Eurasian Plate. One result of these movements was the development of a series of anticlinal welts along the northern side of the Tethyan Geosyncline. Rapid erosion of these elongated, uplifted tracts produced thick sequences of coarse clastics in the intervening troughs. Along the southern margins of the Tethys, conditions were more stable, and relatively undisturbed sequences of carbonates and sandstones were deposited.

Although Australia was predominantly emergent during the Mesozoic, an active geosyncline existed across a tract now occupied by New Zealand. This violently active tectonic zone originated in the Late Paleozoic and reached its orogenic climax during the Jurassic. New Zealand's Cretaceous sediments were for the most part deposited in shallow marginal seas adjacent to the ranges uplifted during the previous period.

The Mesozoic history of the interior regions of Africa is read primarily from continental deposits. Among these, none is more famous than the beds of the Karoo Basin of South Africa. The Karoo sequence includes both Permian and Triassic strata, some of which have yielded splendid collections of the abundant and diverse mammal-like reptiles that once inhabited South Africa.

During the Mesozoic, a largely emergent peninsular India was progressing steadily toward its Eurasian destination. By Cretaceous time, it had approached sufficiently near to cause initial buckling in the eastern Tethys.

The Jurassic was an important geologic period for South America, which during this time began to separate from Africa. The separation continued during the remainder of the Mesozoic, and a new ocean, the South Atlantic, was born. The western margin of the continent formed the leading edge of the South American Plate and was characterized by a single elongate mobile belt—the Andean Geosyncline. Volcanic activity, folding, and intrusions increased in frequency within the mobile belt throughout the Mesozoic.

The orogenies that affected the Americas, Eurasia, and other continents during the Mesozoic did not exhaust the capabilities of these land masses for further tectonic activity. As will be explained in the pages ahead, the geosynclines continued to be the locations of dramatic tectonic events during the Cenozoic Era.

The rocks of the Mesozoic have supplied the world with a variety of important economic resources. These include such fossil fuels as coal, petroleum, and natural gas, as well as nuclear fuels. Orogenic activity along the Mesozoic geosynclinal belts has resulted in the emplacement of such critical metals as copper, zinc, chromite, gold, silver, lead, mercury, and many others. Theories of ore genesis now being formulated suggest that metallic ores emplaced during mountain building along continental margins may represent concentrates of materials on the sea floor that have been conveyed to orogenic belts by sea floor spreading. Once such materials are brought to the plate boundary, favorable physical and chemical conditions must exist locally to provide for concentration of the dispersed metals into ore deposits.

Questions for Review

1. What is the total duration, in years, of the Mesozoic Era? Which period of the Mesozoic had the greatest duration?
2. What sort of geologic happenings or events seem to immediately precede the separation of a land mass into two or more parts? Cite an example from Mesozoic geologic history.
3. Describe the general geologic conditions associ-

ated with the deposition of the Newark Series. Why did normal faulting rather than thrust-faulting occur in this area?
4. What was the probable source area for the sediments of the Morrison Formation of the Rocky Mountain region? What evidence indicates the Morrison was a continental rather than a marine formation?
5. During which period of the Mesozoic were epicontinental seas most extensive? During which period were such incursions most limited?
6. What was the Tethys Sea? In which of today's mountain ranges might one go to examine the now lithified deposits of the Tethys Sea?
7. At what time in geologic history did the modern South Atlantic Ocean begin to form?
8. What nonmetallic mineral resources are derived from Mesozoic strata? What metallic ores occur?
9. What is the paleontologic importance of the Permian-Triassic strata of the Karoo Basin of South Africa?
10. What kind of deformation is particularly characteristic of the Sevier Orogeny?
11. Of all the Gondwanaland segments that separated from Pangaea II, which seems to have drifted over the greatest distance during the Mesozoic?
12. During what geologic period did most of the ranges of the Rocky Mountains experience their major deformation? In what areas are thrust faults prevalent?
13. During what geologic period of the Mesozoic were large volumes of evaporites deposited in the Gulf Coast region? What relationships, if any, do these deposits have to oil entrapment?

Terms to Remember

carnotite
Chinle Formation
Deccan Traps
Décollement
Karoo sequence
Kayenta Formation
Mesocordilleran Highlands
Morrison Formation
Navajo Sandstone
Nevadan Orogeny
Newark Series
Niobrara Formation
ophiolites
rudists
Rocky Mountain Orogeny
Sevier Orogeny
Shinarump Conglomerate
Sonoma Orogeny
Sonoma Terrain
Sundance Sea
Tethys Geosyncline
Wingate Formation

Supplemental Readings and References

Ager, D. V. 1980. *The Geology of Europe.* New York, John Wiley and Sons.

Armstrong, R. L. 1968. The Sevier Orogenic Belt in Nevada and Utah. *Bull. Geol. Soc. Amer.* 79:429–452.

Dietz, R., and Holden, J. C. 1971. The breakup of Pangaea. *In Continents Adrift. Readings from Scientific American.* San Francisco. W. H. Freeman Co.

Hamilton, W. 1981. Plate tectonic mechanism of Laramide deformation. *Contrib. Geology, Univ. of Wyoming* 19(2):87–92.

Hsu, K. J. 1976. Paleoceanography of the Mesozoic Alpine Tethys. *Geol. Soc. Amer.* Special Paper 170, 44 pp.

Huxley, T. 1868. *On a Piece of Chalk* (1967 ed.) New York, Charles Scribner's Sons.

Kummel, B. K. 1970. *History of the Earth* (2d ed.). San Francisco, W. H. Freeman Co.

Kurten, B. 1973. *The Age of Dinosaurs* (2d ed.). New York, McGraw-Hill Book Co.

Manspeizer, W., Putter, J. H., and Cousminer, H. L. 1978. Separation of Morocco and Eastern North America: a Triassic-Liassic stratigraphic record. *Bull. Geol. Soc. Amer.* 89:901–920.

Mudge, M. R., and Earhart, R. L. 1980. The Lewis Thrust Fault and related structures in the disturbed belt of Northwestern Montana. *U.S. Geol. Survey Professional Paper* No. 1174.

Phillips, J. D., and Forsyth, D. 1972. Plate tectonics, paleomagnetism, and the opening of the Atlantic. *Geol. Soc. Amer. Bull.* 22:1579.

Rona, P. A. 1974. Plate tectonics and mineral resources. *In Planet Earth, Readings from Scientific American.* San Francisco, W. H. Freeman Co., pp. 170–180.

Stanley, K. O., Jordan, W. M., Dott, R. H. 1971. New hypothesis of Early Jurassic paleogeography and sediment dispersal for Western U.S. *Bull. Amer. Assoc. Petrol. Geol.* 55(1):10–19.

Tappan, H. 1968. Primary production, isotopes, extinctions, and the atmosphere. *Palaeogeog. Paleocimat. Paleoecol.* 4(3):187–210.

Two small boys view an actual-size model of the carnivorous dinosaur, *Tyrannosaurus rex*, outside the Oak Knoll Museum of Natural History, St. Louis, Missouri. (Photograph by Stephen D. Levin. Statue is one of a group prepared by Sinclair Oil Corporation for the 1964 New York World's Fair.)

13

The Mesozoic Biosphere

Tyrannosaurs, enormous bipedal caricatures of men, would stalk mindlessly across the sites of future cities and go their way down into the dark of geologic time.

Loren Eiseley
The Immense Journey, 1959

PRELUDE

The **biosphere** is the world of life. We can view the biosphere as consisting of all living things and those parts of the continents, oceans, and atmosphere with which living things interact. In this chapter we direct our attention to the biosphere of the Mesozoic Era. As is true today, the distribution, evolution, and abundance of life during the Mesozoic were strongly influenced by climate. Climate, in turn, was affected by the changing locations of continents, major marine transgressions and regressions, and the formation of mountain ranges.

MESOZOIC CLIMATES

Cool climates seem to have characterized many continental areas during the final days of the Paleozoic Era. The vastness of the Pangaea II supercontinent, the upheaval of mountains, general uplifts, and withdrawal on inland seas in many regions were contributing causes of the generally cooler conditions. Gradually, however, the climate warmed, and glaciers in Africa, Australia, Argentina, and India began to melt away as these continents began to drift away from the south pole. In general, the 160 million years of the Mesozoic seemed to have been blessed with warm and rather equable climates.

During the Triassic the continents were still tightly clustered. The paleoequator extended from central Mexico across the northern bulge of Africa (Fig. 13–1). As noted in Chapter 12, the Triassic was a time of general emergence of the continents. Mountains, thrust upward at the end of the Paleozoic, inhibited the flow of moist air into the more centrally located regions, causing widespread arid-

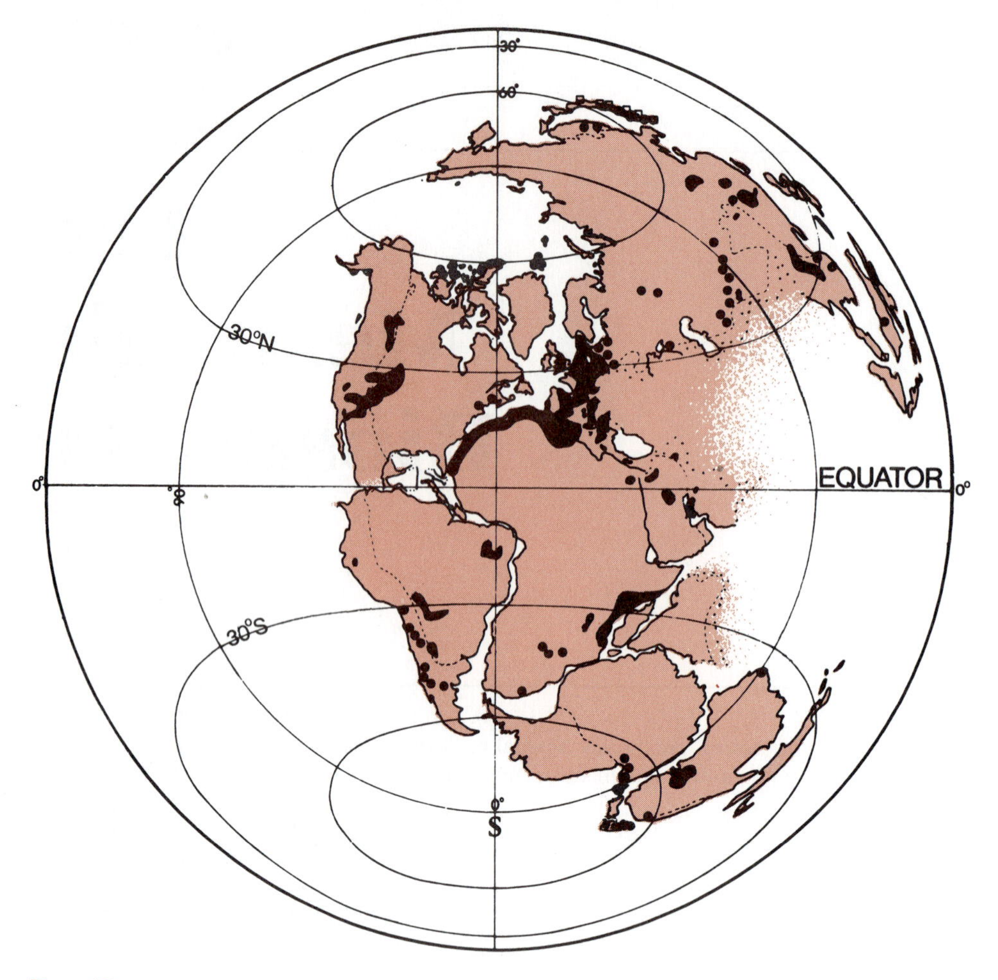

Figure 13–1 Triassic paleogeographic map showing position of equator and land masses, as well as the distribution of evaporites (black areas) and coal (black circles). (From G. E. Drewey, T. S. Ramsey, and A. G. Smith. *J. Geol.*, 82(5):538, 1974.)

ity. Evaporites, dune sandstones, and red beds accumulated at both high and low latitudes and attest to relatively dry and warm conditions.

Reconstructions, based in part on paleomagnetic studies, suggest that during the Jurassic the continents were at the approximate latitudinal positions they occupy today (Fig. 13–2). Marine waters extended northward into a great trough formed by the opening of the Atlantic Ocean, and in many places shallow inland seas spilled out of the deep basins onto the continents. An arm of the Pacific Ocean extended westward as the great Tethys Sea. It is not unreasonable to assume that warm, westward-flowing equatorial currents may have penetrated far into the Tethys. These same equatorial currents, deflected by the east coast of Pangaea II, were shunted to the north along coastal Asia and to the south along northeastern Africa and India. To complete the cycle, the cooled currents may have returned to the equator along the west side of the Americas. The presumed ocean and wind currents, and the rather extensive coverage of seas, both upon and adjacent to continents, perpetuated the mild climates that seem to characterize the Jurassic. Glacial deposits of this age are simply not known, and coal beds occur in Antarctica, India, China, and Canada. Paleobotanical evidence suggests that tropical conditions prevailed over regions that are now in temperate zones. Even dinosaurs lived in temperate areas somewhat north of the present arctic circle.

During most of the Cretaceous, climatic conditions were a continuation of the generally warm and stable conditions that had prevailed during the Jurassic. A remarkably homogeneous flora spread around the world, with subtropical plant families thriving in latitudes 70° from the equator. Coal beds are found on nearly every continent and even

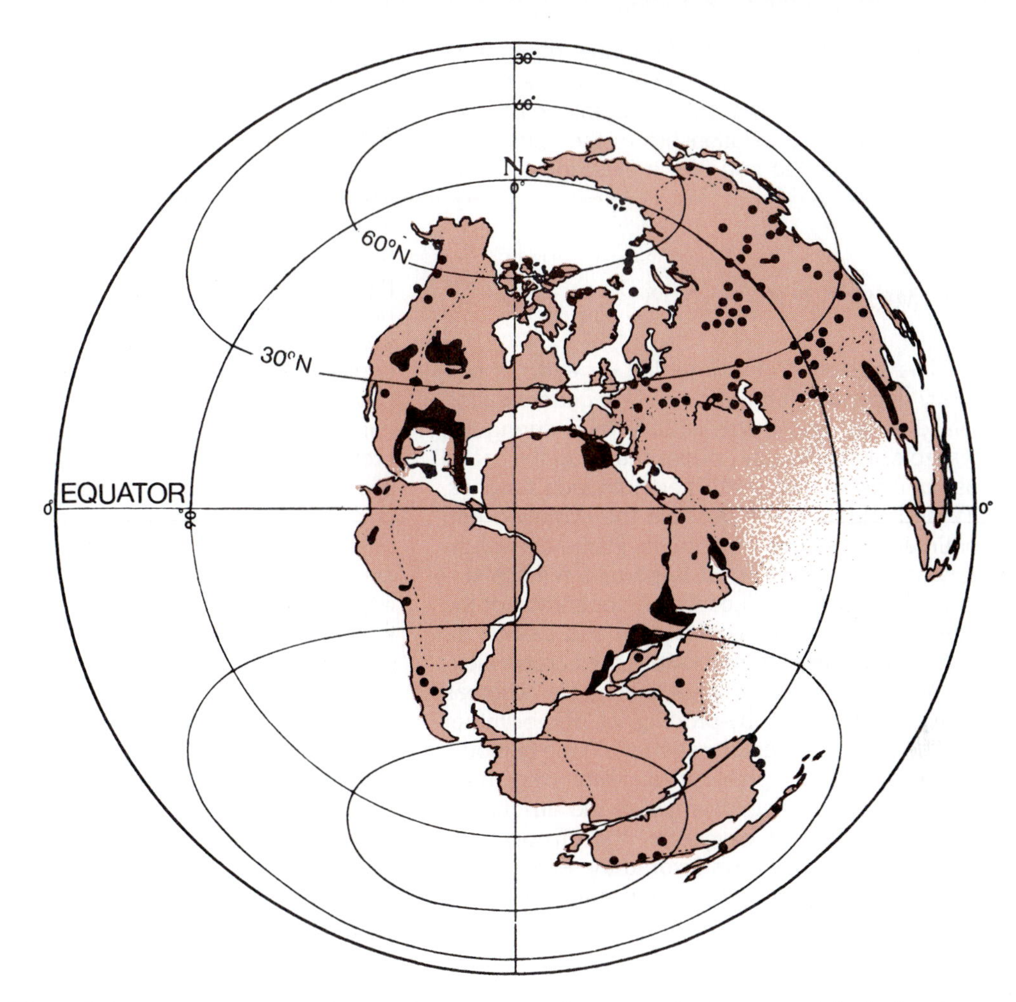

Figure 13–2 Jurassic paleogeographic map showing position of equator and land masses, and distribution of evaporites (black areas) and coal (black circles). (From G. E. Drewey, T. S. Ramsey, and A. G. Smith. *J. Geol.*, 82(5):537, 1974.)

at very high latitudes. However, such pleasant conditions did not persist, and toward the end of the Cretaceous, the climatic pendulum began a slow backward swing toward more rigorous conditions. The south pole was now centrally located in Antarctica, and the north pole was located at the north edge of Ellesmere Island. Europe and North America had moved somewhat farther north, and the widespread inland seas had begun to recede. The worldwide regressions were accompanied in some regions by major orogenic disturbances.

In the Late Cretaceous seas, the golden brown algae known as coccolithophorids spread in enormous profusion. Their calcium carbonate platelets accumulated on the floors of the inland seas and were converted into thick layers of chalk. They are estimated to have been so abundant in the Late Cretaceous that their photosynthetic activity may have produced a carbon dioxide shortage in the atmosphere. Because carbon dioxide provides the earth with a warming "greenhouse effect" it is at least theoretically possible that the great blooms of plankton may have contributed, along with the other environmental conditions, to a temporary chilling of the late Cretaceous world.

There are several lines of evidence that favor the interpretation of a terminal Cretaceous cooling episode. Terrestrial plants provide some clues. The tropics-loving cycads underwent a sharp reduction, and ferns declined in both North America and Eurasia. Hardier plants, such as conifers and angiosperms, extended their realms. Evidence comes also from paleotemperature studies based on the oxygen isotope method described earlier, in Chapter 4 (see Fig. 4–31). The oxygen isotope ratios obtained from open ocean planktonic calcareous organisms indicate a decline in ocean temperatures beginning about 80 million years ago. If there was indeed a dip in worldwide annual mean temperatures, it might have had a deleterious effect on animal life. For this reason, some paleontologists speculate about its relationship to the extinctions that occurred at the end of the Mesozoic.

THE MESOZOIC FLORA

The existence of animal life on earth is ultimately dependent on plant life. This generalization is valid today, and it was equally valid during the Mesozoic. Then, as now, plants made up the broad base of the food pyramid. Their nutritious starches, oils, and sugars made possible the evolution and continuing existence of animals. Plants are a fundamental part of the most important self-sustaining ecologic system on earth. The operation of the system is dependent upon oxygen and carbon dioxide. Animal respiration provides the carbon dioxide needed for plant photosynthesis, whereas plants supply—as a by-product of photosynthesis—the oxygen needed by animals. In the geologic past, variations in plant productivity may have caused corresponding changes in the amount of carbon dioxide and oxygen in the atmosphere. Such variations may have favored the evolution of some organisms over others and may even have been responsible for the demise of particular groups.

Marine Plants

Because plants that live in the oceans are suspended in water, they do not require the vascular and support systems that characterize land plants. The majority of marine plants are therefore unicellular, although they may grow in impressive colonies and aggregates. The major groups of marine plants are part of that vast realm of floating organisms called plankton; because they are plants, they are further designated **phytoplankton.** The geologic record of the important fossil groups of phytoplankton is shown in Figure 13–3. As indicated on the chart, phytoplankton that did not secrete mineralized coverings predominate in the pre-Mesozoic eras. These include both blue-green and green algae, as well as a group of cellulose-covered unicellular algae of uncertain affinity that are loosely called acritarchs. Of more immediate importance to us are the coccolithophorids, dinoflagellates, silicoflagellates, and diatoms, which were the more abundant phytoplankton groups of the Mesozoic. The expansion of the coccolithophorids and dinoflagellates really began in the Early Jurassic.

Dinoflagellates are frequently encountered as fossils and are important aids in Mesozoic and Cenozoic stratigraphy. From the Jurassic on, they were among the primary producers in the marine food chain. Dinoflagellates (Fig. 13–4) are unicellular organisms having a cellulose cell wall like that in pollen. For propulsion, the cell wall is equipped

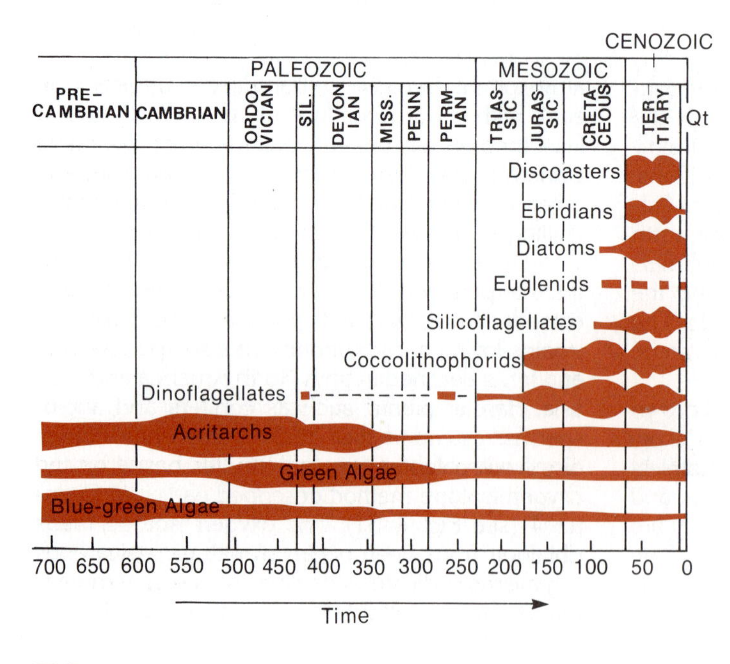

Figure 13–3 Geologic distribution and abundances of phytoplankton. (From H. Tappan and A. R. Loeblich, Jr. Geobiological implications of fossil phytoplankton evolution and time-space distribution. [Symposium of 1966.] *Geol. Soc. Am.*, Special Paper 127:257, 1970.)

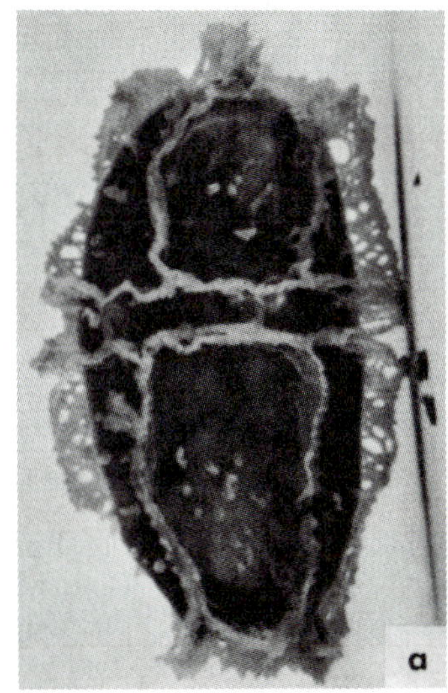
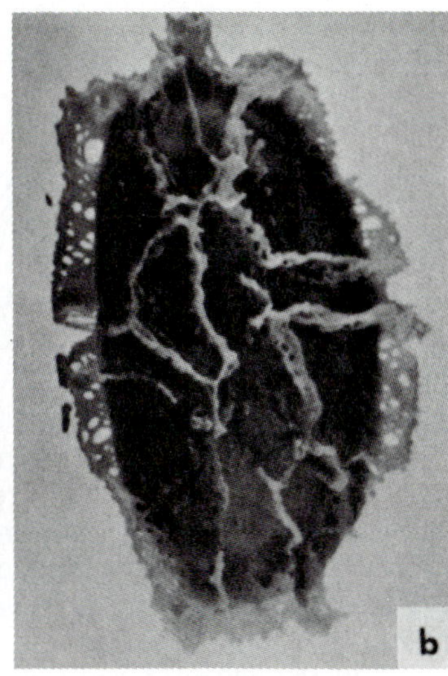

Figure 13–4 Fossil dinoflagellate cyst, *Prionodinium alveolatum*, from the Cretaceous of Alaska. (From H. A. Leffingwell and R. P. Morgan. *J. Paleontol.*, 51(2):292, 1977.)

with two flagella: One is longitudinal and whiplike, and the other is transverse and ribbon-like. During their life cycle, dinoflagellates develop a motile plankton form and a cyst phase that is formed within the motile organism. The cyst form is extremely resistant to decay, and, indeed, only dinoflagellate cysts are known as fossils.

The coccolithophorids also began their expansion during the Early Jurassic. Unlike the dinoflagellates, these calcium carbonate–secreting organisms have a splendid fossil record. Their abundant remains have formed many of the extensive coccolith limestones of the Mesozoic and Early Cenozoic. Even today, they are frequently present in the deep sea sediment known as calcareous ooze. The coccolithophorid organism is one of several varieties of unicellular golden brown algae. These algae deposit calcium carbonate internally on an organic matrix and construct tiny, shieldlike structures called coccoliths. Once formed, the coccoliths move to the surface of the cell and encase it in calcareous armor (Fig. 13–5). Coccoliths measure only between 1 and 16 microns in diameter (1 micron equals 0.001 mm). However, with the aid of the electron microscope, it is possible to discern their intricate construction. With such magnification, coccoliths are seen to consist of one, or sometimes two, superimposed elliptical plates. These plates are concave on one surface so that they can fit closely around the outside of the spherical cell. Each disc or plate is itself composed of still smaller crystalloids, and these may be triangular, rhombic, or variously shaped by the organism. The crystalloids are uniformly arranged, usually in a circular, radial, or spiral plan that is often astonishing in its beauty and precision (see Fig. 2–35). Because they are frequently fossilized, have experienced frequent evolutionary changes through time, and are readily dispersed by oceanic currents, coccoliths have become extremely useful in stratigraphic correlation.

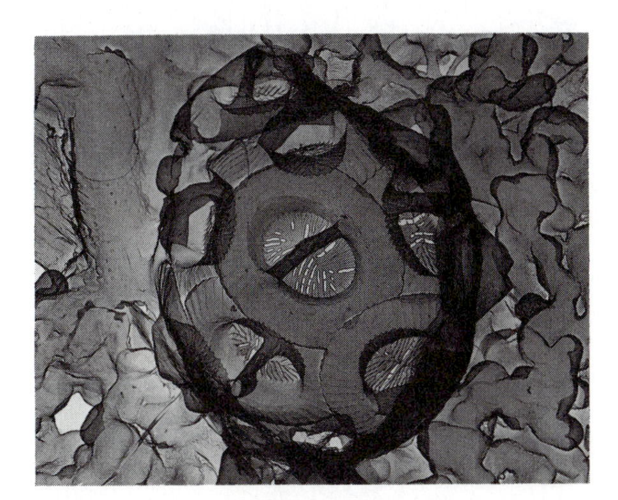

Figure 13–5 Scanning electron micrograph of a coccosphere covered with coccoliths. (From J. M. McCormick and J. V. Thiruvathukal. *Elements of Oceanography*. 2nd ed. Philadelphia, Saunders College Publishing, 1981. Photograph courtesy of S. Honjo.)

The earliest known silicoflagellates (Fig. 13–6) and diatoms appeared in the Middle Cretaceous. Along with other phytoplankton, they experienced a decline at the end of the Cretaceous, and then all groups expanded again into the early epochs of the Cenozoic. The silicoflagellates and diatoms are, along with the coccolithophorids, members of the Phylum Chrysophyta.

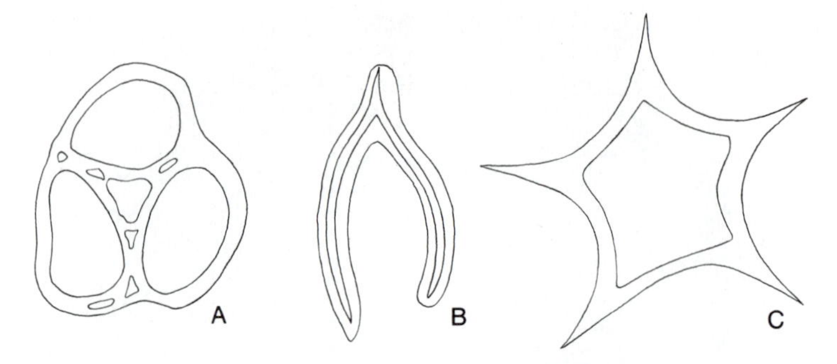

Figure 13–6 Three Mesozoic silicoflagellates. (A) *Corbisema* (Cretaceous-Eocene). (B) *Lyramula* (Cretaceous). (C) *Vallacerta*. (Cretaceous).

As suggested by their somewhat ponderous name, the silicoflagellates are flagella-bearing algae that secrete a siliceous internal "skeleton." Radiating spines characterize most genera, whereas others are roughly stellate. They range in size from 10 to 150 microns.

Like the silicoflagellates, diatoms also secrete siliceous coverings (Fig. 13–7). The covering is called a **frustule.** It is usually composed of an upper and lower part that fit together like a lid on a box. The test may be circular, cylindric, triangular, or a variety of other shapes and is usually quite beautiful. Today, marine diatoms are most prevalent in the cooler regions of the oceans. In the past, a proliferation of diatoms was often associated with volcanic activity. Apparently, the silica supplied to sea water as fine volcanic ash stimulated diatom productivity.

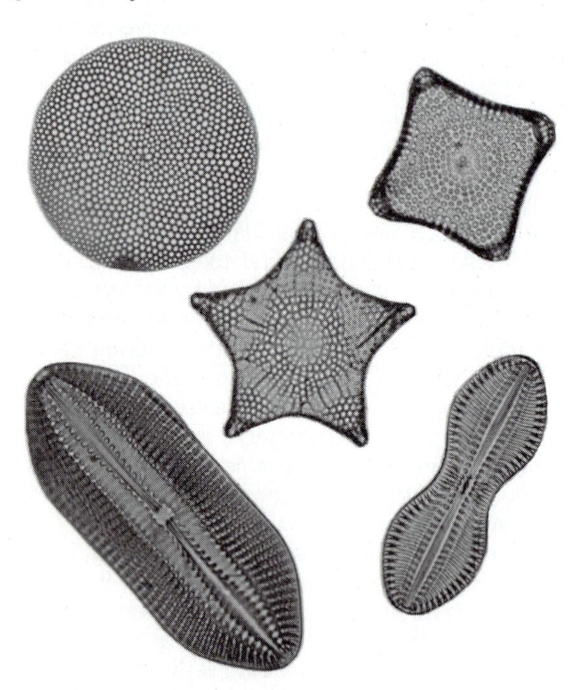

Figure 13–7 Diatoms. These modern diatoms were collected off the coast of Crete. (Photographs courtesy of Naja Mikkelsen, Scripps Institution of Oceanography.)

Terrestrial Plants

For the vertebrate paleontologist, the "Age of Dinosaurs" may seem a splendid way to designate the Jurassic Period. However, a paleobotanist might well argue that the "Age of Cycads" would be equally appropriate. The cycads, more formally known as the Class Cycadophyta (Fig. 13–8), are seed plants in which true flowers have not been developed. Jurassic cycads included tall trees with

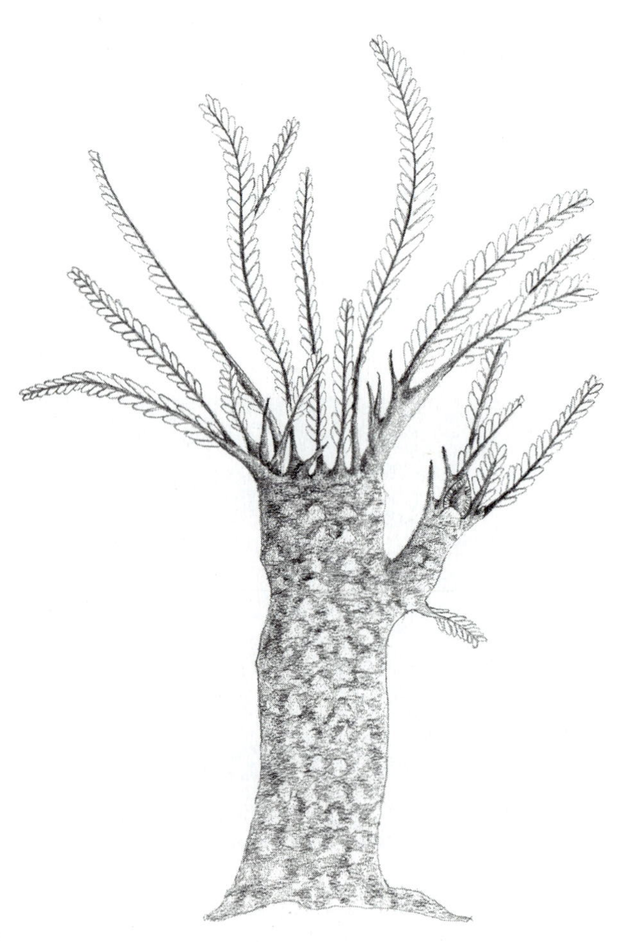

Figure 13–8 Restoration of a cycadophyte *(Williamsonia)* from Jurassic strata of India.

rough, columnar branches marked by the leaf bases of earlier growths and by crowns of leathery pinnate leaves. Actually, the Cycadophyta include two related groups, the cycadeoids ("fossil cycads") and Cycadales ("true cycads"). Although the adult plants in both of these groups were superficially similar, they differed in reproductive structures and in other anatomic details. Cycads experienced a marked decline in the Late Cretaceous, and only a few have survived to the present time. One such survivor is the sago "palm."

In Chapter 11, it was noted that there were three important episodes in the evolution of land plants. The first stage led to the development of the spore-bearing, leafy, treelike plants. The second witnessed the evolution of nonflowering, pollinating seed plants such as cycads, ginkgoes, seed ferns, and conifers. As indicated in Figure 13–9, all but the seed ferns have living representatives, and the conifers are today widespread around the world. The third important episode in plant history is marked by the advent of species that carried enclosed seeds and bore flowers. Such plants are known as **angiosperms.** Angiosperm-like pollen grains obtained from rocks of Cretaceous age provide the earliest evidence of the existence of flowering plants. By Cretaceous time, angiosperms were conspicuously present. Forested areas included stands of birch, maple, walnut, beech, sassafras, poplar, and willow trees (Fig. 13–10). Before the Cretaceous came to a close, angiosperms had surpassed the nonflowering plants in both abundance and diversity. Flowering trees, shrubs, and vines expanded into every corner of the globe and gave a modern appearance to every landscape.

Either directly or indirectly, the evolution of Mesozoic insects, birds, reptiles, and mammals was strongly influenced by this remarkable floral revolution. Angiosperms produced a variety of nuts and fruits that were intended to insure survival of the plant embryo but that also insured the survival of animals that used this bounty as food.

The relationship between insects and flowering plants is well known. The angiosperms, by encouraging insect visits, are able to utilize insects as delivery agents for pollen. This provides a far more efficient means of pollen dispersal than the more random wind pollination, which requires each plant to disperse tens of millions of pollen grains rather than a few dozen. The insect pollinators are en-

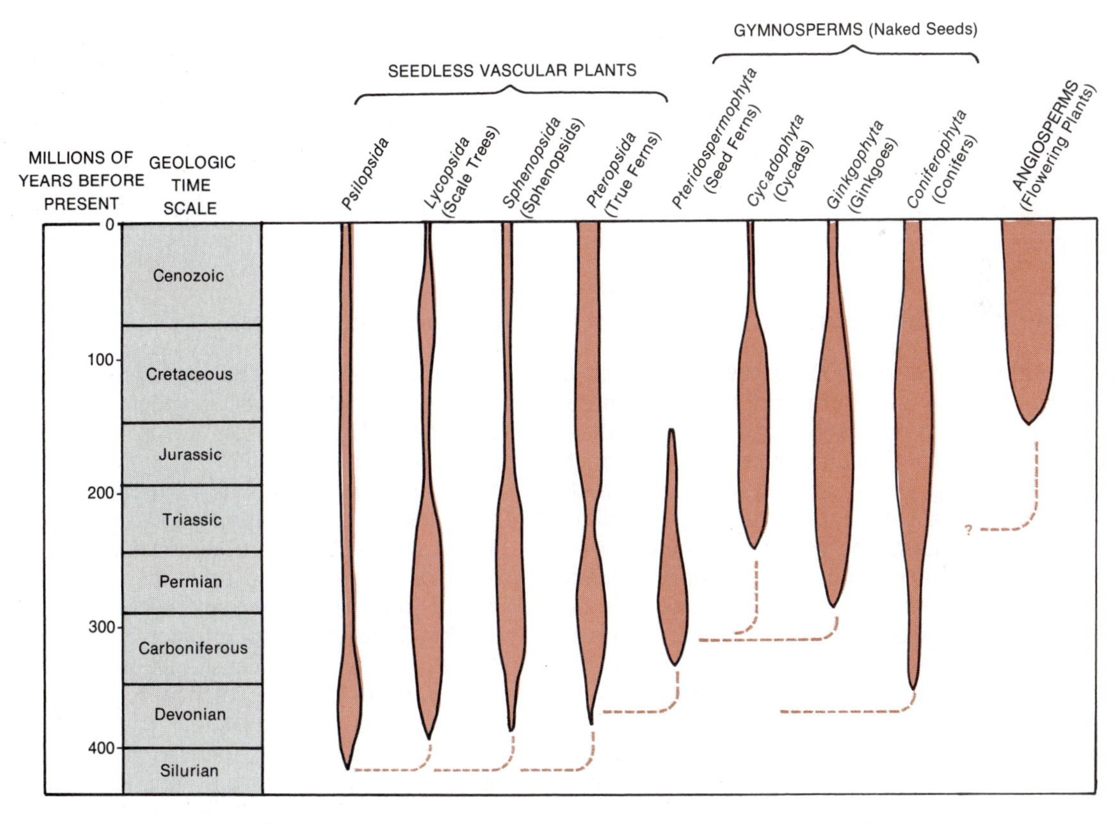

Figure 13–9 Geologic ranges, relative abundances, and evolutionary relationships of vascular land plants.

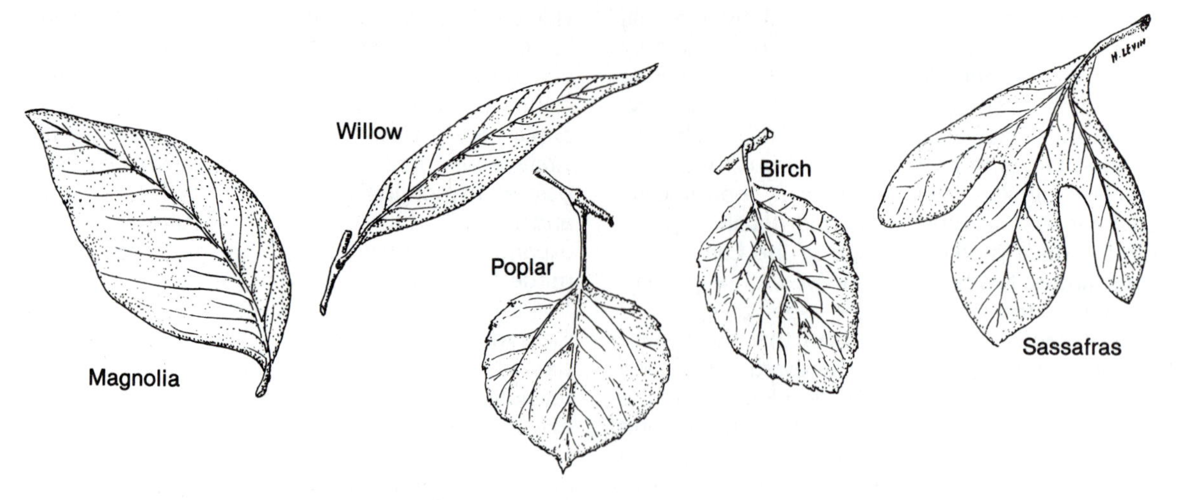

Figure 13–10 Leaves of angiosperms found fossilized in Cretaceous Rocks.

couraged to visit particular flowers in order to obtain the nectar and pollen needed for food. Having had success with a flower of a certain form, color, and scent, they move off to find others of the same kind in order to repeat the favorable experience. In this way, pollen is transported on the hair, legs, and bodies of insects from plant to plant of the same species. The selective competition for efficient pollinators has induced a constantly changing range in variations among both plants and insects. In the angiosperms, the need for each plant to be recognizably different resulted in a spectacular floral variety that has persisted from the Cretaceous down to the present.

THE INVERTEBRATES

Continental Invertebrates

Although paleontologists have assembled an enormous body of information about the marine invertebrates of the Mesozoic, relatively little is known about continental groups. The pulmonate, or air-breathing, snails are rarely found. Freshwater clams and snails are more commonly fossilized. Freshwater crustaceans, especially such groups as **branchiopods** and nonmarine species of ostracods, are frequently found in Mesozoic lake bed deposits. It is reasonable to assume that many varieties of worms existed, but they left few traces. The spiders, millipedes, scorpions, and centipedes that had been abundant in Carboniferous forests undoubtedly were present also in the Mesozoic, although their remains are elusive. Among the insects, flies, mosquitoes, caddis flies, earwigs, wasps, bees, and ants are known from rocks as old as Jurassic. Many of the best fossils are collected from the Solnhofen Limestone, an unusual Jurassic formation in Bavaria. Even the Solnhofen collection is not truly representative, however, because most of the remains are of larger species. This observation has caused paleontologists to assume that smaller insects were eaten by fish or were decomposed.

The fossil record for Cretaceous insects is also sadly inadequate. Insects preserved in amber of Cretaceous age are known from Canada and Alaska. Specimens include bees, wasps, ants, beetles, flies, and mosquitoes. Butterflies, moths, termites, and fleas occur in rocks of the Early Cenozoic but have not yet been found in rocks as old as Cretaceous.

Marine Invertebrates

At the end of the Paleozoic, many families of marine invertebrates either declined or suffered extinction. The Mesozoic resurgence of marine organisms was somewhat tardy, as indicated by the limited nature of Early Triassic faunas. However, after a few groups had become well established, marine life expanded dramatically. The pelecypods (Bivalvia) became increasingly prolific from Middle Triassic on and eventually surpassed the brachiopods in colonization of the sea floor. Among the most successful pelecypods were the oysters (Fig. 13–11), which were represented by such genera as *Gryphaea* (Fig. 13–12) and *Exogyra* (Fig. 13–13). Some members of the oyster group became giants of their kind. Other pelecypods grew conical shells that strikingly resembled the horn corals of the Paleozoic. In these

Figure 13–11 Fossil oysters in the Upper Cretaceous Selma Formation, Lee County, Mississippi.

forms, called **rudists** (see Fig. 12–7), the left valve formed a small lid that closed the open end of the cone. Rudists, which are known from the Jurassic and Cretaceous, formed impressive reef structures.

In those parts of the Mesozoic marine environment where the water was warm, relatively clear, and shallow, corals proliferated. During the Late Jurassic, for example, the Tethyan Geosyncline was the site of major coralline evolution and reef building. The corals of the Mesozoic are termed **scleractinids.** Unlike their Paleozoic predecessors, they did not use calcite to build their skeletal structures but rather the crystallographically different (but compositionally identical) aragonite. As is the case today, scleractinids had rather restrictive environmental requirements. Most could exist only in clear water of normal salinity no deeper than about 50 meters, with temperatures no lower than about 20°C. One reason corals require shallow water is that they have a symbiotic relationship with an alga that lives within the coral polyp and is dependent on sunlight. Corals seem to have lived much far-

Figure 13–12 *Gryphaea.* These specimens were obtained from Jurassic strata in southern England. (Photograph courtesy of Wards Natural Science Establishment, Rochester, N.Y.)

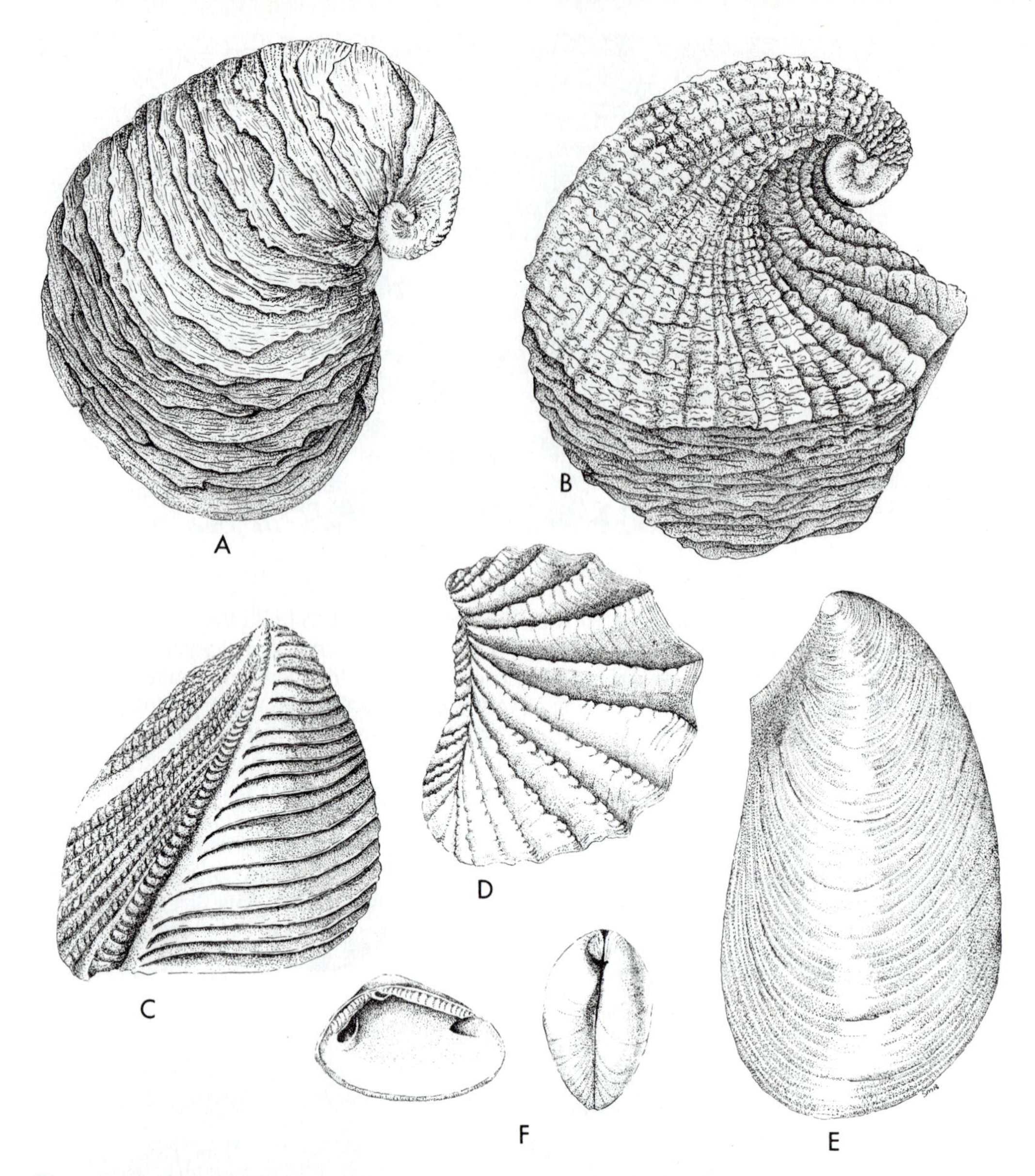

Figure 13–13 Mesozoic pelecypods. (A) *Exogyra ponderosa* (Cretaceous). (B) *Exogyra cancellata* (Cretaceous). (C) *Trigonia thoracica* (Cretaceous). (D) *Trigonia costata* (Jurassic). (E) *Inoceramus labiatus* (Cretaceous). (F) *Nucella percrassa* (interior of right valve and dorsal view).

ther north in the Mesozoic than they do today, suggesting a more northerly location for the paleoequator during that era.

The great reefs of stony corals offered food and shelter to a host of other kinds of oceanic life. Persisting groups of brachiopods clung to the reef structures, as did rudists, algae, bryozoans, and other sedentary creatures. Gastropods grazed ceaselessly along the reef structures, whereas crabs and shrimp scuttled about seeking food in the recesses and cavernous hollows of the reefs. In the quieter lagoonal areas behind the reefs and on the floors of the epicontinental seas, starfish, sea urchins, and crinoids often thrived.

Those cousins of the crinoids, the echinoids, became far more diverse and abundant in the Mesozoic than they had ever been in the preceding era. Some Lower Cretaceous formations in the Gulf Coastal states contain prodigious remains of these spiny creatures. Collectors in Europe prize the silicified echinoids obtained from Cretaceous chalk beds. The "regular" sea urchins were especially

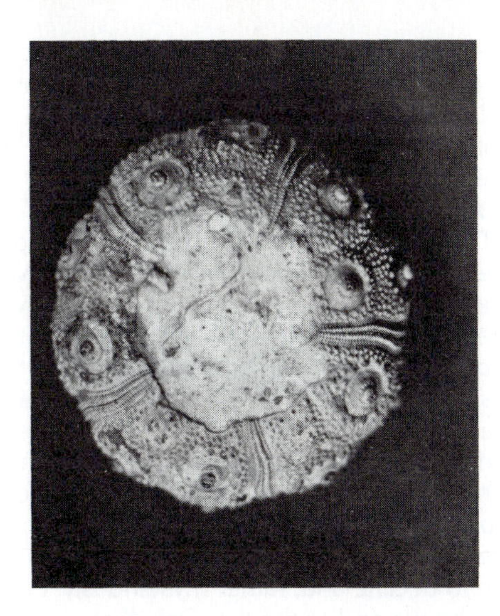

Figure 13–14 A regular echinoid from Jurassic strata in Germany.

numerous. In these forms the symmetry is pentameral, and the shell, or **test,** is spherical (Fig. 13–14). The regular forms were overtaken by the "irregular" echinoids during the Cretaceous (Fig. 13–15). These are mostly flattened, bilateral sea urchins that live as burrowers in the sediment of the sea floor.

If Poseidon were a paleontologist, he would probably have designated the Mesozoic "The Age of Ammonoids." The name would be suitable not only because these cephalopods were so abundant but also because they have proved to be useful in worldwide correlation. Zones developed on the basis of ammonoid guide fossils permit correlation of Mesozoic time-rock units with a level of preci-

Figure 13–15 An "irregular" echinoid from Upper Cretaceous chalk beds near Paris.

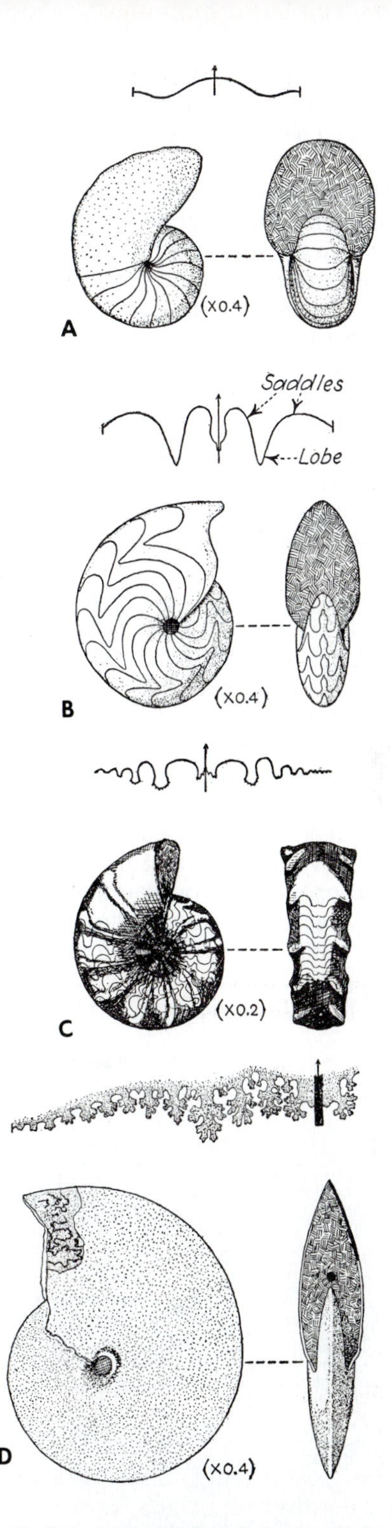

Figure 13–16 Sutures of cephalopod conchs. (A) A nautiloid cephalopod (*arrow* at midventral line pointing toward conch opening). (B) Ammonoid cephalopod with goniatitic sutures. (C) Ammonoid with ceratitic sutures. (D) Ammonoid cephalopod with ammonitic sutures. (From W. H. Twenhofel and R. R. Shrock. *Invertebrate Paleontology*. New York, McGraw-Hill Book Co., 1935. Copyright © 1935 McGraw-Hill Book Co.)

sion surpassing that of radiometric techniques. You may recall that two orders of cephalopods arose during the Paleozoic Era. These were the Nautiloidea, having relatively smooth sutures, and the Ammonoidea, having wrinkled sutures. **Sutures** of cephalopods are lines formed on the inside of the conch where the edge of each chamber's partition, or **septum,** meets the inner wall. Wrinkled sutures are simply a reflection of septa that, like the edges of a pie crust, are fluted. Based on the complexity of the suture patterns (Fig. 13–16), the Ammonoidea can be subdivided into goniatites (see Fig. 11–50), which lived from Silurian through Permian time, ceratites, which were abundant in Permian and Triassic marine areas (Fig. 13–17), and ammonites. Ammonites (Fig. 13–18), although represented in all three Mesozoic periods, were most prolific during the Jurassic and Cretaceous.

Knowledge of the exact suture pattern of ammonoid cephalopods is necessary for their identification and hence their use in correlation. The function of the septal fluting that produced the suture patterns provides an interesting subject for speculation. Most paleontologists favor the theory that, like the corrugated steel panels used in buildings, fluted septa provide greater strength. Comparison studies based on the living cephalopod *Nautilus* have revealed that the gas-filled chambers exert only a slight outward pressure, whereas the water pressure on the outside of the conch wall is considerable. The septal fluting may have helped the animal to withstand the differences in pressure. Proponents of this concept call attention to the fact that ammonoid conchs (unlike nautiloid conchs) tend to thin toward the adult chambers. Apparently, in order to compensate for that thinning, and presumably weaker, conch wall, the septa in many species became more closely spaced and more intricately fluted.

Figure 13–17 An ammonoid cephalopod, *Ceratites* (Triassic), with ceratitic sutures.

The greater variety of Mesozoic ammonoids is a reflection of their success in adapting to a variety of marine environments. They seem to have expanded not only within the shallow epicontinental seas but also in the open oceans. During the Cretaceous, many ammonites became aberrant in shape, so that the normally planispiral forms were joined by species with open spirals, straightened conchs, and even some that coiled in a helicoid fashion, like that of a snail. However, near the end of the Cretaceous, the entire diverse assemblage began to decline and, rather mysteriously, became extinct by the end of the era. Only their close relatives, the nautiloids, managed to survive.

Another group of cephalopods that became particularly common during the Mesozoic were the squidlike belemnites (see Fig. 4–30). The belemnite conch was inside the animal and resembled a cigar in shape. The pointed end was at the rear of the animal, and the forward part was chambered. Most belemnite conchs were less than a half meter in size, but some attained lengths of over 2 meters. A few remarkable specimens from Germany are preserved as thin films of carbon and clearly show the 10 tentacles and body form. X-ray study of these fossils has revealed details of internal anatomy as well. Much like the modern-day squid, the belemnites were probably able to make rapid reverse dashes by jetting water out of a funnel located at the anterior end.

The belemnites were highly successful during the Jurassic and Cretaceous. Triassic belemnites may very well have been the ancestors to the squids, which were also numerous during the Jurassic and Cretaceous. Octopods, because they lack a conch, are inadequately recorded in the Mesozoic. However, their presence is affirmed by an imprint of an octopus found in strata of Late Cretaceous age from Lebanon.

Marine gastropods were also abundant during the Mesozoic. Many are found in sediments that represent old beach deposits. Then, as now, cap-shaped limpets grazed slowly across wave-washed boulders while a variety of snails with tall, helicoid conchs crawled about on the surfaces of shallow reefs. For the most part, the gastropod fauna had a decidedly modern appearance and included many colorful and often beautiful forms that have present-day relatives.

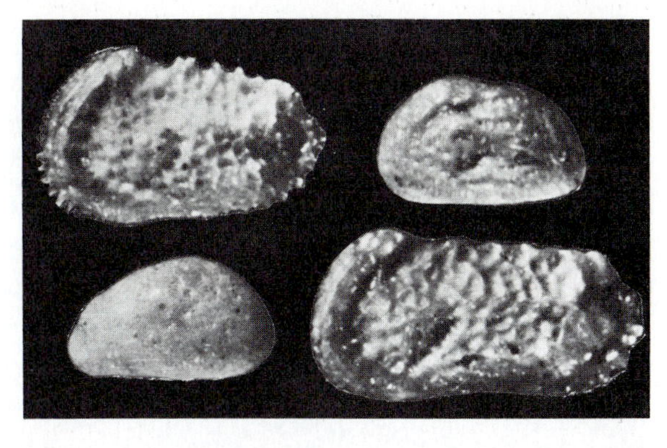

Figure 13–19 **Ostracoda,** 30 ×. (Photograph courtesy of D.J. Echols, Washington University.)

Figure 13–18 Cretaceous ammonoid cephalopods from the Mancos Shale of New Mexico. (A) *Hoplitoides sandovalensis* (vertical diameter of 9.5 cm). (B) Apertural view of another individual of same species as (A). (C) Part of outer whorl of *Tragodesmoceras socorroense* (maximum height of 16 cm). (Photograph courtesy of W. A. Cobban, United States Geological Survey.)

Modern types of marine crustaceans, such as crayfish, lobsters, crabs, shrimp, and ostracods, were abundant by Jurassic time. Of these crustaceans, the ostracods (Fig. 13–19) have become particularly useful to stratigraphers. At some localities, barnacles grew prolifically on rocks and reef structures.

Among the single-celled protozoan animals that crawled or floated about the Mesozoic seas were the radiolarians (Fig. 13–20) and the foraminifers. Radiolarians make their open, delicate, spinose skeletons from opaline silica. In some regions today, radiolarian and diatom skeletal remains accumulate on the sea floor to form extensive deposits of siliceous ooze and in the past have contributed to the formation of chert beds.

The tests of foraminifers are generally more durable than those of radiolarians. The "forams," as they are often called, left an imposing Mesozoic fossil record and one that is of the greatest importance in stratigraphic correlation. They are especially important in the exploration for petroleum. Because of their small size and strong tests, large numbers of foraminifera can be obtained unbroken from the small pieces of rock recovered while drilling for oil. They are then used in tracing stratigraphic units from well to well. Forams are also sensitive indicators of water temperature and salinity and have provided data useful in reconstructing ancient environmental conditions.

Foraminifers were only meagerly represented in the Triassic but began to proliferate in the Jurassic and Cretaceous. Their expansion continued well into the Cenozoic. Nearly all were bottom-dwelling species until Cretaceous time, when plankton groups began their colonization of the upper levels of the ocean in prodigious numbers. Among the planktonic foraminifers, such genera as *Rotalipora* and *Globotruncana* (Fig. 13–21) contributed their empty tests to the calcareous sea floor sedimentary beds, some of which were later lithified into Cretaceous chalks and marls.

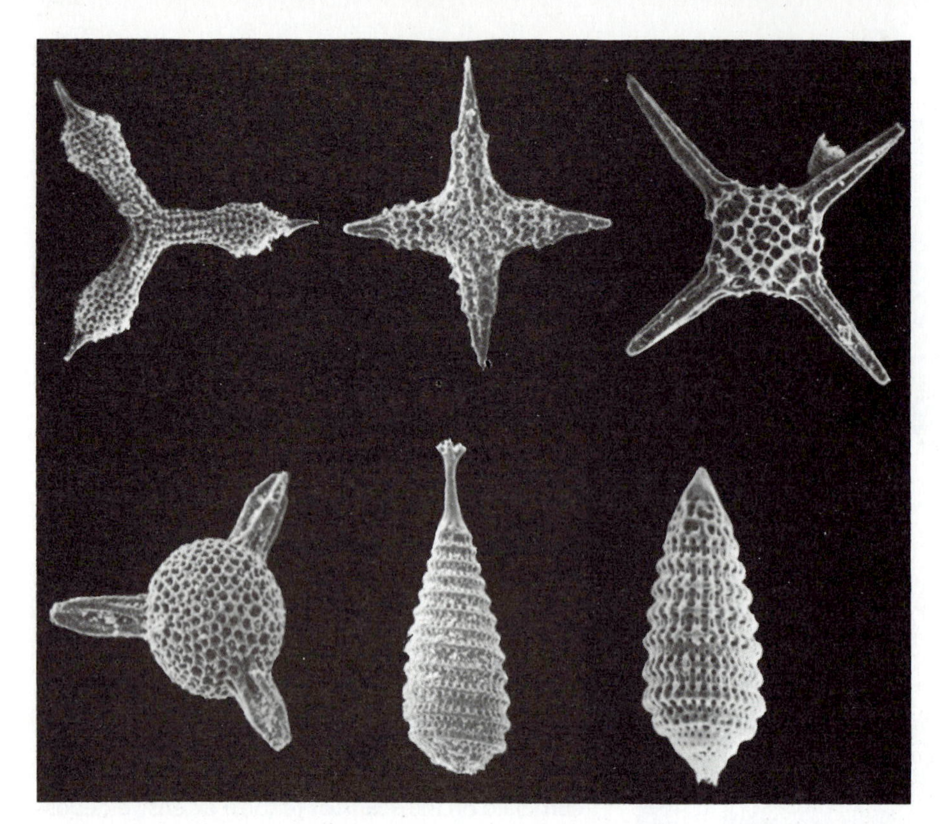

Figure 13–20 Scanning electron micrographs of Jurassic radiolaria from the Coast Ranges of California. Top row from left to right: *Paronaella elegans, Crucella sanfilippoae,* and *Emiluvia antiqua.* Bottom row from left to right: *Tripocyclia blakei, Parvicingula santabarbaraensis,* and *Parvicingula hsui.* (From E. A. Pessagno, Jr. *Micropaleontology,* 23(1): 56–113, 1977, selected from plates 1–12.)

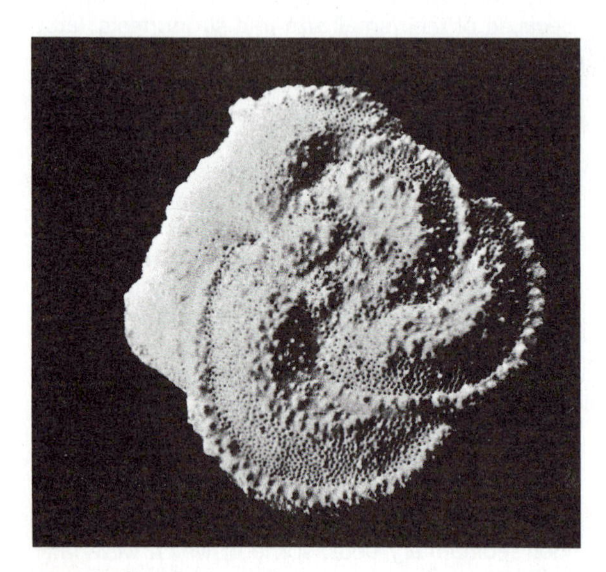

Figure 13–21 The Cretaceous planktonic foraminifer *Globotruncana* (magnification 100×).

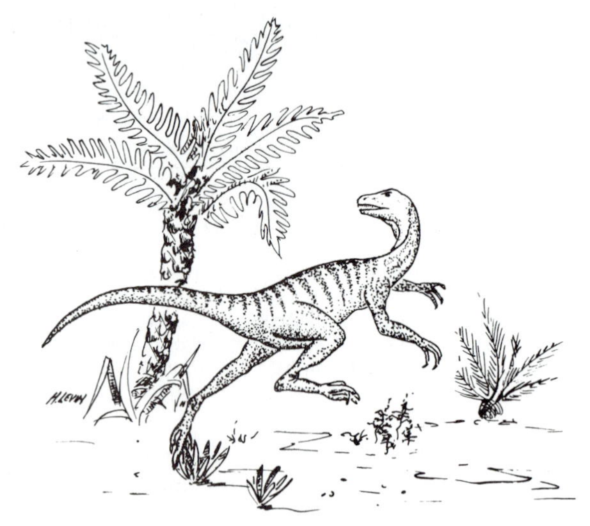

Figure 13–22 *Hesperosuchus* from the Triassic of the southwestern United States.

The Triassic Transition

The general unrest, broad uplifts, and upheavals that occurred during the Carboniferous and Permian periods caused regressions of epicontinental seas, resulted in a variety of continental environments, and generally provided the environmental stimulus needed to maintain the spread and diversification of land vertebrates. Although marine faunas change rather abruptly in passing from the Paleozoic to the Mesozoic, there is considerably more continuity in land faunas. The labyrinthodont amphibians, for example, continued into the Triassic before becoming extinct. The cotylosaurs, or "stem reptiles," also were able to cross the era boundary. Other groups that continued from the Permian were the therapsids. The most progressive of these mammal-like reptiles, the Ictidosauria, succeeded their Permian precursors and lived successfully on into the Late Triassic as contemporaries of primitive mammals.

Many new reptile types appeared in the Triassic. Among these were the ancestors to the first turtles. Triassic turtles were basically similar to their living descendants except that they retained teeth in their jaws. The Triassic was also the geologic period during which various lineages of marine reptiles began to appear. The rhynchocephalians, represented today by the tuatara of New Zealand, were abundant. Most interesting of all, however, were reptiles known as archosaurs. The *Archosauria* are a large and important group of reptiles that include the living crocodiles and the extinct flying reptiles, dinosaurs, and thecodonts. The thecodonts have a distinguished place in vertebrate history, for they were the ancestors of the dinosaurs.

The Thecodonts

Early thecodonts, as exemplified by *Hesperosuchus* (Fig. 13–22) and *Euparkeria* (Fig. 13–23), were small, agile, lightly constructed reptiles with long tails and short forelimbs. They had already developed the unique habit of walking relatively erectly on their hind legs. This bipedal mode was an important innovation. Bipedalism permitted thecodonts to move about more speedily than their sprawling ancestors. Because their forelimbs were not used for support, they could be employed for catching prey; even more important, they could be modified for flight. Thus, the thecodonts were the ancestors not only of dinosaurs but of flying reptiles and birds as well.

Not all thecodonts were nimble, bipedal sprinters; some reverted to a four-footed stance and evolved into either armored land carnivores or large crocodile-like aquatic reptiles called **phytosaurs** (Fig. 13–24). Occasionally in the history of life, initially unlike organisms from separate lineages gradually become more and more similar in form. The once distinctly different groups, in fact, change over many generations so that they are better adapted to a particular environment. The evolutionary process responsible for the trend toward similarity in form is called **convergence.** Phytosaurs and crocodilians are good examples of evolutionary convergence. Indeed, the most visible distinction between the two groups is the position of the nostrils, which are at the end of the snout in crocodiles but were just in front of the eyes in phytosaurs.

The Dinosaurs

Of all the reptiles that now live on this planet or have lived on it in the past, few are more fascinating than dinosaurs (Fig. 13–25). Dinosaurs are the

Figure 13–23 Restoration of the Lower Triassic thecodont, *Euparkeria*.

Figure 13–24 *Rutiodon*, a Triassic phytosaur.

most awesome and familiar of prehistoric beasts. These headliners of the Age of Reptiles compose not one order, but two, each having evolved separately from the thecodonts. The two orders are the Saurischia ("lizard-hipped") and the Ornithischia ("bird-hipped"). As suggested by these names, the arrangement of bones in the hip region provides the criterion for the twofold classification. The reptile pelvis is composed of three bones on each side. The uppermost bone is the **ilium,** which is firmly clamped to the spinal column. The bone extending downward and slightly backward is the ischium. Forward of the **ischium** is the **pubis.** In the saurischians, the arrangement of the three pelvic bones is triradiate, as it was in their thecodont ancestors. However, in the ornithiscians the pubis is swung downward and backward so that it is parallel to the ischium, as in birds (Fig. 13–26).

The earliest dinosaurs were nearly all saurischians. Most were relatively light, nimble, carnivorous bipeds known from fossils discovered in Triassic beds of both South and North America as well as China. *Coelophysis* (Fig. 13–27), for example, was a small, hollow-boned early saurischian found in the Chinle Formation of New Mexico. These birdlike reptiles, called coelurosaurs, continued into the Jurassic and Cretaceous. *Ornithomimus* (Fig. 13–28) must have looked very much like an ostrich, with its long neck, toothless jaws, and small head. Because it lacked teeth, paleontologists speculate that it lived on the eggs laid by its contemporaries; however, it could just as easily have eaten smaller vertebrates. It is possible that *Ornithomimus* and its kin were capable of a certain amount of body temperature regulation—"warm-bloodedness."

Mammals and birds are ideal examples of warm-blooded animals. They are able to maintain a relatively constant body temperature. Like other animals, they produce heat by oxidizing food. When body temperature rises, special physiologic mechanisms regulated by the hypothalamus (part of the brain) help to dissipate the heat. In mammals, these mechanisms include expansion of blood vessels in the skin, perspiring, or (in furry animals) panting. When temperature falls, other mechanisms (such as shuddering and restriction of blood vessels in the skin) minimize heat loss.

Recent studies of dinosaur bone histology have provided clues to body temperature in dinosaurs. In general, the bone of living cold-blooded, or ectothermic, reptiles is rather compact, with a low density of blood vessels and Haversian canals. The bone of warm-blooded animals, such as mammals, is rich in blood vessels and Haversian canals. Examination of dinosaur bones by Robert T. Bakker of Harvard University has revealed that many dinosaurs had blood vessel densities even higher than those in living mammals. Bakker also found evidence of a degree of warm-bloodedness in the predator-prey ratios of dinosaurs. Dinosaur ratios more closely resembled the situation for warm-blooded than for cold-blooded populations. He noted further that the presence of dinosaur remains at locations that were cool during the Mesozoic can be more readily explained if the creatures were, at least in part, endothermic. These new ideas are now being actively examined. Their validity is based on a limited amount of fossil evidence, and thus there is a danger of overinterpretation. With the present information, it does seem that a case can be made for true endothermy, at least among the coelurosaurs.

The larger carnivorous saurischians (including coelurosaurs) are called **theropods.** Of more dramatic dimensions than the coelurosaurs were the

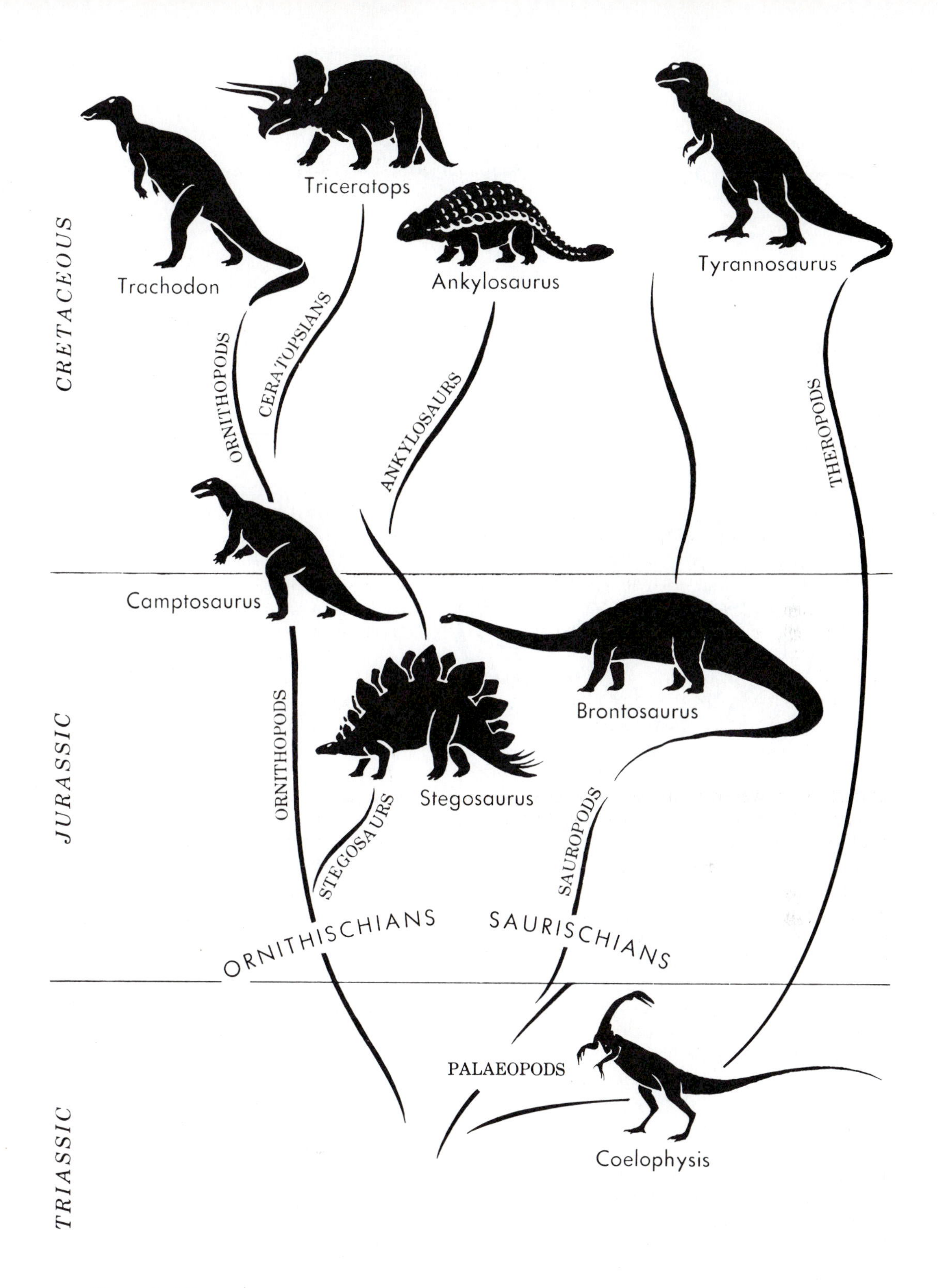

Figure 13–25 Evolution of the dinosaurs. (From E. H. Colbert. *Evolution of the Vertebrates*. New York, John Wiley & Sons, 1969, with permission of the author, artist Lois Darling, and publisher.)

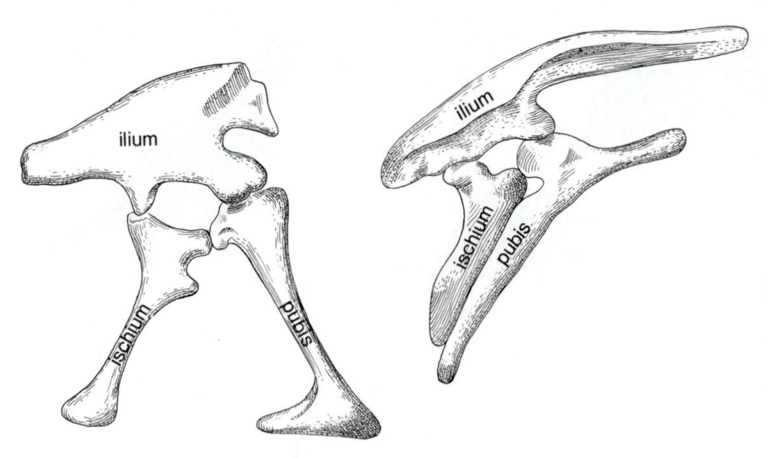

Figure 13–26 Basis for the division of "dinosaurs" into two groups. On the left is the arrangement of pelvic bones in the *Saurischia*, and on the right, the arrangement in *Ornithischia*. (Anterior is toward the right.) These views show one side only; the bones are duplicated on the other side.

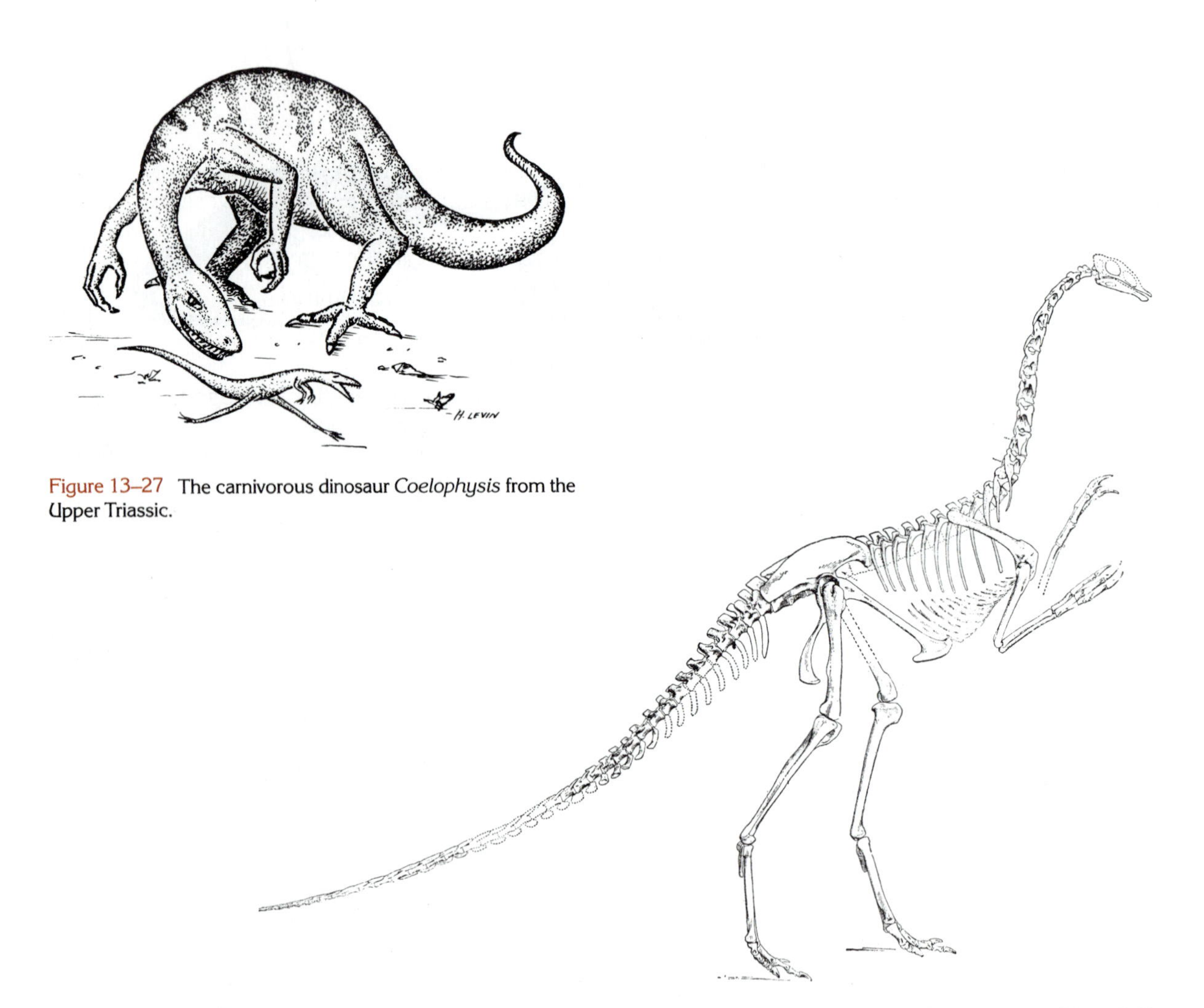

Figure 13–27 The carnivorous dinosaur *Coelophysis* from the Upper Triassic.

Figure 13–28 *Ornithomimus*, an ostrich-like Cretaceous dinosaur that was capable of rapid running. (From H. F. Osborn. Skeletal adaptations of *Ornitholestes, Struthiomimus, Tyrannosaurus. Bull. Am. Mus. Nat. Hist.,* 35:733–777, 1917.)

Figure 13–29 Restoration of a scene in western United States during the Late Jurassic. Plants include cycadophytes, ferns, and horsetails. The large bipedal dinosaur is the carnivorous form *Allosaurus. Stegosaurus* is in the center foreground. The large sauropod, *Diplodocus,* is on the right. In front of *Diplodocus'* right foreleg is the ornithopod *Camptosaurus.* (Courtesy of the Smithsonian Institution.)

large carnosaurians like *Allosaurus* (Fig. 13–29) of the Jurassic and the Cretaceous dinosaurs *Deinonychus* (Fig. 13–30) and *Tyrannosaurus* (Fig. 13–31). The last-named beasts attained lengths of over 13 meters and weighed in excess of 4 metric tons. Carnosaur hind limbs were robust and muscular. Great curved claws for tearing flesh protruded from each of three toes, whereas a nearly functionless fourth toe bore a smaller claw. The forelimbs were, in our view, ridiculously small. As befits an animal that must kill with its jaws and teeth, the head of the carnosaur was large. Doubly serrated teeth, up to 6 inches long, lined the powerful jaws. These were truly spectacular predators.

The saurischian group included herbivorous sauropods as well as flesh-eating theropods. The ancestry of the sauropods can be traced to the Late Triassic "protosauropod" known as *Plateosaurus* (Fig. 13–32). From this smaller, partially bipedal form, the more typical giant sauropods appeared at the beginning of the Jurassic and survived right up to the end of the Cretaceous. They are the ani-

Figure 13–30 The Cretaceous dinosaur *Deinonychus.* This reptile was about 8 ft long. It possessed a large skull, and the margins of the jaws were set with serrated, saber-like teeth.

Figure 13–31 *Tyrannosaurus,* a late Cretaceous carnosaur.

mals that people first think of when they hear the word "dinosaur." The best-known sauropods were enormous long-necked, long-tailed beasts that had returned to the four-legged stance to support their tremendous bulk. A well-known Jurassic representative whose remains have been found in the Morrison Formation of Colorado is *Brontosaurus* (Fig. 13–33), the "thunder beast." This favorite of school children (as well as producers of Hollywood movies) measured almost 20 meters in length and weighed about 30 metric tons. *Diplodocus* (Fig. 13–29), a contemporary, was less bulky and had greater length.

For many years paleontologists have speculated that, even with their massive, pillar-like legs, these largest of land animals could not have supported their own weight continuously. It was therefore surmised that they dwelt in the buoyant waters of lakes and streams. However, this long-held theory of sauropod habits has recently been challenged. According to the new view, brontosaurs roamed through the forests much like modern elephants. For sauropods, there were probably advantages in being large. Great size affords protection and slows changes in body temperature. The ratio of surface area to mass for an animal decreases as

Figure 13–32 *Plateosaurus,* the Late Triassic ancestor of the giant sauropods.

Figure 13–33 The enormous Jurassic sauropod *Brontosaurus*.

size increases. Consequently, the large animal has a proportionately smaller surface for heat loss, and, just as a large pot of water loses its heat more slowly than a small pot, so does the large animal lose its heat more slowly than a small animal.

The other major dinosaur line, the Ornithischia, evolved near the end of the Triassic and thrived throughout the remaining Mesozoic. Ornithischians were plant eaters. The teeth in the forward part of the jaws were replaced by a beak suitable for cropping vegetation. The group included both quadrupedal and bipedal varieties, with the bipedal condition considered more primitive. Even the most advanced quadruped ornithischians had such short fore limbs that their descent from bipedal forms seems certain.

The bipedal group of ornithischians is known as **ornithopods.** Their evolutionary history began in the Triassic with relatively small species that lived primarily on dry land. A representative large Jurassic ornithopod is *Camptosaurus* (Fig. 13–34). This was a bipedal dinosaur of medium size with a heavy tail, short fore limbs, and long hind legs. The articulation of the jaw was arranged to bring the teeth together at the same time, an arrangement frequently seen in herbivores down through the ages. Leaves and stems were cropped by the forward, beaklike part of the jaws and passed backward to the cheek teeth for chopping and chewing. From camptosaurid-like ancestors, the larger Cretaceous ornithopods developed. Among these was *Iguanodon* (Fig. 13–35), one of the first dinosaurs to be scientifically described. *Iguanodon,* sometimes called the "thumbs-up" dinosaur because of the horny spike that substituted for a thumb, is thought to have been a gregarious animal that moved about in "herds." Evidence for this comes from a Belgian coal mine where 29 of these individuals were found together as a result of having fallen into an ancient fissure.

By Cretaceous time, most members of the ornithopod group had moved to lake environments—perhaps to avoid encounters with the ferocious carnosaurs. The most successful of these amphibious reptiles were ornithopods known as trachodonts or hadrosaurs, or more commonly "duck-bill dinosaurs" (Fig. 13–36). Mummified remains of these creatures clearly show the appearance of the skin. The aquatic ornithopods possessed webbed feet and a flattened tail for use in

Figure 13–34 *Camptosaurus,* a small to medium-sized (6 to 8 feet long) herbivorous dinosaur. Although there were no teeth in the beaklike forward part of the jaws, the remainder of the jaws was equipped with a tight mosaic of row upon row of teeth suitable for grinding plant food.

Figure 13–35 *Iguanodon,* a herbivorous ornithischian dinosaur from the Lower Cretaceous of Europe. (Approximately 10 meters long.)

Figure 13–36 The duck-bill dinosaur, *Anatosaurus annectens,* from the Cretaceous Lance Formation of Wyoming. Notice the flattened, toothless "duck bill," long line of cheek teeth, and the compressed swimming tail. (Courtesy of the Smithsonian Institution.)

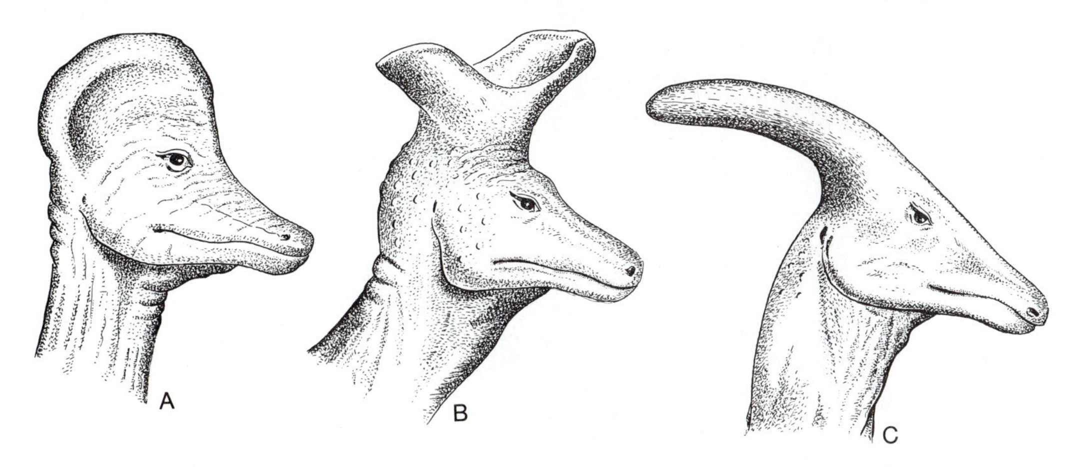

Figure 13–37 Bizarre skull crests developed in hadrosaurs. (A) *Corythosaurus.* (B) *Lambeosaurus.* (C) *Parasaurolophus.*

swimming. From the skeletons that have been unearthed, paleontologists have reconstructed the droll-looking face, in which the forward part of the jaws was often flattened and widened to resemble the bill of a duck. Behind the toothless forward part of the jaw were closely appressed lines of teeth that seem well suited for chewing coarse vegetation.

An interesting peculiarity of some hadrosaurs was the development of bony skull crests containing tubular extensions of the nasal passages (Figs. 13–37 and 13–38). The function of these bizarre structures is of special interest to paleontologists. It has been suggested recently that the cranial crests were used to catch the attention of sexual partners of the same species and could be further employed as vocal resonators for promoting an individual's success in obtaining a breeding partner.

Not all ornithopods had skull crests, and in some the crests lacked nasal tubes. One crestless ornithopod that certainly qualifies as a legitimate "bone head" was *Pachycephalosaurus* (Fig. 13–39). The skull in pachycephalosaurians consisted mostly of solid bone with only a small space to accommodate the unimpressive brain.

The best known of the quadrupedal ornithischians are the **stegosaurs,** (Fig. 13–40). Stegosaurs had two pairs of heavy spikes mounted on

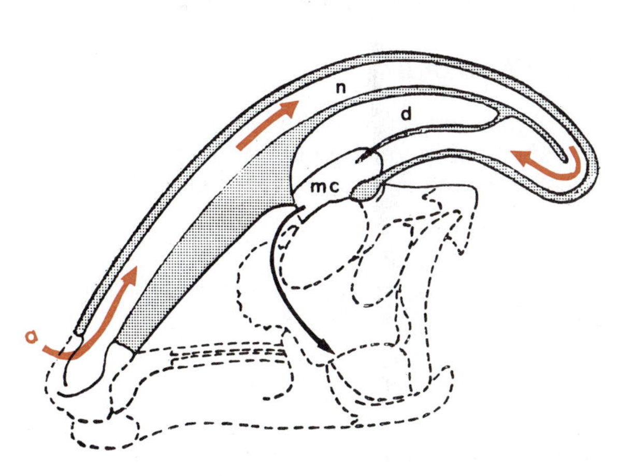

Figure 13–38 Internal structure of the skull crest of *Parasaurolophus cyrtocristatus.* The superficial bone of the left side has been removed to expose the left nasal passage (n). Air enters nostrils at *a*, moves up and around partition in crest, and from there moves down and back to internal openings in the palate. (From J. A. Hopson. The evolution of cranial display structures in hadrosaurian dinosaurs. *Paleobiology*, 1:24, 1975.)

Figure 13–39 Reconstruction of the head of *Pachycephalosaurus,* the "bone-head" dinosaur.

Figure 13–40 *Stegosaurus.*

the tail. These were used for defense. However, their more identifying feature was the double row of alternating pentagonal plates that stood upright along the back. Scientists have debated the purpose of these plates for many years. One theory that is currently being considered is that the plates functioned in the regulation of body temperature by serving as body-heat dissipaters. Indeed, the arrangement, size, and shape of the plates, and their probable rich supply of blood vessels, would favor the thermoregulatory suggestion, although the plates may have served as camouflage or protection as well.

During the Cretaceous, stegosaurs were succeeded by the heavily armored **ankylosaurs** (Fig. 13–41). These bulky, squat ornithiscians were completely covered by closely fitted bony plates that covered the entire length of their 6-meter backsides. A great ball of bone at the end of the tail in *Ankylosaurus* could be used as a bludgeon against an unwary foe.

The fourth group of quadrupedal ornithiscians are the **ceratopsians.** These beasts take their name from horns that grew on the face of all but the earliest forms. Typically, ceratopsians possessed a median horn just above the nostrils, and in some species an additional pair projected from the "forehead." The head was quite large in proportion to the body and displayed a shieldlike bony frill at the back of the skull roof that served as protection for the neck region and as a place of attachment for powerful jaw and neck muscles. All ceratopsians possessed a parrot-like beak (Fig. 13–42). Judging from the scars that mark the shield bones of ceratopsians, they were often attacked by the great carnosaurs. No doubt they frequently emerged the victor. During the Late Cretaceous, ceratopsians moved eastward from Asia across the land connection to North America and inhabited the region that lay to the west of the epicontinental sea.

The Pterosauria

Again and again in the history of life, the descendants of a small group of animals that were initially adapted to a narrow range of ecologic con-

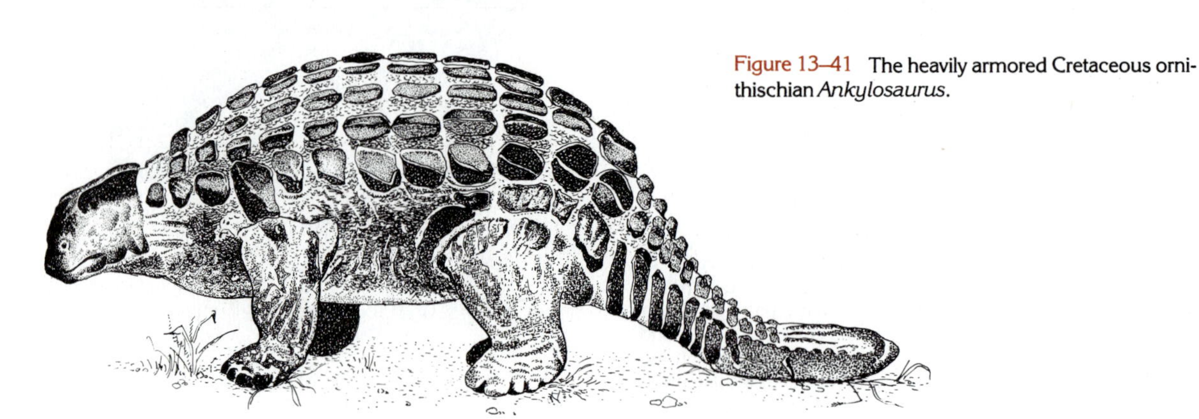

Figure 13–41 The heavily armored Cretaceous ornithischian *Ankylosaurus*.

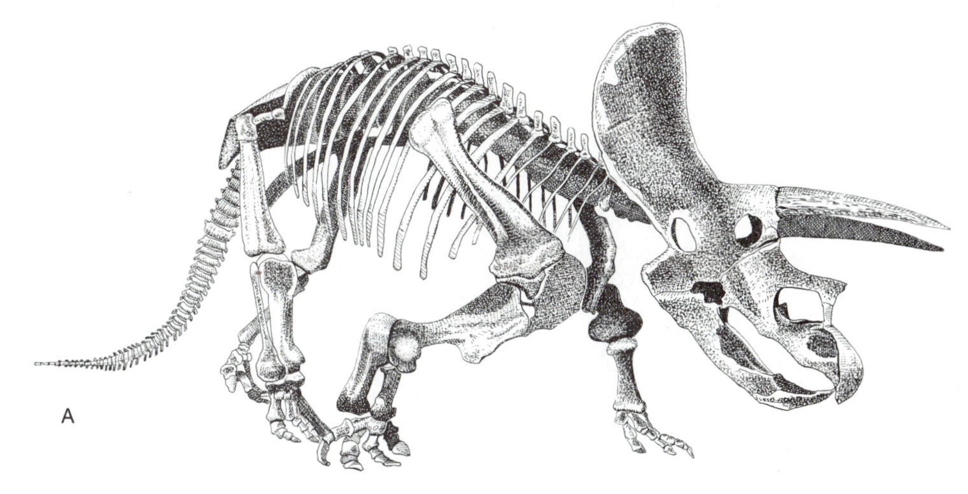

Figure 13–42 To an advancing enemy, *Triceratops* presented a triple-horned threat. (A) Mounted skeleton. (B) Reconstruction. (Courtesy of the Smithsonian Institution.)

ditions have dispersed into a great variety of environments. As they responded through evolutionary processes to the formerly unoccupied living space, the animals changed in ways that made them better suited to their new surroundings. This process, known as **adaptive radiation,** is nicely demonstrated by the Mesozoic reptiles. The adaptive radiation originated with the stem reptiles of the Late Carboniferous and ultimately produced the enormous variety of large and small, herbivorous and carnivorous, dry land and aquatic animals of the Mesozoic. The radiation did not end with terrestrial vertebrates, however, for during the Mesozoic, reptiles invaded the marine environment and even managed to overcome the pervasive tug of gravity and achieve soaring flight. Those reptiles that took to the air were the **pterosaurs.** They were the first flying vertebrates, and, although they may appear a bit graceless, their existence from Early Jurassic until late in the Cretaceous attests to their adaptive success.

The typical pterosaur had a rather large head

Figure 13–43 The Jurassic pterosaur *Rhamphorhynchus*.

and eyes and long jaws, which in most forms were lined with thin, slanted teeth. The bones of the fourth finger were lengthened to help support the wing, whereas the next three fingers were of ordinary length and terminated in claws. The wing was a sail made of skin stretched between the elongate digit, the sides of the body, and the rear limbs. There were two general groups of pterosaurs. The more primitive were the rhamphorhynchoids (Fig. 13–43), which had long tails, and the advanced were tail-less pterodactyloids. The latter group is exemplified by *Pteranodon* (Fig. 13–44), species of which had an astonishing wingspan of over 7 meters. The body of *Pteranodon* was about the size of that of a goose. The skeleton was lightly constructed, as is fitting for an aerial vertebrate. The animals probably glided along much like oversized sea birds, snapping up various sea creatures in their toothless jaws.

Relative to body size, pterosaurs had somewhat larger brains than some of their land-dwelling relatives. Perhaps this was a result of the higher level of nervous control and coordination needed for flight. They may also have had a degree of endothermy or warm-bloodedness. Strong evidence for this theory has been obtained by the Russian paleontologist A. Sharov. Sharov discovered a superbly preserved fossil of a Jurassic flying reptile, which he named *Sordus pilosus.* The name means "hairy devil" and is an appropriate reference to the growth of hair or hairlike feathers that covered the body and limbs. Remarkably, this insu-

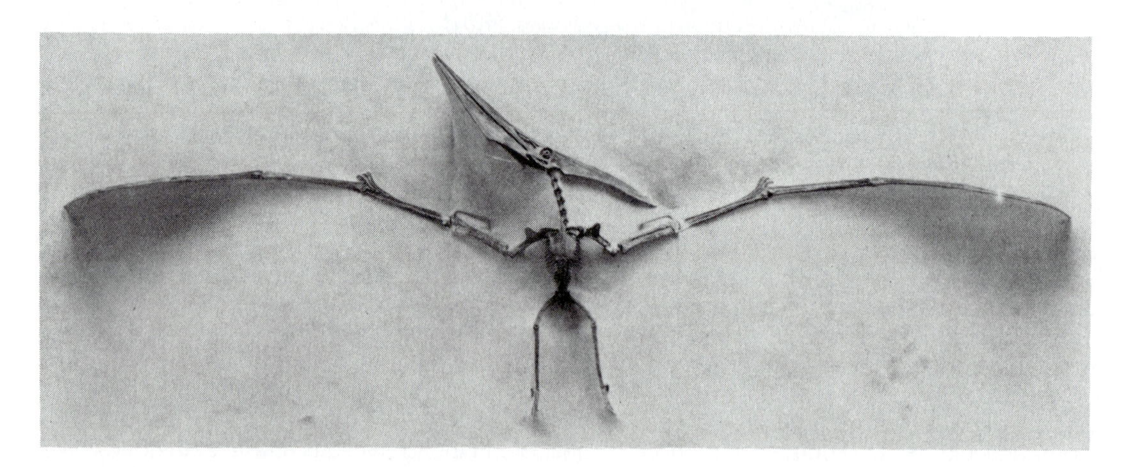

Figure 13–44 Restored and mounted skeleton of the great flying reptile *Pteranodon ingens* from Cretaceous chalk deposits of Kansas.

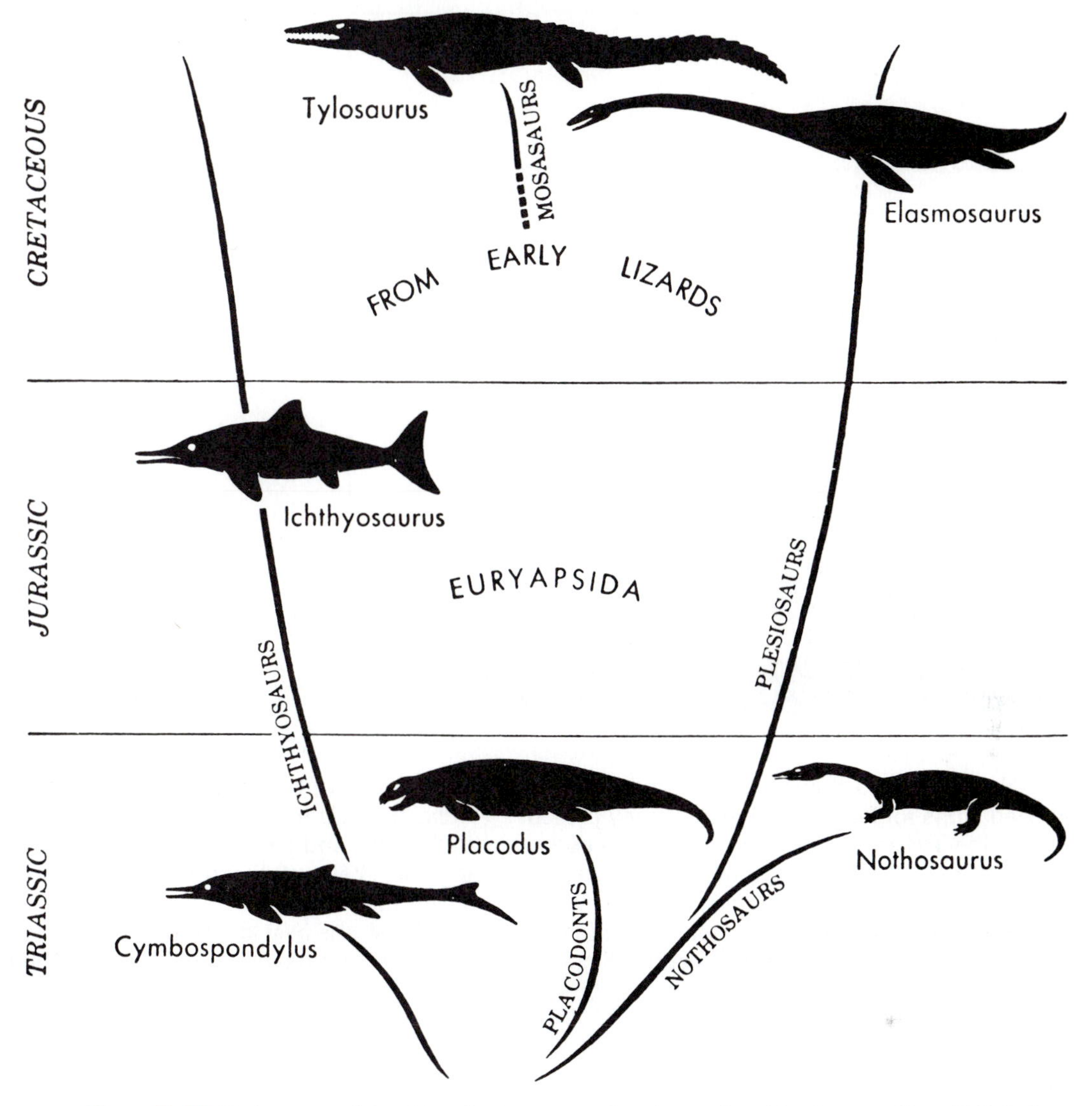

Figure 13–45 Evolutionary relationships of major groups of marine reptiles of the Mesozoic. (From E. H. Colbert. *Evolution of the Vertebrates.* New York, John Wiley & Sons, 1969, with permission of the author, artist Lois Darling, and publisher.)

lating cover is actually preserved in the fine Jurassic lake shales in which the creature was entombed.

A Return to the Sea

The marine habitat is one in which the archosaurs were not notably successful. Only one archosaurian group—the sea crocodiles—were able to invade the oceanic environment. Other reptilian groups, such as the ichthyosaurs, plesiosaurs, mosasaurs, and sea turtles, however, were very successful in their return to the productive seas of the Mesozoic (Fig. 13–45). Many made their diet of the myriads of sharks and bony fishes that had preceded them in populating the seas. The Cretaceous witnessed the modernization of the fish group, culminating in the appearance of the most modern of bony fishes, the **teleosts.**

Not unexpectedly, the invasion of the marine environment required many modifications in form and function. Paddle-shaped limbs and streamlined bodies evolved to allow efficient movement through the water. Because these reptiles were unable to abandon air breathing and reconvert to gills, their lungs were modified for greater efficiency. In those sea-going reptiles that were unable to lay their eggs ashore, reproductive adaptations provided for birth of the young while at sea.

Figure 13–46 Bulky mollusk-eating Triassic marine reptile *Placodus*.

Marine reptiles that depended upon paddle-shaped limbs for locomotion were already present during the Triassic Period. One group, the **nothosaurs,** were just beginning to take on adaptations that would be perfected in their likely descendants, the plesiosaurs. The nothosaurs were joined in the Triassic by a group of mollusk-eating, flippered reptiles known as **placodonts** (Fig. 13–46). These bulky animals had distinctive pavement type teeth in the jaws and palate, which they used for crushing the shells of the marine invertebrates upon which they fed.

By far the best known of the paddle swimmers were the **plesiosaurs** (Fig. 13–47). Their earliest

Figure 13–47 The large Cretaceous plesiosaur *Elasmosaurus*.

remains are found in Jurassic strata. Sometimes nicknamed "swan lizards," they had short, broad bodies and large, many-boned flippers. In some species the neck was extraordinarily long and was terminated by a smallish head. Slender, curved teeth, suitable for ensnaring fish, lined the jaws. *Elasmosaurus,* a well-known Cretaceous plesiosaur, attained an overall length in excess of 12 meters.

In addition to the long-necked types, there were many plesiosaurs characterized by short necks and large heads. It is likely that the short-necked plesiosaurs were aggressive divers. *Kronosaurus,* a giant, short-necked form from the Lower Cretaceous of Australia, had a skull size that probably exceeds that of any known reptile. It measured 3 meters in length.

The most fishlike in form and habit of all the marine reptiles were the Triassic and Jurassic **ichthyosaurs** (Fig. 13–48). In many ways, they were the reptilian counterparts of the toothed whales of our present-day oceans. Ichthyosaurs had fishlike tails in which vertebrae extended downward into the lower lobe, boneless dorsal fins to help prevent sideslip and roll, and paddle limbs for steering and braking. The head was a suitably pointed entering wedge for cutting rapidly through the water. A ring of bony plates surrounded the large eyes and may have helped protect them against water pressure. Clearly, these were active predators with good vision and the ability to move swiftly through water.

The **mosasaurs,** a group of giant monster lizards, were a highly successful group of Cretaceous sea dwellers. A typical mosasaur (Fig. 13–49) looked somewhat like a large moray eel to which four flippers had been added. The creatures propelled themselves through the water by the sculling action of their long, vertically flattened tails and the rhythmic undulations of their long bodies. The lower jaw had an extra hinge at midlength, which greatly increased its flexibility and gape. Mosasaurs were primarily fish eaters, but some frequently dined on large mollusks. Conchs of cephalopods

Figure 13–49 The great marine lizard of the Cretaceous, *Tylosaurus,* a mosasaur.

have been found with puncture marks that precisely match the dental pattern of their mosasaurian foes.

Perhaps less spectacular than the mosasaurs, but far more persevering, were the sea turtles. In this group, also, we find a trend toward increase in size. The Cretaceous turtle *Archelon* (Fig. 13–50), for example, reached a length of nearly 4 meters. As an adaptation to their aquatic habitat, the cara-

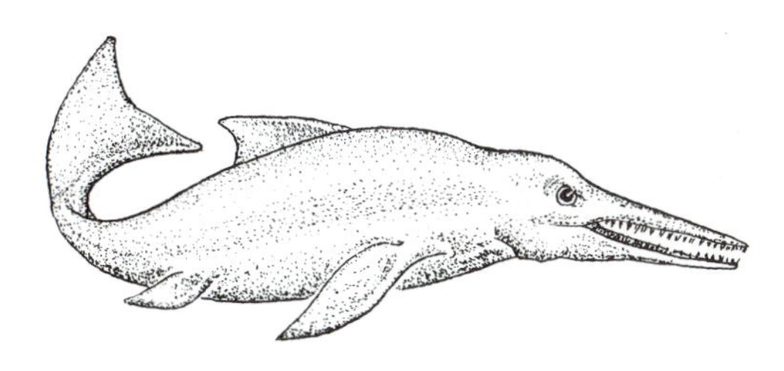

Figure 13–48 Restoration of an ichthyosaur, a marine reptile similar to a modern porpoise in form and habits.

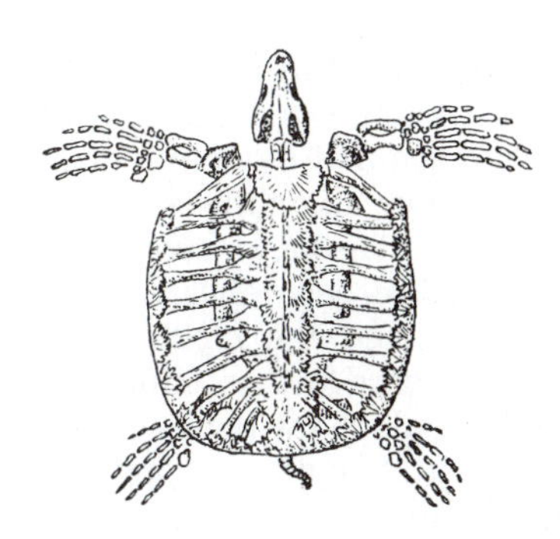

Figure 13–50 Skeleton of the giant Cretaceous marine turtle *Archelon.*

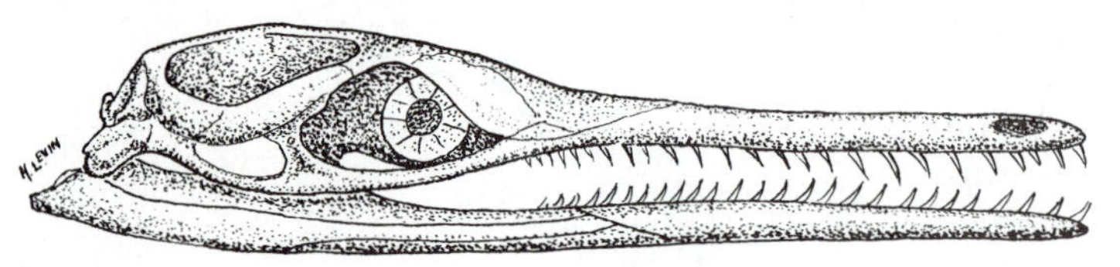

Figure 13–51 Toothy skull of the marine Jurassic crocodile *Geosaurus*. (Length of skull is about 45 cm.)

pace of the marine turtle was greatly reduced and the limbs were modified into broad paddles.

As briefly indicated earlier, the marine crocodiles (Fig. 13–51) were the only members of the archosaurian group that entered the sea. They became relatively common during the Jurassic, but only a few remained by Early Cretaceous. It is quite likely that they did not fare well in competition with the mosasaurs.

The Birds

From the time of Darwin, naturalists have been aware of the structural similarities between birds and reptiles. These similarities have prompted the statement that birds are only glorified reptiles that have gained wings and feathers and lost their teeth. However, such statements depreciate the marvelous attainments of birds, not the least of which are their superior powers of flight and high level of endothermy. Both of these attributes are related to the transformation into feathers of what once were reptilian scales.

There is little doubt that birds descended from Triassic thecodonts, which were already birdlike in their hollow bone structure and bipedal stance. Some thecodonts may even have had thermal insulation. Middle Triassic lake beds in Turkestan have yielded the remains of a small thecodont that was covered with long, overlapping, keeled scales that might very well have served to trap an insulating layer of air next to the body. The scales seem to be an ideal transitional form for the feathery insulation of birds.

The earliest bird thus far discovered, *Archaeopteryx* (Fig. 13–52), is a perfect link between the bipedal thecodonts and modern birds. *Archaeopteryx* was a creature about the size of a crow. With the exception of its distinctly fossilized feathers, its features were still largely reptilian. The jaws bore thecodont teeth, and the creature had a long, lizardlike tail that bore feathers. Unlike the wings of modern birds, in which the bones of the digits coalesce for greater strength, the primitive wings of *Archaeopteryx* retained claw-bearing free fingers for climbing and grasping. The lightweight sternum lacked a keel, indicating that the heavy muscles needed for sustained flight were lacking. Clearly, *Archaeopteryx* was not a vigorous aviator but rather a forest dweller that glided from one tree branch to another.

Small, delicate, hollow-boned animals are not readily preserved, and the fossil record for birds—especially Mesozoic birds—is not good. The Cretaceous provides the next partial glimpse of bird evolution. Toothed birds otherwise resembling terns and gulls are occasionally found in the Cretaceous deposits of the inland chalky seas. Some Cretaceous birds, such as *Hesperornis,* became excellent swimmers. They retained feathers and other birdlike characteristics but lost their wings and relied upon their webbed feet for swimming (see Fig. 12–27). Marine sediments provided the rapid burial needed to preserve these aquatic birds. On land, preservation of Mesozoic birds was rare indeed, and their fossil record is quite inadequate.

The Mammalian Vanguard

While the Mesozoic reptiles were having their heyday, small, furry animals were scurrying about in the undergrowth and unwittingly awaiting their day of supremacy. These shrewlike creatures were the primitive mammals. On the basis of rare and often minuscule remains, they are known from all three systems of the Mesozoic. Among the earliest of the mammals were the **morganucodonts** (Fig. 13–53), jaws, teeth, and skull fragments of which have been recovered in Late Triassic rocks of south Wales. It is evident from these rather scrappy remains that morganucodonts still retained many vestiges of reptilian structures. However, the articulation of the jaw to the skull was mammalian, and the lower jaw was functionally a single bone, as in mammals.

There were at least six additional groups of Mesozoic mammals. Each is recognized primarily on the basis of tooth morphology. Among those known from the Jurassic and Cretaceous are the **docodonts, triconodonts, symmetrodonts, multituberculates** and **pantotherians** (Fig. 13–54). The docodonts had multicusped molar teeth, which suggests that they may have been the stock from which present-day monotremes evolved. Very

Figure 13–52 The ancestral bird, *Archaeopteryx,* from the Jurassic of Bavaria. (A) Skeleton in the limestone in which it was discovered. The specimen is part of the collection of the Berlin Museum of Natural History. (B) Restoration of *Archaeopteryx.* (Courtesy of Berlin Museum of Natural History.)

Figure 13–53 Restoration of *Morganucodon,* an early mammal from the Late Triassic of Wales.

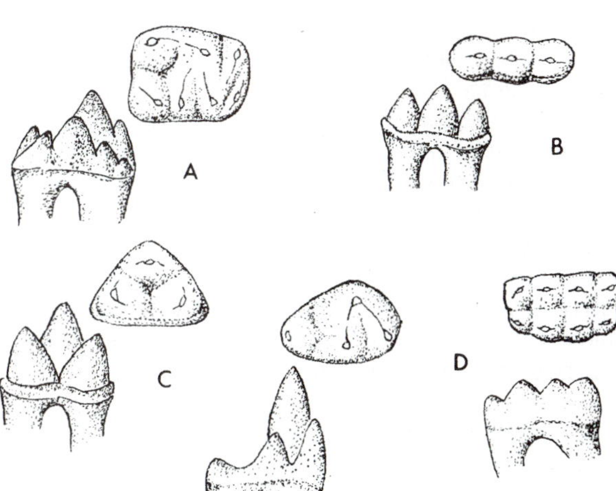

Figure 13–54 Molar teeth of Mesozoic mammals. Side views of lower molars (as viewed from inside the mouth) and top views of the oral surfaces. (A) *Docodonta.* (B) *Triconodonta.* (C) *Symmetrodonta.* (D) *Multituberculata.* (E) *Pantotheria*.

likely, they fed upon insects, as did many of these primitive mammals. The triconodonts can be recognized by cheek teeth in which three cusps are aligned in a row. Some were as large as a cat and may very well have preyed upon smaller vertebrates. Symmetrodonts had molars constructed on a more or less triangular plan. As suggested by their name, multituberculates had teeth with many cusps. They may have been the first entirely herbivorous mammals. Their chisel-like incisors and the gap between the incisors and cheek teeth give them a decidedly rodent-like appearance (Fig. 13–55).

Figure 13–55 The rodent-like multituberculate *Taeniolabis*.

From an evolutionary point of view, the pantotheres are the most important of the mammalian vanguard, for in the Late Cretaceous they gave rise to the marsupials and placentals. Marsupials, for example, have an asymmetrically triangular pattern of cusps that was evidently inherited from the pantotheres. In general, pantotheres were small, ratlike animals with long, slender, toothy jaws. It is likely that they preyed not only upon insects but also upon small lizards and mammals.

For the mammals, the Mesozoic was a time of evolutionary experimentation. For 100 million years, they effectively and unobtrusively lived among the great reptiles while simultaneously improving their nervous, circulatory, and reproductive anatomy. Equipped with an exceptionally reliable system for control of body temperature, they were better able to survive in cooler climates than many of their cold-blooded contemporaries. As the reptile population declined near the end of the era, mammals quickly expanded into the many habitats vacated by the saurians.

A SECOND TIME OF CRISIS

Just as the end of the Paleozoic was a time of crisis for animal life, so also was the conclusion of the Mesozoic. Primarily on land but also at sea, gradual extinction overtook many seemingly secure groups of vertebrates and invertebrates. In the seas, the ichthyosaurs, plesiosaurs, and mosasaurs perished. The ammonoid cephalopods and their close relatives, the belemnites, as well as the rudistid pelecypods, disappeared. Entire families of echinoids, bryozoans, planktonic foraminifers, and calcareous phytoplankton became extinct. On land the most noticeable losses were among the great clans of reptiles. Gone forever were the magnificent dinosaurs and soaring pterosaurs. Turtles, snakes, lizards, crocodiles, and the New Zealand reptile *Sphenodon* (See Fig. 15–16) (the "tuatara lizard") are the only reptiles that survived the great biologic crash (Fig. 13–56). Altogether, in the Late Cretaceous, extinctions eliminated about a fourth of all known families of animals.

The question of what caused the decimation in animal life at the end of the Mesozoic continues to intrigue paleontologists. Scores of theories, some scientific and many preposterous, have been offered. Those that have the most credibility attempt to explain simultaneous extinctions of both marine and terrestrial animals and seek a single or related sequence of events as a cause. In general, the theories tend to fall into two broad categories. The first of these relies on some sort of extraterrestrial interference, such as an influx of abnormally high amounts of cosmic radiation, which might either destroy organisms or damage their reproductive capabilities. Some proponents of this concept feel that reversals of the earth's magnetic field may have temporarily eliminated protection from cosmic radiation. Unfortunately, the theory has some aspects that are difficult to defend. It does not explain why marine organisms, which could descend a few meters and escape radiation damage, were decimated at levels far in excess of those among unprotected land plants. Also, several known periods of polar reversals do not correlate to episodes of mass extinction. In addition, the Mesozoic extinctions spanned a time interval far greater than that encompassed by documented magnetic reversals.

Another possibility for radiation damage to the Cretaceous biosphere might have been the arrival of a blast of lethal rays from a nearby supernova. Supernovas are colossal stellar explosions that radiate about as much energy as 10 billion suns and would affect the surfaces of planets located over 100 light years away (1 light year being about 8,898,000,000,000 km). Astronomers estimate that, on the average, a supernova may occur within 50 light years of the earth every 70 million years. If such an event did take place, there seems little doubt that plant and animal communities would be

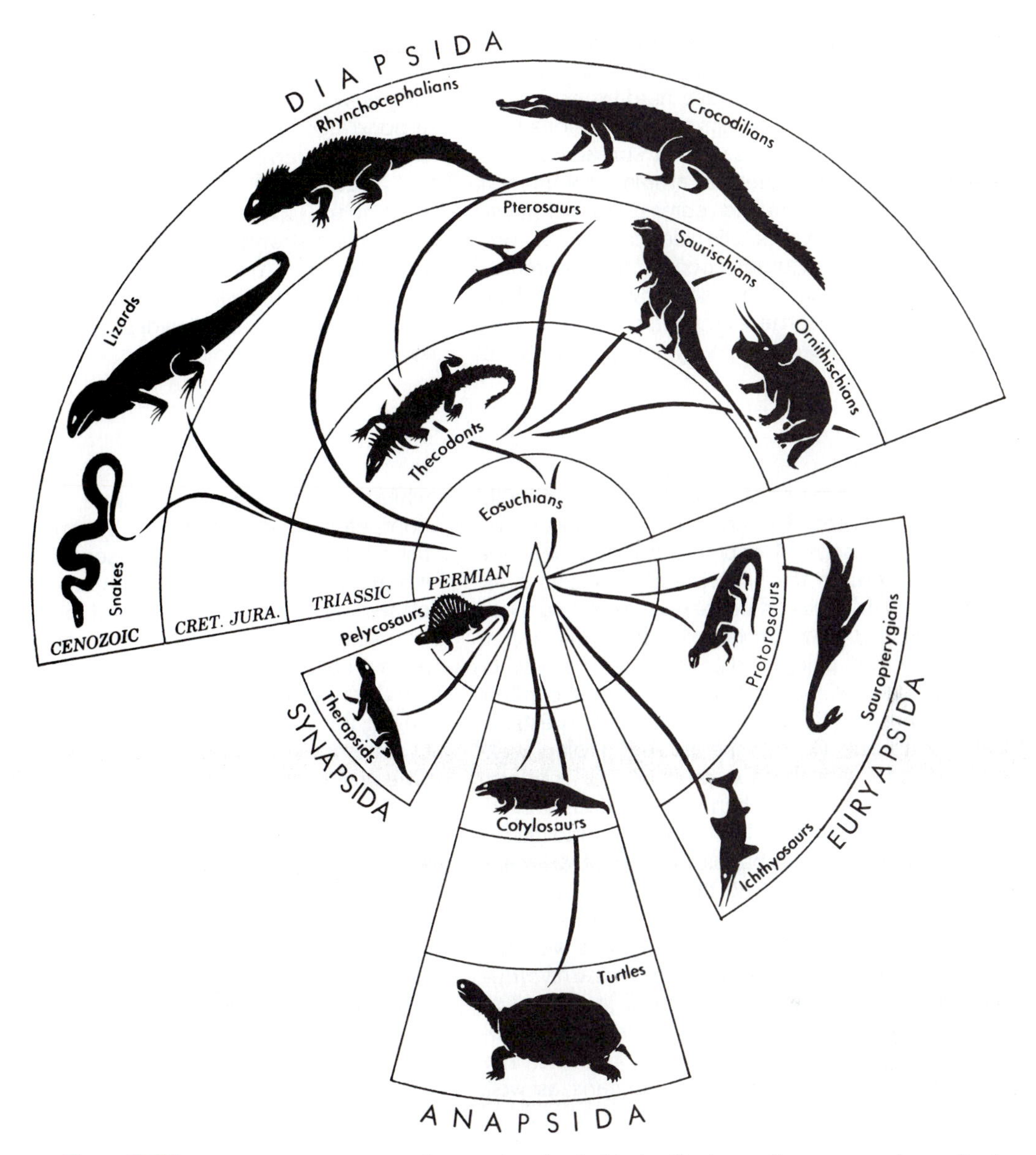

Figure 13–56 Evolution and general classification of reptiles. In this classification reptiles are grouped according to the position and number of temporal openings in the skull. Anapsida have no openings. Diapsida have two, Synapsida have one located low on the skull, and Euryapsida have one located higher on the skull. (From E. H. Colbert. *Evolution of the Vertebrates*. New York, John Wiley & Sons, 1969, with permission of the author, artist Lois Darling, and publisher.)

severely affected. For the present, however, we must await firm evidence that such a stellar explosion did occur nearby about 65 million years ago.

The other category of extinction hypotheses calls upon phenomena that are intrinsic to the earth itself. One concept within this second group is of particular interest in that it explains both marine and terrestrial extinctions in a manner that is ecologically reasonable and that correlates with geologic events of the Late Mesozoic. As indicated in the previous chapter, the Late Cretaceous was a time of extensive marine transgressions. Several lines of evidence indicate that these marine invasions were the result of displacement of ocean water by rising midoceanic ridges. The uplift of the ridges might have been a consequence of acceleration in the rate of sea floor spreading. Whatever the cause, the enormous expanses of inland seas on most of the continents helped to moderate and stabilize world climate. Times were good for living things, which experienced remarkable increases in variety and abundance. And then this Mesozoic "Garden of Eden" experienced hard times. The inland seas regressed. Perhaps the regressions were caused by a slowing of sea floor spreading rates; whatever the cause, there is ample geologic evidence that seas did indeed regress. The result was an episode of harsher climatic conditions, prolonged drought, general cooling, and increased seasonality. The highly diverse forms of animal life, however, had become adapted to the previous, more stable environment. Many lineages could not adjust to the changes and were exterminated. Their demise affected associated organisms in varying degrees and resulted in a wave of extinctions among ecologically dependent species higher in the food pyramid.

Much of the latter theory is speculation. However, there is now a large body of evidence that indicates a strong correlation between episodes of extinction and major withdrawals of epicontinental seas. The association of those withdrawals with plate tectonic theory provides a new twist to an idea long favored by geologists.

Summary

Climates of the Mesozoic were in general mild and equable except for occasional intervals of aridity and an episode of cooler conditions near the end of the era. In the widespread Mesozoic seas, coccoliths and diatoms flourished, as did such invertebrate groups as ammonoids, belemnites, oysters and other pelecypods, echinoderms, corals, and foraminifers. On land, seed ferns and conifers were common in Triassic and Jurassic forests. In the Cretaceous Period, the flowering plants expanded, and with them a multitude of modern-looking insects. The changing composition of plant populations was matched by innovations among terrestrial vertebrates. The most dramatic of these changes involved the evolution of dinosaurs, pterosaurs, and birds from the small bipedal thecodonts of the Triassic. The dinosaurs were the ruling reptiles of the Mesozoic. Both carnivorous and herbivorous varieties occupied a variety of habitats. The pterosaurs and certain of the dinosaurs may have been partially endothermic. The reptilian dynasty extended to the oceans as well, where ichthyosaurs, plesiosaurs, and mosasaurs competed successfully with the most modern of fishes, the teleosts. The Mesozoic is also noteworthy as the era during which two new vertebrate classes appear: mammals and birds. The birds, with true feathers for insulation and flight, evolved from reptilian ancestors and first appear in rocks of the Jurassic System. Mammals made their debut during the Triassic. In general, most of these primitive mammals remained small and inconspicuous. One group, the pantotheres, gave rise to marsupial and placental mammals during the Cretaceous.

Like the Paleozoic, the Mesozoic Era closed with an episode of extinctions. Many groups of both marine and terrestrial reptiles succumbed, as did the ammonoids and belemnites. Several of the major taxonomic classes survived, but entire families within those classes were exterminated. Still other groups of Mesozoic organisms experienced rapid evolutionary development and thereby were able to keep pace with environmental changes.

Questions for Review

1. What are coccoliths? In what way are they involved in the formation of chalk layers? When were they particularly abundant?
2. What are diatoms? How do they differ in composition and morphology from coccolithophorids?
3. In what way might prodigious phytoplankton productivity affect global climate and atmospheric composition?
4. How did the flora of terrestrial plants during the Cretaceous differ from Jurassic floras?
5. In general, how did the marine invertebrates of the Mesozoic Era differ from those of the preceding Paleozoic?
6. How did Mesozoic ammonoids differ from nautiloids? What attributes of the ammonoids resulted in their having special value as guide or index fossils?
7. What are foraminifers? What is there about these organisms that has resulted in their extensive use by petroleum geologists in the correlation of subsurface strata?
8. What clues might one hope to find in attempting to establish that a particular group of Mesozoic reptiles were endothermic ("warm blooded")?
9. What two major classes of vertebrates appear for the first time during the Mesozoic Era?
10. What is meant by evolutionary convergence? Cite an example.
11. Formulate a single sentence description of the following Mesozoic animals.

a. Pterosaurs
b. Ichthyosaurs
c. Sauropods
d. Ornithischians
e. Teleosts
f. Theropods
g. Ornithopods
h. Mosasaurs
i. Morganucodonts
j. Saurischians

12. What is the evolutionary importance of the following Mesozoic vertebrates?
    a. Thecodonts
    b. *Archaeopteryx*
    c. Pantotheres
13. What reptilian groups managed to survive the wave of extinctions that occurred at the end of the Cretaceous Period?
14. What attributes already present in Jurassic and Cretaceous mammals contributed to their survival during the biologic crisis at the conclusion of the Mesozoic?
15. How might the extent of continental coverage by inland seas influence rates of extinction among marine invertebrates? How might worldwide transgressions be related to plate tectonic theory?

Terms to Remember

adaptive radiation
angiosperm
ankylosaur
biosphere
ceratopsian
convergence
docodont
frustule
ilium
ischium
mosasaur
morganucodont
multituberculate
nothosaur
pantotherian
phytoplankton
phytosaur
plesiosaur
pterosaur
pubis
rudist
scleractinid coral
septum(a)
stegosaur
suture
symmetrodont
teleost
test
theropod
triconodont

Supplemental Readings and References

Coombs, W. P. 1975. Sauropod habits and habitats. *Palaeogeog. Palaeoclimatol. Palaeoecol.* 17:1–33.

Farlow, J. O., Thompson, C. V., and Rasner, D. E. 1976. Plates of the dinosaur *Stegosaurus*: forced convection heat loss fins? *Science* 192(4244): 1123–1125.

Gartner, S., and McGuirk, P. 1979. Terminal Cretaceous extinction, scenario for catastrophe. *Science* 206:1272–1276.

Hopson, J. A. 1975. The evolution of cranial display structures in hadrosaurian dinosaurs. *Paleobiology* 1:21–43.

Langston, W., Jr. 1981. Pterosaurs. *Sci. Amer.* 244(2):122–137.

Lipps, J. H. 1970. Plankton evolution. *Evolution* 24(1):1–22.

Phillips, J. D., and Forsyth, D. 1972. Plate tectonics, paleomagnetism, and the opening of the Atlantic: *Geol. Soc. Amer. Bull.* 82:1579.

Russell, D. A. 1979. The enigma of the extinction of the dinosaurs. *Ann. Rev. Earth Planet. Sci.* 7:163–182.

Geologic team explores the frozen banks of the Anaktuvuk River in Alaska. This region is still in the grip of the Ice Age. (Courtesy Sinclair Oil Corporation and American Petroleum Institute Photo Library.)

14

The Cenozoic Era

Many an aeon moulded earth before her highest, man, was born,
Many an aeon too may pass when earth is manless and forlorn,
Earth so huge and yet so bounded—pools of salt and plots of land—
Shallow skin of green and azure—chains of mountains, grains of sand!

A. Tennyson, *Locksley Hall Sixty Years After*, 1866

PRELUDE

The Cenozoic is the final era of geologic time. It is the era in which the continents acquired their present form. The landscapes and life of the modern world are products of Cenozoic events.

There are two sets of terms used in dividing the 65 million years of the Cenozoic (Table 14–1). In a scheme long used by most geologists, the era is separated into a Tertiary and a Quaternary Period. Some European geologists, however, prefer a time table that divides the era into Paleogene and Neogene periods, arguing that this represents a more equal division of the epochs and also a more natural way of dividing Cenozoic rocks of Europe. The Paris Basin is the type area for most of the Cenozoic epochs. In that area, a major unconformity representing a marine regression does indeed serve nicely as a boundary between the Paleogene and Neogene systems. However, because predominant usage still seems to favor the old scheme, it will be employed here.

Certainly one general characteristic of the Cenozoic Era was a vigorous amount of tectonic plate motion and sea floor spreading. It has been estimated that approximately 50 per cent of the present ocean floor has been renewed along mid-oceanic ridges during the past 65 million years. Much of this new ocean floor was emplaced in the expanding Atlantic and Indian Oceans. As this wid-

Table 14–1 GEOCHRONOLOGIC TERMINOLOGY USED FOR DIVISIONS OF THE CENOZOIC ERA

ERA	PERIOD		EPOCH
CENOZOIC	QUATERNARY	NEOGENE	RECENT
			PLEISTOCENE
	TERTIARY		PLIOCENE
			MIOCENE
		PALEOGENE	OLIGOCENE
			EOCENE
			PALEOCENE

ening was in progress, the Americas were drifting westward. California came into contact with the northward moving Pacific Plate and thereby produced the San Andreas Fault system. South America moved firmly against the Andean Trench and actually bent and displaced it. Orogenic and volcanic activity was vigorous along the western "backbone" of the Americas and resulted in the eventual formation of a connecting volcanic causeway between the continents that we now call the Isthmus of Panama. The North Atlantic rift extended to the north and eventually separated Greenland from Scandinavia and destroyed the connection between Europe and North America. Far to the south, Australia separated from Antarctica (Fig. 14–1) and drifted northward to its present location. Meanwhile, Antarctica moved to its south polar position. During the Cenozoic, a branch of the Indian Ocean rift split Arabia away from Africa and in the process created the Gulf of Aden and the Red Sea. However, the most dramatic crustal event of the era must surely have been the collision of Africa and India with Eurasia. This magnificent smashup transformed the Tethys Geosyncline into great mountain ranges, not the least of which are the Alps and Himalayas.

The interior regions of continents stood relatively high during the Cenozoic, and as a result, marine transgressions were quite limited. Climatic zones were more sharply defined than in earlier

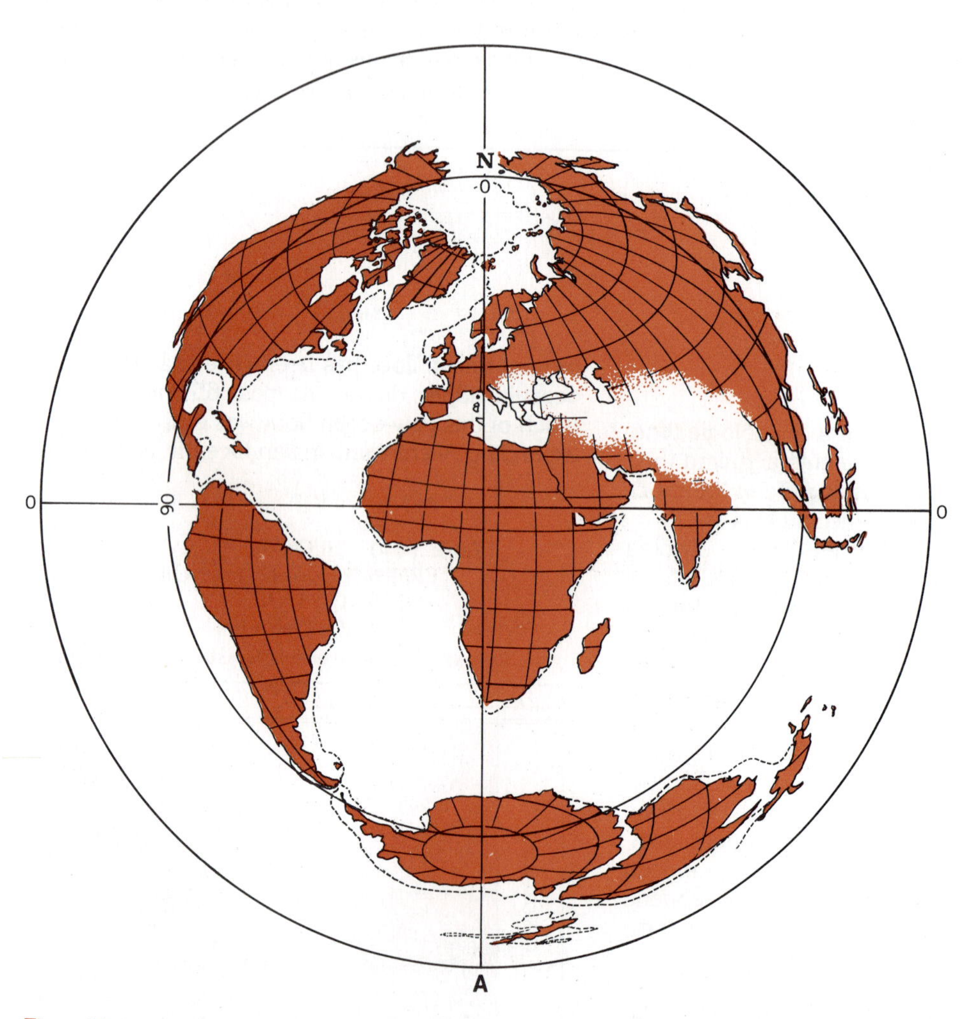

Figure 14–1 Location of major land masses during the Eocene (A), about 50±5 million years ago, and the world as it is today (B) plotted on the same (Lambert Equal Area) projection. (From J. C. Briden, G. E. Drewey, and A. G. Smith. Phanerozoic equal area world maps, *J. Geol.,* 82:556, 558, 1974.)

Illustration continued on opposite page

eras. A cooling trend that was to culminate in the Pleistocene ice age is clearly indicated by paleobotanical studies.

BEFORE THE ICE AGE

North America

The East

Eastern North America witnessed little in the way of geologic cataclysms during the Cenozoic. Erosion continued quietly in the Appalachians. Periodically, the lands would be beveled by erosion almost to sea level, and then broad, gentle uplifts would revitalize streams and lead to the erosional sculpturing of a new generation of graceful ridges and valleys.

The initial episode of erosion of the Appalachians was already underway during the Cretaceous, so that near the beginning of the Cenozoic nearly all of the mountain belt had been reduced to a low, undulating plain that we call the **Schooley Peneplain.** Subsequently, the region was gently arched and uplifted. Streams, rejuvenated by the uplift, began a new cycle or erosion that ultimately produced broad, level lowlands on those areas underlain by weaker formations. These lowlands were interrupted by ridges composed of more resistant formations. The name **Harrisburg erosional surface** has been applied to these lowlands developed on softer rocks. The sequence of uplift and erosion that resulted in the Harrisburg surface was repeated once again to produce a third surface of more local extent called the **Somerville surface.** Nor has the process ended, for slight, sporadic uplifts apparently continue in the Appalachians down to the present day.

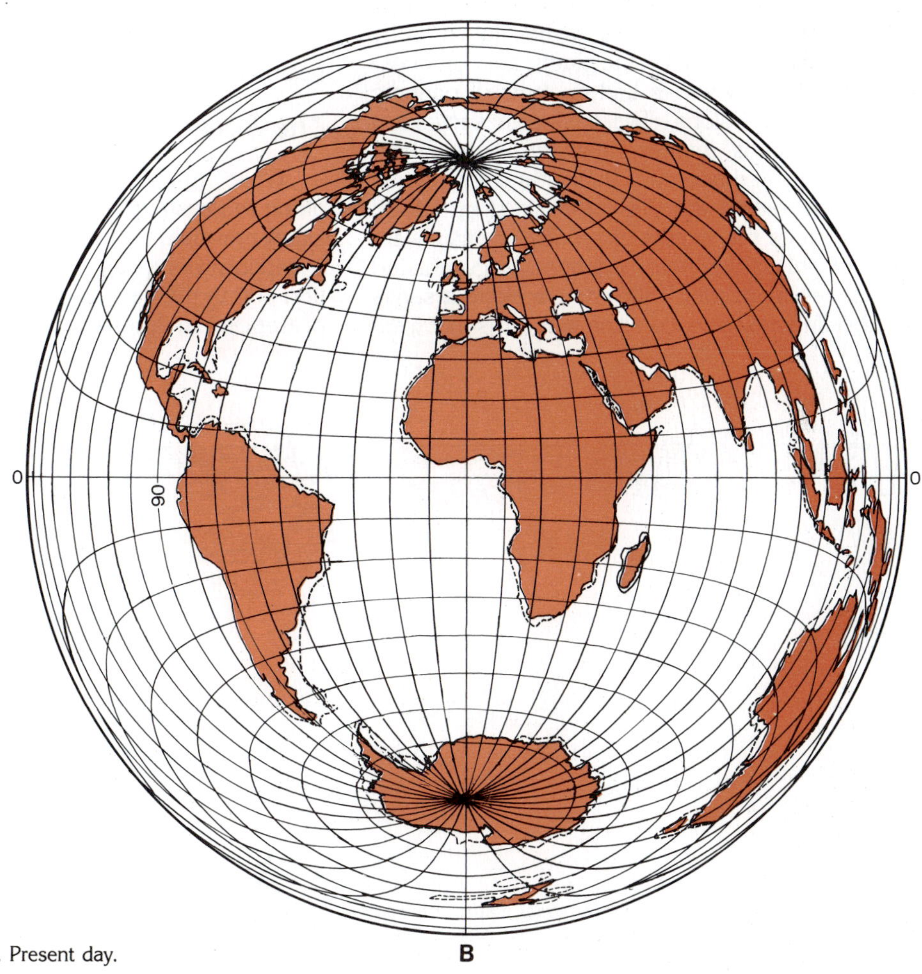

Figure 14–1 *Continued,* Present day.

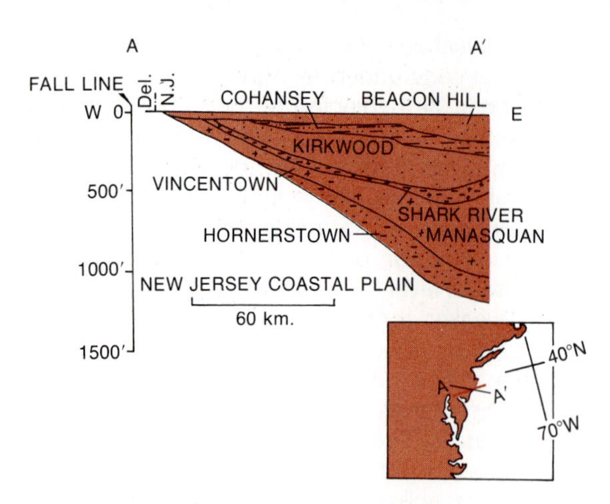

Figure 14–2 Cross-section of Tertiary strata across the New Jersey Coastal Plain. (From *Stratigraphic Atlas of North and Central America.* Courtesy of Shell Oil Company.)

The raising of the Appalachian belt was accompanied by a gentle tilting of the coastal plains and adjacent parts of the continental shelf. Clastic sediments were brought out of the highlands by streams and deposited on the plains, and, on occasion, reworked by waves and currents of periodic marine transgressions. The Cenozoic section of marine rocks is thinner near the Appalachian source areas and becomes thicker and less clastic toward the region that lies offshore (Fig. 14–2). Southward, in the vicinity of Florida, clastics were less available. Carbonate sediments accumulated to thicknesses of over 2500 meters along a series of subsiding elongate, coralline platforms that were probably much like the Bahama Banks (Fig. 14–3). In the final ages of the Tertiary, uplift along the northern end of this tract raised the land area of Florida above the waves.

Gulf Coast

The best record of Cenozoic marine strata is found in the Gulf Coastal Plain (Fig. 14–4). Altogether, eight major transgressions and regressions are recorded in this region. The Paleocene transgression brought marine waters as far north as southern Illinois. Frequently, during marine regressions, nearshore deltaic sands were deposited above offshore shales. The resulting interfingering of permeable sands and impermeable clays provided ideal conditions for the entrapment of oil and gas.

"A wedge of sediments that thickens seaward" is a particularly suitable description for the Cenozoic formations of the Gulf Coast. Geophysical measurements suggest that the thickness of Tertiary sediments beneath the northern border of the Gulf may exceed 6000 meters. The area must have been subsiding as it received this great mass of sediment.

The Rocky Mountains and High Plains

While marine sedimentation was underway along the eastern coastal regions, terrestrial deposition prevailed in the Rocky Mountain region. The exception was a single area of marine sedimentation during the Paleocene. The strata recording the presence of this seaway are found in western North

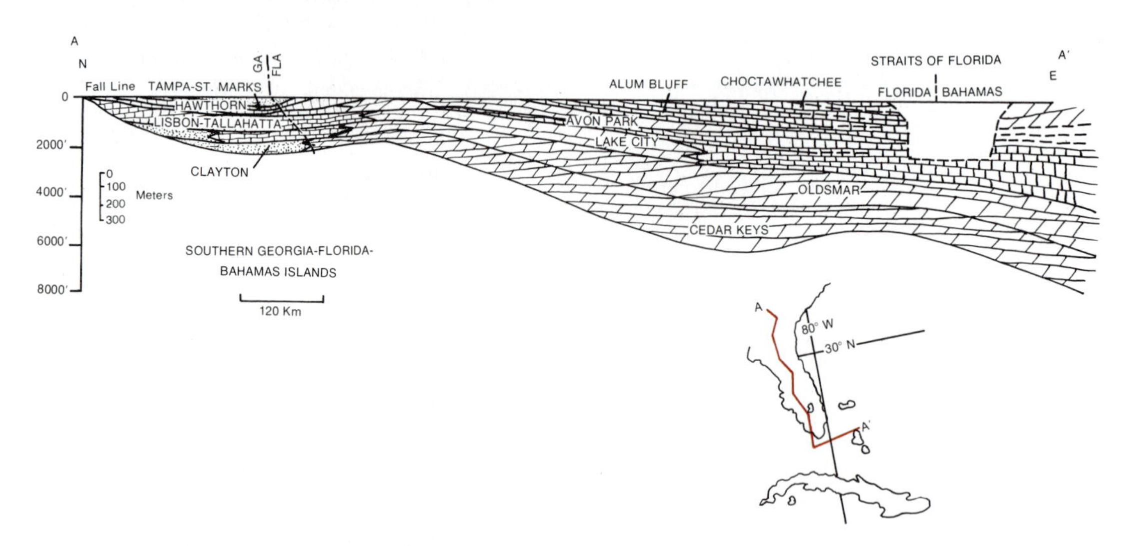

Figure 14–3 Cross-section of Tertiary strata across the trend indicated on the location map from southern Georgia to the Bahamas. (From *Stratigraphic Atlas of North and Central America.* Courtesy of Shell Oil Company.)

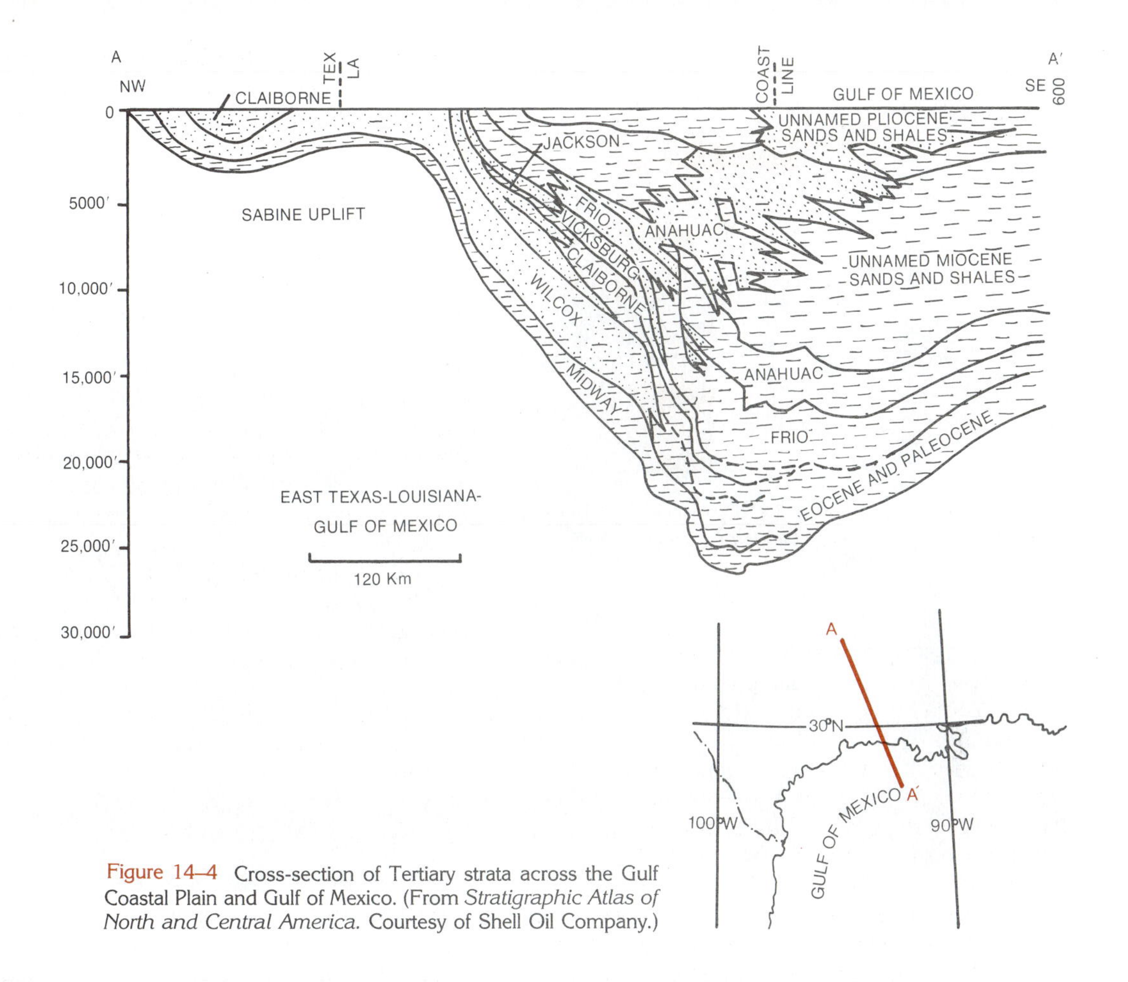

Figure 14–4 Cross-section of Tertiary strata across the Gulf Coastal Plain and Gulf of Mexico. (From *Stratigraphic Atlas of North and Central America.* Courtesy of Shell Oil Company.)

Dakota and consist of dark shales containing over 150 species of invertebrates. There is no evidence of a connection between this sea and either the Gulf of Mexico or the Arctic, and it is therefore believed to be a vestige of the more extensive Late Cretaceous epicontinental sea.

The Late Cretaceous and Early Tertiary phases of deformation described in Chapter 12 were largely responsible for the major structural features of the Western Cordillera. However, the present topography of this region is due primarily to erosion following uplifts that began during the Miocene. As always, erosion acting upon the tilted and folded hard and soft layers was the final factor in shaping the landscape. Following Late Cretaceous orogenesis, erosional debris was trapped in lowlands and intermontane basins, and these terrestrial sediments provide the record of the earlier Tertiary epochs. Later uplifts and erosion resulted in the spectacular relief of the Rockies and caused the detritus of the older basins and newer uplands to be spread over the plains that lay to the east. The result was the creation of that vast apron of nonmarine Oligocene through Pliocene sands, shales, and lignites that underlies the western high plains. Beds of volcanic ash interspersed within the Tertiary sections of the Western Cordillera attest to periodic episodes of igneous pyrotechnics.

For the most part, however, lower Tertiary strata are fluvial in origin. Typical of these deposits are the brilliantly colored layers of the Wasatch Group, which are spectacularly sculptured by erosion in Utah's Bryce Canyon (Fig. 14–5). These beds have yielded the bones of the world's first horse *Hyracotherium* (formerly known as *Eohippus*). Not unexpectedly, basins within the Cordillera received the waters of streams coming off the highlands and filled up to form lake systems.

One especially notable region of lacustrine deposition was the Green River Basin (Figs. 14–6 and 14–7) of southwestern Wyoming and northeastern Colorado. Over 600 meters of fine, evenly laminated shales were deposited in this basin as it slowly subsided. The laminations take the form of varves. Each varve consists of a thin, dark winter layer and a lighter colored summer layer. By count-

Figure 14–5 The pink cliffs of Bryce Canyon, Bryce Canyon National Park, Utah. The cliffs are sculptured by erosion from the Cedar Breaks Formation of Early Paleocene Age. (Photograph courtesy of United States National Park Service.)

ing the varves, it has been determined that over 6½ million years were required to deposit the Green River sediments. Within the fine shales are well-preserved fishes (Fig. 14–8), insects, and plant fragments. In addition, the shales are rich in waxy hydrocarbons. For this reason, they are called **oil shales** and can be processed to yield up to 200 liters of petroleum per ton. The Wasatch Sandstone, also of Eocene age, underlies and interfingers laterally with the Green River beds. Several oil fields in Wyoming derive their oil from the Wasatch Formation.

During the Oligocene Epoch, explosive volcanic activity blanketed the region that today includes Yellowstone Park and the San Juan Mountains with layers of volcanic dust and ash. However, for paleontologists, the more interesting rocks of

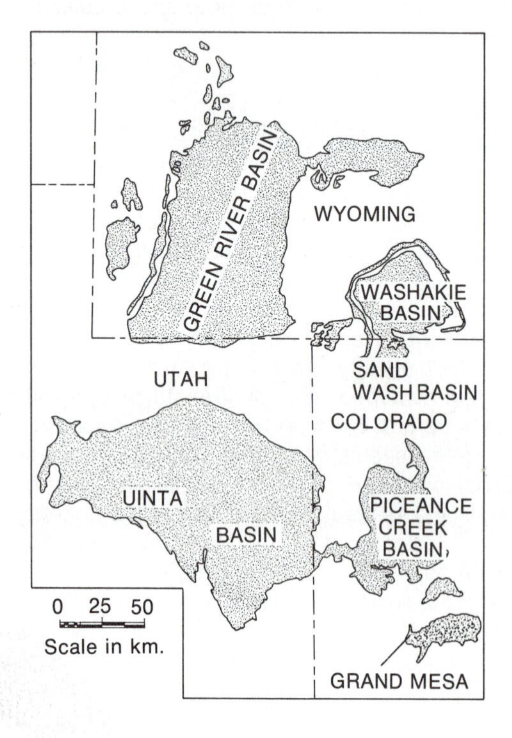

Figure 14–6 Location of major oil shale deposits in the United States. (Adapted from G. L. Cook. Oil shale—an impending energy source. *J. Oil Tech.*, November 24, 1972, pp. 1325–1330.)

Figure 14–7 The Green River Formation exposed in a road cut about 10 mi west of Soldier Summit, Wyoming. The exposure here consists of dark shales and thin, lighter colored limestones. (Photograph courtesy of J. C. Brice.)

this epoch are floodplain deposits of the White River Formation. This famous formation contains entire skeletons of Tertiary mammals in unexcelled number, variety, and condition of preservation. If one recalls the newsreels of recent floods, it is not difficult to understand how Oligocene mammals became buried in the sediment of overflowing streams. The clays, silts, and ash beds of the White River Formation are the sediments from which the Badlands of South Dakota have been sculptured.

Another particularly interesting occurrence of Oligocene beds is exposed near Florissant, Colorado. Explosive volcanic activity in this area produced a great deal of ash, which settled to the bottom of a neighboring lake, burying thousands of insects (Fig. 14–9), leaves, fish, and even a few birds. The Florissant flora, known not only from leaves but also from spores and pollen, indicates subhumid conditions for the region and elevations of between 300 and 900 meters.

The Miocene Epoch was another time of frequent volcanic activity and continued sedimentation in basins. Widespread beds of volcanic ash and numerous lava flows in the central and southern Rockies attest to the vigor of this igneous activity. The well-known gold deposits at Cripple Creek, Colorado, are mined from veins associated with a Miocene volcano. Regional uplift of the Rockies also began in the Miocene and was accompanied by increased rates of erosion. Great volumes of terrigenous detritus eroded from uplifted areas, filled intermontane basins, and spread eastward, where the detritus contributed to the construction of the Great Plains (Fig. 14–10). Some of these crustal movements continued throughout the remaining epochs of the Cenozoic and brought some of the highest peaks of the Rockies to spectacular elevations.

As indicated earlier, the present Rocky Mountain topography is largely the result of the uplift and erosional sculpturing that began in the Miocene. Much of the erosional detritus was spread on top of

Figure 14–8 The Eocene fish *Priscacara peali* from the Green River Formation of Wyoming. (Photograph courtesy of Wards Natural Science Establishment, Rochester, N.Y.)

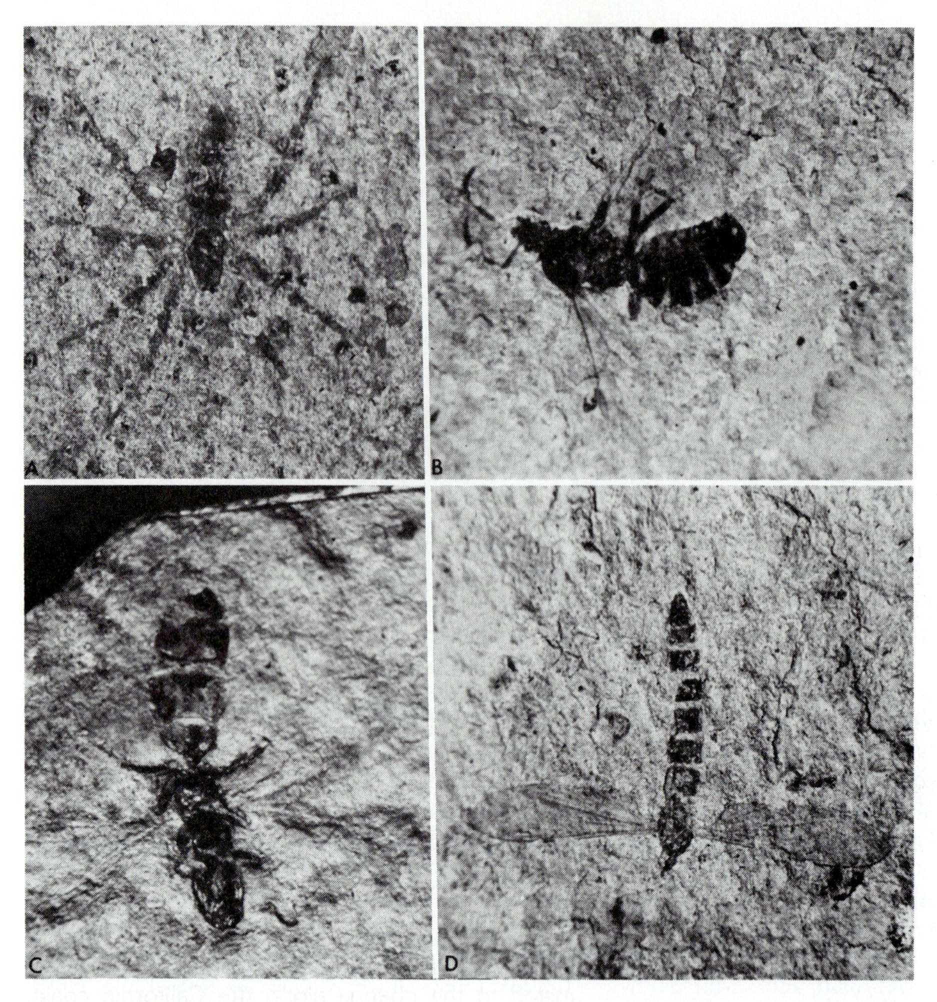

Figure 14–9 Fossil insects from the Oligocene tuff beds near Florissant, Colorado. (Tuff is consolidated volcanic ash.) (From the Washington University Collection.)

White River and equivalent beds over extensive areas of South Dakota and Nebraska. The general character of the sedimentary layers does not change in the overlying Pliocene. For example, the Pliocene Ogallala Group consists of slightly consolidated sands, gravels, silts and clays also derived from the erosion of the Rockies. Remains of plants and animals in some of these sediments indicate that the Pliocene was somewhat cooler and drier than had been the Miocene. The stage was being readied for the coming of the great ice age.

At particular localities west of the Great Plains, Tertiary geologic events produced some of the most striking scenic attractions in the world. Normal faulting and volcanism accompanying Late Tertiary uplifts were responsible for some of these features. In northwestern Wyoming, the lofty Teton Range (Fig. 14–11) was elevated along great normal faults, with displacements of up to 6000 meters. The magnificent east face of the Tetons is a fault scarp that rises nearly 2½ km to an altitude of over 4000 meters.

Basin and Range Province

A larger scale physiographic feature that is the result of Late Tertiary faulting is the Basin and Range Province of Nevada, Arizona, New Mexico, and southern California (Fig. 14–12). This region had been folded and overthrust during the Mesozoic and existed as a structural arch during most of pre-Miocene time. Then, beginning in the Miocene, the arch subsided between great normal faults that developed on both the west and the east sides. Similar faults with general north-south alignment developed in the interior of the region. The uplifted

Figure 14–10 The Great Plains in central Kansas. (Courtesy of Kansas Geological Survey. Photograph by Jim Enyeart.)

blocks formed linear mountain ranges that became sources of sediment for the adjacent downdropped basins (Fig. 14–13).

Subsequent geologic history in the region is recorded in the coarse clastics washed out of the mountains. These Miocene to recent sediments frequently blocked the passes and caused lakes to develop. Miocene sediments often include the salty layers formed when these lakes evaporated.

The cause of all the large-scale tensional faulting is still being debated by geologists. On the basis of paleomagnetic evidence, some geologists believe that uparching and faulting in this region occurred when westward-moving North America overrode part of the oceanic plate and spreading center (East Pacific Rise) that was being subducted along the coast of California. When the subducted spreading center reached the region beneath eastern Nevada and western Utah, it caused uplift and stretching of the crust in east and west directions. Recently, however, this idea has fallen into disfavor because of a lack of evidence of progressive eastward deformation that should accompany the passage of a spreading center beneath the sole of a continent. As an alternative theory, many geologists now believe that the normal faulting in the Basin and Range is simply the way the crust adjusted to the change along the California coast, when oblique shearing of the edge of the continent during the Early Miocene replaced the earlier subduction zone. Yet another model proposes extension and uplift of the crust from the remnants of an oceanic plate that had been carried beneath the Basin and Range region by an earlier episode of subduction. When subduction ceased, the oceanic slab may have formed a partially molten buoyant mass that pressed upward against the overlying crust and caused tensional faulting and escape of lava along fault and fracture zones. Finally, there is the possibility that the crustal extension and tensional faulting is related to convectional movements beneath the continental plate similar to those that cause the breaking apart of continents.

The Colorado Plateau

One of the most magnificent regions of uplift in the American West is the Colorado Plateau (Figs. 14–14 and 14–15). Somehow this block of crust had remained undeformed during the Rocky

Figure 14–11 The Teton Range in Winter. Mount Moran is the highest peak on the right. (Photograph courtesy of Wyoming Travel Commission.)

Mountain Orogeny, for its Paleozoic and Mesozoic rocks are relatively flat-lying. The region formed a buttress around the perimeter of which folding and faulting produced circumventing highlands. The plateau was repeatedly raised during Early to Middle Pliocene time (about 5 to 10 million years ago). Steep faults cracked the Plateau during its rise and provided avenues for the upward escape of volcanic materials. The San Francisco Mountains near Flagstaff, Arizona, are a group of impressive recent volcanoes and cinder cones built up above the level surface of the plateau.

The best-known feature resulting from the linked processes of uplift and erosion on the Colorado Plateau is the Grand Canyon of the Colorado River (Fig. 14–16). This awe-inspiring monument to the forces of erosion has by now reached a maximum depth of over 2600 meters, and the river has penetrated deep into crystalline Precambrian basement rocks.

Columbia Plateau and Cascades

Unlike the Colorado Plateau, which is constructed of layered sedimentary rocks, the Columbia Plateau in the northwestern corner of the United States has been built by volcanic activity (Fig. 14–17). During the Late Tertiary and Quaternary, low-silica lavas erupted along deep fissures in the region. The liquid rock spread out and buried over 200,000 sq mi of existing topography under layer after layer of lava. In some places, these low-viscosity lavas flowed over a distance of 170 km from their source. Their combined thickness exceeds 2800 meters.

West of the Columbia Plateau lies an uplifted belt that was also the site of extensive volcanic activity. Here, however, the outpourings of more viscous lavas resulted in the mountains of the Cascade Range. Volcanism in this region began about 4 million years ago.

Figure 14–12 (A) View of part of the Basin and Range Province of southeastern California. Paiute Mountains west of Essex, California, are in the distance. (B) Nopah Range of the Basin and Range Province. (Photograph courtesy of J. C. Brice.)

The fact that it has continued down to the present was made dramatically obvious to Americans by the eruption of Mount St. Helens on May 18, 1980 (Fig. 14–18). The volcano exploded with a force equivalent to 50 million tons of TNT and with a roar heard over 300 km away. A great turbulent cloud of hot gases, steam, pulverized rock, and ash burst from the north side of the mountain, rose 20,000 meters into the atmosphere, and began to drift slowly toward the east. An estimated 1 cu km of air-borne ash and other rock debris from the explosion blocked out the sun's light and caused automatic street lights to switch on in towns hundreds of kilometers downwind from the rumbling volcano. The ash fell like a dismal gray snow, blanketing streets, dangerously burdening the roofs of buildings, choking the engines of vehicles, and covering the leaves of trees and crops. Hot gas and ash from the volcano melted part of St. Helens' snow and ice cap, and the resulting meltwater mixed with ash and formed mudflows, which surged down the mountain slopes at speeds as great as 80 km per hour. In the nearby town of Toutle, the mudflows destroyed 123

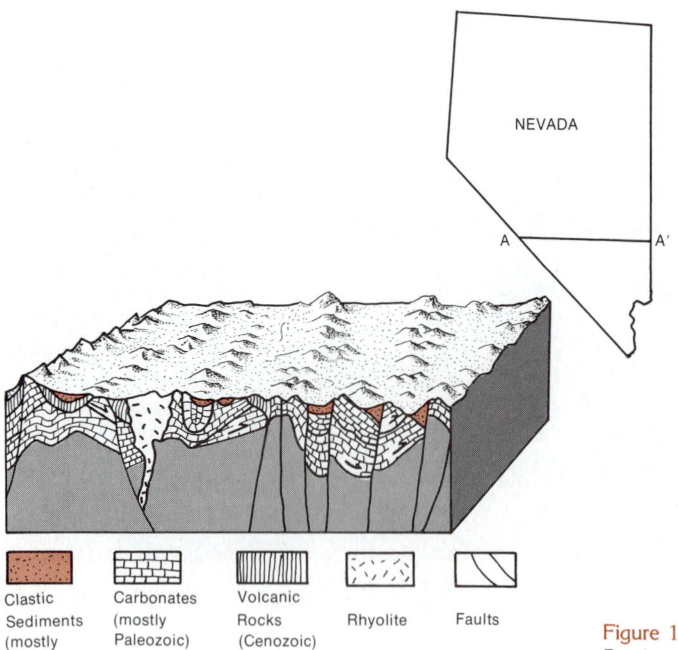

Figure 14–13 Geologic section across the Basin and Range Province along line A——A′ in southern Nevada.

homes. In addition, 22 people camping or working nearby were killed by heat, gases, or burial under the downpour of ash.

The recent activity at Mount St. Helens and the older eruptions that gave us the volcanic peaks of the Cascades are surface manifestations of an ongoing encounter between two of the earth's lithospheric plates. One of these is the North American Plate, and the other is the small Gorda Plate of the eastern Pacific, which moves eastward on a collision course toward the coasts of Oregon and Washington. The Gorda Plate plunges beneath the coastline, and molten rock generated as the plate moves downward rises to supply lava to the volcanoes.

Mount St. Helens is not the only famous volcanic mountain in the Cascades, nor is it the only peak having periodic eruptions. As recently as 1914, Mount Lassen extruded lava and ash for a year and then exploded violently on May 19, 1915, producing a *nuée ardente* that roared down the mountainside, destroying everything in its path. Other well-known Cascade peaks include Mount Baker, Mount Hood, and Mount Rainier. The last

Figure 14–14 Prominent butte in Monument Valley, Colorado Plateau, southern Utah. (Photograph courtesy of J. C. Brice.)

Figure 14–15 Canyon of the Dolores River, Colorado Plateau south of Grand Junction, Colorado. (Photograph courtesy of J. C. Brice.)

Figure 14–16 The Grand Canyon of the Colorado River. View across O'Neill Butte to Bright Angel Creek. (Photograph courtesy of National Park Service.)

Figure 14–17 Lava flows exposed on either side of Summer Falls, Columbia Plateau. (Photograph courtesy of United States Department of the Interior, Bureau of Reclamation, Columbia Basin Project.)

Figure 14–18 Aerial view of erupting Mount St. Helens volcano taken at about 11:30 AM on May 18, 1980. Clouds of steam and ash are being blown toward the northeast from the prominent plume which reached about 20,000 meters altitude. The white linear features are logging roads, and the dark patches are stands of mature trees. This May 18 eruption produced an amount of ash roughly equivalent to that ejected during the AD 79 eruption of Mount Vesuvius that buried Pompeii. (Photograph by Austin Post, courtesy of the U.S. Geological Survey.)

Figure 14–19 A view of the west wall of Crater Lake, with Wizard Island rising from the caldera. (Photograph courtesy of National Park Service, Crater Lake National Park.)

major eruption of Mount Rainier was about 2000 years ago, although minor disturbances occurred frequently in the late 1800's. Crater Lake in Oregon (Fig. 14–19) was formed when the top of a volcanic cone collapsed into its subsiding lava column to form a caldera. Mount Mazama, the volcano from which Crater Lake is derived, had been active until late in the Pleistocene.

Sierra Nevada Mountains and California

South of the Cascades are the great Sierra Nevada Mountains. The rocks of these ranges had been compressed and intruded during the Jurassic Nevadan Orogeny. For most of the Tertiary, the peaks were steadily reduced by erosion until their granitic basement lay exposed at the surface. Then, in Pliocene time and continuing into the Pleistocene, the entire Sierra Nevadan block was raised along normal faults bounding its eastern side and tilted westward. Its high eastern front was lifted an astonishing 4000 meters. The depressed western side formed the California trough. As these movements were underway, rejuvenated streams and powerful valley glaciers began to sculpture the magnificent landscapes of the present-day Sierras (Fig. 14–20). The raising of the Sierras was probably the result of isostatic adjustment as the great block of granitic crust sought flotational equilibrium with denser underlying materials. Such adjustment may have been impossible earlier because of compression generated along the West Coast subduction zone.

Deposits of marginal marine embayments characterize the region west of the Sierras. Early in the Cenozoic, islands formed as a result of folding along the present sites of the coast ranges. Between islands and peninsulas, seas entered low-lying areas of the coast ranges and Great Valley of California. Fine-grained clastics and siliceous deposits, such as cherts and diatomites, are common in marine Cenozoic sections in California. During the Miocene, folding and uplift caused marine regressions in many areas, and by the end of the Tertiary, seas were restricted to a narrow tract along the western edge of California. Of course, the final chapter in the tectonic history of California has

Figure 14–20 Yosemite Valley owes it grandeur to glacial erosion. the massive granites that form the valley walls were emplaced during the Nevadan Orogeny.

not yet been written. It may indeed be the most dramatic chapter of all, for California lies uneasily along the juncture of the active American and Pacific tectonic plates.

There are impressive sections of Tertiary rocks far to the north in Alaska and the arctic islands of Canada. Coal found associated with these strata, along with fossil spores and pollen, indicate that a temperate or even warm-temperate climate prevailed in these northern lands. Along the coastal mountains and Aleutian chain, Tertiary sands and shales are interspersed with layers of pyroclastics and lava flows. Then, as now, Alaska was nervously astir.

A Change in West Coast Tectonics

As described in Chapter 12, the western edge of North America during most of the Cenozoic was the site of an eastward-dipping subduction zone. This subduction was in one way or another responsible for the batholiths, compressional structures, volcanism, and metamorphism that accompanied Mesozoic and Early Tertiary orogenies. The oceanic plate that was being fed into the subduction zone has been named the *Farallon Plate.* During the Cenozoic, the Farallon Plate was being consumed at the subduction zone faster than it was receiving additions at its spreading center. As a result, most of the Farallon Plate and part of the East Pacific Rise that generated it was gobbled up at the subduction zone along the western edge of California (Fig. 14–21). Today, the small Juan de Fuca Plate near Oregon and Washington and the Cocos Plate off the coast of Mexico are the remnants of the once more extensive Farallon Plate.

With the loss of Farallon Plate near California, the North American Plate was brought into direct contact with the Pacific Plate, and a new set of plate motions came into operation. Before making contact with the West Coast, the Pacific Plate had been moving toward the northeast. As a result, when contact was made, the Pacific Plate did not plunge under the continental margin but rather slipped along laterally, giving rise to the San Andreas Fault. No longer did the California sector of the West Coast have an Andean type of subduction zone (Fig. 14–22). Strike-slip movements now characterized the western margin of the United States.

An important result of the shearing and wrenching of the West Coast was the tearing away of Baja California from the mainland of Mexico about 5 million years ago. Because it was located west of the San Andreas Fault system, the Baja area was sheared from the American Plate and now accompanies the northward movement of the Pacific Plate.

South America

Extensive deformation, metamorphism, and emplacement of granitic masses had characterized the Andean Belt during the Cretaceous. Folding and volcanism continued to be widespread during the Cenozoic as well and was especially intense during the Miocene Epoch. Subsequently, the Andean highlands were eroded to low relief and then in Late Pliocene uplifted. Erosion of the uplifted surface is responsible for the present relief and topography of the Andes. Cenozoic instability in the Andes has been related to the continued movement of the floor of the South Pacific Ocean be-

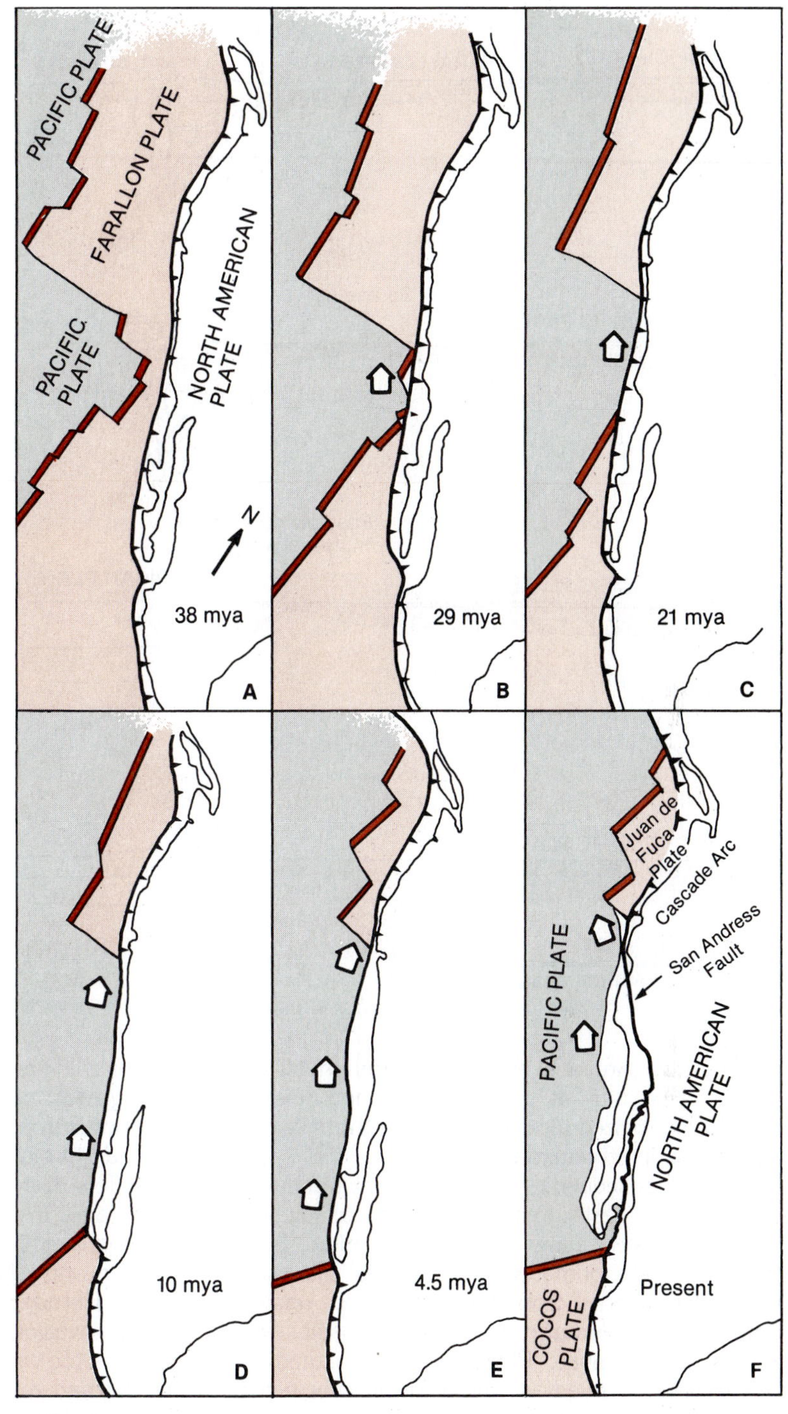

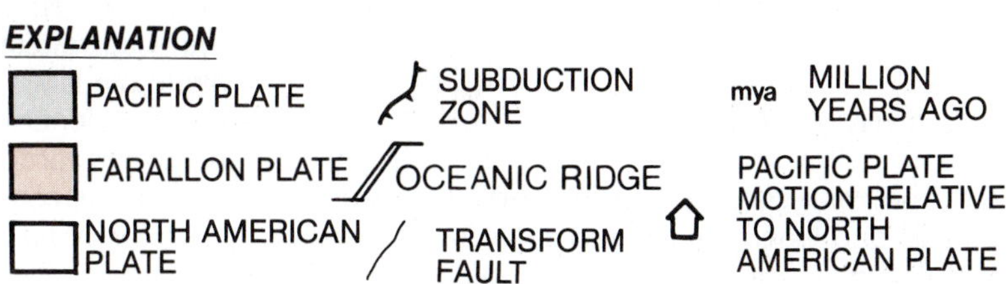

Figure 14–21 Schematic model of the interaction of the Pacific-Farallon and North American Plates for six time intervals during the Cenozoic. Note how the Farallon Plate was largely subducted by Late Cenozoic, leaving only remnants to the north (Juan de Fuca Plate) and to the south (Cocos Plate). The San Andreas and associated faults were caused by right lateral movements beginning about 29 million years ago. (Adapted from T. H. Nilsen. Introduction to Late Mesozoic and Cenozoic sedimentation and tectonics in California. San Joaquin Geological Society Short Course No. 3, 1977. Data source T. Atwater. *Geol. Soc. Amer. Bull.,* 81(12):3513–3536, 1970.)

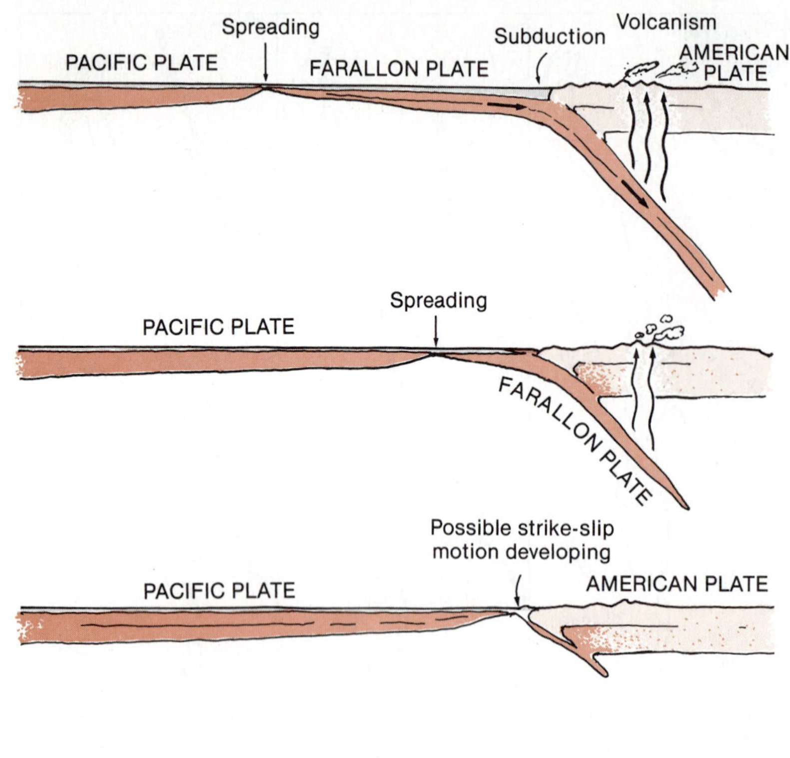

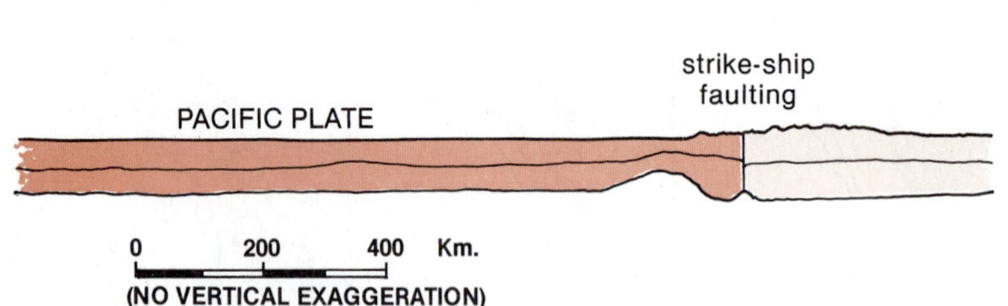

Figure 14–22 Sequence of cross-sections of California and its offshore area illustrating the subduction of the Farallon Plate. In this model, the Pacific Plate is considered fixed as the spreading center (East Pacific Rise) encounters the continental margin. (After T. Atwater. Implications of plate tectonics for the tectonic evolution of western North America. *Bull. Geol. Soc. Amer.* 81(12):3513–3536, 1970.)

neath South America. An oceanic trench off the Pacific coast and an eastwardly inclined seismic zone characterized by deep focus earthquakes attest to this interpretation.

As a consequence of the frequent upheavals that beset the west side of South America, epicontinental seas were either marginal or limited. Sedimentation was primarily terrestrial, as silt, sand, and gravel eroded from the Andes was washed down into the Amazon and Orinoco Basins and across the Pampas. Riftlike basins created by block-faulting served as collecting sites for prodigious thicknesses of clastic sediments.

The Tethyan Realm

The conversion of a major seaway separating Eurasia and Gondwanaland to a spectacular array of mountains and plateaus must be considered one of the greatest events of the Cenozoic Era. During most of the era, the Tethys Gulf lay to the south of Europe. Its waters spilled onto the northern margin of Africa as an epicontinental sea. Foraminifers of the genus *Nummulites* (Fig. 14–23) proliferated in this sea. In Egypt and Libya, their remains contributed to the formation of thick layers of nummulitic limestone. To the north of the Tethys, Europe was only intermittently subjected to marine invasions. The most extensive of these incursions took place during the Oligocene Epoch and formed a wide swath across eastern Europe from the Tethys to the North Sea. Subsequently, the seas retreated except in the Tethys itself and along the lands bordering the North Sea and western France. Over much of central Europe, the Cenozoic record is read from extensive lake and stream deposits.

Rumblings prophetic of the impending tectonic storm began in the Eocene with large-scale

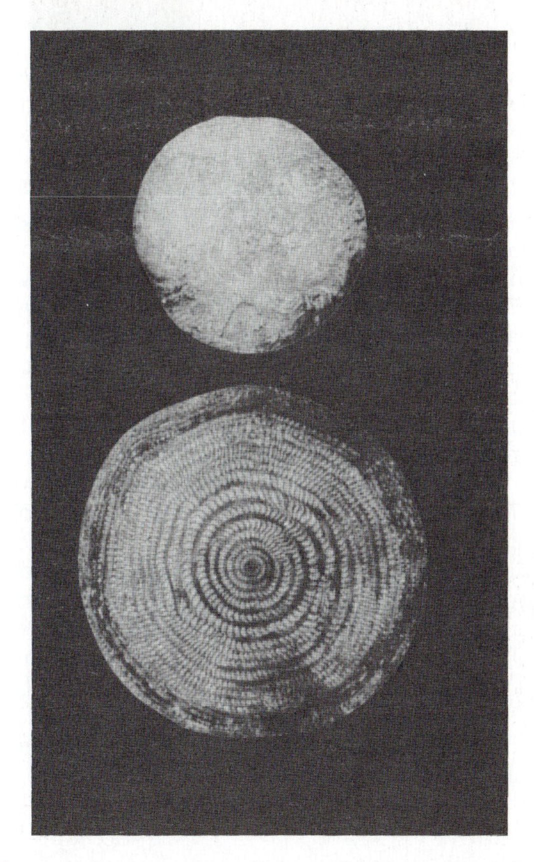

Figure 14–23 *Nummulites.* The upper specimen is complete, and the lower is a thin section that has been enlarged to show internal complexity. Actual diameter of specimen is 16 mm.

folding and thrust-faulting. It was at approximately this time that the northward-moving African block first encountered the western underside of Europe and crumpled the strata that now compose the Pyrenees and Atlas mountains. Then, with a sort of scissor-like movement, the Alpine region began to be squeezed. At first, marine conditions persisted between the emerging folds. Siliceous shales, cherts, and poorly sorted sandstones accumulated between elongate submarine bands.

By Oligocene time, compression from the south caused enormous recumbent folds to rise as mountain arcs out of the old seaway and to slide forward over the lands that once lay to the north of the Tethys. Great folds were cut along their undersides by thrust faults and pushed on top of one another as spectacular monuments to the forces involved in plate collisions (Fig. 14–24). North of these contorted and rising structures lay a topographic depression that received the piedmont deposits eroded from the mountains. These terrestrial clastics, termed **molasse** by European geologists, resemble similar deposits swept eastward from the Rocky Mountains during the Cenozoic.

Even after the Oligocene, the compressions continued, and during the Pliocene, further thrusts from the south carried the older folded belts northward over the molasse deposits and crumpled the Jura folds, which now form the northern front of the Alps. The thrusting was followed by spasmodic regional uplifts that continue even to the present day.

That part of the Tethys Geosyncline that extended far to the east of the Alps experienced its own paroxysm of mountain making during the Cenozoic. Volcanism, folding, thrusting, and emplacement of massive granitic intrusions began early in the era and increased markedly during the Miocene. Great elongate tracts of the geosyncline's floor were squeezed into folds and thrust southward. Much of the early deformation occurred at or near sea level, but ultimately the geosynclinal section was forced up and above the level of the waves. A broad, subsiding trough formed along the northern edge of newly arrived peninsular India. On this lowland were spread over 5000 meters of stream and lake deposits. These sediments contain an important fossil record of Cenozoic mammals and plants. In the two final epochs of the era, regional (epeirogenic) uplifts brought the plateaus and ranges to lofty elevations and caused the retreat of those marginal seas that still remained. The uplift may very well have been caused by the con-

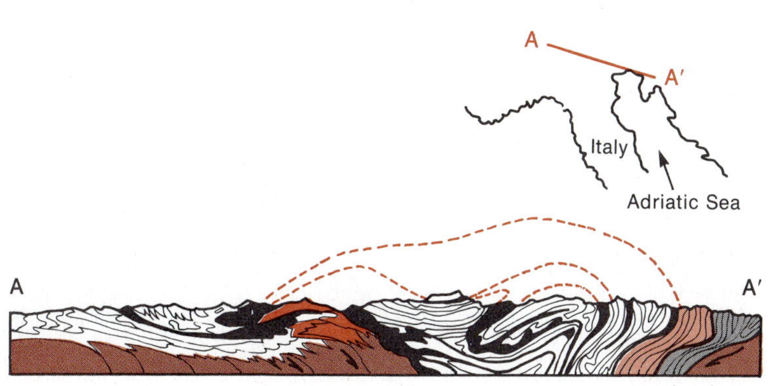

Figure 14–24 The complexity of Alpine deformation is evident in this section across northern Italy.

tinued movement of the Indian block northward beneath the southern edge of the Asian plate.

The history of the Tethys provides yet another remarkable example of how a region of the earth's crust can undergo a complete changeabout in form. The once quiet seaway that bordered Eurasia on the south was transformed into a structurally complex region of rugged highlands that includes the Alps (Fig. 14–25), Atlas, Apennines, Carpathians, Caucasus, Pamirs, and Himalayas. The Mediterranean, Black, and Caspian seas are the only surviving parts of the great Tethys gulf.

Important geologic events were also occurring in regions north of the Tethys during the Cenozoic. In the early part of the era, lavas were extruded in Scotland, Ireland, Spitzbergen, Greenland, and Baffin Island. In Ireland, these lavas are vertically jointed and form the famous Devil's Causeway. The volcanic activity appears to have accompanied the separation of Greenland and Europe. Far to the east, in Mongolia and China, epeirogenic uplift occurred during the time that the Himalayas were being deformed. Lakes and swamps were plentiful across Asia, and some extended over vast regions. Meandering streams built wide floodplains. Important Cenozoic mammalian remains have been unearthed in these Asian fluvial deposits.

Africa

The southern margin of the Tethys Sea formed the upper boundary of Africa during most of the Cenozoic. The region had a much quieter tectonic history than did the northern Tethys. In Libya and Egypt, the formations are flat or only moderately folded carbonates; the only severely folded mountain ranges are in western North Africa, which experienced pulses of orogeny through the Cenozoic.

The larger part of Africa that lay south of the Tethys was generally emergent during the Cenozoic. It was a time in which erosion prevailed. The most conspicuous changes that occurred were those associated with the rift valleys and uplands along the east side of the continent. During the Late Cenozoic, much of eastern Africa was arched upward nearly 3000 meters above sea level. Fracturing and faulting occurred across the crest of the arch. The famous **rift valleys** formed as narrow slivers of crust that had slipped downward between great fault blocks. As the faulting continued, volcanoes developed along the fault trends. Mount Kenya and Mount Kilimanjaro are two of the most notable of these volcanoes, but there are many others as well. In addition to the volcanoes, a series of elongate lakes (Fig. 14–26) formed within the downfaulted blocks. Today, Lake Nyasa and Lake Tanganyika are splendid examples of these fault-controlled inland water bodies. This region of lakes, volcanoes, and fault-controlled ranges is well known for its exceptional scenic grandeur.

The Western Pacific

Most of Australia was comparatively stable during the Cenozoic Era. Marine deposits are restricted to peripheral basins. Near the center of the continent, coal-bearing terrestrial clastics are

Figure 14–25 A view of the Swiss Alps near St. Moritz. A prominent valley glacier is in the center of the photograph.

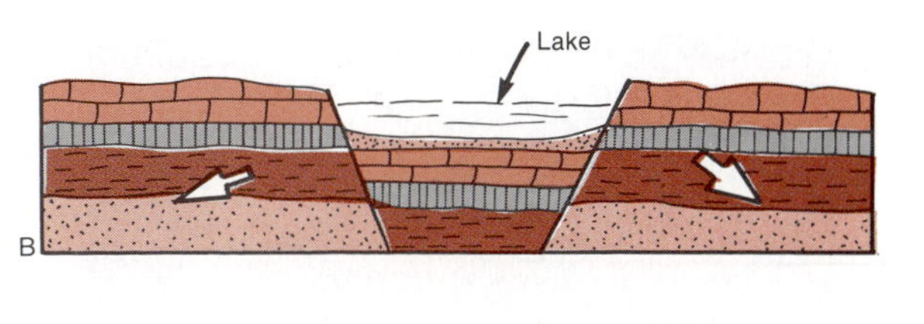

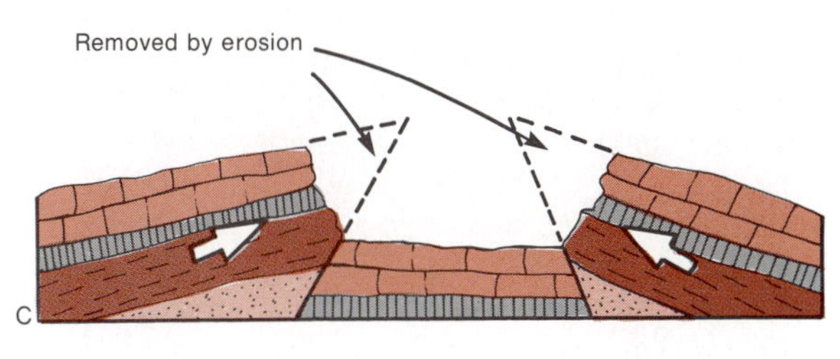

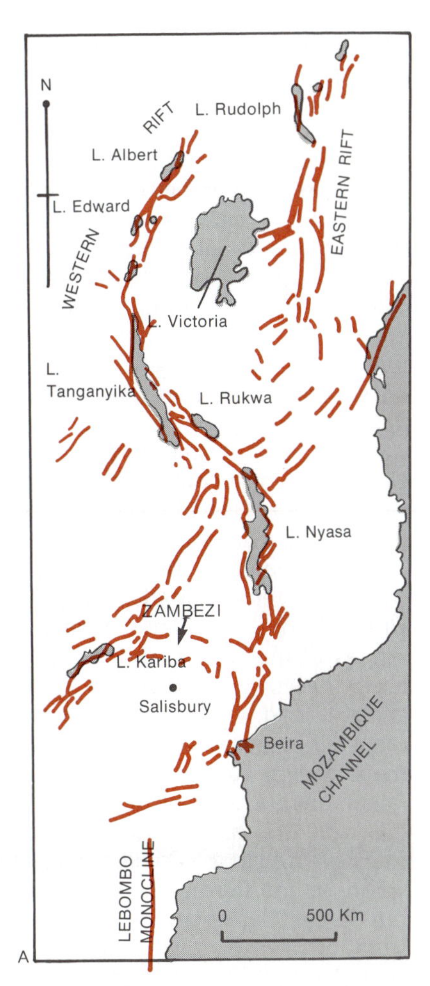

Figure 14–26 (A) East African rift valleys and associated lakes. (B) Schematic illustration of the formation of rift valleys from the action of tensional forces in the crust. (C) Example of how compressional forces may also result in the formation of rift valleys. (A, From H. H. Read and J. Watson. *Introduction to Geology.* London, Macmillan & Co., 1975, p. 237, Figure 8.3.)

found. However, the general quiet that prevailed over most of the continent was not characteristic of the extreme southeastern border, where Paleocene and Eocene volcanism was intense. The fireworks were apparently associated with the separation of Antarctica from Australia an estimated 50 or 60 million years ago.

The Cenozoic history of New Zealand and the region north of Australia was also marked by exceptional instability. Throughout the islands that extend from the Aleutians down through Japan, the Philippines, Indonesia, and into New Zealand, folding, volcanism and spectacular upheavals were common occurrences. Tropical rains and storms caused rapid erosion and deposition of coarse sediments. Deep sedimentary basins received the erosional detritus and frequently spread it seaward as vast deltas.

The enormous volumes of Cenozoic andesites and basalts found in New Zealand clearly indicate its participation in the instability. Indeed, a belt of active volcanoes persists today on the north island. Late in the Cenozoic, vertical movements raised a great fault segment an incredible 18 km along the east side of the north island.

The crustal activity that characterized the western Pacific during the Cenozoic has continued to the present. If one were to speculate about the most active region of the globe during the upcoming "Post-Cenozoic," the sinuous arcs from the Kamachatka Peninsula through Japan and down into New Guinea would be likely candidates for the title "Most Likely to be Deformed." (The Caribbean region might qualify as a second choice.)

Antarctica

One of the most interesting of the inferences that have resulted from the study of Cenozoic rocks of Antarctica is that this now frigid land had a genial or semi-tropical climate throughout the early part of the era. It was not until the Miocene that snow and ice began to accumulate and spread

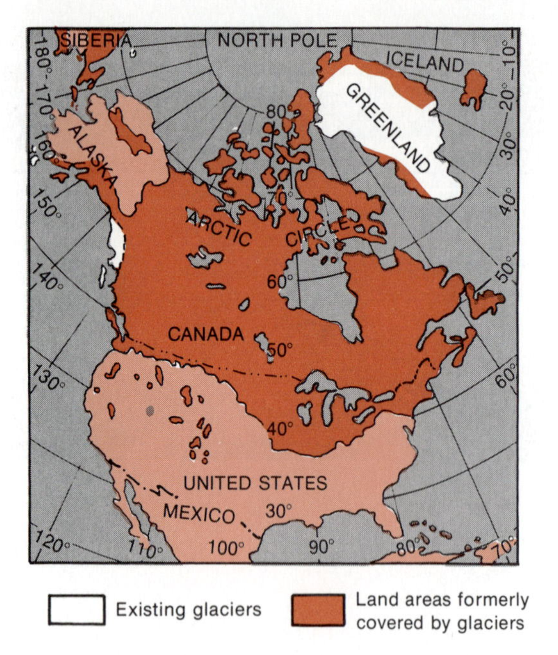

Figure 14–27 Maximum area of glacial coverage in North America. (From United States Geological Survey Pamphlet, *The Great Ice Age*. United States Government Printing Office, 1969, O-357–128).

across the land mass. Geosynclinal sediments and volcanics accumulated along the Pacific side of Antarctica. The instability of west Antarctica can be inferred to be the result of the eastward movement of an oceanic plate against and beneath the continent.

THE GREAT ICE AGE

The final two epochs of the Cenozoic Era are the Pleistocene and Holocene (or Recent). They represent only about 2 million years or so. However, for us they are an exceedingly significant interval. In the first place, it was during the Pleistocene that primates of our own species appeared and rose to a dominant position. Second, the Pleistocene was the epoch in which over 40 million cu km of snow and ice were dumped on about a third of the land surface of the globe (Figs. 14–27 and 14–28). Such an extensive cover of ice and snow had profound effects not only upon the glaciated terrains themselves (Fig. 14–29) but also upon regions at great distances from the ice fronts. Climatic zones in the northern hemisphere were shifted southward, and arctic conditions prevailed across northern Europe and the United States. Mountains and highlands in the Cordilleran and Eurasian ranges were sculptured by spectacular mountain glaciers. While the snow and ice accumulated and spread in higher latitudes, rainfall increased in the lower latitudes, with generally beneficial effects on plant and animal life. Even as late as the beginning of the Holocene, presently arid regions in north and east Africa were well watered, fertile, and populated by nomadic tribes. Peoples of the Middle and Late Pleistocene seemed to thrive by hunting along the fringes of the continental glaciers, where game was abundant and the meat kept longer with less danger of spoilage. Animal furs provided warm clothing, and following the discovery of fire, the caves were warm against the arctic winds.

In addition to the glaciations, the Pleistocene was a time of recurring crustal unrest. Volcanoes were active in New Mexico, Arizona, Idaho, Mexico, Iceland, Spitzbergen, and the Pacific borders of both North and South America.

Pleistocene crustal uplifts occurred in the Grand Tetons, the Sierra Nevada Mountains, and ranges of the central and northern Rocky Mountains. Pulses of uplift during the Pleistocene also characterized the Alps, the Himalayas, and the ranges that lay between. All of this crustal and climatic activity suggests that the Pleistocene was a rather unique time in geologic history. However, this is not the case, for widespread glaciations have also occurred in the Precambrian, late in the Ordovician, in the Permo-Carboniferous, and possibly during the Oligocene and Pliocene as well. Indeed, study of relatively accessible Pleistocene and Recent glacial deposits has greatly improved our abil-

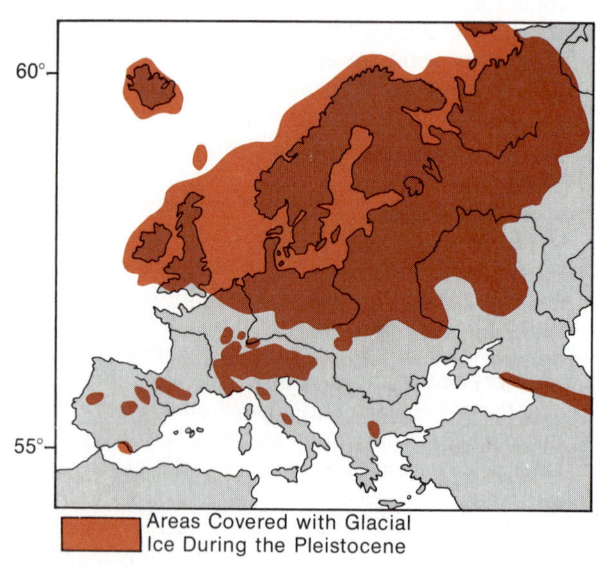

Figure 14–28 Areal extent of major glaciers in Europe during the Pleistocene Epoch.

Figure 14–29 Glaciated Canadian Shield north of Montreal. (Photograph courtesy of Royal Canadian Air Force.)

ity to interpret the vestiges of more ancient glaciations. Carbon and oxygen isotope analyses of foraminifers from deep sea cores have provided evidence of several episodes of ocean water cooling that may not necessarily have been associated with glaciation on land.

Pleistocene and Holocene Chronology

It is a popular, but biostratigraphically untenable, conception that the Pleistocene Epoch began with a sudden worldwide onslaught of frigid climates and that the lower boundary of the Pleistocene Series is easily recognized by sedimentologic and paleontologic clues to that frigidity. However, as is the case with the other epochs of the Cenozoic, the Pleistocene was defined by Charles Lyell in 1839 according to the proportion of extinct to living species of mollusk shells in the layers of sediment. For example, strata containing 90 to 100 per cent of present-day mollusks were designated as belonging to the Pleistocene. Thus the definition of what should be designated Pleistocene is straightforward, but it is no simple matter at all to find suitably fossiliferous sediments in various parts of the globe that can be confidently correlated to Lyell's type section in eastern Sicily.

At the present time, the most widely accepted figure for the beginning of the Pleistocene Epoch is 1.6 million years ago. However, actual glaciation (as indicated by glacial sediments) may not have begun until about a million years ago. Also, the low temperatures and abundant precipitation required for extensive continental glaciation occurred at different times at different places. As a result, debate will doubtless continue over the precise actual age for the beginning of the Pleistocene Epoch.

In those places where radiometric dating is not feasible, biostratigraphers still use fossils for correlating Pleistocene beds. In some cases, a horizon correlative to the standard section can be recognized by the earliest appearance of certain cooler water mollusks and foraminifers. When studying cores of deep sea sediments, one can frequently recognize the basal Pleistocene oozes by the extinction point of fossils called **discoasters.** Discoasters (Fig. 14–30) are calcareous, often star-shaped fossils believed to have been produced by golden brown algae related to coccoliths. In continental deposits, the fossil remains of the modern horse (*Equus*), the first true elephants, and particular species of other vertebrates are used to recognize the deposits of the Lower Pleistocene.

Pleistocene deposits can be related to four distinct glacial stages, which are separated by three interglacial stages (Fig. 14–31). In North America, each glacial stage is named after a state in which the deposits are well exposed. Interglacial stages in North America are named for localities where interglacial soils and other deposits are well displayed. Glacial and interglacial stages followed one another at intervals that lasted between 30,000 and 100,000 years.

The end of the Pleistocene (and beginning of the Holocene) is drawn at the time of melting of the most recent ice sheets and the concomitant rise in sea level. This would mean that the Pleistocene ended about 8000 years ago. However, there is some argument about the date of this boundary. Some geologists prefer to set the Pleistocene-

Figure 14–30 Photographs of discoasters taken with the scanning electron microscope.

NORTH AMERICA	ALPINE REGION
WISCONSIN	Würm
Sangamon	Riss-Würm
ILLINOIAN	Riss
Yarmouth	Mindel-Riss
KANSAN	Mindel
Aftonian	Gunz-Mindel
NEBRASKAN	Gunz
Pre-Nebraskan	Pre-Gunz

A

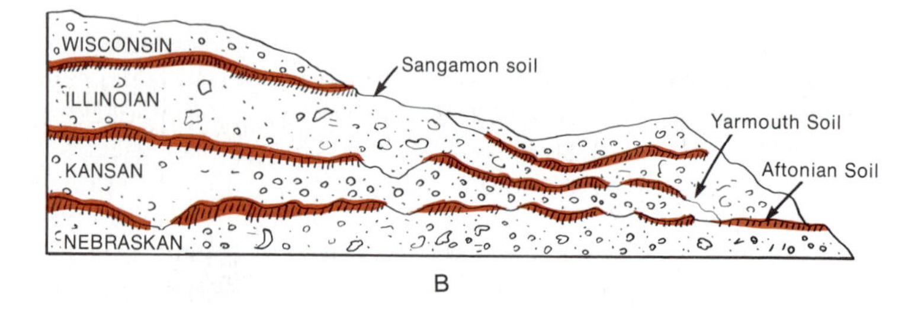

Figure 14–31 Standard Pleistocene nomenclature for glacial (all capital letters) and interglacial stages are shown in A. B is an idealized cross-section showing a succession of deposits of glacial stages and interglacially developed soils. (A section this complete would be a rare occurrence.)

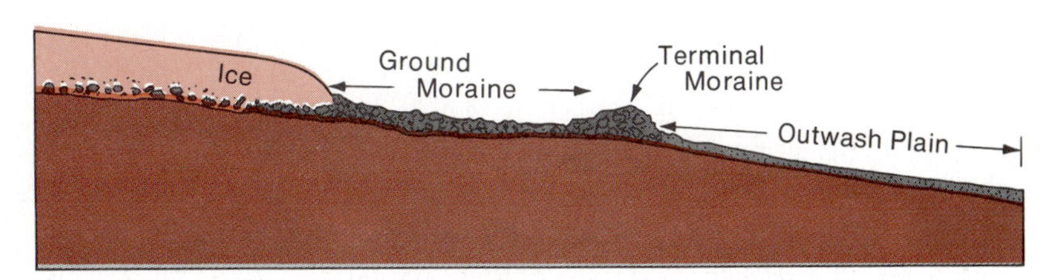

Figure 14–32 Terminal moraine, ground moraine, and outwash plain.

Holocene boundary at the midpoint in warming of the oceans, in which case the ice age would have ended between 11,000 and 12,000 years ago.

Stratigraphy of Terrestrial Pleistocene Deposits

Because they are frequently chaotic mixtures of coarse materials, glacial deposits are difficult to correlate. Moving glaciers may carry away loose material and leave a surface of scoured and polished bedrock. However, if the ice does not advance for a time, a terminal moraine may be built up in front of the glacier. As the ice melts away, the terminal moraine (Fig. 14–32) may be left behind to provide a record of the former extent of the glaciers. As more of the ice sheet is wasted away, a widespread ground moraine is left behind. Morainal deposits may consist of **till,** an unsorted mixture of particles from clay to boulder size that is dropped directly by the ice (Fig. 14–33). **Stratified drift,** on the other hand, has been washed by meltwater, is better sorted and has many of the characteristics of stream deposits. In either case, however, the stratigrapher is faced with similar-looking masses of coarse clastics that are woefully deficient in index fossils. As a result, he or she must employ several substantiating criteria in making correlations. The degree to which one sheet of glacial debris has been dissected by streams may indicate that it is older or younger than another sheet. The depth of oxidation and amount of chemical weathering may also provide some esti-

Figure 14–33 Till deposit from an end moraine west of Madison, Wisconsin. (Photograph courtesy of Professor Robert F. Black.)

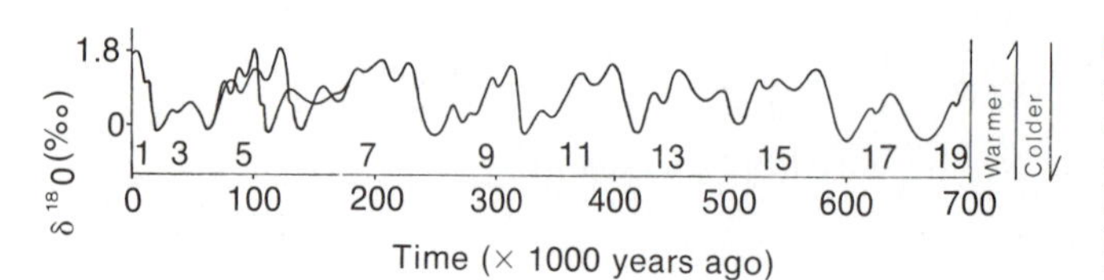

Figure 14–34 Generalized paleotemperature curves for the Caribbean during the past 700,000 years. Time data from radiometric dating; temperature data from oxygen isotope analyses of deep sea cores. (From C. Emiliani and N. J. Shackleton. The Brunhes Epoch: Isotope temperatures and geochronology. *Science,* 183(4124):513, 1974, Fig. 4.)

mate of the relative age of interglacial soils. Careful examination of fossil pollen grains in thick exposures of bog or lake deposits often clearly reflects fluctuations of climate and can be used to mark times of glacial advance and retreat. Varved clays deposited in lakes near the glaciers can sometimes be correlated to similar sediments of other lakes and in addition may provide an estimate of the time required for deposition of the entire thickness of lake deposits. However, the most accurate means of dating and correlating Pleistocene sediments is to extract pieces of wood, bone, or peat and determine their age by radiocarbon dating techniques. Unfortunately, even this method has its limitations, the most significant of which is the relatively short half-life (5570 years) of carbon 14. This limits the method to materials less than 50,000 years old and thus restricts the use of the carbon 14 technique to deposits of the last glacial stage.

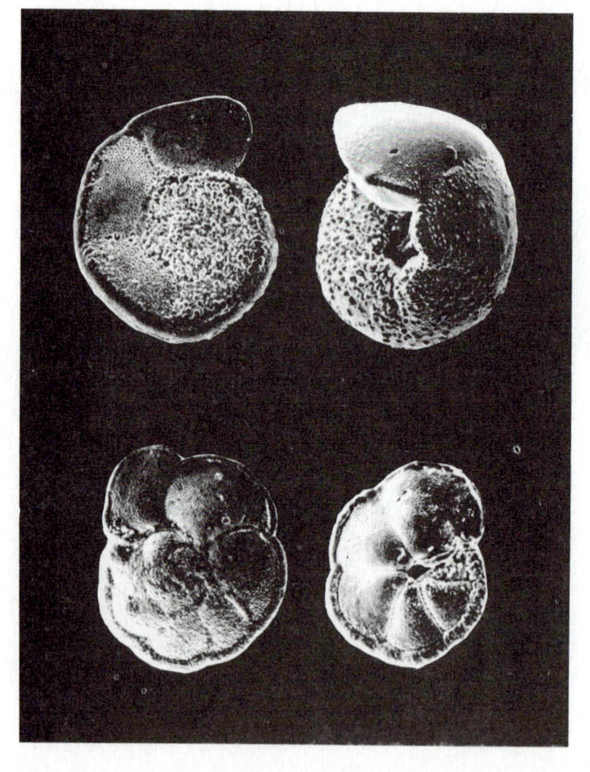

Figure 14–35 Two well-known species of planktonic foraminifers used in correlating deep sea sediments. At the top are views of both sides of *Globorotalia truncatulinoides.* Below are views of opposite sides of *Globorotalia menardii.* (Magnification of all specimens is 50×.)

Pleistocene Deep Sea Sediments

Deep sea sediments can be very useful in Pleistocene geochronology. The deep ocean basins are sites of a relatively continuous sedimentary record; the column of sediments contain abundant fossil remains; and the deposits can be dated by relating them to paleomagnetic data and to radiometric isotopes having short half-lives.

Continuous sections of deep sea sediments can be obtained by means of piston-coring devices. These tools provide cores over 15 meters long. The cores can then be subjected to various kinds of analyses. One approach is to use the oxygen-isotope ratios of calcareous foraminifers within the cored sediment to determine the water temperatures in which the foraminifers once lived. Such temperatures plotted against depth would indicate variation in temperature with time (Fig. 14–34). Indications of cooler conditions are then correlated with glacial stages and a decline in worldwide sea level.

Another way to use deep sea sediments in Pleistocene chronology is to plot at each level in the core the relative abundance of species of foraminifers that are known to be especially sensitive to temperature. For example, the tropical species *Globorotalia menardii* (Fig. 14–35) is alternately present or absent within Pleistocene cores from the equatorial Atlantic. The absence of the species in part of a core is taken to indicate an episode of cooler climates and glaciation. Carbon 14 dates in the upper part of such cores permit one to determine the rates of sedimentation, which is assumed to be uniform for the lower portions.

Another foraminifer, *Globorotalia truncatulinoides,* also permits one to recognize alternate cooling and warming of the oceans (Fig. 14–36). This species is trochospirally coiled, which means that it coils in a spiral with all the whorls visible on one side but only the last whorl visible on the other side. Individuals that coil to the right dominate in warmer water, whereas left-coiled individuals prefer colder water. During glacial advances, when ocean

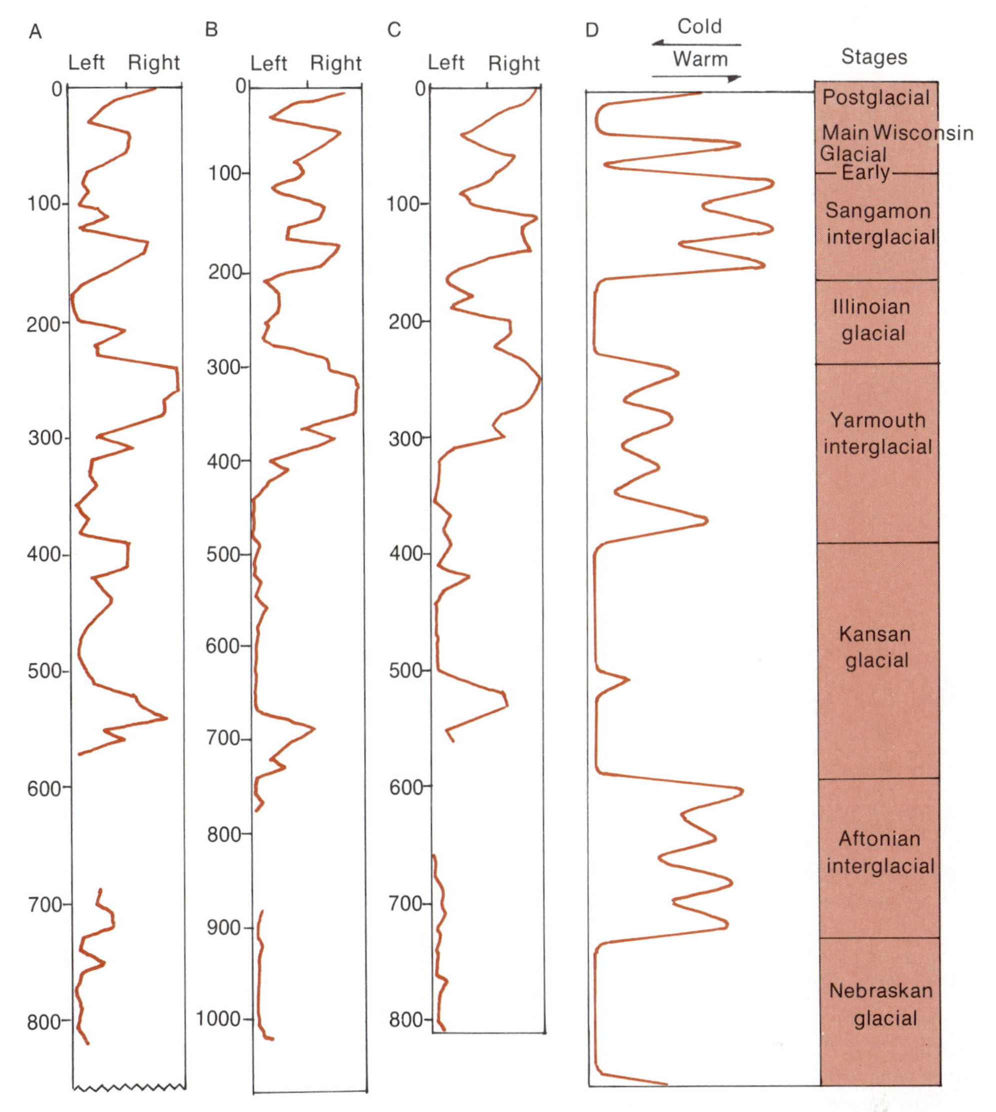

Figure 14–36 A, B and C are curves depicting percentages of right and left coiled *Globorotalia truncatulinoides*. The scale runs from 100 per cent right-coiling at the right hand margins; left-coiling forms prefer cooler water. Numbers to the left of the columns are core depths in centimeters. D is a generalized climate curve based on variation in the frequency of *Globorotalia menardii* complex in 10 cores. Pleistocene divisions provide correlation. (From D. B. Ericson and E. Wollin. Pleistocene chronology and deep sea cores. *Science,* 162(3859):1233, December, 1968, Figs. 6 and 7.)

temperatures declined, the right-coiled populations of *G. truncatulinoides* in middle and low latitudes were replaced by populations in which left coiling predominated. The record of such changes is clearly apparent in the deep sea cores.

Remanent magnetism in some deep sea sediments is sometimes sufficiently strong and stable that a particular section of the core can be correlated to a magnetic reversal documented on land in volcanic rocks of known age. However, the analysis is likely to be in error if the original orientation of detrital magnetic grains has been disturbed by burrowing organisms.

The Effects of Pleistocene Glaciations

Earlier it was noted that during maximum glacial coverage, over 40 million cu km of ice and snow lay upon the continents, equivalent to a tremendous amount of water; removal of that water from the oceans had a multitude of effects on the

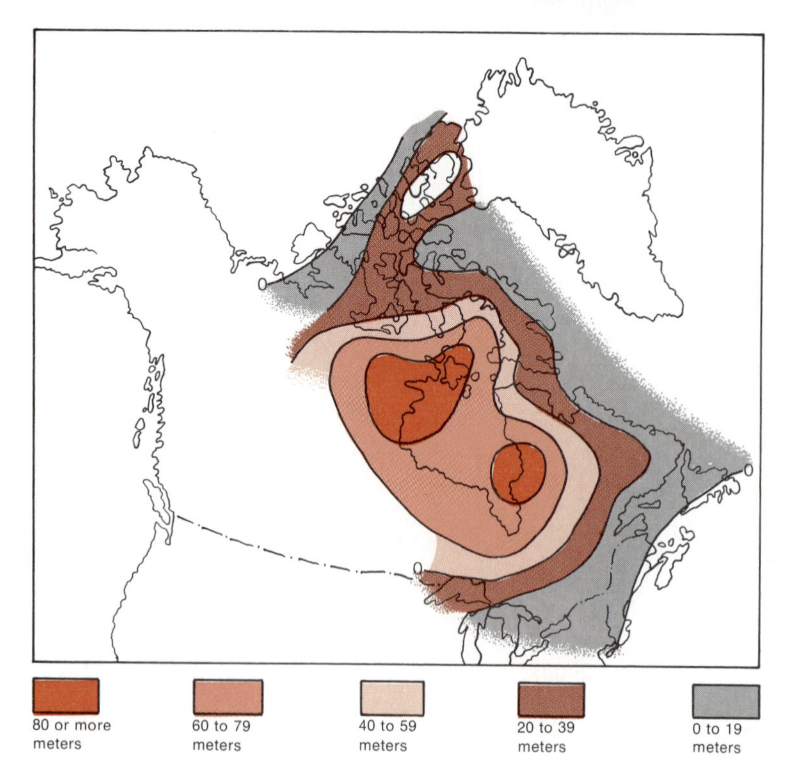

Figure 14–37 Postglacial uplift of North America determined by measuring elevation of marine sediments 6000 years old. (Simplified and adapted from J. T. Andrews. The pattern and interpretation of restrained, postglacial and residual rebound in the area of Hudson Bay. In *Earth Science Symposium on Hudson Bay,* Ottawa, 1968: Canadian Geological Survey Paper 68–53, p. 53, 1969.)

environment. It has been estimated that sea level may have dropped at least 75 meters during maximum ice coverage. Extensive tracts of the present continental shelves became dry land and were covered with forests and grasslands. The British Isles were joined to Europe, and a land bridge stretched from Siberia to Alaska. During interglacial stages, marine waters returned to the low coastal areas, drowning the flora and forcing terrestrial animals inland.

The glaciers had a direct impact on the erosion of lands and the creation of glacial land forms. The great weight of the ice depressed the crust of the earth over large parts of the glaciated area, in some places to a level of 200 to 300 meters below the preglacial position. With the removal of the last ice sheet, the downwarped areas of the crust gradually began to return to their former positions. The rebound is dramatically apparent in parts of the Baltic, the Arctic, and the Great Lakes region of North America, where former coastal features are now elevated high above sea level (Fig. 14–37).

As the great continental glaciers advanced, they obliterated old drainage channels and caused streams to erode new channels. These dislocations are especially evident in the north-central United States. Prior to the ice age, the northern segment of the Missouri River drained northward into Hudson Bay, and the northern part of the Ohio River flowed northeastward into the Gulf of St. Lawrence. The lower Ohio drained into a preglacial stream named the Teays River. Today, geologists recognize the former location of the Teays by a thick linear trend of sands and gravels. Those parts of the Missouri and Ohio Rivers that once flowed toward the north were turned aside by the ice sheet and forced to flow along the fringe of the glacier until they found a southward outlet. The present trend of the Missouri and Ohio Rivers approximates the margin of the most southerly advance of the ice. The equilibrium of streams was also affected, for with glacial advances there was a lowering of sea level and thus an increase in stream gradients near coastlines. A reverse effect occurred with wasting away of the ice sheets.

Prior to the Pleistocene, there were no Great Lakes in North America. The present floors of these water bodies were lowlands. Glaciers moved into these lowlands and scoured them deeper. As the glaciers retreated, their meltwaters collected in the vacated depressions (Fig. 14–38). Niagara Falls, between Lake Erie and Lake Ontario, came into existence when the retreating ice of the Wisconsin glacial stage uncovered an escarpment formed by a southwardly tilted resistant strata. Water from the Niagara River tumbled over the edge of the escarpment, which is supported by the resistant Lockport Limestone. Weak shales beneath the limestone are continually undermined, causing southward retreat of the falls.

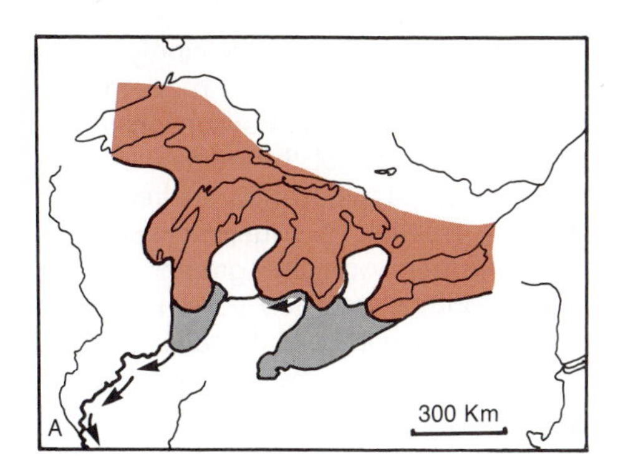

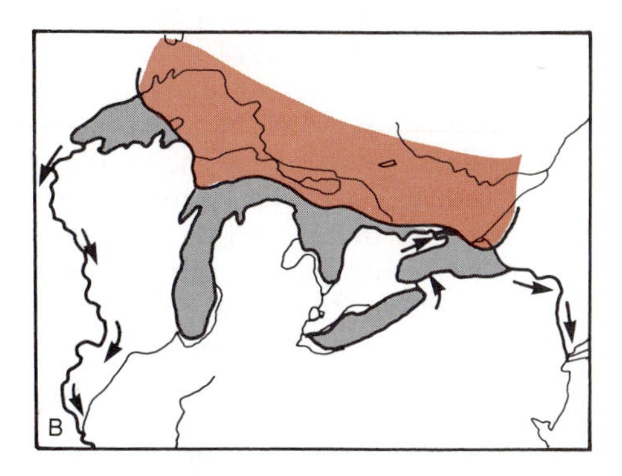

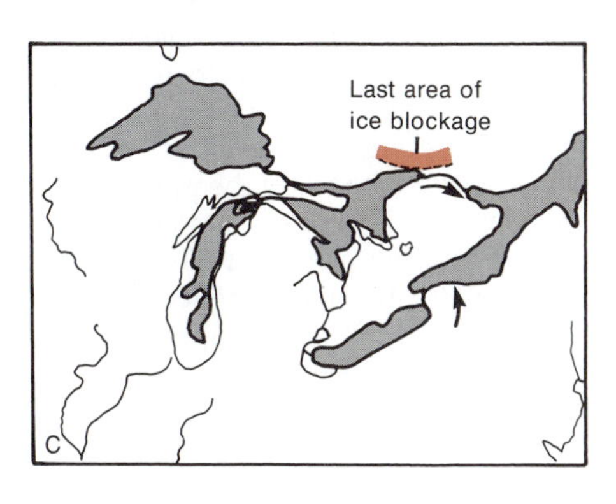

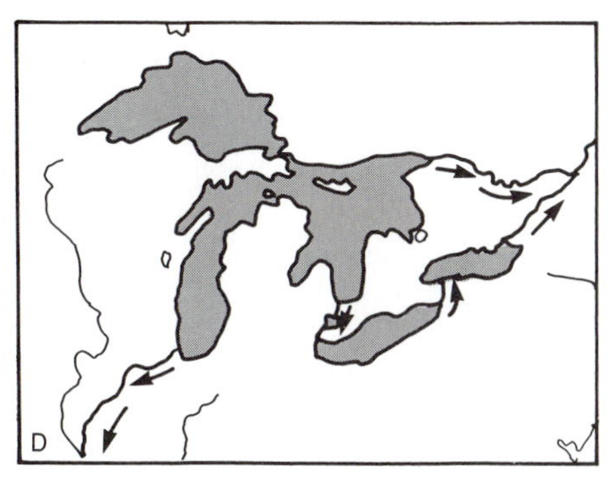

Figure 14–38 Four stages in the development of the Great Lakes as the ice of the last glacial advance moved away. (After J. L. Hough. *Geology of the Great Lakes.* Urbana, University of Illinois Press, 1958, Figs. 56, 69, 73, and 74.)

Another large system of ice-dammed lakes covered a vast area of North Dakota, Minnesota, Manitoba, and Saskatchewan. The largest of these lakes has been named Lake Agassiz in honor of Jean Louis Rodolphe Agassiz, the great French naturalist who initially insisted on the existence of the ice age. Today, rich wheatlands extend across what was once the floor of the lake.

Other lakes developed during the Pleistocene, occupying basins that were not near ice sheets. These were formed as a consequence of the greatly increased precipitation and runoff that characterized regions south of the glaciers. Such water bodies are called pluvial lakes (L. *pluvia,* rain). Pluvial lakes were particularly numerous in the northern part of the Basin and Range Province of North America, where faulting produced more than 140 closed basins. So-called pluvial intervals, when lakes were most extensive, were generally synchronous with glacial stages, whereas during interglacial stages many lakes shrank to small saline remnants or even dried out completely. Lake Bonneville in Utah was such a lake. It once covered over 50,000 sq km and was as deep as 300 meters in some places. Parts of Lake Bonneville persist today as Great Salt Lake, Utah Lake, and Sevier Lake.

A rather spectacular event associated with Pleistocene lake formation occurred in the northwestern corner of the United States. Lobes of the southwardly advancing ice sheet repeatedly blocked the Clark Fork River, and the impounded water formed a long, narrow lake extending diagonally across part of western Montana. The freshwater body, called Lake Missoula, contained an estimated 2000 cu km of water. With the recession of the glacier, the ice dam broke, and tremendous floods of water rushed out catastrophically across more than 38,000 sq km of what are now appropriately termed the channeled scablands (Fig. 14–39).

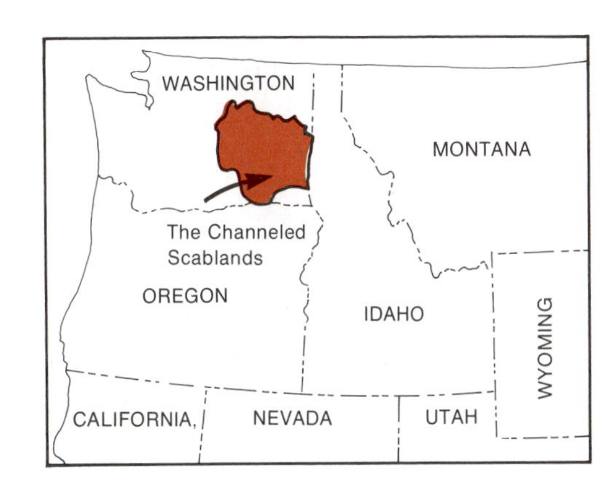

Figure 14–39 Location of the channeled scablands.

The glacial conditions of the Pleistocene also had an effect upon soils. In many northern areas, fertile topsoil was stripped off the bedrock and transported to more southerly regions, which are now among the world's most productive farmlands. Because of the flow of dense, cold air coming off the glaciers, winds were strong and persistent. Fine-grained glacial sediments that had been spread across outwash plains and floodplains were picked up and transported by the wind and then deposited as thick layers of windblown silt called **loess.** Such deposits blanket large areas of the Missouri River Valley, central Europe, and northern China.

Cause of Pleistocene Climatic Conditions

The results of oxygen isotope research indicate that world climates grew progressively cooler from Middle Cenozoic to the Pleistocene. The culmination of this trend was not a single sudden plunge into frigidity but rather an oscillation of glacial and interglacial stages. Any theory that adequately explains glaciation must consider not only the long-term decline in worldwide temperatures but the oscillations as well. Further, the theory must include reasons for the ideal combination of temperature and precipitation required for the buildup of continental glaciers. Although geologists, physicists, and meteorologists have been speculating about the cause of the ice age for over a hundred years, no single causative agent has been found. Indeed, it appears likely that Pleistocene climatic conditions came about as a result of several simultaneously occurring factors.

One particularly interesting theory of ice age origin was developed by M. Milankovitch, a Yugoslavian astronomer who convincingly calculated that irregularities of the earth's movement could cause climatic cycles. Milankovitch drew attention to the fact that the tilt of the earth's axis with respect to the plane of the orbit varies between about 22° and 24°. This tilt, of course, causes the seasons, and, furthermore, the greater the tilt, the greater the contrast between summer and winter temperatures at a particular latitude. Second, the axis also has a wobbling motion called precession (Fig. 14–40). In addition, the earth's orbit around the sun is an ellipse that varies in its eccentricity between known limits. According to the Milankovitch theory, a combination of these astronomic factors would periodically result in longer, colder winters and shorter, warmer summers at particular latitudes. The coldest point in the cycle would occur every 40,000 years or so. The timing of glaciations as predicted from the Milankovitch calculations correspond rather well to indications of cooling obtained from oxygen-isotope and foraminiferal analyses of deep sea cores. However, there is one problem with the theory. If the Milankovitch effect has been in existence for billions of years, why haven't there been Pleistocene-like glaciations continuously down through geologic time? Apparently, other factors must also be involved.

One such factor might be a variation in the amount of solar energy reflected from the earth back into space rather than being absorbed. The fraction of solar energy reflected back into space is termed the earth's **albedo.** At the present time, it is about 33 per cent. Theorists suggest that by the end of the Cenozoic, when continents were fully emergent, and hence highly reflective, temperatures may have been lowered enough for the Milankovitch effect to begin to operate. This is not an unreasonable suggestion, for only a 1 per cent lessening of retained solar energy could lead to as

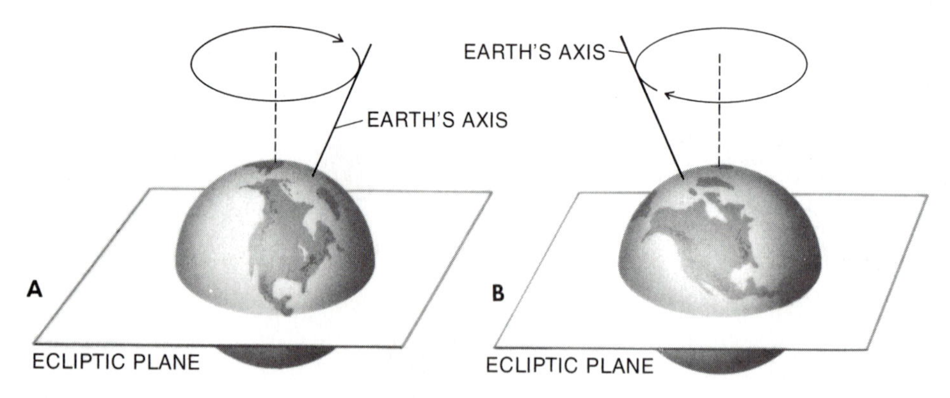

Figure 14–40 Two positions in the precession of the earth's axis. (From J. Pasachoff. *Contemporary Astronomy.* 2nd ed. Philadelphia, Saunders College Publishing, 1981.)

much as an 8°C drop in average surface temperatures. This would be sufficient to trigger a glacial advance if ample precipitation were available over continental areas. Still other geologists speculate that absorption of solar energy was hindered by cloud cover, volcanic ash, and dust in the atmosphere or fluctuations in carbon dioxide. A decrease in carbon dioxide content would cause a corresponding decrease in the warmth-gathering "greenhouse effect," for example. On the other hand, it is also possible that a buildup of CO_2 might trigger glaciation, for as warming occurred, there might be more rapid evaporation and an increase in highly reflective cloud cover.

Yet another proposal, developed by the geophysicists M. Ewing and W. Donn, stresses the importance of ample precipitation in building continental glaciers and draws on the interrelationships of ocean currents and pole positions to supply the necessary moisture. Although there are areas of uncertainty in the Ewing and Donn theory, it does account for both the cooling of the atmosphere that led to the glacial epoch and the oscillating advance and retreat of ice sheets. Several conditions may have had a bearing on the initiation of the ice age. One was the drifting of continental areas over or near the poles, where the cold would not be dissipated by the general movement of ocean waters. For example, when the north pole became established in the nearly enclosed Arctic Ocean Basin, the heat exchange system of ocean currents was impaired, and polar regions became increasingly cooler. For a period of time, the Arctic Ocean supplied moist air for snowfall on surrounding land areas. Additional moisture shunted northward from the Atlantic and Pacific further nourished the ice sheet. Ultimately, the great blanket of ice and snow increased albedo to the point at which temperatures fell so low that the surface of the Arctic Ocean froze and evaporation declined. Deprived of precipitation, the ice sheets began to melt away. The climate warmed, and the Arctic Ocean became ice-free once again. The world entered an interglacial stage, but only temporarily, for once again the original conditions for glacial growth were established. Snow and ice began to accumulate once more, and another stage of glaciation commenced. If the theory is valid, then we may be in an interglacial stage today, and climates will continue to oscillate between glacial and nonglacial until the relationship of the continents and poles is changed in some way.

Although scientists are as yet not ready to formulate a complete and unified theory for the cause of the Pleistocene's multiple glaciations, it does appear that they may be very near the answer. Few deny that the Milankovitch effect, albedo, ocean currents, and pole positions may have played a role in creating Pleistocene climatic conditions. The problem is to determine the relative importance of these factors and how they relate to one another. If Pleistocene-like climatic cycles are fundamentally the result of changes in the earth's orbital geometry, then one can make some interesting speculations about our future climate. Calculations indicate, for example, that the long-term climatic trend over the next 20,000 years is toward extensive northern hemisphere glaciation and, of course, cooler conditions. What effect man's activities, such as the burning of coal, oil, and gas, might have on this trend is uncertain.

CENOZOIC CLIMATES

The worldwide cooling that culminated in the great ice age actually had begun during the Early Cenozoic. However, the decrease in temperatures was not entirely uniform. There were warming trends during the late Paleocene and Eocene, as indicated by fossils of palm trees and crocodiles found in Minnesota, in Germany, and near London. Trees that are today characteristic of moist temperature zones thrived in Alaska, Spitzbergen, and Greenland. Corals grew in regions 10° to 20° north of their present optimum habitat. Cooling resumed during the Oligocene, and coral reefs began a slow retreat toward the equator. On land, temperate and tropical forests were also displaced toward lower latitudes. Another pulse of warmer conditions occurred during the Miocene, after which climates grew steadily cooler, as if in anticipation of the approaching Pleistocene glaciations.

Although the Pleistocene Epoch clearly experienced the greatest glaciations of the era, ice accumulated on a more limited scale at other times during the Cenozoic. Deep sea cores taken near Antarctica indicate that an ice sheet already existed there during the Eocene, and terrestrial glacial deposits have been found in Miocene and Pliocene strata of Antarctica as well. Ice-rafted glacial debris of Miocene age has been observed in deep sea cores from the Bering Sea and from the north Pacific. Evidence for Pliocene glaciations has also been found in the Sierra Nevada Mountains, Iceland, South America, and Russia.

MINERAL RESOURCES OF THE CENOZOIC

Most of the known accumulations of **oil** have been found in Tertiary reservoir rocks. For example, it has been estimated that nearly half of the sum of past production; and probable unproduced reserves of oil in the United States lies within Tertiary strata and that the upper half of the system accounts for more than 30 per cent of all the petroleum ever discovered on the earth. Important Paleocene reservoirs occur in Libya and beneath the North Sea. Petroleum trapped in Eocene sandstones and permeable limestones are being exploited in Texas, Louisiana, Iraq, Russia, Pakistan, and Australia. The largest reserves of oil shale (Fig. 14–41) in the world are within the Eocene Green River Formation of the western United States. Rocks of Oligocene age yield oil in Europe, Burma, the United Stated Gulf Coast, and California. However, of all the Tertiary epochs, the Miocene can be considered the most petroliferous. Miocene reservoir sandstones yield oil on every continent except Australia. They are extraordinarily productive in California, the offshore areas of Texas and Louisiana, and the Middle East. Offshore oil fields in the Gulf Coast and California also tap oil held in rocks of Pliocene age.

Coal also occurs in Cenozoic rocks. Most Cenozoic coal is of the lignitic or sub-bituminous variety and is of little value. However, Paleocene coal of the Fort Union Group has been mined in the Dakotas, Wyoming, and Montana. Scattered coal beds, mostly of Eocene age, are exploited along the Pacific Coast.

During the early Cenozoic, **gold-bearing** Jurassic quartz veins of the Mother Lode region of California were exposed by erosion. Gold particles eroded from the quartz were frequently concentrated in gravel deposits (Fig. 14–42). Since the middle 1800's, this placer gold has been gathered by dredging and hydraulic mining. About 60 per cent of the world's supply of **tin** is also obtained from Cenozoic stream deposits.

In addition to placer gold and tin, valuable deposits of **copper, silver, lead, mercury,** and **zinc** are frequently found in association with Tertiary intrusive rocks of western North America and the Andes and in Tertiary orogenic belts along the western margins of the Pacific Ocean. Similar Cenozoic deposits were developed in southern Europe and Asia as a consequence of the deformation of the Tethys geosyncline.

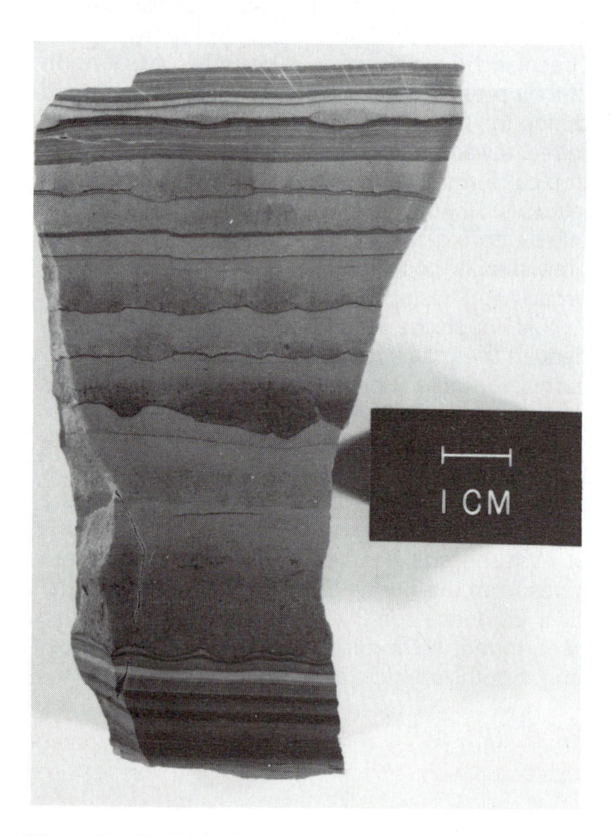

Figure 14–41 Eocene oil shale from northeastern Colorado. (Courtesy of United States Geological Survey.)

Figure 14–42 Typical placer gold from the North Trinity River of Trinity County, California. (Courtesy of United States Geological Survey and P.E. Holt.)

Nonmetallic resources are also exploited from Cenozoic rocks. **Diatomite,** the white, porous rock formed from the siliceous coverings of diatoms, is mined in great quantities from Miocene strata in California. Cenozoic **building stone, clay, phosphates, sulfur, salt,** and **gypsum** are quarried in North America, Europe, the Middle East, and Asia.

Summary

A compilation of the great historical events of the Cenozoic would certainly include the birth of the Alpine-Himalayan mountain system, the drift of continents to their present locations, the uplift of mountains around the perimeter of the Pacific, the development of an ice age, and the evolution of *Homo sapiens.* For most of the duration of the Cenozoic, the continents stood at relatively high elevations; epicontinental and geosynclinal seas were of limited extent.

Along the east side of North America, the Appalachian chain was subjected to cycles of erosion and uplift that served to sculpture the surface to its present topography. Along the Atlantic and Gulf Coastal Plains, repeated transgressions and regressions reworked the clastics transported there from the interior. Florida was a shallow, limestone-forming, submarine bank until uplifted late in the era. In the Gulf of Mexico, Cenozoic deposits accumulated in thicknesses in excess of 9000 meters as the depositional basin was continuously downwarped.

In western North America, compressional forces that had produced many of the structures of the Cordillera ceased, and in their place vertical crustal adjustments occurred. The erosional debris worn from the highland areas filled the basins between ranges and was then spread eastward to form the clastic wedge of the Great Plains. Lakes formed in some basins and served as collecting sites for oil shales. Crustal movements late in the Tertiary caused the elevation of the Sierra Nevada along a great fault and produced the Basin and Range Province. Intense volcanic activity accompanied these crustal movements. Volcanism continues today along sectors of the West Coast of North America that lie adjacent to subduction zones, as, for example, in the Cascades, which are fed by lavas from the subducting Gordo Plate. Also in the Far West, the Farallon Plate and its spreading center were subducted along the California sector, causing a shift from subduction tectonics to strike-slip tectonics and creating the San Andreas Fault system.

In the Tethys region, there was intense deformation during the Cenozoic as major tectonic plates moved against one another. Orogenies that accompanied the compressions and uplifts along the southern margin of the Tethys Geosyncline resulted in the Atlas Mountains, whereas along the northern border, the Alps, Carpathians, Apennines, Pyrenees, and Himalayas were constructed from the sediments that had accumulated in the geosyncline. The Black, Caspian, and Aral Seas were left as remnants of the Tethys Sea. Orogenic activity was also prevalent over much of the Middle East and Far East and along the western borders of the Pacific. Africa witnessed Cenozoic marine deposition along its northern border. However, the most dramatic African occurrences were the continuing development of the great rift valleys that extend along the eastern side of the continent.

It is apparent from fossil evidence and oxygen isotope studies that the earth's climate was on a somewhat fluctuating cooling trend during the Cenozoic. The culmination of that trend was the Pleistocene ice age, which began about 1.6 million years ago. Variations in climate during the Pleistocene caused the vast ice sheets that had formed on the northern continents to alternately advance and recede. Altogether, there were four major advances, with minor fluctuations before and within each. The ice, at times attaining thicknesses of 3000 meters, depressed the earth's crust, smoothed and rounded the landscape, deranged drainage patterns, indirectly altered climates of adjacent regions, and was responsible for the development of numerous lakes, not the least of which are the Great Lakes. As the ice alternately accumulated and melted, sea level was caused to rise and fall with profound effects on low-lying coastal areas. It is a possibility that today we live in an interglacial stage and that some thousands of years from now the ice will again accumulate in Canada and northern Europe and advance southward.

Although a wealth of gold, tin, copper, silver, and other metals is found in deposits or associated with intrusions of Cenozoic rock, the greatest resource of the era has been petroleum. Indeed, most of the world's petroleum has been found in strata that range from 1 to 60 million years of age.

Questions for Review

1. Briefly describe the manner in which each of the major physiographic features below developed:
   a. The mountains of the Basin and Range Province
   b. The Great Plains
   c. The Columbia and Snake River Plateau
   d. The Red Sea
   e. The Teton Range
   f. The Cascade Range
2. What Eurasian mountain ranges resulted from the compression and upheaval of large areas of the Tethys Geosyncline?
3. What epochs of the Cenozoic Era are included within the Tertiary? What epochs compose the Paleogene?
4. The Cenozoic stratigraphic record for the Gulf Coastal region indicates that there were eight major transgressions and regressions. What characteristics of a stratigraphic column might indicate that it was deposited during a transgression? A regression?
5. What evidence indicates that the Gulf of Mexico experienced steady subsidence during the Cenozoic?
6. What is the paleontologic and economic importance of the Green River Formation of Wyoming and Colorado?
7. In terms of plate tectonics, why did the western borders of the Americas experience folding and volcanisms throughout the Cenozoic?
8. Account for the frequent juxtaposition of older, often chaotic, marine deposits lying *above* younger molasse deposits in areas of the Alpine-Himalayan belt.
9. What is the origin of such east African lakes as Nyasa and Tanganyika?
10. What is the difference in texture and mode of origin between stratified drift and till?
11. What are the advantages and limitations of the carbon 14 isotopic dating method for determining the age of Pleistocene deposits?
12. What, in brief, is the "Milankovitch effect"? Why is it unlikely that this effect alone could be the cause of Pleistocene glaciations?
13. What is meant by *albedo*? What circumstances might cause an increase in albedo and possibly the initiation of an ice age?
14. Why were extensive deposits of loess formed in the Pleistocene, but not in earlier Cenozoic epochs? Most of the particles in loess are within the range from $1/16$ to $1/32$ mm, and very little sand or clay is present. How do you account for this characteristic of loess?
15. Why is the Great Salt Lake of Utah so salty?
16. What might be the effect upon humans if all the ice present today as continental ice sheets were to undergo relatively rapid melting?
17. In a 30-meter marine core that had penetrated the entire Pleistocene Series, how might you differentiate the sediment that had accumulated during an interglacial stage from that of a glacial stage?
18. What land bridge, important in the migration of ice age mammals, resulted from glacial lowering of sea level?
19. What is the explanation for the gradual rise in land elevations that has occurred within historic time around Hudson Bay, the Great Lakes, and the Baltic Sea?
20. What changes in movement of tectonic plates along the western border of the United States were responsible for the development of the San Andreas Fault?

Terms to Remember

albedo
discoaster
Farallon Plate
Harrisburg erosional surface
loess
Milankovitch effect
molasse
Neogene
nuée ardente
oil shale
Paleogene
pyroclastics
rift valley
Schooley Peneplain
Somerville erosional surface
stratified drift
till

Supplementary Readings and References

Atwater, T. 1970. Implications of plate tectonics for the Cenozoic evolution of western North America. *Bull. Geol. Soc. Amer.* 81:3513–3536.

Continental Tectonics. 1980. Studies in Geophysics, National Academy of Sciences, Washington, D.C.

Curtis, B. F. (ed.) 1975. Cenozoic history of the southern Rocky Mountains. *Geol. Soc. Amer.* Memoir 144:1–279.

Ericson, D. B., and Wollin, G. 1964. *The Deep and the Past.* New York, Alfred R. Knopf, Inc.

Ewing, M., and Donn, W. L. 1966. A theory of ice ages III. *Science* 152:1706.

Girdler, R. W. (ed.) 1972. *East African Rifts.* Amsterdam, Elsevier.

Hays, J. D., Imbrie, J., and Shackleton, N. J. 1976. Variations in the earth's orbit: Pacemaker of the ice ages. *Science,* 194(No. 4270):1121–1132.

Leopold, E. B. 1969. Late Cenozoic palynology. *In* Tschudy, R. H., and Scott, R. A. (eds). *Aspects of Palynology.* New York, Wiley Interscience.

Match, C. L. 1976. *North America and the Great Ice Age.* New York, McGraw-Hill.

Matthews, R. K. 1974. *Dynamic Stratigraphy.* Englewood Cliffs, N.J., Prentice-Hall.

Nilsen, T. H. 1977. Introduction to Late Mesozoic and Cenozoic sedimentation and tectonics in California. *San Joaquin Geol. Soc. Short Course* No. 3, pp. 7–19.

Read, H. H., and Watson, J. 1975. *Introduction to Geology.* Vol. 2, *Earth History.* Part 2, Later stages of earth history. New York, John Wiley.

Rutten, M. G. 1969. *The Geology of Western Europe.* Amsterdam, Elsevier.

A landscape during the Eocene Epoch. The six-horned bulky animals at the right are *Uintatherium*. Several early horses (Orohippus) are in the left foreground. (Courtesy Chicago Museum of Natural History. Painting by C. R. Knight.)

15

Life of the Cenozoic

> During the Cenozoic Era or the Age of Mammals, the vegetation was of modern cast and life conditions in the main were similar to those of today, although there is much evidence of a gradual elevation of nearly all lands with a consequent increase in aridity and diminution of moisture-loving vegetation.
>
> Richard Swann Lull, *Organic Evolution*, 1924

PRELUDE

Because biologic developments of the Cenozoic have occurred so recently and because Cenozoic fossils are topmost in the stratigraphic column, we have more facts about this era than about the far lengthier ones that preceded it. Armed with this more adequate body of data, paleontologists are better able to trace the relationship between biologic evolution and the environmental changes of the Cenozoic. Continental fragmentation clearly stimulated biologic diversification and resulted in distinctive faunal radiations on separated land masses. Among the many interesting evolutionary developments during the Cenozoic, none seems more fascinating than the changes experienced by the primate group, which by Late Tertiary had produced species considered to be the direct ancestors of humans. In the Pleistocene Epoch, our own species, *Homo sapiens,* appeared.

PLANTS

The Response of Mammals to the Spread of Prairies

Although they are the most recent group to evolve, the flowering plants are now the most widespread of all vascular land plants. Angiosperm floras did not explode upon the lands until Cretaceous time (Fig. 15–1). There were few spectacular floral innovations during the Cenozoic. Rather, this was a time of steady progress toward the development of today's complex plant populations. The Miocene is particularly noteworthy as the epoch during which grasses appeared and grassy plains and prairies spread widely over the lands. In response to the proliferation of this particular kind of forage, grazing mammals began their remarkable evolution.

Numerous evolutionary modifications among herbivorous mammals can be correlated to the development of extensive grasslands. Especially evident were changes in the dentition of grazing mammals. Grasses are abrasive materials, and because they grow close to the ground, they are often coated with fine particles of soil. To compensate for the wear that results from chewing grasses, the major groups of herbivores evolved high-crowned cheek teeth that continued to grow at the roots during part of the animals' lives. To provide space for these high-crowned teeth, the overall length of the face in front of the eyes was increased. Enamel, the most resistant tooth material, became folded, so that when the tooth was worn, a complex system of enamel ridges was formed on the grinding surface (see Fig. 4–12). In general, incisors were gradually aligned into a curved arc for nipping and chopping grasses.

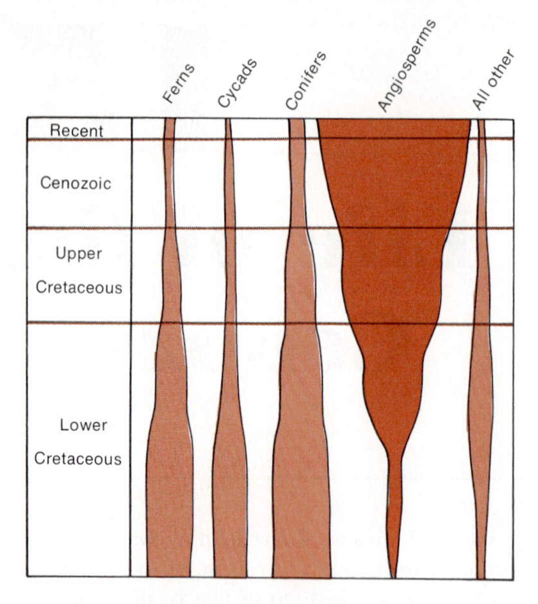

Figure 15–1 Relative proportions of genera in plant groups of Cretaceous floras.

In the wide open plains environment, it was more difficult to escape detection by predators. As a result, many herbivores evolved modifications that permitted speedy flight from their enemies. Limb and foot bones were lengthened, strengthened, and redesigned by the forces of selection to prevent strain-producing rotation and permit rapid fore-and-aft motion. To achieve greater speed, the ankle was elevated, and, like sprinters, animals ran on their toes. In many forms, side toes were gradually lost. Hoofs developed as a unique adaptation for protecting the toe bones while the animals ran across hard prairie sod. Herbivores called ungulates evolved the four chambered stomach in response to selection pressures favoring improved digestive mechanisms for breaking down the tough grassy materials. The response of mammals to the spread of grasses provides a fine example of how the environment influences the course of evolution.

Marine Phytoplankton

As noted in Chapter 14, entire families of phytoplankton groups experienced extinction at the end of the Cretaceous. Only a few species in each major group survived and continued into the Tertiary. However, the survivors were able to take advantage of decreased competitive pressures and rapidly became diversified. In general, peaks in diversity of species were reached in the Eocene and Miocene. A decrease in the diversity has been recorded in the intervening Oligocene. Diatoms (Fig. 15–2), dinoflagellates (Fig. 15–3), and coccolithophores (Fig. 15–4) provided the most abundant populations of Cenozoic marine phytoplankton.

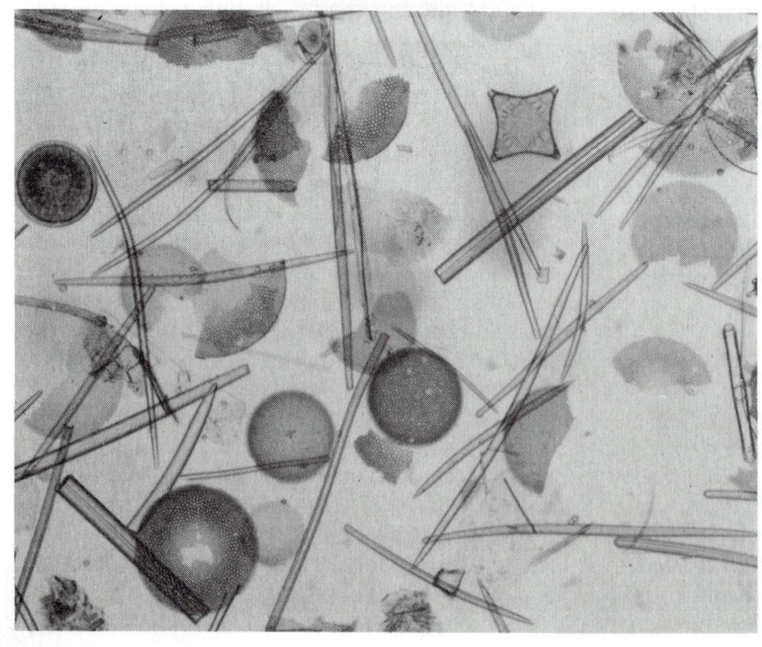

Figure 15–2 Recent diatoms from Asteri, Crete. (Courtesy of Naja Mikkelsen, Scripps Institution of Oceanography.)

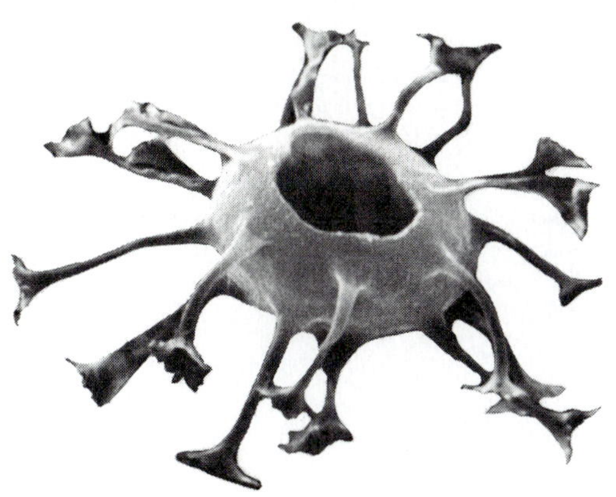

Figure 15–3 Dinoflagellate as seen with the aid of the scanning electron microscope. From Paleocene sedimentary rocks of Alabama. (Courtesy Standard Oil Company of California; photographer, W. Steinkraus.)

Figure 15–4 (A) Photograph of transmission electron microscope image of *Coccolithus.* (B) Photomicrograph of Eocene discoaster *(Discoaster lodoensis)* as it appears when viewed with an optical microscope. (C) Same species viewed with the aid of a scanning electron microscope.

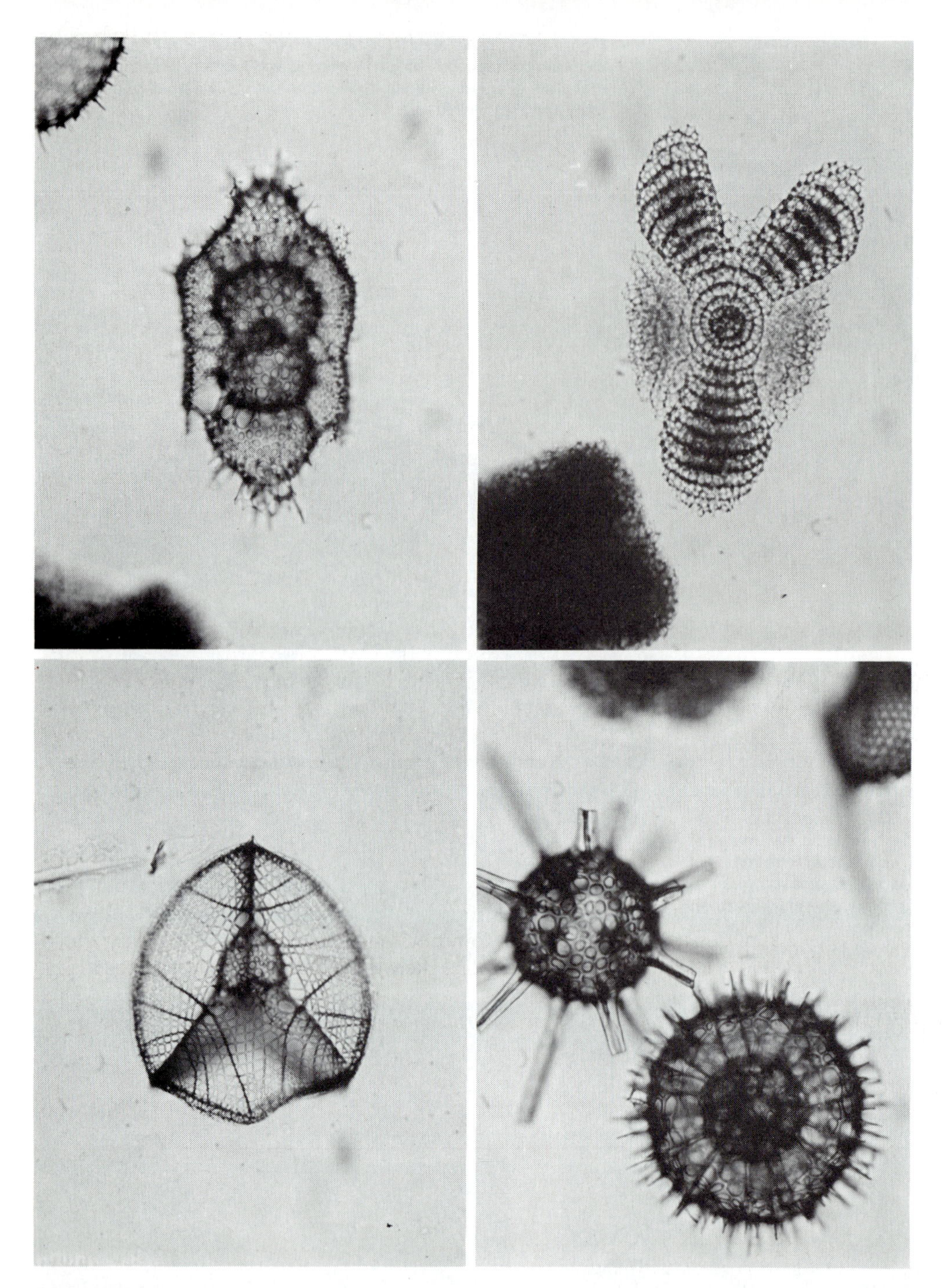

Figure 15–5 Quaternary radiolarians from the tropical Pacific Ocean. (Courtesy of Annika Sanfilippo, Scripps Institution of Oceanography.)

INVERTEBRATES

The invertebrate animals of the Cenozoic seas included dense populations of foraminifers, radiolarians (Fig. 15–5), mollusks of all classes, corals, bryozoans, and echinoids. The combined fauna had a decidedly modern aspect. Such once successful groups as ammonites and rudistid cephalopods were no longer present. No new major groups of invertebrates appeared during the Cenozoic.

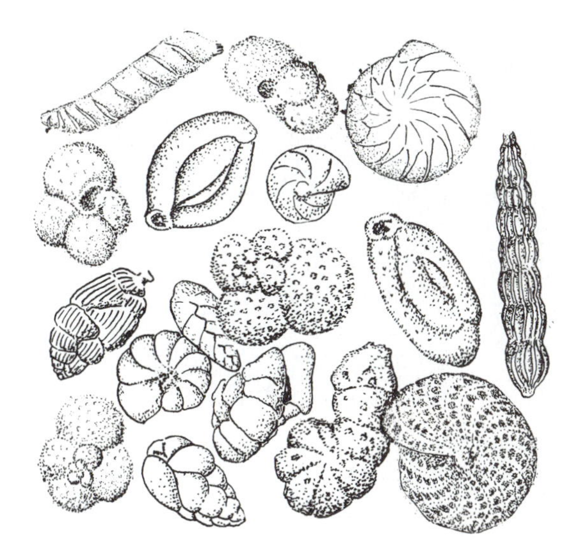

Figure 15–6 Cenozoic foraminifers. (Magnification 50 ×.)

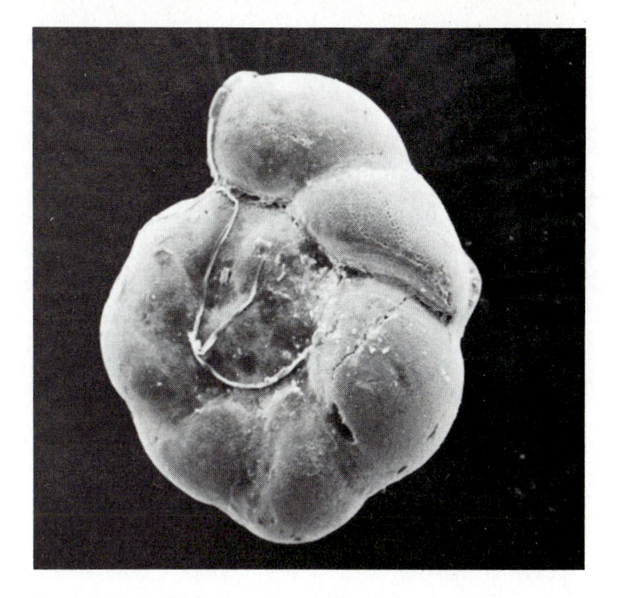

Figure 15–7 Electron micrograph of the rotalid foraminifer *Valvulineria* from the Miocene Monterey Formation of California. (Diameter is about 0.35 mm.)

Protozoans

The enormously prolific and varied Cenozoic foraminifers (Fig. 15–6) included both the exceptionally large and the more persistent smaller forms. The smaller foraminifers were dominated by numerous genera of rotalids (Fig. 15–7), miliolids (Fig. 15–8), and the planktonic globigerinids and globorotalids (Fig. 15–9). Among the larger genera, coin-shaped nummulites (Fig. 14–23) thrived in the clear Tethys Sea as well as in warm water areas of the western Atlantic. The tests of these organisms have accumulated to form thick beds of nummulitic limestone. The ancient Egyptians quarried this rock and used it to construct the Pyramids of Gizeh. Even the famous Sphinx was carved from a large residual block of nummulitic limestone. The incredible numbers and variety of foraminifers have permitted their extensive use in correlating Cenozoic strata in oil fields of the Gulf Coast, California, Venezuela, the East Indies, and the Near East.

Mollusks

Cenozoic shells of mollusks look very much like those likely to be found along coastlines today.

Figure 15–8 Electron micrograph of the miliolid foraminifer *Quinqueloculina*. (Height of specimen is 0.83 mm.)

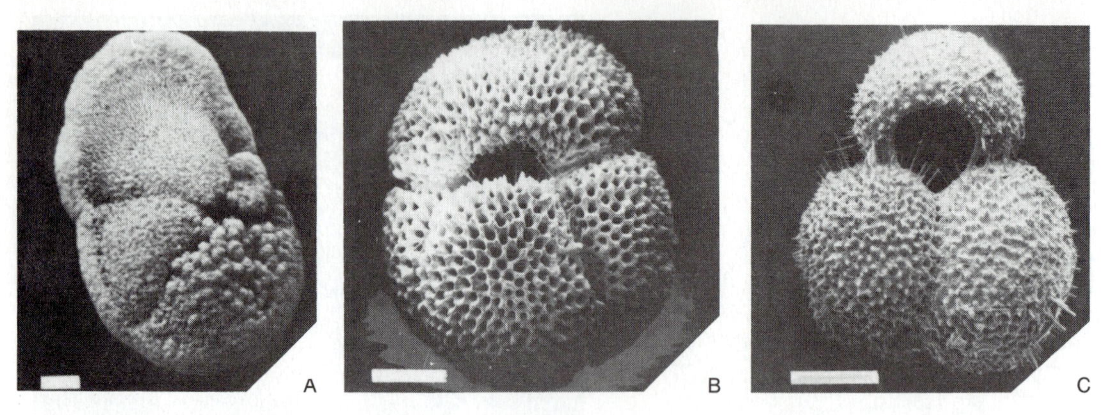

Figure 15–9 Present-day planktonic foraminifers. (A) *Globorotalia tumida;* (B) *Globigerinoides conglobatus;* (C) *Globigerinoides rubra.* (Courtesy of C. G. Adelseck, Jr., and W. H. Berger, Scripps Institution of Oceanography.)

Figure 15–10 Tertiary pelecypods and gastropods. (A) *Venus;* (B) *Glycymeris;* (C) *Pecten;* (D) *Turritella;* (E) *Ecphora;* (F) *Arca;* (G) *Fissurella;* (H) *Conus.*

Pelecypods and gastropods were the dominant classes of Cenozoic mollusks (Fig. 15–10). Their range of adaptation was truly outstanding. Arcoids, mytiloids, pectinoids, cardioids, veneroids, and oysters were particularly abundant pelecypods. Although the climax of cephalopod evolution had passed with the demise of the ammonites, nautiloids similar to the modern pearly nautilus lived in Cenozoic seas, as they do today. Shell-less cephalopods, such as squid, octopi, and cuttlefish, were also well represented in the marine environment, although their fossil record is rather sparse.

Corals

Corals (Fig. 15–11) grew extensively in the warmer waters of the Cenozoic oceans. Fossil solitary corals are commonly found in shallow-water deposits of Early Tertiary age in Europe and the United States Gulf Coastal region. Reef corals were most extensively developed in parts of the Tethyan belt, West Indies, Caribbean, and Indo-Pacific regions. Careful comparison of coral species in Cenozoic rocks on either side of the Isthmus of Panama has given geologists clues to when the Atlantic and Pacific Ocean were connected across this present-day barrier. Continuous deposition of reef limestones occurred throughout the Cenozoic in atolls of the open Pacific. **Atolls** are ringlike coral reefs that grow around a central lagoon. In 1842, Charles Darwin brilliantly discerned the relationship between atolls and volcanic islands. In his **subsidence theory,** he proposed that, because of their great weight, volcanic islands subside slowly. Furthermore, such subsidence is usually slow enough to permit the corals that grow around the fringe of the sinking island to grow upward and maintain their optimum habitat near the ocean's surface. Eventually, the central island would be submerged, but the encircling, living reef would persist (Fig. 15–12). Several deep borings into atolls have validated Darwin's theory. The boring at Eniwetok encountered the basaltic summit of the volcano at depths of about 1200 meters. Eocene reef structures were directly above these igneous rocks.

There is an alternate possibility for atoll formation. It is possible that corals might build **fringing reefs** around the perimeter of volcanic islands that had been erosionally truncated during an interglacial period of low sea level. If the sea level subsequently rose (as it did when the great Pleistocene ice sheets began to melt), then the corals would strive to build the reef upward in order to stay at their optimum living depth. It seems not unreasonable that particular atolls may be primarily the result of subsidence of the island, whereas others may have been triggered by a rise in sea level. Whatever their developmental history, atolls are one indication of the healthy state of Cenozoic reef corals, as well as the multitude of other creatures that contributed to the growth and biology of the reefs.

Echinoderms, Bryozoans, Crustaceans, and Brachiopods

Many other invertebrates continue successfully through the Cenozoic. Echinoderms, for the most part free-moving types, were particularly prolific. Members of the Phylum Bryozoa are common in Tertiary rocks and are still very abundant in many parts of the ocean. It was during this final era that the modern crustaceans became firmly estab-

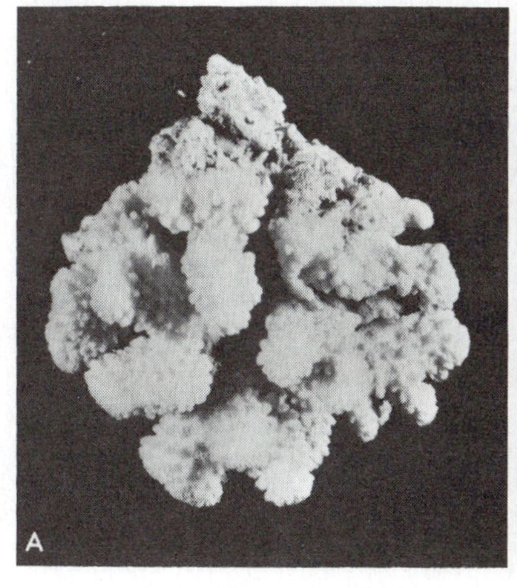

Figure 15–11 Two common scleractinid corals. *Acropora* (A) is one of the most ubiquitous and successful of all reef corals. It originated during the Eocene. *Fungia* (B) is the familiar flat coral from the Indo-Pacific. The genus is known from Miocene to Recent.

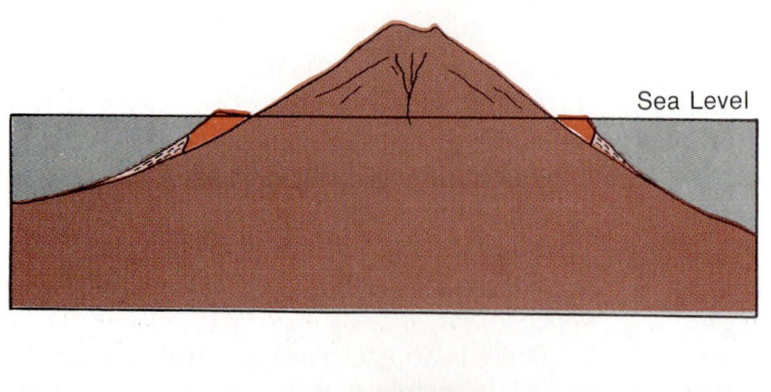

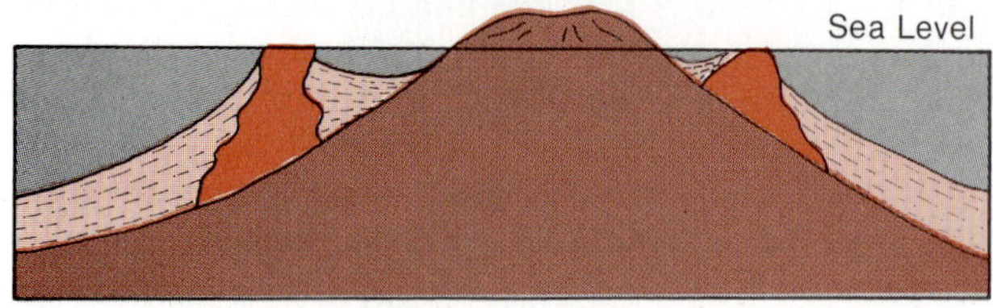

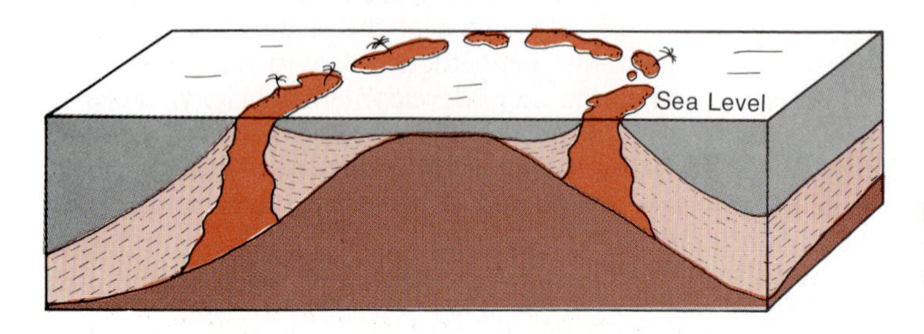

Figure 15–12 Three stages in the development of an atoll caused by gradual submergence of a volcano and upward growth of reefs along its margin.

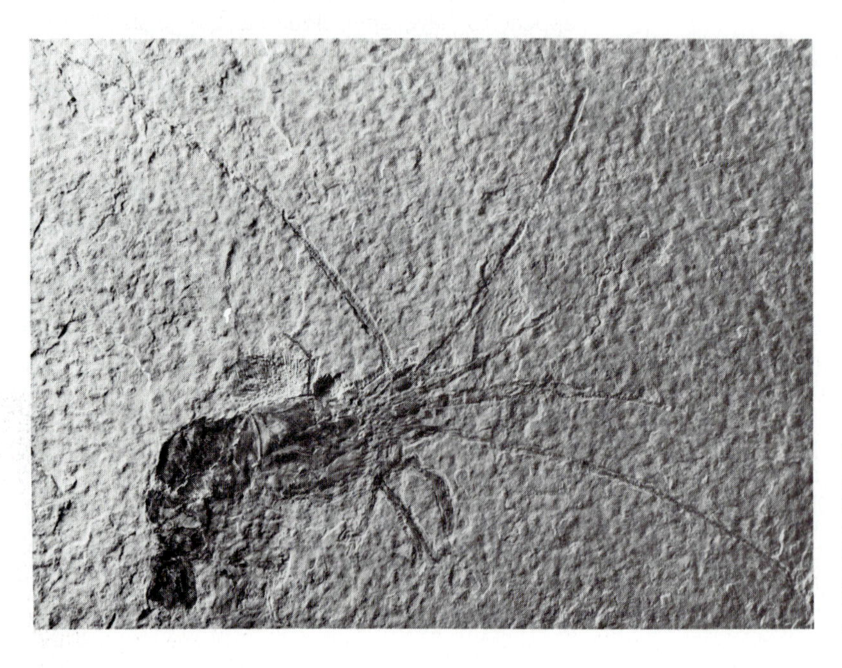

Figure 15–13 The freshwater shrimp *Bechleja rostrata* from the Eocene Green River Formation. (Photograph courtesy of Rodney M. Feldmann. From M. Feldmann, et al. Decapod fauna of the Green River Formation (Eocene) of Wyoming. *J. Paleontol.* 55(4):788–799, 1981.)

lished both in freshwater bodies (Fig. 15–13) and in the oceans. Perhaps the least important of the Cenozoic invertebrates were the brachiopods. Fewer than 60 genera survive today. They consist mostly of terebratulids, rhynchonellids, and inarticulates such as *Lingula* (Fig. 15–14).

Figure 15–14 *Lingula,* a persistent primitive type of inarticulate brachiopod with a thin shell of proteinaceous material and a long, fleshy siphuncle.

VERTEBRATES

Fishes and Amphibians

Those bony fishes that have evolved to the highest level of ossification and skeletal perfection are the teleosts. Teleosts achieved an enormous range of adaptive radiation during the Cenozoic. That radiation included such forms as perches, bass, snappers, seahorses, sailfishes, barracudas, swordfishes, flounders, and others too numerous to mention. The Green River strata of Wyoming are well known for large numbers of beautifully preserved Eocene freshwater fishes (Fig. 15–15). In addition to the bony fishes, sharks were at least as common in the Tertiary as they are today. *Carcharodon,* an exceptionally large shark, had teeth as big as a man's hand and from head to tail extended over 20 meters.

Amphibians throughout the Cenozoic closely resembled modern forms. Frogs, toads, and salamanders were relatively abundant. The first frogs were Triassic in age. By Jurassic, they had become completely modern in appearance and continued almost unchanged over a span of more than 200 million years.

Reptiles

By the beginning of the Cenozoic, the dinosaurs had disappeared from the lands, as had the flying reptiles from the air and several groups of marine reptiles from the seas. The reptiles that have managed to survive and continue to the present day are the turtles, crocodilians, lizards and snakes, and the lizard-like "tuatara" (Fig. 15–16), which lives on islands off the coast of New Zealand. For the most part, the survivors are not significantly more primitive than their predecessors, nor do they appear to have specializations that insured their survival. They were simply very well suited to their places in the biosphere, and their particular niche was not successfully invaded by other vertebrates that came along in the Cenozoic.

Figure 15–15 Fossil fish *(Phareodus testis)* from the shale beds deposited in Green River Lake during Eocene time. (Courtesy of the Smithsonian Institution.)

Figure 15–16 *Sphenodon,* or, as it is called in New Zealand, Tuatara.

Birds

The Cenozoic fossil record for birds is extremely poor. Such animals are rarely preserved. The rather fragmentary evidence available suggests that birds were toothless and essentially like those of today. The avian fauna included a rich variety of woodland, aquatic, predatory, and flightless birds.

The fossil record for the large flightless birds is somewhat better. *Diatryma* (Fig. 15–17), an Eocene representative, stood more than 2 meters tall. In New Zealand, huge moas lived until relatively recent time. Some were over 3 meters tall and laid eggs with a 2-gallon capacity. Perhaps the most famous of all Cenozoic ground birds was the "dodo" *(Didus ineptus),* which lived on an island east of Madagascar until around 1700, when they

Figure 15–17 *Diatryma* from the Eocene (Wasatch Formation) of Wyoming. This large, flightless bird was about 7 ft tall.

were exterminated by sailors searching for provisions. The moas of New Zealand were also exterminated by humans. The Maori tribespeople killed them for food. The African ostrich, the South American rhea, and the emus and cassowaries of Australia are surviving flightless land birds.

Mammals

The Cenozoic Era witnessed the spectacular rise of the mammals (Fig. 15–18). However, as we have noted earlier, the evolution of mammalian traits was already underway among the therapsids of the Permo-Triassic. The Karoo beds of Africa, for example, contain bones of "near mammals" that had almost made the transition from reptile to mammal.

Just as birds are easily recognized by the possession of feathers, so are mammals by their possession of hair. It is sometimes only present as a few whiskers, but it is present. Like feathers, hair functions as an insulating layer. Mammals are, of course, also recognized by mammary glands. However, neither body hair nor mammary glands are particularly helpful to the paleontologist, who must recognize mammalian remains on the basis of skeletal characteristics. Among these, the lower jaw is particularly useful. It consists of a single bone, the dentary, on either side. In reptiles and birds, there are always several. Another mammalian trait is the bony mechanism that conveys sound across the middle ear. In mammals, there is a chain of three little bones rather than a single one, as in lower tetrapods. Two of the ossicles, the incus and malleus, were derived from bones of the reptilian jaw. Unfortunately, these delicate ossicles are so small that they are rarely found. Typically, mammals have seven cervical, or neck, vertebrae, regardless of the length of the neck. The skull is usually recognized by the expanded brain case, and teeth are nearly always of different kinds, serving different functions in eating.

Paleontologists speculate that more rigorous climatic conditions during the Permo-Triassic favored selection of the mammalian traits of warm-bloodedness and postnatal care. All evidence indicates that the first mammals were diminutive creatures, and small animals lose heat rapidly. To compensate for this heat loss, warm coats and the ability to find ample food were required. If the earliest mammals still laid eggs, as do primitive mammalian monotremes today, then the newly hatched young might also have had to cope with the heat problem. They were very likely kept warm by snuggling next to the soft fur that may have formed an "incubation patch" on the female. Perhaps at the same time, they were nourished by secretions from glands that preceded the development of true mammae. Unfortunately, fossils provide few clues to the reproductive characteristics of early mammals, and theories will continue to rely heavily on inference.

Earliest Mammals

The mammalian fossil record begins with rare finds of tiny teeth, jaws, and fragments of skulls; the oldest remains are Late Triassic in age. Largely on the basis of tooth morphology, these fossils are separated into categories bearing such names as **morganucodonts, docodonts, triconodonts, symmetrodonts, pantotheres,** and **multituberculates.** The oldest remains are of morganucodonts from Upper Triassic beds of Great Britain and from recent discoveries in South Africa (see Fig. 13–54). All of the groups just named lived during the Mesozoic, and the rodent-like multituberculates managed to cross the era boundary and live until Eocene time. The docodonts seem the probable ancestors of modern monotremes. However, of all the Mesozoic groups, the pantotheres were of the greatest importance, for during the Cretaceous, they gave rise to both the marsupial and the placental mammals.

The stage was thus set for the endless procession of mammals that evolved and spread to all corners of the planet during the Cenozoic. The absence of dinosaurs and other groups of previously numerous reptiles favored a vigorous mammalian radiation. Evolutionary experimentation among Mesozoic ancestors had provided Cenozoic mammals with more efficient nervous and reproductive systems, greater speed and agility, reliable systems of body terperature control, larger brains, and levels of animal intelligence that far exceeded those of earlier classes of vertebrates. Armed with these attributes, mammals quickly expanded into the habitats vacated by the reptiles and found new pathways of adaptation as well (Table 15–1). By Paleocene time, 18 mammalian orders had appeared, and throughout the Early Tertiary, there was an intricate interplay of appearances of new groups and extinctions of older groups that was ultimately to produce more than 30 taxonomic orders of mammals. The list is far too long to treat each group, but a brief review of some of the more interesting forms may convey insight into the spectacular variety among Cenozoic mammals.

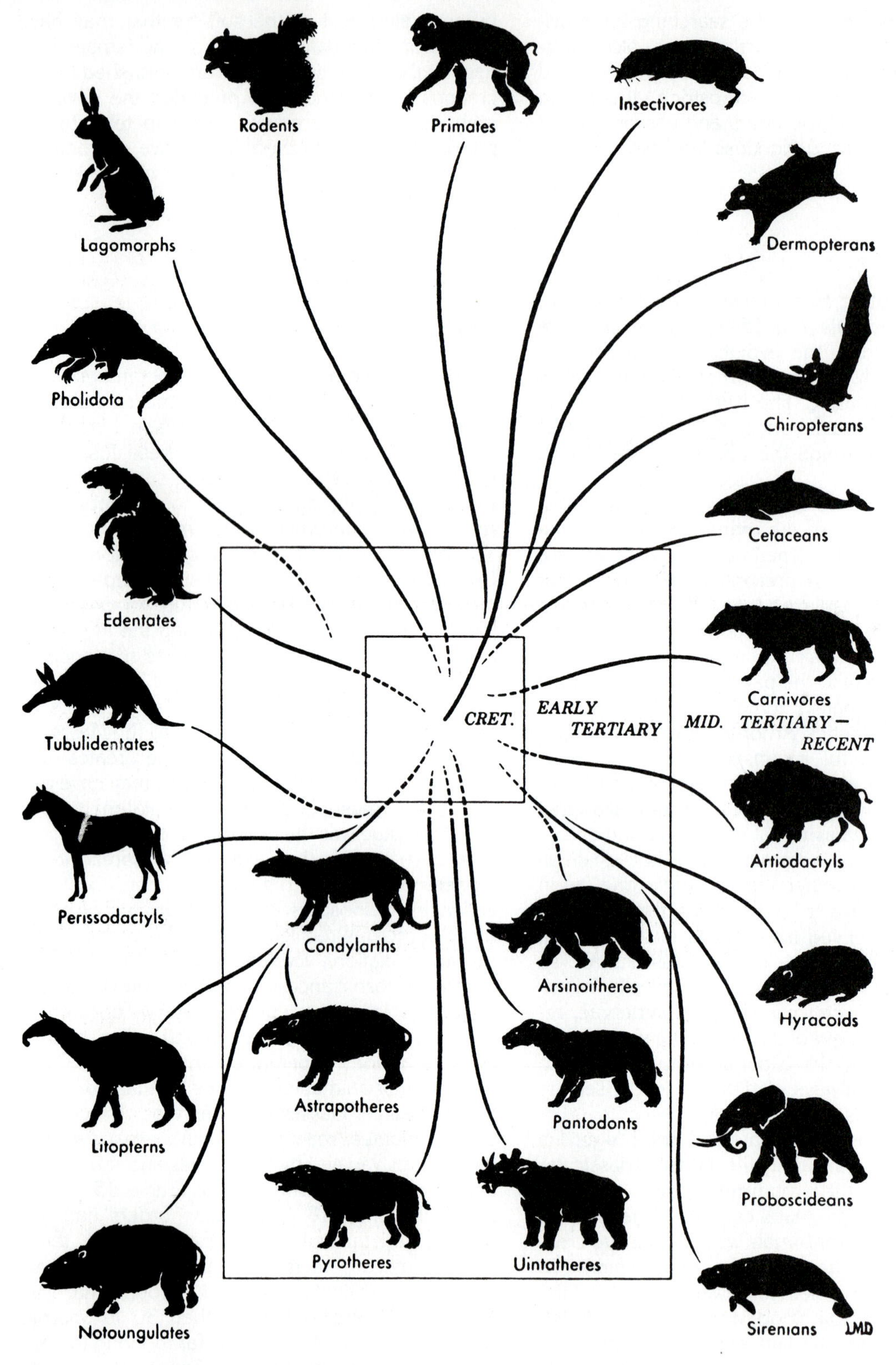

Figure 15–18 General relationships of the major orders of placental mammals. (From E. H. Colbert. *Evolution of the Vertebrates.* New York, John Wiley & Sons, 1969. With permission of the author, artist Lois Darling, and publisher.)

Table 15–1 EVOLUTIONARY MODIFICATIONS OF MAMMALS FOR A VARIETY OF HABITATS*

Habitat	Limbs	Teeth	Other Features
Primitive walking (ambulatory)	Limbs generalized; feet flat on ground	Incisors for grasping; canines for piercing; premolars for cutting; molars for grinding	Tongue, stomach, and intestine generalized; cecum small
Running herbivores (cursorial) horse and cow	Limbs elongate, toes elongate and reduced in number; hoof of horse is one toenail; in cow, two toenails	Incisors for nipping grass; cheek teeth heavy and ridged to withstand wear of grass	Stomach (cow) or intestine (horse) large and complex, cecum long
Running carnivores (cursorial) wolf, cheetah	Limbs elongate; often walk on toes; claws often long and sharp	Canines well-developed; premolars sharp for cutting flesh	Stomach large, intestine short
Heavy body elephant	Bones in limbs massive; flat-jointed; toes in circle around pad	Cheek teeth massive and ridged for grinding plant materials	Nose elongated into trunk to reach food on ground or in trees
Tree-climbing (arborial) tree squirrel, monkey	Limbs elongate; wide range of motion; toes long, flexible	Cheek teeth usually smooth-surfaced and flattened for crushing	Tail sometimes prehensile; diet usually fruit or vegetation
Diet of small, colonial insects anteaters, spiny echidna	Claws enlarged for digging after insects.	Reduced or absent	Tongue long, sticky, and extensile
Burrowing (fossorial) moles, marsupial moles various rodents	Limbs short and stout; forefeet often enlarged, shovel-like; claws long and sharp.	Teeth various, depending upon diet (vegetation or insects and worms)	Pinna of ears usually short or absent; fur short; tail short
Aquatic whales, seals, sea cows	Limbs shortened; feet enlarged and paddle-shaped or absent.	Often reduced to simple, conelike structures, sometimes absent	Tail often modified into swimming organ
Volant flying phalangers, flying squirrels	Limbs extendible to sides to hold gliding membrane outstretched	Various but usually show herbivorous adaptations	Tail often flattened for use in balance
Flying bats	Fore limbs and especially fingers of fore limbs elongate to support wing	Various for diets of insects, fruits and nectar, or other specific foods	Tongue, stomach, and intestine various, correlated with diet

*From E. L. Cockrum and W. J. McCauley. *Zoology*. Philadelphia, W. B. Saunders Co., 1965.

Figure 15–19 Two present-day monotremes; the duck-bill platypus of Australia and Tasmania (A) and the spiny anteater of Australia (B).

Monotremes

The most primitive of all living mammals are called **monotremes.** These relics of an older time still lay eggs in the reptilian manner. However, unlike reptiles, monotremes have primitive mammae and provide their newly hatched offspring with nourishment, if only for a brief period. Unfortunately, the fossil record for monotremes begins in Pleistocene sediments, and we know nothing of their early evolution.

The platypus of Australia and Tasmania and two species of spiny anteaters of New Guinea and Australia (Fig. 15–19) are the only three species of monotremes still in existence.

Marsupials

Marsupials are mammals that nurture their young in a special pouch, or **marsupium.** Today, they are represented by such animals as kangaroos, wallabies, wombats, phalangers, bandicoots, koalas of Australia, and opossums of the New World (Fig. 15–20). Compared with the placentals, it is apparent that they are a dwindling group.

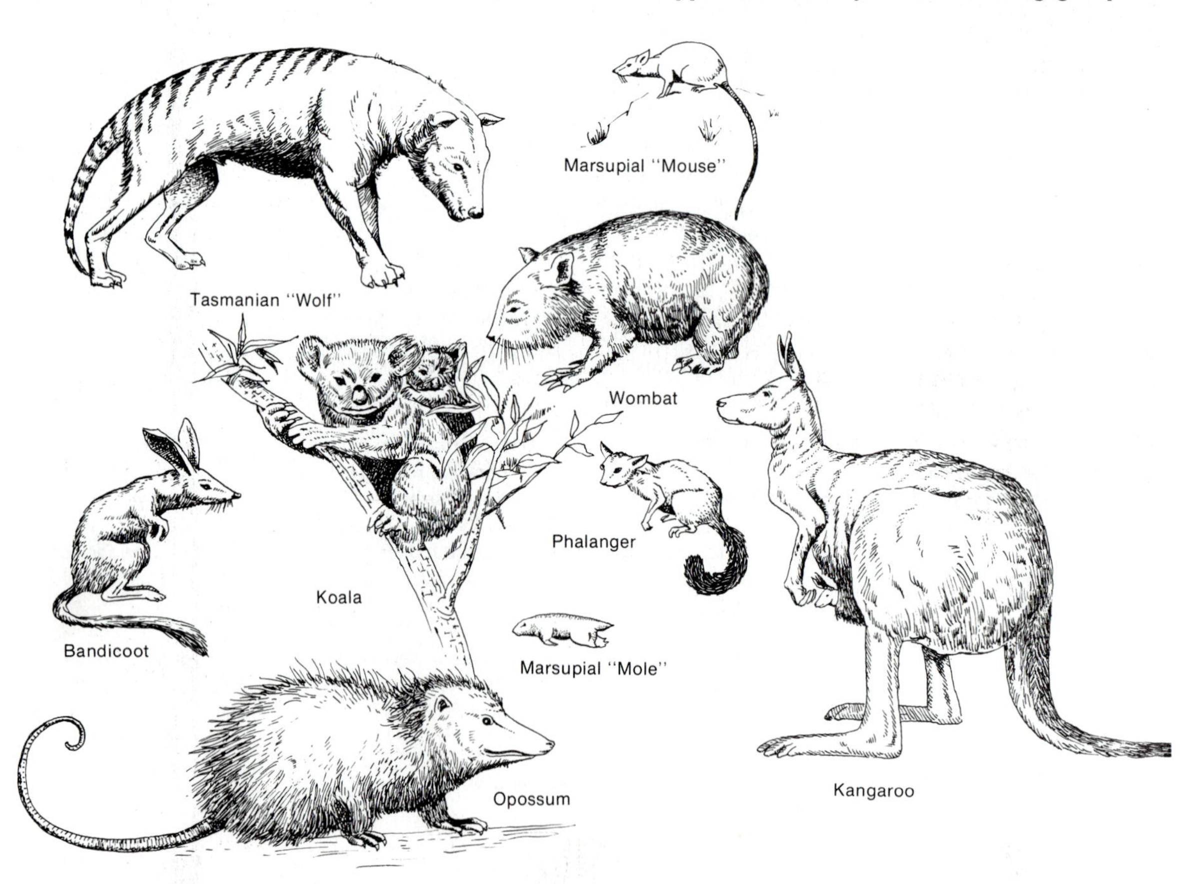

Figure 15–20 Modification of marsupials for various special habitats is evident in their diversity.

Figure 15–21 The marsupial saber-tooth *Thylacosmilos.*

Marsupials had their greatest success in Australia and South America. Both continents were more or less isolated during most of the Tertiary, and marsupials were able to evolve without excessive competition from placentals. Although a few Middle-Tertiary Australian mammal sites have recently been found, the fossil record for Australian marsupials is not good until Pleistocene. An older and more complete record exists in South America, where marsupials sprang from opossum-like ancestors and produced an array of mostly flesh-eating types. *Thylacosmilus* (Fig. 15–21), a large Pliocene form, had an uncanny resemblance to the North American sabre-toothed cats. The South American marsupials fared poorly in competition with placental immigrants from the north, when the connection between North and South America was established in the Early Pleistocene.

Placental Mammals

Placental mammals appear during the Cretaceous as small, unspecialized insectivores. Modern moles are members of the **Insectivora,** but the tiny shrew (Fig. 15–22) is more representative of the kind of animal from which other orders of Cenozoic placentals evolved. The descendants of the insectivores include the edentates, bats, rodents, flesh-eating mammals, a host of herbivores, and varied marine mammals.

Armadillos, tree sloths, and South American anteaters are **edentates** that have not become extinct. Fossil species of armadillos have been found in rocks as old as Paleocene. Among the extinct edentates are the the glyptodonts, which survived up until the Pleistocene. *Glyptodon* (Fig. 15–23) was a walking fortress with a spike-covered knob on its tail for bludgeoning pesky predators. Quite unlike the glyptodonts were the ungainly Cenozoic ground sloths. This group of edentates includes truly colossal Pleistocene beasts such as *Megatherium* (Fig. 15–24).

The **rodents** have been exceptionally successful Cenozoic mammals. Today, they probably outnumber all other mammals and have invaded every habitat of the earth. Their diversification has produced burrowers such as the marmot, the partially aquatic muskrat and beaver, the desert-dwelling jerboas and kangaroo rats, and the arboreal squirrels. The fossil record for rodents begins in the Paleocene with squirrel-like forms, such as *Paramys* (Fig. 15–25), and the horned rodent *Ceratogaulus.*

Beavers are well represented among fossil rodents. During the Pleistocene, beavers, along with many other mammals, grew to exceptional sizes. Some were as large as modern grizzly bears. The other "gnawers and nibblers" of the Cenozoic that are not classified as rodents are the rabbits, hares, and pikas. The fossil record for these so-called **lagomorphs** begins with skeletal fragments of a primitive rabbit found in Paleocene beds of Mongolia.

Figure 15–22 The tree shrew. (Length of body is 3 cm.)

Figure 15–23 Restoration of a scene during the Mid-Pleistocene in Argentina. The heavily armored animals are glyptodonts. On the left is a giant ground sloth. (Copyright Field Museum of Natural History. Courtesy of Chicago Museum of Natural History; painting by C. R. Knight.)

Figure 15–24 The great ground sloth *Megatherium,* which lived during the Pleistocene and was nearly as large as a present-day elephant. (Courtesy and copyright, Chicago Museum of Natural History; painting by C. R. Knight.)

Figure 15–25 The Paleocene rodent *Paramys,* which resembled a large squirrel in size and form.

The history of flesh-eating placentals began in the Cretaceous with the advent of small, often weasel-like animals called creodonts (Fig. 15–26). They were soon joined by members of the **Order Carnivora,** which evolved independently from the **Creodonta.** It appears that, toward the end of Eocene time, with the coming of speedier and more progressive plant eaters, the creodonts gradually lost ground to carnivores. Most became extinct by the end of the Eocene, although one hyena-like group persisted until Pliocene. The expansion of the Carnivora accelerated during the Late Tertiary, producing a host of familiar flesh eaters, including bears, raccoons, weasels, genets, hyenas, dogs, and cats (Fig. 15–27). Wild dogs were doing very well on earth long before humans came along to make companions of them. The modern genus *Canis,* in the form of the dire wolf, was much in evidence during the Pleistocene. The most famous of the extinct cats are the so-called stabbing cats, exemplified by *Smilodon* (Fig. 15–28). This robust flesh eater preyed upon the larger herbivorous animals of the Pleistocene. A second line of felid evolution produced the biting cats, which resembled modern leopards and pumas. The skeletons of these carnivores indicate they were strong, speedy, and agile predators.

As is true of carnivores today, Cenozoic carnivores were an essential element in the evolutionary process. To survive, they had to equal or better the speed and cunning of the herbivores on which they fed. The herbivores, in turn, responded to the carnivore threat by evolving adaptations for greater speed and defense. Then, as now, predators were not villains but necessary constitutents of the total biologic scheme. They were able to cull out the weak, deformed, or sickly animals and thereby help to counteract the effects of degenerative mutation and to prevent overpopulation.

Not all the Carnivora of the Cenozoic were land dwellers. Descendants of sea-dwelling carnivores, such as seals, sea lions, and walruses, gather their food in or at the edge of the sea. As indicated by their sharp, pointed teeth, seals and sea lions eat fish. The walrus has (in addition to tusks) broad, flat teeth for crushing the shells of mollusks. Fossils of marine carnivores older than Pliocene have not been found. The marine carnivores, or **pinnipeds,** probably descended from semiaquatic mammals somewhat similar to present-day otters, but the transitional forms are as yet undiscovered.

In adaptation to the marine environment, **cetaceans** (whales and porpoises) have made the most complete adjustment. Whales first appear quite suddenly about 50 million years ago as already fully developed oceanic creatures. For this reason, paleontologists suspect that cetaceans underwent an extraordinarily rapid evolution. The first whales are exemplified by *Basilosaurus* (Fig. 15–29). The structure of the skull in this whale provides some evidence that the line was derived from early Cenozoic terrestrial flesh eaters. Modern whales arose from the group that included *Basilosaurus* and divided into two lineages: the toothed whales and the "whalebone" whales. Toothed whales, which made their debut during the Oligocene, today include porpoises, killer whales, and sperm whales like Melville's famous Moby Dick. The titanic blue whale, right whale, and Greenland whale are all representatives of the second cetacean group. These plankton-feeding giants first appeared in the Miocene. Instead of teeth, they possess ridges of hardened skin that extend downward in rows from the roof of the mouth. The ridges are fringed with hair that serves to entangle the tiny invertebrates on which these cetaceans feed—thus, the paradox of the largest of all amimals feed-

Figure 15–26 A characteristic predator of the Eocene, *Oxyaena,* a creodont.

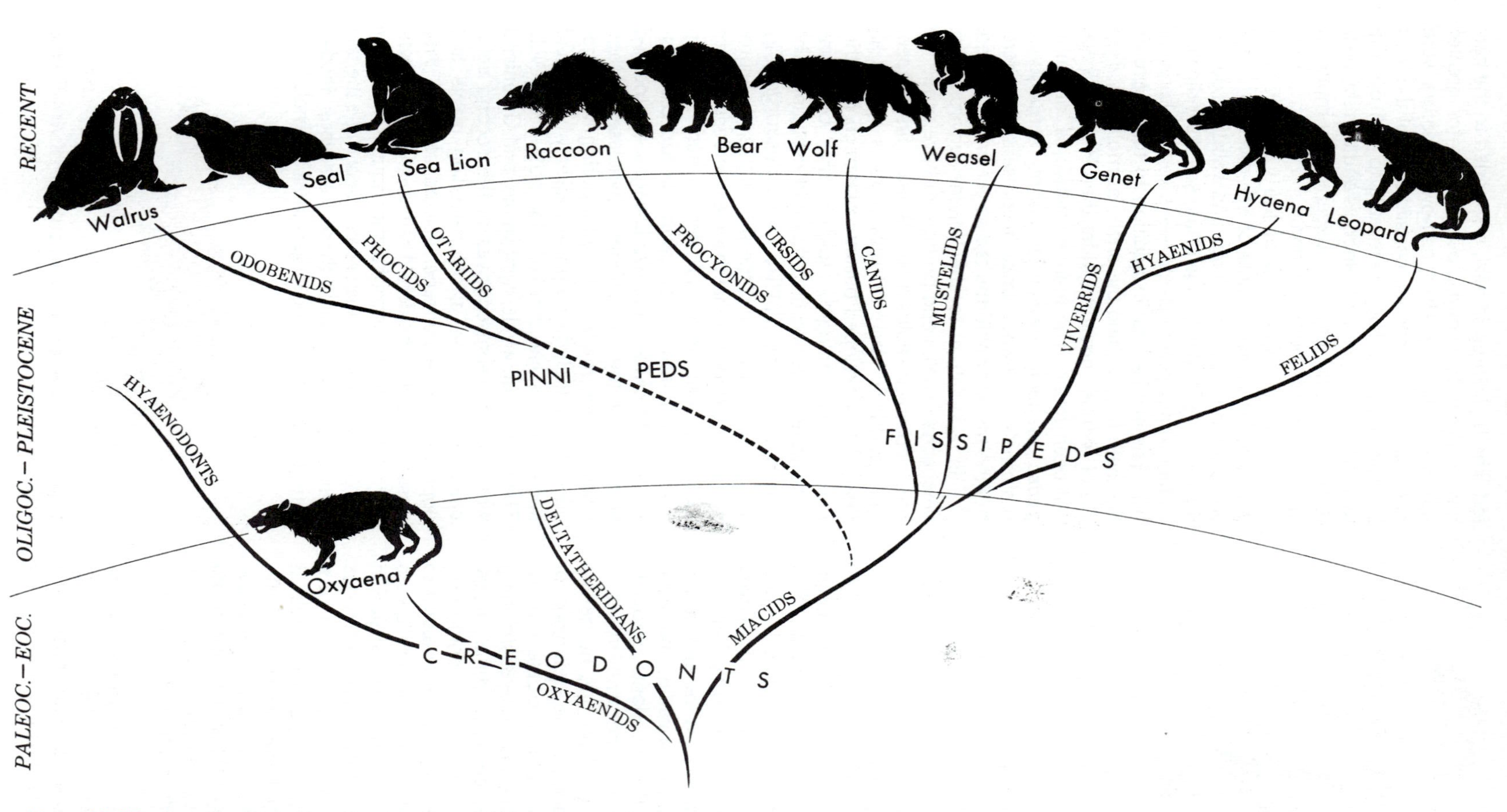

Figure 15–27 General relationships of major groups of flesh-eating placental mammals. (From E. H. Colbert. *Evolution of the Vertebrates.* New York, John Wiley & Sons, 1969. With permission of the author, artist Lois Darling, and publisher.)

Figure 15–28 Restoration of a Pleistocene water hole near the present site of Los Angeles, California. On the right is a saber-tooth cat *(Smilodon)*. The wolf *Aenocyon* observes on the left. Large carrion-eating birds were attracted to the site by animals trapped in asphalt seeps beneath the water and elsewhere. (Courtesy and copyright, Chicago Museum of Natural History; painting by C. R. Knight.)

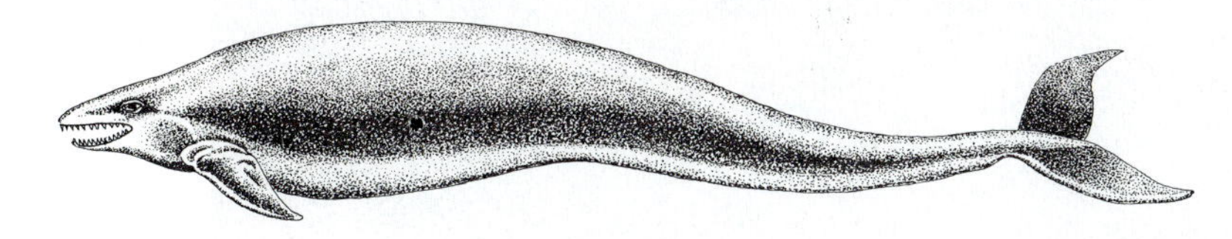

Figure 15–29 The Eocene whale *Basilosaurus.* The tendency toward increase in size was already evident in this whale, which was over 20 meters long.

ing on the smallest. The great blue whales far exceed in size even the largest dinosaurs, for some have attained lengths of 30 meters and weights of over 135 metric tons.

The largest category of Cenozoic planteaters is that comprising the **ungulates.** Simply defined, ungulates are animals that walk upon hoofs and feed upon plants. The earliest ungulates were members of a group called condylarths. *Phenacodus* (Fig. 15–30), a representative condylarth, had simple primitive teeth and five toes, each terminated with a small hoof. The animal walked mostly in a flat-footed fashion, for these ungulates had not yet evolved the ability to walk or run on their toes for greater speed. This type of stance, termed **plantigrade,** was characteristic of such Early Tertiary ungulates as *Coryphodon* (Fig. 15–31) and *Uintatherium* (Fig. 15–32). Primitive, plantigrade, and relatively small-brained plant eaters are often referred to as **archaic ungulates.**

An interesting sidelight to the early history of ungulates was the development of a separate and distinctive group in South America. During the Paleocene, a few condylarths managed to wander southward across the Isthmus of Panama before it was severed. Afterward, they experienced an evolu-

Figure 15–30 *Phenacodus,* an early hoofed mammal of the group known as condylarths. *Phenacodus* was somewhat larger than a sheep.

Figure 15–31 *Coryphodon,* an early Eocene ungulate. (This animal was about 1.3 meter tall at the shoulder.)

Figure 15–32 Eocene mammals. *Uintatherium* (A) is the large six-horned and tusked animal in the upper right. Other animals on this restoration are: (B) small, fleet rhinoceros *Hyrachus;* (C) *Trogosus,* a gnawing-toothed mammal; (D) *Mesonyx,* a hyena-like flesh-eater; (E) *Stylinodon,* a gnawing-toothed mammal; (F) three early members of the horse lineage *(Orohippus)*; (G) a saber-tooth mammal, *Machaeroides;* (H) *Patriofelis,* an early carnivore; (I) *Palaeosyops,* an early titanothere. Restorations are based on skeletal remains from the Middle Eocene Bridger Formation of Wyoming. (Courtesy of the Smithsonian Institution; mural by J. H. Matternes.)

tionary radiation that resulted in such strange-looking animals as litopterns and heavy-bodied notungulates (Fig. 15–33).

The modern, and more familiar, ungulates fall into two categories: the **perissodactyls** and **artiodactyls.** The perissodactyls include the modern horses, tapirs, and rhinoceroses and the extinct **chalicotheres** and **titanotheres** (Fig. 15–34). They seem to have originated from condylarth ancestors and reached the peak of their evolutionary history during the Miocene Epoch. Since that time, they have declined steadily.

Perissodactyls have certain distinctive characteristics. For example, the number of toes on each foot is usually odd, and the axis of the foot, along which the weight of the body is primarily supported, lies through the third, or middle, toe. There is a tendency toward reduction of the lateral toes. In the modern horse, only the single, central toe remains. They are clearly digitigrade, meaning that they run or walk on their toes in order to attain a longer stride and greater speed.

The oldest known perissodactyl, *Hyracotherium* (Fig. 15–35), has been found in Late Paleocene and Eocene strata of both North America and England. Its ancestors were probably condylarths. Familiarly known as *"Eohippus," Hyracotherium* has the distinction of being the earliest member of the horse family. This little "dawn horse" was not much larger than a fox. It had four hoofed toes on the front feet and three on the hind feet. The back curved like that of a condylarth, and the dentition was primitive, with canines still present and the premolar teeth not similar to the molars, as in more recent perissodactyls. In addition, the molars were bluntly cusped for browsing and

Figure 15–33 Two South American ungulates. *Top,* The litoptern *Macrauchenia. Bottom, Toxodon,* a large notungulate of the Pleistocene. The initial discovery of *Toxodon* was made by Charles Darwin, who discovered a partial skeleton of the beast in the bank of an Argentine stream.

had not yet developed the ridged oral surface and high crowns that characterize modern grazers.

From ancestors like *Hyracotherium,* the horse family proceeded to evolve, at least until Miocene, in a rather straightforward manner (see Fig. 4–11). The skeletal remains clearly show progressive increases in size, length of legs, height of crowns of cheek teeth, and brain size. The curved back became straightened, and the middle toe was strengthened and emphasized at the expense of lateral toes. The premolars came to resemble the molars, and their grinding surfaces developed increasingly complicated patterns of resistant enamel ridges. *Mesohippus,* an Oligocene horse about twice the height of *Hyracotherium,* showed the beginnings of many of these trends but still retained relatively low-crowned teeth, best adapted for browsing.

With the spread of grass during the Miocene, the family tree of horses became more complicated as more conservative members stayed behind in the forests, whereas others took advantage of the more open environment by making suitable adaptations for grazing and running across the open plains. These prairie dwellers were the ancestors of the modern horse. *Merychippus* (Fig. 15–36), a horse in the vanguard of this new group, ran on a single middle toe; the side toes were much reduced and of little use. The head was deeper and the face longer to accommodate the long cheek teeth.

In the Pliocene, two groups of horses arose from the general *Merychippus* base. The more

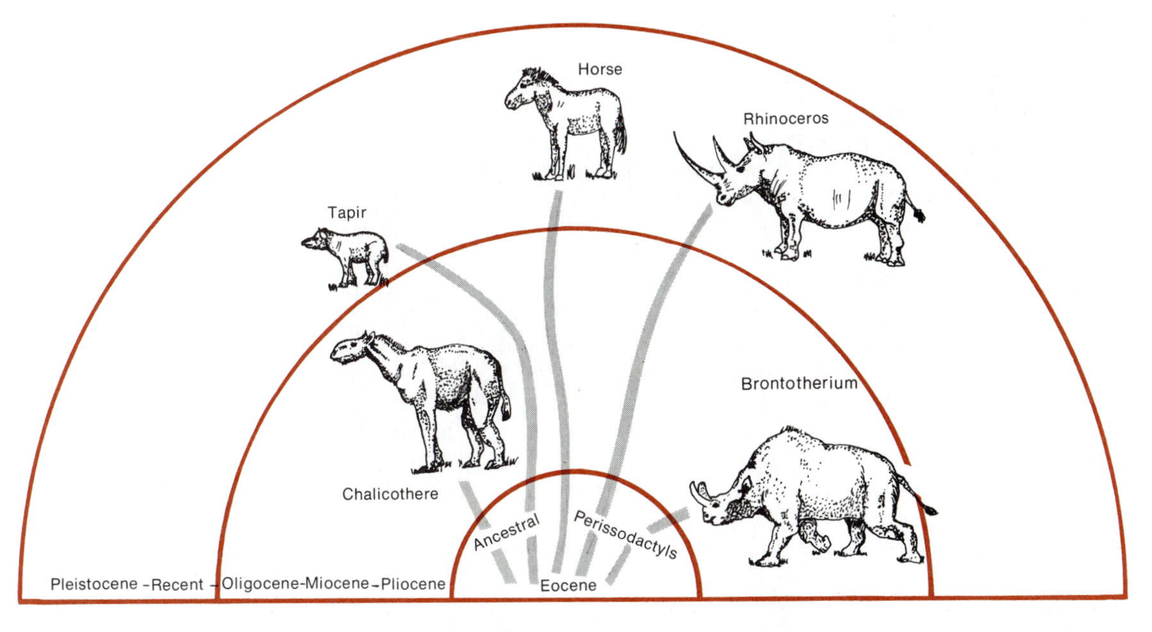

Figure 15–34 Generalized diagram of the evolution of perissodactyls.

Figure 15–35 The "Early Eocene" horse *Hyracotherium,* more popularly known as "Eohippus," the dawn horse. The animal measured only a half meter in length.

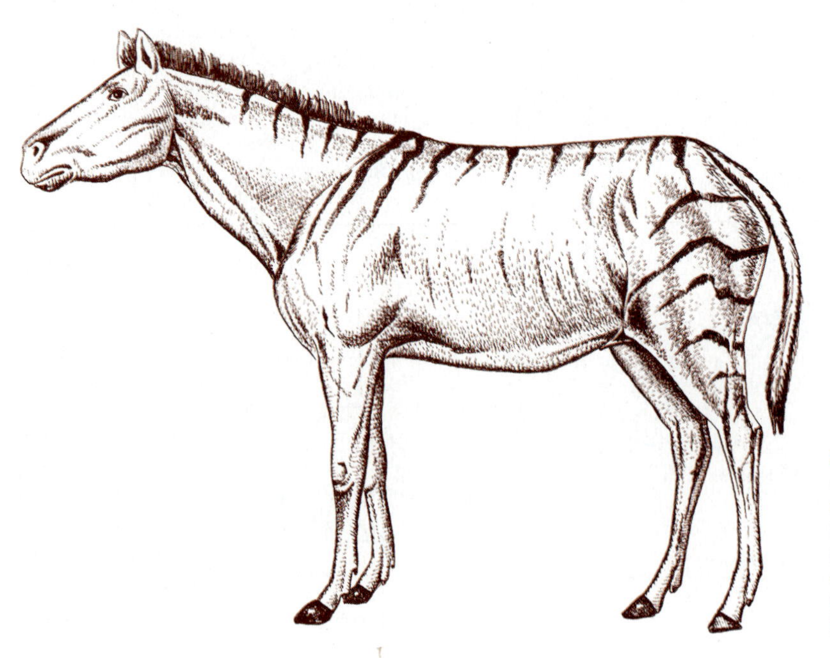

Figure 15–36 *Merychippus,* a Miocene three-toed horse about the size of a small pony. Although the feet retained three toes, the lateral toes were much reduced and of little use. *Merychippus* walked on the single middle toe, the end of which bore a rounded hoof.

Figure 15–37 The giant Oligocene to early Miocene hornless rhinoceros *Baluchitherium.*

conservative group, typified by *Hipparion,* retained three-toed feet. They found their way from North America into Europe and survived until the beginning of the Pleistocene. The more progressive branch of the horse family tree produced *Pliohippus,* a completely one-toed form with lateral digits reduced to mere splints that were concealed beneath the skin. The descendants of *Pliohippus* also evolved along two separate branches. The horses of one branch, represented by *Hippidium,* migrated into South America over the Panamanian land bridge. The other branch led to *Equus,* the genus to which all modern horses belong.

During the Pleistocene, several different species of horses thrived. Some, such as *Equus occidentalis,* were the size of small ponies, but others, such as *Equus giganteus,* were larger than the great draft horses of today. They spread across the length of North America, Eurasia, and Africa. Only a few thousand years ago, horses suffered extinction in North America. (The continent was restocked with the progeny of domestic horses that had escaped from the Spanish explorers in the sixteenth century.) The cause of the extinction of horses in North America is somewhat of a mystery. Some believe it was brought on by contagious disease, whereas others speculate that the cause was overkill by prehistoric human hunters.

Among the other surviving perissodactyls are the tapirs and rhinoceroses. Tapirs retain the primitive condition of four toes on the forefeet and three on the rear, as well as low-crowned teeth. They are forest-dwelling, leaf-eating animals whose fossil record begins in the Oligocene. More impressive to humans are the rhinoceroses, which began their evolution during the Eocene as small, swift-running creatures and eventually produced such giants as *Baluchitherium* (Fig. 15–37). *Baluchitherium* is the largest land mammal thus far discovered. It stood 5 meters tall at the shoulders.

The perissodactyls that did not survive into modern times were the **titanotheres** and **chalicotheres.** The most familiar titanothere of the Cenozoic was ponderous *Brontotherium* (Fig. 15–38), readily remembered because of the pair of hornlike processes that grew over the snout. Chalicotheres differed from all other Cenozoic perissodactyls in having three claws rather than hoofs on their feet. These odd creatures (Fig. 15–39A) were rather similar to a horse in the appearance of the head and torso. The fore legs were longer than the hind legs, and the back sloped rearward. They lived from Eocene into the Pleistocene, at which time they became extinct.

The even-toed ungulates, or artiodactyls, have been far more successful than the perissodactyls in terms of survival, variety, and abundance (Fig. 15–40). Modern artiodactyls include pigs, deer, hippos, goats, sheep, cattle, and camels. The center of artiodactyl evolution was not North America, as had been the case with the odd-toed ungulates, but rather Eurasia and Africa. They were already present during the Eocene and by Late Tertiary clearly achieved numerical and varietal superiority over other herbivores.

Most artiodactyls have an even number of toes on each foot, either four or two. Most of the weight of the animal is carried on the two middle toes, which form an identical pair and thus a "cloven," or "split," hoof. Unlike advanced perissodactyls, the molar and premolar teeth are dissimilar. Among the Tertiary artiodactyls that became extinct, the **oreodonts** and **entelodonts** are particularly interesting. Oreodonts (Fig. 15–41H) were short, rather stocky grazers that roamed the grassy plains of North America in enormous numbers. Entelodonts

Figure 15–38 The titanothere *Brontotherium.*

(Fig. 15–41D), some of which were as large as buffalo, were repulsive-looking, hoglike beasts. Their trademarks were curious bony processes that grew along the sides of the skull and jaws.

The major radiation of artiodactyls began in the Eocene and ultimately produced the three major surviving categories: swine, camels, and ruminants. Of these, the swine family has probably remained the most primitive. They have kept their four toes, even though most of the weight is carried by the two middle digits. The swine group includes both pigs and the more lightly constructed and primarily South American peccaries. Hippopotami are the only modern amphibious artiodactyls. They are relatives of the pig family, having arisen from a group of Miocene piglike animals called anthracotheres.

People are often surprised to learn that much of the evolutionary development of the camel lineage—**Tylopoda**—occurred in North America. The geologic history of camels and llamas began in Eocene time with tiny creatures about the same size as little *"Eohippus."* As they evolved through the Tertiary, they lost their side toes and increased

Figure 15–39 Mural depicting an assemblage of Lower Miocene mammals. (A) The chalicothere *Moropus;* (B) The small artiodactyl *Merychyus;* (C) *Daphaenodon,* a large, wolflike dog; (D) *Parahippus,* a three-toed horse; (E) *Syndyoceras,* an antelope-like animal; (F) *Dinohyus,* a giant, piglike artiodactyl; (G) *Oxydactylus,* a long-legged camel; (H) *Stenomylus,* a small camel. (Courtesy of the Smithsonian Institution; mural by J. H. Matternes.)

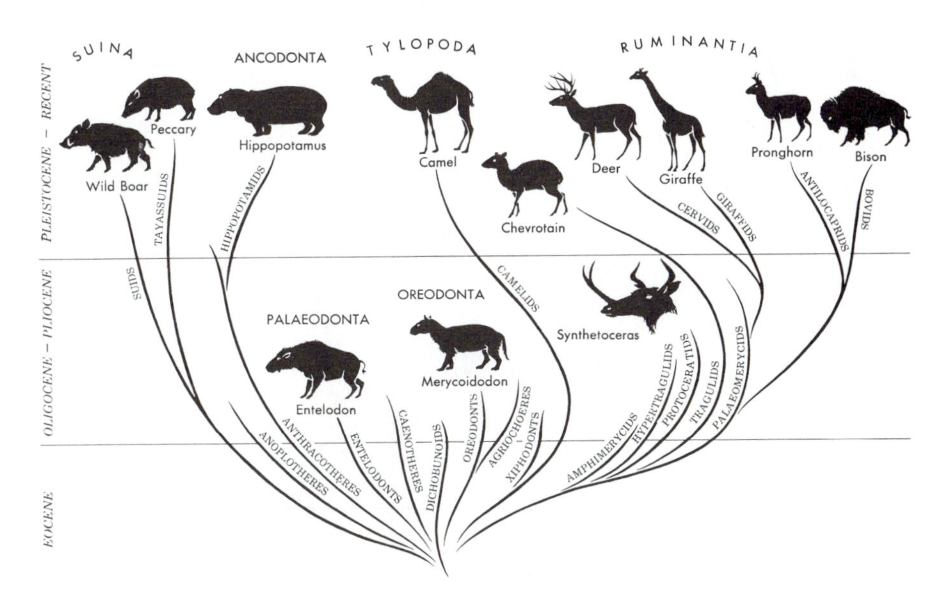

Figure 15–40 The evolutionary radiation of the even-toed ungulates, or artiodactyls. (From E. H. Colbert. *Evolution of the Vertebrates.* New York, John Wiley & Sons, 1969, with permission of the author, artist Lois Darling, and the publisher.)

Figure 15–41 A restoration of Oligocene mammals based primarily on skeletal remains from the White River Formation of South Dakota and Nebraska. The flora is based on nearly contemporaneous plant fossils from the Florissant beds of Colorado. (A) *Trigonias,* an early rhinoceros; (B) *Mesohippus,* a three-toed horse; (C) *Aepinacodon,* a remote relative of the hippopotamus; (D) *Archaeotherium,* an entelodont; (E) *Protoceras,* a horned ruminant; (F) *Hyracodon,* small, fleet rhinoceros; (G) the giant titanothere *Brontotherium;* (H) creodonts named *Merycoidon;* (I) *Hyaenodon,* an Oligocene carnivore; (J) *Poëbrotherium,* an ancestral camel. (Courtesy of The Smithsonian Institution; mural by J. H. Matternes.)

Figure 15–42 *Alticamelus,* with exceptionally long neck (the head was 3 meters above the ground) and legs. The animal was evidently adapted for browsing in tall trees.

the length of their legs and neck. The long-necked *Alticamelus* (Fig. 15–42), of the Pliocene probably browsed on leaves in much the same manner as the giraffe. By Pleistocene time, there were numerous modern-looking camels and llamas in North America. The llamas moved southward and took up residence in the highlands and plains of South America. Camels migrated across the Bering Land Bridge to Eurasia and Africa, where they established themselves in arid regions. Those that were left behind in North America mysteriously suffered extermination.

The **ruminants** are the most varied and abundant of modern-day artiodactyls. This group takes their name from the *rumen,* or first of four compartments in their multichambered stomach. For the most part, they are cud chewers. The earliest ruminants were small, delicate, four-toed animals

Figure 15–43 *Archaeomeryx,* a small animal about the size of the Oriental chevrotain, or "mouse deer" (about 25 cm tall), represents the kind of artiodactyl from which ruminants evolved.

called **tragulids.** They are represented today by the skinny-legged and timid "mouse deer" of Africa and Asia. *Archaeomeryx* (Fig. 15–43), whose remains are found in Eocene strata of Mongolia, was such an animal. Later lines of tragulids sometimes developed peculiar assortments of horns, exemplified by *Synthetoceras* (Fig. 15–47B). This Pliocene tragulid had not only a pair of horns over the eyes but also a Y-shaped horn on the snout.

Ruminants such as sheep, cattle, giraffes, and deer are called **pecorans.** Deer are primarily browsers that have made the forests their principal habitat ever since their first appearance in the Oligocene. The early stages of deer evolution are exemplified by *Blastomeryx,* a small, hornless creature that lived in North America during the Miocene. Subsequent evolution involved increases in size and development of antlers. The culmination of this trend is represented by the Pleistocene *Megaloceros,* whose antlers measured more than 3 meters from tip to tip (Fig. 15–44).

The giraffe family probably branched off from

Figure 15–44 *Megaloceros,* the giant Irish "elk," was actually a deer whose remains are frequently found in the peat bogs of Pleistocene Age in Ireland. (Courtesy and copyright, The Chicago Museum of Natural History; painting by C. R. Knight.)

Figure 15–45 The first proboscideans (members of the group that includes elephants) were moeritheres, named after this typical genus *Moeritherium*, which lived in Egypt during the Late Eocene.

the deer lineage sometime during the Miocene and became specialized in browsing on leaves of trees that grew rather sparsely in areas where most other herbivores were grazers. *Palaeotragus*, the ancestral giraffe, looked much like the okapi of Africa. From this general form, the obvious evolutionary trends emphasized leg and neck elongation.

Of course, the bovids—cattle, sheep, and goats—are presently the most numerous of ruminants. Miocene strata provide the earliest fossil record of bovids. The bison are particularly interesting to Americans. Seven species of bison lived in North America during the Pleistocene, some of which were truly giants with horns that measure 2 meters from end to end.

A discussion of Cenozoic mammals would not be complete without a brief mention of elephants and their older relatives, the mammoths and mastodons. Because these animals have a trunk, they are collectively termed **proboscideans.** The trunk, of course, is the elephant's principal means of bringing food to its mouth. Other animals as tall as elephants reach food on the ground easily because of their long necks. However, proboscideans, with short, muscular necks to support their massive heads, have evolved their own unique anatomic solution to food gathering. Paleontologists are able to follow the development of the trunk in early proboscideans by noting the position of the external nasal openings at the front of the skull. Those openings recede toward the rear of the skull in sequential stages of trunk development. Another elephantine trademark is the tusks, which evolved by elongation of the second pair of incisors.

The fossil record for proboscideans begins with a trunkless, tapir-like animal named *Moeritherium* (Fig. 15–45). Bones of this ancestral proboscidean have been found in Lower Eocene and Early Oligocene beds near Lake Moeris is Egypt. It is likely the moeritheres arose from Early Tertiary condylarths. From an ancestry represented by the moeritheres, proboscideans separated into two branches. One led toward a group of Miocene and Pliocene animals called **dinotheres.** The tusks of dinotheres were distinctive in that they were present only in the lower jaws and curved downward and backward, an orientation presumably useful for uprooting plants and digging for roots and tubers.

The other branch of the proboscidean family tree produced mastodons and elephants. *Palaeomastodon* (Fig. 15–46), in the Oligocene of North

Figure 15–46 *Palaeomastodon*, one of the earliest of the mastodon lineage.

Figure 15–47 A variety of Early Pliocene mammals. (A) *Amebelodon,* the shovel-tusked mastodon; (B) *Synthetoceras;* (C) *Cranioceras;* (D) *Merycodus,* an extinct prong-horned antelope; (E) *Epigoulis,* a burrowing, horned rodent; (F) *Neohipparion,* a Pliocene horse; (G) the giant camel *Megatylopus* and smaller *Procamelus.* (H) *Prosthennops,* an extinct peccary; and (I) the short-faced canid *Osteoborus.* (Courtesy of the Smithsonian Institution; mural by J. H. Matternes.)

Africa, was probably representative of the group from which true mastodons evolved. The face was long, the trunk was short, and the second incisors in both the upper and lower jaws had developed into tusks. By Miocene time, a larger mastodon named *Gomphotherium* had found its way into North America by way of the Bering Isthmus. Subsequent proboscidean evolution produced a variety of long-jawed mastodons. *Trilophodon* from the Pliocene had lower jaws almost 2 meters long. The most bizarre mastodon was the "shovel-tusker" *Amebelodon* (Fig. 15–47A) of the Pliocene. The tusks of the lower jaws in these beasts were flattened and formed two sides of a broad, scooplike ivory shovel. Mastodons with shorter jaws and longer trunks are best represented by species of *Mammut* (Fig. 15–48), which were common throughout North America during the great ice age and survived until comparatively recent times. *Mammut* was heavier but not as tall as the African elephants. It had the sharply crested molars typical of mastodons, a stocky build, and great curving tusks nearly 3 meters long.

True elephants and mammoths evolved from mastodons in the Old World. Their line of evolution can be traced backward to a Miocene animal named *Stegolophodon.* Next on the evolutionary scene was *Stegodon,* a Pliocene form with a short, tuskless lower jaw and molar teeth already containing the numerous cross-ridges that characterize true elephants. Such teeth are more suitable for a diet of abrasive grasses. Further along in the evolution of the elephant, the trunk grew longer, the height of the skull increased enormously, and the jaws became so short that they could accommodate no more than eight grinding teeth at one time. As the molars of elephants gradually wear down, new teeth form in the posterior region of the jaws and push forward until the older, worn teeth break out. The last molars appear when the animals are 40 or 50 years old.

Mammoth is a term loosely applied to the ice-age elephants of North America, Europe, and Africa. They were a magnificent group of animals that included the famous woolly mammoths (Fig. 15–49) drawn by our own ancestors on the walls of

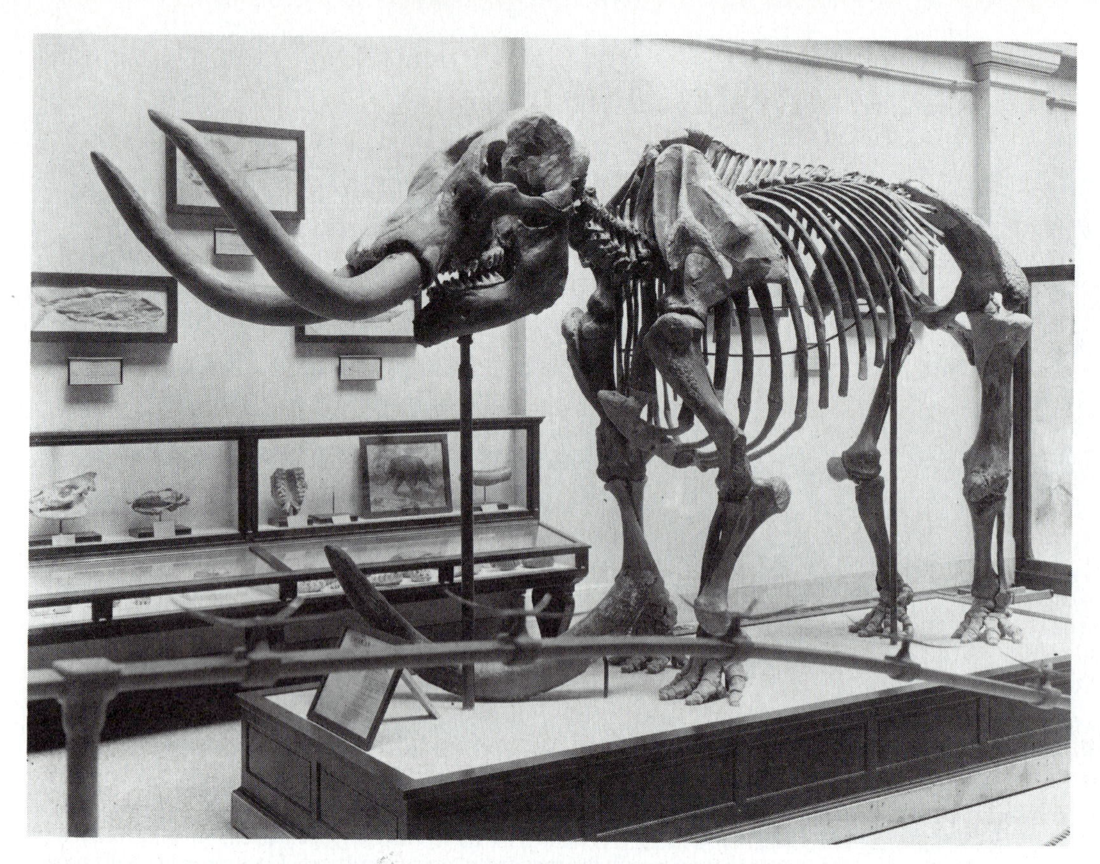

Figure 15–48 Mounted skeleton of the American mastodon *Mammut americanus* found in a Pleistocene swamp deposit in Indiana. (Courtesy of the Smithsonian Institution.)

Figure 15–49 The woolly mammoth. In the Late Pleistocene these magnificent animals lived along the borders of the continental glaciers. Their remains have been found frozen in the tundra of northern Siberia. (Courtesy and copyright, The Chicago Museum of Natural History; painting by C. R. Knight.)

their caves. The great "imperial mammoth" reached heights of 4½ meters and ranged widely across California, Mexico, and Texas. The Columbian elephant had immense spiral tusks that in older individuals overlapped at the tips, thus becoming useless for digging purposes. Many of the Late Pleistocene proboscideans were hunted by humans. Then, about 8000 years ago, all but two genera of elephants became extinct. Their continued survival, as is the case with many wild animals, will depend upon our good judgment and that of the governments we support.

THE DEMISE OF THE PLEISTOCENE GIANTS

At the time of maximum continental glaciation (approximately 11,000 years ago), the northern hemisphere supported an abundant and varied fauna of large mammals, comparable to that which existed in Africa south of the Sahara several decades ago. The fauna included giant beavers, mammoths, mastodons, elk, most species of perissodactyls, many even-toed forms, and ground sloths. From all available evidence, most of these great beasts maintained their numbers quite well during the most severe episodes of glaciation but experienced rapid decline and extinction in the period around 8000 years ago, when the climate became milder. What might have been the cause for the extermination of these large Pleistocene mammals? Explanations tend to fall into two categories: In one, theories attempt to explain the extinctions as being related to human overkill; a second interpretation seeks to relate the losses to climatically controlled environmental causes.

The overkill concept is based on the inference that early humans had developed the ability to hunt in highly organized social groups and skillfully bring down large numbers of big animals on open prairies and tundras. Human predators may have killed then, as they do now, in excess of their needs. Unlike such predators as wolves, early human hunters probably did not seek out the weak or sick to kill but brought down the best animals in the herds. By decimating particular species of herbivores, the other, nonhuman predators would have necessarily also suffered. In support of this idea, paleontologists note that most Pleistocene extinctions involved large terrestrial animals. Marine genera, protected by the sea from human predation, continued to thrive. Small mammals also survived; although frequently hunted, they were difficult to exterminate because of their more rapid breeding rate, their greater numbers, and the probability that they were killed only one at a time. The big, gregarious herbivores were easily accessible and reproduced more slowly, and entire herds could be driven off cliffs or into ravines, thus providing opportunities for slaughter of hundreds at the same time.

Another line of evidence favoring overkill stresses the observation that an early wave of extinction began in Africa and southern Asia, where predatory humans first became prevalent. The better known extinctions of the colder north did not occur until much later, when humans had moved into this region after having first devised the means to better clothe and shelter themselves.

The overkill concept has not, however, gone unchallenged. Many paleontologists feel that human populations were too small and that nomadic tribes shifted too frequently to account for the extermination of entire species of mammals. Also, it cannot be demonstrated that the primitive peoples who live at stone age levels today have endangered the survival of species of animals they hunt. In some cases, animals that were probably not desirable for food, such as the cave lion, became extinct; more desirable forms, such as certain species of deer, also disappeared completely. Such arguments have led some paleontologists to consider favorably theories that stipulate that changes in climate operated selectively against large terrestrial animals.

It is possible that, as the climate was warming up following the final retreat of the ice, parts of the temperate belt became excessively arid, producing severe stress on the large animals, which make the biggest demands on their environment. Also, North American herbivores could not freely move north and south over great distances to escape unfavorable conditions (as could, for example, those in Africa).

Unfortunately, theories that relate Pleistocene extinctions to climatic alteration are difficult to validate. Fossilized pollen (as clues to climate) cannot usually be precisely correlated to the demise of particular species of mammals. Whatever the cause of the exterminations, it is apparent that they coincide with the expansion of ancient human tribes and with a change from cool and moist to warmer and dryer climates over large continental tracts.

Summary

The Cenozoic was a time of gradual approach to the conditions of the present. The flowering plants continued to expand and diversify, with the most important single occurrence being the Miocene proliferation of grasses as extensive prairies and savannahs. Among the marine invertebrates, large discoidal foraminifera like nummulites were characteristic of areas in the Tethys Sea and the tropical west Atlantic. Smaller planktonic foraminifers proliferated and underwent a marked diversification during the Cenozoic. Gastropods and pelecypods were the dominant higher marine invertebrates. Reef corals grew extensively across tracts of shallow warm water in the Tethys belt, the Caribbean, the West Indies, and around the many volcanic islands of the Pacific.

The reptile dynasty had collapsed at the end of the Mesozoic. Only turtles, crocodiles, lizards, snakes, and rhynchocephalians lived through the Cenozoic and survived to modern times. Although the fossil record for birds is less than adequate, it is likely that most modern orders were present by Eocene time. Nearly all the continents seem to have had one or more species of large, flightless, flesh-eating birds during the Cenozoic.

With considerable justification, the Cenozoic has been called the Age of Mammals. During the Cenozoic, mammals expanded rapidly into the multitude of potential habitats vacated by the reptiles. From ancestral shrewlike creatures, either directly or indirectly, the higher orders of placental mammals evolved. In the vanguard were small-brained archaic ungulates, such as amblypods and condylarths, as well as primitive flesh-eating creodonts.

Following this Early Cenozoic radiation, a second wave of placentals began to evolve in the middle part of the era, and most of the modern mammalian orders became established. This adaptive radiation was as remarkable as had been that of the earlier Mesozoic reptiles. Mammals in the form of bats conquered the air, whereas others, such as whales, seals, and walruses, returned to the sea, where the evolution of vertebrates had begun long ago.

The Carnivora diversified and became dependent upon the multitude of advanced ungulates that populated forests and grasslands. Ungulates included two great categories: the perissodactyls and artiodactyls. The former include horses, rhinoceroses, and tapirs, as well as the extinct titanotheres and chalicotheres. Among artiodactyls, or even-toed ungulates, the large, piglike entelodonts and smaller oreodonts failed to survive beyond the Oligocene. However, living artiodactyls

Figure 15–50 An assemblage of animals that lived in Central Alaska about 12,000 years ago, during the Late Pleistocene. Fossil remains of this period are abundant and indicate a fauna in which grazing animals predominated, with the remaining fauna composed of browsers and predators. (Courtesy of The Smithsonian Institution; mural by J. H. Matternes.)

include swine, camels, hippopotami, and a diverse host of ruminants. As all of these ungulates evolved during the Cenozoic, they underwent specialization of teeth for grazing and limbs for running across relatively open plains. Grass tends to wear down teeth rapidly. To compensate, grass-eaters evolved teeth with complicated ridges of hard enamel and exceptionally high crowns.

Similar changes also occurred in members of the proboscidean line. The earliest proboscideans are recorded as fossils in Eocene beds of Egypt. Their radiation produced an impressive array of variously tusked and specialized mastodons and mammoths. The remains of those great beasts are frequently found in Pleistocene bog deposits in the eastern United States, Alaska, and Siberia.

The Pleistocene was a time of splendid diversity among mammals (Fig. 15–50). The fauna was probably more varied and impressive than that of Africa a century ago. The Pleistocene was an age of giants, in which colossal ground sloths, beavers over 2 meters tall, giant dire wolves, bison, and tall, rangy mammoths and woolly rhinoceroses ranged far and wide across the continents. One might think an assemblage of beasts that had proved their stamina against the difficulties of an ice age would persist for at least as long as did the dinosaurs. Yet, widespread extermination and extinction was the dominant biologic theme at the end of the Pleistocene. The cause of the extinctions is not known, although paleontologists suspect that, either directly or indirectly, the advent of humans may have had a bearing on the decline of some of these animals.

Questions for Review

1. What was the effect of geographic isolation on the evolution and distribution of Cenozoic mammals?
2. Which groups of marine phytoplankton proliferated during the Cenozoic?
3. How did the cephalopod faunas of the Mesozoic differ from those of the Cenozoic?
4. What depth of water is preferred by reef corals? What is the relationship of this preference to the origin of coral atolls?
5. Which group of bony fishes is particularly characteristic of the Cenozoic?
6. During what epoch of the Cenozoic did grasslands first become extensive? What effect did this have upon adaptations developed by herbivorous mammals?
7. What is the importance of the Order Insectivora in the evolutionary history of placental mammals?
8. What general evolutionary changes are apparent in the Cenozoic history of horses? What environmental conditions stimulated these changes?
9. Discuss the premise that the adaptive radiation of Cenozoic mammals was at least as successful as that of Mesozoic reptiles.
10. What groups of reptiles managed to survive the exterminations of the Late Mesozoic and survive today?
11. How might a paleontologist determine that a fossil lower jaw containing a full complement of teeth belonged to a mammal rather than a reptile?
12. What possible advantages may be inherent in the ruminant type of digestion? Name three living groups of ruminants.
13. How is it possible to trace the development of the trunk, or proboscis, in the skeletal remains of ancient proboscideans?
14. What is the evolutionary importance of the primitive Jurassic mammals known as pantotheres?
15. Account for the reductions and extinctions in marsupialian groups in South America during the Early Pleistocene.

Terms to Remember

archaic ungulate
artiodactyl
cetacean
chalicothere
dinothere
docodont
entelodont
lagomorph
mammoth
marsupium
monotreme
morganucodont
multituberculate
oreodont
pantothere
pecoran
perissodactyl
pinneped
plantigrade
proboscidean
ruminant
symmetrodont
teleost
titanothere
tragulid
tricondont
ungulate

Supplementary Readings and References

Colbert, E. H. 1980. *Evolution of the Vertebrates* (3d ed.). New York, John Wiley & Sons.

Halstead, L. B. 1968. *The Pattern of Vertebrate Evolution.* San Francisco, W. H. Freeman Co.

Kürten, B. 1972. *The Age of Mammals.* New York, Columbia University Press.

Match, C. L. 1976. *North America and the Great Ice Age.* New York, McGraw-Hill.

Martin, P. S., and Wright, H. E. 1967. *Pleistocene Extinctions.* New Haven, Yale University Press.

Pearson, R. 1964. *Animals and Plants of the Cenozoic Era.* London, Butterworths Press.

Scott, W. B. 1937. *A History of Land Mammals in the Western Hemisphere.* New York, Macmillan.

Zappler, G., and Zappler, L. 1970. *The World After the Dinosaurs.* Garden City, N.Y., Natural History Press.

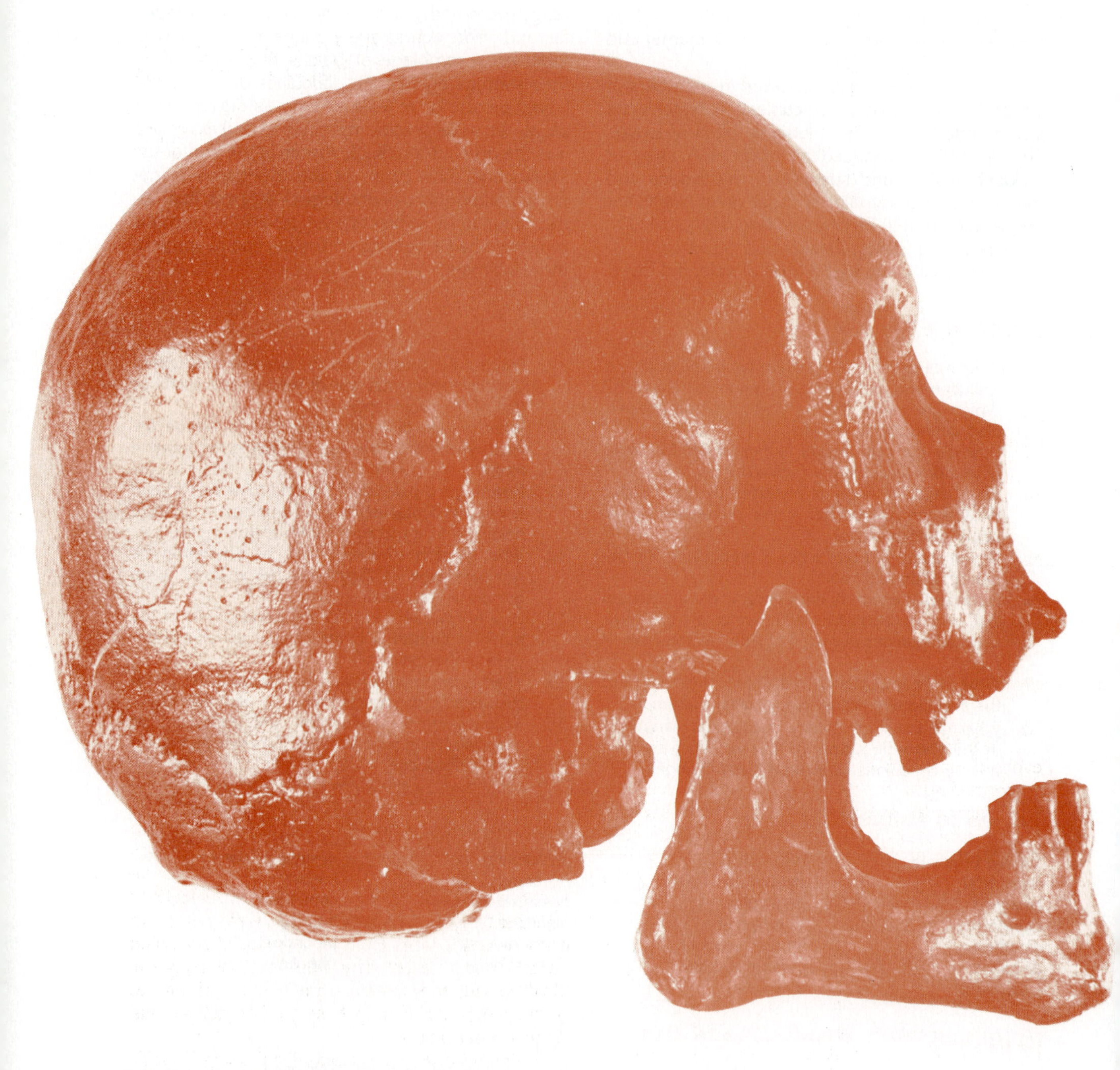

Reproduction of the skull of Tepexpan Man. A nearly complete skeleton was recovered in 1947 from a Late Pleistocene gravel deposit north of Mexico City. Deposits thought to be of equivalent age have yielded the remains of mammoths and are thought to be about 11,000 years old. (Courtesy of Oak Knoll Museum of Natural History, St. Louis, Missouri.)

16 Human Origins

It walked on its hind feet, like something out of the vanished Age of Reptiles.
The mark of the trees was in its body and hands.
It was venturing late into a world dominated by fleet runners and swift killers.
By all the biological laws this gangling, ill-armed beast should have perished, but
you who read these lines are its descendant.

Loren Eiseley, 1971, *The Night Country*.
New York: Charles Scribner's Sons.

PRELUDE

The late epochs of the Cenozoic witnessed the evolution and proliferation of a mammal capable of controlling and shaping his own environment. That mammal, of course, was our own species, *Homo sapiens.* Unlike previous vertebrates, this remarkable creature profoundly changed the surface of the planet, modified environment in ways both beneficial and destructive, and so manipulated the populations of other creatures as to change the entire biosphere. For these reasons, it is fitting that the final chapter in our history deal with the evolution of humankind.

PRIMATES

Primate Characteristics

What traits should an animal have in order to qualify as a primate? It seems that this question cannot be answered easily, for primates have remained structurally generalized compared with most other orders of placental mammals. They retain the primitive number of five digits, have teeth that are not specialized for dealing with either grain or flesh, and have never developed hoofs, horns, trunks, or antlers. In the course of their evolution from shrewlike insectivores, the principal changes they have experienced have involved progressive enlargement of the brain and certain modifications of the hand, foot, and thorax that were related to their life in the trees and their manner of obtaining food. These adaptations were not insignificant, however, for they enabled the human primate to shape a mode of life qualitatively different from that of any other animal.

The fictional pig Snowball, in George Orwell's book *Animal Farm,* remarked that "the distinguishing mark of man is the hand, the instrument with which he does all his mischief." Although this is clearly a biased point of view, it is correct in its assessment of the importance of the primate grasping hand with its opposable thumb (Fig. 16–1). This characteristic not only permitted primates a firm grip on their perches but also allowed them to grasp, release, and manipulate food and other objects. The forearms retained their primitive mobility, in which rotation of the ulna and radius upon one another permitted the hands to be reversed in position.

The development of the grasping, mobile hand was accompanied by improvement in visual attributes. The eyes of primates became positioned toward the front of the face so that there was considerable overlap of both fields of vision, resulting in an improved ability to judge distances. It would seem that grasping hands and feet and good binocular vision are obvious adaptations for an animal that leaps from branch to branch and seeks its food from precarious boughs. Yet many tree-dwelling mammals, such as squirrels, civets, and opossums, lack the short face, close-set eyes, and opposable digits and get along very well. For this reason, anthropologists have recently suggested that the visual attributes of primates originated as predatory adaptations that allowed early insectivorous primates to gauge accurately the distance to insect prey without movement of the head. By being able to grasp narrow supports securely with its feet, the animal was able to use both of its mobile hands to catch the prey. Claws, which are utilized by animals like squirrels in moving about on relatively wide branches, would not be as advantageous for climbing about on thin boughs and vines and might not provide a sufficiently secure hold for the quick catch of a moving insect. Thus, it seems that binocular vision and the grasping hand may not have been originally related to arboreal locomotion alone, but to the cautious, well-controlled, manual capture of visually located prey in the insect-rich canopy of tropical forests.

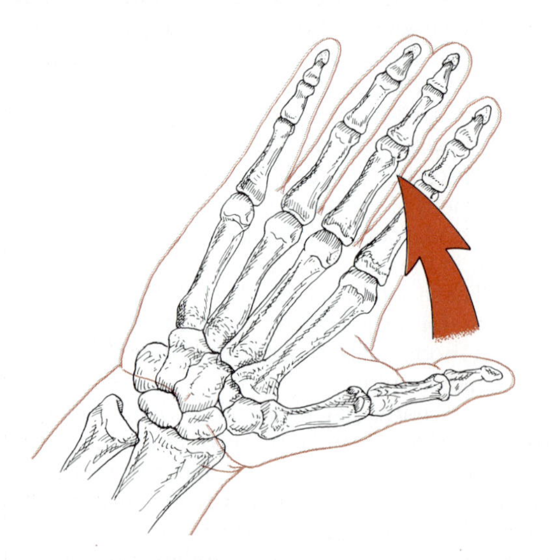

Figure 16–1 The right hand of a human (palm up). The human hand is not used in locomotion and can be used to manipulate small objects between the fingers and thumb.

Other evolutionary modifications of primates were related to changes in the eyes and limbs. To protect the eyes from bulging jaw muscles, a postorbital bar developed behind the eye orbits. As the eyes were positioned more closely together, the snout was reduced, so that the face became flatter. In response to the branchiating habit (swinging from branches), fore limbs and hind limbs diverged in form and function, and a general predisposition toward upright posture developed.

Modern Primates

The primates are divided into the suborders *Prosimii* (Fig. 16–2), or tree shrews, lemurs, lorises, and tarsiers, and the suborder *Anthropoidea* (Fig. 16–3), which includes monkeys, apes, and humans (Table 16–1). The *Prosimii* are more primitive. The tree shrews, for example, possess clawed feet and a long muzzle with eyes in the lateral positions. The lemurs, which are confined largely to the Island of Madagascar, still have a long snout and lateral eyes, although they are positioned more to the front than are those in the shrew. Some of the digits are clawed, whereas others are equipped with flattened nails. In the East Indian tarsier, the face is relatively flat. As befits a nocturnal animal, the eyes are large and positioned toward the front, so that vision is very nearly stereoscopic. The fingers and toes terminate in nails rather than claws.

In the more progressive Anthropoidea, the trends initiated by prosimians are further developed. Monkeys are the more primitive members of the Anthropoidea. They are grouped into New World and Old World forms. The *Ceboidea,* or New World monkeys, are an early branch and are not involved in the eventual evolution of humans. Included among the ceboids are the howler, spider, and squirrel monkeys as well as the familiar little capuchin, or organ-grinder, monkey. New World monkeys can be identified by their flattish faces, widely separated nostrils, and prehensile tails. Most are small in comparison to their Old World cousins.

The more advanced Old World monkeys, or *Cercopithecoidea,* are widely distributed in tropical regions of Africa and Asia. They include the familiar macaque, or rhesus monkey of laboratories and zoos, the Barbary ape of Gibraltar, langurs, baboons, and mandrills. In this group of monkeys, the nostrils are close together and directed downward (as in humans), and the tail is not prehensile.

The anthropoid apes are tail-less primates. Modern species probably evolved from the same ancestral stock that produced humans. However,

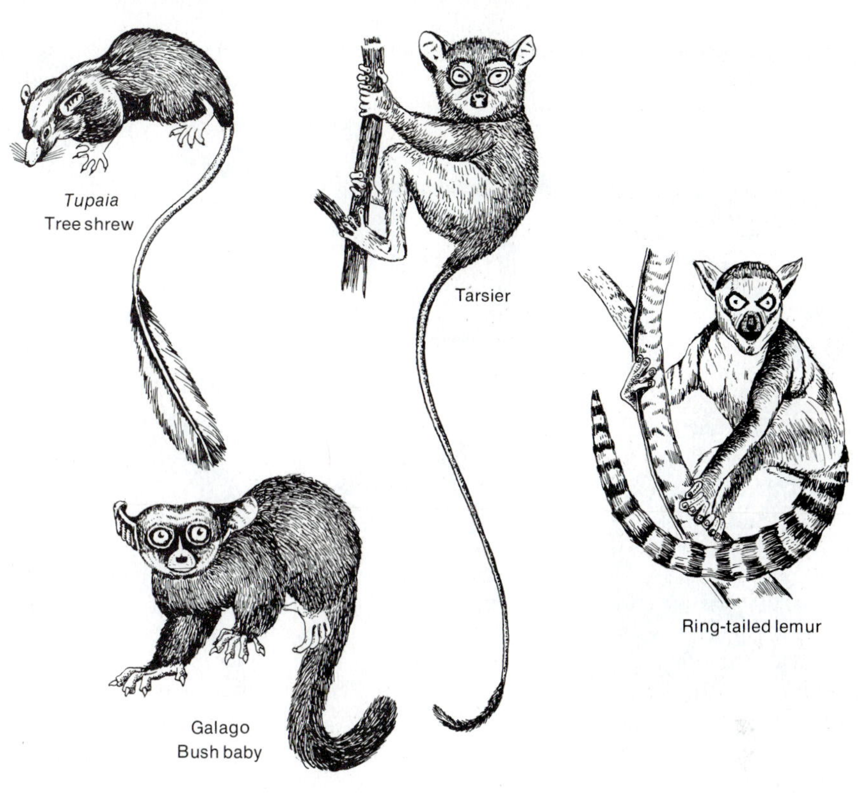

Figure 16–2 Representative prosimian primates. (After W. E. Le Gros Clark. *History of the Primates.* London, British Museum [Natural History], 1950.)

Table 16–1 A SIMPLIFIED CLASSIFICATION OF THE ORDER PRIMATES

ORDER	SUBORDER	SUPERFAMILY	Family	COMMON NAMES OF REPRESENTATIVE FORMS
Primates	Prosimii	TUPAIOIDEA		Tree shrew
		LEMUROIDEA		Lemur
		LORISOIDEA		Bush baby, Slender loris
		TARSIOIDEA		Tarsier
	Anthropoidea	CEBOIDEA New world monkeys		Howler monkey, Spider monkey Capuchin, Common marmoset, Pinche monkey
		CERCOPITHECOIDEA Old world monkeys		Macaque, Baboon, Wanderloo, Common langur, Proboscis monkey
		HOMINOIDEA	HYLOBATIDAE	Gibbon, Siamang
			PONGIDAE	Orangutan, Chimpanzee, Gorilla
			HOMINIDAE	Humans

Figure 16–3 Representative types of anthropoid primates. (From E. L. Cockrum, W. J. McCauley, and N. A. Younggren. *Biology*. Philadelphia, Saunders College Publishing, 1966, p. 652.)

the divergence probably took place back in the Middle Tertiary. The more primitive branch of the tail-less apes is composed of the *Hylobatidae,* best represented by those long-armed acrobats, the gibbons. Orangutans, chimpanzees, and gorillas are grouped within the Pongidae. The red-haired orangutan, or "wild man," of Borneo and Sumatra is considerably larger than the gibbon and has a larger brain as well. Chimpanzees and gorillas are similar in many ways and tend to be less pronounced in their arboreal adaptations. Gorillas are essentially ground dwellers and spend only a small fraction of their time in the trees.

THE PROSIMIAN VANGUARD

The fossil record for primates begins with the appearance of a creature named *Purgatorius,* known only from a few teeth discovered in the Hell

Creek Formation at Purgatory Hill in Montana. These finds indicate that the earliest primates were contemporaries of the last of the dinosaurs, at least in the tropical latest Cretaceous environments of North America. The fossil record improves somewhat in the early epochs of the Cenozoic. *Plesiadapis* (Fig. 16–4), found in Paleocene beds of both the United States and France, has the distinction of being the only genus of primate, other than that of man, that has inhabited both the Old World and the New World. The presence of this **prosimian** on the now widely separated continents is one of many clues indicating that they were not yet completely separated by the widening Atlantic at this time. *Plesiadapis* was a rather distinctive and specialized primate and represents a sterile offshoot of the primate family tree. The incisors were rodent like and were separated from the cheek teeth by a toothless gap, or **diastema.**

Figure 16–4 The Paleocene rodent-like prosimian *Plesiadapis*.

Figure 16–5 The Eocene lemur *Notharctus.*

Fingers and toes terminated in claws rather than nails.

General trends in prosimian evolution during the Eocene Epoch involved reduction in the length of the muzzle, increase in brain size, and shifting of the eye orbits to a more forward position. These trends are apparent in the well-known Eocene genera *Notharctus* (Fig. 16–5) and *Smilodoectes,* both of which were quite modern-looking lemurs. Prosimian populations during the Late Paleocene and Eocene also included tarsiers. *Tetonius* was an Eocene form with the tarsioid traits of closely spaced large eyes and shortened muzzle already quite evident.

Both tarsiers and lemurs were abundant and widely dispersed on northern hemisphere continents during the Eocene. However, with the advent of cooler Oligocene climates, they virtually deserted North America, whereas in the eastern hemisphere they were forced southward into the warmer latitudes of Asia, Africa, and the East Indies. Surviving prosimians are much reduced in variety and number. The prevailing opinion is that many have been replaced by the monkeys.

THE EARLY ANTHROPOIDS

The next step in the primate evolutionary advance was the appearance of the first **anthropoids.** Discoveries in the Fayum district of Egypt (Fig. 16–6) have provided a wealth of information about the early anthropoids. The fossils there occur in several Late Oligocene horizons and include well over 100 specimens. Although many of the skull fragments and teeth retain subtle vestiges of a prosimian ancestry, none of the remains are from prosimians. They are the fossils of primates clearly recognizable as relatives of either apes or Old World monkeys. The Fayum specimen named *Oligopithecus,* for example, appears to have been quite close to the lineage that gave rise to the Old World monkeys. Another Fayum genus, *Propliopithecus,* may be near the evolutionary branch that led to living **pongids** and man. It is apparent from the study of these and other Fayum discoveries that the prosimian-anthropoid transition had taken place by Late Oligocene time.

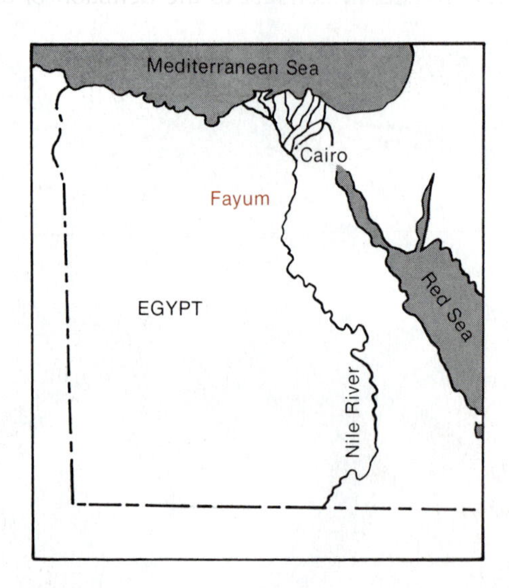

Figure 16–6 Location of the Fayum region of Egypt, noted for the discoveries of Late Oligocene anthropoids and other contemporary mammals.

Fossils of the earliest known New World monkeys are known from Late Oligocene and Early Miocene strata of South America. New World monkeys, although they appear superficially similar to some Old World forms, evolved quite independently from prosimian ancestors and without genetic contact with their Old World cousins.

Because evolution is a continuing process, with each animal truly a transitional link between older and younger species, it is a difficult and often arbitrary task to assign the fragment of a fossil jaw or a tooth to either the monkey or the ape category. One of many clues used in attempting to make this distinction is the pattern of cusps on certain molar teeth. In Old World monkeys, specific molars have four cusps. Among apes (and humans), these same molars have five cusps, with an intervening Y-shaped trough. Molars belonging to *Propliopithecus* and *Aegyptopithecus* (another Fayum fossil) display this "5-Y" pattern (Fig. 16–7), and this is one reason anthropologists consider these genera possibly very early ancestors to Miocene apes. It seems safe to state that the Oligocene Epoch may have witnessed not only the transition from prosimians to anthropoids but also the differentiation of apes from monkeys.

The Miocene record of primate evolution is known chiefly from deposits in East Africa. These beds are about 10 million years younger than those at Fayum. So large an interval without fossils leaves a serious gap in the history of Cenozoic primates. Among the new players that come on the scene during the Miocene is the gibbon-like *Pliopithecus.* Indeed, this form may have been among the group from which modern long-armed gibbons evolved.

Another important group of primates that roamed Southern Europe, Asia, and Africa both in Miocene and Pliocene time were the dryopithecines. The **dryopithecines** (Fig. 16–8) varied in size and appearance and apparently lived predominantly on the ground in open savannah coun-

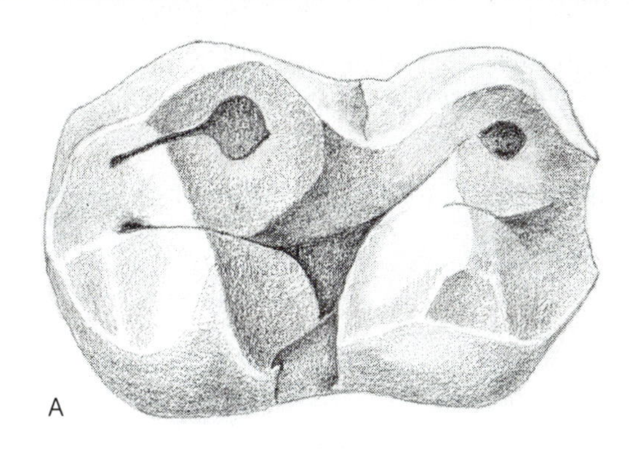

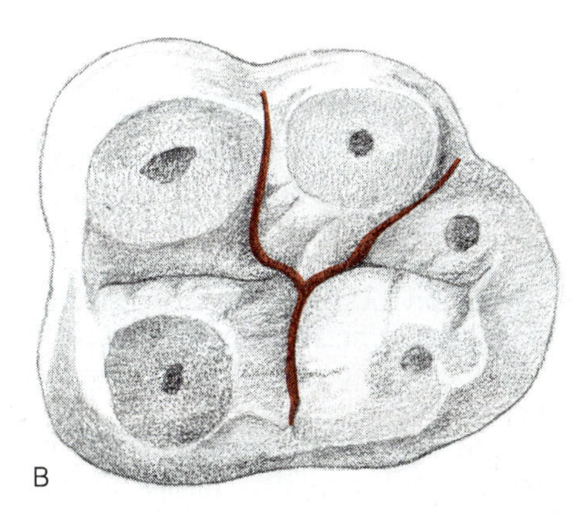

Figure 16–7 General pattern of cusps on the molars of Old World monkeys. *A* illustrates a cusp at each corner. The "lazy Y-5" pattern shown in *B* is characterized by a Y-shaped depression that separates five cusps. *A* is the lower molar of a baboon, *B* is that of a chimpanzee.

try. The group includes the well-known **Proconsul** (*Dryopithecus*), first found in the 1930's on an island in Lake Victoria. Anthropologists regard the dryopithecines as true apes, and yet the creatures also displayed monkey-like traits and even some hints of manlike traits. They had, for example, the "5-Y" molar pattern of apes and early humans. Could this mean that dryopithecines were close to the human line of descent? Apparently not, for the hard palate that roofs the mouth was flat in dryopithecines, and the jaws were U-shaped, with parallel rows of teeth. These are traits characteristic of apes rather than humans. The dryopithecines went on to diversify during the Miocene and Pliocene. They disappeared before the end of the Pliocene, presumably by evolving into modern types of great apes.

The search for the earliest creature that might be representative of the line leading to humans was ultimately successful. While excavating in the Siwalik Hills of India in the 1930's, Professor G. E. Lewis of Yale found part of an upper jaw containing a few teeth that appeared more human-like than apelike in appearance. He named his discovery *Ramapithecus*. Additional finds of ramapithecines were made by the husband and wife team of Louis and Mary Leakey in East Africa. From generally scrappy remains there emerged a picture of a Late Miocene to Early Pliocene hominid that most anthropologists tentatively agree was derived from dryopithecine ancestry and that can be considered ancestral to modern humans. The adult animal was about the size of a chimpanzee, and even the fragmentary remains indicate that the face was foreshortened. In contrast to the dentition of apes (Fig. 16–9), the incisors were of modest size, the

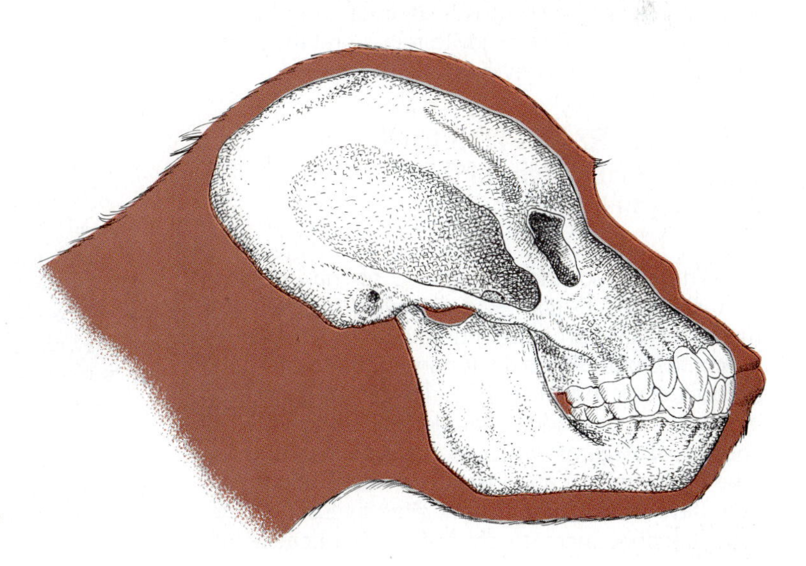

Figure 16–8 Skull of Proconsul, from Lake Victoria, Kenya.

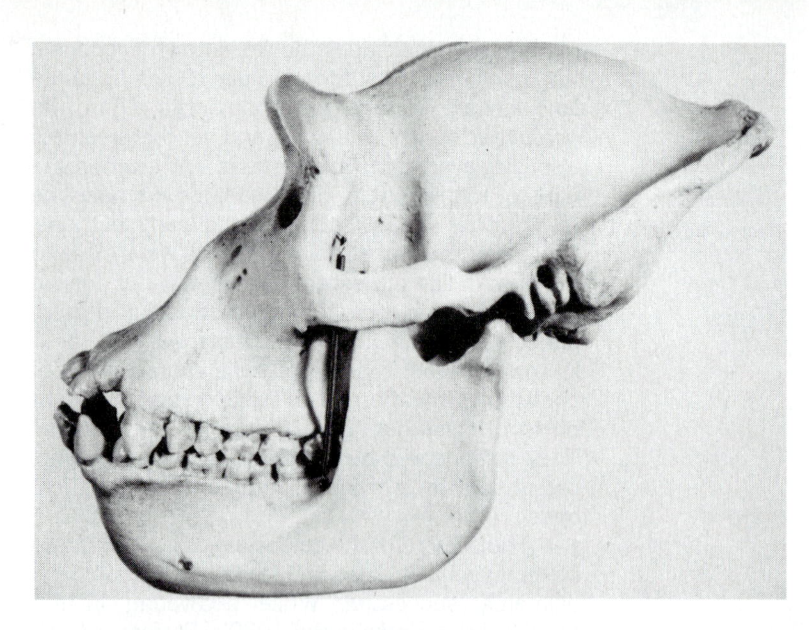

Figure 16–9 Gorilla skull. (Photograph courtesy of David G. Gantt.)

canines were not much larger than the premolars, and the teeth were arranged in a broad curve, with the widest part at the back (Fig. 16–10). It was a human-like dental arch, and the ratio in size of front to rear teeth was similar to that in humans. Lacking even large canines for defense, these small and vulnerable protohumans may have used sticks, stones, and the limb bones of larger animals to defend themselves.

The East African and Egyptian fossil beds clearly indicate that primate evolution in its early stages was not a case of orderly sequential change along a single trend. The family tree of the primates was a complex of parallel and diverging branches. For this reason, the attempt to trace the ascent of humans up through the many bifurcations and dead ends is an exciting but difficult task. One must be ready to reinterpret the story each time new fossil evidence is uncovered.

THE AUSTRALOPITHECINE STAGE AND THE EMERGENCE OF HOMINIDS

The story of the emergence of man begins with Raymond Dart's 1924 discovery of fossil remains of an immature primate in a limestone quarry in South Africa. Dart named the fossil *Australopithecus africanus* (Fig. 16–11). In succeeding years, many additional skeletal fragments of species of *Australopithecus* have been found in Late Pliocene or Early Pleistocene deposits of Africa, Java, and China. Fossil sites in East Africa have become increasingly important because of the veritable bonanza of hominid bones they have yielded. The new collections of fossils have given paleoanthropologists an unsurpassed record of human evolution over the past 3 million years. Some of the East African sites, like Olduvai Gorge, are now famous as a result of the lifelong research programs of Mary Leakey and the late Louis Leakey. Their efforts, and recently those of their son Richard, have provided fossils of relatively heavy-bodied, robust australopithecines as well as

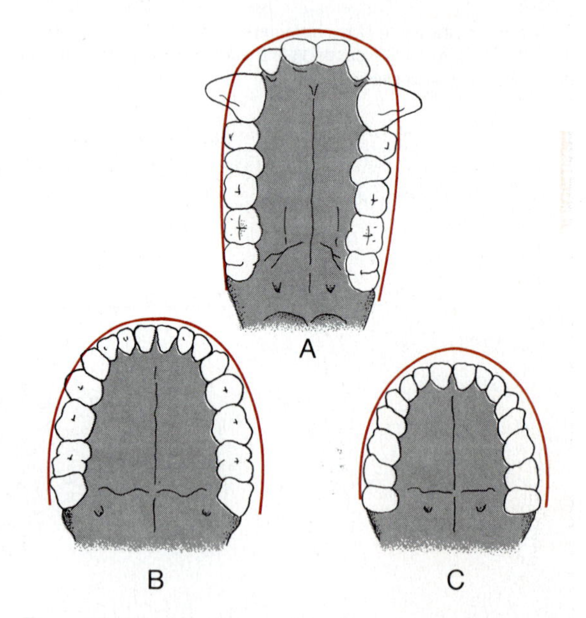

Figure 16–10 The palates and dentition of a gorilla (*A*), reconstructed *Ramapithecus (B)*, and human (*C*). The rather square U-shape of the gorilla's dental arcade differs from the more rounded (human-like) dental row of *Ramapithecus*.

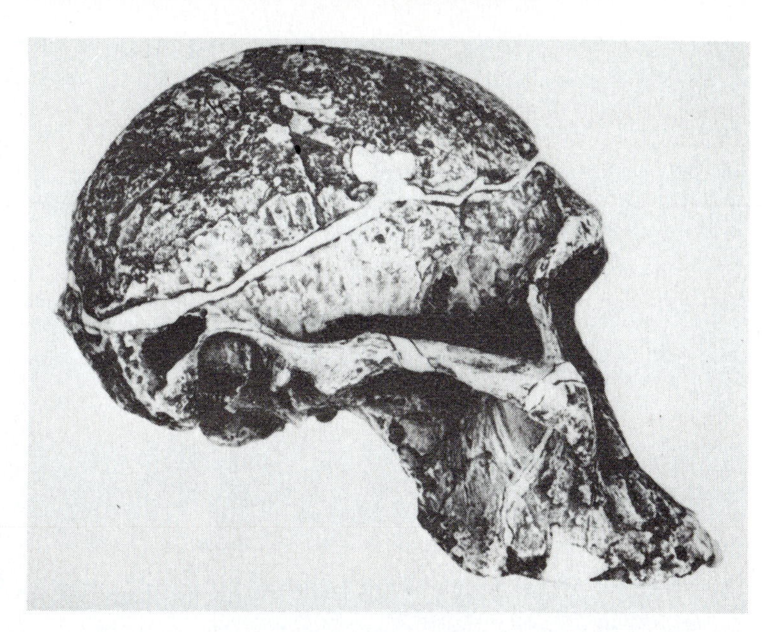

Figure 16–11 *Australopithecus africanus* from the Transvaal of South Africa. (Photograph of Wenner-Gren Foundation replica by David G. Gantt.)

lighter, so-called "gracile" types. Among the latter are remains of creatures that may well belong within our own genus.

East Africa during the Cenozoic experienced a lively history of volcanic activity. As a result, many of the fossil sites have interspersed layers of volcanic ash and lava flows. This fact has been of great benefit to the paleoanthropologists, for samples of ash can often be dated by the potassium-argon method. Dates from two succeeding ash beds would then provide a close estimate of the age of fossils found in the intervening layers of sediment.

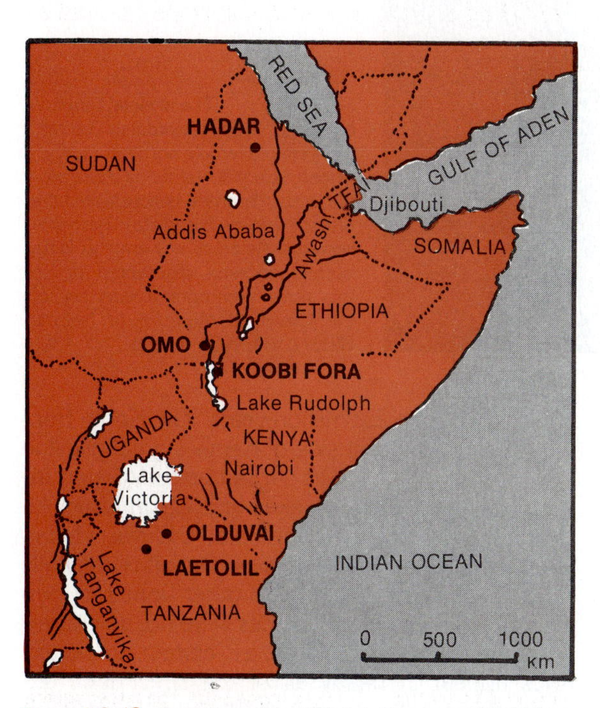

Figure 16–12 Location of the Olduvai, Omo, and Koobi Fora (East Rudoph) paleontologic sites.

Among the oldest undisputed hominids of East Africa are those discovered at Laetolil in Tanzania and Hadar in Ethiopia (Fig. 16–12). These localities have yielded remains of human ancestors that roamed Africa between 3 and 4 million years ago. Here have been found the earliest hominids. Among them, none is more famous than the skeleton of a young female discovered by Donald C. Johanson in 1974. The individual, informally dubbed "Lucy," is the most complete Pliocene hominid thus far discovered (Fig. 16–13). Lucy was about 3.5 to 4.0 feet tall (about 1 meter), and she was fully bipedal. The fact that she walked erect was amply demonstrated by her bones as well as those of her contemporaries, also found at Hadar. Further evidence of bipedalism came from the discovery of footprints in layers of solidified ash at Laetolil. Although not very large, Lucy was well muscled and had powerful arms that were slightly longer relative to body size than in ourselves. Her head was about as big as a softball and contained a brain comparable in size to that of a chimpanzee. The shape of the skull was more like an ape than a human, with large, forward-thrusting jaws and no chin.

Lucy, whose scientific name is *Australopithecus afarensis,* was not a solitary wanderer into the Hadar area. The site has provided bones of between 35 and 65 other individuals, all within the 2.6

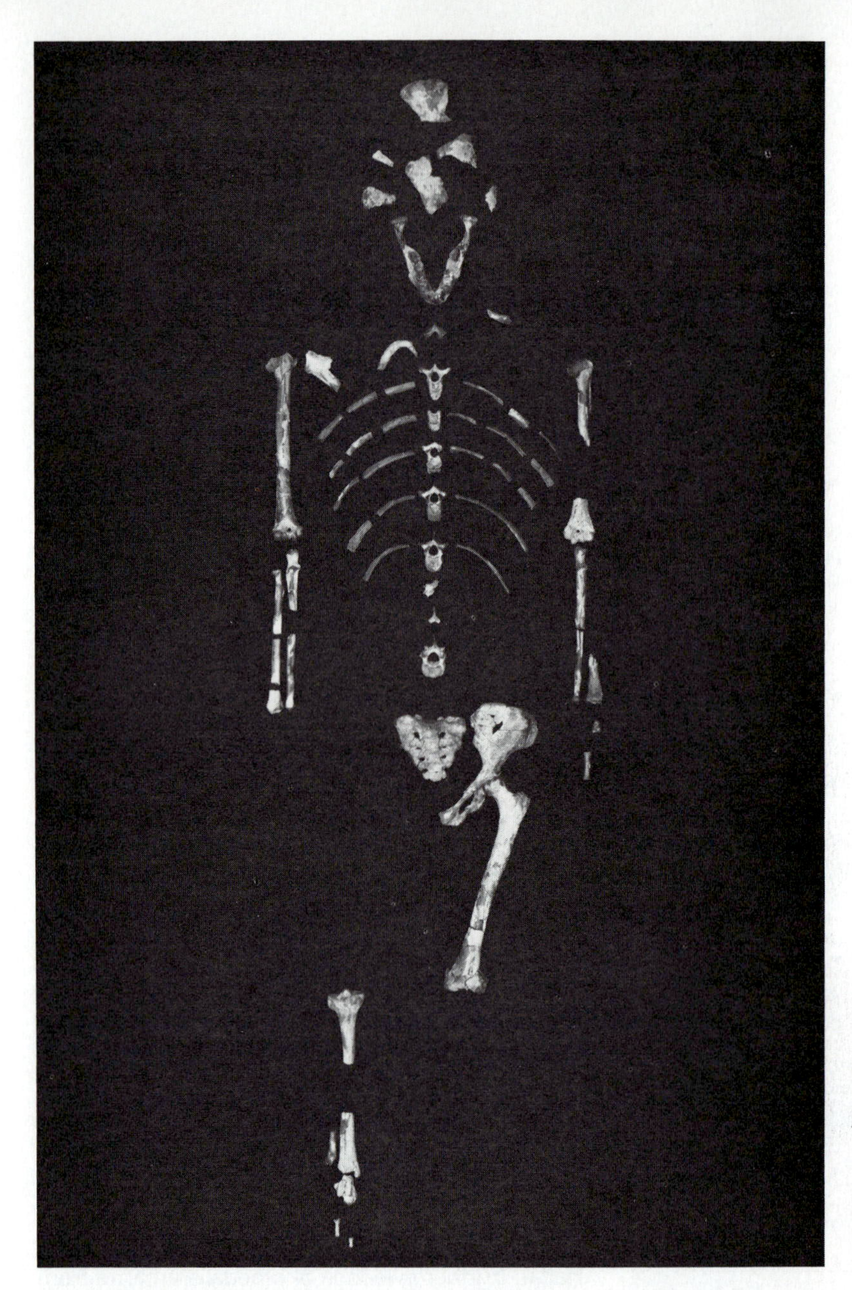

Figure 16–13 The skeletal remains of "Lucy," an approximately 3.5 million year old female hominid. "Lucy," whose scientific name is *Australopithecus afarensis,* is considered the oldest, most complete, best-preserved skeleton of any erect walking human ancestor discovered up to this time. (Photograph courtesy of the Cleveland Museum of Natural History, with permission.)

to 3.3-million-year-old Hadar Formation. Although debate about the precise evolutionary position of these fossils is currently lively, many paleoanthropologists consider Lucy and her kin the direct ancestors of the genus *Homo,* as well as later species of *Australopithecus* (Fig. 16–14).

Two other East African fossil localities that have yielded species of both *Homo* and *Australopithecus* are the Omo and Koobi Fora sites. The Omo digs are located along the Omo River in remote southwestern Ethiopia. Here one finds a nearly continuous section of sediments and volcanics that span 2.2 million years. The Koobi Fora locality is on the east side of Kenya's great alkaline Lake Rudolf (Fig. 16–15). It is currently being investigated by Richard Leakey and Glynn Isaac. Altogether, the Omo, Koobi Fora, Olduvai, Hadar, and Laetolil localities have given us sufficient skulls (Fig. 16–16), jaws, teeth, and other bones to indicate that the australopithecines were not an entirely homogeneous group. However, certain characteristics appear in nearly all of the specimens unearthed. From the structure of their pelvic girdles, it is known that they stood upright in a fashion more

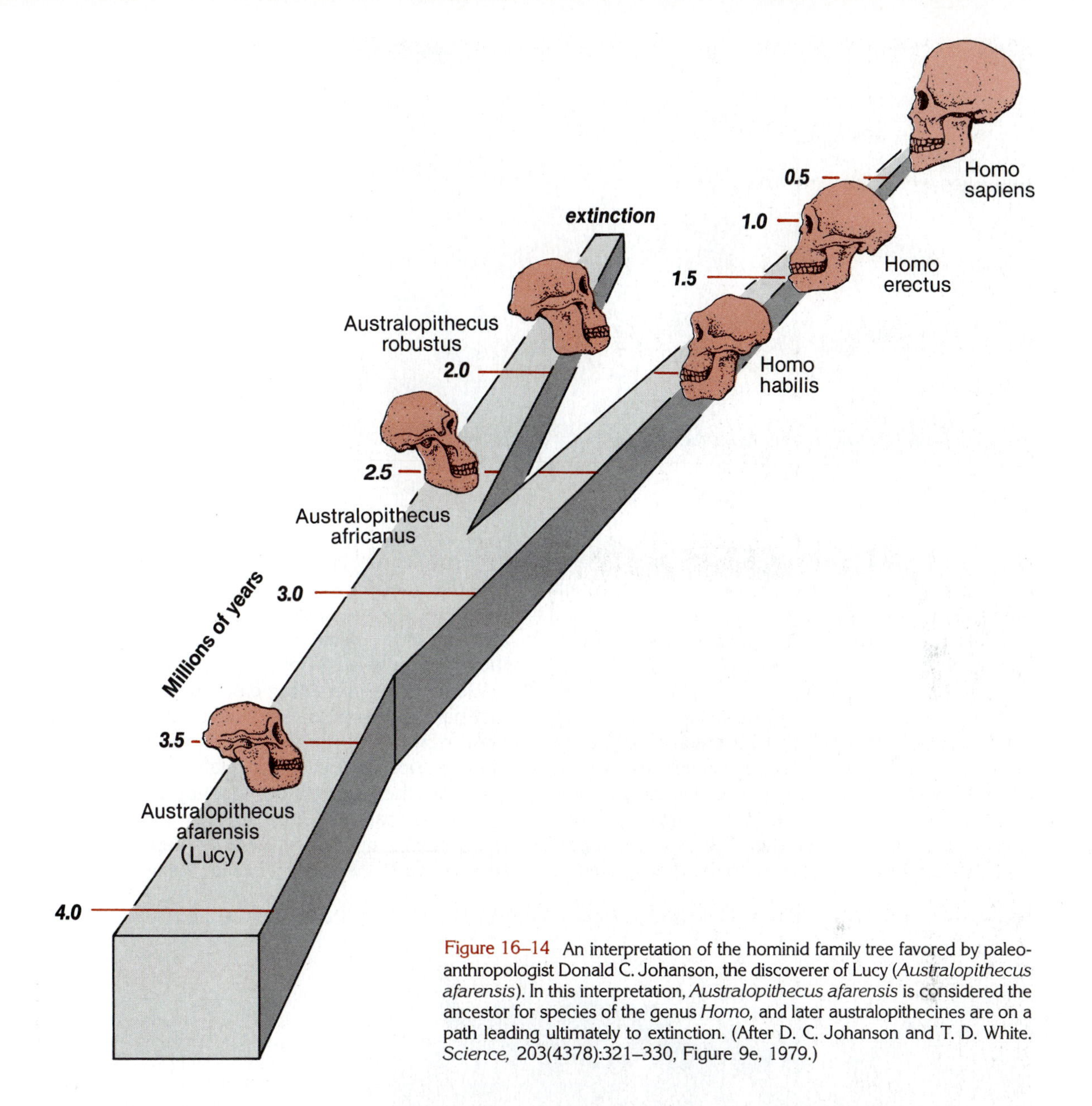

Figure 16–14 An interpretation of the hominid family tree favored by paleoanthropologist Donald C. Johanson, the discoverer of Lucy (*Australopithecus afarensis*). In this interpretation, *Australopithecus afarensis* is considered the ancestor for species of the genus *Homo,* and later australopithecines are on a path leading ultimately to extinction. (After D. C. Johanson and T. D. White. *Science,* 203(4378):321–330, Figure 9e, 1979.)

human-like than apelike. Australopithecine dentition was essentially human, although the teeth were more robust than in modern humans. In fact, the teeth of *Australopithecus* resembled those of *Ramapithecus.* In contrast to these human-like characteristics, australopithecine cranial capacity was comparable to that of modern large apes, and reached a volume of 600 or 700 cc. However, even with a brain only about half the size of *Homo sapiens,* they were able to make several kinds of crude implements from horn, teeth, and bone.

Recent investigations by anthropologists working in Kenya and Ethiopia have provided evidence that there may have been several manlike lineages coexisting in Africa between 1 and 3 million years ago. One of these lineages included *Australopithecus africans* and the more recent *Australopithecus habilis* (some prefer "*Homo habilis*") from Olduvai Gorge. The possible existence of a second lineage became apparent in 1972, when a team led by Richard Leakey unearthed a skull at the Koobi Fora site in beds that are about 2.9 million years old. The skull, assigned the code name KNM-ER-1470, had a cranial capacity of about 800 cc, much larger than that of any *Australopithecus* and close to the cranial capacity of later hominids already assigned to the genus *Homo.* Thus, large-brained human-like

Figure 16–15 An East Rudoph archaeologic site discovered in 1969 by Dr. Kay Behrensmeyer. Dr. Glynn Isaac (dark shirt), coleader of the Koobi Fora Research Project, provides on-site instruction. (Photograph courtesy of Dr. Kay Behrensmeyer.)

creatures may have been present on earth much earlier than previously suspected.

Yet another group of primates at the australopithecine stage had smaller brains but were larger than the *Australopithecus africanus* line. This group is exemplified by *Paranthropus* and may have evolved into a still heavier creature appropriately named *Gigantopithecus.* Such species as *Australopithecus boisei* (once called *Zinjanthropus boisei*) and *Australopithecus robustus* (formerly *Paranthropus robustus* [Fig. 16–17]) apparently represent sterile side branches that failed to give rise to any known descendant. Careful study of the teeth of some of these larger australopithecines suggest that, unlike *Australopithecus africanus,* they may have been primarily herbivorous.

From the preceding discussion, it is apparent that paleoanthropologists have established a rather large number of names for various australopithecine groups that display relatively minor differences. Some students do not support the recognition of so many different species and believe that many forms now designated as separate species may prove to be local races or subspecies.

Figure 16–16 The first relatively complete skull of *Australopithecus* to be discovered at the East Rudoph (now known as Koobi Fora) lies on the examination table at the fossil site. (Courtesy of Dr. Kay Behrensmeyer.)

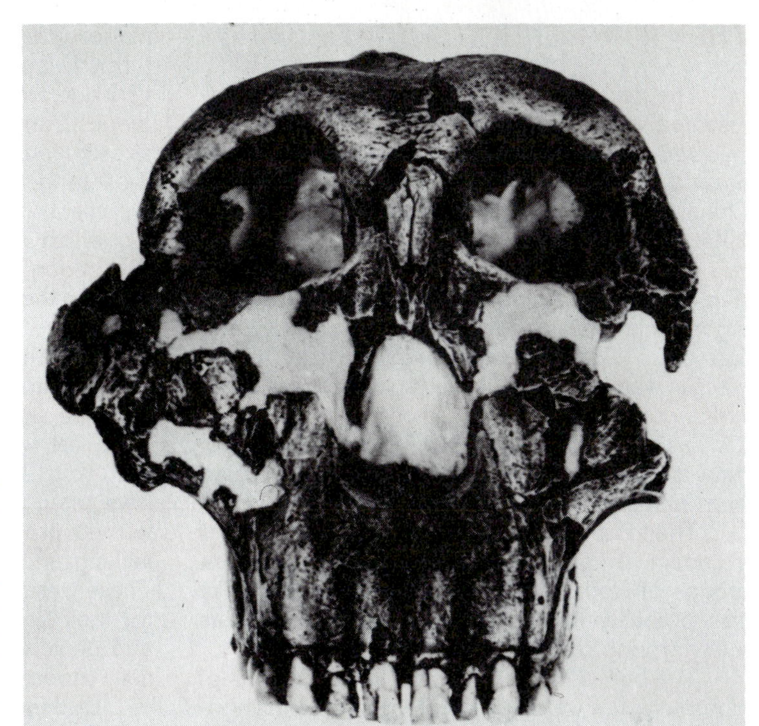

Figure 16–17 *Australopithecus robustus.* Because of the observation that forms like this one had larger molars and premolars than did *Australopithecus africanus,* some paleoanthropologists believed it to have been primarily a vegetarian, as well as taxonomically distinct from *A. africanus.* Recent interpretations favor inclusion of these creatures under the generic name *Australopithecus.* This specimen was derived from Bed I on the south side of the main ravine at Olduvai Gorge. (Photograph of Wenner-Gren Foundation replica by David G. Gantt.)

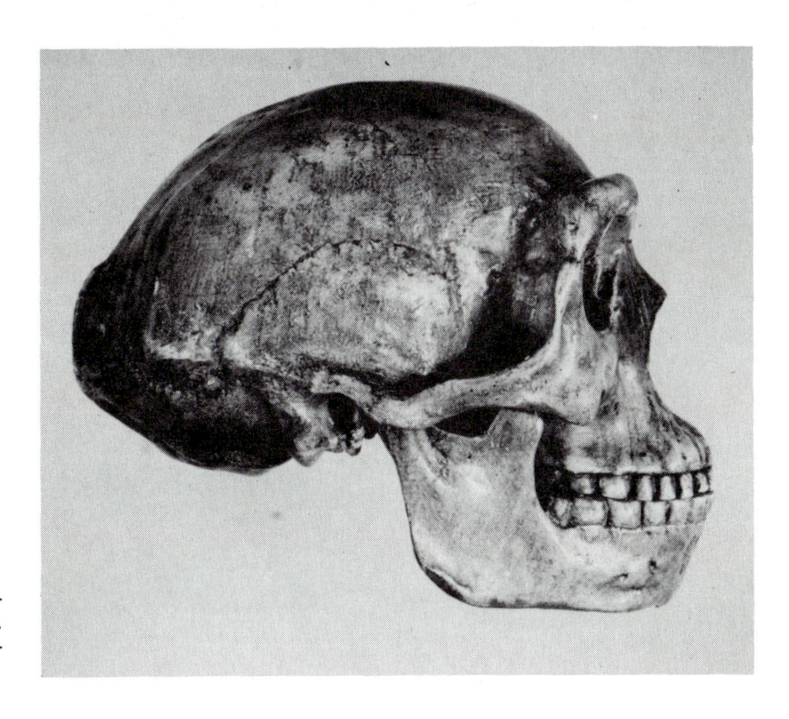

Figure 16–18 Replica of *Homo erectus,* previously known as *Sinanthropus pekinensis.* (Photograph of Wenner-Gren Foundation replica by David G. Gantt.)

THE HOMO ERECTUS STAGE

The next stage in homonoid evolution is represented by *Homo erectus,* formerly known as *Pithecanthropus erectus* from Java and *Sinanthropus pekinensis* (Fig. 16–18) from China. Another important locality was added in 1963, when Dr. Louis Leakey discovered a skull cap of *Homo erectus* in Bed II of Olduvai Gorge (Fig. 16–19). Bed II had been dated by potassium-argon techniques and found to be around 3/4 million years old. These finds indicate that *Homo erectus* was probably widely dispersed in Asia and Africa during the Middle Pleistocene. *Homo erectus* has been considered the first true species of man and is generally regarded as having evolved from a species of *Australopithecus.*

The fossil evidence already available indicates a certain amount of variability between different groups of *Homo erectus.* However, the variability was probably no greater than that which exists today among members of our own species.

The bones of the axial skeleton and limbs of *Homo erectus* were quite similar to those of modern humans. In this regard, these hominids were more advanced than the australopithecines. For example, the pelvic bones of *Australopithecus* indicate that, although they were fully bipedal, they walked with feet turned outward in a sort of half-running, rolling gait. *Homo erectus,* on the other hand, was an excellent walker.

Although the postcranial skeleton of *Homo erectus* was quite modern, the skull was not. Cranial capacity ranged from about 775 cc in earlier forms to nearly 1300 cc in specimens of more recent age. Brain capacity in modern *Homo sapiens* ranges from 1200 to 1500 cc, and thus the brain size of *Homo erectus* overlaps at least the lowermost range of modern peoples. Pithecanthropines represent a stage in the evolution of hominids during which relatively rapid increases in brain size had begun. No doubt the expansion of the brain involved the reshaping not only of the cranium but also of the birth canal. Selection would very likely have favored large pelvic inlets to accommodate fetuses with larger brains.

The skull of *Homo erectus* was massive and rather flat, with heavy "supraorbital ridges" over the eyes (Fig. 16–19). The forehead sloped, and the jaws jutted forward at the tooth line in a condition termed **prognathous.** A definite jutting chin was lacking, and very likely the nose was broad and flat. These were rather primitive traits. However, except for their being somewhat more robust, the teeth and dental arcade in *Homo erectus* were essentially modern.

Evidence exists that *Homo erectus* made good use of his larger brain. From the bones of other animals found at living sites, it is clear that these hominids were good hunters. Of more importance are indications that they had discovered the use of fire. They were also skilled at making simple implements of flint and chert (Fig. 16–20), such as axes and scrapers, the former being equipped with wooden handles. Some also appear to have engaged in cannibalism, perhaps not for

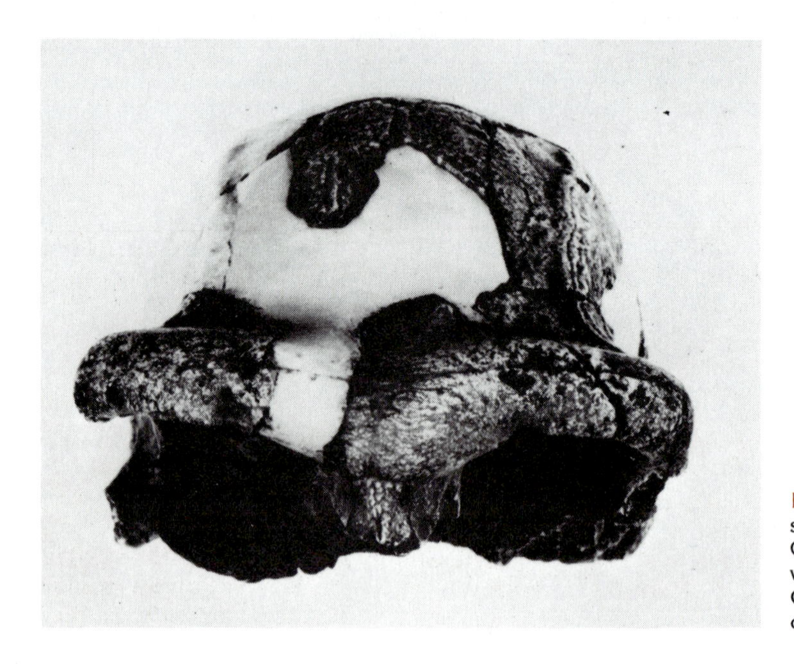

Figure 16–19 Upper portion of skull of a specimen of *Homo erectus* from Olduvai Gorge. This specimen was found associated with stone implements of the type known as Chellan. (Photograph of Wenner-Gren Foundation replica by David G. Gantt.)

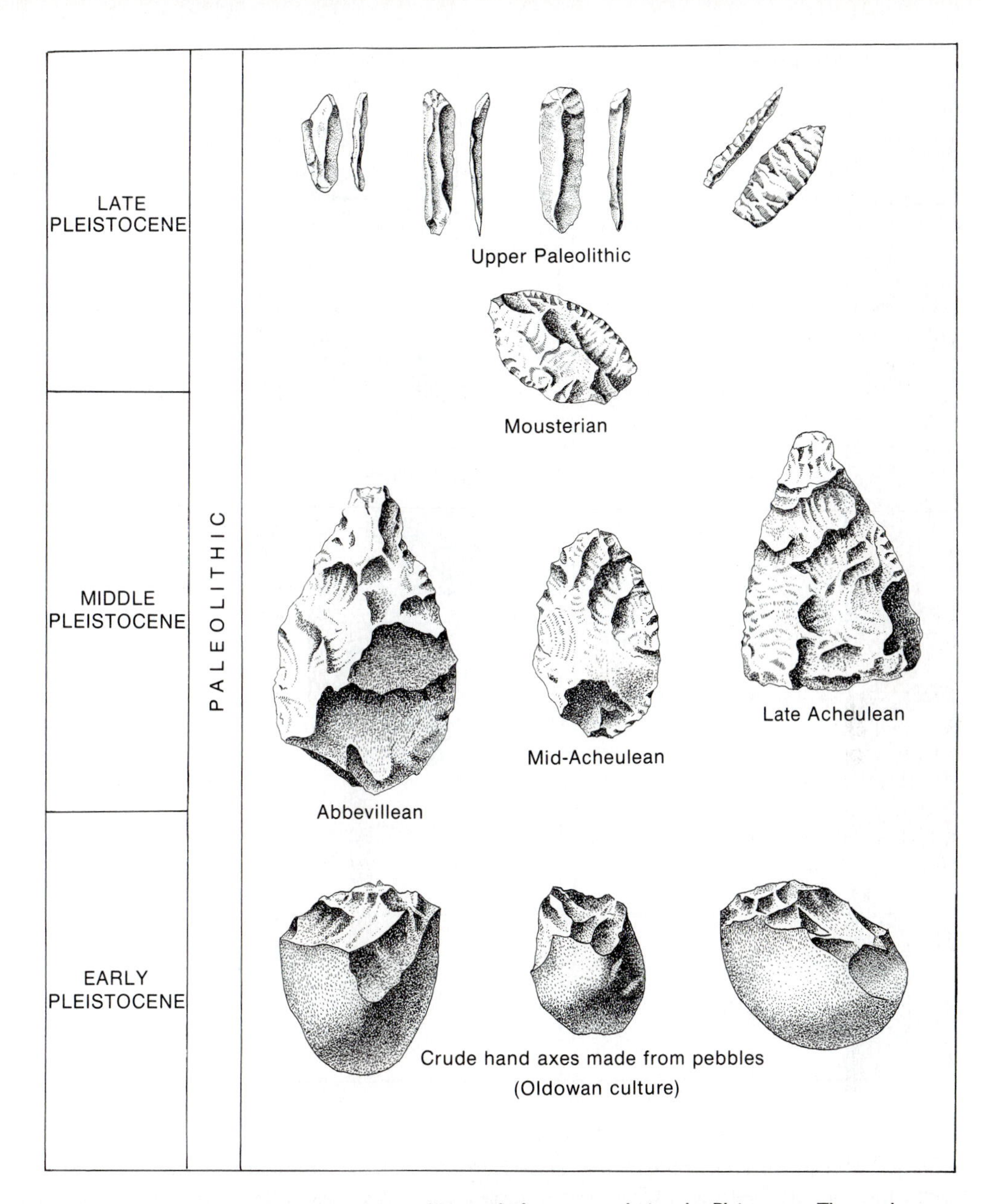

Figure 16–20 Progressive improvement in making tools from stone during the Pleistocene. The crude stone tools of the Early Pleistocene were produced by Australopithecines. *Homo erectus* produced the better shaped tools of the Middle Pleistocene. The Upper Paleolithic tools included carefully chipped blades and points. The next stage, not shown here, would be the Neolithic, characterized by refined and polished tools of many kinds.

food but as part of a ritual system. Anthropologists do not know if *Homo erectus* spoke a distinctive language, wore clothes, or built dwellings. There is vague evidence that they hunted together in bands. What appear to be pithecanthropine slaughter sites have been found in Europe. Unfortunately, these sites did not yield a single bone of *Homo erectus*.

HOMO SAPIENS

The Neanderthals

From the *Homo erectus* stage of the Middle Pleistocene (Fig. 16–21), it is only a short step to Late Pleistocene *Homo neanderthalensis* (Fig.

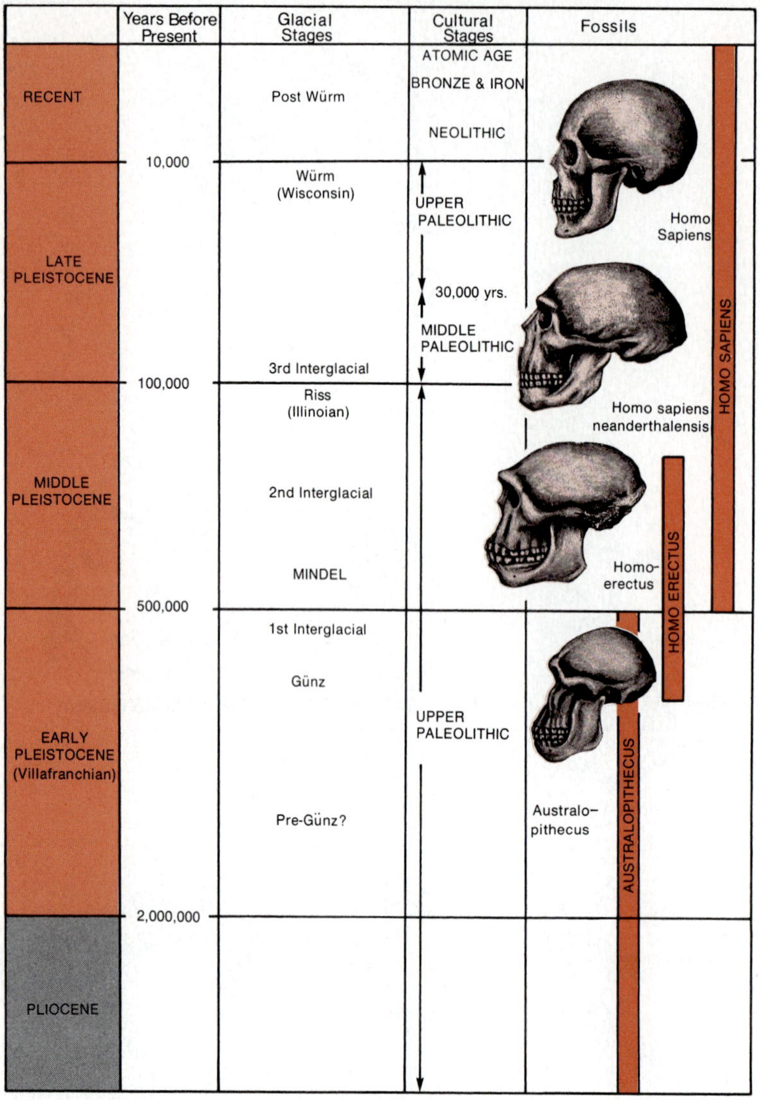

	Years Before Present	Glacial Stages	Cultural Stages	Fossils
RECENT		Post Würm	ATOMIC AGE BRONZE & IRON NEOLITHIC	Homo Sapiens
LATE PLEISTOCENE	10,000	Würm (Wisconsin) 3rd Interglacial	UPPER PALEOLITHIC 30,000 yrs. MIDDLE PALEOLITHIC	Homo sapiens neanderthalensis
MIDDLE PLEISTOCENE	100,000	Riss (Illinoian) 2nd Interglacial MINDEL		Homo-erectus
EARLY PLEISTOCENE (Villafranchian)	500,000	1st Interglacial Günz Pre-Günz?	UPPER PALEOLITHIC	Australo-pithecus
PLIOCENE	2,000,000			

HOMO SAPIENS · HOMO ERECTUS · AUSTRALOPITHECUS

Figure 16–21 Chronologic chart of Pleistocene fossil humans.

16–22). The initial specimen of this early human was found in the Neander Valley near Dusseldorf, Germany. Subsequently, sufficient additional fossils have been found to indicate that the **Neanderthal** people ranged across the entire expanse of the Old World. With their heavy brow ridges and prognathous, chinless jaws, Neanderthals have become the very personification of the "cave man." Indeed, the face seems a brutish carryover from Middle Pleistocene. However, in other features the neanderthaloids were quite modern. Below the neck, their skeletons matched our own, and their brain size equalled or exceeded that of present humans. Thus, the once popular depiction of Neanderthal as a bent-kneed, flat-footed, intoed, bull-necked brute with a curved back is false. Re-examination of the skeleton on which the original restoration had been based revealed that it was that of an elder individual, between 60 and 70 years old, who had been severely afflicted with osteoarthritis.

Most of the so-called classic Neanderthals were sturdy men of small stature that had adapted to life in the cold climates near the edge of the ice sheet. The relatively short limbs and bulky torso may have been an advantage in helping them to conserve body heat. Neanderthals preferred to live in caves and successfully hunted many contemporary cold-tolerant mammals, including cave bears, mammoths, woolly rhinoceroses, reindeer, bison, and fierce ancestors of modern cattle known as **aurochs.** They were not at all devoid of culture and

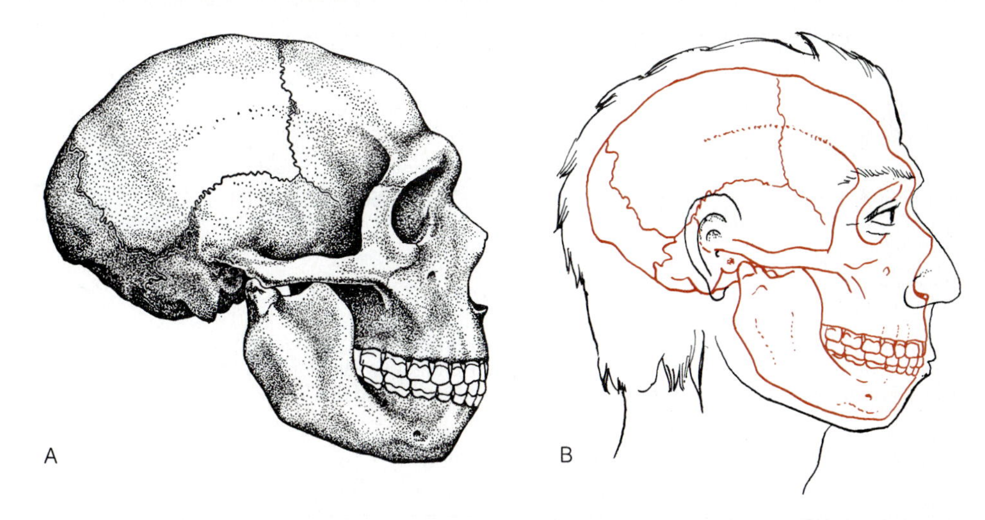

Figure 16–22 *Homo neanderthalensis.* (A) Lateral view of skull; (B) Partial restoration.

manufactured a variety of stone spear points, scrapers, borers, knives, and saw-edged tools. In addition, Neanderthals made ample use of fire and apparently could ignite one at will in the hearths they excavated in the floors of caves. There is evidence that they also constructed shelters of skins, sticks, and bones in areas where caves may not have been available. Perhaps as a hunting ritual, these people killed cave bears and stacked the skulls carefully in chests constructed of stones. That they also pondered the nature of death and believed in an afterlife is indicated by their custom of burying artifacts along with the dead.

Cro-Magnon

About 34,000 years ago, during the fourth glacial stage, humans closely resembling modern Europeans moved into regions inhabited by the Neanderthals and in a short span of time completely replaced them, probably by tribal warfare and by competition for hunting grounds. The new breed, formerly designated *Homo sapiens sapiens* but informally dubbed **Cro-Magnon** (Fig. 16–23), were mostly taller than their predecessors, had a more vertical brow, and had a decided projection to the chin. In short, Cro-Magnon's bones were

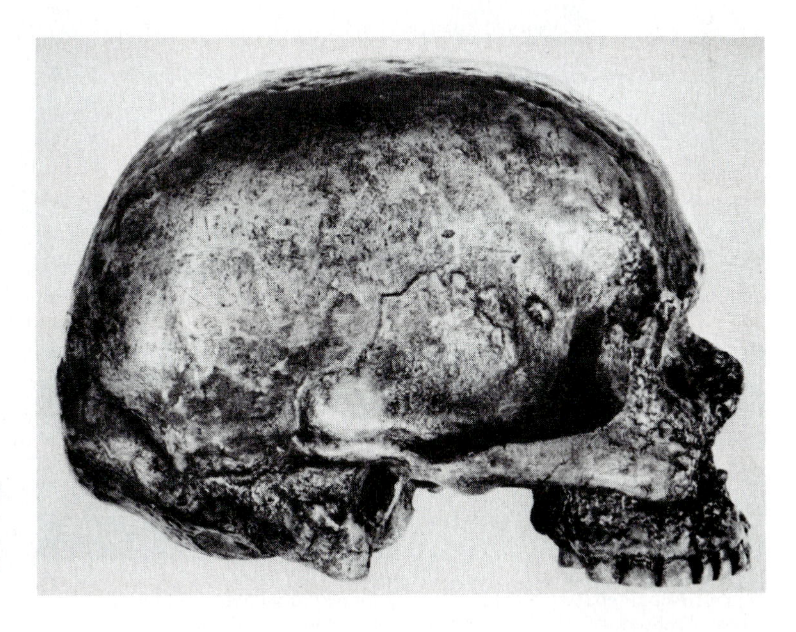

Figure 16–23 Lateral view of the skull of Cro-Magnon (*Homo sapiens*). Note absence of heavy supraorbital ridges. (Photograph of Wenner-Gren Foundation replica by David G. Gantt.)

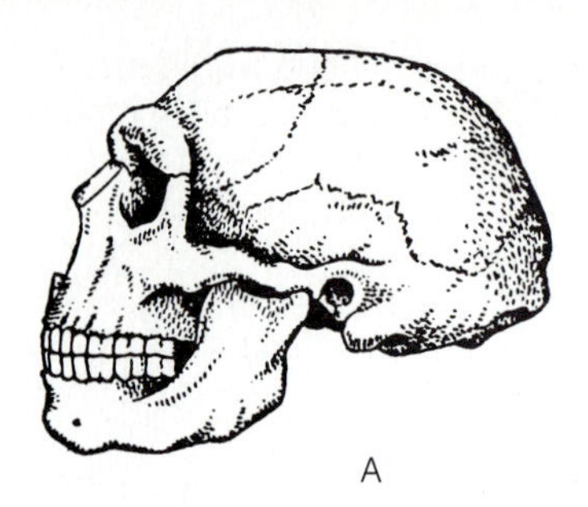

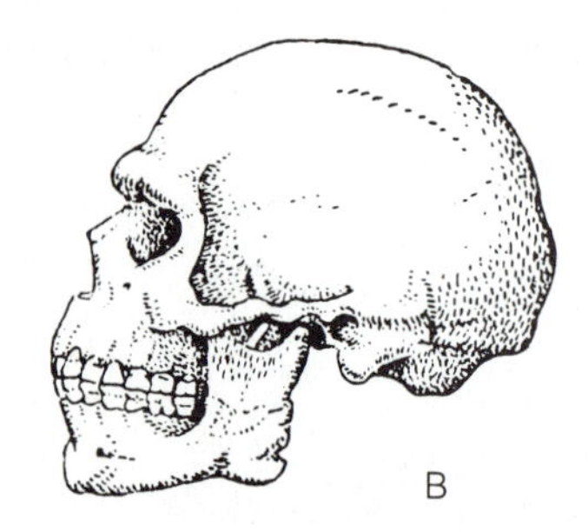

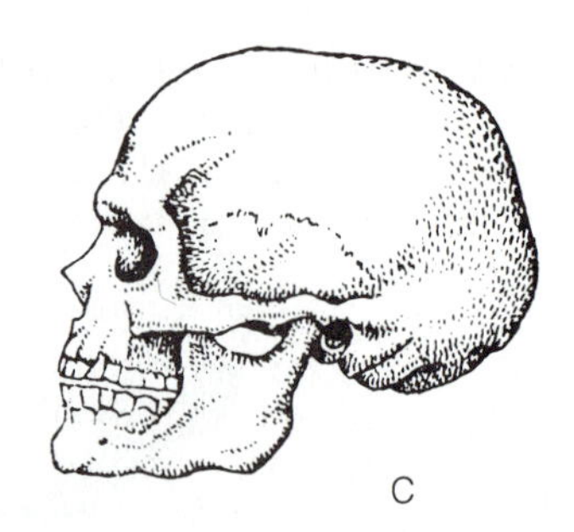

Figure 16–24 Comparison of the skulls of *Homo neanderthalensis.* (A) An intermediate form; (B) excavated from the rock shelter of Skhūl on the slopes of Mount Carmel in Palestine; and (C) Cro-Magnon. The Skhūl population was apparently intermediate between classic Neanderthal and Cro-Magnon.

modern, and anthropologists have recognized definite Cro-Magnon skull types among today's western and northern Europeans, as well as North Africans and native Canary Islanders.

The modern types of *Homo sapiens* did not appear suddenly on the evolutionary scene. Changes in dentition and supporting facial architecture were not an "overnight" occurrence. Transitional forms between Neanderthal and Cro-Magnon do exist (Fig. 16–24) and include fossil skulls from Palestine, South Africa, Germany, and Czechoslovakia. These forms vary among themselves but generally show less robust brow ridges, the beginnings of the modern chin, and less pronounced muscle markings than are evident in typical Neanderthal specimens. The transitional forms seem to have evolved outside Europe and, after they had reached the Cro-Magnon stage, migrated into Europe and Asia during a temporary regression of the ice sheets about 35,000 years ago.

The cultural traditions of the Neanderthals appear to have been continued and further devel-

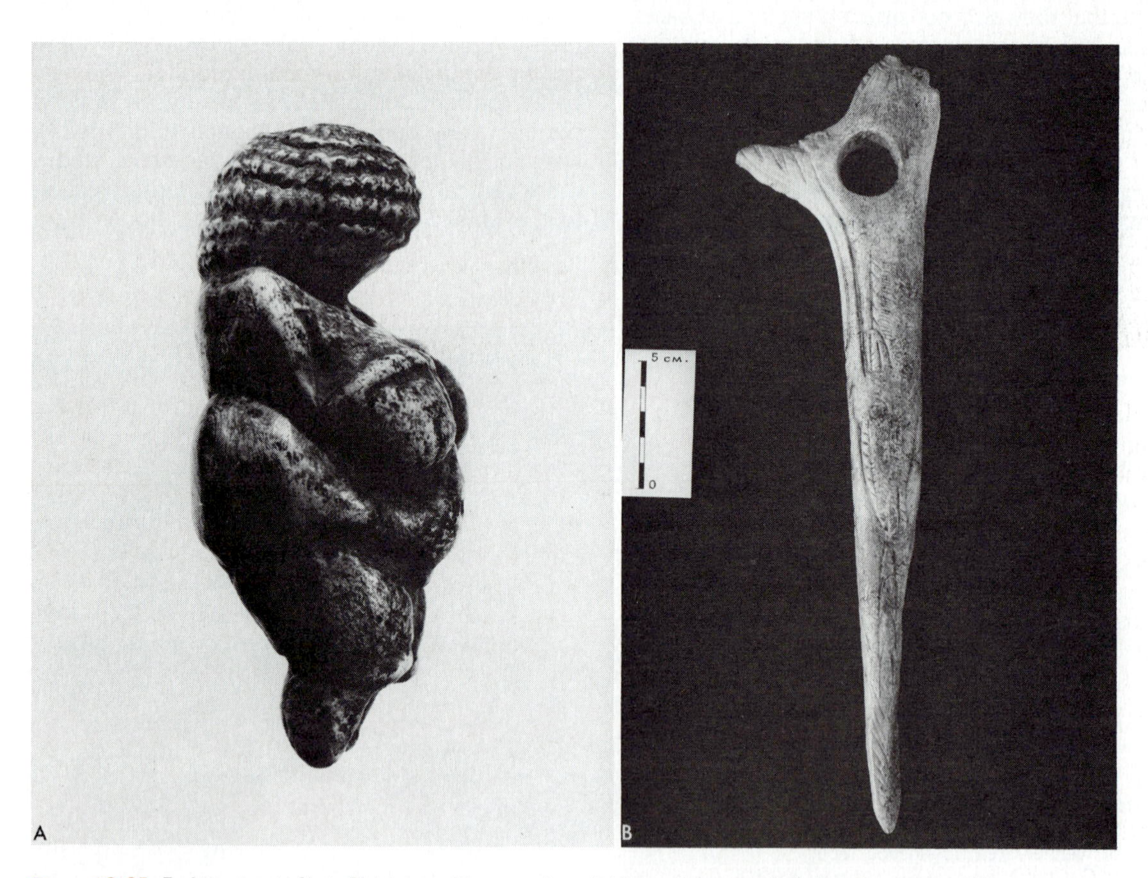

Figure 16–25 Prehistoric art of Late Pleistocene *Homo sapiens.* (A) Venus figure originally found in Austria ("The Venus of Willendorf"). (B) Thong-stropper used either to work hide thongs or to straighten arrow shafts. Tool is made from an antler. (Courtesy of Musée de l'Homme, Paris, France.)

oped by Cro-Magnon. The variety and perfection of stone and bone tools were increased. Finely crafted spear points, awls, needles, scrapers, and other tools are found in caves once inhabited by these people. These caves also retain splendid paintings and drawings on their walls and ceilings. Carvings and sculptures of obese women (Fig. 16–25), probably used in fertility rites, were produced from fragments of bone or ivory, as were small, elegant engravings and statues of mammoths and horses. Cro-Magnon people enjoyed wearing body ornaments and frequently fashioned necklaces from pieces of ivory, shells, or teeth. Burial of the dead became a truly elaborate affair. Hunters were buried with their weapons and children with their ornaments. This concern for an afterlife, and the sense of self-awareness that resulted in art and complex ritual, suggests that at least the beginning of the age of the philosopher had arrived.

Through most of his early history, *Homo sapiens* was a wandering hunter and gatherer of wild edible plants. However, about 10,000 to 15,000 years ago, near the beginning of the Holocene Epoch, tribes began to domesticate animals and cultivate plants. They learned to grind their tools to unprecedented perfection and to make utensils of fired clay. With more reliable sources of food, permanent settlements were developed, and individuals, spared the continuous demands of searching for food, were able to build and improve their cultures. Languages improved, and symbols developed into forms of writing. The era of recorded history began.

Early Humans in the Americas

The fossil record for hominids in the New World is discontinuous and often ambiguous. More often than is desirable, anthropologists must trace human presence on the basis of fragments of tools or weapons rather than actual skeletal remains. These clues, however, suggest that humans wandered into Alaska during the late phase of the Wisconsin glacial stage, when the Bering Straits provided a dry land bridge. It is likely that tribal groups seeking new hunting areas then drifted southeastward across the less mountainous parts of Alaska, Canada, and western United States. Precisely when these migrations occurred is difficult to determine and often is hotly debated. Very crude flints, tentatively reported to be artifacts, have been found in alluvial fan sediments in California and may be up to 40,000 years old. Charred bones of a dwarf mammoth believed to have been deliberately roasted in a campfire have been found at Santa Rosa Island off the coast of California. Carbon 14 dating indicates that the bones are about 28,000 years old. At Tule Springs, Nevada, fire pits determined to be 24,000 years old have been found in questionable association with flaked flints. In Mexico, an obsidian blade was found under a log that was dated at about 23,000 years.

There are a number of more recent sites that yield indirect evidence of **paleoindians.** Projectile points are found in definite association with extinct Late Pleistocene elephants and bison in the western United States. Sites near Clovis and Sandia, New Mexico, have yielded the distinctive tools of two separate cultures that may have lasted from about 13,000 to 11,000 years ago. A somewhat younger group of paleoindians lived from about 11,000 to 9000 years ago. This group has been named the Folsom Culture. Folsom people manufactured short, finely flaked, and fluted points that were mounted on shafts. These flints have been found in kill sites in association with extinct species of bison.

The search for skeletal parts of early man in America has provided fewer discoveries than has the search for his artifacts. However, there have been a few significant finds. A skull unearthed near Laguna Beach, California, has been dated by radiocarbon techniques as between 15,680 and 18,620 years old. Charred human bones from a cave on Marmes Ranch in southeastern Washington state have been dated at 11,500 to 13,000 years old. A projectile point, a bone needle, and the roasted bones of rabbits, deer, and elk have been found in association with the remains of **"Marmes Man."** The oldest known human remains from South America consist of a portion of a jaw and teeth from Guitarrero Cave in the north part of Peru. These remains are believed to be about 12,600 years old. Other Pleistocene diggings in Florida, Mississippi, California, Texas, Venezuela, and Argentina clearly indicate the presence of humans in the Americas prior to 11,000 years ago.

Summary

During the early epochs of the Cenozoic, the main groups of placental mammals were differentiated. One group, the probable descendants of primitive insectivores, took to the trees of dense Early Cenozoic forests, where food was plentiful and natural enemies few. The earliest of these primates have been named prosimians, literally "premonkeys." They were widespread during the Paleocene and Eocene but decreased during the Oligocene Epoch. One reason for that decline among prosimians was the appearance and differentiation of monkeys and apes. Looking back over Early Cenozoic primate evolution, we can see the beginnings of the transformation of snouty, squirrel-like prosimians into progressive anthropoids. The head was reshaped and rounded, the muzzle was lost, and the face was flattened. The eyes were moved forward to provide good binocular vision, not only for quickly estimating the distance of nearby moving prey but also for accurately gauging the distance to the next bough. Manual dexterity became important for hunting, feeding, and gripping branches. Fore limbs with clawed digits became arms and hands equipped with an opposable thumb and an improved sense of touch. Swinging from branch to branch by the arms paved the way toward the erect posture that would be exploited at a later stage in primate history.

During the Miocene, a group of apes had appeared that may be the ancestors of both modern apes and the hominid line. They are called dryopithecines. At the end of the Miocene, a creature called *Ramapithecus* arose, very likely from dryopithecine stock. This progressive animal, with rounded dental arcade and human-like teeth, was clearly a human ancestor.

The dryopithecines are in many ways the evolutionary products of changing Late Cenozoic environments. Dense, jungle-like forests that had cloaked entire regions of Africa were being replaced by grassy, parklike habitats called savannahs. The semierect dryopithecines and later the ramapithecines were quick to respond with a restructuring of the axial skeleton, pelvic girdle, and hind feet. With this improved, erect stability, the hands were now completely freed for manipulative purposes. Probably these creatures took advantage of the protection inherent in unified group action and may have had the ability to communicate, at least to the same degree as do modern apes. All of these functions led ultimately to greater development of the cerebral cortex, the part of the brain associated with dexterity, memory, learning, and thought. Great reproductive advantage must have accrued to those species that made the best use of their brains. Although growth in brain size was slow until the end of the Pliocene, it accelerated phenomenally during the Pleistocene.

The australopithecines, or southern ape men, appeared on earth in the Pliocene about 3.5 to 4.0 million years ago. Their cranial capacity was about 600 cc, only about a third as large as that of humans. However, their dentition closely resembled that of modern people. Australopithecines were upright in stature and used crude tools. There is no evidence that they had discovered the use of fire. Structurally, the australopithecines were close to the grade of primate evolution that might warrant their being considered members of the genus *Homo*. In fact, they seem to have been either the ancestors or the contemporaries of larger-brained creatures currently being designated as species of *Homo*.

The next stage of human evolution witnessed the appearance of *Homo erectus*, about 500,000 to a million years ago. These erect, walking primitive men had a brain capacity that ranged from 775 to 1300 cc. By Middle Pleistocene, *Homo erectus* had spread widely through Europe and Africa and had progressively improved his tool-making abilities. These people also knew the use of fire. Fire was of immense importance to Pleistocene humans. Because of the light that fire provides, the length of time during which they could effectively accomplish tasks was increased. Fire provided protection during the night and permitted *Homo erectus* to occupy cave shelters with less danger from great bears and other predators. (This alone has been a boon to anthropologists, in that it decreases the number of places that need be investigated for fossils.) Most important, of course, is that fire provided warmth, thus enabling an animal that evolved in the tropics to extend its range into cold climates.

The Late Pleistocene was marked by the appearance of the very distinctive and famous Neanderthal people. Remains of the Neanderthals have been discovered in dozens of European sites, as well as the Middle East and Africa. The Neanderthal skull was more massive than our own, had a less prominent forehead and chin, and had more prominent ridges above the eyes. The dozens of skulls already collected indicate that the Neanderthal peoples varied as widely in characteristics as do modern human populations. Neanderthals had brains as large as or larger than modern humans. Their tools were more finely constructed than those of *Homo erectus*, they fashioned objects for aesthetic purposes, and they had a religious concern about afterlife.

Modern *Homo sapiens* appeared during the second interglacial stage of the Pleistocene. They continue down to the present day in the form of you and me. An early group, known as Cro-Magnon, replaced the Neanderthals in Europe and continued their cultural traditions. Ever finer tools and weapons in bone and flint were manufactured by these peoples. They exhibited remarkable artistic skills, painting pictures of the animals vital to their survival on the walls of their caves and carving opulent Venus-type statues to encourage fertility.

Humans came to the Americas at a rather late

date. However, artifacts and sparse skeletal remains indicate their presence somewhat over 20,000 years ago. Their spear heads are sometimes found among the remains of the extinct large mammals that they hunted.

Epilogue

One of the many lessons of historical geology is that the advent of *Homo sapiens* represents one recent and momentary event along the sinuous, branching, 500-million-year evolution of vertebrates. We humans are linked by similarities of structure and body chemistry to lungfish struggling out of stagnating Devonian pools, to the small, shrewlike precursors of the primate order, and to the axe-carrying hunters that labored along the margins of great ice sheets. The descendants of those and other Pleistocene hunters have emerged as the dominant species of higher life presently on this planet. Like other animals, *Homo sapiens* has been shaped by the combined powers of genetic change and environment. However, our species has quickly become a pervasive force in modifying the very physical and biologic environment from which it is derived. Humanity has come to rely heavily on science to improve its lot but has had great difficulty in finding ways to manage the resulting technology and to control the burgeoning problems arising from too many humans and too few resources. Survival of *Homo sapiens* will depend largely on how wisely these problems are solved. If the species fails, there may be other vertebrates to continue along different evolutionary pathways.

Questions for Review

1. Early arboreal primates were characterized by close-set eyes and grasping extremities. What function, other than simply moving about in the trees, might these adaptations have performed?
2. In general, what modifications occur in the skulls of primates in the course of evolution from a tree shrew–like primate to *Homo sapiens*?
3. By what route and at what stage in the Pleistocene did *Homo sapiens* enter the New World?
4. During what geologic period do we find the earliest known remains of primates?
5. What physical characteristics of flint and chert result in their being useful in the manufacture of spear points and axes by early humans?
6. What is the evolutionary importance of the Miocene primate *Ramapithecus*?
7. What advantage is there in a fossil site that consists of fossil-bearing sedimentary layers interspersed with layers of volcanic ash or lava flows?
8. What distinctly human-like traits were possessed by *Australopithecus*? What apelike characteristics did this creature retain?
9. With the advent of pithecanthropines, the trend toward increase in brain size was initiated. What related anatomic changes must have been synchronous with increases in cranial capacity?
10. What features of Cro-Magnon peoples differentiate them from Neanderthals?

Terms to Remember

anthropoid
auroch
australopithecine
Cro-Magnon
diastema
dryopithecine
"Marmes Man"
Neanderthal
paleoindian
pongid
Proconsul
prognathous
prosimian

Supplemental Readings and References

Brace, C. L. 1967. *The Stages of Human Evolution.* Englewood Cliffs, N.J., Prentice-Hall.

Campbell, B. G. 1966. *Human Evolution.* Chicago, Aldine Publishing Co.

Cartmill, M. 1974. Rethinking primate origins. *Science* 184:436–443.

Early Man in America. 1973. *Readings from Scientific American.* San Francisco, W. H. Freeman Co.

Howell, F. C. 1969. Remains of Hominidae from Pliocene/Pleistocene formations in the Lower Omo Basin, Ethiopia. *Nature* 223:1234.

Isaac, G. L., Leakey, R. E. F., and Behrensmeyer, A. K. 1971. Archeological traces of early hominid activities east of Lake Rudolf, Kenya. *Science* 173:1129–1133.

Johanson, D. C., and Edey, M. A. 1981. *Lucy, the Beginnings of Humankind.* New York, Simon and Schuster.

Johanson, D. C., and White, T. D. 1979. A systematic assessment of early African hominids. *Science* 203(4378):321–330.

Le Gros Clark, W. E. 1964. The Fossil Evidence for Human Evolution (2d ed.). Chicago, University of Chicago Press.

Leakey, R. E. F. 1976. Hominids in Africa. *American Scientist* 64:174–178.

Loomis, W. G. 1967. Skin pigmentation and vitamin D synthesis in man. *Science* 157:501–506.

Pilbeam, D. 1972. *The Ascent of Man.* New York, Macmillan.

Rosen, S. I. 1974. *Introduction to the Primates.* Englewood Cliffs, N.J., Prentice-Hall.

Simons, E. L., 1972. *Primate Evolution.* New York, Macmillan Co.

Appendix A
A Classification of Living Things

Early students of biology found it convenient to divide all organisms into two great realms or kingdoms, designated the Animalia and Plantae. However, by the late nineteenth century, many biologists suggested that perhaps a third kingdom—the Protista—should be established for certain single-celled organisms that seemed to be neither plant nor animal. Even this was considered inadequate by some biologists, for, as this three-kingdom classification was gaining adherents, evidence was accumulating that clearly indicated the need for still further major groupings. Studies of unicellular organisms had revealed two quite different forms of cell structures. There were, for example, prokaryotic unicellular organisms, like bacteria and blue-green algae, that lacked a cell nucleus and possessed other traits that set them clearly apart from eukaryotic unicellular organisms, which had a true nucleus, well-defined chromosomes, and organelles. Any classification that sought to categorize organisms according to similarity of origin and fundamental differences could not ignore the contrast between eukaryotes and prokaryotes. Finally, it was recognized that the fungi deserved special taxonomic consideration also, for they are dependent upon a supply of organic molecules in their environment, as are animals, yet they absorb their food through cell membranes, as do plants. The fungi appear to have had an evolutionary radiation quite distinct from the other major groups.

In an effort to account for these differences and better represent the evoutionary relationships of organisms, R. H. Whittaker of Cornell University proposed a five-kingdom system of classification.* In this classification, plants, fungi, and animals are regarded as distinct in terms of having specialized for different modes of nutrition, photosynthesis, absorption, and ingestion. The Protista is composed of unicellular eukaryotic organisms, and the Monera include the simplest of organisms that are inferred to be similar to the primitive forms of life from which other kingdoms evolved. A modified version of Whittaker's classification, which is steadily gaining proponents, is presented here.

KINGDOM MONERA .... Prokaryotes

PHYLUM CYANOPHYTA .... Blue-green algae
PHYLUM MYXOBACTERIAE .... Unicellular or filamentous gliding bacteria
PHYLUM SCHIZOPHYTA. .... True bacteria
PHYLUM ACTINOMYCOTA .... Certain branching, filamentous bacteria
PHYLUM SPIROCHAETAE .... Spirochetes

*Whittaker, R. H. 1969. New concepts of kingdoms of organisms. *Science* 163:150–160.

KINGDOM PROTISTA

KINGDOM PROTISTA	Solitary or colonial unicellular eukaryotic organisms that do not form tissues
PHYLUM EUGLENOPHYTA	Euglenoid organisms
PHYLUM XANTHOPHYTA	Yellow-green algae
PHYLUM CHRYSOPHYTA	Golden brown algae, diatoms, and coccolithophorids
PHYLUM PYRROPHYTA	Dinoflagellates and cryptomonads
PHYLUM HYPHOCHYTRIDIOMYCOTA	Hypochytrids
PHYLUM PLASMODIOPHOROMYCOTA	Plasmodiophores
PHYLUM SPOROZOA	Sporozoans (parasitic protists)
PHYLUM CNIDOSPORIDIA	Cnidosporidians
PHYLUM ZOOMASTIGINA	Animal flagellates, protozoa that have whiplike cytoplasmic protrusions (flagellae)
PHYLUM SARCODINA	Rhizopods, protozoa with pseudopodia for locomotion
PHYLUM CILIOPHORA	Ciliates and suctorians, movement accomplished by beating of cilia (adult suctorians are attached to objects)

KINGDOM PLANTAE

DIVISION* RHODOPHYTA	Red algae, usually marine, multicellular
DIVISION PHAEOPHYTA	Brown algae, multicellular often with large bodies, as in seaweeds and kelps
DIVISION CHLOROPHYTA	Green algae
DIVISION CHAROPHYTA	Stoneworts
DIVISION BRYOPHYTA	Liverworts, mosses, and hornworts
DIVISION PSILOPHYTA	Extinct leafless, rootless, vascular plants
DIVISION LYCOPODOPHYTA	Club mosses, with simple vascular systems and small leaves, including scale trees of Paleozoic (lycopsids)
DIVISION EQUISETOPHYTA (ARTHROPHYTA)	Horsetails, scouring rushes, including sphenopsids such as *Calamites* and *Annularia* of the Late Paleozoic
DIVISION POLYPODIOPHYTA	The true ferns or pteropsids
DIVISION PINOPHYTA	The "gymnosperms" include conifers, cycads, and many evergreen plants; no true flowers
Class Lyginopteriodopsida	The seed ferns, known from fossils of the Late Paleozoic and including such forms as *Neuropteris* and *Glossopteris.*
Class Bennettitopsida	The extinct cycadeoids
Class Cycadopsida	Cycads
Class Ginkgoopsida	Ginkgoes
Class Pinopsida	Conifers, as well as the extinct *Cordaites*

*In botany, it is conventional to use the terms "division," "subdivision," and so on in place of "phylum" and "subphylum."

Class Gnetopsida	Certain climbing shrubs and small tropical trees
DIVISION MAGNOLIOPHYTA	Flowering plants or "angiosperms"; seeds enclosed in an ovary
Class Magnoliopsida	Dicotyledonous plants; embryos with two cotyledons or seed leaves; leaves with netlike veins
Class Liliopsida	Monocotyledonous plants; embryos with only one seed leaf; leaves with parallel veins

KINGDOM FUNGI

DIVISION MYXOMYCOPHYTA	Slime molds
DIVISION EUMYCOPHYTA	True fungi

KINGDOM ANIMALIA

PHYLUM MESOZOA	Mesozoans
PHYLUM PORIFERA	Sponges; includes forms with calcareous spicules, siliceous spicules, and proteinaceous spicules; may also include the extinct stromatoporoids
PHYLUM ARCHAEOCYATHA	Extinct spongelike organisms
PHYLUM COELENTERATA (CNIDARIA)	Jellyfishes, corals; radially symmetric, aquatic, with body wall of two layers of cells, in the outer of which are stinging cells
Class Hydrozoa	Hydra-like animals
Class Scyphozoa	True jellyfishes
Class Anthozoa	Corals and sea anemones
Subclass Zoantharia	Hexacorals of modern seas
Subclass Rugosa	Paleozoic tetracorals
Subclass Tabulata	Paleozoic tabulate corals
PHYLUM CTENOPHORA	Modern comb jellies or "sea walnuts." Not known as fossils
PHYLUM PLATYHELMINTHES	Flatworms
PHYLUM NEMERTEA	Proboscis worms
PHYLUM NEMATODA	Roundworms
PHYLUM ACANTHOCEPHALA	Hook-headed worms
PHYLUM NEMATOMORPHA	Horsehair worms
PHYLUM ROTIFERA	Small, wormlike animals with a circle of cilia on the head
PHYLUM GASTROTRICHA	Small, wormlike animals resembling rotifers but lacking circle of cilia
PHYLUM BRYOZOA	The bryozoans, sometimes considered two phyla: *Entoprocta* and *Ectoprocta*
PHYLUM BRACHIOPODA	Marine animals with two parts (valves) to their shell (dorsal and ventral)

Class Inarticulata	Primitive brachiopods having phosphatic or chitinous valves, lacking hinge
Class Articulata	Advanced calcareous brachiopods with valves that are hinged
PHYLUM PHORONIDA	Wormlike marine animals that secrete and live within a leathery tube
PHYLUM ANNELIDA	The segmented worms
PHYLUM ONYCOPHORA	Rare tropical animals considered intermediate between annelids and arthropods
PHYLUM ARTHROPODA	Segmented animals with jointed appendages
SUBPHYLUM TRILOBITA	Trilobites, common marine arthropods of the Paleozoic Era
SUBPHYLUM CHELICERATA	
Class Xiphosura	Horseshoe crabs
Class Eurypterida	Eurypterids
Class Pycnogonida	Sea spiders
Class Arachnida	Scorpions, spiders, ticks, and mites
SUBPHYLUM CRUSTACEA	Lobsters, crabs, barnacles, ostracods
SUBPHYLUM LABIATA	
Class Chilopoda	Centipedes
Class Diplopoda	Millipedes
Class Insecta	The 24 orders of insects
PHYLUM MOLLUSCA	Unsegmented, soft-bodied animals, usually protected by a shell
Class Amphineura	Chitons, marine forms with shell composed of eight segments
Class Scaphopoda	Tusk shells, curved tubular shells open at both ends
Class Gastropoda	Snails, abalones, asymmetric animals with single spiral conch or no shell
Class Pelecypoda	Shells of two valves (right and left); includes clams, mussels, oysters, scallops
Class Cephalopoda	Marine animals with tentacles around head and well-developed eyes and nervous system
ORDER NAUTILOIDEA	Nautiloids; cephalopods with simple suture lines
ORDER AMMONOIDEA	Ammonoids; cephalopods with complexly folded sutural lines
ORDER BELEMNOIDEA	Belemnites
ORDER SEPIOIDEA	Cuttlefishes
ORDER TEUTHOIDEA	Squids
ORDER OCTOPODA	Octopi
PHYLUM POGONOPHORA	Beard worms
PHYLUM CHAETOGNATHA	Arrow worms
PHYLUM ECHINODERMATA	Marine animals that are radially symmetric as adults (bilateral as larvae), have calcareous, spine-bearing plates and unique water vascular system
Class Asteroidea	Starfishes
Class Ophiurodea	Brittle stars and serpent stars

Class Echinoidea . . . . Sea urchins and sand dollars
Class Holothuroidea. . . . . Sea cucumbers
Class Crinoidea . . . . Sea lilies and feather stars
Class Blastoidea. . . . . The extinct blastoids of the Paleozoic

PHYLUM HEMICHORDATA (PROTOCHORDATA) . . . . The acorn worms; larval forms resemble echinoderm larva; adults have anterior proboscis connected by collar to wormlike body

PHYLUM CHORDATA . . . . Bilaterally symmetric animals with notochord, dorsal hollow neural tube, and gill clefts in the pharynx

SUBPHYLUM UROCHORDATA . . . . Sea squirts or tunicates; larval forms have notochord in tail region

SUBPHYLUM CEPHALOCHORDATA . . Branchiostoma ("Amphioxus"); small marine animals with fishlike bodies and notochord

SUBPHYLUM VERTEBRATA . . . . Animals with a backbone of vertebrae, definite head, ventrally located heart, and well-developed sense organs

Class Agnatha. . . . . Living lampreys and hagfish as well as extinct ostracoderms; agnatha lack jaws
Class Acanthodii. . . . . Primitive, extinct, spiny fishes with jaws
Class Placodermi . . . . Primitive, often armored, Paleozoic jawed fishes
Class Chondrichthyes . . . . Sharks, rays, skates, and chimaeras
Class Osteichthyes. . . . . The bony fishes
Subclass Actinopterygii . . . . Ray-finned fishes
Subclass Sarcopterygii . . . . Lobe-finned, air-breathing fishes
ORDER CROSSOPTERGII . . . . Lobe-finned fishes, ancestors of amphibians
ORDER DIPNOI . . . . Lungfishes
Class Amphibia . . . . Amphibians, the earth's earliest land-dwelling vertebrates; include extinct Labyrinthodontia of Late Paleozoic and Triassic
Class Reptilia . . . . Reptiles, reproducing with use of amniotic eggs
Subclass Anapsida . . . . Turtles, as well as the extinct aquatic mesosaurs and terrestrial stem reptiles called cotylosaurs
Subclass Synapsida . . . . Mammal-like reptiles, including sailback forms and therapsids
Subclass Euryapsida . . . . Extinct, generally marine reptiles, including plesiosaurs, ichthyosaurs, and placodonts
Subclass Diapsida . . . . Reptilian group that includes the extinct dinosaurs and crocodilians, lizards, snakes, and the modern tuatara
Class Aves. . . . . Birds; warm-blooded, feathered, and typically winged animals; primitive forms with reptilian teeth, modern forms toothless

Class Mammalia.................	Warm-blooded animals with hair covering; females with mammary glands that secrete milk for nourishing young
Subclass Eutheria...............	Primitive, extinct Triassic and Jurassic mammals
Subclass Prototheria.............	Monotremes such as the duck-billed platypus and spiny anteater. Egg-laying mammals
Subclass Allotheria..............	Extinct early mammals with multi-cusped teeth; multituberculates
Subclass Metatheria.............	Pouched mammals or marsupials
Subclass Eutheria...............	The placental mammals; young develop within uterus of female, obtain moisture via the placenta
ORDER INSECTIVORA	Primitive insect-eating mammals, including moles and shrews
ORDER DERMOPTERA............	The colugo
ORDER CHIROPTERA.............	Bats
ORDER PRIMATES...............	Lemurs, tarsiers, monkeys, apes, humans
ORDER EDENTATA	Living armadillos, anteaters, tree sloths; extinct glyptodonts and ground sloths
ORDER RODENTIA	Squirrels, mice, rats, beavers, porcupines
ORDER LAGOMORPHA............	Hares, rabbits, pikas
ORDER CETACEA...............	Whales and porpoises
ORDER CREODONTA	Extinct, ancient carnivorous placentals
ORDER CARNIVORA	Modern carnivorous placentals, including dogs, cats, bears, hyenas, seals, sea lions, walruses
ORDER CONDYLARTHA	Extinct ancestral hoofed placentals (ancestral ungulates)
ORDER AMBLYPODA.............	Extinct primitive ungulates
ORDER TUBULIDENTATA	Aardvarks
ORDER PHOLIDOTA	Pangolins
ORDER PERISSODACTYLA	Odd-toed hoofed mammals, including living horses, rhinoceroses, and tapirs; and extinct titanotheres and chalicotheres
ORDER ARTIODACTYLA	Even-toed hoofed animals, including living antelopes, cattle, deer, giraffes, camels, llamas, hippos, and pigs; and extinct entelodonts and oreodonts
ORDER PROBOSCIDEA	Elephants and extinct mastodons and mammoths
ORDER SIRENIA................	Sea cows

Appendix B

Formation Correlation Charts For Representative Sections

The following charts may be used to ascertain the approximate equivalence of some of the better known rock formations deposited during the Phanerozoic Era. The vertical scale does not indicate thickness but rather is an approximation of time. Time-rock terms are indicated on either side of the chart. The diagonal lines indicate formations missing by nondeposition or erosion. Fm = formation (generally including two or more lithologic types); Ss = sandstone; Sh = shale; Ls = limestone; Dol = dolostone; Qtzite = quartzite; Congl = conglomerate; Grp = group (two or more formations); Mem = member. Solid horizontal lines between the formations indicate well-established boundaries, whereas broken lines indicate that formation boundary is not firmly established.

CAMBRIAN

SERIES	STAGES	Southern Appalachian	Missouri, Ozark Region	Arbuckle, Wichita Mtns., Okla.	Sawback Range, British Columbia	House Range, Utah	Nopah Range, Calif.	Northern Wales	STAGES	SERIES
CROIXAN	Trempealeauan	Copper Ridge Dol	Eminence Fm; Potosi Dol	Butterfly Dol; Signal Mtn Fm	Lyell Fm	Notch Peak Fm		Lingula Flags: Dolgelly Beds	Trempealeauan	UPPER
CROIXAN	Franconian	Copper Ridge Dol	Elvins Grp: Doe Run Fm; Derby Fm; Davis Fm	Royer Dol; Fort Sill Lms; Honey Creek Fm	Lyell Fm	Notch Peak Fm; Orr Fm	Nopah Fm	Lingula Flags: Festiniog Beds	Franconian	UPPER
CROIXAN	Dresbachian	Conasauga Fm	Bonneterre Dol; Lamotte Ss	Reagan Ss	Sullivan Fm	Weeks Fm		Lingula Flags: Maentwrog Beds	Dresbachian	UPPER
ALBERTAN	Stages not defined	Conasauga Fm			Eldon Dol; Stephen Fm; Cathedral Dol; Ptarmigan Fm	Marjum Fm; Wheeler Sh; Swasey Ls; Whirlwind Fm; Dome Ls; Chisolm Sh; Howell Ls; Tatow Fm	Cornfield Sprs. Fm; Bonanza King Fm; Cadiz Fm	Menevian Beds: Clogan Shales; Cefn Goch Grit; Harlech Grits	Stages not defined	MIDDLE
WAUCOBAN	Stages not defined	Rome Fm; Shady Dol; Weisner Qtzite			Mount Whyte Fm; St. Piran Ss; Fort Mtn Ss	Pioche Fm; Prospect Mtn. Qtzite	Wood Canyon Fm; Stirling Qtzite	Harlech Grits	Stages not defined	LOWER

ORDOVICIAN

SERIES	STAGES	Taconic Area of Western Vt.	Northwestern New York	Virginia	Missouri	Oklahoma	Colorado	Utah	GREAT BRITAIN SERIES
CINCINNATIAN	Gamachian								ASHGILL
	Richmondian		Queenston Sh (Redbeds)	Juanita Fm	Girardeau Ls Orchard Ck Sh Thebes Ss Maquoketa Sh Fernvale Ls	Polk Ck Sh	Fremont Ls	Fish Haven Dol	
	Mays-villian		Oswego Ss						
	Edenian		Lorraine Grp Utica Sh	Martinsburg Fm					
CHAMPLAINIAN	Mohawkian: Trentonian	Snake Hill Sh Rysedorf Congl	Trenton Ls		Kimmswick Ls Decorah Grp	Big Fork Chert	Harding Ss		CARADOC
	Mohawkian: Black River	Normanskill Sh	Black River Ls	Bays Fm Edinburg Fm	Plattin Grp				
	Chazyan			Lincolnshire Ls Whistle Ck Sh New Market Ls	Joachim Ls St. Peter Ss Everton Fm	Womble Sh		Swan Peak Qtzite	LLANDEILO LLANVIRN
CANADIAN	Stages not yet established	Deepkill Sh Shaghticoke Sh		Knox Dol (upper beds)	Smithville Fm Powell Fm Cotter Dol Jefferson City Grp Roubidoux Fm Gasconade Dol	Blakely Ss Mazarn Sh Crystal Mtn Ss	Manitou Ls	Garden City Ls	ARENIG TREMADOC

SILURIAN

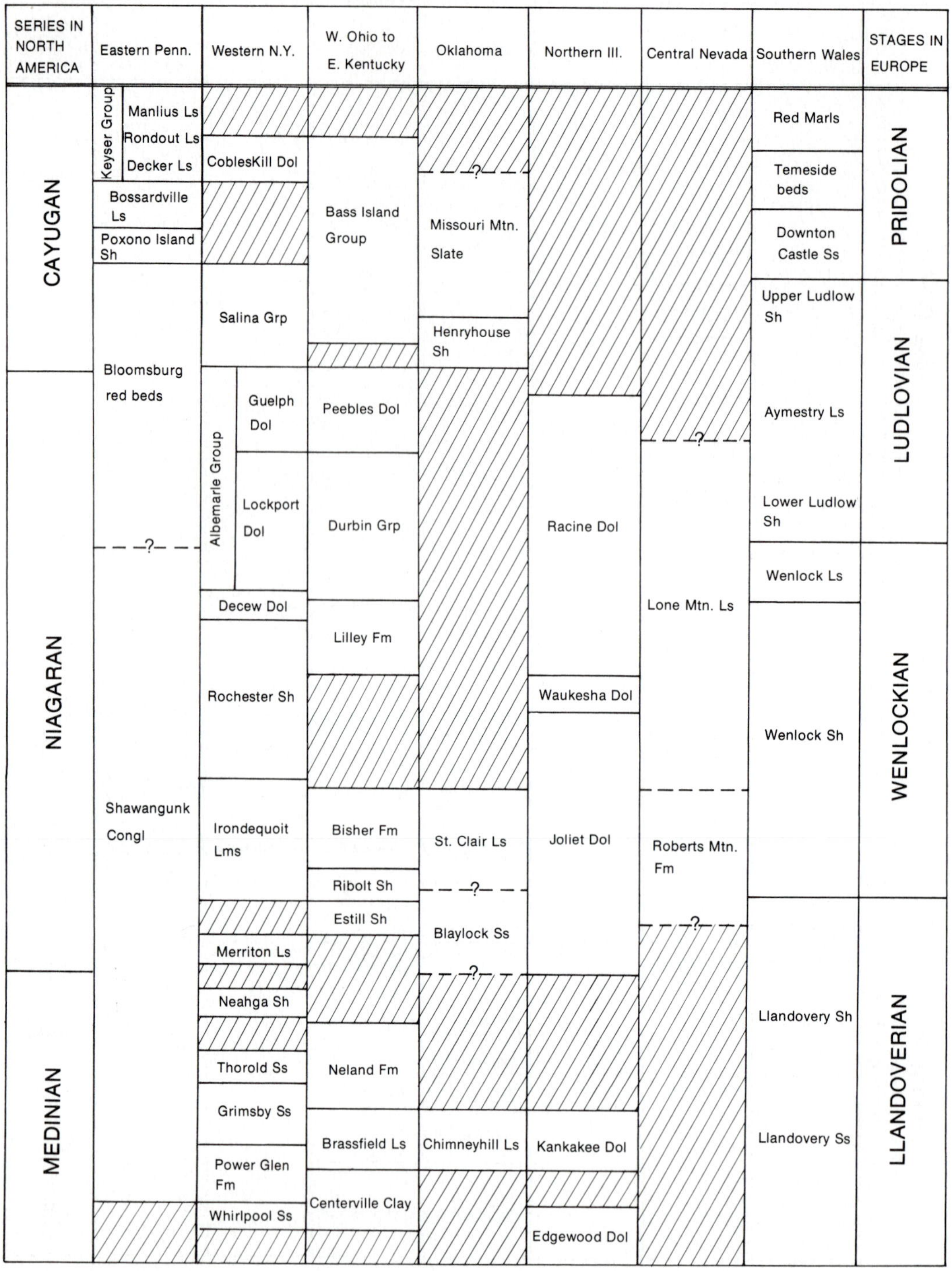

SERIES IN NORTH AMERICA	Eastern Penn.	Western N.Y.	W. Ohio to E. Kentucky	Oklahoma	Northern Ill.	Central Nevada	Southern Wales	STAGES IN EUROPE
CAYUGAN NIAGARAN MEDINIAN	Keyser Group: Manlius Ls, Rondout Ls, Decker Ls Bossardville Ls Poxono Island Sh Bloomsburg red beds Shawangunk Congl	Cobleskill Dol Salina Grp Albemarle Group: Guelph Dol, Lockport Dol Decew Dol Rochester Sh Irondequoit Lms Merriton Ls Neahga Sh Thorold Ss Grimsby Ss Power Glen Fm Whirlpool Ss	Bass Island Group Peebles Dol Durbin Grp Lilley Fm Bisher Fm Ribolt Sh Estill Sh Neland Fm Brassfield Ls Centerville Clay	Missouri Mtn. Slate Henryhouse Sh St. Clair Ls Blaylock Ss Chimneyhill Ls	Racine Dol Waukesha Dol Joliet Dol Kankakee Dol Edgewood Dol	Lone Mtn. Ls Roberts Mtn. Fm	Red Marls Temeside beds Downton Castle Ss Upper Ludlow Sh Aymestry Ls Lower Ludlow Sh Wenlock Ls Wenlock Sh Llandovery Sh Llandovery Ss	PRIDOLIAN LUDLOVIAN WENLOCKIAN LLANDOVERIAN

DEVONIAN

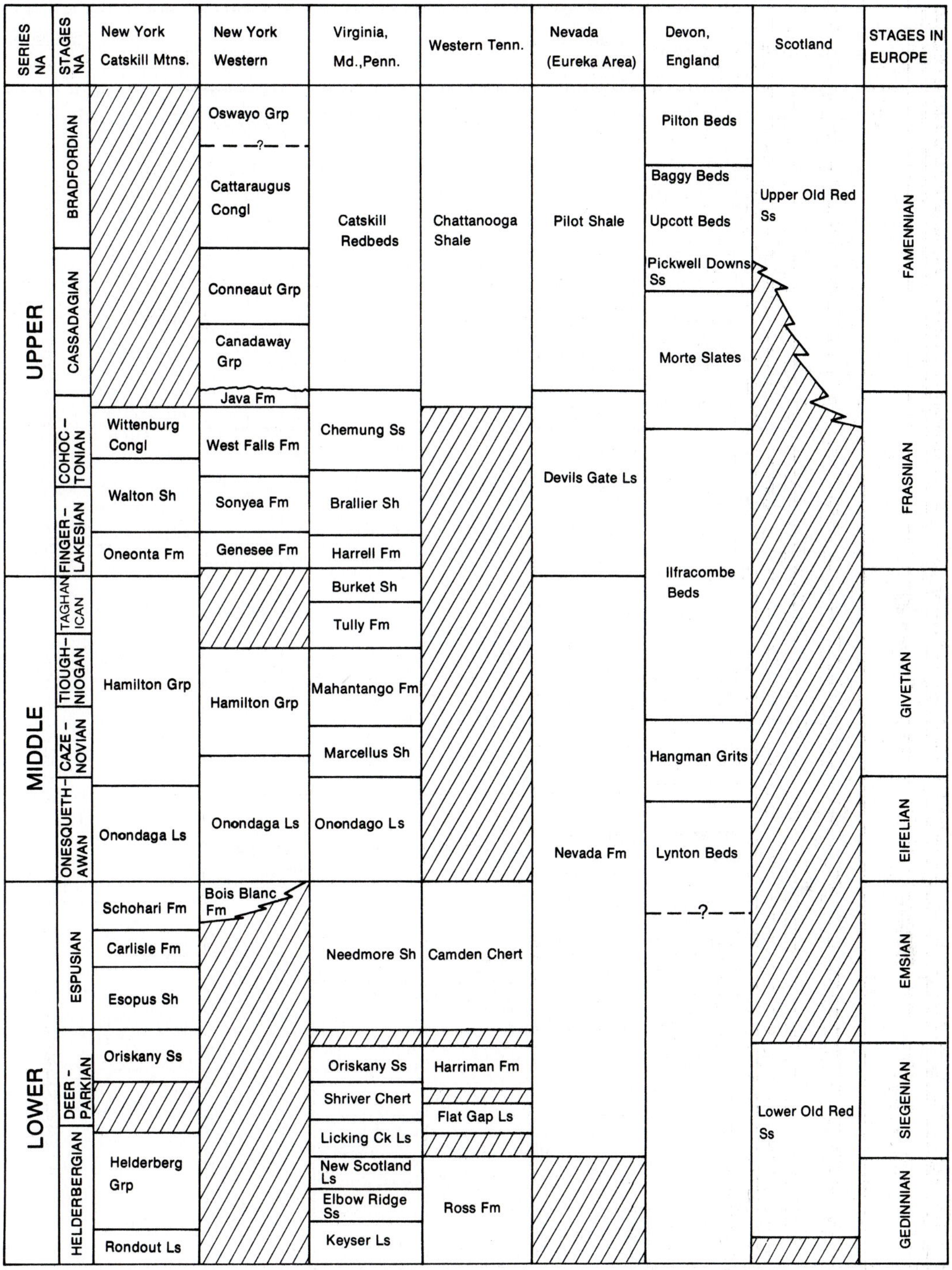

SERIES NA	STAGES NA	New York Catskill Mtns.	New York Western	Virginia, Md.,Penn.	Western Tenn.	Nevada (Eureka Area)	Devon, England	Scotland	STAGES IN EUROPE
UPPER	BRADFORDIAN		Oswayo Grp — —?— — Cattaraugus Congl	Catskill Redbeds	Chattanooga Shale	Pilot Shale	Pilton Beds Baggy Beds Upcott Beds Pickwell Downs Ss	Upper Old Red Ss	FAMENNIAN
	CASSADAGIAN		Conneaut Grp Canadaway Grp Java Fm				Morte Slates		
	COHOC-TONIAN	Wittenburg Congl Walton Sh	West Falls Fm	Chemung Ss		Devils Gate Ls			FRASNIAN
	FINGER-LAKESIAN	Oneonta Fm	Sonyea Fm Genesee Fm	Brallier Sh Harrell Fm					
MIDDLE	TAGHAN-ICAN	Hamilton Grp		Burket Sh Tully Fm			Ilfracombe Beds		GIVETIAN
	TIOUGH-NIOGAN		Hamilton Grp	Mahantango Fm					
	CAZE-NOVIAN			Marcellus Sh			Hangman Grits		
	ONESQUETH-AWAN	Onondaga Ls	Onondaga Ls	Onondago Ls		Nevada Fm	Lynton Beds		EIFELIAN
LOWER	ESPUSIAN	Schohari Fm Carlisle Fm Esopus Sh	Bois Blanc Fm	Needmore Sh	Camden Chert		— —?— —		EMSIAN
	DEER-PARKIAN	Oriskany Ss		Oriskany Ss Shriver Chert	Harriman Fm Flat Gap Ls			Lower Old Red Ss	SIEGENIAN
	HELDERBERGIAN	Helderberg Grp Rondout Ls		Licking Ck Ls New Scotland Ls Elbow Ridge Ss Keyser Ls	Ross Fm				GEDINNIAN

MISSISSIPPIAN

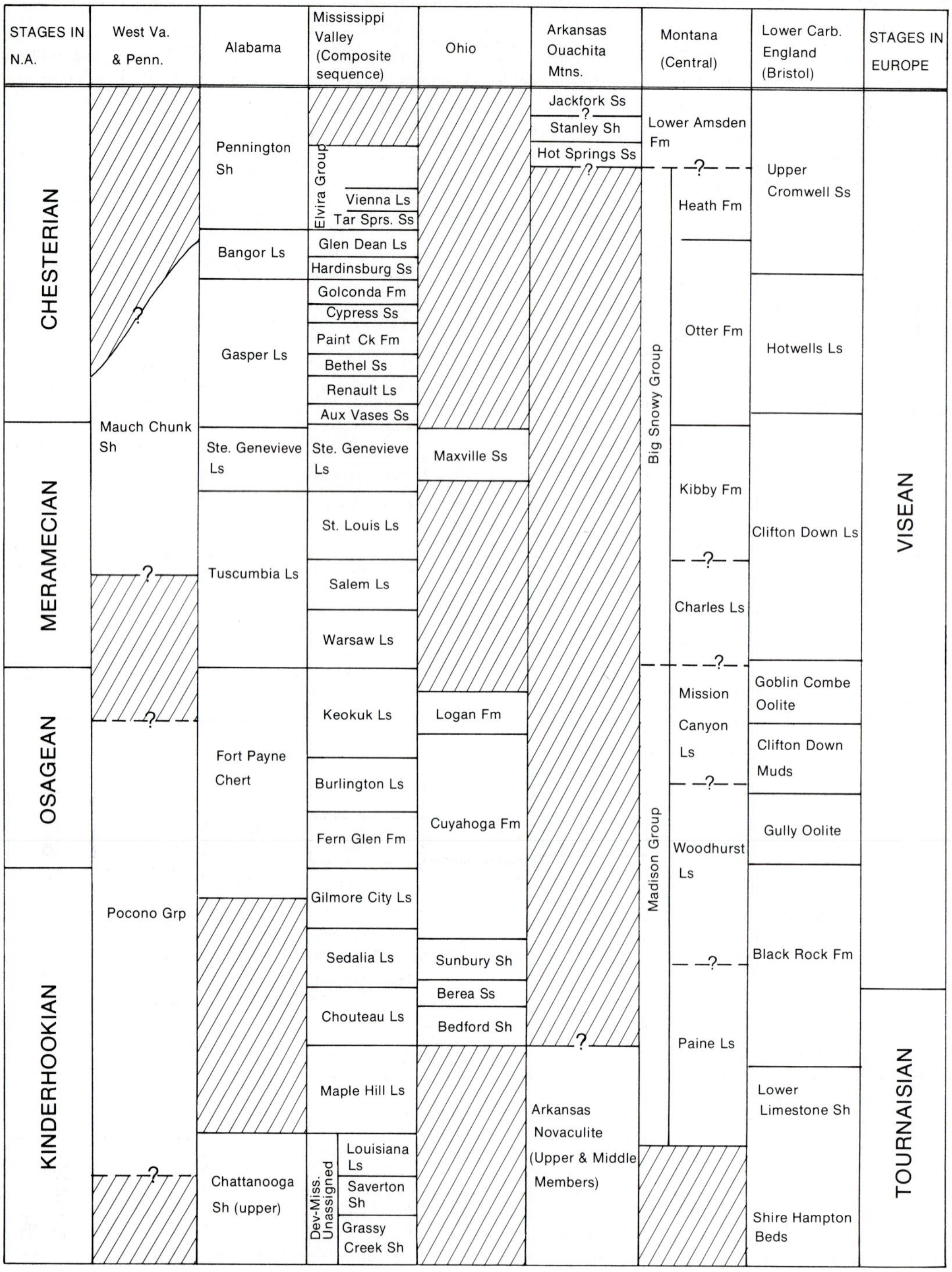

STAGES IN N.A.	West Va. & Penn.	Alabama	Mississippi Valley (Composite sequence)	Ohio	Arkansas Ouachita Mtns.	Montana (Central)	Lower Carb. England (Bristol)	STAGES IN EUROPE
CHESTERIAN	? Mauch Chunk Sh	Pennington Sh Bangor Ls Gasper Ls	Elvira Group: Vienna Ls, Tar Sprs. Ss Glen Dean Ls Hardinsburg Ss Golconda Fm Cypress Ss Paint Ck Fm Bethel Ss Renault Ls Aux Vases Ss		Jackfork Ss ? Stanley Sh Hot Springs Ss ?	Lower Amsden Fm ? Big Snowy Group: Heath Fm Otter Fm	Upper Cromwell Ss Hotwells Ls	VISEAN
MERAMECIAN	Mauch Chunk Sh ?	Ste. Genevieve Ls Tuscumbia Ls	Ste. Genevieve Ls St. Louis Ls Salem Ls Warsaw Ls	Maxville Ss		Kibby Fm ? Charles Ls ?	Clifton Down Ls	VISEAN
OSAGEAN	? Pocono Grp	Fort Payne Chert	Keokuk Ls Burlington Ls Fern Glen Fm	Logan Fm Cuyahoga Fm		Madison Group: Mission Canyon Ls ? Woodhurst Ls	Goblin Combe Oolite Clifton Down Muds Gully Oolite	VISEAN
KINDERHOOKIAN	Pocono Grp ?	Chattanooga Sh (upper)	Gilmore City Ls Sedalia Ls Chouteau Ls Maple Hill Ls Dev-Miss. Unassigned: Louisiana Ls, Saverton Sh, Grassy Creek Sh	Cuyahoga Fm Sunbury Sh Berea Ss Bedford Sh	? Arkansas Novaculite (Upper & Middle Members)	? Paine Ls	Black Rock Fm Lower Limestone Sh Shire Hampton Beds	VISEAN / TOURNAISIAN

PENNSYLVANIAN

Stages in N.A.	Pennsylvania	West Virginia	Arkansas (Ouachita Mtns.)	Southern Illinois	Missouri	Colorado	Northeast England (Upper Carboniferous)	Stages in Europe
VIRGILIAN		Monongahela Series			Waubaunsee Grp Shawnee Grp Douglas Grp			STEPHANIAN
MISSOURIAN	Conemaugh Series	Conemaugh Series		McLeansboro Fm	Pedee Grp Lansing Grp Kansas City Grp Pleasanton Grp			
DESMOINESIAN	Allegheny Series	Allegheny Series		Carbondale Fm	Marmaton Grp Cherokee Grp	Fountain Fm	Upper Coal Measures Middle Coal Measures	WESTPHALIAN
ATOKAN	Pottsville Series	Pottsville Series	Atoka Fm	Tradewater Fm	Riverton Fm Burgner Fm McLouth Fm ? Cheltenham Fm ?		Lower Coal Measures	
MORROWAN			Johns Valley Sh Jackfork Ss Stanley Sh Hot Springs Ss	Caseyville Fm	Hale Fm	Glen Eyrie Sh	Millstone Grit	NAMUR-IAN

PERMIAN

Stages in N.A.	Ohio and W. Va.	Western Texas	Arizona	Wyoming	South Africa	Germany	Russia (Ural Mtns.)	Stages in Europe
OCHOAN		Dewey Lake Fm Rustler Fm Salado Fm Castile Fm			Beaufort Series	Zechstein Fm	Tartarian	TARTARIAN (in part)
GUADALUPIAN		Bell Canyon Fm Cherry Canyon Fm Brushy Canyon Fm		? Phosphoria Fm		Upper Rotliegend	Kazanian Kungurian	KAZANIAN KUNGURIAN
LEONARDIAN		Bone Spring Ls	Kaibab Ls Toroweap Fm Coconino Ss Hermit Sh	?	Ecca Series	Lower Rotliegend	Artinskian	ARTINSKIAN
WOLFCAMPIAN	Dunkard Grp	Heuco Ls	Supai Fm		Dwyka Series		Sakmarian	SAKMARIAN ASSELIAN

TRIASSIC

NA Series	Stages	Connecticut Valley	Pennsylvania (York Co.)	Wyoming	Utah-Idaho	Colo.-S. Dak.	Arizona	Nevada	Germany (Type Section)
UPPER	Rhaetian	Newark Grp: Portland Arkose		Nugget Ss	Ankareh Fm	Dolores Fm (Redbeds)	Chinle Fm	Gabbs Fm	Keuper: Rhät
	Norian	Newark Grp: Meriden Fm	Newark Grp: Gettysburg Fm	Chugwater Fm: Popo Agie Mem	Deadman Ls	Dolores Fm (Redbeds)	Chinle Fm	Gabbs Fm	Keuper: Gyps-keuper
	Carnian	Newark Grp: New Haven Arkose	Newark Grp: New Oxford Fm	Chugwater Fm: Crow Mtn. Ss; Alcova Ls	Higman Grit (Congl)		Shinarump Congl	Luning Fm	Keuper: Letten-kohle
MIDDLE	Ladinian			Chugwater Fm				Grantsville Fm	Muschelkalk
	Anisian			Chugwater Fm				Excelsior Fm	Muschelkalk
LOWER	Sythian			Chugwater Fm: Red Peak Mem; Dinwoody Fm	Thaynes Fm: Woodside Ss Mem, Portneuf Ls Mem; Woodside Redbeds; Dinwoody Fm		Moenkopi Fm	Candelaria Fm	Bunter

JURASSIC

SERIES	STAGES	Gulf of Mexico Region	Utah	Idaho	Wyoming	California Coast Ranges	Alberta Canada	England
UPPER	Portlandian	Cotton Valley Grp: Shuler Fm				Knoxville Fm	Basal Ss of Kootenay Fm	Purbeck Beds; Portland Beds
UPPER	Kimmeridgian	Bossier Fm	Morrison Fm		Morrison Fm	Franciscan Fm	Fernie Grp: "Passage Beds"	Kimmeridge Clay
UPPER	Oxfordian	Buckner Fm; Smackover Fm	Summerville Fm; Curtis Fm; Entrada Ss	Stump Ss; Preuss Ss				Corallian Beds; Oxford Clay
MIDDLE	Callovian	Eagle Mills Fm			Sundance Fm		"Green Beds"; (Shale); Gryphaea Bed	Kellaways Beds
MIDDLE	Bathonian		Carmel Fm	Twin Creek Ls	Gypsum Springs Fm		"Corbula Munda Bed"	Cornbrash Beds; Great Oolite
MIDDLE	Bajocian						Lille Mem?; Sandstones	Inferior Oolite
LOWER	Toarcian						Sandstones; Base Congl	Upper Lias
LOWER	Pliensbachian		Navajo Ss	Nugget Ss				Middle Lias
LOWER	Sinemurian							Lower Lias
LOWER	Hettangian		Kayenta Fm					

CRETACEOUS

Series	Stages	New Jersey	Alabama (Tuscaloosa)	Texas	Kansas (Wallace Co.)	Colorado	Montana	England
UPPER CRETACEOUS	Maastrichtian	Monmouth Grp	Prairie Bluff; Ripley Fm	Navarro Grp: Escondido Fm	?	Animas Fm; McDermott Fm	Hell Ck Fm; Fox Hills Ss; Bear Paw Sh	Upper Chalk
	Campanian	Matawan Grp: Wenonah Sand; Marshall-town Fm; English-town Sand; Woodbury Clay; Merchant-ville Clay	Selma Chalk	Navarro Grp: Corsicana Marl; Taylor Marl	Pierre Sh	Kirkland Fm; Fruitland Fm; Picture Cliffs Ss; Lewis Sh; Mesaverde Grp	Bear Paw Sh; Judith River Fm; Claggett Sh	Upper Chalk
	Santonian	Merchant-ville Clay; ?; Magothy Fm	Selma Chalk; ?	Austin Chalk	?; ?	Mancos Sh	Eagle Ss; Telegraph Ck Fm	Upper Chalk
	Coniacian	Magothy Fm; ?	Eutaw Fm	Austin Chalk	Niobrara Fm; ?	Mancos Sh	Colorado Sh	Upper Chalk
	Turonian	?	?; ?	Eagle Ford Shale	Carlile Sh; Greenhorn Ls	Mancos Sh	Colorado Sh	Middle Chalk
	Cenomanian	Raritan Fm; ?	Tuscaloosa Fm	Eagle Ford Shale; Woodbine Ss	Graneros Sh; Dakota Ss	Dakota Ss	Colorado Sh	Lower Chalk
LOWER CRETACEOUS	Albian			Washita Grp; Fredericksburg Grp	Kiowa Sh; Cheyenne Ss	Dakota Ss; ?	Colorado Sh	Upper Greensand
	Aptian			Trinity Grp		Burro Canyon Fm	Kootenai Fm	Lower Greensand
	Neocomian: Barremian							Wealden Beds
	Neocomian: Hauterivian							Wealden Beds
	Neocomian: Valanginian							Wealden Beds
	Neocomian: Berriasian							Wealden Beds

CENOZOIC

"SYSTEM"	SERIES	STAGES	Gulf Coast	Atlantic Coast	Colorado Wyoming	S. Dakota Nebraska	California (S. Great Valley)	Western Washington	Paris Basin
Quaternary	Rec	Unnamed Recent	Recent Stream and Lake Deps	Pamlico Fm	Recent Sediments	Recent Alluvial Deposits	Recent Alluvial Deposits	Recent Alluvium	Recent Alluvial Deposits
Quaternary	Pleistocene	Wisconsinan	Prairie Fm	Talbot Fm		Pleistocene Sand & Clay	Local Terrace Deposits	Marine & River Terrace Deposits	
Quaternary	Pleistocene	Sangamon	Montgomery Fm	Wicomico Fm					
Quaternary	Pleistocene	Illinoian	Bentley Fm	Sunderland Fm					
Quaternary	Pleistocene	Yarmouth							
Quaternary	Pleistocene	Kansan	Williana Fm	Brandywine Fm			Tulare Fm		
Quaternary	Pleistocene	Aftonian							
Quaternary	Pleistocene	Nebraskan							
Tertiary	Pliocene	Astian	Citronelle Fm	Wacamaw Fm		Ogallala Grp	San Joaquin Fm Etchegoin Fm		
Tertiary	Pliocene	Piacenzan				?	Jacalitos Fm ?		
Tertiary	Miocene	Pontian	Choctawatchee Fm	Yorktown Fm		Sheep Ck Fm	Reef Ridge Sh ?		
Tertiary	Miocene	Sarmatian					McClure Sh		
Tertiary	Miocene	Tortonian		St. Mary's Fm		? Marsland Fm	?		?
Tertiary	Miocene	Helvetian	Shoal River Fm	Choptank Fm		?	Temblor Fm	Astoria Fm	Touraine Marls
Tertiary	Miocene	Burdigalian	Oak Grove Ss Chipola Fm	Calvert Fm Hawthorn Fm		Arikaree Grp	?		? Orléanais Sands
Tertiary	Miocene	Aquitainian	Catahoula Ss	Trent Marl		?		Twin Rivers Fm	
Tertiary	Oligocene	Chattian		Flint River		White River Series		Blakely Fm	Beauce Ls
Tertiary	Oligocene	Rupelian	Vicksburg Ls		Florissant Fm			Lincoln Fm	Stampian Ss
Tertiary	Oligocene	Tongrian			White River Grp	?			Sannosian Ss
Tertiary	Eocene	Ludian	Jackson Fm		Duchesne River Fm		Tumey Ss		"Marls"
Tertiary	Eocene	Bartonian		Castle Hayne Marl	Uinta Fm		Kreyenhagen Sh	Cowlitz Fm	Cresnes and Beauchamp Sands
Tertiary	Eocene	Auversian	Yegua Fm Claiborne Grp: Cook Mtn. Fm Sparta Ss	Shark River Marl	Bridger Fm		?		
Tertiary	Eocene	Lutetian	Claiborne Grp: Mt. Selma Fm				? Domingine Fm		Provins Ls
Tertiary	Eocene	Cuisian	Claiborne Grp: Carrizo Ss Wilcox Grp	Manasquan Marl	Green River Fm		Yokut Ss		
Tertiary	Eocene	Ypresian		Aquia Fm	Cedar Breaks Fm		Lodo Fm	?	? Cuise Sands
Tertiary	Paleocene	Thanetian	Midway Grp: Wills Pt. Fm	Clayton Ls	Fort Union Fm	Fort Union Fm		Metachosin Volcanics	Plastic Clay Bracheux Sands
Tertiary	Paleocene	Montian	Midway Grp: Kincaid Fm				?	?	Meudon Ls
Tertiary	Paleocene	Danian							

Glossary

For terms not included in this glossary, students may wish to consult the Dictionary of Geologic Terms, *published in 1976 by Anchor/Doubleday under the direction of the American Geological Institute. A more comprehensive reference is the* Glossary of Geology, *edited by R.L. Bates and J.A. Jackson, 2d ed., Falls Church, Virginia, American Geological Institute.*

A

Absaroka Sequence—a sequence of Permian-Pennsylvanian sediments bounded both above and below by a regional unconformity and recording an episode of marine transgression over an eroded surface, full flood level of inundation, and regression from the craton.

Absolute geologic age—the actual age, expressed in years, of a geologic material or event.

Acadian Orogeny—an episode of mountain building in the northern Appalachians during the Devonian Period.

Acanthodians—the earliest known vertebrates (fishes) with movable, well-developed lower jaw, or mandible; hence, the first jawed fishes.

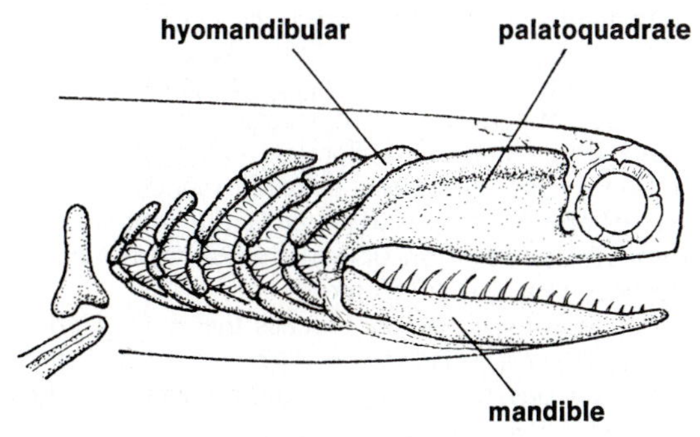

Partially reconstructed skeleton of head and gill region of an acanthodian indicating origin of jaws from gill arches.

Actualism—the uniform operation of actual causes in changing the earth. Actualism regards the laws of nature as invariant and responsible for changes that have occurred through time.

Adaptation—a modification of an organism that better fits it for existence in its present environment or enables it to live in a somewhat different environment.

Adaptive radiation—the diversity that develops among species as each adapts to a different set of environmental conditions.

Adenosine diphosphate (ADP)—a product formed in the hydrolysis of adenosine triphosphate that is accompanied by release of energy and organic phosphate.

Adenosine triphosphate (ATP)—a compound that occurs in all cells and that serves as a source of energy for physiologic reactions such as muscle contraction.

Aerobic organism—an organism that uses oxygen in carrying out respiratory processes.

Age—the time represented by the time-stratigraphic unit called a stage. (Informally, may indicate any time span in geologic history, as "age of cycads.")

Agnatha—the jawless vertebrates, including extinct ostracoderms and living lampreys and hagfishes.

Algae—any of a large group of simple plants (thallophyta) that contain chlorophyll and are capable of photosynthesis.

Allegheny Orogeny—the Late Paleozoic episodes of mountain building along the present trend of the Appalachian Mountains.

Alluvium—unconsolidated, poorly sorted detrital sediments ranging from clay to gravel sizes and characteristically fluvial in origin.

Alpha particle—a particle, equivalent to the nucleus of a helium atom, emitted from an atomic nucleus during radioactive decay.

Alpine Orogeny—in general, the sequence of crustal disturbances beginning in the Middle Mesozoic and continuing into the Miocene that resulted in the geologic structures of the Alps.

Amino acids—nitrogenous hydrocarbons that serve as the building blocks of proteins and are thus essential to all living things.

Ammonites—Ammonoid cephalopods having more complex sutural patterns than either ceratites or goniatites.

Ammonoids—an extinct group of cephalopods, with coiled, chambered conch and having septa with crenulated margins.

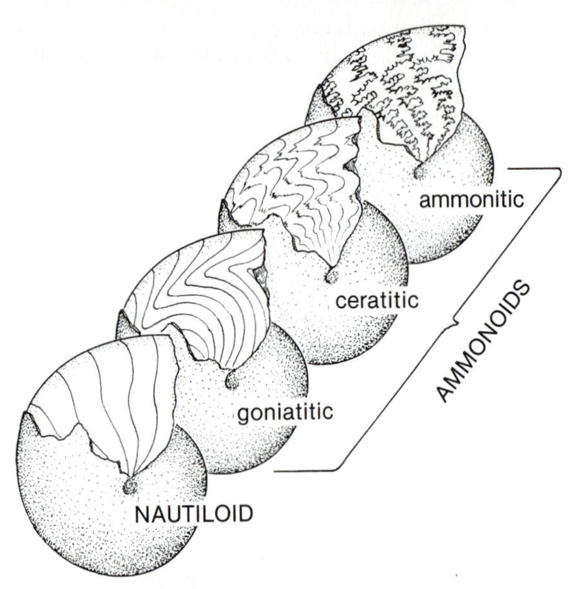

Suture patterns in cephalopods

Amniotic egg—that type of egg produced by reptiles, birds, and monotremes. In this type, the developing embryo is maintained and protected by an elaborate arrangement of shell membranes, yolk, sac, amnion, and allantois.

Amphibians—"cold-blooded" vertebrates that utilize gills for respiration in the early life stages but that have air-breathing lungs as adults.

Amphibole—a ferromagnesian silicate mineral that occurs commonly in igneous and metamorphic rocks.

Anaerobic organism—an organism that does not require oxygen for respiration, but, rather, makes use of processes such as fermentation in order to obtain its energy.

Andesite—a volcanic rock that in chemical composition is intermediate between basalt and granite.

Angiosperms—an advanced group of plants having floral reproductive structures and seeds in a closed ovary. The "flowering plants."

Anthropoidea—the suborder of primates that includes monkeys, apes, and humans.

Anticline—a geologic structure in which strata are bent into an upfold or arch.

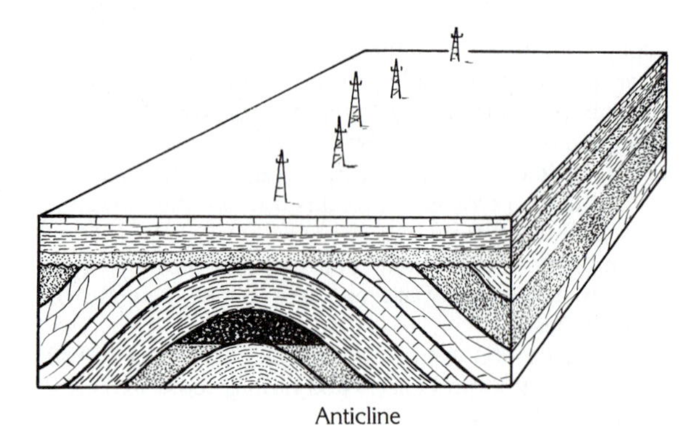
Anticline

Antler Orogeny—a Late Devonian and Mississippian episode of mountain building involving folding and thrusting along a belt across central Nevada to southwestern Alberta.

Archaean—the older of a two-part division of the Precambrian. Same as Archaeozoic, which preceded the Proterozoic.

Archaeocyatha—a group of extinct marine organisms having double, perforated, calcareous, conical to cylindric walls. Archaeocyathids lived during the Cambrian.

Archosaurs—advanced reptiles of a group called diapsids, which includes thecodonts, "dinosaurs," pterosaurs, and crocodiles.

Arcoids—a group of pelecypods exemplified by species of *Arca.*

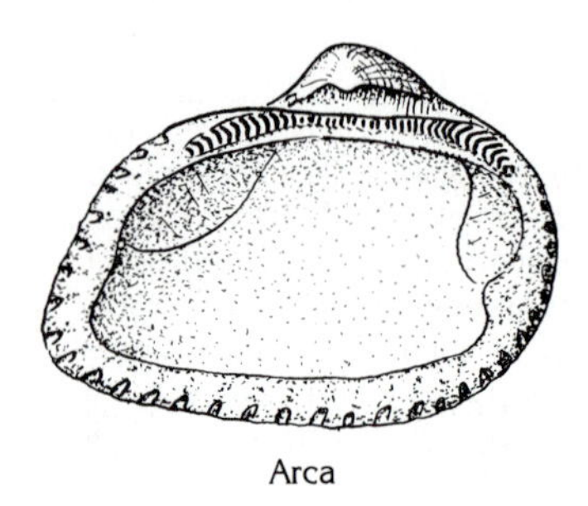
Arca

Artiodactyl—hoofed mammals that typically have two or four toes on each foot.

Asteroid—one of numerous relatively small planetary bodies (less than 800 km in diameter) revolving around the sun in orbits lying between those of Mars and Jupiter.

Asthenosphere—a zone between 50 and 250 km below the surface of the earth where the shock waves of earthquakes travel at much reduced speeds, perhaps because of less rigidity. The asthenosphere may be a zone where convective flow of material may occur.

Atoll—a ringlike island or a series of islands formed by corals and calcareous algae around a central lagoon.

Atom—the smallest divisible unit retaining the characteristics of a specific element.

Atomic fission—a nuclear process that occurs when a heavy nucleus splits into two or more lighter nuclei, simultaneously liberating a considerable amount of energy.

Atomic fusion—a nuclear process that occurs when two light nuclei unite to form a heavier one. In the process, a large amount of energy is released.

Atomic mass—a quantity essentially equivalent to the number of neutrons plus the number of protons in an atomic nucleus.

Atomic number—the number of protons in the nuclei of atoms of a particular element. (An element is thus a substance in which all of the atoms have the same atomic number.)

Australopithecines—a general term applied rather loosely to Pliocene and Early Pleistocene primates whose skeletal characteristics place them between typically apelike individuals and those more obviously human.

Autotroph—an organism that uses an external source of energy to produce organic nutrients from simple inorganic chemicals.

B

Basin—a depressed area that serves as a catchment area for sediments (basin of deposition). A structural basin is an area in which strata slope inward toward a central location. Structural basins tend to experience periodic downsinking and thus receive a thicker and more complete sequence of sediments than do adjacent areas.

Belemnites—members of the molluscan Class Cephalopoda, having straight internal shells.

Benioff Seismic Zone—an inclined zone along which frequent earthquake activity occurs and that marks the location of the plunging-forward edge of a lithospheric plate during subduction.

Benthic—a bottom-dwelling organism.

Bentonite—a layer of clay, presumably formed by the alteration of volcanic ash.

Beta particle—a charged particle, essentially equivalent to an electron, emitted from an atomic nucleus during radioactive disintegration.

Bivalvia—a class of the Phylum Mollusca also known as the Class Pelecypoda. The term Pelecypoda is preferred in this text so that the pelecypods are not confused with other "bivalves," such as brachiopods and ostracods.

Blastoids—sessile (attached) Paleozoic echinoderms having a stem and an attached cup or calyx composed of relatively few plates.

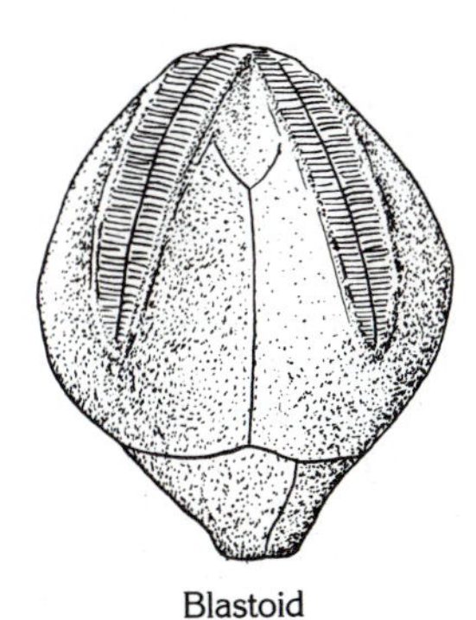

Blastoid

Brachiating—swinging from branch to branch and tree to tree by using the limbs, as among monkeys.

Brachiopod—bivalved (double-shelled) marine invertebrates. They were particularly common and widespread during the Paleozoic and persist in fewer numbers today.

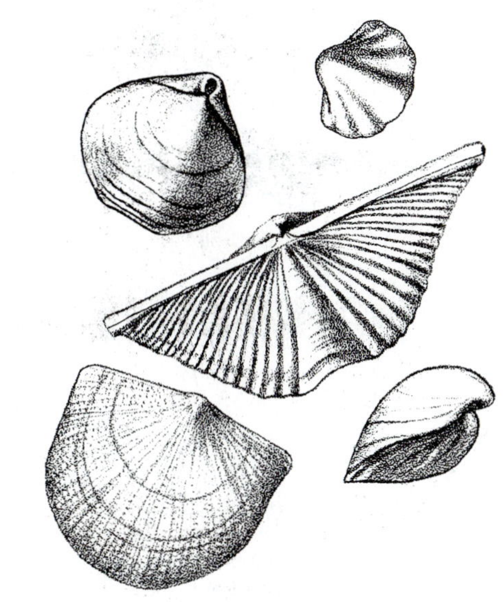

Brachiopods

Breccia—a clastic sedimentary rock composed largely of angular fragments of granule size or larger.

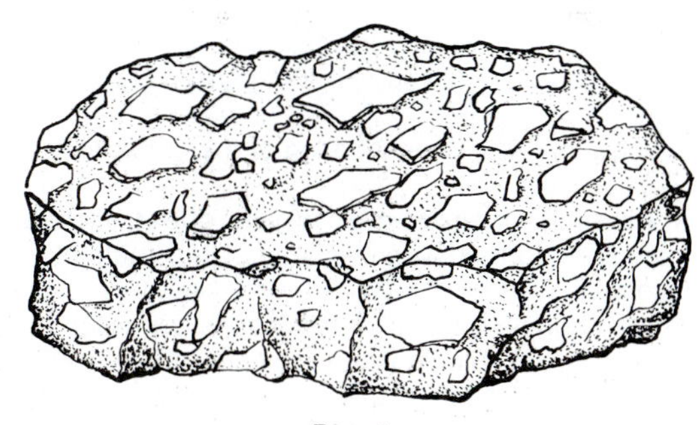

Breccia

Breeder reactor (nuclear)—an atomic reactor that uses uranium 238 but that creates additional fuel by producing more fissionable material than it consumes.

Bryozoa—a phylum of attached and incrusting colonial marine invertebrates.

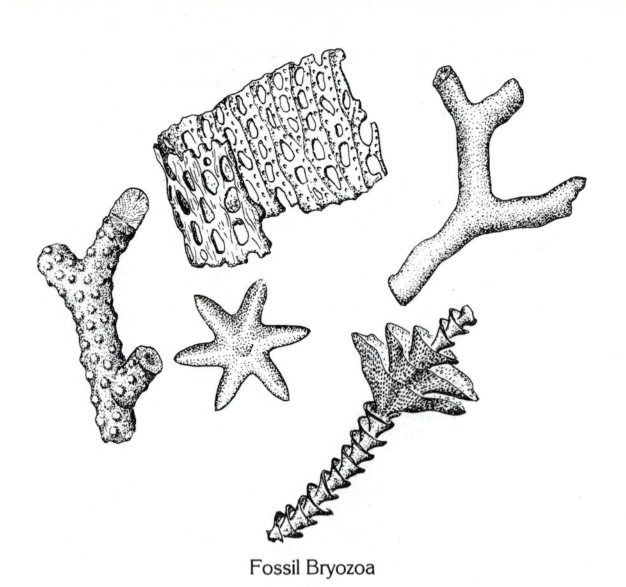
Fossil Bryozoa

Burgess Shale fauna—a beautifully preserved fossil fauna of soft-bodied Cambrian animals discovered in 1910 by Charles Walcott in Kicking Horse Pass, Alberta.

C

Caledonian Orogeny—a major episode of mountain building affecting the Caledonian Geosyncline that was more or less continuous at various locations along the geosyncline from Late Cambrian to Middle Devonian time.

Carbon 14—a radioactive isotope of carbon with an atomic mass of 14. Carbon 14 is frequently used in determining the age of materials less than about 50,000 years old.

Carbonate—a general term for a chemical compound formed when carbon dioxide dissolved in water combines with oxides of calcium, magnesium, potassium, sodium, and iron. The most common carbonate minerals are calcite (which forms the carbonate rock limestone) and dolomite (which is a constituent of dolostone).

Carbonization—the concentration of carbon during fossilization.

Cast (natural)—a replica of an organic object, such as a fossil shell, formed when sediment fills a mold of that object.

Cataract Group—a group of formations deposited during the Lower Silurian and including the Whirlpool Sandstone, Manitoulin Dolomite, Cabot Head Shale, Dyer Bay Limestone, and Wingfield Shale of southern Ontario. The Cataract Group is correlative with the Albion Group of western New York.

Catskill Delta—a buildup of Middle and Upper Devonian clastic sediments as a broad, complex clastic wedge derived from the erosion of highland areas formed largely during the Acadian Orogeny.

Ceboidea—the New World monkeys, characterized by prehensile tails, and including the capuchin, marmoset, and howler monkeys.

Centrifugal force—the apparent outward force experienced by an object moving in a circular path. Centrifugal force is a manifestation of inertia, the tendency of moving things to travel in straight lines.

Ceratites—one of the three larger groups of ammonoid cephalopods having sutural complexity intermediate between goniatites and ammonites.

Ceratopsians—the quadrupedal ornithischian dinosaurs characterized by the development of prominent horns on the head.

Cercopithecoidea—the Old World monkeys (Asia, southern Europe, Africa), including macaques, guenons, langurs, baboons, and mandrills.

Cetaceans—the group of marine mammals that includes whales and porpoises.

Chalicothere—extinct perissodactyls having robust claws rather than hoofs.

Chalk—A soft, white, fine-grained variety of limestone composed largely of the calcium carbonate skeletal remains of marine microplankton.

Chert—a dense, hard sedimentary rock or mineral composed of submicrocrystalline quartz. Unless colored by impurities, chert is white, as opposed to flint, which is dark or black.

Chlorophyll—the catalyst that makes possible the reaction of water and carbon dioxide in green plants to produce carbohydrates. Photosynthesis is the reaction.

Choanichthyes—that group of fishes that includes both the dipnoans (lungfishes with weak pelvic and pectoral fins) and crossopterygians (lungfishes with stout lobe-fins).

Chondrichthyes—the broad category of fishes with cartilaginous skeletons that is exemplified by sharks, skates, and rays.

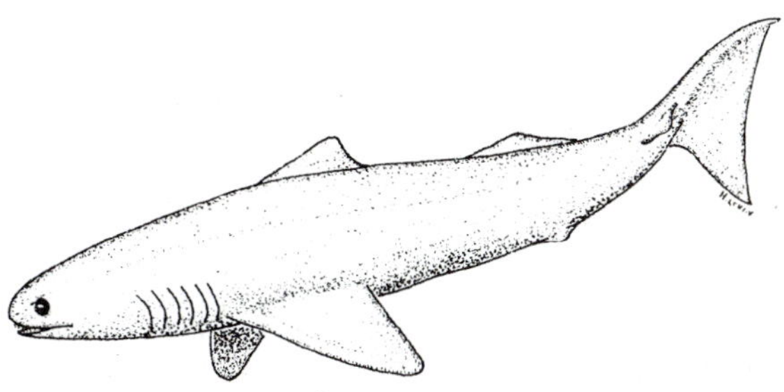

The Devonian chondrichthyan (shark) *Cladoselache*

Chondrites—stony meteorites that contain rounded silicate grains or chondrules. Chondrules are believed to have formed by crystallization of liquid silicate droplets.

Chromosome—threadlike microscopic bodies composed of chromatin. Chromosomes appear in the nucleus of the cell at the time of cell division. They contain the genes. The number of chromosomes is normally constant for a particular species.

Chromosphere—one of the concentric shells of the sun, lying above the photosphere and telescopically visible as a thin, brilliant red rim around the edge of the sun for a second or so at the beginning and end of a solar eclipse.

Clastic texture—texture that characterizes a rock made up of fragmental grains such as sand, silt, or parts of fossils. Conglomerates, sandstones, and siltstones are clastic rocks; the individual clastic grains are termed *clasts*.

Clastic wedge—an extensive accumulation of largely clastic sediments deposited adjacent to uplifted areas. Sediments in the wedge become finer and the section becomes thinner in a direction away from the upland source areas. The Queenstone and Catskill "deltas" are examples of clastic wedges.

Coccolithophorids—marine, planktonic, biflagellate, golden brown algae that typically secrete coverings of discoidal calcareous platelets called *coccoliths*.

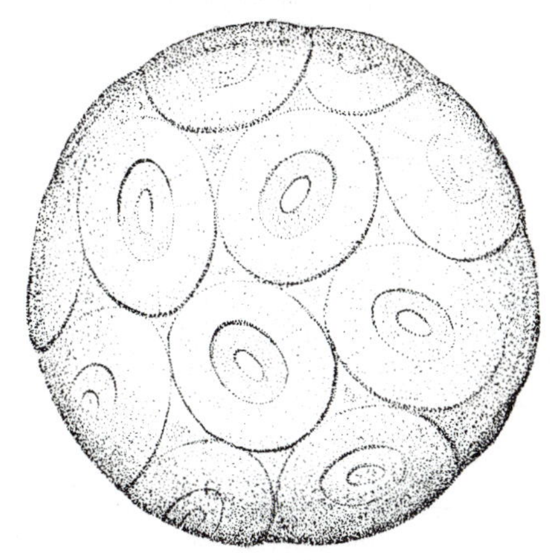

Coccolithophorid

Colorado Mountains—a group of areas uplifted in Pennsylvanian time in Colorado. Sometimes inappropriately termed the "ancestral Rockies."

Conodonts—small, toothlike fossils composed of calcium phosphate and found in rocks ranging from Cambrian to Triassic in age. Although the precise nature of the conodont-producing organism has not been determined, this uncertainty does not detract from their usefulness as guide fossils.

Contact metamorphism—the compositional and textural changes in a rock that result from heat and pressure emanating from an adjacent igneous intrusion.

Convergence (in evolution)—the process by which similarity of form or structure arises among different organisms as a result of their becoming adapted to similar habitats.

Cordaites—a primitive order of treelike plants with long, bladelike leaves and clusters of naked seeds. Cordaites were in some ways intermediate in evolutionary stage between seed ferns and conifers.

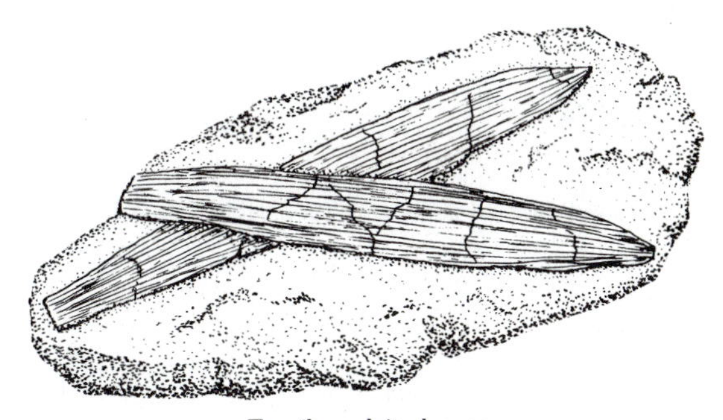

Fossil cordaite leaves

Cordilleran Geosyncline—a great north-south trending geosyncline located along the western margin of the North American continent.

Core—central part of the earth that lies beneath the mantle.

Coriolis Effect—the deflection of winds and water currents to the right in the northern hemisphere and to the left in the southern hemisphere as a consequence of the earth's rotation.

Cosmic rays—extremely high-energy particles, mostly protons, which move through the galaxy and frequently strike the earth's atmosphere.

Cotylosaurs—the earliest and most primitive of Paleozoic reptiles.

Craton—the long-stable region of a continent, commonly with Precambrian rocks either at the surface or only thinly covered with younger sedimentary rocks.

Creodonta—primitive, early, flesh-eating placental mammals.

Crinoids—stalked echinoderms with a calyx composed of regularly arranged plates from which radiate arms for gathering food.

Cross-bedding (cross-stratification)—an arrangement of laminae or thin beds transverse to the planes of stratification. The inclined laminae are usually at inclinations of less than 30° and may be either straight or concave.

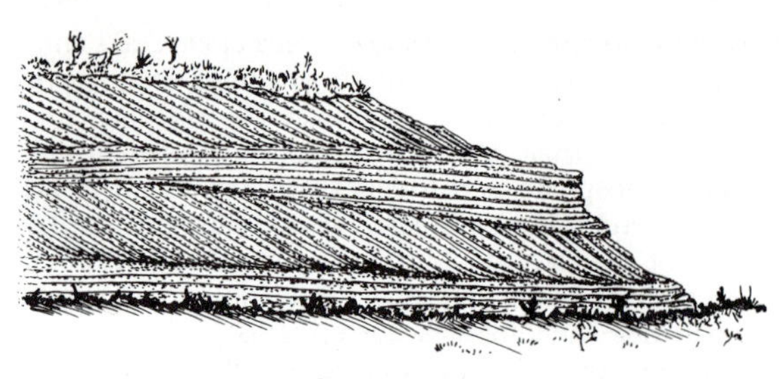

Cross-bedding

Crossopterygii—that group of choanichthyan fishes ancestral to earliest amphibians and characterized by stout pectoral and pelvic fins as well as lungs.

The Devonian crossopteryian *Eusthenopteron*

Crust (earth)—the outer part of the lithosphere; it averages about 32 km in thickness.

Crustacea—a Class of the Phylum Arthropoda that includes such well-known living animals as lobsters and crayfishes.

Cryptozoic—the span of geologic time that preceded the Cambrian Period, about 570 million years ago. Also known as Precambrian.

Curie temperature—the temperature at which a cooling mineral acquires permanent magnetic properties that record the surrounding magnetic-field orientation and strength at the time of cooling. Magnetic properties of a mineral above the Curie temperature will change as the surrounding field changes. Below the Curie temperature, magnetic characteristics of the mineral will not alter.

Cycadales—a group of seed plants that were especially common during the Mesozoic and were characterized by palmlike leaves and coarsely textured trunks marked by numerous leaf scars.

Cyclothem—a vertical succession of sedimentary units reflecting environmental events that occurred in a constant order. Cyclothems are particularly characteristic of the Pennsylvanian System.

Cystoids—attached echinoderms with generally irregular arrangement and number of plates in the calyx and perforated by pores or slits.

D

Deccan traps—a thick sequence (3200 meters) of Late Cretaceous basaltic lava flows that cover about 500,000 sq km of Peninsular India.

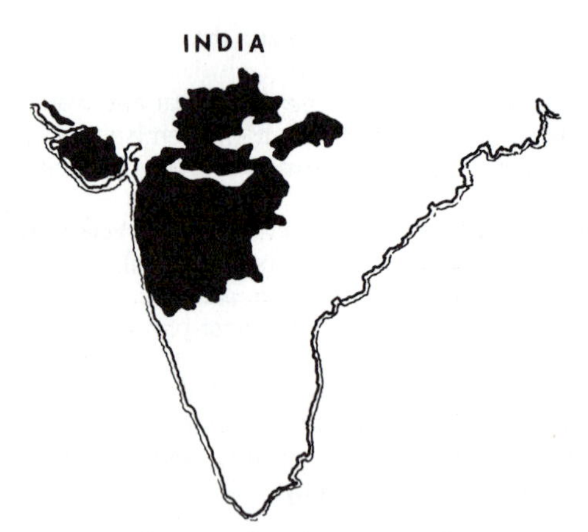

Deccan traps (outcrop map)

Décollement—feature of stratified rocks in which upper formations may become "unstuck" from lower formations, deform, and slide thousands of meters over underlying beds.

Diatoms—microscopic golden brown algae (chrysophytes) that secrete a delicate siliceous frustule (shell).

Diatom

Differentiation—the process by which a planet becomes internally zoned as heavy materials sink toward its center and light materials accumulate near the surface.

Dinoflagellates—unicellular marine algae usually having two flagella and a cellulose wall.

Diploid cells—cells having two sets of chromosomes that form pairs, as in somatic cells.

Dipnoi—an order of lungfishes with weak pectoral and pelvic fins; not considered ancestral to land vertebrates.

Disconformity—a variety of unconformity in which bedding planes above and below the plane of erosion or nondeposition are parallel.

DNA—the nucleic acid found chiefly in the nucleus of cells that functions in the transfer of genetic characteristics and in protein synthesis.

Docodonts—a group of small, primitive Upper Jurassic mammals possibly ancestral to the living monotremes.

Dolostone—a carbonate sedimentary rock that contains more than 50 per cent of the mineral dolomite, $CaMg(CO_3)_2$.

Dome—an upfold in rocks having the general configuration of an inverted bowl. Strata in a dome dip outward and downward in all directions from a central area. An example is the Ozark Dome.

Dryopithecine—in general, a group of lightly built primates that lived during the Miocene and Pliocene in mostly open savannah country and that includes *Dryopithecus,* a form considered to be in the line leading to apes, and *Ramapithecus,* considered by some to be in the ancestry of humans.

Dynamothermal (regional) metamorphism—metamorphism that has occurred over a wide region, caused by deep burial and high temperatures associated with pressures resulting from overburden and orogeny.

E

Echinoderms—the large group (Phylum Echinodermata) of marine invertebrates characterized by prominent pentamerous symmetry and skeleton frequently constructed of calcite elements and including spines. Cystoids, blastoids, crinoids, echinoids, and asteroids are examples of echinoderms.

Edentata—an order of placental mammals that includes extinct ground sloths and glyptodonts, as well as living armadillos, tree sloths, and South American anteaters.

Elastic limit—the greatest amount of stress that can be applied to a body without causing permanent strain.

Endemic population—the native fauna of any particular region.

Entelodonts—a group of extinct artiodactyls bearing a superficial resemblance to giant wild boars.

Entelodont

Eocambrian—name applied to the poorly fossiliferous sequence of sedimentary rocks that lie generally above Precambrian basement rocks but below readily identifiable fossiliferous Cambrian strata.

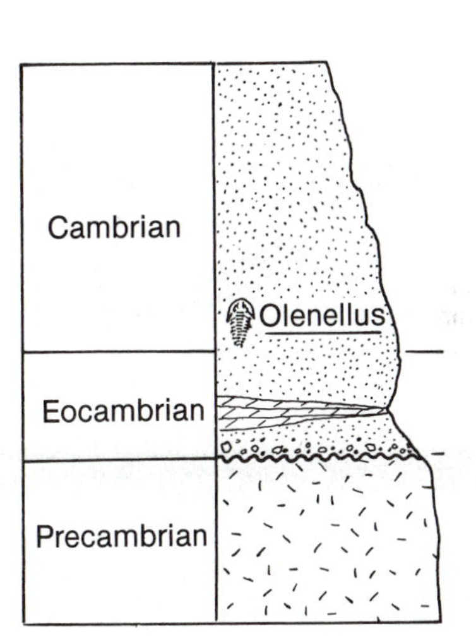

Eon—a major division of the geologic time scale. Two eons are recognized by some geologists: a Crytozoic Eon and a Phanerozoic Eon. The term is also sometimes used to denote a span of 1 billion years.

Epifaunal organisms—organisms living *on,* as distinct from *in,* a particular body of sediment or another organism.

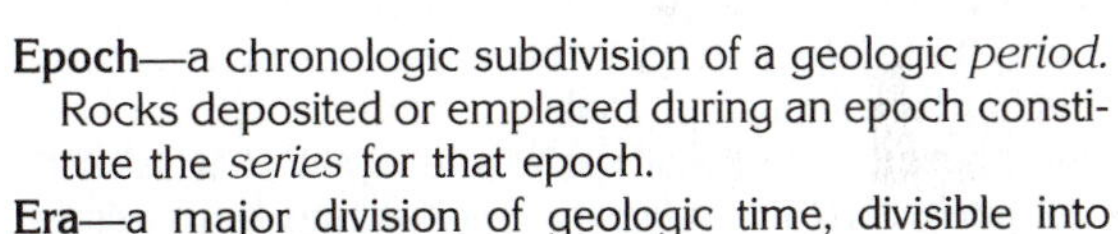

Epoch—a chronologic subdivision of a geologic *period.* Rocks deposited or emplaced during an epoch constitute the *series* for that epoch.

Era—a major division of geologic time, divisible into geologic periods.

Eukaryote—a type of living cell containing a true nucleus, enclosed within a nuclear membrane, and having well-defined chromosomes and cell organelles.

Eugeosyncline—the oceanward side of a geosyncline, characterized by great thickness of poorly sorted clastics, siliceous sediments, and volcanics.

Eurypterids—aquatic arthropods of the Paleozoic, superficially resembling scorpions, and probably carnivorous.

Eustatic—pertaining to worldwide, simultaneous changes in sea level, such as might result from change in the volume of continental glaciers.

Evaporites—sediments precipitated from a water solution as a result of the evaporation of that water. Evaporite minerals include anhydrite gypsum ($CaSO_4$) and halite ($NaCl$).

Evolution—the continuous genetic adaptation of organisms or species to the environment.

F

Facies—a particular aspect of sedimentary rocks that is a direct consequence of sedimentation in a particular depositional environment.

Fault—a fracture in the earth's crust along which rocks on one side have been displaced relative to rocks on the other side.

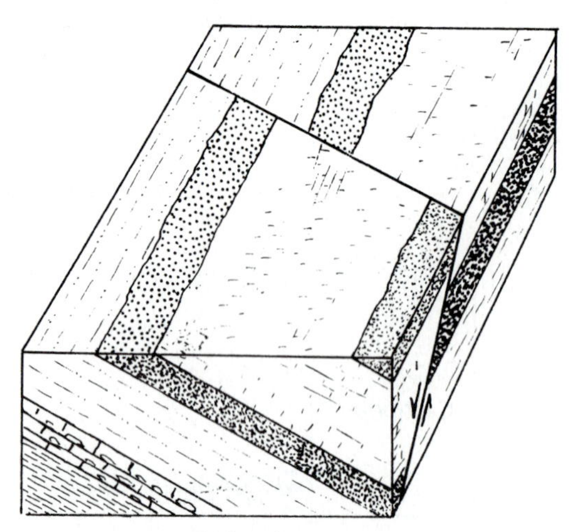

Fault in dipping strata

Fermentation—the partial breakdown of organic compounds by an organism in the absence of oxygen. The final product of fermentation is alcohol or lactic acid.

Filter-feeders (organisms)—an animal that obtains its food, which usually consists of small particles or organisms, by filtering it from the water.

Fissility—that property of rocks that causes them to split into thin slabs parallel to bedding. Fissility is particularly characteristic of shale.

Flood basalts—regionally extensive layers of basalt that originated as low-viscosity lava pouring from fissure eruptions. The lavas of the Columbia Plateau and the Deccan Plateau are flood basalts.

Fluvial—pertaining to sediments or other geologic features formed by streams.

Flysch—term describing thick sequences of rapidly deposited, poorly sorted marine clastics.

Focus (earthquake)—the location at which rock rupture occurs so as to generate seismic waves.

Foliation—a textural feature especially characteristic of metamorphic rocks in which laminae develop by growth or realignment of minerals in parallel orientation.

Foraminifers—an order of mostly marine, unicellular protozoans that secrete tests (shells) that are usually composed of calcium carbonate.

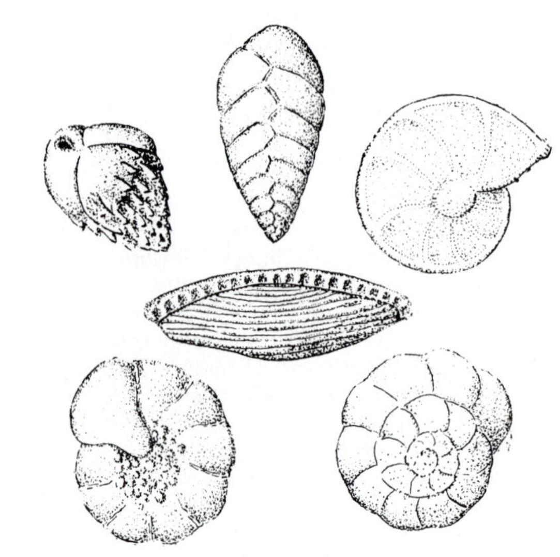

Foraminifers

Formation—a mappable, lithologically distinct body of rock having recognizable contacts with adjacent rock units.

Fossil—the remains or indications of an organism that lived in the geologic past.

Fractional crystallization—the separation of components of a cooling magma by sequential formation of particular mineral crystals at progressively lower temperatures.

Franklinian Geosyncline—the most northerly geosyncline of North America, extending from northern Greenland to Prince Patrick Island, a distance of over 2200 km.

Fusulinids—primarily spindle-shaped foraminifers with a calcareous, coiled test divided into a complex of numerous chambers. Fusulinids were particularly abundant during the Pennsylvanian and Permian Periods.

G

Galaxy—an aggregate of stars and planets, separated from other such aggregates by distances greater than those between member stars.

Gamete—either of two cells (male or female) that must unite in sexual reproduction to initiate the development of a new individual.

Gamma rays—very high-frequency electromagnetic waves.

Garnet—a family of aluminosilicates of iron and calcium that are particularly characteristic of metamorphic rocks.

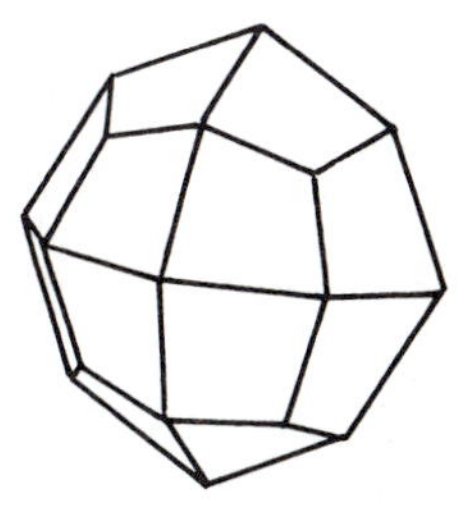

Garnet crystal

Gene—the unit of heredity transmitted in the chromosome.

Genus—the major subdivision of a taxonomic family or subfamily of plants or animals, usually consisting of more than one species.

Geochronology—the study of time as applied to earth and planetary history.

Geologic range—the geologic time span between the first and last appearance of an organism.

Geosyncline—a large troughlike or linear basin in which great thicknesses of sediment may accumulate over an extended span of geologic time. Geosynclines may ultimately be compressed into major mountain systems.

Glacier—a large mass of ice, formed by the recrystallization of snow, that flows slowly under the influence of gravity.

Glauconite—a green clay mineral frequently found in marine sandstones and believed to have formed at the site of deposition.

Glossopteris flora—an assemblage of fossil plants prevalent in rocks of Late Paleozoic and Early Triassic age in South Africa, India, Australia, and South America. The flora takes its name from the seed fern *Glossopteris.*

Gondwanaland—the great Permo-Carboniferous southern hemisphere continent, comprising the assembled present continents of South Africa, India, Australia, Africa-Arabia, and Antarctica.

Goniatites—one of the three large groups of ammonoid cephalopods with sutures forming a pattern of simple lobes and saddles and thus not as complex as either the ceratites or the ammonites.

Gowganda Conglomerate—an apparent tillite of the Canadian Shield. The Gowganda rests upon a surface of older rock that appears to have been polished by glacial action.

Granitization—the formation of granite from previously existing rocks by metamorphic processes. Granitization does not require the origin of granite from a molten state.

Graptolite facies—a Paleozoic sedimentary facies composed of dark shales and fine-grained clastics that contain the abundant remains of graptolites and that are associated with volcanic rocks.

Graptolites—extinct, colonial, marine invertebrates considered to be protochordates. Graptolites range from the Late Cambrian to the Mississippian.

Gravity anomaly—the differences between the observed value of gravity at any point on the earth and the calculated theoretic value.

Greenhouse effect—a process in which incoming solar radiation that is absorbed and reradiated cannot escape back into space because the earth's atmosphere is not transparent to the reradiated energy, which is in the form of infrared radiation (heat).

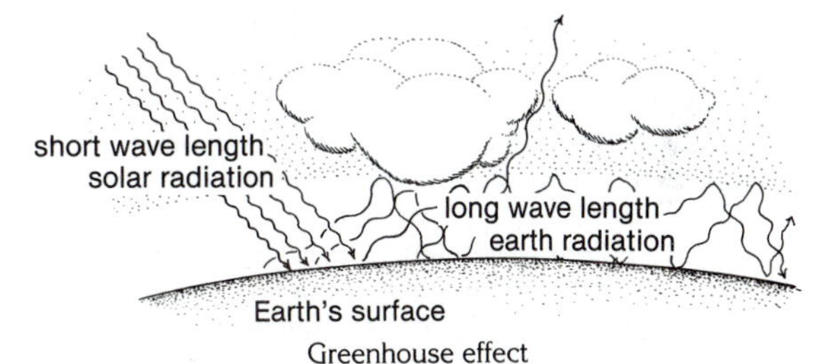

Greenhouse effect

Greenstones—low-grade metamorphic rocks containing abundant chlorite, epidote, and biotite and developed by metamorphism of basaltic extrusive igneous rocks. Great linear outcrops of greenstones are termed greenstone belts and are thought to mark the locations of former volcanic island arcs.

Guide fossil—fossil with a wide geographic distribution but narrow stratigraphic range and thus useful in correlating strata and for age determination.

Gutenberg discontinuity—the boundary separating the mantle of the earth from the core below. The Gutenberg discontinuity lies about 2900 km below the surface.

Gymnosperms—an informal designation for flowerless seed plants in which seeds are not enclosed (hence "naked seeds"). Examples are conifers and cyads.

H

Hadrosaurs—the ornithischian "duck-bill dinosaurs" of the Cretaceous.

Half-life—the time in which one half of an original amount of a radioactive species decays to daughter products.

Haploid cell—one having a single set of chromosomes, as in gametes (see *diploid*).

Hercynian Orogeny—a major Late Paleozoic orogenic episode in Europe that formed the ancient Hercynian Mountains. Today, only the eroded stumps of these

Hadrosaur

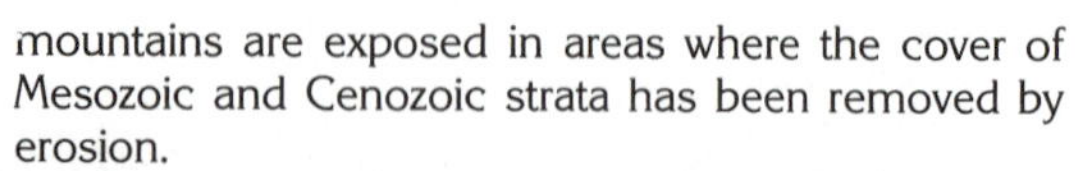

mountains are exposed in areas where the cover of Mesozoic and Cenozoic strata has been removed by erosion.

Heterotroph—an organism that depends upon an external source of organic substances for its nutrition and energy.

Holocene—a term sometimes used to designate the period of time since the last major episode of glaciation. The term is equivalent to "Recent."

Homologous organs—organs having structural and developmental similarities due to genetic relationship.

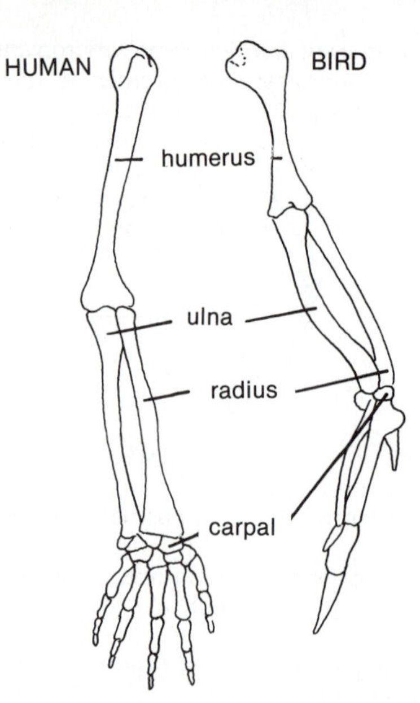

Homologous bones in forelimbs of human and bird

Hydrosphere—the water and water vapor present at the surface of the earth, including oceans, rivers, seas, lakes, and rivers.

Hylobatidae—a group of persistently arboreal, small apes that are exemplified by the gibbons and siamangs.

Hyomandibular bone (in fishes)—the modified upper bone of the hyoid arch, which functions as a connecting element between the jaws and the brain case in certain fishes.

I

Ichthyosaurs—highly specialized marine reptiles of the Mesozoic, recognized by their fishlike form.

Ictidosauria—a group of extinct mammal-like reptiles or therapsids whose skeletal characteristics are considered to be very close to those of mammals.

Igneous rock—a rock formed by the cooling and solidification of magma or lava.

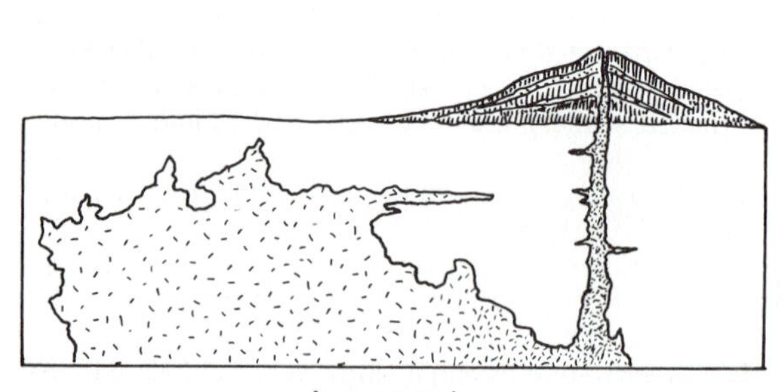

Igneous rocks

Infaunal organisms—organisms that live and feed within bottom sediments.

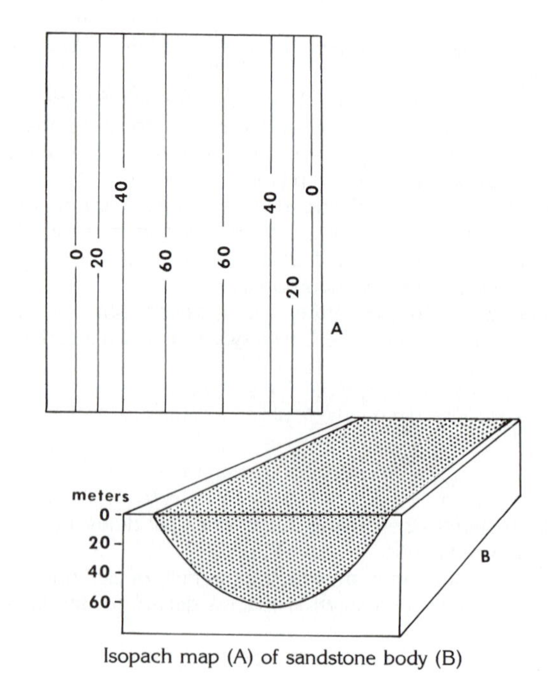

Isopach map (A) of sandstone body (B)

Ion—an atom that, because of electron transfers, has excess positive or negative charges.

Island arcs—chains of islands, arranged in arcuate trends on the surface of the earth. The arcs are sites of volcanic and earthquake activity and are usually bordered by deep oceanic trenches on the convex side.

Isopachous map—a map depicting the thickness of a sedimentary unit.

Isotope—atoms of an element that have the same number of protons in the nucleus, the same atomic number, and the same chemical properties but that have different atomic masses because they have different numbers of neutrons in their nuclei.

K

Karoo "System"—a sequence of Permian to Lower Jurassic rocks, primarily continental formations, which outcrop in Africa and are approximately equivalent to the Gondwana "System" of Peninsular India.

Kaskaskia sequence—a sequence of Devonian-Mississippian sediments, bounded above and below by regional unconformities and recording an episode of transgression followed by full flooding of a large part of the craton and by subsequent regression.

L

Lacustrine—relating to lakes, as in lacustrine sediments (lake sediments).

Lagomorph—the order of small placental mammals that includes rabbits, hares, and pikas.

Lagomorph (pika)

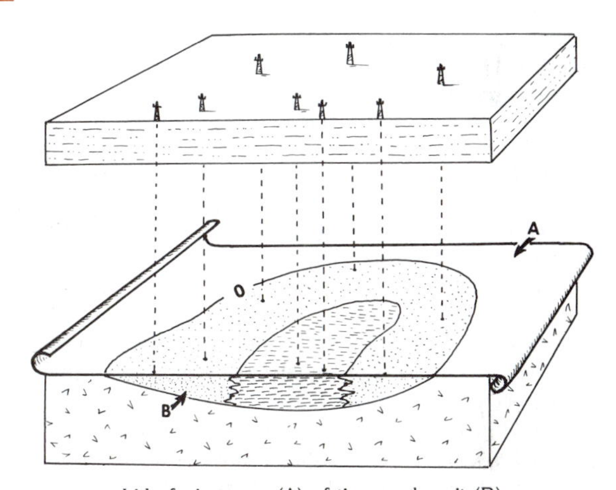

Lithofacies map (A) of time-rock unit (B)

Laramide Orogeny—in general, those pulses of mountain building that were frequent in Late Cretaceous time and were in large part responsible for producing many of the structures of the Rocky Mountains.

Lateral fault—a fault in which the movement is largely horizontal and in the direction of the trend of the fault plane. Sometimes also called a *strike-slip* fault.

Laurasia—a hypothetic "supercontinent" composed of what is now Europe, Asia, Greenland, and North America.

Limestone—a sedimentary rock consisting mainly of calcium carbonate.

Lithofacies map—a map that shows the areal variation in lithologic attributes of a stratigraphic unit.

Lithosphere—the outer shell of the earth, lying above the asthenosphere and comprising the crust and upper mantle.

Litoptern—South American ungulates whose evolutionary history somewhat paralleled that of horses and camels.

Logan's Line—a zone of thrust-faulting produced during the Taconic Orogeny that extends from the west coast of Newfoundland along the trend of the St. Lawrence River to near Quebec and southward along Vermont's western border. (Named after the pioneer Canadian geologist Sir William Logan.)

Lophophore—an organ, located adjacent to the mouth of brachiopods and bryozoans, which bears ciliated tentacles and has as its primary function the capture of food articles.

Low-velocity zone—an interior zone of the earth characterized by lower seismic-wave velocities than the region immediately above it.

Lunar maria—low-lying, dark lunar plains filled with volcanic rocks rich in iron and magnesium.

Lycopsids—leafy plants with simple, closely spaced leaves bearing sporangia on their upper surfaces. They are represented by living club mosses and vast numbers of extinct Late Paleozoic "scale trees."

M

Magnetic declination—the horizontal angle between "true," or geographic, north and magnetic north, as indicated by the compass needle. Declination is the result of the earth's magnetic axis being inclined with respect to the earth's rotational axis.

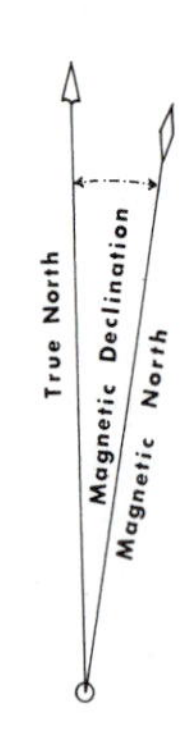

Magnetic declination

Magnetic inclination—the angle between the magnetic lines of force for the earth and the earth's surface; sometimes called "dip." Magnetic inclination can be demonstrated by observing a freely suspended magnetic needle. The needle will lie parallel to the earth's surface at the equator but is increasingly inclined toward the vertical as the needle is moved toward the magnetic poles.

Mammoth—the name commonly applied to extinct elephants of the Pleistocene Epoch.

Marsupials—mammals of the Order Marsupialia. Female marsupials bear mammary glands, and carry their immature young in a stomach pouch.

Mass spectrometer—an instrument for separating ions of different mass but equal charge and measuring their relative quantities.

Mastodon—the group of extinct proboscideans (elephantoids), early forms of which were characterized by long jaws, tusks in both jaws, and low-crowned teeth.

Mazatzal Orogeny—a Precambrian episode of mountain building named for an area in south-central Arizona where clues to crustal unrest in the form of folds, thrust faults, and intrusions are especially evident.

Medusa—the free-swimming, umbrella-shaped jellyfish form of the Phylum Coelenterata.

Meiosis—that kind of nuclear division, usually involving two successive cell divisions, that results in daughter cells having one half the number of chromosomes that were in the original cell.

Mélange—a body of intricately folded, faulted, and severely metamorphosed rocks, examples of which can be seen in the Franciscan rocks of California.

Mesosphere—a zone of the earth's mantle where pressures are sufficient to impart greater strength and rigidity to the rock.

Metamorphic rock—a rock formed from a previously existing rock by subjecting the parent rock to high temperature and pressure but without melting.

Metamorphism—the transformation of previously existing rocks into new types by the action of heat, pressure, and chemical solutions. Metamorphism usually takes place at depth in the roots of mountain chains or adjacent to large intrusive igneous bodies.

Metazoa—all multicellular animals whose cells become differentiated to form tissues (all animals except Protozoa).

Meteorites—metallic or stony bodies from interplanetary space that have passed through the earth's atmosphere and impact on the earth's surface.

Meteors—sometimes called "shooting stars," meteors are generally small particles of solid material from interplanetary space that may approach close enough to the earth to be drawn into the earth's atmosphere, where they are heated to incandescence. Most disintegrate, but a few land on the surface of the earth as *meteorites*.

Micrite—a texture in carbonate rocks that, when viewed microscopically, appears as murky, fine-grained calcium carbonate. Micrite is believed to develop from fine carbonate mud or ooze.

Milankovitch effect—the hypothetic long-term effect on world climate caused by three known components of earth motion. The combination of these components provides a possible explanation for repeated glacial to interglacial climatic swings.

Miliolids—a group of foraminifers with smooth, imperforate test walls and chambers arranged in various planes around a vertical axis. Miliolids are common in shallow marine areas.

Mineral—a naturally occurring element or compound formed by inorganic processes that has a definite chemical composition or range of compositions, as well as distinctive properties and form that reflect its characteristic atomic structure.

Miogeosyncline—the continentward side of a geosyncline, characterized by accumulations of shale, limestones, and sandstones derived from the continent and generally lacking in volcanics.

Mitosis—the method of cell division by means of which each of the two daughter nuclei receives exactly the same complement of chromosomes as had existed in the parent nucleus.

Mobile belt—an elongate region of the earth's crust characterized by exceptional earthquake and volcanic activity, tectonic instability, and periodic mountain building.

Mohorovičić discontinuity—a plane that marks the boundary separating the crust of the earth from the underlying mantle. The "moho," as it is sometimes called, is at a depth of about 70 km below the surface of continents and 6 to 14 km below the floor of the oceans.

Molasse—accumulations of primarily nonmarine, relatively light-colored, irregularly bedded conglomerates, shales, coal seams, and cross-bedded sandstones that are deposited subsequent to major orogenic events.

Mosasaur

Mold—an impression, or imprint, of an organism or part of an organism in the enclosing sediment.

Mollusk—any member of the invertebrate Phylum Mollusca, including pelecypods, cephalopods, gastropods, scaphopods, and chitons.

Monoplacophorans—primitive marine molluscans with simple, cap-shaped shells.

Monotremes—the egg-laying mammals.

Morganucodonts—early mammals found in Triassic beds of Europe and Asia and characterized by their small size and their retention of certain reptilian osteologic traits.

Mosasaurs—large marine lizards of the Late Cretaceous.

Multituberculates—an early group (Jurassic) of Mesozoic mammals with tooth cusps in longitudinal rows and with other dental characteristics that suggest they may have been the earliest herbivorous mammals.

Mutation—a stable and inheritable change in a gene.

Mytiloids—pelecypods having rather triangular shells that are most commonly identical but in some forms unequal. The edible mussel *Mytilus* is a representative form.

Nappe—a large mass of rocks that have been moved a considerable distance over underlying formations by overthrusting, recumbent folding, or both.

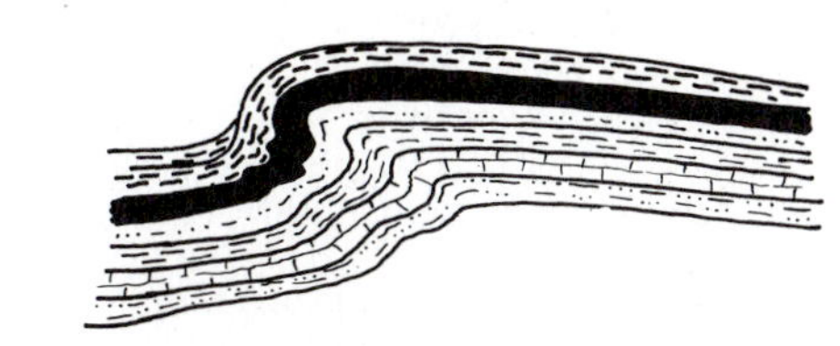

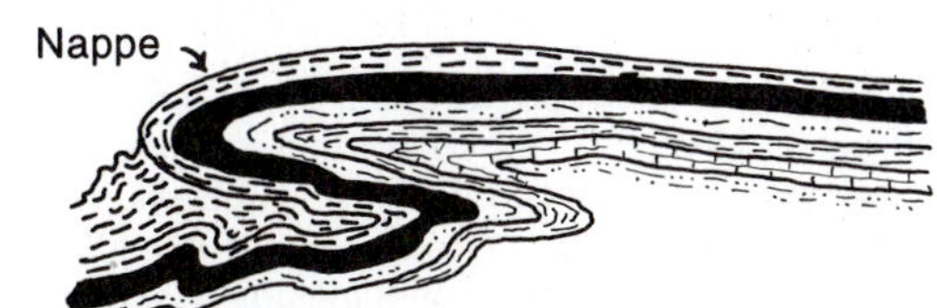

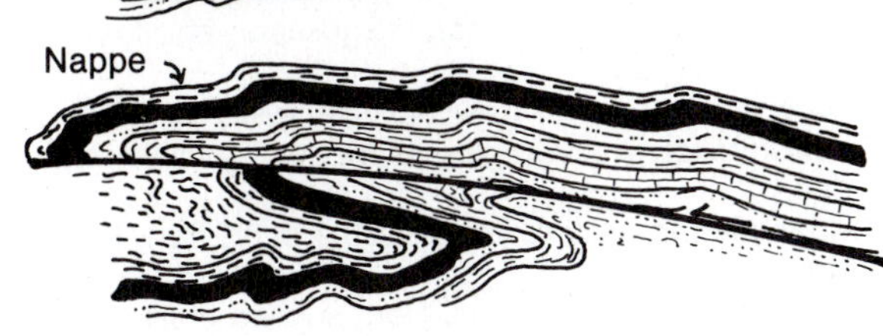

Nappe development

Nektonic—pertaining to swimming organisms.

Neogene—a subdivision of the Cenozoic that encompasses the Miocene, Pliocene, Pleistocene, and Holocene Epochs.

Neutron—electrically neutral (uncharged) particles of matter existing along with protons in the atomic nucleus of all elements exxcept the mass 1 isotope of hydrogen.

Nevadan Orogeny—in general, those pulses of mountain building, intrusion, and metamorphism that were most frequent during the Jurassic and Early Cretaceous along the western part of the Cordilleran Geosyncline.

Newark Series—a series of Late Triassic, nonmarine red beds (shales, sandstones, and conglomerates), lava flows, and intrusions located within downfaulted basins from Nova Scotia to South Carolina.

New World monkeys—monkeys whose habitat is today confined to South America. They are thoroughly arboreal in habit and have prehensile tails by which they hang and swing from tree limbs.

Nonconformity—an unconformity developed between sedimentary rocks and older plutons or massive metamorphic rocks that had been exposed to erosion before the overlying sedimentary rocks were deposited.

Normal fault—a fault in which hanging wall appears to have moved downward relative to the footwall; normally occurring in areas of crustal tension.

Nothosaurs—relatively small Early Mesozoic sauropterygians that were replaced during the Jurassic by the plesiosaurs.

Notochord—a rod-shaped cord of cartilage cells forming the primary axial structure of the chordate body. In vertebrates, the notochord is present in the embryo and is later supplanted by the vertebral column.

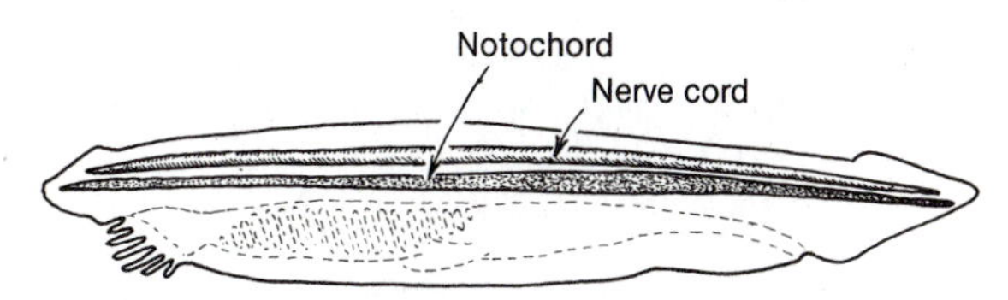

Notochord and dorsal nerve cord in the living lancelet

Notungulates—a group of ungulates that diversified in South America and persisted until Plio-Pleistocene time.

Novaculites—a term applied originally to rocks suitable for whetstones and, in America, to white chert found in Arkansas. Now applied to very tough, uniformly grained cherts composed of microcrystalline quartz.

Nuclear fission tracks—submicroscopic "tunnels" in minerals produced when high-energy particles from the nucleus of uranium are forcibly ejected during spontaneous fission.

Nucleic acid—any of a group of organic acids that control hereditary processes within cells and make possible the manufacture of proteins from the amino acids ingested by the cells as food.

Nuclides—the different weight configurations of an element caused by atoms of that element having differing numbers of neutrons. Nuclides or isotopes of an element differ in number of neutrons but not in chemical properties.

Nummulites—large, coined-shaped foraminifers, especially common in Tertiary limestones.

O

Offlap—a sequence of sediments resulting from a marine regression and characterized by an upward progression from offshore marine sediments (often limestones) to shales and finally sandstones (above which will follow an unconformity).

Oil shale—a dark-colored shale rich in organic material that can be heated to liberate gaseous hydrocarbons.

Old World monkeys—monkeys of Asia, Africa, and southern Europe. Old World monkeys include the macaque, guenons, langurs, baboons, and mandrills.

Onlap—a sequence of sediments resulting from a marine transgression. Normally, the sequence begins with a conglomerate or sandstone deposited over an erosional unconformity and is followed upward in the vertical section by progressively more offshore sediments.

Oölites—spherical carbonate and sand particles composed of concentric laminae of microscopic crystals of calcium carbonate.

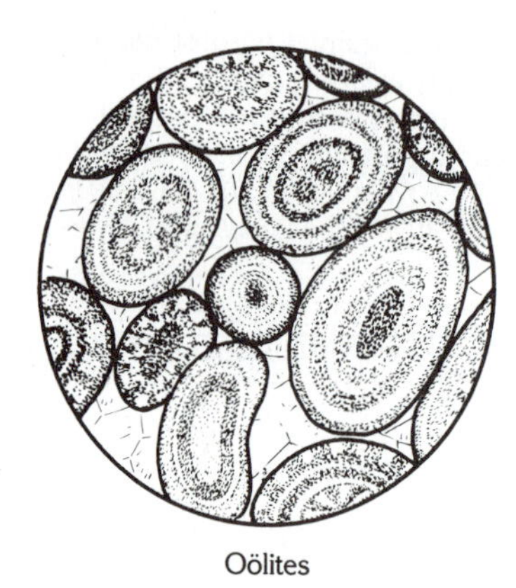

Oölites

Ophiolite suite—an assemblage of rock formations frequently found in great mountain belts and characterized by an upward sequence of ultramafic rock, basaltic rocks, and deep sea sediments. Ophiolite belts are believed to be preserved fragments of vanished ocean basins.

Oreodonts—North American artiodactyls of the Middle and Late Tertiary.

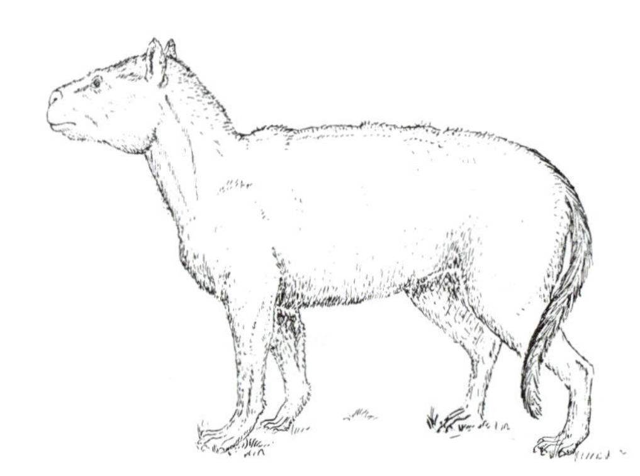

The oreodont *Merycoidodon*

Orogenic belt—great linear tracts of deformed rocks, primarily developed near continental margins by compressional forces accompanying mountain building.

Ornithischia—an order of dinosaurs characterized by birdlike pelvic structures and including such herbivores as the ornithopods, stegosaurs, ankylosaurs, and ceratopsians.

Orogeny—the process by which great systems of mountains are formed. (Orogenesis means mountain building.)

Ostracoderms—extinct jawless fishes of the Early Paleozoic.

Ostracods—small, bivalved, bean-shaped crustaceans.

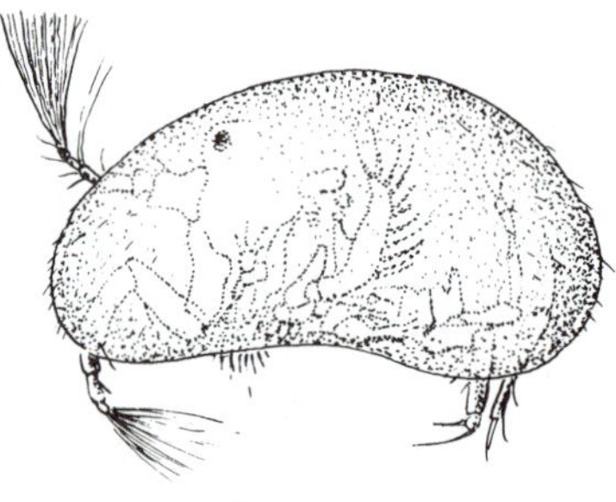

Ostracod

Ostreoids—the "oyster family" of pelecypods.

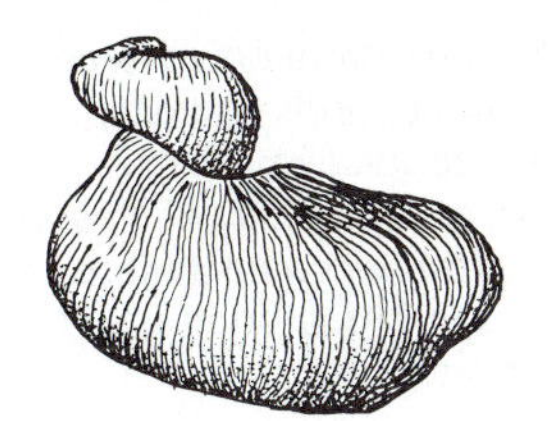
The ostreoid *Exogyra* (Jur.-Cret.)

Outcrop—an area where specific rock units occur at the surface, even though covered by soil and not exposed.

P

Paleoecology—the study of the relationships of ancient organisms to their environment.

Paleogene—a subdivision of the Cenozoic that encompasses the Paleocene, Eocene, and Oligocene Epochs.

Paleogeography—the geography as it existed at some time in the geologic past.

Paleolatitudes—the latitude that once existed across a particular region at a particular time in the geologic past.

Paleomagnetism—the earth's magnetic field and magnetic properties in the geologic past. Studies of paleomagnetism are helpful in determining positions of continents and magnetic poles.

Paleontology—the study of all ancient forms of life, their interaction, and their evolution.

Pangaea I—in Alfred Wegener's theory of continental drift, the supercontinent that included all present major continental masses.

Panthalassa—the great universal ocean that surrounded the supercontinent Pangaea prior to its breakup.

Pantotheres—a group of Jurassic mammals with dentition similar to that of primitive representatives of later marsupial and placental mammals and thus thought to be ancestral to these groups.

Paraconformity—a rather obscure unconformity in which no erosional surface is discernible, and in which beds above and below the break are parallel.

Partial melting—process by which a rock subjected to high temperature and pressure is partially melted, and the liquid fraction moved to another location. Partial melting results from the variations in melting points of different minerals in the original rock mass.

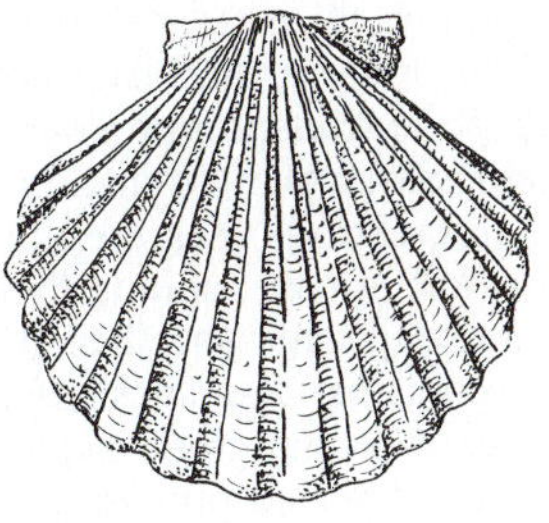
Pecten

Pectenoids—the pectens are pelecypods exemplified by the scallops of commerce. They have generally subcircular shells and straight hinge lines.

Pelycosaurs—early mammal-like reptiles exemplified by the sail-back animals of the Permian Period.

Period—a subdivision of an era.

Perissodactyl—progressive, hoofed mammals, characteristically having an odd number of toes on the hind feet and, usually, on the front feet as well.

Permineralization—a manner of fossilization in which voids in an organic structure (such as bone) are filled with mineral matter.

Petrification—the process of converting organic structures, such as bone, shell, or wood, into a stony substance, such as calcium carbonate or silica.

Phanerozoic—the eon of geologic time during which the earth has been populated by abundant and diverse life. The Phanerozoic Eon followed the Cryptozoic Eon and is divided into Paleozoic, Mesozoic and Cenozoic Eras.

Phosphorite—a sediment composed largely of calcium phosphate.

Photosphere—a relatively thin gaseous layer on the sun that emits nearly all the light that the sun radiates into space.

Photosynthesis—the process of synthesizing carbohydrates from carbon dioxide and water, utilizing the radiant energy of light captured by the chlorophyll in plant cells.

Phytoplankton—microscopic marine planktonic plants, most of which are various forms of algae.

Phytosaurs—extinct aquatic crocodile-like thecodonts.

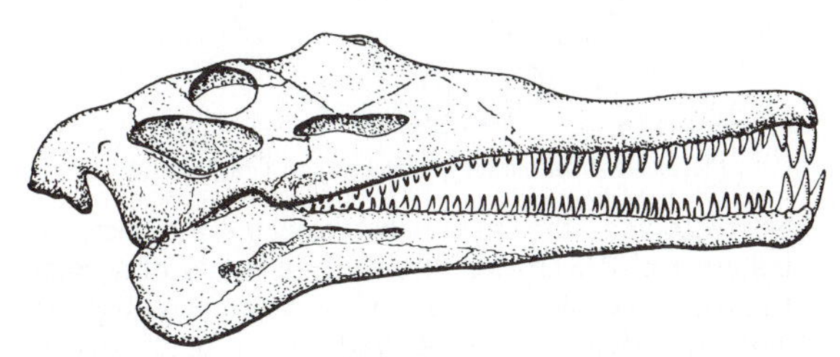
Phytosaur skull

Pillow lava—type of lava that is extruded under water and in which many pillow-shaped lobes break through the chilled surface of the flow and solidify. (Resembles a pile of pillows.)

Pinnipeds—marine carnivores such as seals, sea lions, and walruses.

Placer deposit—an accumulation of sediment rich in a valuable mineral or metal that has been concentrated because of its greater density.

Placoderms—extinct primitive jawed fishes of the Paleozoic Era.

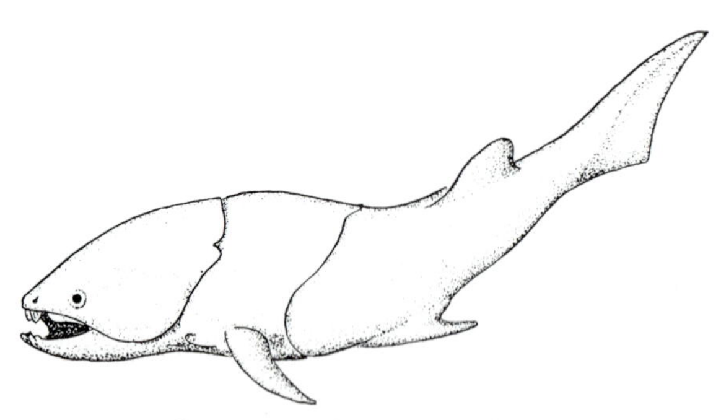

Coccosteus, a Devonian placoderm

Placodonts—extinct walrus-like, marine reptiles that fed principally upon shellfish.

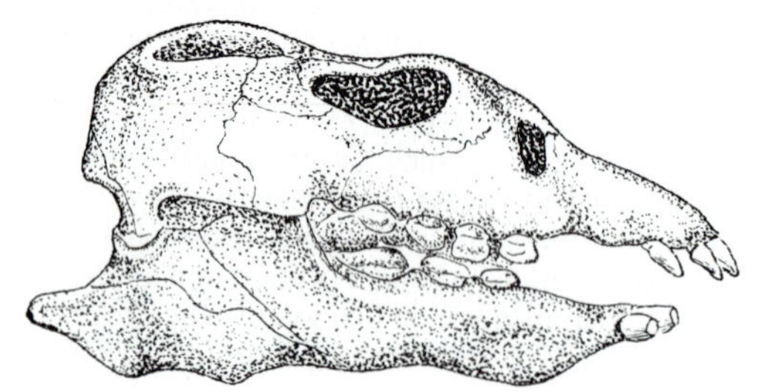

Skull of the placodont *Placodus.* Note blunt teeth for crushing shellfish

Plankton—minute, free-floating aquatic organisms.

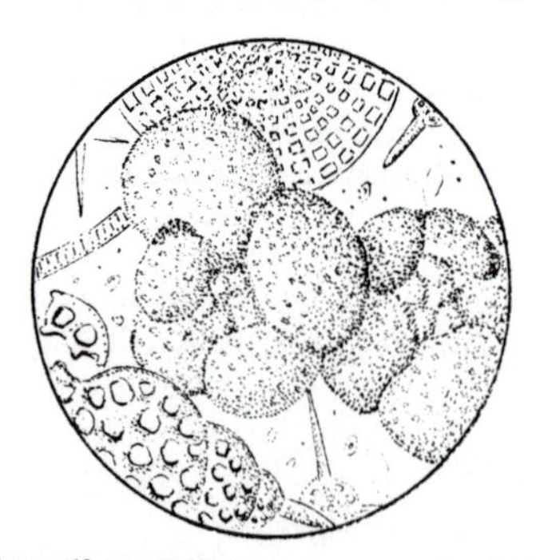

Plankton (foraminifers, diatoms, and radiolaria)

Plate tectonics—the theory that explains the tectonic behavior of the crust of the earth in terms of several moving plates that are formed by volcanic activity at oceanic ridges and destroyed along great ocean trenches.

Platform—that part of a craton covered thinly by layered sedimentary rocks and characterized by relatively stable tectonic conditions.

Plesiosaurs—the group of extinct Mesozoic marine reptiles (sauropterygians) characterized by large, paddle-shaped limbs and broad bodies, with either very long or relatively short necks.

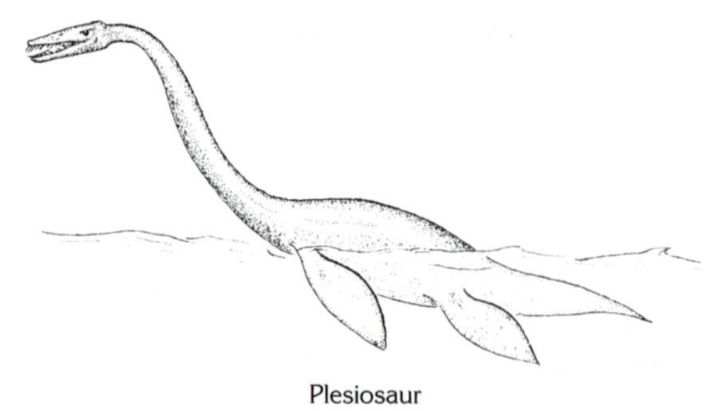

Plesiosaur

Pluvial lake—a lake formed in an earlier climate, when rainfall was greater than at present.

Polyp—the hydra-like form of some coelenterates in which the mouth and tentacles are at the top of the body.

Porphyry—a textural term used to describe an igneous rock in which some of the crystals, called phenocrysts, are distinctly larger than others.

Porphyry

Primary earthquake waves—seismic waves that are propagated through solid rock as a train of compressions and dilations.

Principle of cross-cutting relations—principle that states that geologic features such as faults, veins, and dikes must be younger than the rocks or features they cut across.

Principle of biologic succession—the principle that states that the observed sequence of life forms has changed continuously through time, so that the total aspect of life (as recognized by fossil evidence) for a particular segment of time is distinct and different from that of life of earlier and later times.

Principle of temporal transgression—the principle that stipulates that sediments of advancing (transgressing) or retreating (regressing) seas are not necessarily of the same geologic age throughout their lateral extent.

Proboscidea—the elephants and their progenitors.

Prokaryotes—organisms that lack membrane-bound nuclei (such as the blue-green algae and bacteria).

Prosimii—the less advanced primates, such as lemurs, tarsiers, and tree shrews.

Proteinoids—extra-large organic molecules containing most of the 20 amino acids of proteins and produced in laboratory conditions simulating those found in nature.

Proteins—giant molecules containing carbon, hydrogen, oxygen, nitrogen, and usually sulfur and phosphorus; composed of chains of amino acids and present in all living cells.

Proton—an elemental particle found in the nuclei of all atoms that has a positive electric charge and a mass similar to that of a neutron.

Pyroclastics—fragments of volcanic debris that have usually been fragmented during eruptions.

Pyroxene Group—a group of silicate minerals having single-chain tetrahedral structure.

Q

Queenston Delta—a clastic wedge of red beds shed westward from highlands elevated in the course of the Taconic Orogeny.

R

Radioactive decay—the spontaneous emission of a particle from the atomic nucleus, thereby transforming the atom from one element to another.

Radiolaria—protozoa that secrete a delicate, often beautifully filigreed skeleton of opaline silica.

Recumbent fold—a fold in which the axial plane is essentially horizontal—a fold that has been turned over by compressional forces so that it lies on its side.

Red beds—prevailingly red, usually clastic sedimentary deposits.

Red shift—a phenomenon apparent in telescopic analyses of the spectra of distant galaxies, in which their spectra are shifted to the red by an amount proportional to the distance of the galaxy from the earth.

Regolith—any solid materials, such as rock fragments or soil, lying on top of bedrock.

Regression—a general term signifying that a shoreline has moved toward the center of a marine basin. Regression may be caused by tectonic emergence of the land, eustatic lowering of sea level, or prograding of sediments, as in deltaic build-outs.

Relative geologic age—the placing of an event in a time sequence without regard to the absolute age in years.

Replacement—a fossilization process in which the original skeletal substance is replaced after burial by inorganically precipitated mineral matter.

Reverse fault—a fault formed by compression in which the hanging wall appears to move up relative to the foot wall.

Rhynchonellids—a group of brachiopods having pronounced beaks, accordion-like plications, and triangular outline.

Rift valley—a valley formed by faulting, usually involving a central fault block that moves downward in relation to adjacent blocks.

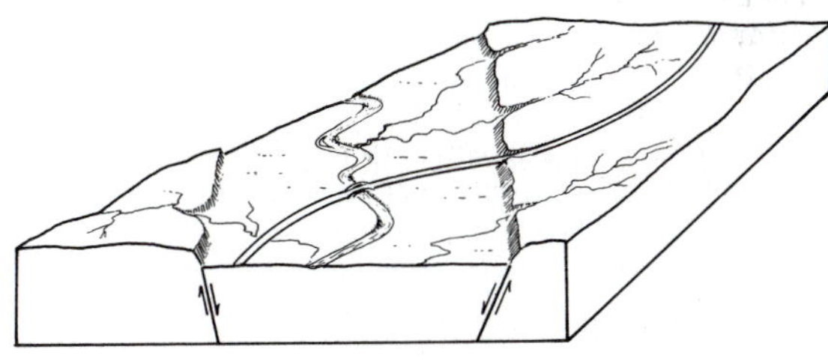

Rift valley

Rudists—peculiarly specialized Mesozoic pelecypods often having one valve in the shape of a horn coral, covered by the other valve in the form of a lid.

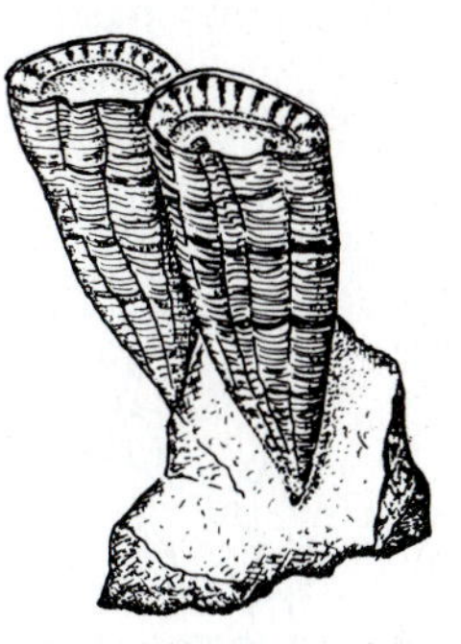

Rudistoid pelecypods

Rugosa—the large group of solitary and colonial Paleozoic horn corals.

Ruminant—a herbivorous, cud-chewing ungulate.

Salt dome—a structural dome in sedimentary strata resulting from the upward flow of a large body of salt.

Sauk sequence—a sequence of Late Precambrian to Ordovician sediments bounded both above and below by a regional unconformity and recording an episode of marine transgression, followed by full flooding of a large part of the craton and ending with a regression from the craton.

Saurischia—an order of dinosaurs with triradiate pelvic structures, including both the gigantic herbivorous sauropods and the carnivorous theropods.

Scleractinid coral—coral belonging to the Order Scleractinia, which includes most modern and post-Paleozoic corals.

Sea floor spreading—the process by which new sea floor crust is produced along midoceanic ridges (divergence zones) and slowly conveyed away from the ridges.

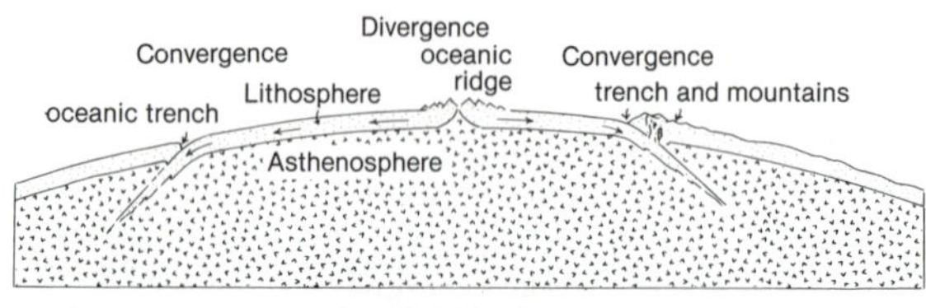

Sea floor spreading

Secondary earthquake wave—a seismic wave in which the direction of vibration of wave energy is at right angles to the direction the wave travels.

Sedimentary rock—a rock that has formed as a result of the consolidation (lithification) of accumulations of sediment.

Sessile—bottom-dwelling aquatic animals that live continuously in one place.

Series—the time-rock term representing the rocks deposited or emplaced during a geologic epoch. Series are subdivisions of systems.

Shelly facies—in general, miogeosynclinal sedimentary deposits consisting primarily of carbonate rocks containing the abundant fossil remains of marine invertebrates.

Silicoflagellates—unicellular, tiny, flagellate marine algae that secrete an internal skeleton composed of opaline silica.

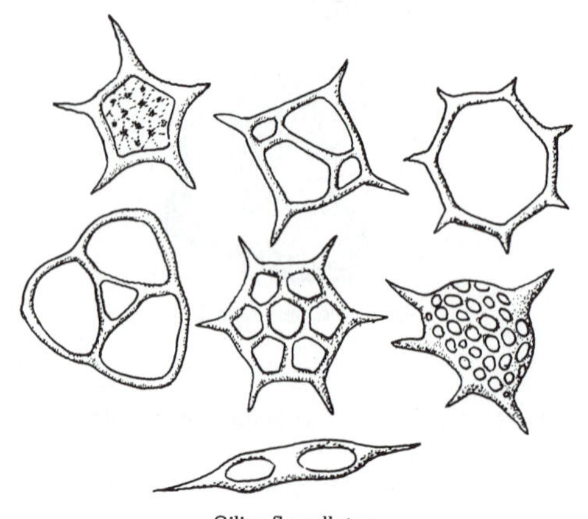

Silicoflagellates

Silicon tetrahedron—an atomic structure in silicates consisting of a centrally located silicon atom linked to four oxygen atoms placed symmetrically around the silicon at the corners of a tetrahedron.

Somona Orogeny—mid-Permian orogenic movements, the structural effects of which are most evident in western Nevada.

Sorting—a measure of the uniformity of the sizes of particles in a sediment or sedimentary rock.

Spar carbonate—as viewed microscopically, the clear, crystalline carbonate that has been deposited in a carbonate rock as a cement between clasts or has developed by recrystallization of clasts.

Species—a unit of taxonomic classification of organisms. In another sense, a species is a population of individuals that are similar in structural and functional characteristics and that in nature breed only with one another.

Sphenodon (Tuatara)—large, lizard-like reptiles that have persisted from Triassic to the present and now inhabit islands off the coast of New Zealand.

Sphenopsids—a group of spore-bearing plants that were particularly common during the Late Paleozoic and were characterized by articulated stems with leaves borne in whorls at the nodes. Only one genus, *Equisetum*, survives.

Spiracle—in cartilaginous fishes, a modified gill opening through which water enters the pharynx.

Spontaneous fission—spontaneous fragmentation of an atom into two or more lighter atoms and nuclear particles.

Spores—a usually asexual reproductive body, such as occurs in bacteria, ferns, and mosses.

Stage—the time-rock unit equivalent to an age. A stage is a subdivision of a series.

Stapes—the innermost of the small bones in the middle ear cavity of mammals; also recognized in amphibians and reptiles.

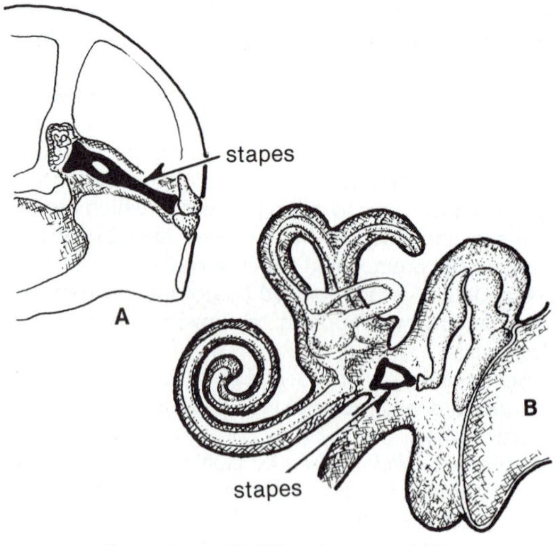

Stapes in reptile (A) and mammal (B)

Strain—deformation of a rock mass in response to stress.

Stratification—the layering in sedimentary rocks that results from changes in texture color or rock type from one bed to the next.

Stratified drift—deposits of glacial clastics that have been sorted and stratified by the action of meltwater.

Stromatolites—distinctly laminated accumulations of calcium carbonate having rounded, branching, or frondose shape and believed to form as a result of the metabolic activity of marine algae. They are usually found in the high intertidal to low supratidal zones.

Stromatoporoids—an extinct group of reef-building organisms now believed to have affinities with the Porifera and noted for the large, often laminated masses constructed by the colonies.

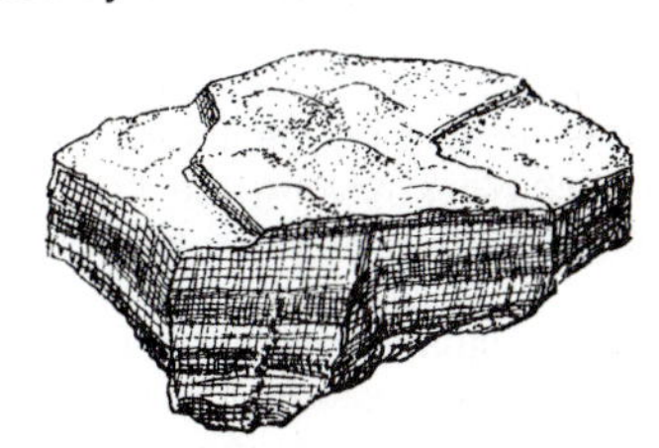

Stromatopora

Subaerial—formed or existing at or near a sediment surface significantly above sea level.

Subduction zone—an inclined planar zone, defined by high frequency of earthquakes, that is thought to locate the descending leading edge of a moving oceanic plate.

Sublittoral zone—the marine bottom environment that extends from low tide seaward to the edge of the continental shelf.

Surface earthquake waves—seismic or earthquake waves that move only about the surface of the earth.

Symmetrodonts—a group of primitive Mesozoic mammals characterized by a symmetric triangular arrangement of cusps on cheek teeth.

Syncline—a geologic structure in which strata are bent into a downfold.

System—the time-rock unit representing rock deposited or emplaced during a geologic period.

T

Taconic Orogeny—a major episode of orogeny that affected the Appalachian Geosyncline in Ordovician time. The northern and Newfoundland Appalachians were the most severely deformed during this orogeny.

Taxon (pl. taxa)—any unit in the taxonomic classification, such as a phylum, class, order, or family.

Taxonomy—the science of naming, describing, and classifying organisms.

Tectonics—the structural behavior of a region of the earth's crust.

Teleosts—the teleosts are the most advanced of the bony fishes and are characterized by thin, rounded scales, completely bony internal skeleton, and symmetric tail. Teleosts range from Cretaceous to Recent.

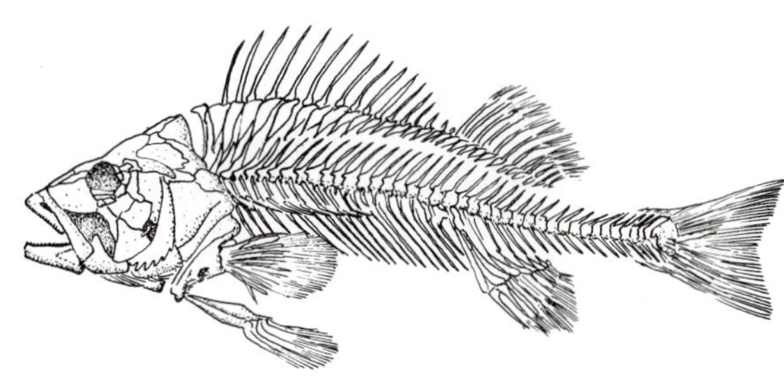

Teleost fish

Terebratulids—a group of Silurian to Recent, mostly smooth-shelled brachiopods having a loop-shaped attachment for the lophophore. Terebratulids were most abundant during the Jurassic and Cenozoic.

Tethys Geosyncline—a great east-west trending seaway lying between Laurasia and Gondwana during Paleozoic and Mesozoic time and from which arose the Alpine-Himalayan Mountain ranges.

Thecodonts—an order of primarily Triassic reptiles considered to be the ancestral archosaurians.

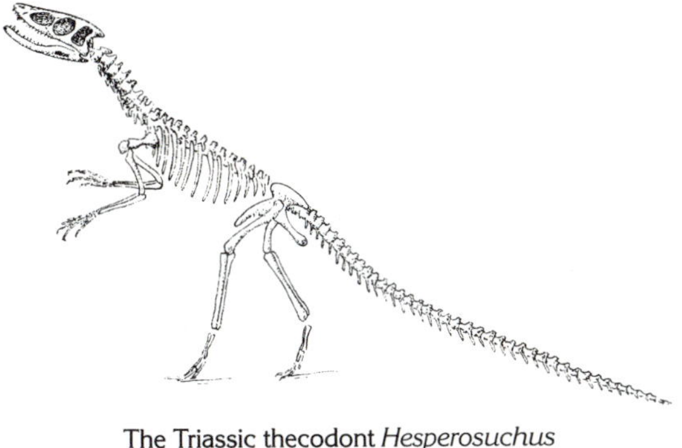

The Triassic thecodont *Hesperosuchus*

Therapsids—an order of advanced mammal-like reptiles.

Cynognathus, a Permian therapsid

Thermal plume—a "hot spot" in the upper mantle believed to exist where a huge column of upwelling magma lies in a fixed position under the lithosphere. Thermal plumes are thought to cause vulcanism in the overlying lithosphere.

Thermoremanent magnetism—permanent magnetization acquired by igneous rocks as they cool past the Curie point while in the earth's magnetic field.

Theropods—the carnivorous saurischian dinosaurs.

The small Jurassic theropod *Ornitholestes*

Thrust fault—a low-angle reverse fault, with inclination of fault plane generally less than 45°.

Till—unconsolidated, unsorted, unstratified glacial debris.

Tillite—unsorted glacial drift (till) that has been converted into solid rock.

Time-rock unit (chronostratigraphic unit)—the rocks formed during a particular unit of geologic time. The Cambrian System, for example, is the time-rock unit for the segment of time known as the Cambrian Period.

Tippecanoe sequence—a sequence of Ordovician to Early Devonian sediments bounded above and below by regional unconformities and recording an episode of marine transgression, followed by full flooding of a large region of the craton, and subsequent regression.

Titanotheres—large, extinct perissodactyls (odd-toed ungulates) that attained the peak of their evolutionary development during the Oligocene. *Brontotherium* is a widely known titanothere.

Trace fossils—tracks, trails, burrows, and other markings made in now lithified sediments by ancient animals.

Transcontinental Arch—an elongate, uplifted region extending from Arizona northeastward toward Lake Superior. During the Cambrian, the arch was emergent, as indicated by the observation that Cambrian marine sediments are located on either side of the arch but are missing above it. (Along the crest of the arch, Post-Cambrian rocks rest upon Pre-Cambrian basement.)

Transform fault—a strike-slip fault bounded at each end by an area of crustal spreading that tends to be more or less perpendicular to the trace of the fault.

Triconodonts—a group of primitive Mesozoic mammals recognized primarily by the arrangement of three principal cheek tooth cusps in a longitudinal row.

Trilobites—Paleozoic marine arthropods of the Class Crustacea, characterized by longitudinal and transverse division of the carapace into three parts, or lobes. Trilobites were especially abundant during the Early Paleozoic.

Tuff—volcanic ash that has become consolidated into rock.

Turbidites—sediment deposited from a turbidity current and characterized by graded bedding and moderate-to-poor sorting.

Turbidity current—a mass of moving water that is denser than surrounding water because of its content of suspended sediment and that flows along slopes of the sea floor as a result of that higher density.

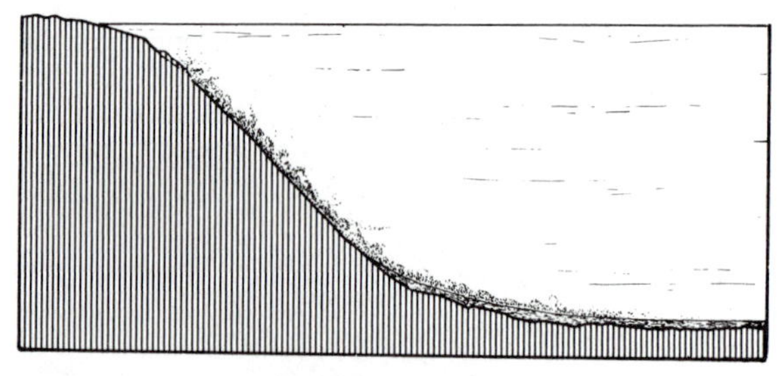

Turbidity Current

Tylopod—the artiodactyl group to which camels and llamas belong.

Alticamelus, a Pliocene tylopod

U

Unconformity—a surface separating an overlying younger rock formation from an underlying formation and representing an episode of erosion or nondeposition. Because unconformities represent a lack of continuity in deposition, they are gaps in the geologic record.

Ungulate—four-legged mammals whose toes bear hoofs.

Uniformitarianism—a general principle that suggests that the past history of the earth can be interpreted and deciphered in terms of what is known about present natural laws.

V

Vagile—bottom dwelling aquatic animal capable of locomotion.

Varve—a thin sedimentary layer or pair of layers that represent the depositional record of a single year.

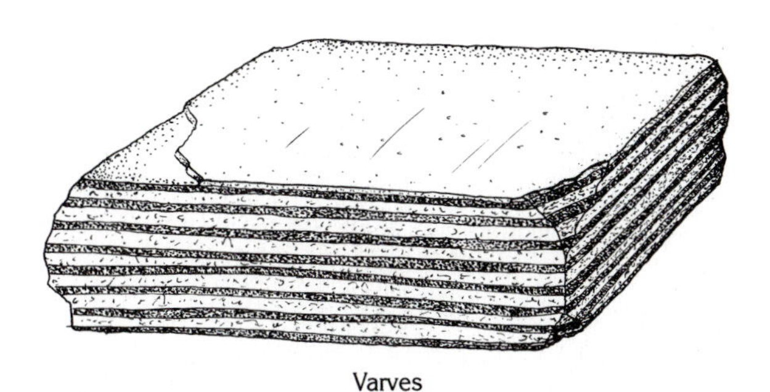

Varves

Vascular plants—plants, including all higher land plants, that have a system of vessels and ducts for distributing moisture and nutrients.

Veneroids—a group of pelecypods exemplified by the common clam, *Venus.*

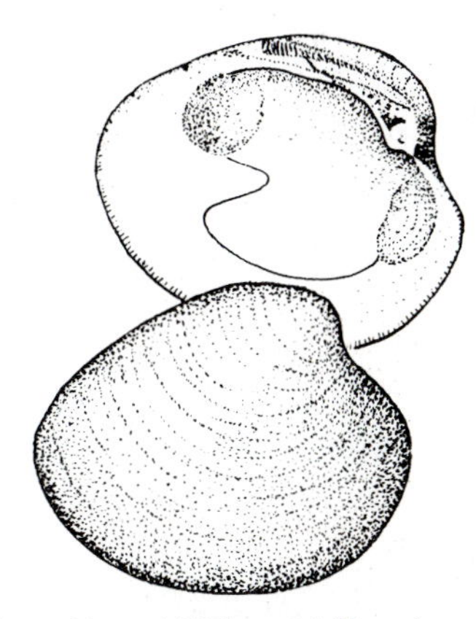

Veneroid Pelecypod (Venus)

Vestigial organ—an organ that is useless, small, or degenerate but that represents a structure that was more fully developed or functional in an ancestral organism.

Vindelician Arch—a highland area believed to have formed a barrier separating the Germanic and Alpine depositional areas during the Triassic and Jurassic.

Williston Basin—a large structural basin extending from South Dakota and Montana northward into Canada; well known for the petroliferous Devonian formations deposited therein.

Zechstein Sea—an arm of the Atlantic that extended across part of northern Europe during the Late Permian and in which were deposited several hundred meters of evaporites, including the well-known potassium salts of Germany.

Zone—a bed or group of beds distinguished by a particular fossil content and frequently named after the fossil or fossils it contains.

Zygote—the cell formed by the union of two gametes. Thus, a zygote is a fertilized egg.

In this index, definitions are indicated by boldface page numbers. Italicized page numbers indicate illustrations. A small *t* following a page number indicates a table. The index includes formation names in the correlation charts, for which the abbreviations Fm. for Formation, Grp. for Group, Ss. for Sandstone, Sh. for Shale, and Ls. for Limestone are used. Terms in the glossary are not included.

A

B

C

D

E

F

J

K

L

M

N

O

T

U

V

W

X

Z